To my father, without whose encouragement
this would not have been possible.

About the Author

Ansel C. Ugural is Research Professor of Mechanical Engineering at New Jersey Institute of Technology. He has been a National Science Foundation (NSF) fellow and taught at the University of Wisconsin. Dr. Ugural has held faculty positions at Fairleigh Dickinson University, where he served for two decades as professor and chairman of the mechanical engineering department. He has considerable and diverse industrial experience in both full-time and consulting capacities as a design, development, and research engineer.

Professor Ugural received his M.S. in Mechanical Engineering and Ph.D. in Engineering Mechanics from the University of Wisconsin–Madison. He has been a member of the American Society of Mechanical Engineers and the American Society of Engineering Education. Dr. Ugural is listed in *Who's Who in Engineering*.

He is the author of several books, including *Stresses in Plates and Shells* (McGraw-Hill, 1999) and *Mechanics of Materials* (McGraw-Hill, 1991) with editions in Korean (1992) and Chinese (1994). Dr. Ugural is also the coauthor (with S. K. Fenster) of *Advanced Strength and Applied Elasticity* (Prentice Hall, 2003). In addition, he has published numerous articles in trade and professional journals.

Mechanical Design
An Integrated Approach

Ansel C. Ugural
New Jersey Institute of Technology

 Higher Education

Boston   Burr Ridge, IL   Dubuque, IA   Madison, WI   New York   San Francisco   St. Louis
Bangkok   Bogotá   Caracas   Kuala Lumpur   Lisbon   London   Madrid   Mexico City
Milan   Montreal   New Delhi   Santiago   Seoul   Singapore   Sydney   Taipei   Toronto

The McGraw·Hill Companies

Higher Education

MECHANICAL DESIGN: AN INTEGRATED APPROACH

Published by McGraw-Hill, a business unit of The McGraw-Hill Companies, Inc., 1221 Avenue of the Americas, New York, NY 10020. Copyright © 2004 by The McGraw-Hill Companies, Inc. All rights reserved. No part of this publication may be reproduced or distributed in any form or by any means, or stored in a database or retrieval system, without the prior written consent of The McGraw-Hill Companies, Inc., including, but not limited to, in any network or other electronic storage or transmission, or broadcast for distance learning.

Some ancillaries, including electronic and print components, may not be available to customers outside the United States.

This book is printed on acid-free paper.

International 2 3 4 5 6 7 8 9 0 DOC/DOC 0 9 8 7
Domestic   2 3 4 5 6 7 8 9 0 DOC/DOC 0 9 8 7

ISBN 0–07–242155–X
ISBN 0–07–121516–6 (ISE)

Publisher: *Elizabeth A. Jones*
Sponsoring editor: *Jonathan Plant*
Administrative assistant: *Rory Stein*
Marketing manager: *Sarah Martin*
Lead project manager: *Jill R. Peter*
Production supervisor: *Sherry L. Kane*
Media project manager: *Jodi K. Banowetz*
Senior media technology producer: *Phillip Meek*
Designer: *David W. Hash*
Cover/interior designer: *Rokusek Design*
Lead photo research coordinator: *Carrie K. Burger*
Compositor: *Interactive Composition Corporation*
Typeface: *10/12 Times*
Printer: *R. R. Donnelley Crawfordsville, IN*

Library of Congress Cataloging-in-Publication Data

Ugural, A. C.
   Mechanical design : an integrated approach / Ansel C. Ugural. — 1st ed.
      p.   cm.
   Includes bibliographical references and index.
   ISBN 0–07–242155–X
   1. Machine design.   2. Structural design.   I. Title.

TJ230 .U45   2004
621.8'15—dc21
                                              2002038093
                                              CIP

INTERNATIONAL EDITION ISBN 0–07–121516–6
Copyright © 2004. Exclusive rights by The McGraw-Hill Companies, Inc., for manufacture and export. This book cannot be re-exported from the country to which it is sold by McGraw-Hill. The International Edition is not available in North America.

www.mhhe.com

Mechanical Design

An Integrated Approach

McGraw-Hill Series in Mechanical Engineering

Average properties of common engineering materials*

SI Units

Material	Density, Mg/m³	Ultimate strength, MPa			Yield strength,† MPa		Modulus of elasticity, GPa	Modulus of rigidity, GPa	Coefficient of thermal expansion, $10^{-6}/°C$	Elongation in 50 mm, %	Poisson's ratio
		Tension	Compression**	Shear	Tension	Shear					
Steel											
Structural, ASTM-A36	7.86	400	—	—	250	145	200	79	11.7	30	0.27–0.3
High strength, ASTM-A242	7.86	480	—	—	345	210	200	79	11.7	21	
Stainless (302), cold rolled	7.92	860	—	—	520	—	190	73	17.3	12	
Cast iron											0.2–0.3
Gray, ASTM A-48	7.2	170	650	240	—	—	70	28	12.1	0.5	
Malleable, ASTM A-47	7.3	340	620	330	230	—	165	64	12.1	10	0.3
Wrought iron	7.7	350	—	240	210	130	190	70	12.1	35	0.33
Aluminum											
Alloy 2014-T6	2.8	480	—	290	410	220	72	28	23	13	0.34
Alloy 6061-T6	2.71	300	—	185	260	140	70	26	23.6	17	
Brass, yellow											
Cold rolled	8.47	540	—	300	435	250	105	39	20	8	0.34
Annealed	8.47	330	—	220	105	65	105	39	20	60	
Bronze, cold rolled (510)	8.86	560	—	—	520	275	110	41	17.8	10	0.34
Copper, hard drawn	8.86	380	—	165	260	160	120	40	16.8	4	0.33
Magnesium alloys	1.8	140–340	—	—	80–280	—	45	17	27	2–20	0.35
Nickel	8.08	310–760	—	—	140–620	—	210	80	13	2–50	0.31
Titanium alloys	4.4	900–970	—	—	760–900	—	100–120	39–44	8–10	10	0.33
Zinc alloys	6.6	280–390	—	—	210–320	—	83	31	27	1–10	0.33
Concrete											0.1–0.2
Medium strength	2.32	—	28	—	—	—	24	—	10	—	
High strength	2.32	—	40	—	—	—	30	—	10	—	
Timber‡ (air dry)											
Douglas fir	0.54	—	55	7.6	—	—	12	—	4	—	
Southern pine	0.58	—	60	10	—	—	11	—	4	—	
Glass, 98% silica	2.19	—	50	—	—	—	65	28	80	—	0.2–0.27
Graphite	0.77	20	240	35	—	—	70	—	7	—	
Rubber	0.91	14	—	—	—	—	—	—	162	600	0.45–0.5

*Properties may vary widely with changes in composition, heat treatment, and method of manufacture.

**For ductile metals the compression strength is assumed to be the same as that in tension.

† Offset of 0.2%.

‡ Loaded parallel to the grain.

Brief Table of Contents

Contents

Appendix C
STRESS CONCENTRATION FACTORS   774

Appendix D
SOLUTION OF THE STRESS
CUBIC EQUATION   781

ANSWERS TO SELECTED PROBLEMS   783

Index   791

Preface

INTRODUCTION

This text developed from classroom notes prepared in connection with junior-senior undergraduate and first-year graduate courses in Mechanical Design, Machine Design, Mechanical Engineering Design, and Engineering Design and Analysis. Many more topics are covered in this book than in any other text on the subject. In addition to its applicability to *mechanical engineering,* and to some extent, aerospace, agricultural, and nuclear engineering and applied engineering mechanics curricula, I have endeavored to make the book useful to *practicing engineers*. The text offers a simple, comprehensive, and methodical presentation of the basic concepts and principles in the design and analysis of machine and structural components. The coverage presumes a knowledge of mechanics of materials and material properties. However, topics that are particularly significant to understanding the subject are reviewed as they are taken up. Special effort has been made to have a book that is as self-explanatory as possible, thus reducing the work of the instructor.

The presentation of the material in this book strikes a balance between the theory necessary to gain insight into mechanics and design methods. I therefore attempt to stress those aspects of theory and application that prepare a student for more advanced study or professional practice in design. Above all, I made an effort to provide a visual interpretation of the equations and present the material in a form useful to a diverse audience. The analysis presented should facilitate the use of computers and programmable calculators. The commonality of the analytical methods needed to design a wide variety of elements and the use of computer-aided engineering as an approach to the design are emphasized.

Mechanical Design: An Integrated Approach provides unlimited opportunities for the use of computer graphics. Computer solutions are usually preferred, because evaluation of the design changes, and "what-if" analysis requires only a few keystrokes. Hence, many examples, case studies, and problems in the text are solved with the aid of a computer. Generally, solid modeling serves as a design tool that can be used to create finite element (F.E.) models for analysis and dynamic simulation. Instructors may use a simple PC-based F.E. program to give the students exposure to the method applied to stress concentration and axisymmetrically loaded and plane stress problems. The website for the text (see Supplements) allows the user to treat problems more realistically, shows tabular and graphical trends, and demonstrates the elements of good computational practice. The text is *independent* of any software package.

Traditional analysis in design, based on the methods of mechanics of materials, is given full treatment. In some instances, the methods of the applied theory of elasticity are employed. The role of the theory of elasticity in this book is threefold: It places limitations on the application of the mechanics of materials theory, it is used as the basis of finite element formulation, and it provides exact solutions when configurations of loading and component shape are simple. Plates, shells, and structural members are discussed to enable the readers to solve real-life problems and understand interactive case studies. Website addresses of component and equipment manufacturers and open-ended web problems are given in numerous chapters to provide the reader access to additional information on those topics. Also presented is finite element analysis (FEA) in computer-aided design. The foregoing unified methods of analysis give the reader opportunity to expand his or her ability to perform design process in a more realistic setting. The book attempts to fill what I believe to be a void in the world of textbooks.

AN INTEGRATED APPROACH

There are two major parts in this text. The fundamentals of loading, stress, strain, materials, deflection, stiffness, buckling, fracture mechanics, failure criteria, and fatigue phenomena are treated first. This is followed by applications to machine, miscellaneous mechanical, and structural components. Both parts attempt to provide an integrated approach that links together a variety of topics by means of case studies. Some chapters and sections in the text are also carefully integrated through cross referencing. Throughout the book, case studies provide numerous machine or component projects. They mostly present different aspects of the same design or analysis problem in successive chapters. For instance, defining the loading in a device in Chapter 1; calculating the stresses, deflections, or factor of safety, due to the loading throughout in Part I; and presenting the design contents in Part II. Case Study 1-1 on the winch crane is one such example.

Attention is given to the presentation of the fundamentals and necessary emprical information required to formulate design problems. Important principles and applications are illustrated with numerical examples, and a broad range of practical problems is provided for solution by the student. This volume offers numerous worked out examples and case studies, aspects of which are presented in several sections of the book; more than 500 problem sets, most are drawn from engineering practice; and a multitude of formulas and tabulations from which design calculations can be made. Most problems can be readily modified for in-class tests. References are listed at the end of each chapter. Answers to selected problems are given at the end of the book.

A sign convention consistent with vector mechanics is used throughout for loads, internal forces (with the exception of the shear in beams), and stresses. This convention has been carefully chosen to conform to that used in most classical mechanics of materials, elasticity, and engineering design texts as well as to that most often employed in the numerical analysis of complex machines and structures. Both the international system of units (SI) and the U.S. customary system of units are used; but since in practice the former is replacing the latter, this book places a greater emphasis on SI units.

TEXT ARRANGEMENT

A glance at the table of contents shows the topics covered and the way in which they are organized. Because of the extensive subdivision into a variety of topics and use of alternative design and analysis methods, the text should provide flexibility in the choice of assignments to cover courses of varying length and content. A discussion of the design process and an overview of the material included in the book is given in Sections 1.1

through 1.4. Most chapters are substantially self-contained. Hence, the order of presentation can be smoothly altered to meet an instructor's preference. It is suggested, however, that Chapters 1 and 2 should be studied first. The sections marked with an asterisk (*) deal with special or advanced topics. These are optional for a basic course in design and can be deleted without destroying the continuity of the book.

This text attempts to provide synthesis and analysis that cuts through the clutter and saves readers time. Every effort has been made to eliminate errors. I hope I've maintained a clarity of presentation, as much simplicity as the subject permits, unpretentious depth, an effort to encourage intuitive understanding, and a shunning of the irrelevant. In this context, emphasis is placed on the use of fundamentals to build students' understanding and ability to solve more complex problems throughout.

MEDIA SUPPLEMENTS

The book is accompanied by a comprehensive **Online Learning Center** website (OLC) for instructors and students. The OLC includes: Book, author, and errata pages; password protected text problem solutions for instructors; PowerPoint slides of book illustrations; MATLAB examples and simulations for machine design (by Sid Wang); "FEPC" a 2-D finite element program students can use to do basic finite element explorations (by Charles Knight). An accompanying Finite Element Primer on the site explains the use of FEPC, and specific applications of FEA to mechanical design; interactive Fundamentals of Engineering (FE) exam questions for machine design (by Edward Anderson); and chapter overviews for students.

SUPPLEMENTS

A **Solutions Manual** is available upon request to instructors in printed form (as well as on th eOLC website). Written and class tested by the author, it features complete solutions to problems in the text.

ACKNOWLEDGMENTS

To acknowledge everyone who contributed to this book in some manner is clearly impossible. A major debt is, however, owed to reviewers who offered constructive suggestions and made detailed comments. These include the following: D. Beale, Auburn University; D. M. McStravick, Rice University; T. R. Grimm, Michigan Technological University; R. E. Dippery, Kettering University; Yuen-Cjen Yong, California Polytechnic University–San Luis Obispo; A. Shih, North Carolina State University; J. D. Gibson, Rose Hulman Institute of Technology; R. Paasch, Oregon State University; J. P. H. Steele, Colorado School of Mines; C. Nuckolls, University of Central Florida; D. Logan, University of Wisconsin–Platteville; E. Conley, New Mexico State University; L. Dabaghian, California State University–Sacramento; E. R. Mijares, California State University–Long Beach; T. Kozik, Texas A&M University; C. Crane, University of Florida; B. Bahram, North Carolina State University; A. Mishra, Auburn University; M. Ramasubramanian, North Carolina State University; S. Yurgartis, Clarkson University; M. Corley, Louisiana Tech.

University; R. Rowlands, University of Wisconsin; B. Hyman, University of Washington; and G. R. Pennock, Purdue University. P. Brackin, Rose Hulman Institute of Technology checked the accuracy of all problems and examples. E. Kenyon checked the accuracy of solutions to chapter problems, case studies, and selected examples. H. Kountorous and R. Sodhi, New Jersey Institute of Technology, read Chapter 1 and offered valuable perspectives on some case studies based upon student design projects. To all of the foregoing colleagues I am pleased to express my gratitude for their invaluable attention and advice.

Editors Jonathan Plant and Amy Hill made extraordinary efforts in developing the book. Also, production was managed skillfully by Jill Peter and handled efficiently by the staff of McGraw-Hill and Interactive Composition Corporation. Thanks are due for their professional help. Final checking of problems and examples in the text, and typing of several drafts of the Solutions Manual were done expertly by my former student, Dr. Youngjin Chung; I am grateful to him for his hard work. Lastly, I deeply appreciate the understanding and encouragement of my wife, Nora, daughter, Aileen, and son, Errol, who also assisted with computer work during preparation of the text.

Ansel C. Ugural
Holmdel, N.J.

Abbreviations

all	allowable
avg	average
Bhn	Brinell hardness number
CD	cold drawn
CCW	counterclockwise
cr	critical
CW	clockwise
ft	foot, feet
fpm	foot per minute
HD	hard drawn
hp	horsepower
hr	hour
H.T.	heat treated
Hz	hertz (cycles per second)
ID	inside diameter
in.	inch, inches
ipm	inch per minute
ips	inch per second
J	Joule
kip	kilopound (1000 lb)
kips	kilopounds
kg	kilogram(s)
ksi	kips per square inch (10^3 psi)
kW	kilowatt
log	common logarithm (base 10)
lb	pound(s)
ln	Naperian natural logarithm

max	maximum
m	meter
min	minimum
mph	miles per hour
m/s	meter per second
N	newton
N.A.	neutral axis
OD	outside diameter
OQ&T	oil quenched and tempered
OT	oil tempered
Pa	pascal
psi	pounds per square inch
Q&T	quenched and tempered
R_C	Rockwell hardness, C scale
rad	radian
req	required
res	residual
rpm	revolutions per minute
rps	revolutions per second
s	second
SI	system of international units
st	static
SUS	Saybolt universal seconds
SUV	Saybolt universal viscosity
VI	Viscosity index
W	watt
WQ&T	water quenched and tempered

Symbols

See Sections 11.2, 11.4, 11.9, 11.11, 12.3, 12.5, 12.6, 12.8, and 12.9 for some gearing symbols.

ROMAN LETTERS

A	amplitude ratio, area, coefficient, cross-sectional area
A_f	final cross-sectional area
A_o	original cross-sectional area
A_e	effective area of clamped parts, projected area
A_t	tensile stress area, tensile stress area of the thread
a	acceleration, crack depth, distance, radius, radius of contact area of two spheres
B	coefficient
b	distance, width of beam, band, or belt; radius
C	basic dynamic load rating, bolted-joint constant, centroid, constant, heat transfer coefficient, specific heat, spring index
C_c	limiting value of column slenderness ratio
C_f	surface finish factor
C_r	reliability factor
C_s	basic static load rating, size factor
c	distance from neutral axis to the extreme fiber, radial clearance, center distance
D	diameter, mean coil diameter, plate flexural rigidity $[Et^3/12(1-\nu^2)]$
d	diameter, distance, pitch diameter, wire diameter
d_{avg}	average diameter
d_c	collar (or bearing) diameter
d_m	mean diameter
d_p	pitch diameter
d_r	root diameter
E	modulus of elasticity
E_k	kinetic energy
E_b	modulus of elasticity for the bolt
E_p	modulus of elasticity for clamped parts, potential energy
e	dilatation, distance, eccentricity, efficiency

F	force, tension
F_a	axial force, actuating force
F_b	bolt axial force
F_c	centrifugal force
F_d	dynamic load
F_i	initial tensile force or preload
F_n	normal force
F_p	clamping force for the parts, proof load
F_r	radial force
F_t	tangential force
F_u	ultimate force
f	coefficient of friction, frequency
f_c	collar (or bearing) coefficient of friction
f_n	natural frequency
G	modulus of rigidity
g	acceleration due to gravity
H	time rate of heat dissipation, power
H_B	Brinell hardness number (Bhn)
H_V	Vickers hardness number
h	cone height, distance, section depth, height of fall, weld size, film thickness
h_f	final length, free length
h_0	minimum film thickness
h_s	solid height
I	moment of inertia
I_e	equivalent moment of inertia of the spring coil
J	polar moment of inertia, factor
K	bulk modulus of elasticity, constant, impact factor, stress intensity factor, system stiffness
K_f	fatigue stress concentration factor
K_c	fracture toughness
K_r	a life adjustment factor
K_s	service factor, shock factor, direct shear factor for the helical spring
K_t	theoretical or geometric stress concentration factor
K_w	Wahl factor
k	buckling load factor for the plate, constant, element stiffness, spring index or stiffness

k_b	stiffness for the bolt	R	radius, reaction force, reliability, stress ratio
k_p	stiffness for the clamped parts	R_B	Rockwell hardness in B scale
		R_C	Rockwell hardness in C scale
L	grip, length, lead	r	aspect ratio of the plate, radial distance, radius, radius of gyration
L_e	equivalent length of the column		
L_f	final length	r_{avg}	average radius
L_o	original length	r_i	inner radius
L_{10}	rating life	r_o	outer radius
L_5	rating life for reliability greater than 90%		
l	direction cosine, length	S	section modulus, Saybolt viscometer measurement in seconds, Sommerfeld number, strength
M	moment	S_e	endurance limit of mechanical part
M_a	alternating moment	S_e'	endurance limit of specimen
M_f	moment of friction forces	S_n'	endurance strength of specimen
M_m	mean moment	S_{es}	endurance limit in shear
M_n	moment of normal forces	S_n	endurance strength of mechanical part
m	direction cosine, mass, module, mass	S_f	fracture strength
		S_p	proof strength, proportional limit strength
N	normal force, number of friction planes, number of teeth, fatigue life or cycles to failure	S_y	yield strength in tension
		S_{ys}	yield strength in shear
N_a	number of active spring coils	S_u	ultimate strength in tension
N_{cr}	critical load of the plate	S_{uc}	ultimate strength in compression
N_t	total number of spring coils	S_{us}	ultimate strength in shear
N_θ	hoop force	s	distance, sample standard deviation
N_ϕ	meridianal force		
n	constant, direction cosine, factor of safety, modular ratio, number, number of threads, rotational speed	T	temperature, tension, torque
		T_a	alternating torque
		T_d	torque to lower the load
n_{cr}	critical rotational speed	T_m	mean torque
		T_f	friction torque
P	force, concentrated load, axial load, equivalent radial load for a roller bearing, radial load per unit projected area	T_o	torque of overhauling
		T_t	transition temperature
		T_u	torque to lift the load
P_a	alternating load	t	temperature, distance, thickness, time
P_{all}	allowable load	t_a	temperature of surrounding air
P_{cr}	critical load of the column or helical spring	t_o	average oil film temperature
P_m	mean load	U	strain energy, journal surface velocity
p	pitch, pressure, probability	U_o	strain energy density
p_{all}	allowable pressure	U_{ov}	dilatational strain energy density
p_i	internal pressure	U_{od}	distortional strain energy density
p_o	outside or external pressure	U_r	modulus of resilience
p_0	maximum contact pressure	U_t	modulus of toughness
p_{max}	maximum pressure	U^*	complementary energy
p_{min}	minimum pressure	U_o^*	complementary energy density
$p(x)$	probability or frequency function	u	radial displacement, fluid flow velocity
Q	first moment of area, imaginary force, volume, flow rate	V	linear velocity, a rotational factor, shear force, volume
Q_s	side leakage rate	V_s	sliding velocity
q	notch sensitivity factor, shear flow	v	displacement, linear velocity

W	work, load, weight
w	distance, unit load, deflection, displacement
X	a radial factor
Y	Lewis form factor based on diametral pitch or module, a thrust factor
y	distance from the neutral axis, Lewis form factor based on circular pitch, quantity
$\bar{y}$	distance locating the neutral axis
Z	curved beam factor, section modulus
z	number of standard deviations

GREEK LETTERS

α	angle, angular acceleration, coefficient, coefficient of thermal expansion, cone angle, form factor for shear, thread angle
α_n	thread angle measured in the normal plane
β	angle, coefficient, half-included angle of the V belt
γ	included angle of the disk clutch or brake, pitch angle of the sprocket, shear strain, weight per unit volume; γ_{xy}, γ_{yz}, and γ_{xz} are shear strains in the xy, yz, and xz planes
γ_{max}	maximum shear strain
Δ	gap, material parameter in computing contact stress
δ	deflection, displacement, elongation, radial interference or shrinking allowance, a virtual infinitesimally small quantity
δ_{max}	maximum or dynamic deflection
δ_s	solid deflection
δ_{st}	static deflection
δ_w	working deflection
ϵ	eccentricity ratio
ε	normal strain; ε_x, ε_y, and ε_z are normal strains in the x, y, and z directions
ε_f	normal strain at fracture
ε_t	true normal strain
ε_u	ultimate strain
η	absolute viscosity or viscosity

θ	angle, angular displacement, slope
θ_p	angle to a principal plane or to a principal axis
θ_s	angle to a plane of maximum shear
λ	lead angle, helix angle, material constant
μ	population mean
ν	kinematic viscosity, Poisson's ratio
Π	potential energy function
ρ	mass density
σ	normal stress; σ_x, σ_y, and σ_z are normal stresses in the x, y, and z planes, standard deviation
σ_a	alternating stress
σ_{all}	allowable stress
σ_{cr}	critical stress
σ_e	equivalent stress
σ_{ea}	equivalent alternating stress
σ_{em}	equivalent mean stress
σ_{max}	maximum normal stress
σ_{min}	minimum normal stress
σ_{nom}	nominal stress
σ_{oct}	octahedral normal stress
σ_{res}	residual stress
τ	shear stress; τ_{xy}, τ_{yz}, and τ_{xz} are shear stresses perpendicular to the x, y, and z axes and parallel to the y, z, and x axes
τ_{avg}	average shear stress
τ_{all}	allowable shear stress
τ_d	direct shear stress
τ_{oct}	octahedral shear stress
τ_{max}	maximum shear stress
τ_{min}	minimum shear stress
τ_{nom}	nominal shear stress
τ_t	torsional shear stress
ϕ	angle, angle giving the position of minimum film thickness, pressure angle, angle of twist, angle of wrap
ϕ_{max}	position of maximum film pressure
ψ	helix angle, spiral angle
ω	angular velocity, angular frequency ($\omega = 2\pi f$)
ω_n	natural angular frequency

PART

I

FUNDAMENTALS

Outline

INTRODUCTION TO DESIGN

Outline

1.1   SCOPE OF TREATMENT

As an applied science, engineering uses scientific knowledge to achieve a specific objective. The mechanism by which a requirement is converted to a meaningful and functional plan is called a *design*. The design is an innovative, iterative, and decision-making process. This book deals with the analysis and design of *machine elements* and basic *structural members* that compose the system or assembly. Typical truss, frame, plate, and shell-like structures also are considered. The purpose and scope of this text may be summarized as follows: It presents a body of knowledge that will be useful in component design for performance, strength, and durability; provides treatments of "design to meet strength requirements" of members and other aspects of design involving prediction of the displacements and buckling of a given component under prescribed loading; presents classical and numerical methods amenable to electronic digital computers for the analysis and design of members and structural assemblies; presents many examples, case studies, and problems of various types to provide an opportunity for the reader to develop competence and confidence in applying the available design formulas and deriving new equations as required.

The text consists of two parts. Part I focuses on fundamental principles and methods, a synthesis of stress analysis and materials engineering, that forms the cornerstone of the subject and has to be studied carefully. It begins with a discussion of basic concepts in design and analysis and definitions relating to properties of a variety of engineering materials. Detailed equilibrium and energy methods of analysis for determining stresses and deformations in variously loaded members, design of bars and beams, buckling, failure criteria, and reliability are presented in this part. A thorough grasp of these topics will prove of great value in attacking new and complex problems. Part II is devoted mostly to mechanical component design. The fundamentals are applied to specific machine elements such as shafts, bearings, gears, belts, chains, clutches, brakes, and springs and typical design situations that arise in the selection and application of these members and others. Power screws; threaded fasteners; bolted, riveted, and welded connections; adhesive bonding; and axisymmetrically loaded components are also considered in some detail. In conclusion, introductory finite element analysis in design is covered.

The full understanding of both terminology in statics and principles of mechanics is an essential prerequisite to the analysis and design of machines and structures. Design methods for members are founded on the methods of mechanics of materials; and the theory of elasticity is used or referred to in design of certain elements. The objective of this chapter is to provide the reader the basic definitions and process of the design, load analysis, and the concepts of solid mechanics in a condensed form. Selected references provide readily available sources where additional analysis and design information can be obtained.

1.2   ENGINEERING DESIGN

Design is the formulation of a plan to satisfy a particular need, real or imaginary. *Engineering design* can be defined as the process of applying science and engineering methods to prescribe a component or a system in sufficient detail to permit its realization. A system constitutes several different elements arranged to work together as a whole. Design is thus the essence, art, and intent of engineering. *Design function* refers to the process in which mathematics, computers, and graphics are used to produce a plan.

Mechanical design means the design of components and systems of a mechanical nature—machines, structures, devices, and instruments. For the most part, mechanical design utilizes the stress analysis methods and materials engineering. A *machine* is an apparatus consisting of interrelated elements or a device that modifies force or motion. *Machine design* is the art of planning or devising new or improved machines to accomplish specific purpose. Although *structural design* is most directly associated with civil engineering, it interacts with any engineering discipline that requires a structural system or member. As noted earlier, the topic of machine design is the main focus of this text.

The ultimate goal in a mechanical design process is to size and shape the elements and choose appropriate materials and manufacturing processes so that the resulting system can be expected to perform its intended function without failure. An *optimum design* is the best solution to a design problem within prescribed constraints. Of course, such a design depends on a seemingly limitless number of variables. When faced with many possible choices, a designer may make various design decisions based on experience, reducing the problem to that with one or few variables.

Generally, it is assumed that a good design meets performance, aesthetics, and cost goals. Another attribute of a good design is robustness, a resistance to quality loss or deviation from desired performance. Knowledge from the entire engineering curricula goes into formulating a good design. Communications is as significant as technology. Basically, the means of communication are in written, oral, and graphical forms. The first fundamental canon in the *Code of Ethics for Engineers* [1] states that, "Engineers shall hold paramount the safety, health, and welfare of the public in the performance of their professional duties." Therefore, engineers must design products that are safe during their intended use for the life of the products. Product safety implies that the product will protect humans from injury, prevent property damage, and prevent harm to the environment.

A plan for satisfying a need often includes preparation of individual preliminary design. Each *preliminary design* involves a thorough consideration of the loads and actions that the structure or machine has to support. For each case, a mechanical analysis is necessary. *Design decisions,* or choosing reasonable values of the factors, is important in the design process. As a designer gains more experience, decisions are reached more readily.

1.3 THE DESIGN PROCESS

The *process* of *design* is basically an exercise in creativity. The complete process may be outlined by design flow diagrams with feedback loops [2–12]. Figure 1.1 shows some aspects of such a diagram. In this section, we discuss the *phases of design* common to all disciplines in the field of engineering design. Most engineering designs involve safety, ecological, and societal considerations. It is a challenge to the engineer to recognize all of these in proper proportion. Fundamental actions proposed for the design process are establishing need as a design problem to be solved, understanding the problem, generating and evaluating possible solutions, and deciding on the best solution.

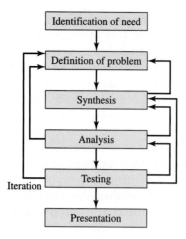

Figure 1.1   Design process.

PHASES OF DESIGN

The design process is independent of the product and is based on the concept of product life cycle. To understand fully all that must be considered in the process of design, here we explain the characteristics of each phase of Figure 1.1. Note that, the process is neither exhaustive nor rigid and will probably be modified to suit individual problems.

Identification of Need

The design process begins with a recognition of a need, real or imagined, and a decision to do something about it. For example, present equipment may require improving durability, efficiency, weight, speed, or cost. New equipment may be needed to perform an automated function, such as computation, assembly, or servicing. The identification aspect of design can have origin in any number of sources. Customer reports on the product function and quality may force a redesign. Business and industrial competition constantly force the need for new or improved apparatus, processes, and machinery designs. Numerous other sources of needs give rise to contemporary design problems.

Definition of the Problem

This phase in design conceives the mechanisms and arrangements that will perform the needed function. For this, a broad knowledge of members is desirable, because new equipment ordinarily consists of new members, perhaps with changes in size and material. All specifications, that is, all forms of input and output quantities, must be carefully spelled out. Often, this area is also labeled *design and performance requirements*. The specifications also include the definitions of the member to be manufactured, the cost, the range of the operating temperature, expected life, and the reliability. A *standard* is a set of specifications for parts, materials, or processes intended to achieve uniformity, efficiency, and a specified quality. A *code* is a set of specifications for the analysis, design, manufacture, and construction of something. The purpose of a code is to achieve a specified degree of safety, efficiency, and performance or quality. All organizations and technical societies (listed in Section 1.6) have established specifications for standards and safety or design codes.

Once the specifications have been prepared, relevant design information is collected to make a *feasibility study*. The purpose of this study is to verify the possible success or failure of a proposal both from the technical and economic standpoints. Frequently, as a result of this study, changes are made in the specifications and requirements of the project. The designer often considers the engineering feasibility of various alternative proposals. When some idea as to the amount of space needed or available for a project has been determined, to-scale layout drawings may be started.

Synthesis

The synthesis (putting together) of the solution represents perhaps the most challenging and interesting part of design. Frequently termed the *ideation and invention phase,* it is where the largest possible number of creative solutions is originated. The philosophy, functionality, and uniqueness of the product are determined during synthesis. In this step, the designer combines separate parts to form a complex whole of various new and old ideas and concepts to produce an overall new idea or concept.

Analysis

Synthesis and analysis are the main stages that constitute the design process. Analysis has as its objective satisfactory performance, as well as durability with minimum weight and competitive cost. Synthesis cannot take place without both analysis or resolution and optimization, because the product under design must be analyzed to determine whether the performance complies with the specifications. If the design fails, the synthesis procedure must begin again. After synthesizing several components of a system, we analyze what effect this has on the remaining parts of the system. It is now necessary to draw the layouts, providing details, and make the supporting calculations that will ultimately result in a prototype design. The designer must specify the dimensions, select the components and materials, and consider the manufacturing, cost, reliability, serviceability, and safety.

Testing

At this juncture, the working design is first fabricated as a *prototype*. Product evaluation is the final proof of a successful design and usually involves testing a prototype in a laboratory or on a computer that provides the analysis database. More often computer prototypes are utilized because they are less expensive and faster to generate. By evaluation we discover whether the design really satisfies the need and other desirable features. Subsequent to many *iterations* (i.e., repetitions or returns to a previous state), the process ends with the vital step of communicating the design to others.

Presentation

The designer must be able to understand the need and describe a design graphically, verbally, and in writing. This is the presentation of the plans for satisfying the need. A successful presentation is of utmost importance as the final step in the design process. Drawings are utilized to produce blueprints to be passed to the manufacturing process. A number of references are available on the design process for those seeking a more-thorough discussion [2, 3].

It is interesting to note that individual parts should be designed to be easily fabricated, assembled, and constructed. The goal of the *manufacturing process* is to construct the designed component or system. The process planning attempts to determine the most effective sequence to produce the component. The produced parts are inspected and must pass certain quality control or assurance requirements. Components surviving inspection are assembled, packaged, labeled, and shipped to customers. The features of a product that attract consumers and how the product is presented to the marketplace are significant functions in the success of a product. Marketing is a crucial last stage of the manufacturing process. Market feedback is very important in enhancing products. These feedback loops are usually incorporated into the first stage of a design process [13]. Many disciplines are involved in product development. Therefore, design engineers need to be familiar with other disciplines, at least from a communications standpoint, to integrate them into the design process.

1.4   DESIGN ANALYSIS

The objective of the design analysis is, of course, to attempt to predict the stress or deformation in the component so that it may safely carry the loads that will be imposed on it. The analysis begins with an attempt to put the conceptual design in the context of the abstracted engineering sciences to evaluate the performance of the expected product. This constitutes design modeling and simulation.

THE ENGINEERING MODELING

Geometric modeling is the method of choice for obtaining the data necessary for failure analysis early in design process. Creating a useful engineering model of a design is probably the most difficult, challenging part of the whole process. It is the responsibility of the designer to ensure the adequacy of a chosen geometric model to a particular design. If structure is simple enough, theoretical solutions for basic configurations may be adequate for obtaining the stresses involved. For more complicated structures, finite-element models not only can estimate the stresses but also utilize them to evaluate the failure criteria for each element in a member.

We note that the geometric model chosen and subsequent calculations made merely approximate reality. Assumptions and limitations, such as linearity and material homogeneity, are used in developing the model. The choice of a geometric model depends directly on the kind of analysis to be performed. Design testing and evaluation may require changing the geometric model before finalizing it. When the final design is achieved, the drafting and detailing of the models start, followed by documentation and production of final drawings.

RATIONAL DESIGN PROCEDURE

The rational design procedure to meet the *strength requirements* of a load-carrying member attempts to take the results of fundamental tests, such as tension, compression, and fatigue, and apply them to all complicated and involved situations encountered in present-day structures and machines. However, not all topics in design have a firm analytical base from which to work. In those cases, we must depend on a semi-rational or empirical approach to solving a problem or selecting a design component. In addition, details related to actual service loads and various factors, discussed in Section 8.7, have a marked influence on the strength and useful life of a component. The static design of axially loaded members, beams, and torsion bars are treated by the rational procedure in Chapters 3 and 9. Suffice it to say that complete design solutions are not unique, and often trial and error is required to find the best solution.

METHODS OF ANALYSIS

Design methods are based on the mechanics of materials theory generally used in this text. Axisymmetrically loaded mechanical components are analyzed by methods of the elasticity theory in Chapter 16. The former approach employs assumptions based on experimental evidence along with engineering experience to make a reasonable solution of the practical problem possible. The latter approach concerns itself largely with more mathematical analysis of the "exact" stress distribution on a loaded body [14–16]. The difference between the two methods of analysis is further discussed at the end of Section 3.18.

Note that solutions based on the mechanics of materials give average stresses at a section. Since, at concentrated forces and abrupt changes in cross section, irregular local stresses (and strains) arise, only at distance about equal to the depth of the member from such disturbances are the stresses in agreement with the mechanics of materials. This is due to *Saint-Venant's Principle:* The stress of a member at points away from points of load application may be obtained on the basis of a statically equivalent loading system; that is, the manner of force application on stresses is significant only in the vicinity of the region where the force is applied. This is also valid for the disturbances caused by the changes in the cross section. The mechanics of materials approach is therefore best suited for relatively slender members.

The complete analysis of a given component subjected to prescribed loads by the method of equilibrium requires consideration of three conditions. These *basic principles of analysis* can be summarized as follows:

1. *Statics.* The equations of equilibrium must be satisfied.
2. *Deformations.* Stress-strain or force deformation relations (e.g., Hooke's law) must apply to the behavior of the material.
3. *Geometry.* The conditions of compatibility of deformations must be satisfied; that is, each deformed part of the member must fit together with adjacent parts.

Solutions based on these requirements must satisfy the boundary conditions. Note that it is not always necessary to execute the analysis in this exact order. Applications of the foregoing procedure are illustrated in the problems involving mechanical components as the subject unfolds. Alternatively, stress and deformation can also be analyzed using the energy methods. The roles of both methods are twofold. They can provide solutions of acceptable accuracy, where configurations of loading and member are regular, and they can be employed as a basis of the numerical methods, for more complex problems.

1.5 PROBLEM FORMULATION AND COMPUTATION

The discussion of Section 1.3 shows that synthesis and analysis are the "two faces" of design. They are opposites but symbiotic. These are the phases of the mechanical design process addressed in this book. Most examples, case studies, and problems are set up so the identification of need, specifications, and feasibility phases already have been defined. As noted previously, this text is concerned with the fundamentals involved and mostly with the application to specific mechanical components. The machine and structural members chosen are widely used and will be somewhat familiar to the reader. The emphasis in treating these components is on the methods and procedures used.

Solving Mechanical Component Problems

Ever-increasing industrial demand for more sophisticated machines and structures calls for a good grasp of the concepts of analysis and design and a notable degree of ingenuity. Fundamentally, design is the process of problem solving. It is very important to formulate a mechanical element problem and its solution accurately. This requires consideration of physical and its related mathematical situations. The reader may find the following format helpful in problem formulation and solution:

1. Definition of the problem by stating the given quantities and appropriate assumptions.
2. Preliminary design decisions and sketches.
3. Mathematical models, detailed design analysis, and evaluation.
4. Documentation of the results and comment about the solution.

We illustrate most of these steps in the solution of the sample problems throughout the text.

This book provides the student the ideas and information necessary for understanding the mechanical analysis and design and encourages the creative process based on that understanding. It is significant that the reader visualize the nature of the quantities being computed. Complete, carefully drawn, free-body diagrams facilitate visualizations and we provide these, knowing that the subject matter can be mastered best by solving practical problems. It should also be pointed out that the relatively simple form of many equations usually results from simplifying assumptions made with respect to the deformation and load patterns in their derivation. Designers must be aware of such restrictions.

Computational Tools for Design Problems

A wide variety of computational tools can be used to perform design calculations with success. A high-quality scientific calculator may be the best tool for solving most of the problems in this book. General purpose analysis tools such as spreadsheets and equation solvers have particular merit for certain computational tasks. These mathematical software packages include MATLAB, TK Solver, and MathCAD. The tools have the advantage of allowing the user to document and save completed work in a detailed form. Computer-aided design (CAD) software may be used throughout the design process [13, 17–22], but, it supports the analysis stages of design more than conceptual phases.

The computer-aided drafting software packages can produce realistic three-dimensional representations of a member or solid models. The CAD software allows the designer to visualize without costly models, iterations, or prototypes. Most CAD systems provide an interface to one or more finite element analysis (FEA) or boundary element analysis (BEA) programs. They permit direct transfer of the model's geometry to an FEA or BEA package for analysis of stress and vibration as well as fluid and thermal analysis. However, usually these analyses of design problems require the use of special purpose programs. The finite element analysis techniques are briefly discussed in Chapter 17.

As noted earlier, the Online Learning Center website available with the text contains a simplified, two-dimensional finite element analysis software program ("FEPC"), and MATLAB simulations for mechanical design. The computer-based software may be used as a tool to assist students with design projects and lengthy homework assignments. However, computer output providing analysis results must not be accepted on faith alone; the designer must always check computer solutions. It is necessary that fundamentals of analysis and design be thoroughly understood.

1.6 FACTOR OF SAFETY AND DESIGN CODES

It is sometimes difficult to determine accurately the various factors involved in the phases of design of machines and structures. An important area of uncertainty is related to the assumptions made in the stress and deformation analysis. An equally significant item is the nature of failure. If failure is caused by ductile yielding, the consequences are likely to be less severe than if caused by brittle fracture. In addition, a design must take into account such matters as the following: types of service loads; variations in the properties of the material; whether failure is gradual or sudden; the consequences of failure (minor damage or catastrophe); and human safety and economics.

DEFINITIONS

Engineers employ a safety factor to ensure against the foregoing unknown uncertainties involving strength and loading. This factor is used to provide assurance that the load applied to a member does not exceed the largest load it can carry. The factor of safety, n, is the ratio of the maximum load that produces failure of the member to the load allowed under service conditions:

$$n = \frac{\text{failure load}}{\text{allowable load}} \tag{1.1}$$

The allowable load is also referred to as the *service load* or *working load*. The preceding represents the basic definition of the factor of safety. This ratio must always be greater than unity, $n > 1$. Since the allowable service load is a known quantity, the usual design procedure is to multiply this by the safety factor to obtain the failure load. Then, the member is designed so that it can just sustain the maximum load at failure.

A common method of design is to use a safety factor with respect to strength of the member. In most situations, a linear relationship exists between the load and the stress produced by the load. Then, the factor of safety may also be defined as

$$n = \frac{\text{material strength}}{\text{allowable stress}} \tag{1.2}$$

In this equation, the materials strength represents either static or dynamic properties. Obviously, if loading is static, the material strength is either the yield strength or the ultimate strength. For fatigue loading, the material strength is based on the endurance limit, discussed in Chapter 8. The allowable stress is also called the *applied stress, working stress,* or *design stress*. It represents the required strength. The foregoing definitions of the factor of safety are used for all types of member and loading conditions (e.g., axial, bending, shear). Inasmuch as there may be more than one potential mode of failure for any component, we can have more than one value for the factor of safety. The smallest value of n for any member is of the greatest concern, because this predicts the most likely mode of failure.

SELECTION OF A FACTOR OF SAFETY

Modern engineering design gives a rational accounting for all factors possible, leaving relatively few items of uncertainty to be covered by a factor of safety. The following numerical values of factor of safety are presented as a guide. They are abstracted from a list by J. P. Vidosic [23]. These safety factors are based on the yield strength S_y or endurance limit S_e of a *ductile material*. When they are used with a *brittle material* and the ultimate strength S_u, the factors must be approximately doubled.

1. $n = 1.25$ to 1.5 for exceptionally reliable materials used under controllable conditions and subjected to loads and stresses that can be determined with certainty. Used almost invariably where low weight is a particularly important consideration.

2. $n = 1.5$ to 2 for well-known materials under reasonably constant environmental conditions, subjected to loads and stresses that can be determined readily.

3. $n = 2$ to 2.5 for average materials operated in ordinary environments and subjected to loads and stresses that can be determined.

4. $n = 2.5$ to 4 for less-tried (or 3 to 4 for untried) materials under average conditions of environment, load, and stress.

5. $n = 3$ to 4 also for better-known materials used in uncertain environments or subjected to uncertain stresses.

Where higher factors of safety might appear desirable, a more-thorough analysis of the problem should be undertaken before deciding on their use. In the field of aeronautical engineering, in which it is necessary to reduce the weight of the structures as much as possible, the term factor of safety is replaced by the *margin of safety:*

$$\frac{\text{ultimate load}}{\text{design load}} - 1$$

In the nuclear reactor industries the safety factor is of prime importance in the face of many unknown effects and hence the factor of safety may be as high as 5. The value of factor of safety is selected by the designer on the basis of experience and judgment.

The simplicity of Eqs. (1.1) and (1.2) sometimes mask their importance. A large number of problems requiring their use occur in practice. The employment of a factor of safety in design is a reliable, time-proven approach. When properly applied, sound and safe designs are obtained. We note that the factor of safety method to safe design is based on rules of thumb, experience, and testing. In this approach, the strengths used are always the *minimum* expected values. A concept closely related to safety factor is termed *reliability*. It is the statistical measure of the probability that a member will not fail in use. In the reliability method of design, the goal is to achieve a reasonable likelihood of survival under the loading conditions during the intended design life. For this purpose, mean strength and load distributions are determined and then these two are related to achieve an acceptable safety margin. Reliability is discussed in Chapter 7.

DESIGN AND SAFETY CODES

Numerous engineering societies and organizations publish standards and codes for specific areas of engineering design. Most are merely recommendations, but some have the force of law. For the majority of applications, relevant factors of safety are found in various construction and manufacturing codes; for instance, the ASME Pressure Vessel Codes. Factors of safety are usually embodied into computer programs for the design of specific members. Building codes are legislated throughout this country and often deal with publicly accessible structures (e.g., elevators and escalators). Underwriters Laboratories (UL) have developed its standards for testing consumer products. When a product passes their tests, it may be labeled *listed UL*. States and local towns have codes as well, relating mostly to fire prevention and building standards.

It is clear that, where human safety is involved, high values of safety factor are justified. However, members should not be overdesigned to the point of making them unnecessarily costly, heavy, bulky, or wasteful of resources. The designer and stress analyst must be apprehensive of the codes and standards, lest their work lead to inadequacies.

The following is a partial list of societies and organizations* that have established specifications for standards and safety or design codes.

AA	Aluminum Association
AFBMA	Anti-Friction Bearing Manufacturing Association
AGMA	American Gear Manufacturing Association
AIAA	American Institute of Aeronautics and Astronautics
AISC	American Institute of Steel Construction
AISI	American Iron and Steel Institute
ANSI	American National Standards Institute
API	American Petroleum Institute
ASCE	American Society of Civil Engineers
ASLE	American Society of Lubrication Engineers
ASM	American Society of Metals
ASME	American Society of Mechanical Engineers
ASTM	American Society for Testing and Materials
AWS	American Welding Society
NASA	National Aeronautics and Space Administration
NIST	National Institute for Standards and Technology
IFI	Industrial Fasteners Institute
ISO	International Standards Organization
SAE	Society of Automotive Engineers
SEM	Society for Experimental Mechanics
SESA	Society for Experimental Stress Analysis
SPE	Society of Plastic Engineers

*The addresses and data on their publications can be obtained in any technical library or from a designated website; for example, for specific titles of ANSI standards, see www.ansi.org.

1.7   UNITS AND CONVERSION

The units of the physical quantities employed in engineering calculations are of major significance. The most recent universal system is the International System of Units (SI). The U.S. customary units have long been used by engineers in this country. Both systems of units, reviewed briefly here, are used in this text. However, greater emphasis is placed on the SI units, in line with international conventions. Some of the fundamental quantities in SI and the U.S. customary systems of units are listed in Table 1.1. For further details, see, for example, Reference 24.

We observe from the table that, in SI, force F is a derived quantity (obtained by multiplying the mass m by the acceleration a, in accordance with Newton's second law, $F = ma$). However, in the U.S. customary system, the situation is reversed, with mass being the derived quantity. It is found from Newton's second law, as $lb \cdot s^2/ft$, sometimes called the *slug*. Ordinarily, temperature in SI is the degree Celsius (°C). The temperature is expressed in U.S. units by the degree Fahrenheit (°F). Conversion formula between the temperature scales is given by

$$t_c = \frac{5}{9}(t_f - 32) \tag{1.3}$$

where t designates the temperature.

It is sufficiently accurate to assume that the acceleration of gravity, denoted by g, near earth's surface equals

$$g = 9.81 \text{ m/s}^2 \quad (\text{or } 32.2 \text{ ft/s}^2)$$

From Newton's second law it follows that, in SI, the weight W of a body of mass 1 kg is $W = mg = (1 \text{ kg})(9.81 \text{ m/s}^2) = 9.81 \text{ N}$. In the U.S. customary system, the weight is expressed in pounds (lb).

Tables A.1 and A.2 furnish conversion factors and SI prefixes in common usage. The use of prefixes avoids unusually large or small numbers. Note that a dot is to be used to separate units that are multiplied together. Thus, for instance, a newton meter is written $N \cdot m$ and must not be confused with mN, which stands for millinewtons.

Table 1.1   Basic Units

	SI Unit		U.S. Unit	
Quantity	**Name**	**Symbol**	**Name**	**Symbol**
Length	meter	m	foot	ft
Force*	newton	N*	pound force	lb
Time	second	s	second	s
Mass	kilogram	kg	slug	$lb \cdot s^2/ft$
Temperature	degree Celsius	°C	degree Fahrenheit	°F

*Derived unit ($kg \cdot m/s^2$).

1.8 LOAD CLASSIFICATION AND EQUILIBRIUM

External forces, or loads acting on a structure or member, may be classified as surface forces and body forces. A surface force acts at a point or is distributed over a finite area. Body forces are distributed throughout the volume of a member. All forces acting on a body, including the reactive forces caused by supports and the body forces are considered as external forces. Internal forces are the forces holding together the particles forming the member.

A *static load* is applied slowly, gradually increasing from 0 to its maximum value, and thereafter remaining constant. Thus, a static load can be a stationary (that is, unchanging in magnitude, point of application, and direction) force, torque, moment, or a combination of these acting on a member. In contrast, *dynamic loads* may be applied very suddenly, causing vibration of structure, or they may change in magnitude with time. As observed earlier, in SI, force is expressed in newtons (N). But, because the newton is a small quantity, the kilonewton (kN) is often used in practice. The unit of force in the U.S. customary system is pounds (lb) or kilopounds (kips).

CONDITIONS OF EQUILIBRIUM

When a system of forces acting on a body has zero resultant, the body is said to be in equilibrium. Consider the equilibrium of a body in space. The conditions of equilibrium require that the following *equations of statics* need be satisfied

$$\sum F_x = 0 \qquad \sum F_y = 0 \qquad \sum F_z = 0$$
$$\sum M_x = 0 \qquad \sum M_y = 0 \qquad \sum M_z = 0 \tag{1.4}$$

If the forces act on a body in equilibrium in a single (xy) plane, a planar problem, the most common form of the static equilibrium equations are

$$\sum F_x = 0 \qquad \sum F_y = 0 \qquad \sum M_z = 0 \tag{1.5a}$$

By replacing either or both force summations by equivalent moment summations in Eqs. (1.5a), two *alternate* sets of equations can be obtained [15].

When bodies are accelerated, that is, the magnitude or direction of their velocity changes, it is necessary to use Newton's second law to relate the motion of the body with the forces acting on it. The *plane motion* of a body, symmetrical with respect to a plane (xy) and rotating about an axis (z), is defined by

$$\sum F_x = ma_x \qquad \sum F_y = ma_y \qquad \sum M_z = I\alpha \tag{1.5b}$$

in which m is the mass and I the principal centroidal mass moment of inertia about the z axis. The quantities a_x, a_y, and α represent the linear and angular accelerations of mass center about the principal x, y, and z axes, respectively. The preceding relationships express that the system of external forces is equivalent to the system consisting of the inertia forces $(ma_x$ and $ma_y)$ attached at the mass center and the couple moment $I\alpha$. Equations (1.5b) can be written for all the connected members in a two-dimensional system and an entire set solved simultaneously for forces and moments [25, 26].

A structure or system is said to be *statically determinate* if all forces on its members can be obtained by using only the equilibrium conditions; otherwise, the structure is referred to as *statically indeterminate*. The degree of static indeterminacy is equal to the difference between the number of unknown forces and the number of pertinent equilibrium equations. Since any reaction in excess of those that can be found by statics alone is called *redundant;* the number of redundants is the same as the degree of indeterminacy. To effectively study a structure, it is usually necessary to make simplifying idealizations of the structure or the nature of the loads acting on the structure. These permit the construction of a *free-body diagram:* a sketch of the isolated body and all external forces acting on it. When internal forces are of concern, an imaginary cut through the body at the section of interest is displayed, as illustrated in the next section.

MODES OF LOAD TRANSMISSION

Distributed forces within a member can be represented by statically equivalent internal forces, so-called stress-resultants or load resultants. Usually, they are exposed by an imaginary cutting plane containing the centroid C through the member and resolved into components normal and tangential to the cut section. This process of dividing the body into two parts is called the *method of sections*. Figure 1.2a shows only the isolated left part of a slender member. A bar whose least dimension is less than about $\frac{1}{10}$ its length may usually be considered a *slender* member. Note that the sense of moments follows the right-hand screw rule and, for convenience, is often represented by double-headed vectors. In three-dimensional problems, the four modes of load transmission are axial force P (also denoted F or N), shear forces V_y and V_z, torque or twisting moment T, and bending moments M_y and M_z.

In planar problems, we find only three components acting across a section: the axial force P, the shear force V, and the bending moment M (Figure 1.2b). The cross-sectional face, or *plane,* is defined as positive when its outward normal points in a positive coordinate direction and as negative when its *outward* normal points in the negative coordinate direction. According to Newton's third law, the forces and moments acting on the faces at a cut section are equal and opposite. The location in a plane where the largest internal force resultants develops and failure is most likely to occur is called the *critical section*.

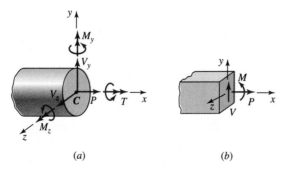

Figure 1.2   Internal forces and moments: (a) the general case; (b) the two-dimensional case.

SIGN CONVENTION

When both the outer normal and the internal force or moment vector component point in a positive (or negative) coordinate direction, the force or moment is defined as positive. Therefore, Figure 1.2 depicts positive internal force and moment components. However, it is common practice for the direction shown in the figure to represent a negative internal shear force. In this text, we use a sign convention for *shear force* in a beam that is contrary to the definition given in Figure 1.2 (see Section 3.6). Note also that the sense of the *reaction* at a support of a structure is arbitrarily assumed; the positive (negative) sign of the answer obtained by the equations of statics will indicate that the assumption is correct (incorrect).

1.9   LOAD ANALYSIS

A *structure* is a unit composed of interconnected members supported in a manner capable of resisting applied forces in static equilibrium. The constituents of such units or systems are bars, beams, plates, and shells, or their combinations. An extensive variety of structures are used in many fields of engineering [27–29]. Structures can be considered in four broad categories: frames, trusses, machines, and thin-walled structures. Adoption of thin-walled structure behavior allows certain simplifying assumptions to be made in the structural analysis (see Section 4.10). The American Society of Civil Engineers (ASCE) lists design loads for buildings and other common structures [30].

Here, we consider load analysis dealing with the assemblies or structures made of several connected members. A *frame* is a structure that always contain at least one multiforce member; that is, a member acted on by three or more forces, which generally are not directed along the member. A *truss* is a special case of a frame, in which all forces are directed along the axis of a member. Machines are similar to frames in that of the elements may be multiforce members. However, as noted earlier, a *machine* is designed to transmit and modify forces (or energy) and always contains moving parts.

The approach used in the load analysis of a pin-jointed structure may be summarized as follows. First, consider the entire structure as a free body and write the equations of static equilibrium. Then, dismember the structure and identify the various members as either two-force (axially loaded) members or multiforce members. Pins are taken to form an integral part of one of the members they connect. Draw the free-body diagram of each member. Clearly, when two-force members are connected to the same member, they are acted on by that member with equal and opposite forces of unknown magnitude but known direction. Finally, the equilibrium equations obtained from the free-body diagrams of the members may be solved to yield various internal forces.

EXAMPLE 1.1 | Determination of Member Forces in a Pin-Connected Frame

The assembly shown in Figure 1.3a, which carries a load of 30 kN, consists of two beams *ABCD* and *CEF*, and one bar *BE*, connected by pins; it is supported by a pin at *A* and a cable *DG*. The dimensions are in meters. Calculate

(a) The components of the forces acting on each member.

(b) The axial force, shear force, and moment acting on the cross section at point *G*.

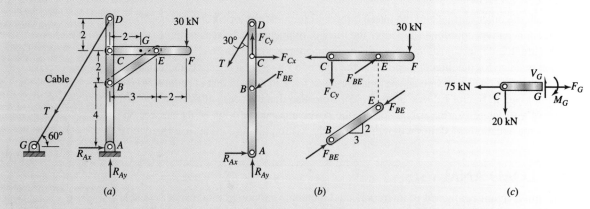

Figure 1.3   Example 1.1. (a) Structural assembly; (b) dismembered structure; (c) portion CG of beam CEF.

Assumptions: Friction forces in the pin joints will be omitted. All forces are coplanar and two dimensional.

Solution: There are two components R_{Ax} and R_{Ay} of the reaction at A and the force T exerted by the cable at D. Therefore, we can compute the reactions by considering the free body of the entire frame (Figure 1.3a):

$$\sum M_A = 30(5) - T\sin 30°(8) = 0 \qquad T = 37.5 \text{ kN}$$

$$\sum F_x = R_{Ax} - 37.5\sin 30° = 0 \qquad R_{Ax} = 18.75 \text{ kN}$$

$$\sum F_y = R_{Ay} - 37.5\cos 30° - 30 = 0 \qquad R_{Ay} = 62.48 \text{ kN}$$

(a) The frame is now dismembered, since only two members are connected at each joint, equal and opposite components or resultants are shown on each member at each joint (Figure 1.3b). We note that BE is a two-force member. Member CEF:

$$\sum M_C = 30(5) - \frac{2}{\sqrt{13}} F_{BE}(3) = 0 \qquad F_{BE} = 90.14 \text{ kN}$$

$$\sum M_E = 30(2) - F_{Cy}(3) = 0 \qquad F_{Cy} = 20 \text{ kN}$$

$$\sum F_x = \frac{3}{\sqrt{13}}(90.14) - F_{Cx} = 0 \qquad F_{Cx} = 75 \text{ kN}$$

Member ABCD: All internal forces have been found. To check the results, using equations of statics we verify that the member $ABCD$ is in equilibrium.

Comment: The positive values obtained means that the directions shown for the force components are correct.

(b) Member *CEF* is cut at point *G*. Choosing the free body of segment *CG* (Figure 1.3c), we have

$$M_G = 20(2) = 40 \text{ kN} \cdot \text{m}, \qquad F_G = 75 \text{ kN}, \qquad V_G = 20 \text{ kN}.$$

Comment: The internal forces at *G* are equivalent to a couple, an axial force, and a shear, acting as shown in the figure.

1.10 CASE STUDIES

A general case study in a component or structure design may go step by step through the request for proposal process, the trade study and configuration development, the detailed design process and construction, and prototype tests. Therefore, a case study presents a product in action. It covers "lessons learned" during the development of a device, such as customer requirements, economics, technology, engineering and manufacturing relationships, goals, and a continuous improvement program. Clearly, a case study brings out derivatives superior to the product currently built [31, 32].

Case studies introduced in select chapters of this text involve situations found in engineering practice. Most of these provide various *mechanical design projects:* the assemblies containing a variety of elements, such as links under combined axial and bending loads; machinery for lifting and moving loads; shafts subjected to bending and torsion simultaneously; gearsets and bearings subject to fluctuating loads; high-speed cutters; winch crane hooks; compression springs, structural connections, and pressure vessels. These case studies show how the design of any one component may be invariably affected by the design of related parts.

A total of 12 case studies are presented throughout the book. We note that, because of space limitations, only certain phases of these studies are briefly discussed. The common website addresses are given in select chapters in the book to provide the reader access to further information on material properties, manufacturer, and description of various mechanical components. The numerical examples to follow are limited to two- and three-dimensional loading situations. A winch crane and a bolt cutter are the systems analyzed.

Case Study 1-1 | WINCH CRANE FRAME LOADING ANALYSIS

A winch crane supported by solid plastic wheels is used for lifting and moving loads in the laboratory or machine shop (Figure 1.4). It has a capacity to lift a load of *P*. The concrete counterbalance weight W_C on the base prevents the crane from tipping forward when it is pushed by a horizontal force *F* at a height *H* from the ground. For safety purposes, the drive system includes a torque limiter coupled to a drum and allows the crane to lift no more than the load *P*. Determine

(a) The design load on the front and rear wheels.

(b) The factor of safety n_t for the crane tipping forward from the loading.

(continued)

Case Study   (CONTINUED)

Figure 1.4   Winch crane.

Given: The geometry of each element is known. The cable and hook are rated at 15 kN, which gives a safety factor of 5. The 85-mm diameter drum is about 20 times the cable diameter. The crane frame carries the load P, counterweight W_C, weights of parts W_i ($i = 1, 2, 3, 4, 5$), and the push force F, as shown in Figure 1.5. The frame is made of $b = 50$ mm and $h = 100$ mm structural steel tubing of $t = 6$ mm thickness with weight w newtons per meter (Table A.4).

Data:

$$P = 3 \text{ kN} \qquad F = 100 \text{ kN} \qquad W_C = 2.7 \text{ kN}$$
$$w \approx 130 \text{ N/m} \qquad a = 0.8 \text{ m} \qquad H = 1 \text{ m}$$
$$L_1 = 1.5 \text{ m} \qquad L_2 = 2 \text{ m} \qquad L_3 = 1 \text{ m}$$
$$L_4 = 0.5 \text{ m} \qquad L_5 = 0.65 \text{ m}$$

Case Study   (CONTINUED)

Figure 1.5   Simple sketch of the winch crane frame shown in Figure 1.4.

and

$$W_1 = wL_1 = 130(1.5) = 195 \text{ N} \qquad W_2 \approx wL_2 = 260 \text{ N}$$

$$W_3 = 130 \text{ N} \qquad W_4 = 65 \text{ N} \qquad W_5 = 84.5 \text{ N}$$

For dimensions and properties of a selected range of frequently used crane members, refer to manufacturers' catalogs.

Assumptions:

1. A line speed of 0.12 meter per second is used, as suggested by several catalogs for lifting. The efficiency of the speed reduction unit or gear box is 95%. The electric motor has 0.5 hp capacity to lift 3 kN load for the preceding line speed and efficiency and includes an internal brake to hold the load when it is inoperative. The gear ratios (see Case Study 11-1) satisfy the drive system requirements.

2. Only the weights of concrete counterbalance and main frame parts are considered. All frame parts are weld connected to one another.

3. Compression forces caused by the cable running along the members is ignored. All forces are static; F is x directed (horizontal) and remaining forces are parallel to the xy plane. Note that the horizontal component of the reaction at B equals $F/2$, not indicated in Figure 1.5.

(continued)

Case Study   (CONCLUDED)

Solution: See Figure 1.5.

(a) Reactional forces R_A and R_B acting on the wheels are determined by applying conditions of equilibrium, $\sum M_z = 0$ at B and $\sum F_y = 0$, to the free-body diagram shown in the figure with $F = 0$. Therefore,

$$R_A = \frac{1}{2}\left(P\frac{L_1}{L_3} + \frac{1}{2}W_1\frac{L_1}{L_3} + \frac{1}{4}W_C + W_3 + \frac{1}{2}W_5\right)$$

$$R_B = -R_A + \frac{1}{2}P + \frac{1}{2}W_1 + W_2 + W_3$$
$$+ \frac{1}{2}W_4 + \frac{1}{2}W_C + \frac{1}{2}W_5 \tag{1.6}$$

Substitution of the given data into the foregoing results in

$$R_A = \frac{1}{2}\left[3000(1.5) + \frac{1}{2}(195)(1.5) + \frac{1}{4}(2700)\right.$$
$$\left. + 130 + \frac{1}{2}(84.5)\right] = 2747 \text{ N}$$

$$R_B = -2747 + \frac{1}{2}(3000) + \frac{1}{2}(195) + 260 + 130$$
$$+ \frac{1}{2}(65) + \frac{1}{2}(2700) + \frac{1}{2}(84.5) = 665 \text{ N}$$

Note that, when the crane is unloaded ($P = 0$), Eqs. (1.6) give

$$R_A = 497 \text{ N} \qquad R_B = 1415 \text{ N}$$

Comment: Design loads on front and rear wheels are 2747 N and 1415 N, respectively.

(b) The factor of safety n_t is applied to tipping loads. The condition $\sum M_z = 0$ at point A:

$$n_t[P(L_1 - L_3) + FH]$$
$$= W_1\left(L_3 - \frac{1}{2}L_1\right) + 2W_2L_3$$
$$+ W_4L_3 + \frac{3}{4}W_C L_3 + \frac{1}{2}(2W_3 + W_5)L_3 \tag{1.7}$$

Introducing given numerical values,

$$n_t[3000(0.5) + 100(1)]$$
$$= 195(0.25) + 2(260)(1) + 65(1)$$
$$+ \frac{3}{4}(2700)(1) + \frac{1}{2}[2(130) + 84.5](1)$$

from which $n_t = 1.77$.

Comments: For the preceding forward tipping analysis, the rear wheels are assumed to be locked and the friction is taken to be sufficiently high to prevent sliding. Side-to-side tipping may be checked similarly.

Case Study 1-2 | BOLT CUTTER LOADING ANALYSIS

Figure 1.6 shows a pin-connected tool in the closed position in the process of gripping its jaws into a bolt. The user provides the input loads between the handles, indicated as the reaction pairs P. Determine the force exerted on the bolt and the pins at joints A, B, and C.

Given: The geometry is known. The data are

$$P = 2 \text{ lb,} \qquad a = 1 \text{ in.,} \qquad b = 3 \text{ in.,}$$
$$c = \tfrac{1}{2} \text{ in.,} \qquad d = 8 \text{ in.,} \qquad e = 1 \text{ in.}$$

Assumptions: Friction forces in the pin joints are omitted. All forces are coplanar, two dimensional, and

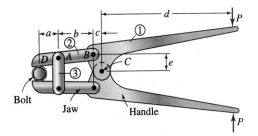

Figure 1.6   Bolt cutter.

Case Study (CONCLUDED)

static. The weights of members are neglected as being insignificant compared to the applied forces.

Solution: The equilibrium conditions are fulfilled by the entire cutter. Let the force between the bolt and the jaw be Q, whose direction is taken to be normal to the surface at contact (point D). Due to the symmetry, only two free-body diagrams shown in Figure 1.7 need be considered. Inasmuch as link 3 is a two-force member, the orientation of force F_A is known. Note also that the force components on the two elements at joint B must be equal and opposite, as shown on the diagrams.

Conditions of equilibrium are applied to Figure 1.7a to give $F_{Bx} = 0$ and

$$\sum F_y = Q - F_A + F_{By} = 0 \qquad F_A = Q + F_{By}$$

$$\sum M_B = Q(4) - F_A(3) = 0 \qquad F_A = \frac{4Q}{3}$$

from which $Q = 3F_{By}$. In a like manner, referring to Figure 1.7b, we obtain

$$\sum F_y = -F_{By} + F_{Cy} - 2 = 0 \qquad F_{Cy} = \frac{Q}{3} + 2$$

$$\sum M_C = F_{Bx}(1) + F_{By}(0.5)$$
$$- 2(8) = 0 \qquad F_{By} = 32 \text{ lb}$$

and $F_{Cx} = 0$. Solving $Q = 3(32) = 96$ lb. The shear forces on the pins at the joints A, B, and C are

$$F_A = 128 \text{ lb}, \quad F_B = F_{By} = 32 \text{ lb}, \quad F_C = F_{Cy} = 34 \text{ lb}$$

Comments: Observe that the high mechanical advantage of the tool transforms the applied load to a large force exerted on the bolt at point D. The handles and jaws are under combined bending and shear forces. Stresses and deflections of the members are taken up in Case Studies 3-2 and 4-2.

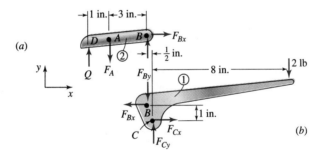

Figure 1.7 Free-body diagrams of bolt cutter shown in Figure 1.6: (a) jaw; (b) handle.

1.11 WORK AND ENERGY

This section provides a brief introduction to the method of work and energy, which is particularly useful in solving problems dealing with buckling design and components subjected to combined loading. All machines or mechanisms consisting of several connected members involve loads and motion that, in combination, represent work and energy. The concept of *work* in mechanics is presented as the product of the magnitudes of the force and displacement vectors and the cosine of the angle between them [25]. The work W done by a constant force F moving through a displacement s in the direction of force can be expressed as

$$W = Fs \tag{1.8}$$

Similarly, the work of a couple of forces or torque T during a rotation θ of the member, such as the wheel, is given by

$$W = T\theta \tag{1.9}$$

The work done by a force, torque, or moment can be regarded as a transfer of energy to the member. In general, the work is stored in a member as potential energy, kinetic energy, internal energy, or any combination of these or dissipated as heat energy. The magnitude of the energy a given component can store is sometimes a significant consideration in mechanical design. Members, when subjected to impact loads, are often chosen based on their capacity to absorb energy. *Kinetic energy E_k* of a member represents the capacity to do work associated with the speed of the member. The kinetic energy of a component in rotational motion may be written

$$E_k = \frac{1}{2}I\omega^2 \tag{1.10}$$

The quantity I is the mass moment of inertia and ω represents the angular velocity or speed. Table A.5 lists mass moments of inertia of common shapes. The work of the force is equal to the change in kinetic energy of the member. This is known as the *principle of work and energy*. Under the action of conservative forces, the sum of the kinetic energy of the member remains constant.

The units of work and energy in SI is the newton meter (N · m), called the joule (J). In the U.S. customary system, work is expressed in foot pounds (ft · lb) and British thermal units (Btu). The unit of energy is the same as that of work. The quantities given in either unit system can be converted quickly to the other system by means of the conversion factors listed in Table A.1. Specific facets are associated with work, energy, and power, as will be illustrated in the analysis and design of various components in the chapters to follow.

EXAMPLE 1.2 | Camshaft Torque Requirement

Figure 1.8   Example 1.2. Camshaft and follower.

A rotating camshaft (Figure 1.8) of an intermittent motion mechanism moves the follower in a direction at right angles to the cam axis. For the position shown, the follower is being moved upward by the lobe of the cam with a force F. A rotation of θ corresponds to a follower motion of s. Determine the average torque T required to turn the camshaft during this interval.

Given: $F = 0.2 \text{ lb},$ $\theta = 8° = 0.14 \text{ rad},$ $s = 0.05 \text{ in}.$

Assumptions: The torque can be considered to be constant during the rotation. The friction forces can be omitted.

Solution: The work done on the camshaft equals to the work done by the follower. Therefore, by Eqs. (1.8) and (1.9), we write

$$T\theta = Fs \tag{1.11}$$

Substituting the given numerical values,

$$T(0.14) = 0.2(0.05) = 0.01 \text{ lb} \cdot \text{in}$$

The foregoing gives $T = 0.071 \text{ lb} \cdot \text{in}$.

Comment: The stress and deflection caused by force F at the contact surface between the cam and follower is considered in Chapter 3.

1.12 POWER

Power is defined as the time rate at which work is done. Note that, in selecting a motor or engine, power is a much more significant criterion than the actual amount of work to be performed. When work involves a force, the rate of energy transfer is the product of the force F and the velocity V at the point of application of the force:

$$\text{Power} = FV \tag{1.12}$$

In the case of a member, such as a shaft rotating with an angular velocity or speed ω in radians per unit time and acted on by a torque T, we have

$$\text{Power} = T\omega \tag{1.13}$$

The mechanical *efficiency,* designated by e, of a machine may be defined as follows:

$$e = \frac{\text{power output}}{\text{power input}} \tag{1.14}$$

Because of energy loses due to friction, the power output is always smaller than the power input. Therefore, machine efficiency is always less than 1. Inasmuch as power is defined as the time rate of doing work, it can be expressed in units of energy and time. Hence, the unit of power in SI is the watt (W), defined as the joule per second (J/s). If U.S. customary units are used, the power should be measured in ft · lb/s or in horsepower (hp).

TRANSMISSION OF POWER BY ROTATING SHAFTS AND WHEELS

The power transmitted by a rotating machine component such as a shaft, flywheel, gear, pulley, or clutch is of keen interest in the study of machines. Consider a circular shaft or disk of radius r subjected to a constant tangential force F. Then, the torque is expressed as $T = Fr$. The velocity at the point of application of the force is V. A relationship between the power, speed, and the torque acting through the shaft is readily found, from first principles, as follows.

In SI units the power transmitted by a shaft is measured by the kilowatt (kW), where 1 kW equals 1000 watts. One watt does the work of one newton-meter per second. The speed n is expressed in revolutions per minute, then the angle through which the shaft rotates equals $2\pi n$ rad/min. Thus, the work done per unit time is $2\pi nT$. This is equal to the power delivered: $2\pi nT/60 = 2\pi nFr/60 = kW(1000)$. Since $V = 2\pi rn/60$, the foregoing may be written as $FV = kW(1000)$. For convenience, power transmitted may be expressed in two forms:

$$kW = \frac{FV}{1000} = \frac{Tn}{9549} \qquad (1.15)$$

where

$T =$ torque (N · m)

$n =$ shaft speed (rpm)

$F =$ tangential force (N)

$V =$ velocity (m/s)

We have one horsepower (hp) equals 0.7457 kW, and the preceding equation may be written

$$hp = \frac{FV}{745.7} = \frac{Tn}{7121} \qquad (1.16)$$

In U.S. customary units, horsepower is defined as a work rate of $550 \times 60 = 33,000$ ft · lb/m. An equation similar to that preceding can be obtained:

$$hp = \frac{FV}{33,000} = \frac{Tn}{63,000} \qquad (1.17)$$

Here, we have $T =$ torque in lb · in., $n =$ shaft speed in rpm, $F =$ tangential force in lb, $V =$ velocity in fpm.

EXAMPLE 1.3 | Power Capacity of Punch Press Flywheel

A high-strength steel flywheel of outer and inner rim diameters d_o and d_i, and length in axial direction of l, rotates at a speed of n (Figure 1.9). It is to be used to punch metal during two-thirds of a revolution of the flywheel. What is the average power available?

(a)   (b)

Figure 1.9   Example 1.3. (a) Punch press flywheel and (b) its cross section.

Given: $d_o = 0.5$ m, $\quad n = 1000$ rpm, $\quad \rho = 7860$ kg/m^3, $\quad$ (Table B.1)

Assumptions:

1. Friction losses are negligible.
2. Flywheel proportions are $d_i = 0.75d_o$ and $l = 0.18d_o$.
3. The inertia contributed by the hub and spokes is omitted: the Flywheel is considered as a rotating ring free to expand.

Solution: Through the use of Eqs. (1.9) and (1.10), we obtain

$$T\theta = \frac{1}{2}I\omega^2 \tag{1.18}$$

where

$$\theta = \frac{2}{3}(2\pi) = 4\pi/3 \text{ rad}$$

$$\omega = 1000(2\pi/60) = 104.7 \text{ rad/s}$$

$$I = \frac{\pi}{32}\left(d_o^4 - d_i^4\right)l\rho \qquad \text{(Table A.5)}$$

$$= \frac{\pi}{32}[(0.5)^4 - (0.375)^4](0.09)(7860) = 2.967 \text{ kg} \cdot \text{m}^2$$

Introducing the given data into Eq. (1.18) and solving, $T = 7765$ N $\cdot$ m. Equation (1.15) is therefore

$$\text{kW} = \frac{Tn}{9549} = \frac{7765(1000)}{9549}$$

$$= 813.2$$

Comment: The braking torque required to stop a similar disk in two-third revolution would have an average value of 7.77 kN $\cdot$ m (see Section 16.5).

1.13 STRESS COMPONENTS

Stress is a concept of paramount importance to a comprehension of solid mechanics. It permits the mechanical behavior of load-carrying members to be described in terms essential to the analyst and designer. Applications of this concept to typical members are discussed in Chapter 3. Consider a member in equilibrium, subject to external forces. Under the action of these forces, internal forces and hence stresses are developed between the parts of the body [15]. In SI units, the stress is measured in newtons per square meter (N/m^2) or pascals. Since the pascal is very small quantity, the megapascal (MPa) is commonly used. Typical prefixes of the SI units are given in Table A.2. When the U.S. customary system is used, stress is expressed in pounds per square inch (psi) or kips per square inch (ksi).

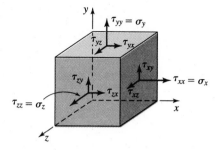

Figure 1.10 Element in three-dimensional stress. (Only stresses acting on the positive faces are shown.)

The three-dimensional state of stress at a point, using three mutually perpendicular planes of a cubic element isolated from a member, can be described by nine *stress components* (Figure 1.10). Note that only three (positive) faces of the cube are actually visible in the figure, and that oppositely directed stresses act on the hidden (negative) faces. Here the stresses are considered to be identical on the mutually parallel faces and uniformly distributed on each face. The general state of stress at a point can be assembled in the form

$$\begin{bmatrix} \tau_{xx} & \tau_{xy} & \tau_{xz} \\ \tau_{yx} & \tau_{yy} & \tau_{yz} \\ \tau_{zx} & \tau_{zy} & \tau_{zz} \end{bmatrix} = \begin{bmatrix} \sigma_x & \tau_{xy} & \tau_{xz} \\ \tau_{yx} & \sigma_y & \tau_{yz} \\ \tau_{zx} & \tau_{zy} & \sigma_z \end{bmatrix} \tag{1.19}$$

This is a matrix presentation of the *stress tensor*. It is a second-rank tensor requiring two indices to identify its elements. (A vector is a tensor of first rank; a scalar is of zero rank.) The double subscript notation is explained as follows: The first subscript denotes the direction of a normal to the face on which the stress component acts; the second designates the direction of the stress. Repetitive subscripts are avoided in this text. Therefore, the *normal stresses* are designated σ_x, σ_y, and σ_z, as shown in Eq. (1.19). In Section 3.17, it is demonstrated rigorously for the *shear stresses* that $\tau_{xy} = \tau_{yx}$, $\tau_{yz} = \tau_{zy}$, and $\tau_{xz} = \tau_{zx}$.

SIGN CONVENTION

When a stress component acts on a positive plane (Figure 1.10) in a positive coordinate direction, the stress component is *positive*. Also a stress component is considered positive when it acts on a negative face in the negative coordinate direction. A stress component is considered negative when it acts on a positive face in a negative coordinate direction (or vice versa). Hence, tensile stresses are always positive and compressive stresses are always negative. The sign convention can also be stated as follows: A stress component is positive if both the outward normal of the plane on which it acts and its direction are in coordinate directions of the same sign; otherwise, it is negative. Figure 1.10 depicts a system of positive normal and shearing stresses. This sign convention for stress, which agrees with that adopted for internal forces and moments, is used throughout the text.

SPECIAL CASES OF STATE OF STRESS

The general state of stress reduces to simpler states of stress commonly encountered in practice. An element subjected to normal stresses σ_1, σ_2, and σ_3, acting in mutually

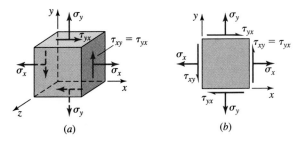

Figure 1.11 (a) Element in plane stress; (b) two-dimensional presentation of plane stress.

perpendicular directions alone with respect to a particular set of coordinates, is said to be in a state of *triaxial stress*. Such a stress can be represented as

$$\begin{bmatrix} \sigma_1 & 0 & 0 \\ 0 & \sigma_2 & 0 \\ 0 & 0 & \sigma_3 \end{bmatrix}$$

The absence of shearing stresses indicates that these stresses are the principal stresses for the element (Section 3.16).

In the case of two-dimensional or *plane stress* only the x and y faces of the element are subjected to stresses ($\sigma_x, \sigma_y, \tau_{xy}$) and all the stresses act parallel to the x and y axes, as shown in Figure 1.11a. Although the three-dimensional aspect of the stress element should not be forgotten, for the sake of convenience, we usually draw only a two-dimensional view of the plane stress element (Figure 1.11b). A thin plate loaded uniformly over the thickness, parallel to the plane of the plate, exemplifies the case of plane stress. When only two normal stresses are present, the state of stress is called *biaxial*.

In *pure shear,* the element is subjected to plane shear stresses acting on the four side faces only; for example, $\sigma_x = \sigma_y = 0$ and τ_{xy} (Figure 1.11b). Typical pure shear occurs over the cross sections and on longitudinal planes of a circular shaft subjected to torsion. Examples include axles and drive shafts in machinery, propeller shafts (Chapter 9), drill rods, torsional pendulums, screwdrivers, steering rods, and torsion bars (Chapter 14). If only one normal stress exists, the one-dimensional stress is referred to as a *uniaxial* tensile or compressive stress.

1.14 NORMAL AND SHEAR STRAINS

In the preceding section, our concern was with the stress within a loaded member. We now turn to deformation caused by the loading, the analysis of which is as important as that of stress. The analysis of deformation requires the description of the concept of strain, that is, the intensity of deformation. As a result of deformation, extension, contraction, or change of shape of a member may occur. To obtain the actual stress distribution within a member, it is necessary to understand the type of deformation occurring in that member. Only small displacements, commonly found in engineering structures, are considered in this text.

Figure 1.12   (a) Deformation of a bar;
(b) distortion of a rectangular plate.

The strains resulting from small deformations are small compared with unity, and their products (higher-order terms) are neglected. The preceding assumption leads to one of the fundamentals of solid mechanics, the *principle of superposition* that applies whenever the quantity (deformation or stress) to be obtained is directly proportional to the applied loads. It permits a complex loading to be replaced by two or more simpler loads and thus renders a problem more amenable to solution, as will be observed repeatedly in the text.

The fundamental concept of normal strain is illustrated by considering the deformation of the homogenous prismatic bar shown in Figure 1.12a. A prismatic bar is a straight bar having constant cross-sectional area throughout its length. The initial length of the member is L. Subsequent to application of the load, the total deformation is δ. Defining the normal strain ε as the unit change in length, we obtain

$$\varepsilon = \frac{\delta}{L}$$

(1.20)

A positive sign designates elongation; a negative sign, contraction. The foregoing state of strain is called *uniaxial strain*. When an unconstrained member undergoes a temperature change ΔT, its dimensions change and a normal strain develops. The uniform thermal strain for a homogeneous and isotropic material is expressed as

$$\varepsilon_t = \alpha \Delta T$$

(1.21)

The coefficient of expansion α is approximately constant over a moderate temperature change. It represents a quantity per degree Celsius (1/°C) when ΔT is measured in °C.

Shear strain is the tangent of the total change in angle taking place between two perpendicular lines in a member during deformation. Inasmuch as the displacements considered are small, we can set the tangent of the angle of distortion equal to the angle. Thus, for a rectangular plate of unit thickness (Figure 1.12b), the shear strain γ measured in radians is defined as

$$\gamma = \frac{\pi}{2} - \beta$$

(1.22)

Here β is the angle between the two rotated edges. The shear strain is positive if the right angle between the reference lines decreases, as shown in the figure; otherwise, the shearing strain is negative. Because normal strain ε is the ratio of the two lengths, it is a dimensionless quantity. The same conclusion applies to shear strain. Strains are also often measured in terms of units mm/mm, in/in, and radians or microradians. For most engineering materials, strains rarely exceed values of 0.002 or 2000 μ in the elastic range. We read this as "2000 micros."

REFERENCES

1. American Society of Mechanical Engineers. *Code of Ethics for Engineers*. New York: ASME, 1997.
2. Ullman, D. G. *The Mechanical Design Process*, 2nd ed. New York: McGraw-Hill, 2002.
3. Nevins, J. L., and D. E. Whitney, eds. *Concurrent Design Products and Processes*. New York: McGraw-Hill, 1989.
4. Shigley, J. E., and C. R. Mischke. *Mechanical Engineering Design,* 6th ed. New York: McGraw-Hill, 2001.
5. Juvinall, R. C., and K. M. Marshek. *Fundamentals of Machine Component Design,* 3rd ed. New York: Wiley, 2000.
6. Norton, R. L. *Machine Design: An Integrated Approach,* 2nd ed. Upper Saddle River, NJ: Prentice Hall, 2000.
7. Spotts, M. F., and T. E. Shoup. *Design of Machine Elements,* 7th ed. Upper Saddle River, NJ: Prentice Hall, 1998.
8. Deutcshman, A. D., W. J. Michels, and W. J. Wilson. *Machine Design Theory and Practice*. New York: Macmillan, 1975.
9. Mott, R. L. *Machine Elements in Mechanical Design*. New York: Macmillan, 1992.
10. Hamrock, B. J., B. Jacobson, and S. R. Schmid. *Fundamentals of Machine Elements*. New York: McGraw-Hill, 1997.
11. Burr, A. H., and J. B. Cheatham. *Mechanical Analysis and Design,* 2nd ed. Upper Saddle River, NJ: Prentice Hall, 1995.
12. Rothbart, H. A., ed. *Mechanical Design and Systems Handbook,* 2nd ed. New York: McGraw-Hill, 1985.
13. Zeid, I. *CAD/CAM Theory and Practice*. New York: McGraw-Hill, 1991.
14. Ugural, A. C., and S. K. Fenster. *Advanced Strength and Applied Elasticity,* 4th ed. Upper Saddle River, NJ: Prentice Hall, 2003.
15. Ugural, A. C. *Mechanics of Materials*. New York: McGraw-Hill, 1991.
16. Timoshenko, S. P., and N. Goodier. *Theory of Elasticity,* 3rd ed. New York: McGraw-Hill, 1970.
17. Wilson, C. E. *Computer Integrated Machine Design*. Upper Saddle River, NJ: Prentice Hall, 1997.
18. Akin, J. E. *Computer-Aided Mechanical Design*. St. Paul, MN: West Publishing, 1989.
19. Pao, Y. C. *Elements of Computer-Aided Design and Manufacturing*. New York: Wiley, 1984.
20. Dimarogonas, A. D. *Computer-Aided Machine Design*. Upper Saddle River, NJ: Prentice Hall, 1988.
21. Dimarogonas, A. D. *Machine Design: A CAD Approach*. New York: Wiley, 2001.
22. Bhonsle, S. B., and K. J. Weinmann. *Mechanical Modeling for Design of Machine Components (TK Integrated)*. Upper Saddle River, NJ: Prentice Hall, 1999.
23. Vidosic, J. P. *Machine Design Projects*. New York: Ronald Press, 1957.
24. *ASTM Standard for Metric Practice*. Publication E 380.86. West Conshohocken, PA: American Society for Testing and Materials, 1916.
25. Beer, E. P., and E. R. Johnston, Jr. *Vector Mechanics for Engineers: Dynamics,* 6th ed. New York: McGraw-Hill, 2000.

26. Norton, R. L. *Design of Machinery,* 3rd ed. New York: McGraw-Hill, 2004.
27. West, H. H. *Fundamentals of Structural Analysis,* 2nd ed. New York: Wiley, 2002.
28. Abraham, L. H. *Structural Design of Missiles and Spacecraft.* New York: McGraw-Hill, 1962.
29. Sack, R. I. *Structural Analysis.* New York: McGraw-Hill, 1986.
30. *Minimum Design Loads for Buildings and Other Structures,* ASCE-7, American Society of Civil Engineers, Reston, VA, 1998.
31. *A Case Study in Aircraft Design: The Boeing 727.* Reston, VA: AIAA, 1978.
32. *Products in Action.* Canonsburg, PA: ANSYS, Inc., available at www.ansys.com.

PROBLEMS

Problems 1.1 through 12 are to be solved using the following assumptions, as needed: Weights of the members are insignificant compared to the applied forces and are neglected, all loads are static, friction in the pin joints is ignored, contact surfaces are smooth, friction in the bearings is neglected, and bearings act as simple supports.

Sections 1.1 through 1.10

1.1 A beam *CAB* with simple supports at *A* and *B* and an overhang *AC* carries loads as shown in Figure P1.1. All forces are coplanar and two dimensional. Determine the shear force and moment acting on the cross sections at the points *D* and *E*.

Figure P1.1

1.2 and **1.3** Two planar pin-connected frames are supported and loaded as shown in Figures P1.2 and P1.3. For each structure, determine the following:

Figure P1.2

Figure P1.3

(a) The components of reactions at B and C.

(b) The axial force, shear force, and moment acting on the cross section at point D.

1.4 The piston, connecting rod, and crank of an engine system are shown in Figure P1.4. Calculate

(a) The torque T required to hold the system in equilibrium.

(b) The normal force in the rod AB.

Given: A force $P = 4$ kips acts as indicated in the figure.

Figure P1.4

1.5 A planar frame is supported and loaded as shown in Figure P1.5. Determine the reaction at hinge B.

Figure P1.5

1.6 A hollow transmission shaft AB is supported at A and E by bearings and loaded as depicted in Figure P1.6. Calculate the following:

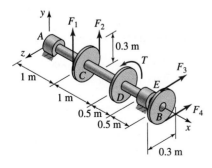

Figure P1.6

(a) The torque T required for equilibrium.

(b) The reactions at the bearings.

Given: $F_1 = 4$ kN, $F_2 = 3$ kN, $F_3 = 5$ kN, $F_4 = 2$ kN.

1.7 Pin-connected members *ADB* and *CD* carry a load W applied by a cable-pulley arrangement, as shown in Figure P1.7. Determine

(a) The components of the reactions at *A* and *C*.

(b) The axial force, shear force, and moment acting on the cross-section at point *G*.

Given: The pulley at *B* has a radius of 150 mm. Load $W = 1.6$ kN.

Figure P1.7

1.8 A bent rod is supported in the *xz* plane by bearings at *B*, *C*, *D* and loaded as shown in Figure P1.8. Dimensions are in millimeters. Calculate moment and shear force in the rod on the cross section at point *E*, for $P_1 = 200$ N and $P_2 = 300$ N.

Figure P1.8

1.9 Redo Problem 1.8, for the case in which $P_1 = 0$ and $P_2 = 400$ N.

1.10 A gear train is used to transmit a torque $T = 150$ N · m from an electric motor to a driven machine (Figure P1.10). Determine the torque acting on driven machine shaft, T_d, required for equilibrium.

1.11 A planar frame formed by joining a bar with a beam with a hinge is loaded as shown in Figure P1.11. Calculate the axial force in the bar *BC*.

1.12 A frame *AB* and a simple beam *CD* are supported as shown in Figure P1.12. A roller fits snugly between the two members at *E*. Determine the reactions at *A* and *C* in terms of load *P*.

Figure P1.10

Figure P1.11

Figure P1.12

Sections 1.11 through 1.14

1.13 The input shaft to a gearbox operates at speed of n_1 and transmits a power of 30 hp. The output power is 27 hp at a speed of n_2. What is the torque of each shaft (in kip · in) and the efficiency of the gearbox?

Given: $n_1 = 1800$ rpm, $n_2 = 425$ rpm.

1.14 A punch press with a flywheel produces N punching strokes per minute. Each stroke provides an average force of F over a stroke of s. The press is driven through a gear reducer by a shaft. Overall efficiency is e. Determine

(a) The power output.

(b) The power transmitted through the shaft.

Given: $N = 150$, $F = 500$ lb, $s = 2.5$ in., $e = 88\%$.

1.15 A rotating ASTM A-48 cast-iron flywheel has outer rim diameter d_o, inner rim diameter d_i, and length in the axial direction of l (Figure 1.9). Calculate the braking energy required in slowing the flywheel from 1200 to 1100 rpm.

Assumption: The hub and spokes add 5% to the inertia of the rim.
Given: $d_o = 400$ mm, $d_i = 0.75d_o$, $l = 0.25d_o$, $\rho = 7200$ kg/m^3 (see Table B.1).

1.16 A hollow cylinder is under an internal pressure that increases its 300-mm inner diameter and 500-mm outer diameter by 0.6 and 0.4 mm, respectively. Calculate

(a) The maximum normal strain in the circumferential direction.

(b) The average normal strain in the radial direction.

1.17 A thin triangular plate ABC is uniformly deformed into a shape $AB'C$, as shown by the dashed lines in Figure P1.17. Determine

(a) The normal strain along the centerline OB.

(b) The normal strain along the edge AB.

(c) The shear strain between the edges AB and BC.

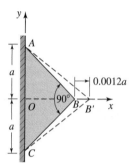

Figure P1.17

1.18 A 200 mm × 250 mm rectangle $ABCD$ is drawn on a thin plate prior to loading. After loading, the rectangle has the dimensions (in millimeters) shown by the dashed lines in Figure P1.18. Calculate, at corner point A,

(a) The normal strains ε_x and ε_y.

(b) The final length of side AD.

Figure P1.18

1.19 A thin rectangular plate, $a = 200$ mm and $b = 150$ mm (Figure P1.19) is acted on by a biaxial tensile loading, resulting in the uniform strains $\varepsilon_x = 1000\ \mu$ and $\varepsilon_y = 800\ \mu$. Determine the change in length of diagonal BD.

Figure P1.19

1.20 As a result of loading, the thin rectangular plate (Figure P1.19) deforms into a parallelogram in which sides AB and CD shorten 0.004 mm and rotate 1000 μ rad counterclockwise, while sides AD and BC elongate 0.006 mm and rotate 200 μ rad clockwise. Determine, at corner point A,

(a) The normal strains ε_x and ε_y and the shear strain γ_{xy}.

(b) The final lengths of sides AB and AD.

Given: $a = 50$ mm, $b = 25$ mm.

1.21 When loaded, the plate of Figure P1.21 deforms into a shape in which diagonal AC elongates 0.2 mm and diagonal BD contracts 0.5 mm while they remain perpendicular. Calculate the average strain components ε_x, ε_y, and γ_{xy}.

Figure P1.21

2

MATERIALS

Outline

2.1 INTRODUCTION

A great variety of materials has been produced and more are being produced in seemingly endless diversification. Material may be crystalline or noncrystalline. A crystalline material is made up of a number of small units called *crystals* or *grains*. Most materials must be processed before they are usable. Table 2.1 gives a general classification of engineering materials. This book is concerned with the macroscopic structural behavior: Properties are based on experiments using samples of materials of appreciable size. It is clear that a macroscopic structure includes a number of elementary particles forming a continuous and homogeneous structure held together by internal forces. The website at www.mathweb.com offers extensive information on materials.

In this chapter, the mechanical behavior, characteristics, treatment, and manufacturing processes of some common materials are briefly discussed. A review of the subject matter presented emphasizes how a viable as well as an economic design can be achieved. Later chapters explore typical material failure modes in more detail. The average properties of selected materials are listed in Table B.1 [1–7]. Unless specified otherwise, we assume in this text that the material is homogeneous and isotropic. With the exception of Section 16.7, our considerations are limited to the behavior of elastic materials. Note that the design of plate and shell-like members, for example, as components of a missile or space vehicle, involve materials having characteristics dependent on environmental conditions [8–10]. We refer to

Table 2.1   Some Commonly Used Engineering Materials

Metallic Materials	
Ferrous Metals	Nonferrous Metals
Cast iron	Aluminum
Malleable iron	Chromium
Wrought iron	Copper
Cast steel	Lead
Plain carbon steel	Magnesium
Steel alloys	Nickel
Stainless steel	Platinum
Tool steel	Silver
Special steels	Tin
Structural steel	Zinc
Nonmetallic Materials	
Carbon and graphite	Plastics
Ceramics	Brick
Cork	Stone
Felt	Elastomer
Glass	Silicon
Concrete	Wood

the ordinary properties of engineering materials in this volume. It is assumed that the reader has had a course in material science [11].

2.2   MATERIAL PROPERTY DEFINITIONS

The mechanical properties are those that indicate how the material is expected to behave when subjected to varying conditions of load and environment. These characteristics are determined by standardized destructive and nondestructive test methods outlined by the American Society for Testing and Materials (ASTM). A thorough understanding of material properties permits the designer to determine the size, shape, and method of manufacturing mechanical components.

Durability denotes the ability of a material to resist destruction over long periods of time. The destructive conditions may be chemical, electrical, thermal, or mechanical in nature or combinations of these conditions. The relative ease with which a material may be machined, or cut with sharp-edged tools, is termed its *machinability. Workability* represents the ability of a material to be formed into required shape. Usually, *malleability* is considered a property that represents the capacity of a material to withstand plastic deformation in compression without fracture. We see in Section 2.10 that *hardness* may represent the ability of a material to resist scratching, abrasion, cutting, or penetration. Frequently, the limitations imposed by the materials are the controlling factors in design. Strength and stiffness are main factors considered in the selection of a material. However, for a particular design, durability, malleability, workability, cost, and hardness of the materials may be equally significant. In considering the cost, attention focuses on not only the initial cost but also the maintenance and replacement costs of the part. Therefore, selecting a material from both its functional and economic standpoints is vitally important.

An elastic material returns to its original dimensions on removal of applied loads. This elastic property is called *elasticity*. Usually, the elastic range includes a region throughout which stress and strain have a linear relationship. The elastic portion ends at a point called the *proportional limit*. Such materials are linearly elastic. In a viscoelastic solid, the state of stress is function not only of the strain but the time rates of change of stress and strain as well. A *plastically* deformed member does not return to its initial size and shape when the load is removed. A homogenous solid displays identical properties throughout. If properties are the same in all directions at a point, the material is *isotropic*. A *composite* material is made up of two or more distinct constituents. A nonisotropic, or *anisotropic,* solid has direction-dependent properties. Simplest among them are that the material properties differ in three mutually perpendicular directions. A material so described is *orthotropic*. Some wood material may be modeled by orthotropic properties. Many manufactured materials are approximated as orthotropic, such as corrugated and rolled metal sheet, plywood, and fiber-reinforced concrete.

The capacity of a material to undergo large strains with no significant increase in stress is called *ductility*. Thus, a ductile material is capable of substantial elongation prior to failure. Such materials include mild steel, nickel, brass, copper, magnesium, lead, and teflon. The converse applies to a *brittle material*. A brittle material exhibits little deformation before rupture; for example, concrete, stone, cast iron, glass, ceramic materials, and many metallic alloys. A member that ruptures is said to *fracture*. Metals with strains at rupture in *excess* of

0.05 inches per inch in the tensile test are sometimes considered to be ductile [12]. Note that, generally, ductile materials fail in *shear,* while brittle materials fail in *tension.* Further details on material property definitions are found in Sections 2.12 and 2.13, where description of metal alloys, the numbering system of steels, and typical nonmetallic materials are included.

2.3 STATIC STRENGTH

In analysis and design, the mechanical behavior of materials under load is of primary importance. Experiments, mainly in tension or compression tests, provide basic information about overall response of specimens to the applied loads in the form of stress-strain diagrams. These curves are used to explain a number of mechanical properties of materials. Data for a stress-strain diagram are usually obtained from a *tensile test.* In such a test, a specimen of the material, usually in the form of a round bar, is mounted in the grips of a testing machine and subjected to tensile loading, applied slowly and steadily or statically at room temperature (Figure 2.1). The ASTM specifies precisely the dimensions and construction of standard tension specimens.

The tensile test procedure consists of applying successive increments of load while taking corresponding electronic extensometer readings of the elongation between the two gage marks (gage length) on the specimen. During an experiment, the change in gage length is noted as a function of the applied load. The specimen is loaded until it finally ruptures. The force necessary to cause rupture is called the *ultimate load.* Figure 2.2 illustrates a steel specimen that has fractured under load and the extensometer attached at the right by two arms to it. Based on the test data, the stress in the specimen is found by dividing the force by the cross-sectional area, and the strain is found by dividing the elongation by the gage length. In this manner, a complete stress-strain diagram, a plot of strain as abscissa and stress as the ordinate, can be obtained for the material. The stress-strain diagrams differ widely for different materials.

STRESS-STRAIN DIAGRAMS FOR DUCTILE MATERIALS

A typical stress-strain plot for a ductile material such as structural or mild steel in tension is shown in Figure 2.3a. Curve *OABCDE* is a conventional or engineering stress-strain diagram. The other curve, *OABCF,* represents the true stress-strain. The *true stress* refers to the load divided by the actual instantaneous cross-sectional area of the bar; the *true strain* is the sum of the elongation increments divided by the corresponding momentary length. Clearly, engineering stress equals the load divided by the initial cross-sectional area; the engineering strain is defined by Eq. (1.20). For most practical purposes, the conventional stress-strain diagram provides satisfactory information for use in design.

Yield Strength

The portion *OA* of the diagram is the elastic range. The linear variation of stress-strain ends at the *proportional limit, S_p,* point A. The lowest stress (point B) at which there is a marked increase in strain without corresponding increase in stress referred to as the *yield point* or *yield strength S_y.* For most cases, in practice, the proportional limit and yield point are assumed to be one: $S_p \approx S_y$. In the region between B and C, the material becomes *perfectly plastic,* meaning that it can deform without an increase in the applied load.

Figure 2.1   Tensile loading machine with automatic data-processing system. (Courtesy of MTS Systems Corp.)

Strain Hardening: Cold Working

The elongation of a mild steel specimen in the yield (or perfect plasticity) region is typically 10 to 20 times the elongation that occurs between the onset of loading and the proportional limit. The portion of the stress-strain curve extending from *A* to the point of fracture (*E*) is the plastic range. In the range *CD*, an increase in stress is required for a

Figure 2.2 A tensile test specimen with extensometer attached; the specimen has fractured. (Courtesy of MTS Systems Corp.)

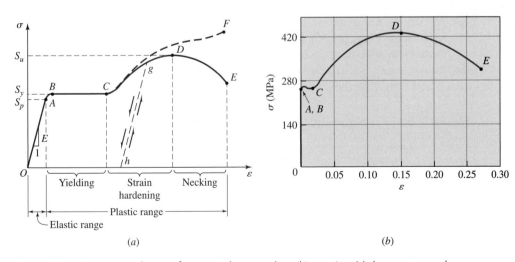

Figure 2.3 Stress-strain diagram for a typical structural steel in tension: (a) drawn not to scale; (b) drawn to scale.

continued increase in strain. This is called the *strain hardening* or *cold working*. If the load is removed at a point *g* in region *CD*, the material returns to no stress at a point *h* along a new line parallel to the line *OA*: A *permanent set Oh* is introduced. If the load is reapplied, the new stress-strain curve is *hg DE*. Note that there is now new yield point (*g*) that is higher than before (point *B*) but reduced ductility. This process can be repeated until the material becomes brittle and fractures.

Ultimate Tensile Strength

The engineering stress diagram for the material when strained beyond *C* displays a typical ultimate stress (point *D*), referred to as the *ultimate* or *tensile strength* S_u. Additional elongation is actually accompanied by a reduction in the stress, corresponding to fracture strength S_f (point *E*) in the figure. Failure at *E* occurs, by separation of the bar into two parts (Figure 2.2), along the *cone*-shaped surface forming an angle of approximately 45° with its axis that corresponds to the planes of maximum shear stress. In the vicinity of the ultimate stress, the reduction of the cross-sectional area or the lateral contraction becomes clearly visible and a pronounced *necking* of the bar occurs in the range *DE*. An examination of the ruptured cross-sectional surface depicts a fibrous structure produced by stretching of the grains of the material.

Interestingly, the standard measures of ductility of a material are defined on the basis of the geometric change of the specimen, as follows:

$$\text{Percent elongation} = \frac{L_f - L_o}{L_o}(100) \tag{2.1}$$

$$\text{Percent reduction in area} = \frac{A_o - A_f}{A_o}(100) \tag{2.2}$$

Here A_o and L_o denote, respectively, the original cross-sectional area and gage length of the specimen. Clearly, the ruptured bar must be pieced together to measure the final gage length L_f. Similarly, the final area A_f is measured at the fracture site where the cross section is minimal. Note that the elongation is not uniform over the length of the specimen but concentrated in the region of necking. Therefore, percent elongation depends on the gage length.

The diagram in Figure 2.3a depicts the general characteristics of the stress-strain diagram for mild steel, but its proportions are *not* realistic. As already noted, the strain between *B* and *C* may be about 15 times the strain between *O* and *A*. Likewise, the strains from *C* to *E* are many times greater than those from *O* to *A*. Figure 2.3b shows a stress-strain curve for mild steel drawn to scale. Clearly, the strains from *O* to *A* are so small that the initial part of the curve appears to be a vertical line.

Offset Yield Strength

Certain materials, such as heat treated steels, magnesium, aluminum, and copper, do not show a distinctive yield point and it is usual to use a yield strength S_y at an arbitrary strain. According to the so-called 0.2% offset method, a line is drawn through a strain of 0.002 (that is 0.2%), parallel to the initial slope at point *O* of the curve, as shown in Figure 2.4. The intersection of this line with the stress-strain curve defines the *offset yield strength* (point *B*). For the materials mentioned in the preceding discussion, the offset yield strength is slightly above the proportional limit.

Figure 2.4
Determination of yield strength by the offset method.

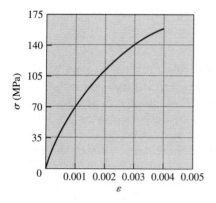

Figure 2.5 Stress-strain diagram for gray cast iron in tension.

STRESS-STRAIN DIAGRAM FOR BRITTLE MATERIALS

The tensile behavior of gray cast iron, a typical brittle material, is shown in Figure 2.5. We observe from the diagram that rupture occurs with no noticeable prior change in the rate of elongation. Therefore, for brittle materials, there is no difference between the ultimate strength and the fracture strength. Also, the strain at the rupture is much smaller for brittle materials than ductile materials. The stress-strain diagrams for brittle materials are characterized by having no well-defined linear region. The fracture of these materials is associated with the tensile stresses. Therefore, a brittle material breaks normal to the axis of the specimen, because this is the plane of maximum tensile stress.

STRESS-STRAIN DIAGRAMS IN COMPRESSION

Compression stress-strain curves, analogous to those in tension, may also be obtained for a variety of materials. Most ductile materials behave approximately the same in tension and compression over the elastic range. For these materials, the yield strength is about the same in tension and compression: $S_y \approx S_{yc}$, where the subscript c denotes compression. But, in the plastic range, the behavior is quite different. Since compression specimens expand instead of necking down, the compressive stress-strain curve continues to rise instead of reaching a maximum and dropping off.

A material having basically equal tensile and compressive strengths is termed an *even material*. For brittle materials, entire compression stress-strain diagram has a shape similar to the shape of the tensile diagram. However, brittle materials usually have characteristic stresses in compression that are much greater than in tension. A material that has different tensile and compressive strengths is referred to as an *uneven material*.

2.4 HOOKE'S LAW AND MODULUS OF ELASTICITY

Most engineering materials have an initial region on the stress-strain curve where the material behaves both elastically and linearly. The linear elasticity is a highly important property of materials. For the straight-line portion of the diagram (Figure 2.3), the stress is directly proportional to the strain. Therefore,

$$\sigma = E\varepsilon \tag{2.3}$$

This relationship between stress and strain for a bar in tension or compression is known as *Hooke's law*. The constant E is called the *modulus of elasticity, elastic modulus,* or *Young's modulus*. Inasmuch as ε is a dimensionless quantity, E has units of σ. In SI units, the elastic modulus is measured in newtons per square meter (or pascals); and in the U.S. customary system of units, it is measured in pounds per square inch (psi).

Equation (2.3) is highly significant in most of the subsequent treatment; the derived formulas are based on this law. We emphasize that Hooke's law is valid only up to the proportional limit of the material. The modulus of elasticity is seen to be the *slope* of the stress-strain curve in the linearly elastic range and is different for various materials. The E represents the stiffness of material in tension or compression. It is obvious that a material has a high elastic modulus value when its deformation in the elastic range is small. Similarly, linear elasticity can be measured in a member subjected to pure shear loading. Referring to Eq. (2.3), we now have

$$\tau = G\gamma \tag{2.4}$$

This is the Hooke's law for shear stress τ and shear strain γ. The constant G is called the *shear modulus of elasticity* or modulus of rigidity of the material, expressed in the same units as E: pascals or psi. The values of E and G for common materials are included in Table B.1.

We note that the slope of the stress-strain curve above the proportional limit is the *tangent modulus* E_t. That is, $E_t = d\sigma/d\varepsilon$. Likewise, the slope of a line from the origin to the point on the stress-strain curve above the proportional limit is known as the *Secant modulus* E_s. Therefore, $E_s = \sigma/\varepsilon$. Below the proportional limit, both E_t and E_s equal E.

In the elastic range, the ratio of the lateral strain to the axial strain is constant and known as *Poisson's ratio:*

$$\nu = -\frac{\text{lateral strain}}{\text{axial strain}} \tag{2.5}$$

Here the minus sign means that the lateral strain is of sense opposite to that of the axial strain.* Figure 2.6 depicts the lateral contraction of a rectangular parallelepiped element of side lengths a, b, and c in tension. Observe that the faces of the element at the origin are assumed fixed in position. The deformations are greatly exaggerated and the final shape of the element is shown by the dashed lines in the figure. The preceding definition is valid only for a uniaxial state of stress. Experiments show that, in most common materials, the values of ν are in the range 0.25 to 0.35. For steels, Poisson's ratio is usually assumed to be 0.3. Extreme cases include $\nu = 0.1$ for some concretes and $\nu = 0.5$ for rubber (Table B.1).

---

*It should be mentioned that there are some solids with a negative Poisson's ratio. These materials become fatter in the cross section when stretched [13].

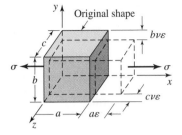

Figure 2.6 Axial elongation and lateral contraction of an element in tension (Poisson's effect).

2.5 GENERALIZED HOOKE'S LAW

For a two- or three-dimensional state of stress, each of the stress components is taken to be a linear function of the components of strain within the linear elastic range. This assumption usually predicts the behavior of engineering materials with good accuracy. In addition, the principle of superposition applies under multiaxial loading, since strains components are small quantities. In the following development, we rely on certain experimental evidence to derive the stress-strain relations for linearly elastic isotropic materials: A normal stress creates no shear strain whatsoever, and shear stress produces only shear strain.

Consider now an element of unit thickness subjected to a biaxial state of stress (Figure 2.7). Under the action of the stress σ_x, not only would the direct strain σ_x/E occur, but a y contraction as well, $-\nu\sigma_x/E$. Likewise, were σ_y to act only, an x contraction $-\nu\sigma_y/E$ and a y strain σ_y/E would result. Therefore, simultaneous action of both stresses σ_x and σ_y results in the following strains in the x and y directions:

$$\varepsilon_x = \frac{\sigma_x}{E} - \nu\frac{\sigma_y}{E} \tag{2.6a}$$

$$\varepsilon_y = \frac{\sigma_y}{E} - \nu\frac{\sigma_x}{E} \tag{2.6b}$$

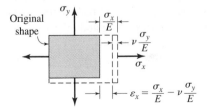

Figure 2.7 Element deformations caused by biaxial stress.

The elastic stress-strain relation, Eq. (2.4), for the state of two-dimensional pure shear, is given by

$$\gamma_{xy} = \frac{\tau_{xy}}{G}$$ (2.6c)

Inversion of Eqs. (2.6) results in the stress-strain relationships of the form

$$\sigma_x = \frac{E}{1 - \nu^2}(\varepsilon_x + \nu\varepsilon_y)$$

$$\sigma_y = \frac{E}{1 - \nu^2}(\varepsilon_y + \nu\varepsilon_x)$$ (2.7)

$$\tau_{xy} = G\gamma_{xy}$$

Equations (2.6) and (2.7) represent *Hooke's law for two-dimensional stress*.

The foregoing procedure is easily extended to a three-dimensional stress state (Figure 1.10). Then, the strain-stress relations, known as the *generalized Hooke's law*, consist of the expressions:

$$\varepsilon_x = \frac{1}{E}[\sigma_x - \nu(\sigma_y + \sigma_z)]$$

$$\varepsilon_y = \frac{1}{E}[\sigma_y - \nu(\sigma_x + \sigma_z)]$$

$$\varepsilon_z = \frac{1}{E}[\sigma_z - \nu(\sigma_x + \sigma_y)]$$ (2.8)

$$\gamma_{xy} = \frac{\tau_{xy}}{G}, \qquad \gamma_{yz} = \frac{\tau_{yz}}{G}, \qquad \gamma_{xz} = \frac{\tau_{xz}}{G}$$

The shear modulus of elasticity G is related to the modulus of elasticity E and Poisson's ratio ν. It can be shown that

$$G = \frac{E}{2(1 + \nu)}$$ (2.9)

So, for an isotropic material, there are only two independent elastic constants. The values of E and G are determined experimentally for a given material and ν can be found from the preceding basic relationship. Since the value of Poisson's ratio for ordinary materials is between 0 and $\frac{1}{2}$, we observe from Eq. (2.9) that G must be between $\frac{1}{3}E$ and $\frac{1}{2}E$.

DILATATION AND BULK MODULUS

The unit change in volume e, the change in volume ΔV per original volume V_o, in elastic materials subjected to stress is defined by

$$e = \frac{\Delta V}{V_o} = \varepsilon_x + \varepsilon_y + \varepsilon_z$$ (2.10)

The shear strains cause no change in volume. The quantity e is also referred to as *dilatation*. Equation (2.10) can be used to calculate the increase or decrease in volume of a member under loading, provided that the strains are known.

Based on the generalized Hooke's law, the dilatation can be found in terms of stresses and material constants. Using Eqs. (2.8), the stress-strain relationships may be expressed as follows

$$\sigma_x = 2G\varepsilon_x + \lambda e \qquad \tau_{xy} = G\gamma_{xy}$$
$$\sigma_y = 2G\varepsilon_y + \lambda e \qquad \tau_{yz} = G\gamma_{yz} \qquad \text{(2.11)}$$
$$\sigma_z = 2G\varepsilon_z + \lambda e \qquad \tau_{xz} = G\gamma_{xz}$$

In the preceding, we have

$$e = \varepsilon_x + \varepsilon_y + \varepsilon_z = \frac{1-2\nu}{E}(\sigma_x + \sigma_y + \sigma_z) \qquad \text{(2.12)}$$

$$\lambda = \frac{\nu E}{(1+\nu)(1-2\nu)} \qquad \text{(2.13)}$$

where λ is an elastic constant.

When an elastic member is subjected to a hydrostatic pressure p, the stresses are $\sigma_x = \sigma_y = \sigma_z = -p$ and $\tau_{xy} = \tau_{yz} = \tau_{xz} = 0$. Then Eq. (2.12) becomes $e = -3(1-2\nu)p/E$. This may be written in the form

$$K = -\frac{p}{e} = \frac{E}{3(1-2\nu)} \qquad \text{(2.14)}$$

The quantity K represents the modulus of volumetric expansion or so-called bulk modulus of elasticity. Equation (2.14) shows that, for incompressible materials ($e = 0$), $\nu = \frac{1}{2}$. For most materials, however, $\nu < \frac{1}{2}$ (Table B.1). Note that, in the perfectly plastic region behavior of a material, no volume change occurs and hence Poisson's ratio may be taken as $\frac{1}{2}$.

Determination of Displacements of a Plate

EXAMPLE 2.1

A steel panel of a device is approximated by a plate of thickness t, width b, length a, subjected to stresses σ_x and σ_y, as shown in Figure 2.8. Calculate

(a) The value of σ_x for which length a remains unchanged.

(b) The final thickness t' and width b'.

(c) The normal strain for the diagonal AC.

Given: $a = 400$ mm, $b = 300$ mm, $t = 6$ mm, $E = 200$ GPa, $\nu = 0.3$, and $\sigma_y = 220$ MPa.

Assumption: The plate is in plane state of stress.

Figure 2.8   Example 2.1. Plate in biaxial stress.

Solution: Inasmuch as the length does not change, we have $\varepsilon_x = 0$. In addition, plane stress $\sigma_z = 0$. Then, Eqs. (2.8a–c) become

$$\sigma_x = \nu\sigma_y \tag{2.15a}$$

$$\varepsilon_y = \frac{1}{E}(\sigma_y - \nu\sigma_x) \tag{2.15b}$$

$$\varepsilon_z = -\frac{\nu}{E}(\sigma_x + \sigma_y)$$

(a)   The given data are carried into Eq. (2.15a) to yield

$$\sigma_x = 0.3(220 \times 10^6) = 66 \text{ MPa}$$

(b)   Through the use of Eqs. (2.15b), we obtain

$$\varepsilon_y = \frac{10^6}{200(10^9)}[220 - 0.3(66)] = 1001\mu$$

$$\varepsilon_z = -\frac{0.3}{200(10^3)}(66 + 220) = -429\mu$$

In the foregoing, a minus sign means a decrease in the thickness. Therefore,

$$t' = t(1 + \varepsilon_z) = 6(0.9996) = 5.998 \text{ mm}$$

$$b' = b(1 + \varepsilon_y) = 300(1.0010) = 300.300 \text{ mm}$$

(c)   The original and final lengths of the diagonal are, respectively,

$$AC = (300^2 + 400^2)^{1/2} = 500 \text{ mm}$$

$$A'C' = (300.300^2 + 400^2)^{1/2} = 500.180 \text{ mm}$$

The normal strain for the diagonal is then

$$\varepsilon_{AC} = \frac{500.180 - 500}{500} = 360\mu$$

Comment: Alternatively, this result may readily be found by using the strain transformation equations, to be discussed in Section 3.11.

2.6 THERMAL STRESS-STRAIN RELATIONS

When displacements of a heated isotropic member is prevented, thermal stresses occur. The effects of such stresses can be severe, particularly since the most adverse thermal environments are frequently associated with design requirements dealing with unusually stringent constraints as to weight and volume. The foregoing is especially true in aerospace and machine design (e.g., engine, power plant, and industrial process) applications.

The total strains are obtained by adding to thermal strains of the type described by Eq. (1.21), the strains owing to the stress resulting from mechanical loads. In doing so, for instance, referring to Eqs. (2.6) for two-dimensional stress:

$$\varepsilon_x = \frac{1}{E}(\sigma_x - \nu\sigma_y) + \alpha T$$

$$\varepsilon_y = \frac{1}{E}(\sigma_y - \nu\sigma_x) + \alpha T \qquad (2.16)$$

$$\gamma_{xy} = \frac{\tau_{xy}}{G}$$

From these equations, we obtain the stress-strain relations as

$$\sigma_x = \frac{E}{1 - \nu^2}(\varepsilon_x + \nu\varepsilon_y) - \frac{E\alpha T}{1 - \nu}$$

$$\sigma_y = \frac{E}{1 - \nu^2}(\varepsilon_y + \nu\varepsilon_x) - \frac{E\alpha T}{1 - \nu} \qquad (2.17)$$

$$\tau_{xy} = G\gamma_{xy}$$

The quantities T and α represent the temperature change and the coefficient of expansion, respectively. Equations for three-dimensional stress may be readily expressed in a like manner.

Note that because free thermal expansion causes no distortion in an isotropic material, the shear strain is uneffected, as shown in the preceding expressions. The differential equations of equilibrium are based on purely mechanical considerations and unchanged for thermoelasticity. The same is true of the strain-displacement relations and hence the conditions of compatibility, which are geometrical in character (see Section 3.18). Thermoelasticity and ordinary elasticity therefore differ only to the extent of Hooke's law. Solutions for the problems in the former are usually harder to obtain than solutions for the problems in the latter.

In statically determinate structures, a uniform temperature change will not cause any stresses, as thermal deformations are permitted to occur freely. On the other hand, a temperature change in a structure supported in a statically indeterminate manner induces stresses in the members. Detailed discussions and illustrations of thermal loads and stresses in components and assemblies are given in Chapters 4 and 16.

2.7 TEMPERATURE AND STRESS-STRAIN PROPERTIES

A large deviation in temperature may cause a change in the properties of a material. In this section, temperature effects on stress-strain properties of materials are considered. Effects of temperature on impact and fatigue strengths are treated in Sections 2.9 and 8.7, respectively.

Another important thermal effect results because most materials expand with an increase in temperature.

SHORT-TIME EFFECTS OF ELEVATED AND LOW TEMPERATURES

For the static short-time testing of metals at elevated temperatures, it is generally found that the ultimate strength, yield strength, and modulus of elasticity are lowered with increasing temperature whereas the ductility increases with temperature. Here *elevated temperatures* refer to the absolute temperatures in excess of about one-third of the melting point absolute temperature of the material. On the contrary, at low temperatures, there is an increase in yield strength, ultimate strength, modulus of elasticity, and hardness and a decrease in ductility for metals. Therefore, when the operating temperatures are lower than the transition temperature, defined in Section 2.9, the possibility arises that a component could fail due to a brittle fracture.

The problem of designing for extreme temperatures is a special one, in that information concerning material properties is not overly abundant. Figure 2.9 depicts the effect of low and high temperatures on the strength of a type 304 stainless steel [14]. The considerable property variations illustrated by these curves are caused by metallurgical changes that take place as the temperature increases or decreases.

LONG-TIME EFFECTS OF ELEVATED TEMPERATURES: CREEP

Most metals under a constant load at elevated temperatures over a long period develop additional strains. This phenomenon is called *creep*. Creep is time dependent because deformation increases with time until a rupture occurs. For some nonferrous metals and a number of nonmetallic materials such as plastics, wood, and concrete, creep may also be produced at low stresses and normal (room) temperatures.

A typical creep curve, for a mild steel specimen in tension at elevated temperatures, consists of three regions or stages (Figure 2.10). In the first region, the material is becoming stronger because of strain hardening, and the creep rate ($d\varepsilon/dt$) decreases continuously. This stage is important if the load duration is short. The second region begins at a

Figure 2.9   Stress-strain diagrams for AISI type 304 stainless steel in tension: (a) at low temperatures; (b) at elevated temperatures.

Figure 2.10   Creep curve for structural steel in tension at high temperatures.

minimum strain rate and remains constant because of the balancing effects of strain harden-ing and annealing. *Annealing* refers to a process involving softening of a metal by heating and slowly cooling, discussed in Section 2.11. The secondary stage is usually the dominant interval of a creep curve. In the third region, the annealing effect predominates, and the deformation occurs at an accelerated creep rate until a rupture results.

When a component is subjected to a steady loading at elevated temperature and for a long period, the creep-rupture strength of the material determines its failure. However, failure at elevated temperatures due to dynamic loading will most likely occur early in the life of the material. Interest in the phenomenon of creep is not confined to possible failure by rupture but includes failure by large deformations that can make equipment inoperative. Therefore, in many designs, creep deformation must be maintained small. However, for some applications and within certain temperatures, stress, and time limits, creep effects need not be considered, and the stress-strain properties determined from static short-time testing are adequate.

2.8 MODULI OF RESILIENCE AND TOUGHNESS

Some machine and structural elements must be designed more on the basis of absorbing energy than withstanding loads. Inasmuch as energy involves both loads and deflections, stress-strain curves are particularly relevant. A detailed discussion of strain energy and its application is found in Chapter 5. Here we limit ourselves to the case of a member in ten-sion to illustrate how the energy-absorbing capacity of a material is determined.

MODULUS OF RESILIENCE

Resilience is the capacity of a material to absorb energy within the elastic range. The mod-ulus of resilience U_r represents the energy absorbed per unit volume of material, or the *strain energy density,* when stressed to the proportional limit. This is equal to the area under the straight-line portion of the stress-strain diagram (Figure 2.11a), where proportional limit S_p and yield strength S_y are taken approximately the same. The value of modulus of resilience, setting $\sigma_x = S_y$ into Eq. (5.1), has the form

$$U_r = \frac{S_y^2}{2E}$$

(2.18)

(a)   (b)

Figure 2.11 Stress-strain diagram: (a) modulus of resilience; (b) modulus of toughness.

where E is the modulus of elasticity. Therefore, resilient materials are those having high strength and low moduli of elasticity.

MODULUS OF TOUGHNESS

Toughness is the capacity of a material to absorb energy without fracture. The modulus of toughness U_t represents the energy absorbed per unit volume of material up to the point of fracture. It is thus equal to the entire area under stress-strain diagram (Figure 2.11b). Expressed mathematically, the modulus of toughness is

$$U_t = \int_0^{\varepsilon_f} \sigma \, d\varepsilon \qquad (2.19)$$

The quantity ε_f is the strain at fracture. Clearly, the toughness of a material is related to its ductility as well as to its ultimate strength. It is often convenient to perform the foregoing integration graphically. A planimeter can be used to determine this area.

Sometimes the modulus of toughness is approximated by representing the area under the stress-strain curve of ductile materials as the average of the yield strength S_y and ultimate strength S_u times the fracture strain. Therefore,

$$U_t = \frac{S_y + S_u}{2} \varepsilon_f \qquad (2.20)$$

For brittle materials (e.g., cast iron), the approximation of the area under the stress-strain curve (Figure 2.5), as given by Eq. (2.20) would be considerably in error. In such cases, the modulus of toughness is occasionally estimated by assuming that the strain-stress curve is a parabola. Then, using Eq. (2.19) with $\varepsilon_f = \varepsilon_u$, the modulus of toughness:

$$U_t = \frac{2}{3} S_u \varepsilon_u \qquad (2.21)$$

in which ε_u is the strain at the ultimate strength.

Toughness is usually associated with the capacity of a material to withstand an impact or shock load. Two common tests, the Charpy and Izod tests, discussed in the next section, determine the impact strength of materials at various temperatures. We observe that toughness obtained from these tests is as dependent on the geometry of the specimen as on the load rate. The units of both the modulus of toughness and modulus of resilience are expressed in joules ($N \cdot m$) per cubic meter (J/m^3) in SI and in in. $\cdot$ lb per cubic inch in the U.S. customary system. These are the same units of stress, so we can also use pascals or psi as the units for U_r and U_t. As an example, consider a structural steel having $S_y = 250$ MPa, $S_u = 400$ MPa, $\varepsilon_f = 0.3$, and $E = 200$ GPa (Table B.1). For this material, by Eqs. (2.18) and (2.20), we have $U_r = 156.25$ kPa and $U_t = 97.5$ MPa, respectively.

Note that fracture toughness is another material property that defines its ability to resist at the tip of a crack. When stress intensity reaches the fracture toughness, a fracture takes place with no warning. The study of this phenomenon is taken up in Section 7.3.

Material Resilience on an Axially Loaded Rod

EXAMPLE 2.2

During the manufacturing process, a prismatic round steel rod must acquire an elastic strain energy of $U_{app} = 200$ in. · lb (Figure 2.12). Determine the required yield strength S_y for a factor of safety of $n = 2.5$ with respect to permanent deformation.

Figure 2.12 Example 2.2.
Prismatic bar in tension.

Given: $E = 30 \times 10^6$ psi, diameter $d = 7/8$ in., length $L = 4$ ft.

Solution: The volume of the member is

$$V = AL = \frac{\pi}{4}(7/8)^2(4 \times 12) = 28.9 \text{ in.}^3$$

The rod should be designed for a strain energy:

$$U = nU_{app} = 2.5(200) = 500 \text{ in.} \cdot \text{lb}$$

The strain energy density is therefore $500/28.9 = 17.3$ lb/in.^{3}.
Through the use of Eq. (2.18), we have

$$U_r = \frac{S_y^2}{2E}; \qquad 17.3 = \frac{S_y^2}{2(30 \times 10^6)}$$

Solving

$$S_y = 32.2 \text{ ksi}$$

Comment: Observe that the factor of safety is applied to the energy load and not to the stress.

2.9 DYNAMIC AND THERMAL EFFECTS: BRITTLE-DUCTILE TRANSITION

A load applied to a structure or machine is called the *impact load,* also referred to as the *shock load,* if the time of application is less than one-third of the lowest natural period of the structure. Otherwise, it is termed the *static load.* Examples of a shock load include rapidly moving loads, such as those caused by a railroad train passing over a bridge, or direct impact loads, such as result from a drop hammer. In machine operation, impact loads are due to gradually increasing clearances that develop between mating parts with progressive wear, for example, steering gears and axle journals of automobiles; sudden application of loads, as occurs during explosion stroke of a combustion engine; and inertia loads, as introduced by high acceleration, such as in a flywheel.

In this section, we discuss the conditions under which metals may manifest a change from ductile to brittle or from brittle to ductile behavior. The matter of *ductile-brittle transition* has important applications where the operating environment includes a wide variation in temperature or when the rate of dynamic loading changes. The stress raisers such as grooves and notches also have a significant effect on the transition from brittle to ductile failure. The *transition temperature* represents roughly the temperature a material's behavior changes from ductile to brittle [10]. While most ferrous metals have well-defined transition temperature, some nonferrous metals do not. Therefore, the width of the temperature range over which the transition from brittle to ductile failure occurs is material dependent.

Let us, to begin with, refer to Figure 2.13a, where yield strength S_y and fracture strength S_f in tension are shown as functions of the temperature of a metal. Note that S_f exhibits only a small decrease with increasing temperature. The point of intersection (C) of the two strength curves in the figure defines the critical or transition temperature, T_t. If, at a given temperature above T_t, the stress is progressively increased, failure will occur by yielding and the fracture curve will never be encountered. Likewise, for a test conducted at $T < T_t$, the yield curve is not intercepted, since failure occurs by fracture. At temperatures close to T_t, the material generally exhibits some yielding prior to a partially brittle fracture.

A transition phenomenon is more commonly examined from the viewpoint of the energy required to fracture a notched or unnotched specimen, the impact toughness rather than the stress (Figure 2.13b). The transition temperature is then defined as the temperature at which there is a sudden decrease in impact toughness. The *Charpy* and *Izod* method notched-bar impact bending *tests* made at various temperatures utilize specimens to determine the impact toughness. In both tests, the specimen is struck by a pendulum released from a fixed height and the energy absorbed is computed from the height of swing (or indicated on a dial) after fracture. We note that the state of stress around the notch is triaxial and nonuniformly distributed throughout the specimen. Notches (and grooves) reduce the

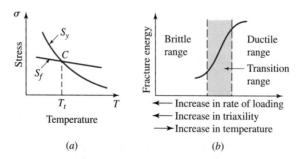

Figure 2.13   Typical transition curves for metals:
(a) variation of yield strength S_y and fracture strength S_f with temperature; (b) effects of loading rate, stress around a notch, and temperature on impact toughness.

energy required to fracture and shift the transition temperature, normally very low, to the range of normal temperatures. This is the reason why most experiments are performed on notched specimens [12].

The foregoing discussion shows that, under certain conditions, a material said to be ductile will behave in a brittle fashion and vice versa. The principal factors governing whether failure occurs by fracture or yielding are summarized as follows (Figure 2.13):

1. *Temperature.* If the temperature increases (exceeds T_t), the resistance to yielding is less than the resistance to fracture ($S_y < S_f$), and the specimen yields. On the contrary, if the temperature decreases (less than T_t), the specimen fractures without yielding.

2. *Loading rate.* Increasing the rate at which the load is applied increases a metal's ability to resist yielding.

3. *Triaxiality.* The effect on the transition of a three-dimensional stress condition around the notch, so-called triaxiality, is similar to that of loading rate.

In addition, other factors may also affect a ductile material to undergo a fracture similar to that of a brittle material. Some of these are fatigue, cyclic loading at normal temperatures (see Section 8.3); creep, long-time static loading at elevated temperatures; severe quenching, in heat treatment, if not followed by tempering; work hardening by sufficient amount of yielding. Internal cavities or voids in casting or forging may have an identical effect.

2.10 HARDNESS

Selection of a material that has good resistance to wear and erosion very much depends on the hardness and the surface condition. Hardness is the ability of a material to resist indentation and scratching. The kind of hardness considered depends on the service requirements to be met. For example, gears, cams, rails, and axles must have a high resistance to indentation. In mineralogy and ceramics, the ability to resist scratching is used as a measure of hardness. The *indentation hardness,* generally used in engineering, is briefly discussed in this section.

Hardness testing is one of the principal methods for ascertaining the suitability of a material for its intended purpose. It is also a valuable inspection tool for maintaining uniformity of quality in heat-treated parts. Indentation hardness tests most often involve one of the three methods: Brinell, Rockwell, or Vickers [12]. The shore scleroscope hardness testing is sometimes employed as well. These nondestructive tests yield a relative numerical measure or scale of hardness, showing how well a material resists indentation.

BRINELL HARDNESS

The Brinell hardness test uses a spherical ball in contact with a flat specimen of the material and subjected to a selected compressive load. Subsequent to removal of the load, the diameter of the indentation is measured with an optical micrometer. The hardness is then defined as the *Brinell hardness number* (Bhn), H_B, which is equal to the applied load divided by the area of the surface of indentation. Tables of hardness values are given in the Standards of the ASTM. The Brinell test is used mainly for materials whose thickness is 6.25 mm or greater. As a rule, the case hardened steels are unsuitable for Brinell testing. The test is used to determine the hardness of a wide variety of materials. The harder is the material, the smaller the indentation and the higher the Brinell number.

ROCKWELL HARDNESS

The Rockwell test uses an indenter (steel ball or diamond cone called *Brale*) pressed into the material. The relationships of the total test force to the depth of indentation provides a measure of Rockwell hardness, which is indicated on a dial gage. Depending on the size of the indenter, the load used, and the material being tested, the Rockwell test furnishes hardness data on various scales. Two common scales, R_B and R_C (i.e., Rockwell *B* and *C*), are frequently used for soft metals (such as mild steel or copper alloys) and hard metals (such as hardened steel or heat-treated alloy steel), respectively. In standard tests, the thickness should be at least 10 times the indentation diameter.

The Rockwell test is simple to perform and the most widely employed method for determining hardness of metals and alloys, ranging from the softest bearing materials to the hardest steels. It can also be used for certain plastics, such as acrylics, acetates, and phenolics. Optical measurements are not required; all readings are direct. Routine testing is usually performed with bench-type Rockwell machines.

VICKERS HARDNESS

The Vickers hardness test is similar to the Brinell test. However, it uses a four-sided inverted diamond pyramid with an apex angle of 136°. The Vickers hardness number (H_V) is the ratio of the impressed load to the square indented area. The Vickers hardness test is of particular value for hard, thin materials where hardness at a spot is required.

SHORE SCLEROSCOPE

The shore scleroscope uses a small diamond-tipped pointer or hammer that is allowed to fall from a fixed height onto the specimen. Hardness is measured by the height of rebound. The method is easy and rapid to apply. However, the results obtained are the least reliable of all machine methods. The hardness of soft plastics and wood felts is measured by this scleroscope.

RELATIONSHIPS AMONG HARDNESS AND ULTIMATE STRENGTH IN TENSION

Figure 2.14 shows the conversion plot between Brinell, Rockwell (*B* and *C*), and the tensile strength of steel. Note that the curves for R_B and R_C are nonlinear and the related values are only approximate. However, the results of the Brinell hardness test have been found to correlate linearly with the tensile strength S_u of most *steels* as follows:

$$S_u = 500 H_B \text{ psi} \tag{2.22}$$

This is indicated by a nearly straight line in the figure. In addition, for stress-relieved (not cold-drawn) steels, the tensile yield strength S_y is given by

$$S_y = 1.05 S_u - 30,000 \text{ psi} \tag{2.23}$$

Figure 2.14 Hardness conversion to ultimate strength in tension of steel.

Substituting Eq. (2.22),

$$S_y = 525H_B - 30,000 \text{ psi} \qquad \text{(2.24)}$$

Formulas (2.22) through (2.24) are estimates and should be used only when definite strain-hardening data are lacking [15, 16].

2.11   PROCESSES TO IMPROVE HARDNESS AND THE STRENGTH OF METALS

A material with metallic properties consisting of two or more elements, one of which is a basic metal, is called an *alloy*. An alloying element is deliberately added to a metal to alter its physical or mechanical properties. For example, the addition of alloying elements to iron results in cast iron and steel. By general usage, the term *metal* is used in a generic sense, often referring to both a simple metal and metallic alloys. Unless specified otherwise, we adhere to this practice.

There are a number of ways to increase the hardness and strength of metals. These include suitably varying the composition or alloying, mechanical treatment, and heat treatment. Various alloys are considered in the next section. Many coatings and surface treatments are also available for materials. Some have the main purpose of inhibiting corrosion while the others are intended to improve surface hardness and wear. We shall here discuss mechanical and heat treating processes.

MECHANICAL TREATMENT

Mechanical forming and hardening consists of hot-working and cold-working processes. A metal can be shaped and formed when it is above a certain temperature, known as the *recrystallization temperature*. Below this temperature the effects of mechanical working are cold worked. On the other hand, in hot working, the material is worked mechanically above its recrystallization temperature. Note that, hot working gives a finer, more uniform grain structure and improves the soundness of the material. However, in general, cold working leaves the part with residual stress on the surface. Thus, resulting mechanical properties in the foregoing processes are quite different.

Cold Working

Cold working, also called strain hardening, is a process of forming the metal usually at a room temperature (see Section 2.3). This results in an increase in hardness and yield strength, with a loss in toughness and ductility (that can be recovered by a heat treatment process termed annealing). Cold working is used to gain hardness on low carbon steels, which cannot be heat treated. Typical of cold-working operations include cold rolling, drawing, turning, grinding, and polishing. As noted previously, the relative ease with which a given material may be machined, or cut with sharp-edged tools, is called its machinability.

The most common and versatile of the cold-working treatments is shot peening. It is widely used with springs, gears, shafts, connecting rods, and many other components. In *shot peening,* the surface is bombarded with high-velocity iron or steel shot (small, spherical pellets) discharged from a rotating wheel or pneumatic nozzle. The process leaves the surface in compression and alters its smoothness. Since fatigue cracks are not known to initiate or propagate in a compression region (see Section 8.1), shot peening has proven very successful in raising the fatigue life of most members. Machine parts made of very high-strength steels (about 1400 MPa), such as springs, are particularly benefited. Shot peening has also been used to reduce the probability of stress corrosion cracking in a turbine rotor and blades.

Hot Working

Hot working reduces the strain hardening of a material but avoids the ductility and toughness loss attributed to cold working. However, hot-rolled metals tend to have greater ductility, lower strength, and a poorer surface finish than cold-worked metals of the identical alloy. Examples of hot-working processes are rolling, forging, hot extrusion, and hot pressing, where the metal is heated sufficiently to make it plastic and easily worked. Forging is an automation of blacksmithing. It uses a series of hammer dies shaped to gradually form the hot metal into the final configuration. Practically any metal can be forged. Extrusion is used mainly for nonferrous metals and it typically uses steel dies.

HEAT TREATMENT

The heat treatment process refers to the controlled heating and subsequent cooling of a metal. It is a complicated process, employed to obtain properties desirable and appropriate for a particular application. For instance, an intended heat treatment may be to strengthen and harden a metal, relieving its internal stresses, hardening its surface only, softening a cold-worked piece, or improving its machineability. The heating is done in the furnace and the maximum temperature must be maintained long enough to refine the grain structure. Cooling is also done in the furnace or an insulated contained. The definitions that follow are concerned with some common heat treating terms [1, 2, 8].

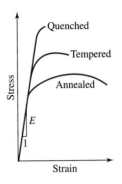

Figure 2.15 Stress-strain diagrams for annealed, quenched, and tempered steel.

Quenching. The rapid cooling of a metal from an elevated temperature by injecting or spraying the metal with a suitable cooling medium, such as oil or water, to increase hardness. The stress-strain curve as a result of quenching a mild steel is depicted in Figure 2.15.

Tempering or drawing. A process of stress relieving and softening by heating then quenching. Figure 2.15 shows a stress-strain curve for a mild steel after tempering.

Annealing. A process involving heating and slowly cooling, applied usually to induce softening and ductility. The quenching and tempering process is reversible by annealing; that is, annealing effectively returns a part to original stress-strain curve (Figure 2.15).

Normalizing. A process that includes annealing, except that the material is heated to a slightly higher temperature than annealing. The result is a somewhat stronger, harder metal than a fully annealed one.

Case hardening or carburizing. A process where the surface layer (or case) is made substantially harder than the metal's interior core. This is done by increasing the carbon content at the surface. Surface hardening by any appropriate method is a desirable hardening treatment for various applications. Some of the more useful case-hardening processes are carburizing, nitriding, cyaniding, induction hardening, and flame hardening. In the induction-hardening process, a metal is quickly heated by an induction coil followed by quenching in oil.

Through hardening. With a sufficiently high carbon content, 0.35–0.50%, the material is quenched and drawn at suitable temperatures to obtain the desired physical properties. Alloy steels will through harden and retain their shape better than plain carbon steels when heat treated. The greater strength and hardness of surface and core for heat-treated material is accompanied by loss of ductility.

2.12 GENERAL PROPERTIES OF METALS

There are various nonferrous and ferrous engineering metals (Table 2.1). This section attempts to provide some general information for the readers to help identify the types of a few selected metal alloys. Appendix B lists common mechanical properties of the foregoing materials.

IRON AND STEEL

Iron is a metal that, in its pure form, has almost no commercial use. The addition of other elements to iron essentially changes its characteristics, resulting in a variety of cast and wrought irons and steel. Cast iron and cast steel result from pouring the metal into molds of the proper form. To make wrought iron and varieties of wrought steel, the metal is cast into a suitable size and shape (e.g., slabs) then hot rolled to form bars, tubing, plate, structural shapes, pipe, nails, wires, and so on. Wrought iron is tough and welds easily. Cast

Table 2.2   Composition of Cast Iron

Carbon	2.00–4.00%
Silicon	0.50–3.00%
Manganese	0.20–1.00%
Phosphorus	0.05–0.80%
Sulfur	0.04–0.15%
Iron	Remainder

steel can also be readily welded. Steel is difficult to cast because it shrinks considerably. Castings are ordinarily inferior to corresponding wrought metals in impact resistance.

CAST IRONS

Cast iron is an iron alloy containing over 2% carbon. Cast irons constitute a whole family of materials. Having such a high carbon content, cast iron is brittle, has a low ductility, and hence cannot be cold worked. While relatively weak in tensile strength, it is very strong in compression. Bronze welding rods are widely used in cast iron that is not easily welded. The common composition of cast iron is furnished in Table 2.2.

The characteristics of cast iron can be altered extensively with the addition of alloying metals (such as copper, silicon, manganese, phosphorus, and sulfur) and proper heat treatment. Cast iron alloys are widely used as crankshafts, cam shafts, and cylinder blocks in engines, gearing, dies, railroad brake shoes, rolling mill rolls, and so on. Cast iron is inexpensive, easily cast, and readily machined. It has superior vibration characteristics and resistance to wear.

Since the physical properties of a cast iron casting are fully affected by its cooling rate during solidification, there are various cast iron types. Gray cast iron is the most widely used form of cast iron. It is common to refer gray cast iron just as *cast iron*. Other basic types of cast iron include malleable cast irons and nodular or ductile cast irons. Nodular cast iron shrinks more than gray cast iron, but its melting temperature is lower than for cast steel. A particlar type of cast iron, called *Meehanite iron,* is made using a patented process by the addition of a calcium-silicon alloy. In practice, cast irons are classified with respect to ultimate strength. In the *ASTM Numbering System* for cast irons, the class number correspond to the minimum ultimate strength. Thus, an ASTM No. 30 cast iron has a minimum tensile strength of 30 ksi (210 MPa). Some average properties and minimum ultimate strengths of cast irons are shown in Tables B.1 and B.2, respectively.

STEELS

Steel is an alloy of carbon containing less than 2% of carbon. Additional alloying elements ease the hardening of steel. Nevertheless, carbon content, almost alone, induces the maximum hardness that can be developed in steel. Steels are used extensively in machine construction. They can be classified as plain carbon steels, alloy steels, high-strength steels, cast steels, stainless steel, tool steel, and special purpose steel.

Table 2.3   Groups and Typical Uses of Plain Carbon Steels

Type	Carbon content (%)	Area of use
Low carbon	0.03–0.25	Plate, sheet, structural parts
Medium carbon	0.30–0.55	Machine parts, crane hooks
High carbon	0.60–1.40	Spring, tools, cutlery

Plain Carbon Steels

These steels contain only carbon, usually less than 1%, as a significant alloying element. Carbon is a potent alloying element, and wide changes in strength and hardness can be obtained by changing the amount of this element. Carbon steel owes its distinct properties chiefly to the carbon it contains. The range of desired characteristics can be further gained by heat treatment. Plain carbon steel is the least expensive steel, manufactured in larger quantities than any other. Table 2.3 summarizes general uses for steels having various levels of carbon content.

Low-carbon or mild steels, also referred to as the *structural steels,* are ductile and thus readily formable. If welded, they do not become brittle. Where a wear-resistant surface is needed, this steel can be case hardened. A minimum 0.30% carbon is necessary to make a heat-treatable steel. Therefore, medium and high-carbon steels can be heat treated to achieve the desired characteristics.

Alloy Steels

There are many effects of any alloy addition to a basic carbon steel. The primary reason is to improve the ease with which steel can be hardened. That is, potential hardness and strength, controlled by the carbon content, can be accomplished with less drastic heat treatment by alloying. When a proper alloy is present in a carbon steel, the metallurgical changes take place during quenching at a faster rate, the cooling effects penetrate deeper, and a large portion of the part is strengthened.

Ordinary alloying elements (in addition to carbon) include, singly or in various combinations, manganese, molybdenum, chromium, vanadium, and nickel. Added to the steel, nickel and chromium also bring significant impact resistance and provide considerable wear as well as corrosion resistance, respectively. A variety of carbon and alloy steels are employed for the construction of machinery and structures.

Stainless Steels

The so-called stainless steels are in widespread use for resisting corrosion and heat-resisting applications. They contain (in addition to carbon) at least 12% chromium as the basic alloying element. Stainless steel is of three types: austenitic (18% chromium, 8% nickel), ferritic (17% chromium), and martensitic (12% chromium). Particularly the austenitic kind of stainless steel polishes to a luster and finish. The mechanical characteristics of various wrough steels are given in Table B.5.

Steel Numbering Systems

Various numbering systems of steels are used. The Society of Automotive Engineers (SAE), the American Iron and Steel Institute (AISI), and the ASTM have devised codes to define the alloying elements and the carbon content in steels. These designations can provide a simple means by which any particular steel can be specified. A brief description of the common systems follows.

The *AISI/SAE numbering system* generally uses a number composed of four digits. The first two digits indicate the principal alloying element. The last two digits give the approximate carbon content, expressed in hundredths of percent. The AISI number for steel is similar to the SAE number, but a letter prefix is included to indicate the process of manufacture (such as A and C for the steels). For instance,

$$\text{AISI C1020} \quad \text{(SAE 1020)} \quad \text{steel}$$

represents a plain carbon steel denoted by the basic number 10, containing 0.20% carbon. In a like manner,

$$\text{AISI A3140} \quad \text{(SAE 3140)} \quad \text{steel}$$

is an alloy steel (with nickel and chromium designated by 31) with 0.40% carbon. We see from the foregoing examples that the first two digits are not so systematic.

The *ASTM numbering system* for steels is based on the ultimate strength. Mostly used specifications are ASTM-A27 mild- to medium-strength carbon-steel castings for general application and A148 high-strength steel castings for structural purposes [1]. Tables B.1 and 14.2 include average properties of a few ASTM steels. Samples of the minimum yield strengths S_y and ultimate strengths S_u for certain ASTM steels are shown in Table 2.4, where Q and T denote quenched and tempered, respectively. Note that the ASTM-A36 is the all purpose carbon grade steel extensively used in building and bridge construction. ASTM-A572 is high-strength low alloy steel, A588 represents atmospheric corrosion resistant high-strength low alloy steel, and A514 is alloy Q&T steel. For complete information on each steel, reference should be made to the appropriate ASTM specification.

The ASTM, the AISI, and the SAE, developed the *Unified Numbering System* (UNS) for metals and alloys. This system also contains cross-reference numbers for other material specifications. The UNS uses a letter prefix to indicate the material (e.g., G for the carbon and alloy steels). The mechanical properties of selected carbon and alloy steels are

Table 2.4   Structural Steel Strengths in the ASTM Numbering System [17]

Steel type	ASTM no.	S_y MPa	(ksi)	S_u MPa	(ksi)	Max. thickness mm	(in.)
Carbon	A36	248	(36)	400	(58)	200	(8)
Low alloy	A572	290	(42)	414	(60)	150	(6)
Stainless	A588	345	(50)	480	(65)	100	(4)
Alloy Q&T	A514	690	(100)	758	(110)	62.5	(2.5)

furnished in Tables B.3 and B.4. The AISI, the SAE, the ASTM, and UNS lists are continuously being revised, and it is necessary to confer the latest edition of a material handbook. We mostly use the AISI/SAE designations for steels.

ALUMINUM AND COPPER ALLOYS

Aluminum alloys are very versatile materials, having good electrical and thermal conductivity as well as light reflectivity. They posses a high strength-to-weight ratio, which can be a very important consideration in the design of, for example, aircraft, missiles, trains, and so on. Aluminum has a high resistance to most corrosive atmospheres because it readily forms a passive oxide surface coating. Lightweight aluminum alloys have extensive applications in manufactured products. Aluminum is readily formed, drawn, stamped, spun, machined, welded, or brazed. The high-strength aluminum alloys have practically the same strength as mild steel.

Numerous aluminum alloys are available in both wrought and cast form. The aluminum *casting alloys* are indicated by three-digit numbers. The *wrought alloys,* shaped by rolling or extruding, use four-digit numbers. Silicon alloys are preferred for casting. Typical wrought aluminum alloys include copper and silicon-magnesium. The comparison and mechanical properties of some typical aluminum alloys are given in Table B.6. The temper of an aluminum alloy is a main factor governing its strength, hardness, and ductility. Temper designation is customarily specified by cold work such as rolling, drawing, or stretching. Other alloys are heat treatable, and their properties can be enhanced considerably by appropriate thermal processing.

Copper alloys are very ductile materials. Copper may be spun, stamped, rolled into a sheet, or drawn into wire and tubing. Owing to its high electrical and thermal conductivity, resistance to corrosion, but relatively low ratio of strength to weight, copper is used extensively in the electrical, telephone, petroleum, and power industries. The most notable copper-base alloys are brass and bronze. *Brass* is a copper-zinc alloy, and *bronze* is composed mainly of copper and tin. Brass and bronze are used in both cast and wrought form. The strength of brass increases with the zinc content. Brass is about equal to copper in corrosion resistance, but bronze is superior to both.

Die castings generally are made from zinc, aluminum, magnesium, and to a lesser extent, brass. They are formed by forcing a molten alloy into metal molds or dies under high pressure. The die cast process is applicable for parts containing very thin sections of intricate forms. Copper and most of its alloys can be fabricated by soldering, welding, or brazing. They can be work strengthened by cold working but cannot be heat treated. The machinability of the brass and bronze is satisfactory. The properties of copper alloys can often be significantly improved by adding small amounts of additional alloying elements (Table B.7).

2.13 GENERAL PROPERTIES OF NONMETALS

The three common categories of nonmetals are of engineering interest: plastics, ceramics, and composites (see Table 2.1). Plastics represent a vast and growing field of synthetics. Sometimes optimum properties can be obtained by combination of dissimilar materials or composites. Ceramics are hard, heat-resistant, brittle materials. Here we briefly discuss the general properties of these materials.

Table 2.5   Selected Plastics

Chemical classification	Trade name
Thermoplastic Materials	
Acetal	Delrin, Celcon
Acrylic	Lucite, Plexiglas
Cellulose acetate	Fibestos, Plastacele
Cellulose nitrate	Celluloid, Nitron
Ethyl cellulose	Gering, Ethocel
Polyamide	Nylon, Zytel
Polycarbonate	Lexan, Merlon
Polyethylene	Polythene, Alathon
Polystyrene	Cerex, Lustrex
Polytetrafluoroethylene	Teflon
Polyvinyl acetate	Gelva, Elvacet
Polyvinyl alcohol	Elvanol, Resistoflex
Polyvinyl chloride	PVC, Boltaron
Polyvinylidene chloride	Saran
Thermosetting Materials	
Epoxy	Araldite, Oxiron
Phenol-formaldehyde	Bakelite, Catalin
Phenol-furfural	Durite
Polyster	Beckosol, Glyptal
Urea-formaldehyde	Beetle, Plaskon

PLASTICS

Plastics are synthetic materials known as *polymers*. They are employed increasingly for structural purposes and thousands of different types are available. Table 2.5 presents several common plastics. The mechanical properties of these materials vary tremendously, with some plastics being brittle and others ductile. When designing with plastics, it is significant to remember that their properties are greatly affected by both change in temperature and the passage of time. Observe from the table that polymers are of two principal classes: thermoplastics and thermosets.

Thermoplastic materials repeatedly soften when heated and harden when cooled. There are also highly elastic flexible materials known as thermoplastic *elastomers*. A common elastomer is a rubber band. Elastomers' industrial applications include belts, hoses, gaskets, seals, machinery mounts, and vibration dampers. *Thermosets* or thermosetting plastics sustain structural change during processing to become permanently insoluble and infusible. Thermoplastic materials may be formed into a variety of shapes by the simple application of heat and pressure, while thermoset plastics can be formed only by cutting or machining.

An examination of tensile stress-strain diagrams at various low temperatures, for instance, a cellulose nitrate and similar ones for other plastics, indicates that the ultimate strength, yield strength, modulus of elasticity, but not ductility increase with the decrease in temperature. These modifications of the stress-strain properties are similar to those found for metals (see Section 2.7). Experiments also show an appreciable decrease in impact toughness with a decrease in temperature for some plastics but not all. A loaded plastic may stretch gradually over a time until it no longer is serviceable. Interestingly, because of its light weight, the strength-to-weight ratio for nylon is about the same as for structural steel. The principal advantage of plastics is their ability to be readily processed, as previously noted; most plastics can be easily molded into complicated shapes. Large elastic deflections allow the design of polymer components that snap together, making assembly fast and inexpensive. Furthermore, many plastics are low-cost materials and display exceptional resistance to wear and corrosive attacks by chemicals.

Fiber reinforcement increases the stiffness, hardness, strength, and resistance to environmental factors and reduces the shrinking of plastics. A glass-reinforced plastic has improved strength by a factor of about 2 or more. Further improvement is gained by carbon reinforcement. The foregoing relatively new materials (with 10–40% carbon) have tensile strengths as high as 280 MPa. Reinforced plastics increasingly are being employed for machine and structural components requiring light weight or high strength-to-weight ratios. Tables B.1 and B.8 show that the ranges of properties that can be obtained with plastics are very large.

CERAMICS

Ceramics are basically compounds of nonmetallic as well as metallic elements, mostly oxides, nitrades, and carbides. Generally, silica and graphite ceramics dominate the industry. However, newer ceramics, often called *technical ceramics,* play a major role in many applications. Glass ceramics are in wide spread usage as electrical, electronic, and laboratory ware. Ceramics have high hardness and brittleness, high compressive but low tensile strengths. They are typically 15 times stronger in compression than in tension. High temperature and chemical resistance, high dielectric strength, and low weight characterize many of these materials. Therefore, attempts are being made to replace customary metals with ceramics in some machine and structural members.

COMPOSITES

As mentioned earlier, a composite material is made up of two or more unique elements. Composites usually consist of high-strength reinforcement material embedded in a surrounding material. They have a relatively large strength-to-weight ratio compared to a homogeneous material and additional other desirable characteristics [18, 19]. Furthermore, there are many situations where different materials are used in combination so that the maximum advantage is gained from each component part. For instance, graphite-reinforced epoxy gets strength from the graphite fibers while the epoxy protects the graphite from oxidation and provides toughness. In this text, the discussions concern isotropic composites like reinforced-concrete beams and multilayer members and filament-wound anisotropic composite cylinders.

REFERENCES

1. Avallone, E. A., and T. Baumeister III, eds. *Mark's Standard Handbook for Mechanical Engineers,* 10th ed. New York: McGraw-Hill, 1996.
2. *Annual Book of ASTM.* Philadelphia: American Society for Testing and Materials. 2001.
3. Ashby, M. J. *Material Selection in Mechanical Design.* Oxford: Pergamon, 1992.
4. Budinsky, K. *Engineering Materials, Properties and Selection.* Upper Saddle River, NJ: Prentice Hall, 1979.
5. Farag, M. M. *Selection of Materials and Manufacturing Processes for Engineering Materials.* Upper Saddle River, NJ: Prentice Hall, 1990.
6. Lewis, G. *Selection of Engineering Materials.* Upper Saddle River, NJ: Prentice Hall, 1990.
7. Dieter, G. E. *Mechanical Metallurgy.* New York: McGraw-Hill, 1976.
8. Miner, D. F., and J. B. Seastone, eds. *Handbook of Engineering Materials.* New York: Wiley, 1955.
9. Abraham, L. H. *Structural Design of Missiles and Spacecraft.* New York: McGraw-Hill, 1962.
10. Faupel, J. H., and F. E. Fisher. *Engineering Design,* 2nd ed. New York: Wiley, 1981.
11. Vlack, L. H. Van. *Elements of Material Science and Engineering,* 6th ed. Reading, MA: Wesley, 1989.
12. Marin, J. *Mechanical Behavior of Engineering Materials.* Upper Saddle River, NJ: Prentice Hall, 1962.
13. Lakes, R. S. "Advances in Negative Poisson's Ratio Materials." *Advanced Materials* 5, no. 4, (1993).
14. *Metals Handbook,* 9th ed., vol. 8, Mechanical Testing. Metals Park, OH: The American Society for Metals (ASM), 1985.
15. Datsko, J. *Material Properties and Manufacturing Processes.* New York: Wiley, 1966.
16. Datsko, J. *Materials in Design and Manufacturing.* Ann Arbor, MI: Malloy, 1977.
17. *Manual of Steel Construction,* 9th ed., 1992, The American Institute of Steel Construction (AISC), Chicago.
18. Jones, R. M. *Mechanics of Composite Materials.* New York: McGraw-Hill, 1975.
19. Ugural, A. C. *Stresses in Plates and Shells,* 2nd ed. New York: McGraw-Hill, 1999.

PROBLEMS

Sections 2.1 through 2.13

2.1 A bar of diameter d, gage length L, is loaded to the proportional limit in a tensile testing machine. A strain gage is placed on the surface of the bar to measure normal strains in the longitudinal direction. With an axial load P, it is found that the bar is elongated $12(10^{-3})$ in. and the diameter reduced $0.24(10^{-3})$ in. Calculate the proportional limit, modulus of elasticity, Poisson's ratio, reduction in area, and percentage elongation.

Given: $d = 0.5$ in., $L = 8$ in., $P = 4$ kips.

2.2 A 15-mm × 15-mm square *ABCD* is drawn on a member prior to loading. After loading, the square becomes the rhombus shown in Figure P2.2. Determine

(*a*) The modulus of elasticity.

(*b*) Poisson's ratio.

(*c*) The shear modulus of elasticity.

Figure P2.2

2.3 A rectangular block of width a, depth b, and length L, is subjected to an axial tensile load P, as shown in Figure P2.3. Subsequent to the loading, dimensions b and L are changed to 1.999 in. and 10.02 in., respectively. Calculate

(a) Poisson's ratio.

(b) The modulus of elasticity.

(c) The final value of the dimension a.

(d) The shear modulus of elasticity.

Given: $a = 3$ in., $b = 2$ in., $L = 10$ in., $P = 100$ kips.

Figure P2.3

2.4 A bar of any given material is subjected to uniform triaxial stresses. Determine the maximum value of Poisson's ratio.

2.W Search the website at www.matweb.com. Review the material property database and select

(a) Four metals with a tensile strength $S_y < 50$ ksi (345 MPa), modulus of elasticity $E > 26 \times 10^6$ psi (179 GPa), Brinell hardness number $H_B < 200$.

(b) Three metals having elongation greater than 15%, $E > 28 \times 10^6$ psi (193 GPa), Poisson's ratio $v < 0.32$.

(c) One metallic alloy that has $E > 30 \times 10^6$ psi (207 GPa) and ultimate strength in compression $S_{uc} = 200$ ksi (1378 MPa).

2.5 The block shown in Figure P2.3 is subjected to an axial load P. Calculate the axial strain.

Given: $P = 25$ kN, $a = 20$ mm, $b = 10$ mm, $L = 100$ mm, $E = 70$ GPa, and $v = 0.3$.
Assumption: The block is constrained against y- and z-directed contractions.

2.6 Consider Figure P2.6 with a bronze block ($E = 100$ GPa, $v = 1/3$) subjected to uniform stresses σ_x, σ_y, and σ_z. Calculate the new dimensions after the loading.

Given: $L = 100$ mm, $a = 50$ mm, and $b = 10$ mm prior to the loading

$$\sigma_x = 150 \text{ MPa}, \sigma_y = -90 \text{ MPa, and } \sigma_z = 0.$$

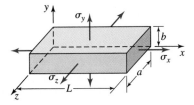

Figure P2.6

2.7   Resolve Problem 2.6 assuming that the block is under a uniform pressure of only $p = 120$ MPa on all its faces.

2.8   Determine the approximate value of the modulus of toughness for a structural steel bar having the stress-strain diagram of Figure 2.3b. What is the permanent elongation of the bar for a 50-mm gage length?

2.9   A strain energy of 9J must be acquired by an 6061-T6 aluminum alloy rod of diameter d and length L, as an axial load is applied (Figure P2.9). Determine the factor of safety n of the rod with respect to permanent deformation.

Given: $d = 5$ mm and $L = 3$ m.

Figure P2.9

2.10   A square steel machine component of sides a by a and length L is to resist an axial energy load of 400 N $\cdot$ m. Determine

(a)   The required yield strength of the steel.

(b)   The corresponding modulus of resilience for the steel.

Given: $a = 50$ mm, $L = 1.5$ m, factor of safety with respect to yielding $n = 1.5$, and $G = 200$ GPa.

2.11   A strain energy of $U_{app} = 150$ in. $\cdot$ lb must be acquired by an ASTM-A36 steel rod of diameter d and length L when the axial load is applied (Figure P2.9). Calculate the diameter d of the rod with a factor of safety n with respect to permanent deformation.

Given: $L = 8$ ft and $n = 5$.

STRESS AND STRAIN

Outline

3.1 INTRODUCTION

This chapter provides a review and insight into the stress and strain analyses. Expressions for both stresses and deflections in mechanical elements are developed throughout the text as the subject unfolds, after examining their function and general geometric behavior. With the exception of Sections 3.12 through 3.18, we employ mechanics of materials approach, simplifying the assumptions related to the deformation pattern so that strain distributions for a cross section of a member can be determined. A fundamental assumption is that *plane sections remain plane*. This hypothesis can be shown to be exact for axially loaded elastic prismatic bars and circular torsion members and for slender beams, plates, and shells subjected to pure bending. The assumption is approximate for other stress analysis problems. Note, however, that there are many cases where applications of the *basic formulas of mechanics of materials*, so-called elementary formulas for stress and displacement, lead to useful results for slender members under any type of loading.

Our coverage presumes a knowledge of mechanics of materials procedures for determining stresses and strains in a homogeneous and an isotropic bar, shaft, and beam. In Sections 3.2 through 3.9, we introduce the basic formulas, the main emphasis being on the underlying assumptions used in their derivations. Next to be treated are the transformation of stress and strain at a point. Then attention focuses on stresses arising from various combinations of fundamental loads applied to members and the stress concentrations. The chapter concludes with discussions on contact stresses in typical members referring to the solutions obtained by the methods of the theory of elasticity and the general states of stress and strain.

In the treatment presented here, the study of complex stress patterns at the supports or locations of concentrated load is not included. According to Saint-Venant's Principle (Section 1.4), the actual stress distribution closely approximates that given by the formulas of the mechanics of materials, except near the restraints and geometric discontinuities in the members. For further details, see texts on solid mechanics and theory of elasticity; for example, References 1 through 3.

3.2 STRESSES IN AXIALLY LOADED MEMBERS

Axially loaded members are structural and machine elements having straight longitudinal axes and supporting only axial forces (tensile or compressive). Figure 3.1a shows a homogeneous prismatic bar loaded by tensile forces P at the ends. To determine the normal stress, we make an imaginary cut (section *a-a*) through the member at right angles to its

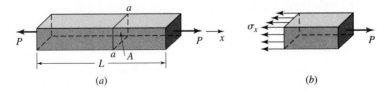

Figure 3.1   Prismatic bar in tension.

axis (x). A free-body diagram of the isolated part is shown in Figure 3.1b. Here the stress is substituted on the cut section as a replacement for the effect of the removed part.

Assuming that the stress has a uniform distribution over the cross section, the equilibrium of the axial forces, the first of Eqs. (1.4), yields $P = \int \sigma_x \, dA$ or $P = A\sigma_x$. The normal stress is therefore

$$\sigma_x = \frac{P}{A} \tag{3.1}$$

where A is the cross-sectional area of the bar. The remaining conditions of Eqs. (1.4) are also satisfied by the stress distribution pattern shown in Figure 3.1b. When the member is being stretched as depicted in the figure, the resulting stress is a uniaxial tensile stress; if the direction of the forces is reversed, the bar is in compression and uniaxial compressive stress occurs. Equation (3.1) is applicable to tension members and chunky, short compression bars. For slender members, the approaches discussed in Chapter 6 must be used.

Stress due to the restriction of thermal expansion or contraction of a body is called *thermal stress, σ_t*. Using Hooke's law and Eq. (1.21), we have

$$\sigma_t = \alpha(\Delta T)E \tag{3.2}$$

The quantity ΔT represents a temperature change. We observe that a high modulus of elasticity E and high coefficient of expansion α for the material increase the stress.

DESIGN OF TENSION MEMBERS

Tension members are found in bridges, roof trusses, bracing systems, and mechanisms. They are used as tie rods, cables, angles, channels, or combinations of these. Of special concern is the design of prismatic tension members for strength under static loading. In this case, a rational design procedure (see Section 1.6) may be briefly described as follows:

1. *Evaluate the mode of possible failure.* Usually the normal stress is taken to be the quantity most closely associated with failure. This assumption applies regardless of the type of failure that may actually occur on a plane of the bar.

2. *Determine the relationships between load and stress.* This important value of the normal stress is defined by $\sigma = P/A$.

3. *Determine the maximum usable value of stress.* The maximum usable value of σ without failure, σ_{max}, is the yield strength S_y or the ultimate strength S_u. Use this value in connection with equation found in step 2, if needed, in any expression of failure criteria, discussed in Chapter 7.

4. *Select the factor of safety.* A safety factor n is applied to σ_{max} to determine the allowable stress $\sigma_{all} = \sigma_{max}/n$. The required cross-sectional area of the member is therefore

$$A = \frac{P}{\sigma_{all}} \tag{3.3}$$

If the bar contains an abrupt change of cross-sectional area, the foregoing procedure is repeated, using a stress concentration factor to find the normal stress (step 2).

EXAMPLE 3.1 | Design of a Hoist

A pin-connected two-bar assembly or hoist is supported and loaded as shown in Figure 3.2a. Determine the cross-sectional area of the round aluminum eyebar AC and the square wood post BC.

Figure 3.2   Example 3.1.

Given: The required load is $P = 50$ kN. The maximum usable stresses in aluminum and wood are 480 and 60 MPa, respectively.

Assumptions: The load acts in the plane of the hoist. Weights of members are insignificant compared to the applied load and omitted. Friction in pin joints and the possibility of member BC buckling are ignored.

Design Decision: Use a factor of safety of $n = 2.4$.

Solution: Members AC and BC carry axial loading. Applying equations of statics to the free-body diagram of Figure 3.2b, we have

$$\sum M_B = -40(2.5) - 30(2.5) + \frac{5}{13}F_A(3.5) = 0 \qquad F_A = 130 \text{ kN}$$

$$\sum M_A = -40(2.5) - 30(6) + \frac{1}{\sqrt{2}}F_B(3.5) = 0 \qquad F_B = 113.1 \text{ kN}$$

Note, as a check, that $\sum F_x = 0$.
   The allowable stress, from design procedure steps 3 and 4,

$$(\sigma_{\text{all}})_{AC} = \frac{480}{2.4} = 200 \text{ MPa}, \qquad (\sigma_{\text{all}})_{BC} = \frac{60}{2.4} = 25 \text{ MPa}$$

By Eq. (3.3), the required cross-sectional areas of the bars,

$$A_{AC} = \frac{130(10^3)}{200} = 650 \text{ mm}^2, \qquad A_{BC} = \frac{113.1(10^3)}{25} = 4524 \text{ mm}^2$$

Comment: A 29-mm diameter aluminum eyebar and a 68 mm × 68 mm wood post should be used.

3.3 DIRECT SHEAR STRESS AND BEARING STRESS

A *shear stress* is produced whenever the applied forces cause one section of a body to tend to slide past its adjacent section. As an example consider the connection shown in Figure 3.3a. This joint consists of a bracket, a clevis, and a pin that passes through holes in the bracket and clevis. The pin resists the shear across the two cross-sectional areas at *b-b* and *c-c*; hence, it is said to be in *double shear*. At each cut section, a shear force V equivalent to $P/2$ (Figure 3.3b) must be developed. Thus, the shear occurs over an area parallel to the applied load. This condition is termed *direct shear*.

The distribution of shear stress τ across a section cannot be taken as uniform. Dividing the total shear force V by the cross-sectional area A over which it acts, we can obtain the average shear stress in the section:

$$\tau_{avg} = \frac{V}{A} \tag{3.4}$$

The average shear stress in the pin of the connection shown in the figure is therefore $\tau_{avg} = (P/2)/(\pi d^2/4) = 2P/\pi d^2$. Direct shear arises in the design of bolts, rivets, welds, glued joints, as well as in pins. In each case, the shear stress is created by a direct action of the forces in trying to cut through the material. Shear stress also arises in an indirect manner when members are subjected to tension, torsion, and bending, as discussed in the following sections.

Note that, under the action of the applied force, the bracket and the clevis press against the pin in bearing and a nonuniform pressure develops against the pin (Figure 3.3b). The

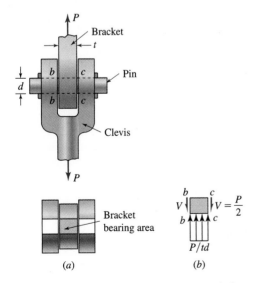

Figure 3.3 (a) A clevis-pin connection, with the bracket bearing area depicted; (b) portion of pin subjected to direct shear stresses and bearing stress.

average value of this pressure is determined by dividing the force P transmitted by the *projected area* A_p of the pin into the bracket (or clevis). This is called the *bearing stress:*

$$\sigma_b = \frac{P}{A_p} \tag{3.5}$$

Therefore, bearing stress in the bracket against the pin is $\sigma_b = P/td$, where t and d represent the thickness of bracket and diameter of the pin, respectively. Similarly, the bearing stress in the clevis against the pin may be obtained.

EXAMPLE 3.2 | Design of a Monoplane Wing Rod

The wing of a monoplane is approximated by a pin-connected structure of beam AD and bar BC, as depicted in Figure 3.4a. Determine

(a)   The shear stress in the pin at hinge C.

(b)   The diameter of the rod BC.

(a)   (b)

Figure 3.4   Example 3.2.

Given: The pin at C has a diameter of 15 mm and is in double shear.

Assumptions: Friction in pin joints is omitted. The air load is distributed uniformly along the span of the wing. Only rod BC is under tension. A round 2014-T6 aluminum alloy bar (see Table B.1) is used for rod BC with an allowable axial stress of 210 MPa.

Solution: Referring to the free-body diagram of the wing ACD (Figure 3.4b),

$$\sum M_A = 36(1.8) - F_{BC}\frac{1}{\sqrt{5}}(2) = 0 \qquad F_{BC} = 72.45 \text{ kN}$$

(a)   Through the use of Eq. (3.4),

$$\tau_{avg} = \frac{F_{BC}}{2A} = \frac{72,450}{2[\pi(0.0075)^2]} = 205 \text{ MPa}$$

(b) Applying Eq. (3.1), we have

$$\sigma_{BC} = \frac{F_{BC}}{A_{BC}}, \qquad 210(10^6) = \frac{72,450}{A_{BC}}$$

Solving

$$A_{BC} = 3.45(10^{-4}) \text{ m}^2 = 345 \text{ mm}^2$$

Hence,

$$345 = \frac{\pi d^2}{4}, \qquad d = 20.96 \text{ mm}$$

Comments: A 21-mm diameter rod should be used. Note that, for steady inverted flight, the rod BC would be a compression member.

3.4 THIN-WALLED PRESSURE VESSELS

Pressure vessels are closed structures that contain liquids or gases under pressure. Common examples include tanks for compressed air, steam boilers, and pressurized water storage tanks. Although pressure vessels exist in a variety of different shapes (see Sections 16.10 through 16.14), only thin-walled cylindrical and spherical vessels are considered here. A vessel having a wall thickness less than about $\frac{1}{10}$ of inner radius is called *thin walled*. For this case, we can take $r_i \approx r_o \approx r$, where r_i, r_o, and r refer to inner, outer, and mean radii, respectively. The contents of the pressure vessel exert internal pressure, which produces small stretching deformations in the membranelike walls of an inflated balloon. In some cases external pressures cause contractions of a vessel wall. With either internal or external pressure, stresses termed *membrane stresses* arise in the vessel walls.

Section 16.11 shows that application of the equilibrium conditions to an appropriate portion of a thin-walled tank suffices to determine membrane stresses. Consider a thin-walled *cylindrical vessel* with closed ends and internal pressure p (Figure 3.5a). The longitudinal or *axial stress* σ_a and circumferential or *tangential stress* σ_θ acting on the side faces

(a)   (b)

Figure 3.5 Thin-walled pressure vessels: (a) cylindrical; and (b) spherical.

of a stress element shown in the figure are principal stresses from Eqs. (16.74):

$$\sigma_a = \frac{pr}{2t} \tag{3.6a}$$

$$\sigma_\theta = \frac{pr}{t} \tag{3.6b}$$

The circumferential strain as a function of the change in radius δ_c is $\varepsilon_\theta = [2\pi(r + \delta_c) - 2\pi r]/2\pi r = \delta_c/r$. Using Hooke's law, we have $\varepsilon_\theta = (\sigma_\theta - \nu\sigma_a)/E$, where ν and E represent Poisson's ratio and modulus of elasticity, respectively. The extension of the radius of the cylinder, $\delta_c = \varepsilon_\theta r$, under the action of the stresses given by Eqs. (3.6) is therefore

$$\delta_c = \frac{pr^2}{2Et}(2 - \nu) \tag{3.7}$$

The tangential stresses σ act in the plane of the wall of a *spherical vessel* and are the same in any section that passes through the center under internal pressure p (Figure 3.5b). *Sphere stress* is given by Eq. (16.71):

$$\sigma = \frac{pr}{2t} \tag{3.8}$$

They are half the magnitude of the tangential stresses of the cylinder. Thus, sphere is an optimum shape for an internally pressurized closed vessel. The radial extension of the sphere, $\delta_s = \varepsilon r$, applying Hooke's law $\varepsilon = (\sigma - \nu\sigma)/E$ is then

$$\delta_s = \frac{pr^2}{2Et}(1 - \nu) \tag{3.9}$$

Note that the stress acting in the radial direction on the wall of a cylinder or sphere varies from $-p$ at the inner surface of the vessel to 0 at the outer surface. For thin-walled vessels, radial stress σ_r is much smaller than the membrane stresses and is usually omitted. The state of stress in the wall of a vessel is therefore considered biaxial. To conclude, we mention that a pressure vessel design is essentially governed by ASME Pressure Vessel Design Codes, discussed in Section 16.13.

Thick-walled cylinders are often used as vessels or pipe lines. Some applications involve air or hydraulic cylinders, gun barrels, and various mechanical components. Equations for exact elastic and plastic stresses and displacements for these members are developed in Chapter 16.* Composite thick-walled cylinders under pressure, thermal, and dynamic loading are discussed in detail. Numerous illustrative examples also are given.

| *Within this chapter, some readers may prefer to study Section 16.3.

Design of Spherical Pressure Vessel

EXAMPLE 3.3

A spherical vessel of radius r is subjected to an internal pressure p. Determine the critical wall thickness t and the corresponding diametral extension.

Assumption: A safety factor n against bursting is used.

Given: $r = 2.5$ ft, $p = 1.5$ ksi, $S_u = 60$ ksi, $E = 30 \times 10^6$ psi, $\nu = 0.3$, $n = 3$.

Solution: We have $r = 2.5 \times 12 = 30$ in. and $\sigma = S_u/n$. Applying Eq. (3.8),

$$t = \frac{pr}{2S_u/n} = \frac{1.5(30)}{2(60/3)} = 1.125 \text{ in.}$$

Then, Eq. (3.9) results in

$$\delta_s = \frac{pr^2(1-\nu)}{2Et} = \frac{1500(30)^2(0.7)}{2(30 \times 10^6)(1.125)} = 0.014 \text{ in.}$$

The diametral extension is therefore $2\delta_s = 0.028$ in.

3.5 STRESS IN MEMBERS IN TORSION

In this section, attention is directed toward stress in prismatic bars subject to equal and opposite end torques. These members are assumed free of end constraints. Both circular and rectangular bars are treated. *Torsion* refers to twisting a structural member when it is loaded by couples that cause rotation about its longitudinal axis. Recall from Section 1.8 that, for convenience, we often show the moment of a couple or torque by a vector in the form of a double-headed arrow.

CIRCULAR CROSS SECTIONS

Torsion of circular bars or shafts produced by a torque T results in a shear stress τ and an angle of twist or angular deformation ϕ, as shown in Figure 3.6a. The *basic assumptions* of the formulations on the torsional loading of a circular prismatic bar are as follows:

1. A plane section perpendicular to the axis of the bar remains plane and undisturbed after the torques are applied.

2. Shear strain γ varies linearly from 0 at the center to a maximum on the outer surface.

3. The material is homogeneous and obeys Hooke's law; hence, the magnitude of the maximum shear angle γ_{max} must be less than the yield angle.

The maximum shear stress occurs at the points most remote from the center of the bar and is designated τ_{max}. For a linear stress variation, at any point at a distance r from center, the shear stress is $\tau = (r/c)\tau_{max}$, where c represents the radius of the bar. On a cross

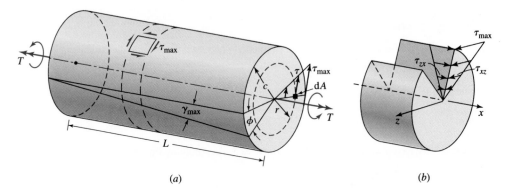

Figure 3.6   (a) Circular bar in pure torsion. (b) Shear stresses on transverse (xz) and axial (zx) planes in a circular shaft segment in torsion.

section of the shaft the resisting torque caused by the stress, distribution must be equal to the applied torque T. Hence,

$$T = \int r\left(\frac{r}{c}\tau_{max}\right)dA$$

The preceding relationship may be written in the form

$$T = \frac{\tau_{max}}{c} \int r^2 \, dA$$

By definition, the polar moment of inertia J of the cross-sectional area is

$$J = \int r^2 \, dA \qquad\qquad\qquad \textbf{(a)}$$

For a solid shaft, $J = \pi c^4/2$. In the case of a circular tube of inner radius b and outer radius c, $J = \pi(c^4 - b^4)/2$.

Shear stress varies with the radius and is largest at the points most remote from the shaft center. This stress distribution leaves the external cylindrical surface of the bar free of stress distribution, as it should. Note that the representation shown in Figure 3.6a is purely schematic. The maximum shear stress on a cross section of a circular shaft, either solid or hollow, is given by the *torsion formula:*

$$\tau_{max} = \frac{Tc}{J} \qquad\qquad\qquad \textbf{(3.10)}$$

The shear stress at distance r from the center of a section is

$$\tau = \frac{Tr}{J} \qquad\qquad\qquad \textbf{(3.11)}$$

The *transverse* shear stress found by Eq. (3.10) or (3.11) is accompanied by an axial shear stress of equal value, that is, $\tau = \tau_{xz} = \tau_{zx}$ (Figure 3.6b), to satisfy the conditions of static equilibrium of an element. Since the shear stress in a solid circular bar is maximum at the outer boundary of the cross section and 0 at the center, most of the material in a solid shaft is stressed significantly below the maximum shear stress level. When weight reduction and savings of material are important, it is advisable to use hollow shafts (see also Example 3.4).

NONCIRCULAR CROSS SECTIONS

In treating torsion of noncircular prismatic bars, cross sections initially plane experience out-of-plane deformation or *warping,* and the first two assumptions stated previously are no longer appropriate. Figure 3.7 depicts the nature of distortion occurring in a rectangular section. The mathematical solution of the problem is complicated. For cases that cannot be conveniently solved by applying the theory of elasticity, the governing equations are used in conjunction with the experimental techniques. The finite element analysis is also very efficient for this purpose. Torsional stress (and displacement) equations for a number of noncircular sections are summarized in references such as [2, 4]. Table 3.1 lists the "exact" solutions of the maximum shear stress and the angle of twist ϕ for a few common cross sections. Note that the values of coefficients α and β depend on the ratio of the side lengths a and b of a rectangular section. For thin sections ($a \gg b$), the values of α and β approach $\frac{1}{3}$.

The following approximate formula for the maximum shear stress in a *rectangular member* is of interest:

$$\tau_{max} = \frac{T}{ab^2}\left(3 + 1.8\frac{b}{a}\right) \tag{3.12}$$

As in Table 3.1, a and b represent the lengths of the long and short sides of a rectangular cross section, respectively. The stress occurs along the centerline of the wider face of the bar. For a *thin section,* where a is much greater than b, the second term may be neglected. Equation (3.12) is also valid for equal-leg angles; these can be considered as two rectangles, each of which is capable of carrying half the torque.

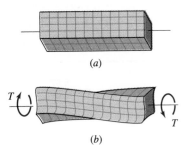

(a)

(b)

Figure 3.7 Rectangular bar (a) before and (b) after a torque is applied.

Table 3.1   Expressions for stress and deformation in some cross-section shapes in torsion

Cross section	Maximum shearing stress	Angle of twist per unit length
Ellipse For circle: $a = b$ (2a wide, 2b high, point A at bottom)	$\tau_A = \dfrac{2T}{\pi ab^2}$	$\phi = \dfrac{(a^2 + b^2)T}{\pi a^3 b^3 G}$
Equilateral triangle (side a, point A at left)	$\tau_A = \dfrac{20T}{a^3}$	$\phi = \dfrac{46.2T}{a^4 G}$
Rectangle (width a, height b, point A at bottom)	$\tau_A = \dfrac{T}{\alpha ab^2}$	$\phi = \dfrac{T}{\beta ab^3 G}$
Hollow rectangle (outer width a, height b, wall t, top wall t_1, points A and B)	$\tau_A = \dfrac{T}{2abt_1}$ $\tau_B = \dfrac{T}{2abt}$	$\phi = \dfrac{(at + bt_1)T}{2tt_1 a^2 b^2 G}$
Hollow ellipse For hollow circle: $a = b$ (2a wide, 2b high, wall t, point A)	$\tau_A = \dfrac{T}{2\pi abt}$	$\phi = \dfrac{\sqrt{2(a^2 + b^2)}\,T}{4\pi a^2 b^2 t G}$
Hexagon (dimension a, point A)	$\tau_A = \dfrac{5.7T}{a^3}$	$\phi = \dfrac{8.8T}{a^4 G}$

Rectangle coefficients:

a/b	α	β
1.0	0.208	0.141
1.5	0.231	0.196
2.0	0.246	0.229
2.5	0.256	0.249
3.0	0.267	0.263
4.0	0.282	0.281
5.0	0.292	0.291
10.0	0.312	0.312
∞	0.333	0.333

Torque Transmission Efficiency of Hollow and Solid Shafts | **EXAMPLE 3.4**

A hollow shaft and a solid shaft (Figure 3.8) are twisted about their longitudinal axes with torques T_h and T_s, respectively. Determine the ratio of the largest torques that can be applied to the shafts.

Figure 3.8   Example 3.4.

Given: $c = 1.15b$.

Assumptions: Both shafts are made of the same material with allowable stress and both have the same cross-sectional area.

Solution: The maximum shear stress τ_{max} equals τ_{all}. Since the cross-sectional areas of both shafts are identical, $\pi(c^2 - b^2) = \pi a^2$:

$$a^2 = c^2 - b^2$$

For the hollow shaft, using Eq. (3.10),

$$T_h = \frac{\pi}{2c}(c^4 - b^4)\tau_{all}$$

Likewise, for the solid shaft,

$$T_s = \frac{\pi}{2}a^3\tau_{all}$$

We therefore have

$$\frac{T_h}{T_s} = \frac{c^4 - b^4}{ca^3} = \frac{c^4 - b^4}{c(c^2 - b^2)^{\frac{3}{2}}} \qquad (3.13)$$

Substituting $c = 1.15b$, this quotient gives

$$\frac{T_h}{T_s} = 3.56$$

Comments: The result shows that, hollow shafts are more efficient in transmitting torque than solid shafts. Interestingly, thin shafts are also useful for creating an essentially uniform shear (i.e., $\tau_{min} \approx \tau_{max}$). However, to avoid buckling (see Section 6.2), the wall thickness cannot be excessively thin.

3.6　SHEAR AND MOMENT IN BEAMS

In beams loaded by transverse loads in their planes, only two components of stress resultants occur: the shear force and bending moment. These loading effects are sometimes referred to as *shear* and *moment in beams*. To determine the magnitude and sense of shearing force and bending moment at any section of a beam, the method of sections is applied. The sign conventions adopted for internal forces and moments (see Section 1.8) are associated with the deformations of a member. To illustrate this, consider the positive and negative shear forces V and bending moments M acting on segments of a beam cut out between two cross sections (Figure 3.9). We see that a positive shear force tends to raise the left-hand face relative to the right-hand face of the segment, and a positive bending moment tends to bend the segment concave upward, so it "retains water." Likewise, a positive moment compresses the upper part of the segment and elongates the lower part.

Load, Shear, and Moment Relationships

Consider the free-body diagram of an element of length dx, cut from a loaded beam (Figure 3.10a). Note that the distributed load w per unit length, the shears, and the bending moments are shown as positive (Figure 3.10b). The changes in V and M from position x to $x + dx$ are denoted by dV and dM, respectively. In addition, the resultant of the distributed load ($w\,dx$) is indicated by the dashed line in the figure. Although w is not uniform, this is permissible substitution for a very small distance dx.

Equilibrium of the vertical forces acting on the element of Figure 3.10b, $\sum F_x = 0$, results in $V + w\,dx = V + dV$. Therefore,

$$\frac{dV}{dx} = w \qquad\qquad (3.14a)$$

Figure 3.9　Sign convention for beams: definitions of positive and negative shear and moment.

(a)　　　　　　　　(b)

Figure 3.10　Beam and an element isolated from it.

This states that, at any section of the beam, the slope of the shear curve is equal to w. Integration of Eq. (3.14a) between points A and B on the beam axis gives

$$V_B - V_A = \int_A^B w\,dx = \text{area of load diagram between } A \text{ and } B \qquad \text{(3.14b)}$$

Clearly, Eq. (3.14a) is not valid at the point of application of a concentrated load. Similarly, Eq. (3.14b) cannot be used when concentrated loads are applied between A and B. For equilibrium, the sum of moments about O must also be 0: $\sum M_O = 0$ or $M + dM - (V + dV)\,dx - M = 0$. If second-order differentials are considered as negligible compared with differentials, this yields

$$\frac{dM}{dx} = V \qquad \text{(3.15a)}$$

The foregoing relationship indicates that the slope of the moment curve is equal to V. Therefore the shear force is inseparably linked with a change in the bending moment along the length of the beam. Note that the maximum value of the moment occurs at the point where V (and hence dM/dx) is 0. Integrating Eq. (3.15a) between A and B, we have

$$M_B - M_A = \int_A^B V\,dx = \text{area of shear diagram between } A \text{ and } B \qquad \text{(3.15b)}$$

The differential equations of equilibrium, Eqs. (3.14a) and (3.15a), show that the shear and moment curves, respectively, always are 1 and 2 degrees higher than the load curve. We note that Eq. (3.15a) is not valid at the point of application of a concentrated load. Equation (3.15b) can be used even when concentrated loads act between A and B, but the relation is not valid if a couple is applied at a point between A and B.

Shear and Moment Diagrams

When designing a beam, it is useful to have a graphical visualization of the shear force and moment variations along the length of a beam. A shear diagram is a graph where the shearing force is plotted against the horizontal distance (x) along a beam. Similarly, a graph showing the bending moment plotted against the x axis is the bending-moment diagram. The signs for shear V and moment M follow the general convention defined in Figure 3.9. It is convenient to place the shear and bending moment diagrams directly below the free-body, or load, diagram of the beam. The maximum and other significant values are generally marked on the diagrams.

We use the so-called summation method of constructing shear and moment diagrams. The procedure of this semigraphical approach is as follows:

1. Determine the reactions from free-body diagram of the entire beam.
2. Determine the value of the shear, successively summing from the *left end* of the beam the vertical external forces or using Eq. (3.14b). Draw the shear diagram obtaining the shape from Eq. (3.14a). Plot a positive V upward and a negative V downward.

3. Determine the values of moment, either continuously summing the external moments from the left end of the beam or using Eq. (3.15b), whichever is more appropriate. Draw the moment diagram. The shape of the diagram is obtained from Eq. (3.15a).

A check on the accuracy of the shear and moment diagrams can be made by noting whether or not they close. Closure of these diagrams demonstrates that the sum of the shear forces and moments acting on the beam are 0, as they must be for equilibrium. When any diagram fails to close, you know that there is a construction error or an error in calculation of the reactions. The following example illustrates the procedure.

EXAMPLE 3.5 | **Shear and Moment Diagrams for a Simply Supported Beam by Summation Method**

Draw the shear and moment diagrams for the beam loaded as shown in Figure 3.11a.

Figure 3.11   Example 3.5: (a) An overhanging beam; (b) free-body or load diagram; (c) shear diagram; and (d) moment diagram.

Assumptions: All forces are coplanar and two dimensional.

Solution: Applying the equations of statics to the free-body diagram of the entire beam, we have (Figure 3.11b):

$$R_A = 6.4 \text{ kN}, \qquad R_B = 9.6 \text{ kN}$$

In the *shear diagram* (Figure 3.11c), the shear at end C is $V_C = 0$. Equation (3.14b) yields

$$V_A - V_C = \frac{1}{2}w(1.5) = \frac{1}{2}(-4)(1.5) = -3, \qquad V_A = -3 \text{ kN}$$

the upward force near to the left of A. From C to A, the load increases linearly, hence the shear curve is parabolic, which has a negative and increasing slope. In the regions AD, DB, and BE, the slope of the shear curve is 0 or the shear is constant. At A, the 6.4 kN upward reaction force increases the shear to 3.4 kN. The shear remains constant up to D where it decreases by a 10 kN downward force to -6.6 kN. Likewise, the value of the shear rises to 3 kN at B. No change in the shear occurs until point E, where the downward 3 kN force closes the diagram. The maximum shear $V_{max} = -6.6$ kN occurs in region BD.

In the *moment diagram* (Figure 3.11d), the moment at end C is $M_C = 0$. Equation (3.15b) gives

$$M_A - M_C = -\int_0^{1.5} \left(\frac{1}{2}\frac{4x}{1.5}x\right) dx \qquad M_A = -1.5 \text{ N} \cdot \text{m}$$

$$M_D - M_A = 3.4(1.5) \qquad M_D = 3.6 \text{ kN} \cdot \text{m}$$

$$M_B - M_D = -6.6(1) \qquad M_B = -3 \text{ kN} \cdot \text{m}$$

$$M_E - M_B = 3(1) \qquad M_E = 0$$

Since M_E is known to be 0, a check on the calculations is provided. We find that, from C to A, the diagram takes the shape of a cubic curve concave downward with 0 slope at C. This is in accordance with $dM/dx = V$. Here V, prescribing the slope of the moment diagram, is negative and increases to the right. In the regions AD, DB, and BE, the diagram forms straight lines. The maximum moment, $M_{max} = 3.6$ kN $\cdot$ m, occurs at D.

A procedure identical to the preceding one applies to axially loaded bars and twisted shafts. The applied axial forces and torques are positive if their vectors are in the direction of a positive coordinate axis. When a bar is subjected to loads at several points along its length, the internal axial forces and twisting moments vary from section to section. A graph showing the variation of the axial force along the bar axis is called an *axial-force diagram*. A similar graph for the torque is referred to as a *torque diagram*. We note that the axial force and torque diagrams are *not* used as commonly as shear and moment diagrams.

3.7 STRESSES IN BEAMS

A *beam* is a bar supporting loads applied laterally or transversely to its (longitudinal) axis. This flexure member is commonly used in structures and machines. Examples include the main members supporting floors of buildings, automobile axles, and leaf springs. We see in Sections 4.10 and 4.11 that the following formulas for stresses and deflections of beams can readily be reduced from those of rectangular plates.

ASSUMPTIONS OF BEAM THEORY

The basic assumptions of the technical or engineering *theory for slender beams* are based on geometry of deformation. They can be summarized as follows [1]:

1. The deflection of the beam axis is *small* compared with the depth and span of the beam.

2. The slope of the deflection curve is very small and its square is negligible in comparison with unity.

3. Plane sections through a beam taken normal to its axis remain plane after the beam is subjected to bending. This is the fundamental hypothesis of the flexure theory.

4. The effect of shear stress τ_{xy} on the distribution of bending stress σ_x is omitted. The stress normal to the neutral surface, σ_y, may be disregarded.

A generalization of the preceding presuppositions forms the basis for the theories of plates and shells [5].

When treating the bending problem of beams, it is frequently necessary to distinguish between pure bending and nonuniform bending. The former is the flexure of a beam subjected to a constant bending moment; the latter refers to flexure in the presence of shear forces. We discuss the stresses in beams in both cases of bending.

NORMAL STRESS

Consider a linearly elastic beam having the y axis as a vertical axis of symmetry (Figure 3.12a). Based on assumptions 3 and 4, the normal stress σ_x over the cross section (such as A-B, Figure 3.12b) varies linearly with y and the remaining stress components are 0:

$$\sigma_x = ky \qquad \sigma_y = \tau_{xy} = 0 \qquad \text{(a)}$$

Here k is a constant, and $y = 0$ contains the neutral surface. The intersection of the neutral surface and the cross section locates the *neutral axis* (abbreviated N.A.). Figure 3.12c depicts the linear stress field in the section A-B.

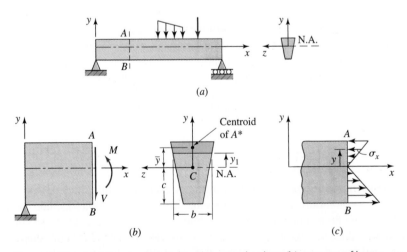

Figure 3.12   (a) A beam subjected to transverse loading; (b) segment of beam; (c) distribution of bending stress in a beam.

Conditions of equilibrium require that the resultant normal force produced by the stresses σ_x be 0 and the moments of the stresses about the axis be equal to the bending moment acting on the section. Hence,

$$\int_A \sigma_x \, dA = 0, \qquad -\int_A (\sigma_x \, dA)y = M \tag{b}$$

in which A represents the cross-sectional area. The negative sign in the second expression indicates that a positive moment M is one that produces compressive (negative) stress at points of positive y. Carrying Eq. (a) into Eqs. (b),

$$k \int_A y \, dA = 0 \tag{c}$$

$$-k \int_A y^2 \, dA = M \tag{d}$$

Since $k = 0$, Eq. (c) shows that the first moment of cross-sectional area about the neutral axis is 0. This requires that the neutral and centroidal axes of the cross section coincide. It should be mentioned that the symmetry of the cross section about the y axis means that the y and z axes are principal centroidal axes. The integral in Eq. (d) defines the moment of inertia, $I = \int y^2 \, dA$, of the cross section about the z axis of the beam cross section. It follows that

$$k = -\frac{M}{I} \tag{e}$$

An expression for the normal stress, known as the elastic *flexure formula* applicable to initially straight beams, can now be written by combining Eqs. (a) and (e):

$$\sigma_x = -\frac{My}{I} \tag{3.16}$$

Here y represents the distance from the neutral axis to the point at which the stress is calculated. It is common practice to recast the flexure formula to yield the maximum normal stress σ_{max} and denote the value of $|y_{max}|$ by c, where c represents the distance from the neutral axis to the outermost fiber of the beam. On this basis, the flexure formula becomes

$$\sigma_{max} = \frac{Mc}{I} = \frac{M}{S} \tag{3.17}$$

The quantity $S = I/c$ is known as the section modulus of the cross-sectional area. Note that the flexure formula also applies to a beam of *unsymmetrical* cross-sectional area, provided I is a principal moment of inertia and M is a moment around a principal axis [1].

Curved Beam of a Rectangular Cross Section

Many machine and structural components loaded as beams, however, are not straight. When beams with initial curvature are subjected to bending moments, the stress distribution is not linear on either side of the neutral axis but increases more rapidly on the inner side. The

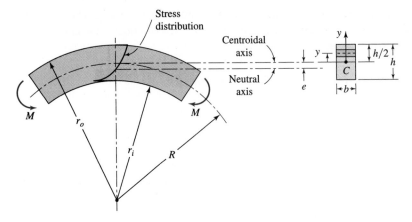

Figure 3.13   Curved bar in pure bending.

flexure and displacement formulas for these axisymmetrically loaded members are developed in the later chapters, using energy, elasticity, or exact, approximate technical theories.*

Here, the general equation for stress in curved members is adapted to the rectangular cross section shown in Figure 3.13. Therefore, for pure bending loads, the normal stress σ in a curved beam of a rectangular cross section, from Eq. (16.55):

$$\sigma = \frac{M}{AR}\left[1 + \frac{y}{Z(R+y)}\right] \tag{3.18}$$

The curved beam factor Z by Table 16.1 is

$$Z = -1 + \frac{R}{h}\ln\frac{r_o}{r_i} \tag{3.19}$$

In the foregoing expressions, we have

A = cross-sectional area

h = depth of beam

R = radius of curvature to the neutral axis

M = bending moment, positive when directed toward the concave side, as shown in the figure

y = distance measured from the neutral axis to the point at which stress is calculated, positive toward the convex side, as indicated in the figure

r_i, r_o = radii of the curvature of the inner and outer surfaces, respectively.

Accordingly, a positive value obtained from Eq. (3.18) means tensile stress.

---

l *Some readers may prefer to study Section 16.8.

The neutral axis shifts toward the center of curvature by distance e from the centroidal axis $(y = 0)$, as shown in Figure 3.13. By Eq. (16.57), we have $e = -ZR/(Z + 1)$. Expression for Z and e for many common cross-sectional shapes can be found referring to Table 16.1. Combined stresses in curved beams is presented in Chapter 16. A detailed comparison of the results obtained by various methods is illustrated in Example 16.7. Deflections of curved members due to bending, shear, and normal loads are discussed in Section 5.6.

SHEAR STRESS

We now consider the distribution of shear stress τ in a beam associated with the shear force V. The vertical shear stress τ_{xy} at any point on the cross section is numerically equal to the horizontal shear stress at the same point (see Section 1.13). Shear stresses as well as the normal stresses are taken to be uniform across the width of the beam. The shear stress $\tau_{xy} = \tau_{yx}$ at any point of a cross section (Figure 3.12b) is given by the shear formula:

$$\tau_{xy} = \frac{VQ}{Ib} \tag{3.20}$$

Here

V = the shearing force at the section

b = the width of the section measured at the point in question

Q = the first moment with respect to the neutral axis of the area A^* beyond the point at which the shear stress is required; that is,

$$Q = \int_{A^*} y \, dA = \bar{y} A^* \tag{3.21}$$

By definition, the area A^* represents the area of the part of the section below the point in question and $\bar{y}$ is the distance from the neutral axis to the centroid of A^*. Clearly, if $\bar{y}$ is measured above the neutral axis, Q represents the *first moment of the area* above the level where the shear stress is to be found. Obviously, shear stress varies in accordance with the shape of the cross section.

Rectangular Cross Section

To ascertain how the shear stress varies, we must examine how Q varies, because V, I, and b are constants for a rectangular cross section. In so doing, we find that the distribution of the shear stress on a cross section of a rectangular beam is parabolic. The stress is 0 at the top and bottom of the section $(y_1 = \pm h/2)$ and has its maximum value at the neutral axis $(y_1 = 0)$ as shown in Figure 3.14. Therefore,

$$\tau_{max} = \frac{V}{Ib} A^* \bar{y} = \frac{V}{(bh^3/12)b} \frac{bh}{2} \frac{h}{4} = \frac{3}{2} \frac{V}{A} \tag{3.22}$$

Figure 3.14   Shear stresses in a beam of rectangular cross section.

where $A = bh$ is the cross-sectional area of a beam having depth h and width b. For *narrow beams* with sides parallel to the y axis, Eq. (3.20) gives solutions in good agreement with the "exact" stress distribution obtained by the methods of the theory of elasticity. Equation (3.22) is particularly useful, since beams of rectangular-sectional form are often employed in practice. Stresses in a wide beam and plate are discussed in Section 4.10 after derivation of the strain-curvature relations.

The shear force acting across the width of the beam per unit length along the beam axis may be found by multiplying τ_{xy} in Eq. (3.22) by b (Figure 3.12b). This quantity is denoted by q, known as the *shear flow*,

$$q = \frac{VQ}{I} \tag{3.23}$$

The foregoing equation is valid for any beam having a cross section that is symmetrical about the y axis. It is very useful in the analysis of *built-up beams*. A beam of this type is fabricated by joining two or more pieces of material. Built-up beams are generally designed on the basis of the assumption that the parts are adequately connected so that the beam acts as a single member. Structural connections are taken up in Chapter 15.

EXAMPLE 3.6   Determining Stresses in a Simply Supported Beam

A simple beam of T-shaped cross section is loaded as shown in Figure 3.15a. Determine

(a) The maximum shear stress.

(b) The shear flow q_j and the shear stress τ_j in the joint between the flange and the web.

(c) The maximum bending stress.

Given: $P = 4$ kN and $L = 3$ m

Assumptions: All forces are coplanar and two dimensional.

Figure 3.15 Example 3.6.

Solution: The distance $\bar{y}$ to the centroid is determined as follows (Figure 3.15b):

$$\bar{y} = \frac{A_1\bar{y}_1 + A_2\bar{y}_2}{A_1 + A_2} = \frac{20(60)70 + 60(20)30}{20(60) + 60(20)} = 50 \text{ mm}$$

The moment of inertia I about the neutral axis is found using the parallel axis theorem:

$$I = \frac{1}{12}(60)(20)^3 + 20(60)(20)^2 + \frac{1}{12}(20)(60)^3 + 20(60)(20)^2 = 136 \times 10^4 \text{ mm}^4$$

The shear and moment diagrams (Figures 3.15c and 3.15d) are drawn using the method of sections.

(a) The maximum shearing stress in the beam occurs at the neutral axis on the cross section supporting the largest shear force V. Hence,

$$Q_{\text{N.A.}} = 50(20)25 = 25 \times 10^3 \text{ mm}^3$$

Since the shear force equals 2 kN on all cross sections of the beam (Figure 3.12c), we have

$$\tau_{\text{max}} = \frac{V_{\text{max}}Q_{\text{N.A.}}}{Ib} = \frac{2 \times 10^3(25 \times 10^{-6})}{136 \times 10^{-8}(0.02)} = 1.84 \text{ MPa}$$

(b) The first moment of the area of the flange about the neutral axis is

$$Q_f = 20(60)20 = 24 \times 10^3 \text{ mm}^3$$

Applying Eqs. (3.23) and (3.20),

$$q_j = \frac{V Q_f}{I} = \frac{2 \times 10^3 (24 \times 10^{-6})}{136 \times 10^{-8}} = 35.3 \text{ kN/m}$$

$$\tau_j = \frac{V Q_f}{I b} = \frac{35.3(10^3)}{0.02} = 1.76 \text{ MPa}$$

(c) The largest moment occurs at midspan, as shown in Figure 3.15d. Therefore, from Eq. (3.19), we obtain

$$\sigma_{\max} = \frac{Mc}{I} = \frac{3 \times 10^3 (0.05)}{136 \times 10^{-8}} = 110.3 \text{ MPa}$$

3.8   DESIGN OF BEAMS

We are here concerned with the elastic design of beams for strength. Beams made of single and two different materials are discussed. We note that some beams must be selected based on allowable deflections. This topic is taken up in Chapters 4 and 5. Occasionally, beam design relies on plastic moment capacity, the so-called limit design [1].

PRISMATIC BEAMS

We select the dimensions of a beam section so that it supports safely applied loads without exceeding the allowable stresses in both flexure and shear. Therefore, the design of the member is controlled by the largest normal and shear stresses developed at the critical section, where the maximum value of the bending moment and shear force occur. Shear and bending-moment diagrams are very helpful for locating these critical sections. In heavily loaded short beams, the design is usually governed by shear stress, while in slender beams, the flexure stress generally predominates. Shearing is more important in wood than steel beams, as wood has relatively low shear strength parallel to the grain.

Application of the rational procedure in design, outlined in Section 3.2, to a beam of ordinary proportions often includes the following steps:

1. It is assumed that failure results from yielding or fracture, and flexure stress is considered to be most closely associated with structural damage.

2. The significant value of bending stress is $\sigma = M_{\max}/S$.

3. The maximum usable value of σ without failure, $\sigma_{\max}$, is the yield strength S_y or the ultimate strength S_u.

4. A factor of safety n is applied to σ_{max} to obtain the allowable stress: $\sigma_{all} = \sigma_{max}/n$. The *required section modulus* of a beam is then

$$S = \frac{M_{max}}{\sigma_{all}} \tag{3.24}$$

There are generally several different beam sizes with the required value of S. We select the one with the lightest weight per unit length or the smallest sectional area from tables of beam properties. When the allowable stress is the same in tension and compression, a doubly symmetric section (i.e., section symmetric about the y and z axes) should be chosen. If σ_{all} is different in tension and compression, a singly symmetric section (for example a T beam) should be selected so that the distances to the extreme fibers in tension and compression are in a ratio nearly the same as the respective σ_{all} ratios.

We now check the *shear-resistance requirement* of beam tentatively selected. After substituting the suitable data for Q, I, b, and V_{max} into Eqs. (3.20), we determine the maximum shear stress in the beam from the formula

$$\tau_{max} = \frac{V_{max} Q}{Ib} \tag{3.25}$$

When the value obtained for τ_{max} is smaller than the allowable shearing stress τ_{all}, the beam is acceptable; otherwise, a stronger beam should be chosen and the process repeated.

Design of a Beam of Doubly Symmetric Section

EXAMPLE 3.7

Select a wide-flange steel beam to support the loads shown in Figure 3.16a.

Given: The allowable bending and shear stresses are 160 and 90 MPa, respectively.

Solution: Shear and bending-moment diagrams (Figures 3.16b and 3.16c) show that $M_{max} = 110$ kN · m and $V_{max} = 40$ kN. Therefore, Eq. (3.24) gives

$$S = \frac{110 \times 10^3}{160(10^6)} = 688 \times 10^3 \text{ mm}^3$$

Using Table A.6, we select the lightest member that has a section modulus larger than this value of S: a 200-mm W beam weighing 71 kg/m ($S = 709 \times 10^3$ mm^3). Since the weight of the beam ($71 \times 9.81 \times 10 = 6.97$ kN) is small compared with the applied load (80 kN), it is neglected.

The approximate or average maximum shear stress in beams with flanges may be obtained by dividing the shear force V by the web area:

$$\tau_{avg} = \frac{V}{A_{web}} = \frac{V}{ht} \tag{3.26}$$

Figure 3.16   Example 3.7.

In this relationship, h and t represent the beam depth and web thickness, respectively. From Table A.6, the area of the web of a $W\ 200 \times 71$ section is $216 \times 10.2 = 2.203(10^3)$ mm^2. We therefore have

$$\tau_{avg} = \frac{40 \times 10^3}{2.203(10^{-3})} = 18.16 \text{ MPa}$$

Comment: Inasmuch as this stress is well within the allowable limit of 90 MPa, the beam is acceptable.

BEAMS OF CONSTANT STRENGTH

When a beam is stressed to a uniform allowable stress, σ_{all}, throughout, then it is clear that the beam material is used to its greatest capacity. For a prescribed material, such a design is of minimum weight. At any cross section, the required section modulus S is given by

$$S = \frac{M}{\sigma_{all}} \tag{3.27}$$

where M presents the bending moment on an arbitrary section. Tapered beams designed in this manner are called *beams of constant strength*. Note that shear stress at those beam locations where the moment is small controls the design.

Beams of uniform strength are exemplified by leaf springs and certain forged or cast machine components (see Section 14.10). For a structural member, fabrication and design constraints make it impractical to produce a beam of constant stress. So, welded cover plates are often used for parts of prismatic beams where the moment is large; for instance, in a bridge girder. If the angle between the sides of a tapered beam is small, the flexure formula allows little error. On the other hand, the results obtained by using the shear stress formula may not be sufficiently accurate for nonprismatic beams. Usually, a modified form of this formula is used for design purposes. The exact distribution in a rectangular wedge is obtained by the theory of elasticity [2].

Design of a Constant Strength Beam

EXAMPLE 3.8

A cantilever beam of uniform strength and rectangular cross section is to support a concentrated load P at the free end (Figure 3.17a). Determine the required cross-sectional area, for two cases: (a) the width b is constant; (b) the height h is constant.

Figure 3.17 Example 3.8.
(a) Uniform strength cantilever;
(b) side view; (c) top view.

Solution:

(a) At a distance x from A, $M = Px$ and $S = bh^2/6$. Through the use of Eq. (3.27), we write

$$\frac{bh^2}{6} = \frac{Px}{\sigma_{all}} \tag{a}$$

Similarly, at a fixed end ($x = L$ and $h = h_1$),

$$\frac{bh_1^2}{6} = \frac{PL}{\sigma_{\text{all}}}$$

Dividing Eq. (a) by the preceeding relationship results in

$$h = h_1 \sqrt{\frac{x}{L}} \qquad \textbf{(b)}$$

Therefore, the depth of the beam varies parabolically from the free end (Figure 3.17b).

(b) Equation (a) now yields

$$b = \left(\frac{6P}{h^2 \sigma_{\text{all}}} \right) x = \frac{b_1}{L} x \qquad \textbf{(c)}$$

Comments: In Eq. (c), the expression in parentheses represents a constant and set equal to b_1/L so that when $x = L$ the width is b_1 (Figure 3.17c). In both cases, obviously the cross section of the beam near end A must be designed to resist the shear force, as shown by the dashed lines in the figure.

COMPOSITE BEAMS

Beams fabricated of two or more materials having different moduli of elasticity are called *composite beams*. The advantage of this type construction is that large quantities of low-modulus material can be used in regions of low stress, and small quantities of high-modulus materials can be used in regions of high stress. Two common examples are wooden beams whose bending strength is bolstered by metal strips, either along its sides or along its top or bottom, and reinforced concrete beams. The assumptions of the technical theory of homogenous beams, discussed in Section 3.7, are valid for a beam of more than one material. We use the common *transformed-section method* to ascertain the stresses in a composite beam. In this approach, the cross section of several materials is transformed into an equivalent cross section of one material in that the resisting forces are the same as on the original section. The flexure formula is then applied to the transformed section.

To demonstrate the method, a typical beam with symmetrical cross section built of two different materials is considered (Figure 3.18a). The moduli of elasticity of materials are denoted by E_1 and E_2. We define the modular ratio, n, as follows

$$n = \frac{E_2}{E_1} \qquad \textbf{(d)}$$

Although $n > 1$ in Eq. (d), the choice is arbitrary; the technique applies well for $n < 1$. The transformed section is composed of only material 1 (Figure 3.18b). The moment of inertia of the entire transformed area about the neutral axis is then denoted by I_t. It can be

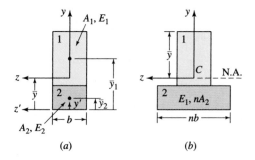

Figure 3.18   Beam of two materials: (a) Cross section; (b) equivalent section.

shown [1] that, the flexure formulas for a composite beam are in the forms

$$\sigma_{x1} = -\frac{My}{I_t}, \qquad \sigma_{x2} = -\frac{nMy}{I_t} \qquad \textbf{(3.28)}$$

where σ_{x1} and σ_{x2} are the stresses in materials 1 and 2, respectively. Obviously, when $E_1 = E_2 = E$, this equation reduces to the flexure formula for a beam of homogeneous material, as expected. The following sample problems illustrate the use of Eqs. (3.28).

Determination of Stress in a Composite Beam   **EXAMPLE 3.9**

A composite beam is made of wood and steel having the cross-sectional dimensions shown in Figure 3.19a. The beam is subjected to a bending moment $M_z = 25$ kN · m. Calculate the maximum stresses in each material.

Given: The modulus of elasticity of wood and steel are $E_w = 10$ GPa and $E_s = 210$ GPa, respectively.

Figure 3.19   Example 3.9: (a) Composite beam and (b) equivalent section.

Solution: The modular ratio $n = E_s/E_w = 21$. We use a transformed section of wood (Figure 3.19b). The centroid and the moment of inertia about the neutral axis of this section are

$$\bar{y} = \frac{150(200)(112) + 3150(12)(6)}{150(200) + 3150(12)} = 52.9 \text{ mm}$$

$$I_t = \frac{1}{12}(150)(200)^3 + 150(200)(59.1)^2 + \frac{1}{12}(3150)(12)^3 + 3150(12)(46.9)^2$$

$$= 288 \times 10^6 \text{ mm}^4$$

The maximum stress in the wood and steel portions are therefore

$$\sigma_{w,\text{max}} = \frac{Mc}{I_t} = \frac{25(10^3)(0.1591)}{288(10^{-6})} = 13.81 \text{ MPa}$$

$$\sigma_{s,\text{max}} = \frac{nMc}{I_t} = \frac{21(25 \times 10^3)(0.0529)}{288(10^{-6})} = 96.43 \text{ MPa}$$

At the juncture of the two parts, we have

$$\sigma_{w,\text{min}} = \frac{Mc}{I_t} = \frac{25(10^3)(0.0409)}{288(10^{-6})} = 3.55 \text{ MPa}$$

$$\sigma_{s,\text{min}} = n(\sigma_{w,\text{min}}) = 21(3.55) = 74.56 \text{ MPa}$$

Stress at any other location may be determined likewise.

EXAMPLE 3.10   Design of Steel Reinforced Concrete Beam

A concrete beam of width b and effective depth d is reinforced with three steel bars of diameter d_s (Figure 3.20a). Note that it is usual to use $a = 50$-mm allowance to protect the steel from corrosion or fire. Determine the maximum stresses in the materials produced by a positive bending moment of $50 \text{ kN} \cdot \text{m}$.

Figure 3.20   Example 3.10. Reinforced concrete beam.

Given: $b = 300$ mm, $d = 380$ mm, and $d_s = 25$ mm.

Assumptions: The modular ratio will be $n = E_s/E_c = 10$. The steel is uniformly stressed. Concrete resists only compression.

Solution: The portion of the cross section located a distance kd above the neutral axis is used in the transformed section (Figure 3.20b). The transformed area of the steel

$$nA_s = 10[3(\pi \times 25^2/4)] = 14{,}726 \text{ mm}^2$$

This is located by a single dimension from the neutral axis to its centroid. The compressive stress in the concrete is taken to vary linearly from the neutral axis. The first moment of the transformed section with respect to the neutral axis must be 0. Therefore,

$$b(kd)\frac{kd}{2} - nA_s(d - kd) = 0$$

from which

$$(kd)^2 + (kd)\frac{2nA_s}{b} - \frac{2nA_s}{b}d = 0 \tag{3.29}$$

Introducing the required numerical values, Eq. (3.29) becomes

$$(kd)^2 + 98.17(kd) - 37.31 \times 10^3 = 0$$

Solving, $kd = 150.2$ mm, and hence $k = 0.395$. The moment of inertia of the transformed cross section about the neutral axis is

$$I_t = \frac{1}{12}(0.3)(0.1502)^3 + 0.3(0.1502)(0.0751)^2 + 0 + 14.73 \times 10^{-3}(0.2298)^2$$

$$= 1116.5 \times 10^{-6} \text{ m}^4$$

The peak compressive stress in the concrete and tensile stress in the steel are

$$\sigma_{c,max} = \frac{Mc}{I_t} = \frac{50 \times 10^3 (0.1502)}{1116.5 \times 10^{-6}} = 6.73 \text{ MPa}$$

$$\sigma_s = \frac{nMc}{I_t} = \frac{10(50 \times 10^3)(0.2298)}{1116.5 \times 10^{-6}} = 102.9 \text{ MPa}$$

These stresses act as shown in Figure 3.20c.

Comments: Often an alternative method of solution is used to estimate readily the stresses in reinforced concrete [6]. We note that, inasmuch as concrete is very weak in tension, the beam depicted in Figure 3.20 would become practically useless, should the bending moments act in the opposite direction. For balanced reinforcement, the beam must be designed so that stresses in concrete and steel are at their allowable levels simultaneously.

3.9   PLANE STRESS

The stresses and strains treated thus far have been found on sections perpendicular to the coordinates used to describe a member. This section deals with the states of stress at points located on inclined planes. In other words, we wish to obtain the stresses acting on the sides of a stress element oriented in any desired direction. This process is termed a *stress transformation*. The discussion that follows is limited to two-dimensional, or plane, stress. A two-dimensional state of stress exists when the stresses are independent of one of the coordinate axes, here taken as z. The plane stress is therefore specified by $\sigma_z = \tau_{yz} = \tau_{xz} = 0$, where σ_x, σ_y, and τ_{xy} have nonzero values. Examples include the stresses arising on inclined sections of an axially loaded bar, a shaft in torsion, a beam with transversely applied force, and a member subjected to more than one load simultaneously.

Consider the stress components σ_x, σ_y, τ_{xy} at a point in a body represented by a two-dimensional stress element (Figure 3.21a). To portray the stresses acting on an inclined section, an infinitesimal wedge is isolated from this element and depicted in Figure 3.21b. The angle θ, locating the x' axis or the unit normal n to the plane AB, is assumed positive when measured from the x axis in a counterclockwise direction. Note that, according to the sign convention (see Section 1.13), the stresses are indicated as positive values. It can be shown that equilibrium of the forces caused by stresses acting on the wedge-shaped element gives the following transformation equations for plane stress [1–3]:

$$\sigma_{x'} = \sigma_x \cos^2 \theta + \sigma_y \sin^2 \theta + 2\tau_{xy} \sin \theta \cos \theta \tag{3.30a}$$

$$\tau_{x'y'} = \tau_{xy}(\cos^2 \theta - \sin^2 \theta) + (\sigma_y - \sigma_x)\sin \theta \cos \theta \tag{3.30b}$$

The stress $\sigma_{y'}$ may readily be obtained by replacing θ in Eq. (3.30a) by $\theta + \pi/2$ (Figure 3.21c). This gives

$$\sigma_{y'} = \sigma_x \sin^2 \theta + \sigma_y \cos^2 \theta - 2\tau_{xy} \sin \theta \cos \theta \tag{3.30c}$$

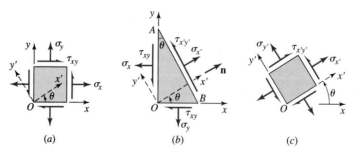

        (a)   (b)   (c)

Figure 3.21   Elements in plane stress.

Using the double-angle relationships, the foregoing equations can be expressed in the following useful alternative form:

$$\sigma_{x'} = \frac{1}{2}(\sigma_x + \sigma_y) + \frac{1}{2}(\sigma_x - \sigma_y)\cos 2\theta + \tau_{xy}\sin 2\theta \qquad (3.31a)$$

$$\tau_{x'y'} = -\frac{1}{2}(\sigma_x - \sigma_y)\sin 2\theta + \tau_{xy}\cos 2\theta \qquad (3.31b)$$

$$\sigma_{y'} = \frac{1}{2}(\sigma_x + \sigma_y) - \frac{1}{2}(\sigma_x - \sigma_y)\cos 2\theta - \tau_{xy}\sin 2\theta \qquad (3.31c)$$

For design purposes, the largest stresses are usually needed. The two perpendicular directions (θ_p' and θ_p'') of planes on which the shear stress vanishes and the normal stress has extreme values can be found from

$$\tan 2\theta_p = \frac{2\tau_{xy}}{\sigma_x - \sigma_y} \qquad (3.32)$$

The angle θ_p defines the orientation of the principal planes (Figure 3.22). The in-plane principal stresses can be obtained by substituting each of the two values of θ_p from Eq. (3.32) into Eqs. (3.31a and c) as follows:

$$\sigma_{\max,\min} = \sigma_{1,2} = \frac{\sigma_x + \sigma_y}{2} \pm \sqrt{\left(\frac{\sigma_x - \sigma_y}{2}\right)^2 + \tau_{xy}^2} \qquad (3.33)$$

The plus sign gives the algebraically larger maximum principal stress σ_1. The minus sign results in the minimum principal stress σ_2. It is necessary to substitute θ_p into Eq. (3.31a) to learn which of the two corresponds to σ_1.

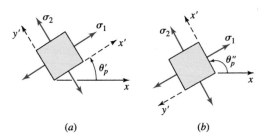

(a) (b)

Figure 3.22 Planes of principal stresses.

EXAMPLE 3.11 | Finding Stresses in a Cylindrical Pressure Vessel Welded along a Helical Seam

Figure 3.23a depicts a cylindrical pressure vessel constructed with a helical weld that makes an angle ψ with the longitudinal axis. Determine

(a) The maximum internal pressure p.

(b) The shear stress in the weld.

Figure 3.23   Example 3.11.

Given: $r = 10$ in., $t = \frac{1}{4}$ in., and $\psi = 55°$. Allowable tensile strength of the weld is 14.5 ksi.

Assumptions: Stresses are at a point A on the wall away from the ends. Vessel is a thin-walled cylinder.

Solution: The principal stresses in axial and tangential directions are, respectively,

$$\sigma_a = \frac{pr}{2t} = \frac{p(10)}{2\left(\frac{1}{4}\right)} = 20p = \sigma_2, \qquad \sigma_\theta = 2\sigma_a = 40p = \sigma_1$$

The state of stress is shown on the element of Figure 3.23b. We take the x' axis perpendicular to the plane of the weld. This axis is rotated $\theta = 35°$ clockwise with respect to the x axis.

(a) Through the use of Eq. (3.31a), the tensile stress in the weld:

$$\sigma_{x'} = \frac{\sigma_2 + \sigma_1}{2} + \frac{\sigma_2 - \sigma_1}{2} \cos 2(-35°)$$

$$= 30p - 10p \cos(-70°) \le 14,500$$

from which $p_{max} = 546$ psi.

(b) Applying Eq. (3.31b), the shear stress in the weld corresponding to the foregoing value of pressure is

$$\tau_{x'y'} = -\frac{\sigma_2 - \sigma_1}{2} \sin 2(-35°)$$

$$= 10p \sin(-70°) = -5.13 \text{ ksi}$$

The answer is presented in Figure 3.23b.

MOHR'S CIRCLE FOR STRESS

Transformation equations for plane stress, Eqs. (3.31), can be represented with σ and τ as coordinate axes in a graphical form known as *Mohr's circle* (Figure 3.24b). This representation is very useful in visualizing the relationships between normal and shear stresses acting on various inclined planes at a point in a stressed member. Also, with the aid of this graphical construction, a quicker solution of stress-transformation problem can be facilitated. The coordinates for point A on the circle correspond to the stresses on the x face or plane of the element shown in Figure 3.24a. Similarly, the coordinates of a point A' on Mohr's circle are to be interpreted representing the stress components $\sigma_{x'}$ and $\tau_{x'y'}$ that act on x' plane. The center is at $(\sigma', 0)$ and the circle radius r equals the length CA. In Mohr's circle representation the normal stresses obey the *sign convention* of Section 1.13. However, for the purposes of *only constructing and reading values* of stress from a Mohr's circle, the shear stresses on the y planes of the element are taken to be positive (as before) but those on the x faces are now negative, Figure 3.24c.

The magnitude of the maximum shear stress is equal to the radius r of the circle. From the geometry of Figure 3.24b, we obtain

$$\tau_{\max} = \sqrt{\left(\frac{\sigma_x - \sigma_y}{2}\right)^2 + \tau_{xy}^2} \qquad (3.34)$$

Mohr's circle shows the planes of maximum shear are always oriented at 45° from planes of principal stress (Figure 3.25). Note that a diagonal of a stress element along which the algebraically larger principal stress acts is called the *shear diagonal*. The maximum shear stress acts toward the shear diagonal. The normal stress occurring on planes of maximum shear stress is

$$\sigma' = \sigma_{\text{avg}} = \frac{1}{2}(\sigma_x + \sigma_y) \qquad (3.35)$$

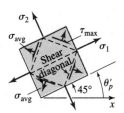

(a)   (b)   (c)

Figure 3.24 (a) Stress element; (b) Mohr's circle of stress; (c) interpretation of positive shear stress.

Figure 3.25 Planes of principal and maximum shear stresses.

It can readily be verified using Mohr's circle that, on any mutually perpendicular planes,

$$I_1 = \sigma_x + \sigma_y = \sigma_{x'} + \sigma_{y'} \qquad I_2 = \sigma_x\sigma_y - \tau_{xy}^2 = \sigma_{x'}\sigma_{y'} - \tau_{x'y'}^2 \qquad (3.36)$$

The quantities I_1 and I_2 are known as two-dimensional *stress invariants,* because they do not change in value when the axes are rotated positions. Equations (3.36) are particularly useful in checking numerical results of stress transformation.

Note that, in the case of triaxial stresses σ_1, σ_2, and σ_3, a Mohr's circle is drawn corresponding to each projection of a three-dimensional element. The three-circle cluster represents Mohr's circle for triaxial stress (see Figure 3.28). The general state of stress at a point is discussed in some detail in the later sections of this chapter. Mohr's circle construction is of fundamental importance because it applies to all (second-rank) tensor quantities; that is, Mohr's circle may be used to determine strains, moments of inertia, and natural frequencies of vibration [7]. It is customary to draw for Mohr's circle only a rough sketch; distances and angles are determined with the help of trigonometry. Mohr's circle provides a convenient means of obtaining the results for the stresses under the following two common loadings.

Axial Loading

In this case, we have $\sigma_x = \sigma_1 = P/A$, $\sigma_y = 0$, and $\tau_{xy} = 0$, where A is the cross-sectional area of the bar. The corresponding points A and B define a circle of radius $r = P/2A$ that passes through the origin of coordinates (Figure 3.26b). Points D and E yield the orientation of the planes of the maximum shear stress (Figure 3.26a), as well as the values of τ_{max} and the corresponding normal stress σ':

$$\tau_{max} = \sigma' = r = \frac{P}{2A} \qquad (a)$$

Observe that the normal stress is either maximum or minimum on planes for which shearing stress is 0.

Torsion

Now we have $\sigma_x = \sigma_y = 0$ and $\tau_{xy} = \tau_{max} = Tc/J$, where J is the polar moment of inertia of cross-sectional area of the bar. Points D and E are located on the τ axis, and Mohr's

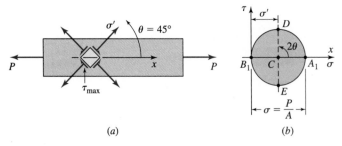

(a)   (b)

Figure 3.26   (a) Maximum shear stress acting on an element of an axially loaded bar; (b) Mohr's circle for uniaxial loading.

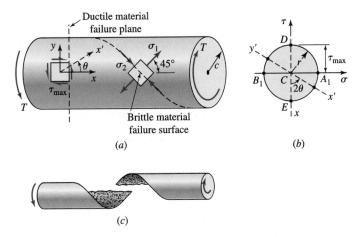

Figure 3.27 (a) Stress acting on a surface element of a twisted shaft; (b) Mohr's circle for torsional loading; (c) brittle material fractured in torsion.

circle is a circle of radius $r = Tc/J$ centered at the origin (Figure 3.27b). Points A_1 and B_1 define the principal stresses:

$$\sigma_{1,2} = \pm r = \pm \frac{Tc}{J} \tag{b}$$

So, it becomes evident that, for a material such as cast iron that is weaker in tension than in shear, failure occurs in tension along a helix indicated by the dashed lines in Figure 3.27a. Fracture of a bar that behaves in a brittle manner in torsion is depicted in Figure 3.27c; ordinary chalk behaves this way. Shafts made of materials weak in shear strength (for example, structural steel) break along a line perpendicular to the axis. Experiments show that a very thin-walled hollow shaft buckles or wrinkles in the direction of maximum compression while, in the direction of maximum tension, tearing occurs.

Stress Analysis of Cylindrical Pressure Vessel Using Mohr's Circle | **EXAMPLE 3.12**

Redo Example 3.11 using Mohr's circle. Also determine maximum in-plane and absolute shear stresses at a point on the wall of the vessel.

Solution: Mohr's circle, Figure 3.28, constructed referring to Figure 3.23 and Example 3.11, describes the state of stress. The x' axis is rotated $2\theta = 70°$ on the circle with respect to x axis.

(a) From the geometry of Figure 3.28, we have $\sigma_{x'} = 30p - 10p \cos 70° \le 14{,}500$. This results in $p_{max} = 546$ psi.

Figure 3.28   Example 3.12.

(b)   For the preceding value of pressure the shear stress in the weld is

$$\tau_{x'y'} = \pm 10(546)\sin 70° = \pm 5.13 \text{ ksi}$$

The largest in-plane shear stresses are given by points D and E on the circle. Hence,

$$\tau = \pm\frac{1}{2}(40p - 20p) = \pm 10(546) = \pm 5.46 \text{ ksi}$$

The third principal stress in the radial direction is 0, $\sigma_3 = 0$. The three principal stress circles are shown in the figure. The absolute maximum shear stresses are associated with points D' and E' on the major principal circle. Therefore,

$$\tau_{\text{max}} = \pm\frac{1}{2}(40p - 0) = \pm 20(546) = \pm 10.92 \text{ ksi}$$

3.10   COMBINED STRESSES

Basic formulas of mechanics of materials for determining the state of stress in elastic members are developed in Sections 3.2 through 3.7. Often these formulas give either a normal stress or a shear stress caused by a single load component being axially, centric, or lying in one plane. Note that each formula leads to stress as directly proportional to the magnitude of the applied load. When a component is acted on simultaneously by two or more loads, causing various internal-force resultants on a section, it is assumed that each load produces the stress as if it were the only load acting on the member. The final or combined stress is then found by superposition of several states of stress. As we see throughout the text, under combined loading, the critical points may not be readily located. Therefore, it may be necessary to examine the stress distribution in some detail.

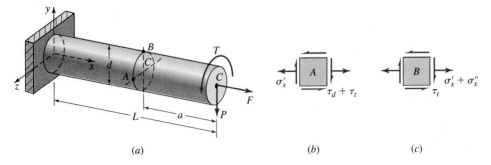

Figure 3.29 Combined stresses owing to torsion, tension, and direct shear.

Consider, for example, a solid circular cantilevered bar subjected to a transverse force P, a torque T, and a centric load F at its free end (Figure 3.29a). Every section experiences an axial force F, torque T, a bending moment M, and a shear force $P = V$. The corresponding stresses may be obtained using the applicable relationships:

$$\sigma'_x = \frac{F}{A}, \qquad \tau_t = -\frac{Tc}{J}, \qquad \sigma''_x = -\frac{Mc}{I}, \qquad \tau_d = -\frac{VQ}{Ib}$$

Here τ_t and τ_d are the torsional and direct shear stresses, respectively. In Figures 3.29b and 3.29c, the stresses shown are those acting on an element B at the top of the bar and on an element A on the side of the bar at the neutral axis. Clearly, B (when located at the support) and A represent the critical points at which most severe stresses occur. The principal stresses and maximum shearing stress at a critical point can now be ascertained as discussed in the preceding section.

The following examples illustrate the general approach to problems involving combined loadings. Any number of critical locations in the components can be analyzed. These either confirm the adequacy of the design or, if the stresses are too large (or too small), indicate the design changes required. This is used in a seemingly endless variety of practical situations, so it is often not worthwhile to develop specific formulas for most design use. We develop design formulas under combined loading of common mechanical components, such as shafts, shrink or press fits, flywheels, and pressure vessels in Chapters 9 and 16.

Determining the Allowable Combined Loading in a Cantilever Bar | **EXAMPLE 3.13**

A round cantilever bar is loaded as shown in Figure 3.29a. Determine the largest value of the load P.

Given: diameter $d = 60$ mm, $T = 0.1P$ N · m, and $F = 10P$ N.

Assumptions: Allowable stresses are 100 MPa in tension and 60 MPa in shear on a section at $a = 120$ mm from the free end.

Solution: The normal stress at all points of the bar is

$$\sigma_x' = \frac{F}{A} = \frac{10P}{\pi(0.03)^2} = 3536.8P \tag{a}$$

The torsional stress at the outer fibers of the bar is

$$\tau_t = -\frac{Tc}{J} = -\frac{0.1P(0.03)}{\pi(0.03)^4/2} = -2357.9P \tag{b}$$

The largest tensile bending stress occurs at point B of the section considered. Therefore, for $a = 120$ mm, we obtain

$$\sigma_x'' = \frac{Mc}{I} = \frac{0.12P(0.03)}{\pi(0.03)^4/4} = 5658.8P$$

Since $Q = A\overline{y} = (\pi c^2/2)(4c/3\pi) = 2c^3/3$ and $b = 2c$, the largest direct shearing stress at point A is

$$\tau_d = -\frac{VQ}{Ib} = -\frac{4V}{3A} = -\frac{4P}{3\pi(0.03)^2} = -471.57P \tag{c}$$

The maximum principal stress and the maximum shearing stress at point A (Figure 3.29b), applying Eqs. (3.33) and (3.34) with $\sigma_y = 0$, Eqs. (a), (b), and (c) are

$$(\sigma_1)_A = \frac{\sigma_x'}{2} + \left[\left(\frac{\sigma_x'}{2}\right)^2 + (\tau_d + \tau_t)^2\right]^{1/2}$$

$$= \frac{3536.8P}{2} + \left[\left(\frac{3536.8P}{2}\right)^2 + (-2829.5P)^2\right]^{1/2}$$

$$= 1768.4P + 3336.7P = 5105.1P$$

$$(\tau_{\max})_A = 3336.7P$$

Likewise, at point B (Figure 3.29c),

$$(\sigma_1)_B = \frac{\sigma_x' + \sigma_x''}{2} + \left[\left(\frac{\sigma_x' + \sigma_x''}{2}\right)^2 + \tau_t^2\right]^{1/2}$$

$$= \frac{9195.6P}{2} + \left[\left(\frac{9195.6P}{2}\right)^2 + (-2357.9P)^2\right]^{1/2}$$

$$= 4597.8P + 5167.2P = 9765P$$

$$(\tau_{\max})_B = 5167.2P$$

It is observed that the stresses at B are more severe than those at A. Inserting the given data into the foregoing, we obtain

$$100(10^6) = 9765P \quad \text{or} \quad P = 10.24 \text{ kN}$$

$$60(10^6) = 5167.2P \quad \text{or} \quad P = 11.61 \text{ kN}$$

Comment: The magnitude of the largest allowable transverse, axial, and torsional loads that can be carried by the bar are $P = 10.24$ kN, $F = 102.4$ kN, and $T = 1.024$ kN · m, respectively.

Determination of Maximum Allowable Pressure in a Pipe under Combined Loading | **EXAMPLE 3.14**

A cylindrical pipe subjected to internal pressure p is simultaneously compressed by an axial load P through the rigid end plates, as shown in Figure 3.30a. Calculate the largest value of p that can be applied to the pipe.

(a)   (b)

Figure 3.30 Example 3.14.

Given: The pipe diameter $d = 120$ mm, thickness $t = 5$ mm, and $P = 60$ kN. Allowable in-plane shear stress in the wall is 80 MPa.

Assumption: The critical stress is at a point on cylinder wall away from the ends.

Solution: The cross-sectional area of this thin-walled shell is $A = \pi dt$. Combined axial and tangential stresses act at a critical point on an element in the wall of the pipe (Figure 3.30b). We have

$$\sigma_x'' = -\frac{P}{\pi dt} = -\frac{60(10^3)}{\pi(0.12 \times 0.005)} = -31.83 \text{ MPa}$$

$$\sigma_x' = \frac{pr}{2t} = \frac{p(60)}{2(5)} = 6p$$

$$\sigma_\theta = \frac{pr}{t} = 12p$$

Applying Eq. (3.34),

$$\tau_{max} = \frac{1}{2}[\sigma_\theta - (\sigma_x' + \sigma_x'')] = \frac{1}{2}[12p - (6p - 31.83)]$$

$$= 3p_{max} + 15.915 \le 80$$

from which

$$p_{max} = 21.36 \text{ MPa}$$

Case Study 3-1 | WINCH CRANE FRAME STRESS ANALYSIS

The frame of a winch crane is represented schematically in Figure 1.5. Determine the maximum stress and the factor of safety against yielding.

Given: The geometry and loading are known from Case Study 1-1. The frame is made of ASTM-A36 structural steel tubing. From Table B.1:

$$S_y = 250 \text{ MPa} \qquad E = 200 \text{ GPa}$$

Assumptions: The loading is static. The displacements of welded joint C are negligibly small, hence part CD of the frame is considered a cantilever beam.

Solution: See Figures 1.5 and 3.31 and Table B.1.

We observe from Figure 1.5 that the maximum bending moment occurs at points B and C and $M_B = M_C$. Since two vertical beams resist moment at B, the critical section is at C of cantilever CD carrying its own weight per unit length w and concentrated load P at the free end (Figure 3.31).

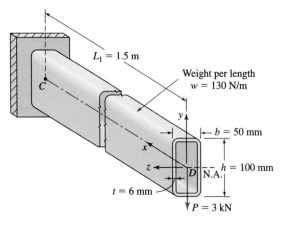

Figure 3.31   Part CD of the crane frame shown in Figure 1.4.

The bending moment M_C and shear force V_C at the cross section through the point C, from static equilibrium, have the following values:

$$M_C = PL_1 + \frac{1}{2}wL_1^2$$

$$= 3000(1.5) + \frac{1}{2}(130)(1.5)^2 = 4646 \text{ N} \cdot \text{m}$$

$$V_C = 3 \text{ kN}$$

The cross-sectional area properties of the tubular beam are

$$A = bh - (b - 2t)(h - 2t)$$

$$= 50 \times 100 - 38 \times 88 = 1.66(10^{-3}) \text{ m}^2$$

$$I = \frac{1}{12}bh^3 - \frac{1}{12}(b - 2t)(h - 2t)^3$$

$$= \frac{1}{12}[(50 \times 100^3) - (38)(88)^3] = 2.01(10^{-6}) \text{ m}^4$$

where I represents the moment of inertia about the neutral axis.

Therefore, the maximum bending stress at point C equals

$$\sigma_C = \frac{Mc}{I} = \frac{4646(0.05)}{2.01(10^{-6})} = 115.6 \text{ MPa}$$

The highest value of the shear stress occurs at the neutral axis. Referring to Figure 3.31, the first moment of the area about the N.A. is

$$Q_{max} = b\left(\frac{h}{2}\right)\left(\frac{h}{4}\right) - (b - 2t)\left(\frac{h}{2} - t\right)\left(\frac{h/2 - t}{2}\right)$$

$$= 50(50)(25) - (38)(44)(22) = 25.716(10^{-6}) \text{ m}^3$$

Hence,

$$\tau_C = \frac{V_C Q_{max}}{Ib}$$

$$= \frac{3000(25.716)}{2.01(2 \times 0.006)} = 3.199 \text{ MPa}$$

Case Study (CONCLUDED)

We obtain the largest principal stress $\sigma_1 = \sigma_{max}$ from Eq. (3.33), which in this case reduces to

$$\sigma_{max} = \frac{\sigma_C}{2} + \sqrt{\left(\frac{\sigma_C}{2}\right)^2 + \tau_C^2}$$

$$= \frac{115.6}{2} + \left[\left(\frac{115.6}{2}\right)^2 + (3.199)^2\right]^{1/2}$$

$$= 115.7 \text{ MPa}$$

The factor of safety against yielding is then

$$n = \frac{S_y}{\sigma_{max}} = \frac{250}{115.7} = 2.16$$

This is satisfactory because the frame is made of average material operated in ordinary environment and subjected to known loads.

Comments: At joint C, as well as at B, a thin (about 6-mm) steel gusset should be added at each side (not shown in the figure). These enlarge the weld area of the joints and help reduce stress in the welds. Case Study 15-2 illustrates the design analysis of the welded joint at C.

Case Study 3-2 | BOLT CUTTER STRESS ANALYSIS

A bolt cutting tool is shown in Figure 1.6. Determine the stresses in the members.

Given: The geometry and forces are known from Case Study 1-2. Material of all parts is AISI 1080 HR steel. Dimensions are in inches. We have

$S_y = 60.9$ ksi (Table B.3), $S_{ys} = 0.5S_y = 30.45$ ksi,

$E = 30 \times 10^6$ psi

Assumptions:

1. The loading is taken to be static. The material is ductile, and stress concentration factors can be disregarded under steady loading.

2. The most likely failure points are in link 3, the hole where pins are inserted, the connecting pins in shear, and jaw 2 in bending.

3. Member 2 can be approximated as a simple beam with an overhang.

Solution: See Figures 1.6 and 3.32.

The largest force on any pin in the assembly is at joint A.

Member 3 is a pin-ended tensile link. The force on a pin is 128 lb, as shown in Figure 3.32a. The normal stress is therefore

$$\sigma = \frac{F_A}{(w_3 - d)t_3} = \frac{128}{\left(\frac{3}{8} - \frac{1}{8}\right)\left(\frac{1}{8}\right)} = 4.096 \text{ ksi}$$

For the bearing stress in the joint A, using Eq. (3.5), we have

$$\sigma_b = \frac{F_A}{dt_3} = \frac{128}{\left(\frac{1}{8}\right)\left(\frac{1}{8}\right)} = 8.192 \text{ ksi}$$

The link and other members have ample material around holes to prevent tearout. The $\frac{1}{8}$-in. diameter pins are in single shear. The worst-case direct shear stress, from Eq. (3.4),

$$\tau = \frac{4F_A}{\pi d^2} = \frac{4(128)}{\pi\left(\frac{1}{8}\right)^2} = 10.43 \text{ ksi}$$

(continued)

Case Study   (CONCLUDED)

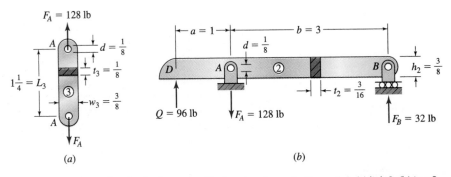

Figure 3.32   Some free-body diagrams of bolt cutter shown in Figure 1.6: (a) link 3; (b) jaw 2.

Member 2, the jaw, is supported and loaded as shown in Figure 3.32b. The moment of inertia of the cross-sectional area is

$$I = \frac{t_2}{12}\left(h_2^3 - d^3\right)$$

$$= \frac{3/16}{12}\left[\left(\frac{3}{8}\right)^3 - \left(\frac{1}{8}\right)^3\right] = 0.793(10^{-3}) \text{ in.}^4$$

The maximum moment, that occurs at point A of the jaw, equals $M = F_B b = 32(3) = 96$ lb · in. The bending stress is then

$$\sigma_C = \frac{Mc}{I} = \frac{96\left(\frac{3}{16}\right)}{0.793 \times 10^{-3}} = 22.7 \text{ ksi}$$

It can readily be shown that, the shear stress is negligibly small in the jaw.

Member 1, the handle, has an irregular geometry and is relatively massive compared to the other components of the assembly. Accurate values of stresses as well as deflections in the handle may be obtained by the finite element analysis.

Comment: The results show that the maximum stresses in members are well under the yield strength of the material.

3.11   PLANE STRAIN

In the case of two-dimensional, or plane, strain, all points in the body before and after the application of the load remain in the same plane. Therefore, in the xy plane the strain components ε_x, ε_y, and γ_{xy} may have nonzero values. The normal and shear strains at a point in a member vary with direction in a way analogous to that for stress. We briefly discuss expressions that give the strains in the inclined directions. These in-plane strain transformation equations are particularly significant in experimental investigations, where strains are measured by means of strain gages. The site at www.measurementsgroup.com includes general information on strain gages as well as instrumentation.

Mathematically, in every respect, the transformation of strain is the same as the stress transformation. It can be shown that [2] *transformation* expressions of stress are converted into strain relationships by substitution:

$$\sigma \to \varepsilon \quad \text{and} \quad \tau \to \gamma/2 \tag{a}$$

These replacements can be made in all the analogous two- and three-dimensional transformation relations. Therefore, the principal strain directions are obtained from Eq. (3.32) in the form, for example,

$$\tan 2\theta_p = \frac{\gamma_{xy}}{\varepsilon_x - \varepsilon_y} \tag{3.37}$$

Using Eq. (3.33), the magnitudes of the in-plane principal strains are

$$\varepsilon_{1,2} = \frac{\varepsilon_x + \varepsilon_y}{2} \pm \sqrt{\left(\frac{\varepsilon_x - \varepsilon_y}{2}\right)^2 + \left(\frac{\gamma_{xy}}{2}\right)^2} \tag{3.38}$$

In a like manner, the in-plane transformation of strain in an arbitrary direction proceeds from Eqs. (3.31):

$$\varepsilon_{x'} = \frac{1}{2}(\varepsilon_x + \varepsilon_y) + \frac{1}{2}(\varepsilon_x - \varepsilon_y)\cos 2\theta + \frac{\gamma_{xy}}{2}\sin 2\theta \tag{3.39a}$$

$$\gamma_{x'y'} = -(\varepsilon_x - \varepsilon_y)\sin 2\theta + \gamma_{xy}\cos 2\theta \tag{3.39b}$$

$$\varepsilon_{y'} = \frac{1}{2}(\varepsilon_x + \varepsilon_y) - \frac{1}{2}(\varepsilon_x - \varepsilon_y)\cos 2\theta - \frac{\gamma_{xy}}{2}\sin 2\theta \tag{3.39c}$$

An expression for the maximum shear strain may also be found from Eq. (3.34). Similarly, the transformation equations of three-dimensional strain may be deduced from the corresponding stress relations, given in Section 3.18.

In *Mohr's circle for strain,* the normal strain ε is plotted on the horizontal axis, positive to the right. The vertical axis is measured in terms of $\gamma/2$. The center of the circle is at $(\varepsilon_x + \varepsilon_y)/2$. When the shear strain is *positive,* the point representing the x axis strain is plotted a distance $\gamma/2$ *below* the axis and vice versa when shear strain is negative. Note that this convention for shearing strain, used *only* in constructing and reading values from Mohr's circle, agrees with the convention used for stress in Section 3.9.

Determination of Principal Strains Using Mohr's Circle | **EXAMPLE 3.15**

It is observed that an element of a structural component elongates 450μ along the x axis, contracts 120μ in the y direction, and distorts through an angle of -360μ (see Section 1.14). Calculate

(a) The principal strains.

(b) The maximum shear strains.

Given: $\varepsilon_x = 450\mu$, $\varepsilon_y = -120\mu$, $\gamma_{xy} = -360\mu$

Assumption: Element is in a state of plane strain.

Solution: A sketch of Mohr's circle is shown in Figure 3.33, constructed by finding the position of point C at $\varepsilon' = (\varepsilon_x + \varepsilon_y)/2 = 165\mu$ on the horizontal axis and of point A at $(\varepsilon_x, -\gamma_{xy}/2) = (450\mu, 180\mu)$ from the origin O.

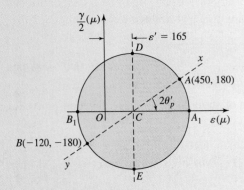

Figure 3.33   Example 3.15.

(a)   The in-plane principal strains are represented by points A and B. Hence,

$$\varepsilon_{1,2} = 165\mu \pm \left[\left(\frac{450 + 120}{2} \right)^2 + (-180)^2 \right]^{1/2}$$

$$\varepsilon_1 = 502\mu \qquad \varepsilon_2 = -172\mu$$

Note, as a check, that $\varepsilon_x + \varepsilon_y = \varepsilon_1 + \varepsilon_2 = 330\mu$. From geometry,

$$\theta'_p = \frac{1}{2} \tan^{-1} \frac{180}{285} = 16.14°$$

It is seen from the circle that θ'_p locates the ε_1 direction.

(b)   The maximum shear strains are given by points D and E. Hence,

$$\gamma_{\max} = \pm(\varepsilon_1 - \varepsilon_2) = \pm 674\mu$$

Comments: Mohr's circle depicts that the axes of maximum shear strain make an angle of 45° with respect to principal axes. In the directions of maximum shear strain, the normal strains are equal to $\varepsilon' = 165\mu$.

3.12 STRESS CONCENTRATION FACTORS

The condition where high localized stresses are produced as a result of an abrupt change in geometry is called the stress concentration. The abrupt change in form or discontinuity occurs in such frequently encountered stress raisers as holes, notches, keyways, threads, grooves, and fillets. Note that the stress concentration is a primary cause of fatigue failure and static failure in brittle materials, discussed in the next section. The formulas of mechanics of materials apply as long as the material remains linearly elastic and shape variations are gradual. In some cases, the stress and accompanying deformation near a discontinuity can be analyzed by applying the theory of elasticity. In those instances that do not yield to analytical methods, it is more usual to rely on experimental techniques or the finite element method (see Case Study 17-4). In fact, much research centers on determining stress concentration effects for combined stress.

A geometric or theoretical *stress concentration factor* K_t is used to relate the maximum stress at the discontinuity to the nominal stress. The factor is defined by

$$K_t = \frac{\sigma_{max}}{\sigma_{nom}} \quad \text{or} \quad K_t = \frac{\tau_{max}}{\tau_{nom}} \tag{3.40}$$

Here the nominal stresses are stresses that would occur if the abrupt change in the cross section did not exist or had no influence on stress distribution. It is important to note that a stress concentration factor is applied to the stress computed for the net or reduced cross section. Stress concentration factors for several types of configuration and loading are available in technical literature [8–13].

The stress concentration factors for a variety of geometries, provided in Appendix C, are useful in the design of machine parts. Curves in the Appendix C figures are plotted on the basis of dimensionless ratios: the shape, but not the size, of the member is involved. Observe that all these graphs indicate the advisability of streamlining junctures and transitions of portions that make up a member; that is, stress concentration can be reduced in intensity by properly proportioning the parts. Large fillet radii help at reentrant corners.

The values shown in Figures C.1, C.2, and C.7 through C.9 are for fillets of radius r that join a part of depth (or diameter) d to the one of larger depth (or diameter) D at a step or shoulder in a member (see Figure 3.34). A full fillet is a 90° arc with radius $r = (D - d_f)/2$. The stress concentration factor decreases with increases in r/d or d/D. Also, results for the axial tension pertain equally to cases of axial compression. However, the stresses obtained are valid only if the loading is not significant relative to that which would cause failure by buckling.

Figure 3.34 A flat bar with fillets and a centric hole under axial loading.

EXAMPLE 3.16   Design of Axially Loaded Thick Plate with a Hole and Fillets

A filleted plate of thickness t supports an axial load P (Figure 3.34). Determine the radius r of the fillets so that the same stress occurs at the hole and the fillets.

Given: $P = 50\,\text{kN}$, $D = 100\,\text{mm}$, $d_f = 66\,\text{mm}$, $d_h = 20\,\text{mm}$, $t = 10\,\text{mm}$

Design Decisions: The plate will be made of a relatively brittle metallic alloy; we must consider stress concentration.

Solution: For the *circular hole*,

$$\frac{d_h}{D} = \frac{20}{100} = 0.2, \qquad A = (D - d_h)t = (100 - 20)10 = 800\ \text{mm}^2$$

Using the lower curve in Figure C.5, we find that $K_t = 2.44$ corresponding to $d_h/D = 0.2$. Hence,

$$\sigma_{\max} = K_t \frac{P}{A} = 2.44\frac{50 \times 10^3}{800(10^{-6})} = 152.5\ \text{MPa}$$

For *fillets*,

$$\sigma_{\max} = K_t \frac{P}{A} = K_t \frac{50 \times 10^3}{660(10^{-6})} = 75.8 K_t\ \text{MPa}$$

The requirement that the maximum stress for the hole and fillets be identical is satisfied by

$$152.5 = 75.8 K_t \quad \text{or} \quad K_t = 2.01$$

From the curve in Figure C.1, for $D/d_f = 100/66 = 1.52$, we find that $r/d_f = 0.12$ corresponding to $K_t = 2.01$. The necessary fillet radius is therefore

$$r = 0.12 \times 66 = 7.9\ \text{mm}$$

3.13   IMPORTANCE OF STRESS CONCENTRATION FACTORS IN DESIGN

Under certain conditions, a normally ductile material behaves in a brittle manner and vice versa. So, for a specific application, the distinction between ductile and brittle materials must be inferred from the discussion of Section 2.9. Also remember that the determination of stress concentration factors is based on the use of Hooke's law.

FATIGUE LOADING

Most engineering materials may fail as a result of propagation of cracks originating at the point of high dynamic stress. The presence of stress concentration in the case of fluctuating (and impact) loading, as found in some machine elements, must be considered, regardless of whether the material response is brittle or ductile. In machine design, then, fatigue stress concentrations are of paramount importance. However, its effect on the nominal stress is not as large, as indicated by the theoretical factors (see Section 8.7).

STATIC LOADING

For static loading, stress concentration is important only for *brittle* material. However, for some brittle materials having internal irregularities, such as cast iron, stress raisers usually have little effect, regardless of the nature of loading. Hence, the use of a stress concentration factor appears to be unnecessary for cast iron. Customarily, stress concentration is ignored in static loading of *ductile* materials. The explanation for this restriction is quite simple. For ductile materials slowly and steadily loaded beyond the yield point, the stress concentration factors decrease to a value approaching unity because of the redistribution of stress around a discontinuity.

To illustrate the foregoing inelastic action, consider the behavior of a mild-steel flat bar that contains a hole and is subjected to a gradually increasing load P (Figure 3.35). When σ_{max} reaches the yield strength S_y, stress distribution in the material is of the form of curve mn, and yielding impends at A. Some fibers are stressed in the plastic range but enough others remain elastic, and the member can carry additional load. We observe that the area under stress distribution curve is equal to the load P. This area increases as overload P increases, and a contained plastic flow occurs in the material [14]. Therefore, with the increase in the value of P, the stress-distribution curve assumes forms such as those shown by line mp and finally mq. That is, the effect of an abrupt change in geometry is nullified and $\sigma_{max} = \sigma_{nom}$, or $K_t = 1$; prior to necking, a nearly *uniform* stress distribution across the net section occurs. Hence, for most practical purposes, the bar containing a hole carries the same static load as the bar with no hole.

The effect of ductility on the strength of the shafts and beams with stress raisers is similar to that of axially loaded bars. That is, localized inelastic deformations enable these members to support high stress concentrations. Interestingly, material ductility introduces a certain element of forgiveness in analysis while producing acceptable design results; for example, rivets can carry equal loads in a riveted connection (see Section 15.13).

When a member is yielded nonuniformly throughout a cross section, *residual stresses* remain in this cross section after the load is removed. An overload produces residual stresses favorable to future loads in the same direction and unfavorable to future loads in the opposite direction. Based on the idealized stress-strain curve, the increase in load

Figure 3.35 Redistribution of stress in a flat bar of mild steel.

capacity in one direction is the same as the decrease in load capacity in the opposite direction. Note that coil springs in compression are good candidates for favorable residual stresses caused by yielding.

3.14   CONTACT STRESS DISTRIBUTIONS

The application of a load over a small area of contact results in unusually high stresses. Situations of this nature are found on a microscopic scale whenever force is transmitted through bodies in contact. The original analysis of elastic contact stresses, by H. Hertz, was published in 1881. In his honor, the stresses at the mating surfaces of curved bodies in compression are called *Hertz contact stresses*. The Hertz problem relates to the stresses owing to the contact surface of a sphere on a plane, a sphere on a sphere, a cylinder on a cylinder, and the like. In addition to rolling bearings, the problem is of importance to cams, push rod mechanisms, locomotive wheels, valve tappets, gear teeth, and pin joints in linkages.

Consider the contact without deflection of two bodies having curved surfaces of different radii (r_1 and r_2), in the vicinity of contact. If a collinear pair of forces (F) presses the bodies together, deflection occurs and the point of contact is replaced by a small area of contact. The first steps taken toward the solution of this problem are the determination of the size and shape of the contact area as well as the distribution of normal pressure acting on the area. The deflections and subsurface stresses resulting from the contact pressure are then evaluated. The following *basic assumptions* are generally made in the solution of the Hertz problem:

1. The contacting bodies are isotropic, homogeneous, and elastic.
2. The contact areas are essentially flat and small relative to the radii of curvature of the undeflected bodies in the vicinity of the interface.
3. The contacting bodies are perfectly smooth, therefore friction forces need not be taken into account.

The foregoing set of presuppositions enables elastic analysis by theory of elasticity. Without going into the rather complex derivations, in this section, we introduce some of the results for both cylinders and spheres. The next section concerns the contact of two bodies of any general curvature. Contact problems of rolling bearings and gear teeth are discussed in the later chapters.*

SPHERICAL AND CYLINDRICAL SURFACES IN CONTACT

Figure 3.36 illustrates the contact area and corresponding stress distribution between two spheres, loaded with force F. Similarly, two parallel cylindrical rollers compressed by forces F is shown in Figure 3.37. We observe from the figures that, in each case, the maximum contact pressure exist on the load axis. The area of contact is defined by dimension a for the spheres and a and L for the cylinders. The relationships between the force of contact F,

*A summary and complete list of references dealing with contact stress problems are given by References [2, 4, 15–17].

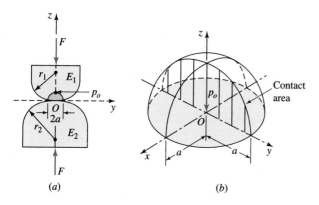

(a)   (b)

Figure 3.36 (a) Spherical surfaces of two members held in contact by force *F*. (b) Contact stress distribution. Note: The contact area is a circle of radius *a*.

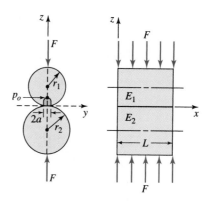

Figure 3.37 Two cylinders held in contact by force *F* uniformly distributed along cylinder length *L*. Note: The contact area is a narrow rectangle of 2*a* × *L*.

maximum pressure p_o, and the *deflection* δ in the point of contact are given in Table 3.2. Obviously, the δ represents the relative displacement of the centers of the two bodies, owing to local deformation. The contact pressure within each sphere or cylinder has a semi-elliptical distribution; it varies from 0 at the side of the contact area to a maximum value p_o at its center, as shown in the figures. For spheres, a is the radius of the circular contact area (πa^2). But, for cylinders, a represents the half-width of the rectangular contact area ($2aL$), where L is the length of the cylinder. Poisson's ratio ν in the formulas is taken as 0.3.

The material along the axis compressed in the z direction tends to expand in the x and y directions. However, the surrounding material does not permit this expansion; hence, the compressive stresses are produced in the x and y directions. The maximum stresses occur along the load axis z, and they are principal stresses (Figure 3.38). These and the resulting maximum shear stresses are given in terms of the maximum contact pressure p_o by the equations to follow [3, 16].

Figure 3.38
Principal stress below the surface along the load axis z.

Two Spheres in Contact (Figure 3.36)

$$\sigma_x = \sigma_y = -p_o \left\{ \left(1 - \frac{z}{a} \tan^{-1} \frac{1}{z/a} \right)(1 + \nu) - \frac{1}{2[1 + (z/a)^2]} \right\} \qquad \textbf{(3.41a)}$$

$$\sigma_z = -\frac{p_o}{1 + (z/a)^2} \qquad \textbf{(3.41b)}$$

Therefore, we have $\tau_{xy} = 0$ and

$$\tau_{yz} = \tau_{xz} = \frac{1}{2}(\sigma_x - \sigma_z) \qquad \textbf{(3.41c)}$$

A plot of these equations is shown in Figure 3.39a.

Table 3.2   Maximum pressure p_o and deflection δ of two bodies in point of contact

Configuration	**Spheres:** $p_o = 1.5\dfrac{F}{\pi a^2}$	**Cylinders:** $p_o = \dfrac{2}{\pi}\dfrac{F}{aL}$
A.	Sphere on a Flat Surface $a = 0.880 \sqrt[3]{Fr_1\Delta}$ $\delta = 0.775 \sqrt[3]{F^2\dfrac{\Delta^2}{r_1}}$	Cylinder on a Flat Surface $a = 1.076 \sqrt{\dfrac{F}{L}r_1\Delta}$ For $E_1 = E_2 = E$: $\delta = \dfrac{0.579F}{EL}\left(\dfrac{1}{3} + \ln\dfrac{2r_1}{a}\right)$
B.	Two Spherical Balls $a = 0.880 \sqrt[3]{F\dfrac{\Delta}{m}}$ $\delta = 0.775 \sqrt[3]{F^2\Delta^2 m}$	Two Cylindrical Rollers $a = 1.076 \sqrt{\dfrac{F\Delta}{Lm}}$
C.	Sphere on a Spherical Seat $a = 0.880 \sqrt[3]{F\dfrac{\Delta}{n}}$ $\delta = 0.775 \sqrt[3]{F^2\Delta^2 n}$	Cylinder on a Cylindrical Seat $a = 1.076 \sqrt{\dfrac{F\Delta}{Ln}}$

Note: $\Delta = \dfrac{1}{E_1} + \dfrac{1}{E_2}$, $\quad m = \dfrac{1}{r_1} + \dfrac{1}{r_2}$, $\quad n = \dfrac{1}{r_1} - \dfrac{1}{r_2}$

where the modulus of elasticity (E) and radius (r) are for the contacting members, 1 and 2. The L represents the length of the cylinder (Figure 3.37). The total force pressing two spheres or cylinders is F.

Two Cylinders in Contact *(Figure 3.37)*

$$\sigma_x = -2vp_o\left[\sqrt{1 + \left(\dfrac{z}{a}\right)^2} - \dfrac{z}{a}\right] \tag{3.42a}$$

$$\sigma_y = -p_o\left\{\left[2 - \dfrac{1}{1 + (z/a)^2}\right]\sqrt{1 + \left(\dfrac{z}{a}\right)^2} - 2\dfrac{z}{a}\right\} \tag{3.42b}$$

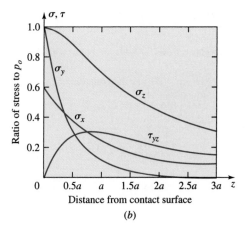

Figure 3.39 Stresses below the surface along the load axis (for $\nu = 0.3$): (a) two spheres; (b) two parallel cylinders. Note: All normal stresses are compressive stresses.

$$\sigma_z = -\frac{p_o}{\sqrt{1 + (z/a)^2}} \tag{3.42c}$$

$$\tau_{xy} = \frac{1}{2}(\sigma_x - \sigma_y), \qquad \tau_{yz} = \frac{1}{2}(\sigma_y - \sigma_z), \qquad \tau_{xz} = \frac{1}{2}(\sigma_x - \sigma_z) \tag{3.42d}$$

Equations (3.42a–3.42c) and the second of Eqs. (3.42d) are plotted in Figure 3.39b. For each case, Figure 3.39 illustrates how principal stresses diminish below the surface. It also shows how the shear stress reaches a maximum value slightly below the surface and diminishes. The maximum shear stresses act on the planes bisecting the planes of maximum and minimum principal stresses.

The subsurface shear stresses is believed to be responsible for the surface fatigue failure of contacting bodies (see Section 8.15). The explanation is that minute cracks originate at the point of maximum shear stress below the surface and propagate to the surface to permit small bits of material to separate from the surface. As already noted, all stresses considered in this section exist along the load axis z. The states of stress off the z axis are not required for design purposes, because the maxima occur on the z axis.

Determining Maximum Contact Pressure between a Cylindrical Rod and a Beam | **EXAMPLE 3.17**

A concentrated load F at the center of a narrow, deep beam is applied through a rod of diameter d laid across the beam width of width b. Determine

(a) The contact area between rod and beam surface.

(b) The maximum contact stress.

Given: $F = 4 \text{ kN},$ $d = 12 \text{ mm},$ $b = 125 \text{ mm}$

Assumptions: Both the beam and the rod are made of steel having $E = 200$ GPa and $\nu = 0.3$.

Solution: We use the equations on the second column of case A in Table 3.2.

(a) Since $E_1 = E_2 = E$ or $\Delta = 2/E$, the half-width of contact area is

$$a = 1.076 \sqrt{\frac{F}{L} r_1 \Delta}$$

$$= 1.076 \sqrt{\frac{4(10^3)}{0.125} \frac{(0.006)2}{200(10^9)}} = 0.0471 \text{ mm}$$

The rectangular contact area equals

$$2aL = 2(0.0471)(125) = 11.775 \text{ mm}^2$$

(b) The maximum contact pressure is therefore

$$p_o = \frac{2}{\pi} \frac{F}{aL} = \frac{2}{\pi} \frac{4(10^3)}{5.888(10^{-6})} = 432.5 \text{ MPa}$$

Case Study 3-3   CAM AND FOLLOWER STRESS ANALYSIS
OF AN INTERMITTENT-MOTION MECHANISM

Figure 3.40 shows a camshaft and follower of an intermittent-motion mechanism. For the position indicated, the cam exerts a force P_{max} on the follower. What are the maximum stress at the contact line between the cam and follower and the deflection?

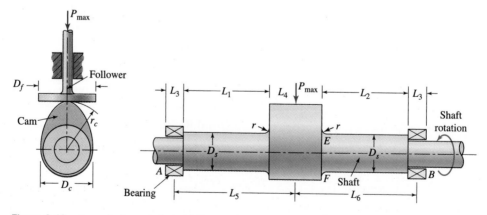

Figure 3.40   Layout of camshaft and follower of an intermittent-motion mechanism.

Case Study (CONCLUDED)

Given: The shapes of the contacting surfaces are known. The material of all parts is AISI 1095, carburized on the surfaces, oil quenched and tempered (Q&T) at 650°C.

Data:

$$P_{max} = 1.6 \text{ kips}, \quad r_c = 1.5 \text{ in.}, \quad D_f = L_4 = 1.5 \text{ in.},$$
$$E = 29 \times 10^6 \text{ psi}, \quad S_y = 80 \text{ ksi},$$

Assumptions: Frictional forces can be neglected. The rotational speed is slow so that the loading is considered static.

Solution: See Figure 3.40, Tables 3.2, B.1, and B.4.
Equations on the second column of case A of Table 3.2 apply. We first determine the half-width a of the contact patch. Since $E_1 = E_2 = E$ and $\Delta = 2/E$, we have

$$a = 1.076 \sqrt{\frac{P_{max}}{L_4} r_c \Delta}$$

Substitution of the given data yield

$$a = 1.076 \left[\frac{1600}{1.5}(1.5) \left(\frac{2}{30 \times 10^6} \right) \right]^{1/2}$$
$$= 11.113(10^{-3}) \text{ in.}$$

The rectangular patch area:

$$2aL = 2(11.113 \times 10^{-3})(1.5) = 33.34(10^{-3}) \text{ in.}^2$$

Maximum contact pressure is then

$$p_o = \frac{2}{\pi} \frac{P_{max}}{aL_4}$$
$$= \frac{2}{\pi} \frac{1600}{(11.113 \times 10^{-3})(1.5)} = 61.11 \text{ ksi}$$

The deflection δ of the cam and follower at the line of contact is obtained as follows

$$\delta = \frac{0.579 P_{max}}{E L_4} \left(\frac{1}{3} + \ln \frac{2r_c}{a} \right)$$

Introducing the numerical values,

$$\delta = \frac{0.579(1600)}{30 \times 10^6 (1.5)} \left(\frac{1}{3} + \ln \frac{2 \times 1.5}{11.113 \times 10^{-3}} \right)$$
$$= 0.122(10^{-3}) \text{ in.}$$

Comments: The contact stress is determined to be less than the yield strength and the design is satisfactory. The calculated deflection between the cam and the follower is very small and does not effect the system performance.

*3.15 MAXIMUM STRESS IN GENERAL CONTACT

In this section, we introduce some formulas for the determination of the maximum contact stress or pressure p_o between the two contacting bodies that have any general curvature [2,15]. Since the radius of curvature of each member in contact is different in every direction, the equations for the stress given here are more complex than those presented in the preceding section. A brief discussion on factors affecting the contact pressure is given in Section 8.15.

Consider two rigid bodies of equal elastic modulus E, compressed by F, as shown in Figure 3.41. The load lies along the axis passing through the centers of the bodies and through the point of contact and is perpendicular to the plane tangent to both bodies at the point of contact. The minimum and maximum radii of curvature of the surface of the upper body are r_1 and r_1'; those of the lower body are r_2 and r_2' at the point of contact. Therefore,

$1/r_1, 1/r_1', 1/r_2,$ and $1/r_2'$ are the principal curvatures. The *sign convention* of the *curvature* is such that it is positive if the corresponding center of curvature is inside the body; if the center of the curvature is outside the body, the curvature is negative. (For instance, in Figure 3.42, r_1, r_1' are positive, while r_2, r_2' are negative.)

Let θ be the angle between the normal planes in which radii r_1 and r_2 lie (Figure 3.41). Subsequent to the loading, the area of contact will be an ellipse with semiaxes a and b. The *maximum contact pressure* is

$$p_o = 1.5 \frac{F}{\pi ab} \qquad (3.43)$$

where

$$a = c_a \sqrt[3]{\frac{Fm}{n}} \qquad b = c_b \sqrt[3]{\frac{Fm}{n}} \qquad (3.44)$$

In these formulas, we have

$$m = \frac{4}{\frac{1}{r_1} + \frac{1}{r_1'} + \frac{1}{r_2} + \frac{1}{r_2'}} \qquad n = \frac{4E}{3(1 - \nu^2)} \qquad (3.45)$$

The constants c_a and c_b are given in Table 3.3 corresponding to the value of α calculated from the formula

$$\cos \alpha = \frac{B}{A} \qquad (3.46)$$

Here

$$A = \frac{2}{m}, \qquad B = \pm \frac{1}{2} \left[\left(\frac{1}{r_1} - \frac{1}{r_1'} \right)^2 + \left(\frac{1}{r_2} - \frac{1}{r_2'} \right)^2 \right.$$
$$\left. + 2 \left(\frac{1}{r_1} - \frac{1}{r_1'} \right) \left(\frac{1}{r_2} - \frac{1}{r_2'} \right) \cos 2\theta \right]^{1/2} \qquad (3.47)$$

The proper sign in B must be chosen so that its values are positive.

Figure 3.41
Curved surfaces of different radii of two bodies compressed by forces F.

Figure 3.42
Contact load in a single-row ball bearing.

Table 3.3   Factors for use in Equation (3.44)

α (degrees)	c_a	c_b	α (degrees)	c_a	c_b
20	3.778	0.408	60	1.486	0.717
30	2.731	0.493	65	1.378	0.759
35	2.397	0.530	70	1.284	0.802
40	2.136	0.567	75	1.202	0.846
45	1.926	0.604	80	1.128	0.893
50	1.754	0.641	85	1.061	0.944
55	1.611	0.678	90	1.000	1.000

Using Eq. (3.43), many problems of practical importance may be solved. These include contact stresses in rolling bearings (Figure 3.42), contact stresses in cam and push-rod mechanisms (see Problem P3.42), and contact stresses between a cylindrical wheel and rail (see Problem P3.44).

Ball Bearing Capacity Analysis   **EXAMPLE 3.18**

A single-row ball bearing supports a radial load F as shown in Figure 3.42. Calculate

(a) The maximum pressure at the contact point between the outer race and a ball.

(b) The factor of safety, if the ultimate strength is the maximum usable stress.

Given: $F = 1.2$ kN, $E = 200$ GPa, $v = 0.3$, and $S_u = 1900$ MPa. Ball diameter is 12 mm; the radius of the groove, 6.2 mm; and the diameter of the outer race is 80 mm.

Assumptions: The basic assumptions listed in Section 3.14 apply. The loading is static.

Solution: See Figure 3.42 and Table 3.3.
For the situation described $r_1 = r_1' = 0.006$ m, $r_2 = -0.0062$ m, and $r_2' = -0.04$ m.

(a) Substituting the given data into Eqs. (3.45) and (3.47), we have

$$m = \frac{4}{\frac{2}{0.006} - \frac{1}{0.0062} - \frac{1}{0.04}} = 0.0272, \qquad n = \frac{4(200 \times 10^9)}{3(0.91)} = 293.0403 \times 10^9$$

$$A = \frac{2}{0.0272} = 73.5294, \quad B = \frac{1}{2}[(0)^2 + (-136.2903)^2 + 2(0)^2]^{1/2} = 68.1452$$

Using Eq. (3.46),

$$\cos \alpha = \pm \frac{68.1452}{73.5294} = 0.9268, \qquad \alpha = 22.06°$$

Corresponding to this value of α, interpolating in Table 3.3, we obtain $c_a = 3.5623$ and $c_b = 0.4255$. The semiaxes of the ellipsoidal contact area are found by using Eq. (3.44):

$$a = 3.5623 \left[\frac{1200 \times 0.0272}{293.0403 \times 10^9} \right]^{1/3} = 1.7140 \text{ mm}$$

$$b = 0.4255 \left[\frac{1200 \times 0.0272}{293.0403 \times 10^9} \right]^{1/3} = 0.2047 \text{ mm}$$

The maximum contact pressure is then

$$p_o = 1.5 \frac{1200}{\pi (1.7140 \times 0.2047)} = 1633 \text{ MPa}$$

(b) Since contact stresses are not linearly related to load F, the safety factor is defined by Eq. (1.1):

$$n = \frac{F_u}{F} \tag{a}$$

in which F_u is the ultimate loading. The maximum principal stress theory of failure gives

$$S_u = \frac{1.5 F_u}{\pi ab} = \frac{1.5 F_u}{\pi c_a c_b \sqrt[3]{(F_u m/n)^2}}$$

This may be written as

$$S_u = \frac{1.5 \sqrt[3]{F_u}}{\pi c_a c_b (m/n)^{2/3}} \tag{3.48}$$

Introducing the numerical values into the preceding expression, we have

$$1900(10^6) = \frac{1.5 \sqrt[3]{F_u}}{\pi(3.5623 \times 0.4255) \left(\frac{0.0272}{293.0403 \times 10^9}\right)^{2/3}}$$

Solving, $F_u = 1891$ N. Equation (a) gives then

$$n = \frac{1891}{1200} = 1.58$$

Comments: In this example, the magnitude of the contact stress obtained is quite large in comparison with the values of the stress usually found in direct tension, bending, and torsion. In all contact problems, three-dimensional compressive stresses occur at the point, and hence a material is capable of resisting higher stress levels.

3.16   THREE-DIMENSIONAL STRESS

In the most general case of three-dimensional stress, an element is subjected to stresses on the orthogonal x, y, and z planes, as shown in Figure 1.10. Consider a tetrahedron, isolated from this element and represented in Figure 3.43. Components of stress on the perpendicular planes (intersecting at the origin O) can be related to the normal and shear stresses on the oblique plane ABC, by using an approach identical to that employed for the two-dimensional state of stress.

Orientation of plane ABC may be defined in terms of the *direction cosines,* associated with the angles between a unit normal $\mathbf{n}$ to the plane and the x, y, z coordinate axes:

$$\cos(\mathbf{n}, x) = l, \qquad \cos(\mathbf{n}, y) = m, \qquad \cos(\mathbf{n}, z) = n \tag{3.49}$$

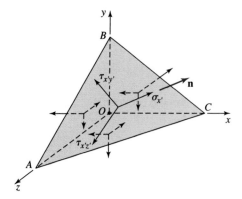

Figure 3.43 Components of stress on a tetrahedron.

The sum of the squares of these quantities is unity:

$$l^2 + m^2 + n^2 = 1 \tag{3.50}$$

Consider now a new coordinate system x', y', z', where x' coincides with $\mathbf{n}$ and y', z' lie on an oblique plane. It can readily be shown that [2] the normal stress acting on the oblique x' plane shown in Figure 3.43 is expressed in the form

$$\sigma_{x'} = \sigma_x l^2 + \sigma_y m^2 + \sigma_z n^2 + 2(\tau_{xy} lm + \tau_{yz} mn + \tau_{xz} ln) \tag{3.51}$$

where l, m, and n are direction cosines of angles between x' and the x, y, z axes, respectively. The shear stresses $\tau_{x'y'}$ and $\tau_{x'z'}$ may be written similarly. The stresses on the three mutually perpendicular planes are required to specify the stress at a point. One of these planes is the oblique (x') plane in question. The other stress components $\sigma_{y'}$, $\sigma_{z'}$, and $\tau_{y'z'}$ are obtained by considering those (y' and z') planes perpendicular to the oblique plane. In so doing, the resulting six expressions represent *transformation equations* for three-dimensional stress.

PRINCIPAL STRESSES IN THREE DIMENSIONS

For the three-dimensional case, three mutually perpendicular planes of zero shear exist; and on these planes, the normal stresses have maximum or minimum values. The foregoing normal stresses are called *principal stresses* σ_1, σ_2, and σ_3. The algebraically largest stress is represented by σ_1 and the smallest by σ_3. Of particular importance are the direction cosines of the plane on which $\sigma_{x'}$ has a maximum value, determined from the equations:

$$\begin{bmatrix} \sigma_x - \sigma_i & \tau_{xy} & \tau_{xz} \\ \tau_{xy} & \sigma_y - \sigma_i & \tau_{yz} \\ \tau_{xz} & \tau_{yz} & \sigma_z - \sigma_i \end{bmatrix} \begin{Bmatrix} l_i \\ m_i \\ n_i \end{Bmatrix} = 0, \qquad (i = 1, 2, 3) \tag{3.52}$$

A nontrivial solution for the direction cosines requires that the characteristic determinant vanishes. Thus

$$\begin{vmatrix} \sigma_x - \sigma_i & \tau_{xy} & \tau_{xz} \\ \tau_{xy} & \sigma_y - \sigma_i & \tau_{yz} \\ \tau_{xz} & \tau_{yz} & \sigma_z - \sigma_i \end{vmatrix} = 0 \qquad \textbf{(3.53)}$$

Expanding Eq. (3.53), we obtain the following stress cubic equation:

$$\sigma_i^3 - I_1\sigma_i^2 + I_2\sigma_i - I_3 = 0 \qquad \textbf{(3.54)}$$

where

$$I_1 = \sigma_x + \sigma_y + \sigma_z$$
$$I_2 = \sigma_x\sigma_y + \sigma_x\sigma_z + \sigma_y\sigma_z - \tau_{xy}^2 - \tau_{yz}^2 - \tau_{xz}^2 \qquad \textbf{(3.55)}$$
$$I_3 = \sigma_x\sigma_y\sigma_z + 2\tau_{xy}\tau_{yz}\tau_{xz} - \sigma_x\tau_{yz}^2 - \sigma_y\tau_{xz}^2 - \sigma_z\tau_{xy}^2$$

The quantities I_1, I_2, and I_3 represent *invariants* of the three-dimensional stress. For a given state of stress, Eq. (3.54) may be solved for its three roots, σ_1, σ_2, and σ_3. Introducing each of these principal stresses into Eq. (3.52) and using $l_i^2 + m_i^2 + n_i^2 = 1$, we can obtain three sets of direction cosines for three principal planes. Note that the direction cosines of the principal stresses are occasionally required to predict the behavior of members. A convenient way of determining the roots of the stress cubic equation and solving for the direction cosines is given in Appendix D.

After obtaining the three-dimensional principal stresses, we can readily determine the maximum shear stresses. Since no shear stress acts on the principal planes, it follows that an element oriented parallel to the principal directions is in a state of triaxial stress (Figure 3.44). Therefore,

$$\tau_{max} = \frac{1}{2}(\sigma_1 - \sigma_3) \qquad \textbf{(3.56)}$$

The maximum shear stress acts on the planes that bisect the planes of the maximum and minimum principal stresses as shown in the figure.

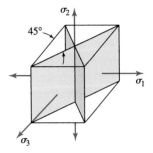

Figure 3.44 Planes of maximum three-dimensional shear stress.

Three-Dimensional State of Stress in a Member

EXAMPLE 3.19

At a critical point in a loaded machine component, the stresses relative to x, y, z coordinate system are given by

$$\begin{bmatrix} 60 & 20 & 20 \\ 20 & 0 & 40 \\ 20 & 40 & 0 \end{bmatrix} \text{MPa} \qquad \text{(a)}$$

Determine the principal stresses $\sigma_1, \sigma_2, \sigma_3$, and the orientation of σ_1 with respect to the original coordinate axes.

Solution: Substitution of Eq. (a) into Eq. (3.54) gives

$$\sigma_i^3 - 60\sigma_i^2 - 2400\sigma_i + 64{,}000 = 0, \qquad (i = 1, 2, 3)$$

The three principal stresses representing the roots of this equation are

$$\sigma_1 = 80 \text{ MPa}, \qquad \sigma_2 = 20 \text{ MPa}, \qquad \sigma_3 = -40 \text{ MPa}$$

Introducing σ_1 into Eq. (3.52), we have

$$\begin{bmatrix} 60 - 80 & 20 & 20 \\ 20 & 0 - 80 & 40 \\ 20 & 40 & 0 - 80 \end{bmatrix} \begin{Bmatrix} l_1 \\ m_1 \\ n_1 \end{Bmatrix} = 0 \qquad \text{(b)}$$

Here l_1, m_1, and n_1 represent the direction cosines for the orientation of the plane on which σ_1 acts.
    It can be shown that only two of Eqs. (b) are independent. From these expressions, together with $l_1^2 + m_1^2 + n_1^2 = 1$, we obtain

$$l_1 = \frac{2}{\sqrt{6}} = 0.8165, \qquad m_1 = \frac{1}{\sqrt{6}} = 0.4082, \qquad n_1 = \frac{1}{\sqrt{6}} = 0.4082$$

The direction cosines for σ_2 and σ_3 are ascertained in a like manner. The foregoing computations may readily be performed by using the formulas given in Appendix D.

SIMPLIFIED TRANSFORMATION FOR THREE-DIMENSIONAL STRESS

Often we need the normal and shear stresses acting on an arbitrary oblique plane of a tetrahedron in terms of the principal stresses acting on perpendicular planes (Figure 3.45). In this case, the x, y, and z coordinate axes are parallel to the principal axes: $\sigma_{x'} = \sigma$, $\sigma_x = \sigma_1$, $\tau_{xy} = \tau_{xz} = 0$, and so on, as depicted in the figure. Let l, m, and n denote the direction cosines of oblique plane ABC. The normal stress σ on the oblique plane, from Eq. (3.51), is

$$\sigma = \sigma_1 l^2 + \sigma_2 m^2 + \sigma_3 n^2 \qquad \text{(3.57a)}$$

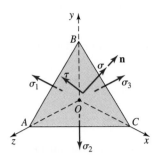

Figure 3.45 Triaxial stress on a tetrahedron.

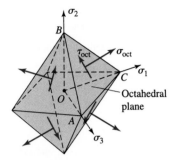

Figure 3.46 Stresses on a octahedron.

It can be verified that, the shear stress τ on this plane may be expressed in the convenient form:

$$\tau = [(\sigma_1 - \sigma_2)^2 l^2 m^2 + (\sigma_2 - \sigma_3)^2 m^2 n^2 + (\sigma_3 - \sigma_1)^2 n^2 l^2]^{1/2} \tag{3.57b}$$

The preceding expressions are the simplified transformation equations for three-dimensional state of stress.

OCTAHEDRAL STRESSES

Let us consider an oblique plane that forms equal angles with each of the principal stresses, represented by face ABC in Figure 3.45 with $OA = OB = OC$. Thus, the normal **n** to this plane has equal direction cosines relative to the principal axes. Inasmuch as $l^2 + m^2 + n^2 = 1$, we have

$$l = m = n = \frac{1}{\sqrt{3}}$$

There are eight such plane or octahedral planes, all of which have the same intensity of normal and shear stresses at a point O (Figure 3.46). Substitution of the preceding equation into Eqs. (3.57) results in, the magnitudes of the *octahedral normal stress* and *octahedral shear stress,* in the following forms:

$$\sigma_{\text{oct}} = \frac{1}{3}(\sigma_1 + \sigma_2 + \sigma_3) \tag{3.58a}$$

$$\tau_{\text{oct}} = \frac{1}{3}[(\sigma_1 - \sigma_2)^2 + (\sigma_2 - \sigma_3)^2 + (\sigma_3 - \sigma_1)^2]^{1/2} \tag{3.58b}$$

Equation (3.58a) indicates that the normal stress acting on an octahedral plane is the mean of the principal stresses. The octahedral stresses play an important role in certain failure criteria, discussed in Sections 5.3 and 7.8.

Determining Principal Stresses Using Mohr's Circle **EXAMPLE 3.20**

Figure 3.47a depicts a point in a loaded machine base subjected to the three-dimensional stresses. Determine at the point

(a) The principal planes and principal stresses.

(b) The maximum shear stress.

(c) The octahedral stresses.

Figure 3.47 Example 3.20.

Solution: We construct Mohr's circle for the transformation of stress in the xy plane as indicated by the solid lines in Figure 3.47b. The radius of the circle is $r = (12.5^2 + 30^2)^{1/2} = 32.5$ MPa.

(a) The principal stresses in the plane are represented by points A and B:

$$\sigma_1 = 47.5 + 32.5 = 80 \text{ MPa}$$

$$\sigma_2 = 47.5 - 32.5 = 15 \text{ MPa}$$

The z faces of the element define one of the principal stresses: $\sigma_3 = -25$ MPa. The planes of the maximum principal stress are defined by θ'_p, the angle through which the element should rotate about the z axis:

$$\theta'_p = \frac{1}{2} \tan^{-1} \frac{30}{12.5} = 33.7°$$

The result is shown on a sketch of the rotated element (Figure 3.47c).

(b) We now draw circles of diameters C_1B_1 and C_1A_1, which correspond, respectively, to the projections in the $y'z'$ and $x'z'$ planes of the element (Figure 3.47b). The maximum shearing stress, the radius of the circle of diameter C_1A_1, is therefore

$$\tau_{max} = \frac{1}{2}(75 + 25) = 50 \text{ MPa}$$

Planes of the maximum shear stress are inclined at 45° with respect to the x' and z faces of the element of Figure 3.47c.

(c) Through the use of Eqs. (3.58), we have

$$\sigma_{oct} = \frac{1}{3}(80 + 15 - 25) = 23.3 \text{ MPa}$$

$$\tau_{oct} = \frac{1}{3}[(80 - 15)^2 + (15 + 25)^2 + (-25 - 80)^2]^{1/2} = 43.3 \text{ MPa}$$

*3.17   VARIATION OF STRESS THROUGHOUT A MEMBER

As noted earlier, the components of stress generally vary from point to point in a loaded member. Such variations of stress, accounted for by the theory of elasticity, are governed by the equations of statics. Satisfying these conditions, the differential equations of equilibrium are obtained. To be physically possible, a stress field must satisfy these equations at every point in a load carrying component.

For the two-dimensional case, the stresses acting on an element of sides dx, dy, and of unit thickness are depicted in Figure 3.48. The body forces per unit volume acting on the element, F_x and F_y, are independent of z, and the component of the body force $F_z = 0$. In general, stresses are functions of the coordinates (x, y). For example, from the lower-left corner to the upper-right corner of the element, one stress component, say, σ_x, changes in value: $\sigma_x + (\partial\sigma_x/\partial x)\, dx$. The components σ_y and τ_{xy} change in a like manner. The stress

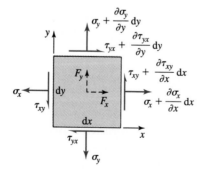

Figure 3.48   Stresses and body forces on an element.

element must satisfy the equilibrium condition $\sum M_z = 0$. Hence,

$$\left(\frac{\partial \sigma_y}{\partial y} \, dx \, dy\right)\frac{dx}{2} - \left(\frac{\partial \sigma_x}{\partial x} \, dx \, dy\right)\frac{dy}{2} + \left(\tau_{xy} + \frac{\partial \tau_{xy}}{\partial x} \, dx\right) dx \, dy$$

$$- \left(\tau_{yx} + \frac{\partial \tau_{yx}}{\partial y} \, dy\right) dx \, dy + F_y \, dx \, dy\frac{dx}{2} - F_x \, dx \, dy\frac{dy}{2} = 0$$

After neglecting the triple products involving dx and dy, this equation results in $\tau_{xy} = \tau_{yx}$. Similarly, for a general state of stress, it can be shown that $\tau_{yz} = \tau_{zy}$ and $\tau_{xz} = \tau_{zx}$. Hence, the shear stresses in mutually perpendicular planes of the element are equal.

The equilibrium condition that x-directed forces must sum to 0, $\sum F_x = 0$. Therefore, referring to Figure 3.48,

$$\left(\sigma_x + \frac{\partial \sigma_x}{\partial x} \, dx\right) dy - \sigma_x \, dy + \left(\tau_{xy} + \frac{\partial \tau_{xy}}{\partial y} \, dy\right) dx - \tau_{xy} \, dx + F_x \, dx \, dy = 0$$

Summation of the forces in the y direction yields an analogous result. After reduction, we obtain the differential equations of equilibrium for a *two-dimensional stress* in the form [2]

$$\frac{\partial \sigma_x}{\partial x} + \frac{\partial \tau_{xy}}{\partial y} + F_x = 0$$

$$\frac{\partial \sigma_y}{\partial y} + \frac{\partial \tau_{xy}}{\partial x} + F_y = 0$$

(3.59a)

In the general case of an element under *three-dimensional stresses*, it can be shown that the differential equations of equilibrium are given by

$$\frac{\partial \sigma_x}{\partial x} + \frac{\partial \tau_{xy}}{\partial y} + \frac{\partial \tau_{xz}}{\partial z} + F_x = 0$$

$$\frac{\partial \sigma_y}{\partial y} + \frac{\partial \tau_{xy}}{\partial x} + \frac{\partial \tau_{yz}}{\partial z} + F_y = 0$$

$$\frac{\partial \sigma_z}{\partial z} + \frac{\partial \tau_{xz}}{\partial x} + \frac{\partial \tau_{yz}}{\partial y} + F_z = 0$$

(3.59b)

Note that, in many practical applications, the weight of the member is only body force. If we take the y axis as upward and designate by ρ the mass density per unit volume of the member and by g the gravitational acceleration, then $F_x = F_z = 0$ and $F_y = -\rho g$ in the foregoing equations.

We observe that two relations of Eqs. (3.59a) involve the three unknowns $(\sigma_x, \sigma_y, \tau_{xy})$ and the three relations of Eqs. (3.59b) contain the six unknown stress components. Therefore, problems in stress analysis are *internally* statically indeterminate. In the mechanics of materials method, this indeterminacy is eliminated by introducing simplifying assumptions regarding the stresses and considering the equilibrium of the finite segments of a load-carrying component.

3.18   THREE-DIMENSIONAL STRAIN

If deformation is distributed uniformly over the original length, the normal strain may be written $\varepsilon_x = \delta/L$, where L and δ are the original length and the change in length of the member, respectively (see Figure 1.12a). However, the strains generally vary from point to point in a member. Hence, the expression for strain must relate to a line of length dx which elongates by an amount du under the axial load. The definition of normal strain is therefore

$$\varepsilon_x = \frac{du}{dx} \tag{3.60}$$

This represents the strain at a point.

   As noted earlier, in the case of two-dimensional or *plane strain,* all points in the body, before and after application of load, remain in the same plane. Therefore, the deformation of an element of dimensions dx, dy, and of unit thickness can contain normal strain (Figure 3.49a) and a shear strain (Figure 3.49b). Note that the partial derivative notation is used, since the displacement u or v is function of x and y. Recalling the basis of Eqs. (3.60) and (1.22), an examination of Figure 3.49 yields

$$\varepsilon_x = \frac{\partial u}{\partial x}, \qquad \varepsilon_y = \frac{\partial v}{\partial y}, \qquad \gamma_{xy} = \frac{\partial v}{\partial x} + \frac{\partial u}{\partial y} \tag{3.61a}$$

Obviously, γ_{xy} is the shear strain between the x and y axes (or y and x axes), hence, $\gamma_{xy} = \gamma_{yx}$. A long prismatic member subjected to a lateral load (e.g., a cylinder under pressure) exemplifies the state of plane strain.

   In an analogous manner, the strains at a point in a rectangular prismatic element of sides dx, dy, and dz are found in terms of the displacements u, v, and w. It can be shown that these *three-dimensional strain* components are ε_x, ε_y, γ_{xy}, and

$$\varepsilon_z = \frac{\partial w}{\partial z}, \qquad \gamma_{yz} = \frac{\partial w}{\partial z} + \frac{\partial v}{\partial z}, \qquad \gamma_{xz} = \frac{\partial w}{\partial x} + \frac{\partial u}{\partial z} \tag{3.61b}$$

where $\gamma_{yz} = \gamma_{zy}$ and $\gamma_{xz} = \gamma_{zx}$. Equations (3.61) represent the components of strain tensor, which is similar to the stress tensor discussed in Section 1.13.

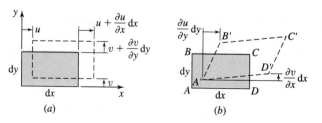

Figure 3.49   Deformations of a two-dimensional element:
(a) normal strain; (b) shear strain.

PROBLEMS IN ELASTICITY

In many problems of practical importance, the stress or strain condition is one of plane stress or plane strain. These two-dimensional problems in elasticity are simpler than those involving three-dimensions. A finite element solution of two-dimensional problems is taken up in Chapter 17. In examining Eqs. (3.61), we see that the six strain components depend linearly on the derivatives of the three displacement components. Therefore, the strains cannot be independent of one another. Six equations, referred to as the *conditions of compatibility*, can be developed showing the relationships among ε_x, ε_y, ε_z, γ_{xy}, γ_{yz}, and γ_{xz} [2]. The number of such equations reduce to one for a two-dimensional problem. The conditions of compatibility assert that the displacements are continuous. Physically, this means that the body must be pieced together.

To conclude, the theory of elasticity is based on the following requirements: strain compatibility, stress equilibrium (Eqs. 3.59), general relationships between the stresses and strains (Eqs. 2.8), and boundary conditions for a given problem. In Chapter 16, we discuss various axisymmetrical problems using the elasticity approaches. In the method of mechanics of materials, simplifying assumptions are made with regard to the distribution of strains in the body as a whole or the finite portion of the member. Thus, the difficult task of solving the conditions of compatibility and the differential equations of equilibrium are avoided.

REFERENCES

1. Ugural, A. C. *Mechanics of Materials*. New York: McGraw-Hill, 1991.
2. Ugural, A. C., and S. K. Fenster. *Advanced Strength and Applied Elasticity*, 4th ed. Upper Saddle River, NJ: Prentice Hall, 2003.
3. Timoshenko, S. P., and J. N. Goodier. *Theory of Elasticity*, 3rd ed. New York: McGraw-Hill, 1970.
4. Young, W. C., and R. C. Budynas. *Roark's Formulas for Stress and Strain*, 7th ed. New York: McGraw-Hill, 2001.
5. Ugural, A. C. *Stresses in Plates and Shells*, 2nd ed. New York: McGraw-Hill, 1999.
6. McCormac, L. C. *Design of Reinforced Concrete*. New York: Harper and Row, 1978.
7. Chen, F. Y. "Mohr's Circle and Its Application in Engineering Design." ASME Paper 76-DET-99, 1976.
8. Peterson, R. E. *Stress Concentration Factors*. New York: Wiley, 1974.
9. Peterson, R. E. *Stress Concentration Design Factors*. New York: Wiley, 1953.
10. Peterson, R. E. "Design Factors for Stress Concentration, Parts 1 to 5." *Machine Design*, February–July 1951.
11. Juvinall, R. C. *Engineering Consideration of Stress, Strain and Strength*. New York: McGraw-Hill, 1967.
12. Norton, R. E. *Machine Design—An Integrated Approach*, 2nd ed. Upper Saddle River, NJ: Prentice Hall, 2000.
13. Juvinall, R. E., and K. M. Marshek. *Fundamentals of Machine Component Design*, 3rd ed. New York: Wiley, 2000.
14. Frocht, M. M. "Photoelastic Studies in Stress Concentration." *Mechanical Engineering*, August 1936, pp. 485–489.

15. Boresi, A. P., and R. J. Schmidt. *Advanced Mechanics of Materials,* 6th ed. New York: Wiley, 2003.

16. Shigley, J. E., and C. R. Mishke. *Mechanical Engineering Design*, 6th ed. New York: McGraw-Hill, 2001.

17. Rothbart, H. A., ed. *Mechanical Design and Systems Handbook,* 2nd ed. New York: McGraw-Hill, 1985.

PROBLEMS

Sections 3.1 through 3.8

3.1 Two plates are fastened by a bolt and nut as shown in Figure P3.1. Calculate

(a) The normal stress in the bolt shank.

(b) The average shear stress in the head of the bolt.

(c) The shear stress in the threads.

(d) The bearing stress between the head of the bolt and the plate.

Assumption: The nut is tightened to produce a tensile load in the shank of the bolt of 10 kips.

Figure P3.1

3.2 A short steel pipe of yield strength S_y is to support an axial compressive load P with factor of safety of n against yielding. Determine the minimum required inside radius a.

Given: $S_y = 280$ MPa, $P = 1.2$ MN, and $n = 2.2$.
Assumption: The thickness t of the pipe is to be one-fourth of its inside radius a.

3.3 The landing gear of an aircraft is depicted in Figure P3.3. What are the required pin diameters at A and B.

Given: Maximum usable stress of 28 ksi in shear.
Assumption: Pins act in double shear.

Figure P3.3

3.4 The frame of Figure P3.4 supports a concentrated load P. Calculate

(a) The normal stress in the member BD if it has a cross-sectional area A_{BD}.

(b) The shearing stress in the pin at A if it has a diameter of 25 mm and is in double shear.

Given: $P = 5$ kN, $A_{BD} = 8 \times 10^3$ mm^2.

Figure P3.4

3.5 Two bars AC and BC are connected by pins to form a structure for supporting a vertical load P at C (Figure P3.5). Determine the angle α if the structure is to be of minimum weight.

Assumption: The normal stresses in both bars are to be the same.

Figure P3.5

3.6 Two beams AC and BD are supported as shown in Figure P3.6. A roller fits snugly between the two beams at point B. Draw the shear and moment diagrams of the lower beam AC.

Figure P3.6

3.7 Design the cross section (determine h) of the simply supported beam loaded at two locations as shown in Figure P3.7.

Assumption: The beam will be made of timber of $\sigma_{all} = 1.8$ ksi and $\tau_{all} = 100$ psi.

Figure P3.7

3.8 A rectangular beam is to be cut from a circular bar of diameter d (Figure P3.8). Determine the dimensions b and h so that the beam will resist the largest bending moment.

Figure P3.8

3.9 The T-beam, whose cross section is shown in Figure P3.9, is subjected to a shear force V. Calculate the maximum shear stress in the web of the beam.

Given: $b = 200$ mm, $t = 15$ mm, $h_1 = 175$ mm, $h_2 = 150$ mm, $V = 22$ kN.

Figure P3.9

3.10 A box beam is made of four 50-mm × 200-mm planks, nailed together as shown in Figure P3.10. Determine the maximum allowable shear force V.

Given: The longitudinal spacing of the nails, $s = 100$ mm; the allowable load per nail, $F = 15$ kN.

50 mm
200 mm
50 mm
|← 200 mm →|

Figure P3.10

3.11 For the beam and loading shown in Figure P3.11, design the cross section of the beam for $\sigma_{all} = 12$ MPa and $\tau_{all} = 810$ kPa.

→| b |←
$2b$
2 kN/m
A
B
1.2 m

Figure P3.11

3.12 Select the S shape of a simply supported 6-m long beam subjected a uniform load of intensity 50 kN/m, for $\sigma_{all} = 170$ MPa and $\tau_{all} = 100$ MPa.

3.13 and **3.14** The beam AB has the rectangular cross section of constant width b and variable depth h (Figures P3.13 and P3.14). Derive an expression for h in terms of x, L, and h_1, as required.

Assumption: The beam is to be of constant strength.

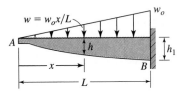

$w = w_o x/L$
w_o
A
h
B
h_1
x
L

w
A
B
h
h_1
x
$L/2$
$L/2$

Figure P3.13   **Figure P3.14**

3.15 A wooden beam 8 in. wide × 12 in. deep is reinforced on both top and bottom by steel plates 0.5 in. thick (Figure P3.15). Calculate the maximum bending moment M about the z axis.

Design Assumptions: The allowable bending stresses in the wood and steel are 1.05 ksi and 18 ksi, respectively. Use $n = E_s/E_w = 20$.

Figure P3.15

3.16 A simply supported beam of span length 8 ft carries a uniformly distributed load of 2.5 kip/ft. Determine the required thickness t of the steel plates.

Given: The cross section of the beam is a hollow box with wood flanges ($E_w = 1.5 \times 10^6$ psi) and steel ($E_s = 30 \times 10^6$ psi), as shown in Figure P3.16.
Assumptions: The allowable stresses are 19 ksi for the steel and 1.1 ksi for the wood.

Figure P3.16

3.17 and **3.18** For the composite beam with cross section as shown (Figures P3.17 and P3.18), determine the maximum permissible value of the bending moment M about the z axis.

Given: $(\sigma_b)_{all} = 120$ MPa $(\sigma_s)_{all} = 140$ MPa

$E_b = 100$ GPa $E_s = 200$ GPa

Figure P3.17

Figure P3.18

3.19 A round brass tube of outside diameter d and an aluminum core of diameter $d/2$ are bonded together to form a composite beam (Figure P3.19). Determine the maximum bending moment M that can be carried by the beam, in terms of E_b, E_s, σ_b, and d, as required. What is the value of M for $E_b = 15 \times 10^6$ psi, $E_a = 10 \times 10^6$ psi, $\sigma_b = 50$ ksi, and $d = 2$ in.?

Design Requirement: The allowable stress in the brass is σ_b.

Figure P3.19

Sections 3.9 through 3.13

3.20 The state of stress at a point in a loaded machine component is represented in Figure P3.20. Determine

(a) The normal and shear stresses acting on the indicated inclined plane a-a.

(b) The principal stresses.

Sketch results on properly oriented elements.

Figure P3.20

3.21 At a point A on the upstream face of a dam (Figure P3.21), the water pressure is -70 kPa and a measured tensile stress parallel to this surface is 30 kPa. Calculate

(a) The stress components σ_x, σ_y, and τ_{xy}.

(b) The maximum shear stress.

Sketch the results on a properly oriented element.

Figure P3.21

3.22 The stress acting uniformly over the sides of a skewed plate is shown in Figure P3.22. Determine

(a)   The stress components on a plane parallel to *a-a*.

(b)   The magnitude and orientation of principal stresses.

Sketch the results on properly oriented elements.

Figure P3.22

3.23 A thin skewed plate is depicted in Figure P3.22. Calculate the change in length of

(a)   The edge *AB*.

(b)   The diagonal *AC*.

Given: $E = 200$ GPa, $v = 0.3$, $AB = 40$ mm, and $BC = 60$ mm.

3.24 The stresses acting uniformly at the edges of a thin skewed plate are shown in Figure P3.24. Determine

(a)   The stress components σ_x, σ_y, and τ_{xy}.

(b)   The maximum principal stresses and their orientations.

Sketch the results on properly oriented elements.

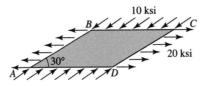

Figure P3.24

3.25 For the thin skewed plate shown in Figure P3.24, determine the change in length of the diagonal *BD*.

Given: $E = 30 \times 10^6$ psi, $v = \frac{1}{4}$, $AB = 2$ in., and $BC = 3$ in.

3.26 The stresses acting uniformly at the edges of a wall panel of a flight structure are depicted in Figure P3.26. Calculate the stress components on planes parallel and perpendicular to *a-a*. Sketch the results on a properly oriented element.

Figure P3.26

3.27 A rectangular plate is subjected to uniformly distributed stresses acting along its edges (Figure P3.27). Determine

(a) The normal and shear stresses on planes parallel and perpendicular to a-a.

(b) The maximum shear stress.

Sketch the results on properly oriented elements.

Figure P3.27

3.28 For the plate shown in Figure P3.27, calculate the change in the diagonals AC and BD.

Given: $E = 210$ GPa, $\nu = 0.3$, $AB = 50$ mm, and $BC = 75$ mm.

3.29 A cylindrical pressure vessel of diameter $d = 3$ ft and wall thickness $t = \frac{1}{8}$ in. is simply supported by two cradles as depicted in Figure P3.29. Calculate, at points A and C on the surface of the vessel,

(a) The principal stresses.

(b) The maximum shear stress.

Given: The vessel and its contents weigh 84 lb per ft of length, and the contents exert a uniform internal pressure of $p = 6$ psi on the vessel.

Figure P3.29

3.30 Redo Problem 3.29, considering point B on the surface of the vessel.

3.31 Calculate and sketch the normal stress acting perpendicular and shear stress acting parallel to the helical weld of the hollow cylinder loaded as depicted in Figure P3.31.

Figure P3.31

3.32 A 40-mm wide × 120-mm deep bracket supports a load of $P = 30$ kN (Figure P3.32). Determine the principal stresses and maximum shear stress at point A. Show the results on a properly oriented element.

Figure P3.32

3.33 A pipe of 120-mm outside diameter and 10-mm thickness is constructed with a helical weld making an angle of 45° with the longitudinal axis, as shown in Figure P3.33. What is the largest torque T that may be applied to the pipe?

Given: Allowable tensile stress in the weld, $\sigma_{all} = 80$ MPa.

Figure P3.33

3.34 The strains at a point on a loaded shell has components $\varepsilon_x = 500\mu$, $\varepsilon_y = 800\mu$, $\varepsilon_z = 0$, and $\gamma_{xy} = 350\mu$. Determine

(a) The principal strains.

(b) The maximum shear stress at the point.

Given: $E = 70$ GPa and $\nu = 0.3$.

3.35 A thin rectangular steel plate shown in Figure P3.35 is acted on by a stress distribution, resulting in the uniform strains $\varepsilon_x = 200\mu$, $\varepsilon_y = 600\mu$, and $\gamma_{xy} = 400\mu$. Calculate

(a) The maximum shear strain.

(b) The change in length of diagonal AC.

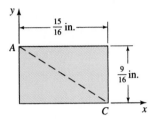

Figure P3.35

3.36 The strains at a point in a loaded bracket has components $\varepsilon_x = 50\mu$, $\varepsilon_y = 250\mu$, and $\gamma_{xy} = -150\mu$. Determine the principal stresses.

Assumptions: The bracket is made of a steel of $E = 210$ GPa and $v = 0.3$.

3.W Review the website at www.measurementsgroup.com. Search and identify

(a) Websites of three strain gage manufacturers.

(b) Three grid configurations of typical foil electrical resistance strain gages.

3.37 A thin-walled cylindrical tank of 500-mm radius and 10-mm wall thickness has a welded seam making an angle of 40° with respect to the axial axis (Figure P3.37). What is the allowable value of p?

Given: The tank carries an internal pressure of p and an axial compressive load of $P = 20\pi$ kN. Assumption: The normal and shear stresses acting simultaneously in the plane of welding are not to exceed 50 and 20 MPa, respectively.

Figure P3.37

3.38 The 15-mm thick metal bar is to support an axial tensile load of 25 kN as shown in Figure P3.38 with a factor of safety of $n = 1.9$ (see Appendix C). Design the bar for minimum allowable width h.

Assumption: The bar is made of a relatively brittle metal having $S_y = 150$ MPa.

Figure P3.38

3.39 Calculate the largest load P that may be carried by a relatively brittle flat bar consisting of two portions, both 12-mm thick, and respectively 30-mm and 45-mm wide, connected by fillets of radius $r = 6$ mm (see Appendix C).

Given: $S_y = 210$ MPa and a factor of safety of $n = 1.5$.

Sections 3.14 through 3.18

3.40 Two identical 300-mm diameter balls of a rolling mill are pressed together with a force of 500 N. Determine

(a) The width of contact.
(b) The maximum contact pressure.
(c) The maximum principal stresses and shear stress in the center of the contact area.

Assumption: Both balls are made of steel of $E = 210$ GPa and $\nu = 0.3$.

3.41 A 14-mm diameter cylindrical roller runs on the inside of a ring of inner diameter 90 mm (see Figure 10.21a). Calculate

(a) . The half-width a of the contact area.

(b) The value of the maximum contact pressure p_o.

Given: The roller load is $F = 200$ kN per meter of axial length.
Assumption: Both roller and ring are made of steel having $E = 210$ GPa and $\nu = 0.3$.

3.42 A spherical-faced (mushroom) follower or valve tappet is operated by a cylindrical cam (Figure P3.42). Determine the maximum contact pressure p_o.

Figure P3.42

Given: $r_2 = r_2' = 10$ in., $r_1 = \frac{3}{8}$ in., and contact force $F = 500$ lb.
Assumptions: Both members are made of steel of $E = 30 \times 10^6$ psi and $v = 0.3$.

3.43 Resolve Problem 3.42, for the case in which the follower is flat faced.

Given: $w = \frac{1}{4}$ in.

3.44 Determine the maximum contact pressure p_o between a wheel of radius $r_1 = 500$ mm and a rail of crown radius of the head $r_2 = 350$ mm (Figure P3.44).

Given: Contact force $F = 5$ kN.
Assumptions: Both wheel and rail are made of steel of $E = 206$ GPa and $v = 0.3$.

Figure P3.44

3.45 Redo Example 3.18 for a double-row ball bearing having $r_1 = r_1' = 5$ mm, $r_2 = -5.2$ mm, $r_2' = -30$ mm, $F = 600$ N, and $S_y = 1500$ MPa.

Assumptions: The remaining data are unchanged. The factor of safety is based on the yield strength.

3.46 At a point in a structural member, stresses with respect to an x, y, z coordinate system are

$$\begin{bmatrix} -10 & 0 & -8 \\ 0 & 2 & 0 \\ -8 & 0 & 2 \end{bmatrix} \text{ksi}$$

Calculate

(a) The magnitude and direction of the maximum principal stress.

(b) The maximum shear stress.

(c) The octahedral stresses.

3.47 The state of stress at a point in a member relative to an x, y, z coordinate system is

$$\begin{bmatrix} 9 & 0 & 0 \\ 0 & 12 & 0 \\ 0 & 0 & -18 \end{bmatrix} \text{ksi}$$

Determine

(a) The maximum shear stress.

(b) The octahedral stresses.

3.48 At a critical point in a loaded component, the stresses with respect to an x, y, z coordinate system are

$$\begin{bmatrix} 42.5 & 0 & 0 \\ 0 & 5.26 & 0 \\ 0 & 0 & -7.82 \end{bmatrix} \text{MPa}$$

Determine the normal stress σ and the shear stress τ on a plane whose outer normal is oriented at angles of 40°, 60°, and 66.2° relative to the x, y, and z axes, respectively.

Deflection and Impact

Outline

4.1   INTRODUCTION

Strength and stiffness are considerations of basic importance to the engineer. The stress level is frequently used as a measure of strength. Stress in members under various loads was discussed in Chapter 3. We now turn to deflection, the analysis of which is as important as that of stress. Moreover, deflections must be considered in the design of statically indeterminate systems, although we are interested only in the forces or stresses.

Stiffness relates to the ability of a part to resist deflection or deformation. Elastic deflection or stiffness, rather than stress, is frequently the controlling factor in the design of a member. The deflection, for example, may have to be kept within limits so that certain clearances between components are maintained. Structures such as machine frames must be extremely rigid to maintain manufacturing accuracy. Most components may require great stiffness to eliminate vibration problems. We begin by developing basic expressions relative to deflection and stiffness of variously loaded members using the equilibrium approaches. Then the integration, superposition and moment-area methods are discussed. Following this, the *impact* or shock loading and bending of plates are treated. The theorems are based upon work-energy concepts, classic methods, and finite element analysis (FEA) for determining the displacement on members or structures are considered in the chapters to follow.

Comparison of various deflection methods shows when one approach is preferred over another and the advantages of each technique. The governing differential equations for beams on integration give the solution for deflection in a problem. However, it is best to limit their application to prismatic beams; otherwise, considerable complexities arise. In practice, the deflection of members subjected to several or complicated loading conditions are often synthesized from simpler loads, using the principle of superposition.

The dual concepts of strain energy and complementary energy provide the basis for some extremely powerful methods of analysis, such as Castigliano's theorem and its various forms. These approaches may be employed very effectively for finding deflection due to applied forces and are not limited at all to linearly elastic structures. Similar problems are treated by the principles of virtual work and minimum potential energy for obtaining deflections or forces caused by any kind of deformation. They are of great importance in the matrix analysis of structures and in finite elements. The moment-area method, a specialized procedure, is particularly convenient if deflection of only a few points on a beam or frame is desired. It can be used to advantage in the solution of statically indeterminate problems as a check. An excellent insight into the kinematics is obtained by applying this technique. The FEA is perfectly general and can be used for the analysis of statically indeterminate as well as determinate, both linear and nonlinear, problems.

4.2   DEFLECTION OF AXIALLY LOADED MEMBERS

Here, we are concerned with the elongation or contraction of slender members under axial loading. The axial stress in these cases is assumed not to exceed the proportional limit of the linearly elastic range of the material. The definitions of normal stress and normal strain and the relationship between the two, given by Hooke's law, are used.

Consider the deformation of a prismatic bar having a cross-sectional area A, length L, and modulus of elasticity E, subjected to an axial load P (see Figure 3.1a). The magnitudes of the axial stress and axial strain at a cross section are found from $\sigma_x = P/A$ and $\varepsilon_x = \sigma_x/E$, respectively. These results are combined with $\varepsilon_x = \delta/L$ and integrated over the length L of the bar to give the following equation for the *deformation* δ of the bar:

$$\delta = \frac{PL}{AE} \tag{4.1}$$

The product AE is known as the *axial rigidity* of the bar. The positive sign indicates elongation. A negative sign would represent contraction. The deformation δ has units of length L. Note that, for tapered bars, the foregoing equation gives results of acceptable accuracy provided the angle between the sides of the rod is no larger than 20° [1].

Most of the force-displacement problems encountered in this book are linear, as in the preceding relationship. The *spring rate,* also known as spring constant, of an axially loaded bar is then

$$k = \frac{P}{\delta} = \frac{AE}{L} \tag{4.2}$$

The units of k are often kilonewtons per meter or pounds per inch. Spring rate, a deformation characteristic, plays a significant role in the design of members.

A change in temperature of ΔT degrees causes a strain $\varepsilon_t = \alpha \Delta T$, defined by Eq. (1.21), where α represents the coefficient thermal expansion. In an elastic body, thermal axial deformation caused by a uniform temperature is therefore

$$\delta_t = \alpha(\Delta T)L \tag{4.3}$$

The thermal strain and deformation usually are positive if the temperature increases and negative if it decreases.

Analysis of a Duplex Structure

EXAMPLE 4.1

A steel rod of cross-sectional area A_s and modulus of elasticity E_s has been placed inside a copper tube of cross-sectional area A_c and modulus of elasticity E_c (Figure 4.1a). Determine the axial

(a)   (b)

Figure 4.1   Example 4.1.

shortening of this system of two members, sometimes called an isotropic duplex structure, when a force P is exerted on the end plate as shown.

Assumptions: Members have the same length L. The end plate is rigid.

Solution: The forces produced in the rod and in the tube are designated by P_s and P_c, respectively.
   Statics.   The equilibrium condition is applied to the free-body of the end plate (Figure 4.1b):

$$P_c + P_s = P \tag{a}$$

This is the only equilibrium equation available, and since it contains two unknowns (P_c and P_s), the structure is statically indeterminate to the first degree (see Section 1.8).
   Deformations.   Through the use of Eq. (4.1), the shortening of the members are

$$\delta_c = \frac{P_c L}{A_c E_c}, \qquad \delta_s = \frac{P_s L}{A_s E_s}$$

Geometry.   Axial deformation of the copper tube is equal to that of the steel rod:

$$\frac{P_c L}{A_c E_c} = \frac{P_s L}{A_s E_s} \tag{b}$$

Solution of Eqs. (a) and (b) gives

$$P_c = \frac{(A_c E_c) P}{A_c E_c + A_s E_s}, \qquad P_s = \frac{(A_s E_s) P}{A_c E_c + A_s E_s} \tag{4.4}$$

The foregoing show that the forces in the members are proportional to the axial rigidities.
   Compressive stresses σ_c in copper and σ_s in steel are found by dividing P_c and P_s by A_c and A_s, respectively. Then, applying Hooke's law together with Eqs. (4.4), we obtain the compressive strain

$$\varepsilon = \frac{P}{A_c E_c + A_s E_s} \tag{4.5}$$

The shortening of the assembly is therefore $\delta = \varepsilon L$.

Comments: Equation (4.5) indicates that the strain equals the applied load divided by the sum of the axial rigidities of the members. Composite duplex structures are treated in Chapter 16.

| **EXAMPLE 4.2** | Analysis of Bolt–Tube Assembly |

In the assembly of the aluminum tube (cross-sectional area A_t, modulus of elasticity E_t, length L_t) and steel bolt (cross-sectional area A_b, modulus of elasticity E_b) shown in Figure 4.2a, the bolt is single threaded, with a 2-mm pitch. If the nut is tightened one-half turn after it has been fitted snugly, calculate the axial forces in the bolt and tubular sleeve.

Figure 4.2   Examples 4.2 and 4.3.

Given: $A_t = 300 \text{ mm}^2$, $E_t = 70 \text{ GPa}$, $L_t = 0.6 \text{ m}$, $A_b = 600 \text{ mm}^2$, and $E_b = 200 \text{ GPa}$.

Solution: The forces in the bolt and in the sleeve are denoted by P_b and P_t, respectively.
   Statics.   The only equilibrium condition available for the free-body of Figure 4.2b gives

$$P_b = P_t$$

That is, the compressive force in the sleeve is equal to the tensile force in the bolt. The problem is therefore statically indeterminate to the first degree.
   Deformations.   Using Eq. (4.1), we write

$$\delta_b = \frac{P_b L_b}{A_b E_b}, \qquad \delta_t = \frac{P_t L_t}{A_t E_t} \tag{c}$$

Here δ_b is the axial extension of the bolt and δ_t represents the axial contraction of the tube.
   Geometry.   The deformations of the bolt and tube must be equal to $\Delta = 0.002/2 = 0.001 \text{ m}$, the movement of the nut on the bolt:

$$\delta_b + \delta_t = \Delta$$

$$\frac{P_b L_b}{A_b E_b} + \frac{P_t L_t}{A_t E_t} = \Delta \tag{4.6}$$

Setting $P_b = P_t$ and $L_b = L_t$, the preceding equation becomes

$$P_b \left(\frac{1}{A_b E_b} + \frac{1}{A_t E_t} \right) = \frac{\Delta}{L_t} \tag{4.7}$$

Introducing the given data, we have

$$P_b \left[\frac{1}{600(200)10^3} + \frac{1}{300(70)10^3} \right] = \frac{0.001}{0.6}$$

Solving, $P_b = 29.8 \text{ kN}$.

EXAMPLE 4.3 | Thermal Stresses in a Bolt-Tube Assembly

Determine the axial forces in the assembly of bolt and tube (Figure 4.2a), after a temperature rise of ΔT.

Given: $\Delta T = 100°C$, $\alpha_b = 11.7 \times 10^{-6}/°C$, and $\alpha_t = 23.2 \times 10^{-6}/°C$.

Assumptions: The data presented in the preceding example remain the same.

Solution: Only force-deformation relations, Eqs. (c), change from Example 4.2. Now the expressions for the *extension* of the bolt and the *contraction* of the sleeve are

$$\delta_b = \frac{P_b L_b}{A_b E_b} + \alpha_b(\Delta T)L_b$$

$$\delta_t = \frac{P_t L_t}{A_t E_t} - \alpha_t(\Delta T)L_t$$

(d)

Note that, in the foregoing, the minus sign indicates a decrease in tube contraction due to the temperature rise.

We have $L_b = L_t$ and $P_b = P_t$. These, carried into $\delta_b + \delta_t = \Delta$, give

$$P_b \left(\frac{1}{A_b E_b} + \frac{1}{A_t E_t} \right) + (\alpha_b - \alpha_t)\Delta T = \frac{\Delta}{L_t}$$

(4.8)

where, as before, Δ is the movement of the nut on the bolt. Substituting the numerical values into Eq. (4.8), we obtain

$$P_b \left[\frac{1}{600(200)10^3} + \frac{1}{300(70)10^3} \right] + (11.7 - 23.2)10^{-6}(100) = \frac{0.001}{0.6}$$

This yields $P_b = 50.3$ kN.

Comment: The final elongation of the bolt and the contraction of the tube can be calculated by substituting the axial force of 50.3 kN into Eqs. (d). Interestingly, when the bolt and tube are made of the same material ($\alpha_b = \alpha_t$), the temperature change does not affect the assembly. That is, the forces obtained in Example 4.2 still hold.

4.3   ANGLE OF TWIST OF BARS

In Section 3.5, the concern was with torsion stress. We now treat angular displacement of twisted prismatic bars or shafts. We assume that the entire bar remains elastic. For most structural materials, the amount of twist is small and hence the member behaves as before.

But in a material such as rubber, where twisting is large, the basic assumptions must be reexamined.

CIRCULAR SECTIONS

Consider a circular prismatic shaft of radius c, length L, and modulus of elasticity in shear G (Figure 3.6). The maximum shear stress τ_{max} and maximum shear strain γ_{max} are related by Hooke's law: $\gamma_{max} = \tau_{max}/G$. Moreover, by the torsion formula, $\tau_{max} = Tc/J$, where J is the polar moment of inertia. Substitution of the latter expression into the former results in $\gamma_{max} = Tc/GJ$. For small deformations, by taking $\tan\gamma_{max} = \gamma_{max}$, we also write $\gamma_{max} = c\phi/L$. These expressions lead to the *angle of twist,* representing the angle through which one end of a cross section of a circular shaft rotates with respect to another:

$$\phi = \frac{TL}{GJ} \tag{4.9}$$

Angle ϕ is measured in radians. The product GJ is called the *torsional rigidity* of the shaft. Equation (4.9) can be used for either solid or hollow bars having circular cross sections. We observe that the *spring rate* of a circular torsion bar is given by

$$k = \frac{T}{\phi} = \frac{GJ}{L} \tag{4.10}$$

Typical units of the k are kilonewton-meters per radian or pound-inches per radian.

Examining Eq. (4.9) implies a method for obtaining the modulus of elasticity in shear G for a given material. A circular prismatic specimen of the material, of known diameter and length, is placed in a torque-testing machine. As the specimen is twisted, increasing the value of the applied torque T, the corresponding values of the angle of twist ϕ between the two ends of the specimen is recorded as a torque-twist diagram. The slope of this curve (T/ϕ) in the linearly elastic region is the quantity GJ/L. From this, the magnitude of G can be calculated.

NONCIRCULAR SECTIONS

As pointed out in Section 3.5, determination of stresses and displacements in noncircular members is a difficult problem and beyond the scope of this book. However, the following angle of twist formula for *rectangular bars* is introduced here for convenience:

$$\phi = \frac{TL}{CG} \tag{4.11}$$

where

$$C = \frac{ab^3}{16} \left[\frac{16}{3} - 3.36\frac{b}{a} \left(1 - \frac{b^4}{12a^4} \right) \right] \tag{4.12}$$

In Eq. (4.12), a and b denote the wider and narrower sides of the rectangular cross section, respectively. Table 3.1 gives the exact solutions of the angle of twist for a number of commonly encountered cross sections [1, 2].

EXAMPLE 4.4 | Determination of Angle of Twist of a Rod with Fixed Ends

A circular brass rod (Figure 4.3a) is fixed at each end and loaded by a torque T at point D. Find the maximum angle of twist.

(a)   (b)

Figure 4.3   Example 4.4.

Given: $a = 20$ in., $b = 40$ in., $d = 1$ in., $T = 500$ lb $\cdot$ in., and $G = 5.6 \times 10^6$ psi.

Solution: The reactions at the end are designated by T_A and T_B.
    Statics.   The only available equation of equilibrium for free-body diagram of Figure 4.3b yields

$$T_A + T_B = T \tag{a}$$

Therefore, the problem is statically indeterminate to the first degree.
    Deformations.   The angle of twist at section D for the left and right segments of the bar are

$$\phi_{AD} = \frac{T_A a}{GJ}, \qquad \phi_{BD} = \frac{T_B b}{GJ}$$

Geometry.   The continuity of the bar at section D requires that

$$\phi_{AD} = \phi_{BD} \quad \text{or} \quad T_A a = T_B b \tag{b}$$

Equations (a) and (b) can be solved simultaneously to obtain

$$T_A = \frac{Tb}{L}, \qquad T_B = \frac{Ta}{L} \tag{4.13}$$

The maximum angle of rotation occurs at section D. Therefore,

$$\phi_{max} = \frac{T_A a}{GJ} = \frac{Tab}{GJL}$$

Substituting the given numerical values into this equation, we have

$$\phi_{max} = \frac{500(20)40}{5.6(10^6)\frac{\pi}{32}(1)^4(60)} = 0.012 \text{ rad} = 0.7°$$

4.4   DEFLECTION OF BEAMS BY INTEGRATION

Beam deflections due to bending are determined from deformations taking place along a span. Analysis of the deflection of beams is based on the assumptions of the beam theory outlined in Section 3.7. As we see in Section 5.5, for slender members, the contribution of shear to deflection is regarded as negligible, since for static bending problems, the shear deflection represents no more than a few percent of the total deflection. Direct integration and superposition methods for determining elastic beam deflection are discussed in the sections to follow.

Governing the differential equations relating the deflection v to the internal bending moment M in a linearly elastic beam whose cross section is symmetrical about the plane (xy) of loading is given by [3]

$$\frac{d^2v}{dx^2} = \frac{M}{EI} \tag{4.14}$$

The quantity EI is called the *flexural rigidity*. The sign convention for applied loading and the internal forces, according to that defined in Section 1.8, is shown in Figure 4.4. The deflection and slope θ (in radians) of the deflection curve are related by the equation

$$\theta = \frac{dv}{dx} = v' \tag{4.15}$$

Positive (and negative) θ, like moments, follow the right-hand rule, as depicted in the figure.

As shown in Section 3.6, internal shear force V, bending moment M, and the load intensity w are connected by Eqs. 3.15 and 3.14. These, combined with Eq. (4.14), give the useful sequence of relationships, for constant EI, in the form:

$$\text{Moment} = M = EI\frac{d^2v}{dx^2} = EIv'' \tag{4.16a}$$

$$\text{Shear} = V = EI\frac{d^3v}{dx^3} = EIv''' \tag{4.16b}$$

$$\text{Load} = w = EI\frac{d^4v}{dx^4} = EIv'''' \tag{4.16c}$$

Figure 4.4   Positive loads and internal forces.

$v(a) = 0$
$\theta(a) = 0$

(a)

$v(a) = 0$
$M(a) = 0$

(b)

$V(a) = 0$
$M(a) = 0$

(c)

$\theta(a) = 0$
$V(a) = 0$

(d)

Figure 4.5   Boundary conditions: (a) fixed end; (b) simply supported end; (c) free end; (d) guided or sliding support.

The deflection v of a beam can be found by solving any one of the foregoing equations by successive integrations. The choice of equation depends on the ease with which an expression of load, shear, or moment can be formulated and individual preference. The approach to solving the deflection problem beginning with Eq. (4.16c) or (4.16b) is known as the *multiple-integration method*. When Eq. (4.16a) is used, because two integrations are required to obtain the v, this is called the *double-integration method*. The constants of the integration are evaluated using the specified conditions on the ends of the beam, that is, the boundary conditions. Frequently encountered conditions that may apply at the ends $(x = a)$ of a beam are shown in Figure 4.5. We see from the figure that the force (static) variables M, V, and the geometric (kinematic) variables v, θ are 0 for common situations.

If the beam has a cross-sectional width b that is large compared to the depth h (i.e., $b \gg h$), the beam is stiffer, and the deflection is less than that determined by Eqs. (4.16) for narrow beams. The large cross-sectional width prevents the lateral expansion and contraction of the material, and the deflection is thereby reduced, as shown in Section 4.9. An improved value for the deflection v of *wide beams* is obtained by multiplying the result given by the equation for a narrow beam by $(1 - v^2)$, where v is Poisson's ratio.

EXAMPLE 4.5 | Finding Beam Deflections by the Double-Integration Method

A simply supported beam is subjected to a concentrated load at a distance a from the left end as shown in Figure 4.6. Develop

(a) The expressions for the elastic curve.

(b) The deflection at point C for the case in which $a = b = L/2$.

Figure 4.6   Example 4.5.

Solution: The reactions are noted in the figure.

(a) The moments for the segments AC and CB of the beam are expressed as

$$M_1 = \frac{Pb}{L}x \qquad\qquad (0 \le x \le a)$$

$$M_2 = \frac{Pb}{L}x - P(x - a) \qquad (a \le x \le L)$$

Double integrations of these equations give the results

For segment AC   *For segment CB*

$$EIv_1'' = \frac{Pb}{L}x \qquad\qquad EIv_2'' = \frac{Pb}{L}x - P(x - a)$$

$$EIv_1' = \frac{Pb}{2L}x^2 + c_1 \qquad EIv_2' = \frac{Pb}{2L}x^2 - \frac{P}{2}(x - a)^2 + c_3$$

$$EIv_1 = \frac{Pb}{6L}x^3 + c_1 x + c_2 \qquad EIv_2 = \frac{Pb}{6L}x^3 - \frac{P}{6}(x - a)^3 + c_3 x + c_4$$

From the boundary and the continuity conditions, we easily obtain

$$v_1'(a) = v_2'(a): \quad c_1 = c_3$$

$$v_1(a) = v_2(a): \quad c_2 = c_4$$

$$v_1(0) = 0: \quad 0 = c_2, \qquad v_2(L) = 0: \quad c_3 = -\frac{Pb}{6L}(L^2 - b^2)$$

The elastic curves for the left- and right-hand segments are therefore

$$v_1 = -\frac{Pbx}{6EIL}(L^2 - b^2 - x^2) \qquad\qquad (0 \le x \le a)$$

$$v_2 = -\frac{Pbx}{6EIL}(L^2 - b^2 - x^2) - \frac{P(x - a)^3}{6EI} \qquad (a \le x \le L)$$

(4.17)

Then, through the use of Eq. (4.15) the slopes for the two parts of the beam can readily be found.

(b) Force P acts at the middle of the beam span ($a = b = L/2$) and hence Eqs. (4.17) result in

$$v_{\max} = v_C = -\frac{PL^3}{48EI}$$

(4.18)

Comments: The minus sign means that the deflection is downward. In this case, the elastic curve is symmetric about the center of the beam.

4.5   BEAM DEFLECTIONS BY SUPERPOSITION

The elastic deflections (and slopes) of beams subjected to simple loads have been solved and are readily available (see Tables A.9 and A.10). In practice, for combined load configurations, the method of superposition may be applied to simplify the analysis and design. The method is valid whenever displacements are linearly proportional to the applied loads. This is the case if Hooke's law holds for the material and deflections are small.

To demonstrate the method, consider the beam of Figure 4.7a, replaced by the beams depicted in Figures 4.7b and 4.7c. At point C, the beam undergoes deflections $(v)_P$ and $(v)_M$, due to P and M, respectively. Hence, the deflection v_C due to combined loading is $v_C = (v_C)_P + (v_C)_M$. From the cases 1 and 2 of Table A.9, we have

$$v_C = \frac{5PL^3}{48EI} + \frac{ML^2}{8EI} \tag{4.19}$$

Similarly, the deflection and the angle of rotation at any point of the beam can be found by the foregoing procedure.

The method of superposition can be effectively applied to obtain deflections or reactions for *statically indeterminate* beams. In these problems, the redundant reactions are considered unknown loads and the corresponding supports are removed or modified accordingly. Next, superposition is employed: The load diagrams are drawn and expressions are written for the deflections produced by the individual loads (both known and unknown); the redundant reactions are computed by satisfying the geometric boundary conditions. Following this, all other reactions can be found from equations of static equilibrium.

The steps described in the preceding paragraph can be made clearer though the illustration of a beam statically indeterminate to the first degree (Figure 4.8a). Reaction R_B is selected as redundant and treated as unknown load by eliminating the support at B. Decomposition of the loads is shown in Figures 4.8b and 4.8c. Deflections due to w and the redundant R_B are (see cases 5 and 8 of Table A.9):

$$(v_B)_w = -\frac{5wL^4}{24EI}, \qquad (v_B)_R = \frac{R_B L^3}{6EI}$$

From geometry of the original beam,

$$v_B = -\frac{5wL^4}{24EI} + \frac{R_B L^3}{6EI} = 0$$

or

$$R_B = \frac{5}{4}wL \tag{4.20}$$

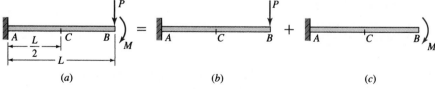

(a)   (b)   (c)

Figure 4.7   Deflections of a cantilevered beam.

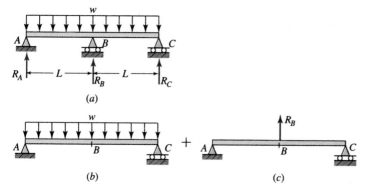

Figure 4.8 Deflections of a two-span continuous beam.

The remaining reactions are $R_A = R_c = 3wL/8$, as determined by applying the equations of equilibrium. Having the reactions available, deflection can be obtained using the method discussed in the preceding section.

Determination of Interaction Force between Two Beams Placed One on Top of the Other

EXAMPLE 4.6

Two simply supported beams are situated one on top of the other as shown in Figure 4.9. The top beam is subjected to a uniformly distributed load of intensity w. Find

(a) The interaction force R at midspan C acting upward on the beam AB and acting downward on beam DE.

(b) The maximum moment and deflection in beam AB.

Figure 4.9 Example 4.6.

Assumption: The beams are supported such a way that they are to deflect by the same amount at the junction *C*.

Solution: Both beams are statically indeterminate to the first degree. We select R as redundant and treat it as an unknown load. Considering the two beams in turn and using the data in Table A.9, the deflections at the center are as follows. For *beam AB*, due to load *w* and owing to R,

$$v_w = \frac{5w(2a)^4}{384EI_a}, \qquad v_R = -\frac{R(2a)^3}{48EI_a}$$

The total downward deflection is therefore

$$v_a = \frac{5wa^4}{24EI_a} - \frac{Ra^3}{6EI_a}$$

For *beam DE*, due to R, the downward deflection is

$$v_b = \frac{Rb^3}{6EI_b}$$

(a)   Equating the two expressions for deflections, $v_a = v_b$, and solving for the interaction force, we have

$$R = \frac{1.25wa}{\alpha} \tag{4.21}$$

where

$$\alpha = 1 + \left(\frac{b}{a}\right)^3 \frac{I_a}{I_b}$$

Comment: If beam *DE* is rigid (i.e., $I_b \to \infty$), then $\alpha = 1$ and $R = 1.25wa$, which is equal to the central reaction for the beam resting on three simple supports.

(b)   Maximum bending moment and deflection in beam *AB* occurring at the center are, respectively,

$$M_C = \frac{1}{2}wa^2 - \frac{1}{2}Ra = \frac{1}{2}wa^2 \left(1 - \frac{1.25}{\alpha}\right) \tag{4.22a}$$

$$v_C = \frac{5w(2a)^4}{384EI_a}\left(1 - \frac{1}{\alpha}\right) = \frac{5wa^4}{24EI_a}\left(1 - \frac{1}{\alpha}\right) \tag{4.22b}$$

Comments: We see from the preceding results that, as beam *DE* is made stiffer by either reducing its span 2*b* or increasing its moment of inertia I_b, the value of α decreases and hence the value of R increases. This decreases the deflection and also reduces the bending moment in the loaded beam *AB*.

Case Study 4-1 | WINCH CRANE FRAME DEFLECTION ANALYSIS

A schematic representation of the frame of a winch crane is shown in Figure 1.5. Determine the deflection under the load using the method of superposition.

Given: The geometry and loading of the critical portion of the frame are known from Case Study 3-1.

Data:

$$E = 200 \text{ GPa}, \qquad I = 2.01(10^{-6}) \text{ m}^4$$

Assumptions: The loading is static. Deflection due to transverse shear is neglected.

Solution: See Figure 3.31 and Table A.9.

When the load P and the weight w of the cantilever depicted in the figure act alone, displacement at D (from

cases 1 and 3 of Table A.9) are $PL_1^3/3EI$ and $wL_1^4/8EI$, respectively. It follows that, the deflection v_D at the free end owing to the combined loading is

$$v_D = -\frac{PL_1^3}{3EI} - \frac{wL_1^4}{8EI}$$

Substituting the given numerical values into the preceding expression, we have

$$v_D = -\frac{1}{200(10^3)(2.01)}\left[\frac{3000(1.5)^3}{3} + \frac{130(1.5)^4}{8}\right]$$

$$= -8.6 \text{ mm}$$

Here minus sign means a downward displacement.

Comment: Since $v_D \ll h/2$, the magnitude of the deflection obtained is well within the acceptable range (see Section 3.7).

Case Study 4-2 | BOLT CUTTER DEFLECTION ANALYSIS

Members 2 and 3 of the bolt cutter shown in Figure 3.32 are critically stressed. Determine the deflections employing the superposition method.

Given: The dimensions (in inches) and loading are known from Case Study 3-2. The parts are made of AISI 1080 HR steel having $E = 30 \times 10^6$ psi.

Assumptions: The loading is static. The member 2 can be approximated as a simple beam with an overhang.

Solution: See Figures 3.32 (repeated here), 4.10 and Table A.9.

Member 3. The elongation of this tensile link (Figure 3.32a) is obtained from Eq. (4.1). So, due to

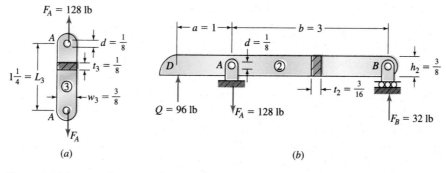

(a)   (b)

Figure 3.32 (repeated) Some free-body diagrams of bolt cutter shown in Figure 1.6: (a) link 3; (b) jaw 2.

(continued)

Case Study   (CONCLUDED)

Figure 4.10   Deflection of simple beam with an overhang.

symmetry in the assembly, the displacement of each end point A is

$$\delta_A = \frac{1}{2}\left(\frac{PL}{AE}\right) = \frac{F_A L_3}{2AE}$$

$$= \frac{128(1.25)}{2\left(\frac{3}{8}\right)\left(\frac{1}{8}\right)(30 \times 10^6)} = 56.9(10^{-6}) \text{ in.}$$

Member 2. This jaw is loaded as shown in Figure 3.32b. The deflection of point D is made up of two parts: a displacement v_1 owing to bending of part DA acting as a cantilever beam and a displacement v_2 caused by the rotation of the beam axis at A (Figure 4.10).

The deflection v_1 at D (by case 1 of Table A.9) is

$$v_1 = \frac{Qa^3}{3EI}$$

The angle θ_A at the support A (from case 7 of Table A.9) is

$$\theta_A = \frac{Mb}{3EI}$$

where $M = Qa$. The displacement v_2 of point D, due to only the rotation at A, is equal to $\theta_A a$, or

$$v_2 = \frac{Qba^2}{3EI}$$

The total deflection of point D, $v_1 + v_2$, is then

$$v_D = \frac{Qa^2}{3EI}(a+b)$$

In the foregoing, we have

$$I = \frac{1}{12}t_2 h_2^3$$

$$= \frac{1}{12}\left(\frac{3}{16}\right)\left(\frac{3}{8}\right)^3 = 0.824(10^{-3}) \text{ in.}^4$$

Substitution of the given data results in

$$v_D = \frac{96(1^2)(1+3)}{3(30 \times 10^6)(0.824 \times 10^{-3})} = 5.18 \times 10^{-3} \text{ in.}$$

Comment: Only very small deflections are allowed in members 2 and 3 to guarantee the proper cutting stroke, and the values found are acceptable.

4.6   BEAM DEFLECTION BY THE MOMENT-AREA METHOD

In this section, we consider a semigraphical technique called the *moment-area method* for determining deflections of beams. The approach uses the relationship between the derivatives of the deflection v and the properties of the area of the bending moment diagram. Usually it gives more rapid solution than integration methods when the deflection and slope at only one point of the beam are required. The moment-area method is particularly effective in the analysis of beams of variable cross sections with uniform or concentrated loading [3–5].

MOMENT-AREA THEOREMS

Two theorems form the basis of the moment-area approach. These principles are developed by considering a segment AB of the deflection curve of a beam under an arbitrary loading. The sketches of the M/EI diagram and greatly exaggerated deflection curve are shown in

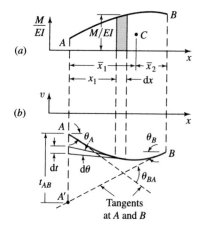

Figure 4.11 Moment-area method:
(a) M/EI diagram; (b) elastic curve.

Figure 4.11. Here M is the bending moment in the beam and EI represents the flexural rigidity. The changes in the angle $d\theta$ of the tangents at the ends of an element of length dx and the bending moment are connected through Eqs. (4.14) and (4.15):

$$d\theta = \frac{M}{EI}\, dx \qquad \text{(a)}$$

The difference in slope between any two points A and B for the beam (Figure 4.11) can be expressed as follows:

$$\theta_{BA} = \theta_B - \theta_A = \int_A^B \frac{M\, dx}{EI} = \left[\text{area of } \frac{M}{EI} \text{ diagram between } A \text{ and } B \right] \qquad \text{(4.23)}$$

This is called the *first moment-area theorem:* The change in angle θ_{BA} between the tangents to the elastic curve at two points A and B equals the area of the M/EI diagram between those points. Note that the angle θ_{BA} and the area of the M/EI diagrams have the same sign. That means a positive (negative) area corresponds to a counterclockwise (clockwise) rotation of the tangent to the elastic curve as we proceed in the x direction. Hence, θ_{BA} shown in Figure 4.11b is positive.

Inasmuch as the deflection of a beam are taken to be small, we see from Figure 4.11b that the vertical distance dt due to the effect of curvature of an element of length dx equals $x\, d\theta$, where $d\theta$ is defined by Eq. (a). Therefore, vertical distance AA', the *tangential deviation* t_{AB} of point A from the tangent at B is

$$t_{AB} = \int_A^B x_1 \frac{M\, dx}{EI} = \left[\text{area of } \frac{M}{EI} \text{ diagram between } A \text{ and } B \right] \bar{x}_1 \qquad \text{(4.24)}$$

in which x is the horizontal distance to the centroid C of the area from A. This is the *second moment-area theorem:* The tangential deviation t_{AB} of point A with respect to the tangent at B equals the moment with respect to A of the area of the M/EI diagram between A and B.

Likewise, we have

$$
t_{BA} = \left[\text{area of } \frac{M}{EI} \text{ diagram between } A \text{ and } B \right] \bar{x}_2 \tag{4.25}
$$

The quantity $\bar{x}_2$ represents the horizontal distance from point B to the centroid C of the area (Figure 4.11a). Note that $t_{AB} \neq t_{BA}$ generally. Also observe from Eqs. (4.24) and (4.25) that the *signs* of t_{AB} and t_{BA} depend on the sign of the bending moments. In many beams, it is obvious whether the beam deflects upward or downward and whether the slope is clockwise or counterclockwise. When this is the case, it is not necessary to follow the sign conventions described for the moment-area method: We calculate the absolute values and find the directions by inspection.

APPLICATION OF THE MOMENT-AREA METHOD

Determination of beam deflections by moment-area theorems is fairly routine. They are equally applicable for rigid frames. In continuous beams, the two sides of a joint are 180° to one another, whereas in rigid frames the sides of a joint often are at 90° to one another. Our discussion is limited to beam problems. A correctly constructed M/EI diagram and a sketch of the elastic curve are always necessary. Table A.3 may be used to obtain the areas and centroidal distances of common shapes. The slopes of points on the beam with respect to one another can be found from Eq. (4.23) and the deflection, using Eq. (4.24) or (4.25). The moment-area procedure is readily used for beams in which the direction of the tangent to the elastic curve at one or more points is known (e.g., cantilevered beams). For computational simplicity, often M/EI diagrams are drawn and the formulations made in terms of the quantity EI; that is, numerical values of EI may be substituted in the final step of the solution.

For a statically determinate beam with various loads or an indeterminate beam, the displacements determined by the moment-area method are usually best found by superposition. This requires a series of diagrams indicating the moment due to each load or reaction drawn on a separate sketch. In this manner, calculations can be simplified, because the areas of the separate M/EI diagrams may be simple geometric forms. When treating statically indeterminate problems, each additional compatibility condition is expressed by a moment-area equation to supplement the equations of statics.

EXAMPLE 4.7 | Shaft Deflection by the Moment-Area Method

A simple shaft carries its own weight of intensity w, as depicted in Figure 4.12a. Determine the slopes at the ends and center deflection.

Assumption: Bearings act as simple supports.

Solution: Inasmuch as the flexural rigidity EI is constant, the M/EI diagram has the same parabolic shape as the bending-moment diagram (Figure 4.12b), where the area properties are taken from Table A.3. The elastic curve is depicted in Figure 4.12c, with the tangent drawn at A.

(a)

(b)

(c)

Figure 4.12 Example 4.7.

The shaft and loading are symmetric about the center C; hence, the tangent to the elastic curve at C is horizontal: $\theta_C = 0$. Therefore, $\theta_{CA} = 0 - \theta_A$ or $\theta_A = -\theta_{CA}$ and

$$A_1 = \frac{2}{3}\left(\frac{L}{2}\right)\left(\frac{wL^2}{8EI}\right) = \frac{wL^3}{24EI}$$

By the first moment-area theorem, $\theta_{CA} = A_1$:

$$\theta_A = -\frac{wL^3}{24EI} = -\theta_B \tag{4.26}$$

The minus sign means that the end A of the beam rotates clockwise, as shown in the figure.

Through the use of the second moment-area theorem, Eq. (4.25),

$$t_{CA} = A_1\left(\frac{3L}{16}\right) = \frac{wL^4}{128EI}$$

in which $t_{CA} = CC'$ and $\theta_A L/2 = C'C''$ (Figure 4.12c). The maximum deflection, $v_{max} = CC''$, is

$$v_{max} = -\frac{wL^3}{24EI}\left(\frac{L}{2}\right) + \frac{wL^4}{128EI} = -\frac{5wL^4}{384EI} \tag{4.27}$$

The minus sign indicates that the deflection is downward.

Comment: Alternatively, the moment of area A_1 about point A, Eq. (4.24), readily gives the numerical value of v_{max}.

EXAMPLE 4.8 | Displacements of a Stepped Cantilevered Beam by the Moment-Area Method

A nonprismatic cantilevered beam with two different moment of inertia carries a concentrated load P at its free end (Figure 4.13a). Find the slope at B and deflection at C.

Figure 4.13   Example 4.8.

Solution: The M/EI diagram is divided conveniently into its component parts, as shown in Figure 4.13b:

$$A_1 = -\frac{PL^2}{8EI}, \qquad A_2 = -\frac{PL^2}{16EI}, \qquad A_3 = -\frac{PL^2}{8EI}$$

The elastic curve is in Figure 4.13c. Inasmuch as $\theta_A = 0$ and $v_A = 0$, we have $\theta_C = \theta_{CA}$, $\theta_B = \theta_{BA}$, $v_C = t_{CA}$, and $v_B = t_{BA}$.

Applying the first moment-area theorem,

$$\theta_B = A_1 + A_2 + A_3 = -\frac{5PL^2}{16EI} \tag{4.28a}$$

The minus sign means that the rotations are clockwise. From the second moment-area theorem,

$$v_C = A_1\left(\frac{L}{4}\right) + A_2\left(\frac{L}{3}\right) = -\frac{5PL^3}{96EI} \tag{4.28b}$$

The minus sign shows that the deflection is downward.

Reactions of a Propped up Cantilever by the Moment-Area Method | **EXAMPLE 4.9**

A propped cantilevered beam is loaded by a concentrated force P acting at the position shown in Figure 4.14a. Determine the reactional forces and moments at the ends of the beam.

(a)

(b)

(c)

Figure 4.14 Example 4.9.

Solution: The reactions indicated in Figure 4.14a shows that the beam is statically indeterminate to the first degree. We select R_B as a redundant (or unknown) load and remove support B (Figure 4.14b). The corresponding M/EI diagram is in Figure 4.14c, with the component areas

$$A_1 = \frac{R_B L^2}{2EI}, \qquad A_2 = -\frac{Pa^2}{2EI}$$

One displacement compatibility condition is required to find the redundant load. Observe that the slope at the fixed end and the deflection at the supported end are 0; the tangent to the elastic curve at A passes through B, or $t_{BA} = 0$. Therefore, by the second moment-area theorem,

$$\frac{R_B L^2}{2EI}\left(\frac{2L}{3}\right) - \frac{Pa^2}{2EI}\left(L - \frac{a}{3}\right) = 0$$

Solving,

$$R_B = \frac{Pa^2}{2L^3}(3L - a) \qquad\qquad \textbf{(4.29)}$$

Comments: The remaining reactions are obtained from equations of statics. Then, the slope and deflection are found as needed by employing the usual moment-area procedure.

4.7   IMPACT LOADING

A moving body striking a structure delivers a suddenly applied dynamic force that is called an *impact* or *shock load*. Details concerning the material behavior under dynamic loading are presented in Section 2.9 and Chapter 8. Although the impact load causes elastic members to vibrate until equilibrium is reestablished, our concern here is with only the influence of impact or shock force on the maximum stress and deformation within the member.

Note that the design of engineering structures subject to suddenly applied loads is complicated by a number of factors, and theoretical considerations generally serve only qualitatively to guide the design [6–8]. In Sections 4.8 and 4.9, typical impact problems are analyzed using the *energy method* of the mechanics of materials theory together with the following common *assumptions:*

1. The displacement is proportional to the loads.
2. The material behaves elastically, and a static stress-strain diagram is also valid under impact.
3. The inertia of the member resisting impact may be neglected.
4. No energy is dissipated because of local deformation at the point of impact or at the supports.

Obviously, the energy approach leads to an approximate value for impact loading. It presupposes that the stresses throughout the impacted member reach peak values at the same time. In a more exact method, the stress at any position is treated as a function of time, and waves of stress are found to sweep through the elastic material at a propagation rate. This wave method gives higher stresses than the energy method. However, the former is more complicated than the latter and not discussed in this text. The reader is directed to references for further information [9–12].

4.8   LONGITUDINAL AND BENDING IMPACT

Here, we determine the stress and deflection caused by linear or longitudinal and bending impact loads. In machinery, the longitudinal impact may take place in linkages, hammer-type power tools, coupling-connected cars, hoisting rope, and helical springs. Examples of bending impact are found in shafts and structural members, such as beams, plates, shells, and vessels.

FREELY FALLING WEIGHT

Consider the free-standing spring of Figure 4.15a, on which is dropped a body of mass m from a height h. Inasmuch as the velocity is 0 initially and 0 again at the instant of maximum deflection of the spring (δ_{max}), the change in kinetic energy of the system is 0. Therefore, the work done by gravity on the body as it falls is equal to the resisting work done by the spring:

$$W(h + \delta_{max}) = \frac{1}{2}k\delta_{max}^2 \qquad \textbf{(4.30)}$$

in which k is the spring constant.

Figure 4.15 (a) Freely falling body. (b) Horizontal moving body.

The deflection corresponding to a static force equal to the weight of the body is simply W/k. This is called the *static* deflection, δ_{st}. The general expression of *maximum dynamic deflection* is, using Eq. (4.30),

$$\delta_{max} = \delta_{st} + \sqrt{(\delta_{st})^2 + 2\delta_{st}h} \qquad \textbf{(4.31a)}$$

This may be written in the form

$$\delta_{max} = \left(1 + \sqrt{1 + \frac{2h}{\delta_{st}}}\right)\delta_{st} = K\delta_{st} \qquad \textbf{(4.31b)}$$

The *impact factor K*, is defined by

$$K = 1 + \sqrt{1 + \frac{2h}{\delta_{st}}} \qquad \textbf{(4.32)}$$

Multiplying the K by W gives an equivalent static, or *maximum dynamic load:*

$$P_{max} = KW \qquad \textbf{(4.33)}$$

To compute the maximum stress and deflection resulting from impact loading, P may be used in the formulas for static loading.

Two extreme situations are clearly of particular interest. When $h \gg \delta_{max}$, the work term, $W\delta_{max}$, in Eq. (4.30) may be neglected, reducing the expression to

$$\delta_{max} = \sqrt{2\delta_{st}h} \qquad \textbf{(4.34)}$$

On the other hand, when $h = 0$, the load is suddenly applied, and Eq. (4.30) reduces to

$$\delta_{max} = 2\delta_{st} \qquad \textbf{(4.35)}$$

HORIZONTALLY MOVING WEIGHT

An analysis similar to the preceding one may be used to develop expressions for the case of a mass ($m = W/g$) in horizontal motion with a velocity v, stopped by an elastic body. In this case, kinetic energy $E_k = mv^2/2$ replaces $W(h + \delta_{max})$, the work done by W, in

Eq. (4.30). By so doing, the maximum dynamic deflection and load are

$$\delta_{max} = \delta_{st}\sqrt{\frac{v^2}{g\delta_{st}}} = \sqrt{\frac{2E_k}{k}} \tag{4.36a}$$

$$P_{max} = m\sqrt{\frac{v^2 g}{\delta_{st}}} = \sqrt{2E_k k} \tag{4.36b}$$

The quantity δ_{st} is the static deflection caused by a horizontal force W. Note that m is measured in kg in SI or lb·s^2/in. in U.S. units. Likewise expressed are v (in m/s or in./s), the gravitational acceleration g (in m/s^2 or in./s^2), and E_k (in N·m or in.·lb).

When the body hits the end of a *prismatic bar* of length L and axial rigidity AE (Figure 4.15b), we have $k = AE/L$ and hence $\delta_{st} = mgL/AE$. Equations (4.36) are therefore

$$\delta_{max} = \sqrt{\frac{mv^2 L}{AE}} \tag{4.37a}$$

$$P_{max} = \sqrt{\frac{mv^2 AE}{L}} \tag{4.37b}$$

The corresponding maximum dynamic compressive stress, taken to be uniform through the bar, is

$$\sigma_{max} = \sqrt{\frac{mv^2 E}{AL}} \tag{4.38}$$

The foregoing shows that the stress can be reduced by increasing the volume AL or decreasing the kinetic energy E_k and the modulus of elasticity E of the member. We note that the stress concentration in the middle of a notched bar would reduce its capacity and tend to promote brittle fracture. This point has been treated in Section 2.9.

EXAMPLE 4.10 | Impact Loading on a Rod

The prismatic rod depicted in Figure 4.16 has length L, diameter d, and modulus of elasticity E. A rubber compression washer of stiffness k and thickness t is installed at the end of the rod.

(a) Calculate the maximum stress in the rod caused by a sliding collar of weight W that drops from a height h onto the washer

(b) Redo part a, with the washer removed.

Given: $L = 5$ ft, $h = 3$ ft, $d = \frac{1}{2}$ in. $t = \frac{1}{4}$ in.
$E = 30 \times 10^6$ psi, $k = 25$ lb/in., $W = 8$ lb

Solution: The cross-sectional area of the rod $A = \pi(1/2)^2/4 = \pi/16 \text{ in.}^2$

(a) For the rod *with* the washer, the static deflection is

$$\delta_{st} = \frac{WL}{AE} + \frac{W}{k}$$

$$= \frac{8(16)(5 \times 12)}{\pi(30 \times 10^6)} + \frac{8}{25} = 0.081 \times 10^{-3} + 0.32 \text{ in.}$$

The maximum dynamic stress, from Eqs. (4.33) and (4.32),

$$\sigma_{max} = \frac{WK}{A}$$

$$= \frac{8 \times 16}{\pi}\left[1 + \sqrt{1 + \frac{2(3 \times 12)}{0.32}}\right] = 653 \text{ psi}$$

(b) In the absence of the washer, this equation results in

$$\sigma_{max} = \frac{8 \times 16}{\pi}\left[1 + \sqrt{1 + \frac{2(3 \times 12 + 0.25)}{0.081 \times 10^{-3}}}\right] = 38.6 \text{ ksi}$$

Comments: The difference in stress for the preceding two solutions is large. This suggests the need for flexible systems for withstanding impact loads. Interestingly, bolts subjected to dynamic loads, such as those used to attach the ends to the tube in pneumatic cylinders, are often designed with long grips (see Section 15.9) to take advantage of the more favorable stress conditions.

Figure 4.16
Example 4.10.

Impact Loading on a Beam

EXAMPLE 4.11

A weight W is dropped a height h, striking at midspan a simply supported steel beam of length L. The beam is of rectangular cross section of width b and depth d (Figure 4.17). Calculate the maximum deflection and maximum stress for these two cases:

(a) The beam is rigidly supported at each end.

(b) The beam is supported at each end by springs.

Figure 4.17   Example 4.11.

Given: $W = 100$ N, $h = 150$ mm, $L = 2$ m, $b = 30$ mm, and $d = 60$ mm.

Assumptions: Modulus of elasticity $E = 200$ GPa and spring rate $k = 200$ kN/m.

Solution: We have $M_{max} = WL/4$ at point C and $I = bd^3/12$. The maximum deflection, due to a static load, is (from case 5 of Table A.9):

$$\delta_{st} = \frac{WL^3}{48EI} = \frac{100(2)^3(12)}{48(200 \times 10^9)(0.03)(0.06)^3} = 0.154 \text{ mm}$$

The maximum static stress equals

$$\sigma_{st} = \frac{M_{max}C}{I} = \frac{100(2)(0.03)(12)}{4(0.03)(0.06)^3} = 2.778 \text{ MPa}$$

(a) The impact factor, using Eq. (4.32),

$$K = 1 + \sqrt{1 + \frac{2(0.15)}{0.154(10^{-3})}} = 45.15$$

Therefore,

$$\delta_{max} = 45.15(0.154) = 6.95 \text{ mm}$$

$$\sigma_{max} = 45.15(2.778) = 125 \text{ MPa}$$

(b) The static deflection of the beam due to its own bending and the deformation of the springs is

$$\delta_{st} = 0.154 + \frac{50}{200} = 0.404 \text{ mm}$$

The impact factor is then

$$K = 1 + \sqrt{1 + \frac{2(0.15)}{0.404(10^{-3})}} = 28.27$$

Hence,

$$\delta_{max} = 28.27(0.404) = 11.42 \text{ mm}$$

$$\sigma_{max} = 28.27(2.778) = 78.53 \text{ MPa}$$

Comments: Comparing the results, we observe that dynamic loading considerably increases deflection and stress in a beam. Also noted is a reduction in stress with increased flexibility, owing to the spring added to the supports. However, the values calculated are probably somewhat higher than the actual quantities, because of our simplifying assumptions 3 and 4.

4.9   TORSIONAL IMPACT

In machinery, torsional impact occurs in rotating shafts of punches and shears, in geared drives, at clutches, brakes, torsional suspension bars, and numerous other components. Here, we discuss the stress and deflection in members subjected to impact torsion. The problem is analyzed by the approximate energy method used in the preceding section. Advantage will be taken of the analogy between linear and torsional systems to readily write the final relationships.

Consider a circular prismatic shaft of flexural rigidity GJ, length L, fixed at one end and subjected to a suddenly applied torque T at the other end (Figure 4.18). The shaft stiffness, from Eq. (4.10), is $k = GJ/L$, where $J = \pi d^4/32$ and d is the diameter. The maximum dynamic angle of twist (in rad), from Eq. (4.36a), is

$$\phi_{max} = \sqrt{\frac{2E_k L}{GJ}} \tag{4.39a}$$

in which E_k is the kinetic energy. Similarly, the maximum dynamic torque, referring to Eq. (4.36b), is

$$T_{max} = \sqrt{\frac{2E_k GJ}{L}} \tag{4.39b}$$

The maximum dynamic shear stress, $\tau_{max} = 16T_{max}/\pi d^3$, is therefore

$$\tau_{max} = 2\sqrt{\frac{E_k G}{AL}} \tag{4.40}$$

Here A represents the cross-sectional area of the shaft.

Recall from Section 1.11 that, for a rotating wheel of constant thickness, the kinetic energy is expressed in the form

$$E_k = \frac{1}{2}I\omega^2 \tag{4.41}$$

Figure 4.18   Bar subjected to impact torsion.

with

$$I = \frac{1}{2}mb^2 \qquad\qquad \text{(4.42a)}$$

$$m = \pi b^2 t \rho \qquad\qquad \text{(4.42b)}$$

In the foregoing, we have

I = mass moment of inertia ($\text{N} \cdot \text{s}^2 \cdot \text{m}$ or $\text{lb} \cdot \text{s}^2 \cdot \text{in.}$)

ω = angular velocity (rad/s)

m = mass (kg or $\text{lb} \cdot \text{s}^2/\text{in.}$)

b = radius

t = thickness

ρ = mass density (kg/m^3 or $\text{lb} \cdot \text{s}^2/\text{in.}^4$).

As before, W and g are the weight and acceleration of gravity, respectively. A detailed treatment of stress and displacement in disk flywheels is given in Section 16.5.

Note that in the case of wheel of variable thickness, the mass moment of inertia may conveniently be obtained from the expression

$$I = mr^2 \qquad\qquad \text{(4.43)}$$

The quantity r is called the *radius of gyration* for the mass. It is hypothetical distance from the wheel center at which the entire mass could be concentrated and still have the same moment of inertia as the original mass.

EXAMPLE 4.12 | Impact Loading on a Shaft

A shaft of diameter d and length L has a flywheel (radius of gyration r, weight W, modulus of rigidity G, yield strength in shear S_{ys}) at one end and runs at a speed of n. If the shaft is instantly stopped at the other end, determine

    (a)　The maximum shaft angle of twist.

    (b)　The maximum shear stress.

Given:　$d = 3$ in., $L = 2.5$ ft, $W = 120$ lb, $r = 10$ in., $n = 150$ rpm

Assumption:　The shaft is made of ASTM-A242 steel. So, by Table B.1, $G = 11.5 \times 10^6$ psi and $S_{ys} = 30$ ksi.

Solution:　The area properties of the shaft are

$$A = \frac{\pi(3)^2}{4} = 7.069 \text{ in.}^2, \qquad J = \frac{\pi(3)^4}{32} = 7.952 \text{ in.}^4$$

The angular velocity equals

$$\omega = n\left(\frac{2\pi}{60}\right) = 150\left(\frac{2\pi}{60}\right) = 5\pi \text{ rad/s}$$

(a) The kinetic energy of the flywheel must be absorbed by the shaft. So, substituting Eq. (4.43) into Eq. (4.41), we have

$$E_k = \frac{W\omega^2 r^2}{2g} \qquad\qquad\qquad \text{(a)}$$

$$= \frac{120(5\pi)^2(10)^2}{2(386)} = 3835 \text{ in} \cdot \text{lb}$$

From Eq. (4.39a),

$$\phi_{max} = \sqrt{\frac{2E_k L}{GJ}} = \left[\frac{2(3835)(2.5 \times 12)}{(11.5 \times 10^6)(7.952)}\right]^{1/2}$$

$$= 0.05 \text{ rad} = 2.87°$$

(b) Through the use of Eq. (4.40),

$$\tau_{max} = 2\sqrt{\frac{E_k G}{AL}} = 2\left[\frac{(3835)(11.5 \times 10^6)}{(7.069)(2.5 \times 12)}\right]^{1/2}$$

$$= 28.84 \text{ ksi}$$

Comment: The stress is within the elastic range, $28.84 < 30$, and hence assumption 2 of Section 4.7 is satisfied.

*4.10 BENDING OF THIN PLATES

A plate is an initially flat structural member with thickness small compared with remaining dimensions. It is usual to divide the plate thickness t into equal halves by a plane parallel to the faces. This plane is called the *midsurface* of the plate. The plate thickness is measured in a direction normal to the midsurface at each point under consideration. Plates of technical importance are usually defined as thin when the ratio of the thickness to the smaller span length is less than $1/20$. Here, we discuss briefly the bending of thin plates. For a detailed treatment of the subject, see [13–15].

BASIC ASSUMPTIONS

Consider a plate before deformation, depicted in Figure 4.19a, where the xy plane coincides with the midsurface and hence the z deflection is 0. When deformation occurs due to external loading, the midsurface at any point x_A, y_A undergoes a deflection w. Referring to

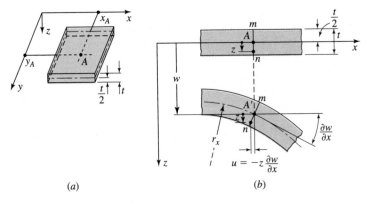

(a)   *(b)*

Figure 4.19   Plate in bending: (a) before deformation; (b) rotation of a plane section due to deflection.

the coordinate system shown, the basic *assumptions* of the small deflection theory of bending for isotropic, homogenous, thin plates may be summarized as follows:

1. The deflection w of the midsurface is small in comparison with the thickness t of the plate; hence, the slope of the deflected surface is much less than unity.

2. Straight lines (such as mn) initially normal to the midsurface remain straight and normal to that plane after bending.

3. No midsurface straining occurs due to bending. This is equivalent to stating that strains γ_{yz}, γ_{xz}, and ε_z are negligible.

4. The component of stress normal to the midsurface, σ_z, may be neglected.

These presuppositions are analogous to those associated with the simple bending theory of beams.

STRAIN-DISPLACEMENT RELATIONS

On the basis of the foregoing assumptions, the strain displacement relations of Eq. (3.61) reduce to

$$\varepsilon_x = \frac{\partial u}{\partial x}, \qquad \varepsilon_y = \frac{\partial v}{\partial y}, \qquad \gamma_{xy} = \frac{\partial v}{\partial x} + \frac{\partial u}{\partial y} \tag{a}$$

and it can be shown that

$$u = -z\frac{\partial w}{\partial x}, \qquad v = -z\frac{\partial w}{\partial y} \tag{b}$$

Combining Eqs. (a) and (b), we have

$$\varepsilon_x = -z\frac{\partial^2 w}{\partial x^2}, \qquad \varepsilon_y = -z\frac{\partial^2 w}{\partial y^2}, \qquad \gamma_{xy} = -2z\frac{\partial^2 w}{\partial x\partial y} \tag{4.44}$$

Because, in small deflection theory, the square for a slope may be considered negligible, the partial derivatives of these relations represent the curvatures of the plate. The *curvatures* at the midsurface in planes parallel to the zx (Figure 4.19b), yz, and xy planes are, respectively,

$$\frac{1}{r_x} = \frac{\partial^2 w}{\partial x^2}, \qquad \frac{1}{r_y} = \frac{\partial^2 w}{\partial y^2}, \qquad \frac{1}{r_{xy}} = \frac{\partial^2 w}{\partial x \partial y} \tag{4.45}$$

The quantity r represents the radius of curvature. Clearly, the curvatures are the rates at which the slopes vary over the plate.

Examining the preceding relationships, we are left to conclude that a circle of curvature can be constructed similarly to Mohr's circle of strain. The curvatures hence transform in the same manner as strains. It can be verified by using Mohr's circle that $(1/r_x) + (1/r_y) = \nabla^2 w$. The sum of the curvatures in perpendicular directions, called the *average curvature,* is invariant with respect to rotation of the coordinate axis. This assertion is valid at any location on the midsurface.

PLATE STRESS, CURVATURE, AND MOMENT RELATIONS

For a thin plate, substituting Eqs. (4.44) into Hooke's law, we obtain

$$\sigma_x = -\frac{Ez}{1-v^2}\left(\frac{\partial^2 w}{\partial x^2} + v\frac{\partial^2 w}{\partial y^2}\right)$$

$$\sigma_y = -\frac{Ez}{1-v^2}\left(\frac{\partial^2 w}{\partial y^2} + v\frac{\partial^2 w}{\partial x^2}\right) \tag{4.46}$$

$$\tau_{xy} = -\frac{Ez}{1+v}\frac{\partial^2 w}{\partial x \partial y}$$

where $\tau_{xy} = \tau_{yx}$. We see from these relations that the stresses are 0 at the midsurface and vary linearly over the thickness of the plate.

The stresses distributed over the side surfaces of the plate, while producing no net force, result in bending and twisting moments. These moment resultants per unit length (in SI units $N \cdot m/m$, or simply N) are designated M_x, M_y, and M_{xy}. With reference to Figure 4.20a,

$$\int_{-t/2}^{t/2} z\sigma_x \, dy \, dz = dy \int_{-t/2}^{t/2} z\sigma_x \, dz = M_x \, dy$$

or

$$M_x = \int_{-t/2}^{t/2} z\sigma_x \, dz \tag{4.47}$$

Here, t is the thickness of the plate. Substituting into this expression the stress σ_x given by Eqs. (4.46), we obtain M_x in terms of curvatures. Expressions involving M_y and

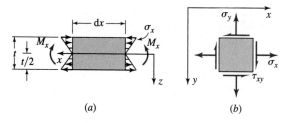

Figure 4.20   (a) Plate segment in pure bending.
(b) Positive stresses on an element in the bottom
half of a plate.

$M_{xy} = M_{yx}$ are derived in a like manner. In so doing, we have

$$M_x = -D\left(\frac{\partial^2 w}{\partial x^2} + v\frac{\partial^2 w}{\partial y^2}\right)$$

$$M_y = -D\left(\frac{\partial^2 w}{\partial y^2} + v\frac{\partial^2 w}{\partial x^2}\right) \qquad \text{(4.48)}$$

$$M_{xy} = -D(1 - v)\frac{\partial^2 w}{\partial x\partial y}$$

The quantity D represents the *flexural rigidity:*

$$D = \frac{Et^3}{12(1 - v^2)} \qquad \text{(4.49)}$$

Interestingly, if a plate element of unit width were free to expand sidewise under the given loading, the flexural rigidity would be $Et^3/12$. The remainder of the plate does not allow this action, however. Because of this, a plate shows *greater stiffness* than a narrow beam by a factor $1/(1 - v^2)$ or about 10% for $v = 0.3$.

According to the *sign convention,* a positive moment is one that results in positive stresses in the positive (bottom) half of the plate (see Section 1.13), as depicted in Figure 4.20b. The maximum stresses occurring on the surface of the plate are obtained by substituting $z = t/2$ into Eqs. (4.46), together with the use of Eqs. (4.48), as

$$\sigma_{x,\max} = \frac{6M_x}{t^2}, \qquad \sigma_{y,\max} = \frac{6M_y}{t^2}, \qquad \tau_{xy,\max} = \frac{6M_{xy}}{t^2} \qquad \text{(4.50)}$$

Since there is a direct correspondence between the moments and stresses, the equation for transforming the stresses should be identical with that used for the moments. Mohr's circle therefore may be applied to moments as well as to stresses.

4.11 DEFLECTION OF PLATES BY INTEGRATION

Consider an element $dx\,dy$ of the plate subject to a uniformly distributed load per unit area p (Figure 4.21). In addition to the moments M_x, M_y, and M_{xy} previously discussed, we now find vertical shear forces Q_x and Q_y (force per unit length) acting on the sides of the element. With the change of location, for example, by dx, one of the moment components, say, M_x, acting on the negative x face, varies in value relative to the positive x face as $M_x + (\partial M_x / \partial x)\,dx$. Treating all components similarly, the state of stress resultants shown in the figure is obtained.

Variations in the moment and force resultants are governed by the conditions of equilibrium. Application of the equations of statics to Figure 4.21 leads to a single differential equation in terms of the moments. This, when combined with Eqs. (4.48), results in [13]

$$\frac{\partial^4 w}{\partial x^4} + 2\frac{\partial^4 w}{\partial x^2 \partial y^2} + \frac{\partial^4 w}{\partial y^4} = \frac{p}{D} \tag{4.51a}$$

or, in concise form,

$$\nabla^4 w = \frac{p}{D} \tag{4.51b}$$

The preceding expression, first derived by Lagrange in 1811, is the governing differential equation for deflection of thin plates.

BOUNDARY CONDITIONS

Two common boundary conditions that apply along the edge at $x = a$ of the rectangular plate with edges parallel to the x and y axes (Figure 4.22) may be expressed as follows.

In the *clamped edge* (Figure 4.22a), both the slope and the deflection must vanish:

$$w = 0, \qquad \frac{\partial w}{\partial x} = 0, \qquad (x = a) \tag{4.52}$$

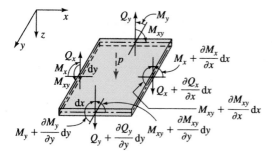

Figure 4.21 Positive moments and shear forces per unit length and distributed loads per unit area on a plate element.

Figure 4.22 Two common boundary conditions: (a) fixed edge; (b) simply supported edge.

In the *simply supported* edge (Figure 4.22b), both the deflection and bending moment are 0. The latter implies that, at edge $x = a$, $\partial w/\partial y = 0$ and $\partial^2 w/\partial y^2 = 0$. Therefore,

$$w = 0, \qquad \frac{\partial^2 w}{\partial x^2} = 0, \qquad (x = a) \tag{4.53}$$

Other typical conditions at the boundaries may be expressed similarly.

To determine the deflection w, we must integrate Eq. (4.51) with the constants of integration dependent on the appropriate boundary conditions. Having the deflection available, the stress (as well as strain and curvature) components are obtained using the formulas derived in the preceding section. This is illustrated in the solution of the following problem.

EXAMPLE 4.13

Determination of Deflection and Stress in a Plate

A long, narrow plate of width b and thickness t, a so-called plate strip, is simply supported at edges $y = 0$ and $y = b$, as depicted in Figure 4.23. The plate carries the loading of the form

$$p(y) = p_o \sin\left(\frac{\pi y}{b}\right) \tag{a}$$

Figure 4.23   Example 4.13.

The quantity p_o represents the load intensity along the x' axis. Let $\nu = 1/3$. Determine

(a)   The equation of the deflection surface and maximum stresses.

(b)   The maximum deflection and stresses for $p_o = 15$ kPa, $b = 0.5$ m, $t = 12.5$ mm, and $G = 210$ GPa.

Solution: Due to symmetry in the loading and end restraints about the x axis, the plate deforms into a cylindrical surface with its generating line parallel to the x' axis. Since, for this situation, $\partial w/\partial x = 0$ and $\partial^2 w/\partial x \partial y = 0$, Eqs. (4.48) reduce to

$$M_x = -\nu D \frac{d^2 w}{dy^2}, \qquad M_y = -D \frac{d^2 w}{dy^2} \tag{4.54}$$

Also Eq. (4.51) becomes

$$\frac{d^4 w}{dy^4} = \frac{p}{D} \tag{4.55}$$

This equation is the same as the wide beam equation, and we conclude that the solution proceeds as in the case of a beam.

(a) Introducing Eq. (a) into Eq. (4.55), integrating, and satisfying the boundary conditions

$$w = 0, \qquad \frac{d^2w}{dy^2} = 0, \qquad (y = 0, \ y = b)$$

we obtain the deflection

$$w = \left(\frac{b}{\pi}\right)^4 \frac{p_o}{D} \sin\left(\frac{\pi y}{b}\right) \tag{4.56}$$

The largest deflection of the plate occurs at $y = b/2$. Therefore,

$$w_{max} = \left(\frac{b}{\pi}\right)^4 \frac{p_o}{D} \tag{4.57}$$

The moments are now readily determined carrying Eq. (4.56) into (4.54). Then, the maximum stresses, occurring at $y = b/2$, are found applying Eqs. (4.50) as

$$\sigma_{y,max} = 0.6 p_o \left(\frac{b}{t}\right)^2, \qquad \sigma_{x,max} = \nu \sigma_{y,max} = 0.2 p_o \left(\frac{b}{t}\right)^2, \qquad \tau_{xy} = 0 \tag{4.58}$$

(b) Substituting the given data into Eq. (4.49), we have

$$D = \frac{210(10^9)(0.0125)^3}{12\left(1 - \frac{1}{9}\right)} = 38,452 \ \text{N·m}$$

Similarly, Eqs. (4.57) and (4.58) lead to

$$w_{max} = \left(\frac{0.5}{\pi}\right)^4 \frac{15(10^3)}{38,452} = 0.25 \ \text{mm}$$

$$\sigma_{y,max} = 0.6(15 \times 10^3)\left(\frac{500}{12.5}\right)^2 = 14.4 \ \text{MPa}$$

$$\sigma_{x,max} = \frac{1}{3}(14.4) = 4.8 \ \text{MPa}$$

Comment: The result, $w_{max}/t = 0.02$, shows that the deflection surface is extremely flat, as is often the case for small deflections.

REFERENCES

1. Ugural, A. C., and S. K. Fenster. *Advanced Strength and Applied Elasticity,* 4th ed. Upper Saddle River, NJ: Prentice Hall, 2003.
2. Young, W. C., and R. C. Budynas. *Roark's Formulas for Stress and Strain,* 7th ed. New York: McGraw-Hill, 2001.
3. Ugural, A. C. *Mechanics of Materials.* New York: McGraw-Hill, 1991.
4. West, H. H. *Fundamentals of Structural Analysis,* 2nd ed. New York: Wiley, 2002.

5. Wang, C. K., and C. G. Salmon. *Introductory Structural Analysis.* Upper Saddle River, NJ: Prentice Hall, 1984.

6. Timoshenko, S. P., and J. N. Goodier. *Theory of Elasticity,* 3rd ed. New York: McGraw-Hill, 1970.

7. Flügge, W., ed. *Handbook of Engineering Mechanics.* New York: McGraw-Hill, 1962, Section 27.3.

8. Harris, C. M. *Shock and Vibration Handbook.* New York: McGraw-Hill, 1988.

9. Burr, A. H., and J. B. Cheatham, *Mechanical Analysis and Design,* 2nd ed. Upper Saddle River, NJ: Prentice Hall, 1995.

10. Goldsmith, W. *Impact.* London: Edward Arnold Ltd., 1960.

11. Kolsky, H. *Stress Waves in Solids.* New York: Dover, 1965.

12. Varley, E., ed. "The Propagation of Shock Waves in Solids." AMD-17 Symposium. New York: ASME, 1976.

13. Ugural, A. C. *Stresses in Plates and Shells,* 2nd ed. New York: McGraw-Hill, 1999.

14. Szilard, R. *Theory and Analysis of Plates.* Upper Saddle River, NJ: Prentice Hall, 1974.

15. Timoshenko, S. P., and S. Woinowsky-Krieger. *Theory of Plates and Shells,* 2nd ed. New York: McGraw-Hill, 1959.

PROBLEMS

Sections 4.1 through 4.6

4.1 A high-strength steel rod of length L, used in control mechanism, must carry a tensile load of P without exceeding its yield strength S_y with a factor of safety n nor stretching more than δ.

(*a*) What is the required diameter of the rod?

(*b*) Calculate the spring rate for the rod.

Given: $P = 10$ kN, $E = 200$ GPa, $S_y = 250$ MPa, $L = 6$ m, $\delta = 5$ mm
Design Decision: The rod will be made of ASTM-A242 steel. Take $n = 1.2$.

4.2 Before loading there is a gap Δ between the wall and the right end of the copper rod of diameter d (Figure P4.2). Calculate the reactions at A and B, after the rod is subjected to an axial load of P.

Given: $\Delta = 0.014$ in., $d = \frac{1}{2}$ in., $P = 8$ kips, $E = 17 \times 10^6$ psi

Figure P4.2

4.3 At room temperature (20°C), a gap Δ exits between the wall and the right end of the bars shown in Figure P4.3. Determine

(*a*) The compressive axial force in the bars after temperature reaches 140°C.

(*b*) The corresponding change in length of the aluminum bar.

Figure P4.3

Given:

$$A_a = 1000 \text{ mm}^2, \qquad E_a = 70 \text{ GPa}, \qquad \alpha_a = 23 \times 10^{-6}/°C, \qquad \Delta = 1 \text{ mm}$$
$$A_s = 500 \text{ mm}^2, \qquad E_s = 210 \text{ GPa}, \qquad \alpha_s = 12 \times 10^{-6}/°C.$$

4.4 Redo Problem 4.3 for the case in which $\Delta = 0$.

4.5 Two steel shafts are connected by gears and subjected to a torque T, as shown in Figure P4.5. Calculate

(a) The angle of rotation in degrees at D.

(b) The maximum shear stress in shaft AB.

Given: $G = 79$ GPa, $T = 500$ N·m, $d_1 = 45$ mm, $d_2 = 35$ mm

Figure P4.5

4.6 Determine the diameter d_1 of shaft AB shown in Figure P4.5, for the case in which the maximum shear stress in each shaft is limited to 150 MPa.

Design Decisions: $d_2 = 65$ mm. The factor of safety against shear is $n = 1.2$.

4.7 A disk is attached to a 40-mm diameter, 0.5-m long steel shaft ($G = 79$ GPa) as depicted in Figure P4.7.

Figure P4.7

Design Requirement: To achieve the desired natural frequency of torsional vibrations, the stiffness of the system is specified such that the disk will rotate $1.5°$ under a torque of $1 \text{ kN} \cdot \text{m}$. How deep ($h$) must a 22-mm diameter hole be drilled to satisfy this requirement?

4.8   A solid round shaft with fixed ends is under a distributed torque of intensity $T(x) = T_1$, lb · in./in., as shown in Figure P4.8. Determine the reactions at the walls.

Figure P4.8

4.9   Resolve Problem 4.8, for the case in which $T(x) = (x/L)T_1$.

4.10   A simply supported beam AB carries a triangularly distributed load of maximum intensity w_o (Figure P4.10).

(a)   Employ the fourth-order differential equation of the deflection to derive the expression for the elastic curve.

(b)   Determine the maximum deflection v_{max}.

Figure P4.10

4.11   A simply supported beam is loaded with a concentrated moment M_o, as shown in Figure P4.11. Derive the equation of the elastic curve for the segment AC of the beam.

Figure P4.11

4.12   A simple beam of wide-flange cross section carries a uniformly distributed load of intensity w (Figure P4.12). Determine the span length L.

Given: $h = 12.5$ in., $E = 30 \times 10^6$ psi
Design Requirements: $\sigma_{max} = 10$ ksi, $v_{max} = \frac{1}{8}$ in.

Figure P4.12

4.13 The overhanging beam *ABC* supports a concentrated load *P* at the free end, as shown in Figure P4.13. For the segment *BC* of the beam,

(a) Derive the equation of the elastic curve.
(b) What is the maximum deflection v_{max}?
(c) Calculate the value of the v_{max} for the following data:

$$I = 5.12 \times 10^6 \text{ mm}^4, \qquad E = 200 \text{ GPa}, \qquad P = 25 \text{ kN}$$

$$L = 2 \text{ m}, \qquad\qquad a = 0.5 \text{ m}$$

Figure P4.13

4.14 Two cantilever beams *AB* and *CD* are supported and loaded as shown in Figure P4.14. What is the interaction force *R* transmitted through the roller that fits snugly between the two beams at point *C*? Use the method of superposition and the deflection formulas of the beams from Table A.9.

Figure P4.14

4.15 Figure P4.15 shows a compound beam with a hinge at point *B*. It is composed of two parts: a beam *BC* simply supported at *C* and a cantilevered beam *AB* fixed at *A*. Apply the superposition method using Table A.9 to determine the deflection v_B at the hinge.

Figure P4.15

4.16 A propped cantilevered beam carries a uniform load of intensity w (Figure P4.16). Determine the reactions at the supports, using the second-order differential equation of the beam deflection.

Figure P4.16

4.17 A fixed-ended beam AB is under a symmetric triangular load of maximum intensity w_o, as shown in Figure P4.17. Determine all reactions, the equation of the elastic curve, and the maximum deflection.

Requirement: Use the second-order differential equation of the deflection.

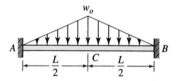

Figure P4.17

4.18 A fixed-ended beam supports a concentrated load P at its midspan (Figure P4.18). Determine all reactions and the equation of the elastic curve.

Figure P4.18

4.19 Redo Problem 4.16, using the method of superposition together with Table A.9.

4.20 A cantilever beam is subjected to a partial loading of intensity w, as shown in Figure P4.20. Use the area moments to determine

(a) The slope at the free end.

(b) The deflection at the free end.

Figure P4.20

4.21 A simple beam with two different moments of inertia is under a center load P, as shown in Figure P4.21. Apply the area moments to find

(*a*) The slope at point B.

(*b*) The maximum deflection.

Figure P4.21

4.22 and **4.23** A simple beam with an overhang and a continuous beam are supported and loaded, as shown in Figures P4.22 and P4.23, respectively. Use the area moments to determine the support reactions.

Figure P4.22

Figure P4.23

Sections 4.7 through 4.11

4.24 The uniform rod AB is made of steel. Collar D moves along the rod and has a speed of $v = 3.5$ m/s as it strikes a small plate attached to end A of the rod (Figure P4.24). Determine the largest allowable weight of the collar.

Given: $S_y = 250$ MPa, $E = 210$ GPa
Design Requirement: A factor of safety of $n = 3$ is used against failure by yielding.

Figure P4.24

4.25 The 20-kg block D is dropped from a height h onto the steel beam AB (Figure P4.25). Determine

(a)   The maximum deflection of the beam.

(b)   The maximum stress in the beam.

Given: $h = 0.5$ m, $E = 210$ GPa

Figure P4.25

4.26 Collar of weight W, depicted in Figure P4.26, is dropped from a height h onto a flange at the end B of the round rod. Determine the W.

Given: $h = 3.5$ ft, $d = 1$ in., $L = 15$ ft, $E = 30 \times 10^6$ psi
Requirement: The maximum stress in the rod is limited to 35 ksi.

Figure P4.26

4.27 The collar of weight W falls a distance h when it comes into contact with end B of the round steel rod (Figure P4.26). Determine diameter d of the rod.

Given: $W = 20$ lb, $h = 4$ ft, $L = 5$ ft, $E = 30 \times 10^6$ psi
Design Requirement: The maximum stress in the rod is not to exceed 18 ksi.

4.28 The collar of weight W falls onto a flange at the bottom of a slender rod (Figure P4.26). Calculate the height h through which the weight W should drop to produce a maximum stress in the rod.

Given: $W = 500$ N, $L = 3$ m, $d = 20$ mm, $E = 170$ GPa
Design Requirement: Maximum stress in the rod is limited to $\sigma_{max} = 350$ MPa.

4.29 Design the shaft (determine the minimum required length L_{min}), described in Example 4.12, so that yielding does not occur. Based on this length and the given impact load, what is the angle of twist?

4.30 The steel shaft and abrasive wheels A and B at the ends of a belt-drive sheave rotates at n rpm (Figure P4.30). If the shaft is suddenly stopped at the wheel A because of jamming, determine

   (a) The maximum angle of twist of the shaft.

   (b) The maximum shear stress in the shaft.

   Given: $D_a = 125$ mm, $\quad D_b = 150$ mm, $\quad d = t = 25$ mm, $\quad L = 0.3$ m,

           $n = 1500$ rpm, $\quad G = 79$ GPa, $\quad S_{ys} = 250$ MPa,

           density of wheels $\rho = 1800$ kg/m^3

   Assumption: Abrasive wheels are considered solid disks.

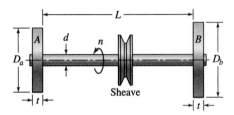

Figure P4.30

4.31 Redo Problem 4.30, for the cease in which the shaft runs at $n = 1200$ rpm and the wheel end B is suddenly stopped because of jamming.

4.32 A rectangular sheet plate of thickness t is bent into circular cylinder of radius r. Determine the diameter D of the cylinder and the maximum moment M_{max} developed in the plate.

Given: $t = \frac{1}{8}$ in., $E = 30 \times 10^6$ psi, $v = 0.3$.
Design Assumption: The maximum stress in the plate is limited to 18 ksi.

4.33 A long, narrow rectangular plate is under a nonuniform loading

$$p = p_o \sin \frac{\pi y}{b}$$

and clamped at edges $y = 0$ and $y = b$ (Figure P4.33). Determine

   (a) An expression for the deflection surface w.

   (b) The maximum bending stress.

   (c) The values of maximum deflection and stress for the data

     $b = 20$ in., $\quad t = 0.4$ in., $\quad E = 10 \times 10^6$ psi, $\quad v = 0.3$, $\quad p_o = 5$ psi

Figure P4.33

4.34 Figure P4.34 depicts a long, narrow rectangular plate with edge $y = 0$ simply supported and edge $y = b$ clamped. The plate is under a uniform load of intensity p_o. Determine

(a) An expression for the deflection surface w.

(b) The maximum bending stress at $y = b$.

(c) The values of maximum deflection at $y = b/2$ and maximum bending stress at $y = b$, based on the data given in Problem 4.33.

Figure P4.34

ENERGY METHODS IN DESIGN

Outline

5.1   INTRODUCTION

As pointed out in Section 1.4, instead of the equilibrium methods, displacements and forces can be ascertained through the use of energy methods. The latter are based on the concept of strain energy, which is of fundamental importance in analysis and design. Application of energy techniques is effective in cases involving members of variable cross sections and problems dealing with elastic stability, trusses, and frames. In particular, strain energy approaches can greatly ease the chore of obtaining the displacement of members under combined loading.

In this chapter, we explore two principal energy methods and illustrate their use with a variety of examples. The first deals with the *finite* deformation experienced by load-carrying components (Sections 5.2 through 5.7). The second, the variational methods, based on a *virtual* variation in stress or displacement, is discussed in the remaining sections. Literature related to the energy approaches is extensive [1–19].

5.2   STRAIN ENERGY

Internal work stored in an elastic body is the internal energy of deformation or the elastic strain energy. It is often convenient to use the quantity, called *strain energy per unit volume* or *strain energy density*. The area under the stress-strain diagram represents the strain energy density, designated U_o, of a tensile specimen (Figure 5.1). Therefore

$$U_o = \int \sigma_x \, d\varepsilon_x \tag{5.1a}$$

The area above the stress-strain curve is termed the *complimentary energy density:*

$$U_o^* = \int \varepsilon_x \, d\varepsilon_x \tag{5.2}$$

Observe from Figure 5.1b that, for a nonlinearly elastic material, these energy densities have different values.

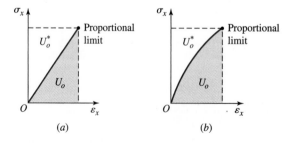

Figure 5.1   Work done by uniaxial stress: (a) linearly elastic material; (b) nonlinearly elastic material.

In the case of a linearly elastic material, from origin up to the proportional limit, substituting σ_x/E for ε_x, we have

$$U_o = \frac{1}{2}\sigma_x\varepsilon_x = \frac{1}{2E}\sigma_x^2 \qquad (5.1b)$$

and the two areas are equal $U_o = U_o^*$, as shown in Figure 5.1a. In SI units, the strain energy density is measured in joules per cubic meter (J/m^3) or in pascals; in U.S. customary units, it is expressed in inch-pounds per cubic inch (in. $\cdot$ lb/in.^{3}) or psi. Similarly, strain energy density for shear stress is given by

$$U_o = \frac{1}{2}\tau_{xy}\gamma_{xy} = \frac{1}{2G}\tau_{xy}^2 \qquad (5.3)$$

When a body is subjected to a general state of stress, the total strain energy density equals simply the sum of the expressions identical to the preceding equations. We have then

$$U_o = \frac{1}{2}(\sigma_x\varepsilon_x + \sigma_y\varepsilon_y + \sigma_z\varepsilon_z + \tau_{xy}\gamma_{xy} + \tau_{yz}\gamma_{yz} + \tau_{xz}\gamma_{xz}) \qquad (5.4)$$

Substitution of generalized Hooke's law into this expression gives the following equation, involving only stresses and elastic constants:

$$U_o = \frac{1}{2E}\left[\sigma_x^2 + \sigma_y^2 + \sigma_z^2 - 2\nu(\sigma_x\sigma_y + \sigma_y\sigma_z + \sigma_x\sigma_z) + \frac{1}{2G}\left(\tau_{xy}^2 + \tau_{yz}^2 + \tau_{xz}^2\right)\right] \qquad (5.5)$$

When the principal axes are used as coordinate axes, the shear stresses are 0. The preceding equation then becomes

$$U_o = \frac{1}{2E}\left[\sigma_1^2 + \sigma_2^2 + \sigma_3^2 - 2\nu(\sigma_1\sigma_2 + \sigma_2\sigma_3 + \sigma_1\sigma_3)\right] \qquad (5.6)$$

in which σ_1, σ_2, and σ_3 are the principal stresses.

The elastic strain energy U stored within an elastic body can be obtained by integrating the strain energy density over the volume V. Thus

$$U = \int_V U_o\,dV = \iiint U_o\,dx\,dy\,dz \qquad (5.7)$$

This equation is convenient in evaluating the strain energy for a number of commonly encountered shapes and loading. It is important to note that the strain energy is a nonlinear (quadratic) function of load or deformation. The principle of superposition therefore is not valid for the strain energy.

5.3   COMPONENTS OF STRAIN ENERGY

The three-dimensional state of stress at a point (Figure 5.2a) may be separated into two parts. The state of stress in Figure 5.2b is associated with the volume changes, so-called dilatations. In the figure σ_m represents the mean stress or the octahedral stress σ_{oct}, defined in Section 3.16. On the other hand, the shape changes, or distortions, are caused by the set of stresses shown in Figure 5.2c.

The dilatational strain-energy density can be obtained through the use of Eq. (5.6) by letting $\sigma_1 = \sigma_2 = \sigma_3 = \sigma_m$. In so doing, we have

$$U_{ov} = \frac{3(1-2\nu)}{2E}(\sigma_m)^2 = \frac{1-2\nu}{6E}(\sigma_1 + \sigma_2 + \sigma_3)^2 \qquad (5.8)$$

The *distortional strain-energy density* is readily found by subtracting the foregoing from Eq. (5.6):

$$U_{od} = \frac{1}{12G}[(\sigma_1 - \sigma_2)^2 + (\sigma_2 - \sigma_3)^2 + (\sigma_3 - \sigma_1)^2] = \frac{3}{4G}\tau_{oct}^2 \qquad (5.9)$$

The quantities G and E are related by Eq. (2.10). The octahedral planes where σ_{oct} and τ_{oct} act are shown in Figure 3.46.

Test results indicate that the dilatational strain energy is ineffective in causing failure by yielding of ductile materials. The energy of distortion is assumed to be completely responsible for material failure by inelastic action. This is discussed further in Chapter 8. Stresses and strains associated with both components of the strain energy are also very useful in describing the plastic deformation [4, 6].

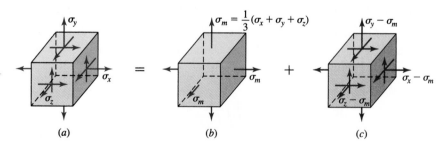

Figure 5.2   (a) Principal stresses, resolved into (b) dilatational stresses and (c) distortional stresses.

Determining the Components of Strain Energy in a Prismatic Bar | **EXAMPLE 5.1**

A structural steel bar having uniform cross-sectional area A carries an axial tensile load P. Find the strain energy density and its components.

Solution: The state of stress is uniaxial tension, $\sigma_x = \sigma = P/A$, and the remaining stress components are 0 (Figure 5.2a). We therefore have the stresses causing volume change $\sigma_m = \sigma/3$ and shape change $\sigma_x - \sigma_m = 2\sigma/3$, $\sigma_y - \sigma_m = \sigma_z - \sigma_m = \sigma/3$ (Figures 5.2b and 5.2c). The strain energy densities for the stresses in Figure 5.2, from Eqs. (5.5), (5.8), and (5.9), are

$$U_o = \frac{\sigma^2}{2E}$$

$$U_{ov} = \frac{(1-2v)\sigma^2}{6E} = \frac{\sigma^2}{12E}$$

$$U_{od} = \frac{(1+v)\sigma^2}{3E} = \frac{5\sigma^2}{12E}$$

Comments: We observe that $U_o = U_{ov} + U_{od}$ and that $5U_{ov} = U_{od}$. That is, to change the shape of a unit volume element subjected to simple tension, five times more energy is absorbed than to change the volume.

5.4   STRAIN ENERGY IN COMMON MEMBERS

Recall from Section 5.2 that the method of superposition is not applicable to strain energy; that is, the effects of several forces or moments on strain energy are not simply additive, as illustrated in Example 5.2. In this section, the following types of loads are considered for the various members of a structure: axial loading, torsion, bending, and shear. Note that the equations derived are restricted to linear material behavior.

AXIALLY LOADED BARS

The normal stress at any transverse section through a bar subjected to an axial load P is $\sigma_x = P/A$, where A represents the cross-sectional area and x is the axial axis (Figure 5.3).

Figure 5.3   Nonprismatic bar with varying axial loading.

Substitution of this and Eq. (5.1) into Eq. (5.7) and setting $dV = A\,dx$, we obtain

$$U_a = \frac{1}{2} \int_0^L \frac{P^2\,dx}{AE} \tag{5.10}$$

For a prismatic bar, subjected to end forces of magnitude P, Eq. (5.10) becomes

$$U_a = \frac{P^2 L}{2AE} \tag{5.11}$$

The quantity E represents the modulus of elasticity and L is the length of the member.

EXAMPLE 5.2

Figure 5.4
Example 5.2.

Determination of Strain Energy Stored in a Bar Due to Combined Loading

A prismatic bar suspended from one end carries, in addition to its own weight, an axial load F (Figure 5.4). Find the strain energy stored in the member.

Solution: The axial load, in an element at a distance x from the fixed end, is expressed as

$$P = \gamma A(L - x) + F$$

Here γ is the weight of the material and A is the cross-sectional area. Introducing this equation into Eq. (5.10),

$$U_a = \int_0^L \frac{[\gamma A(L - x) + F]^2}{2AE}\,dx = \frac{\gamma^2 A L^3}{6E} + \frac{\gamma F L^2}{2E} + \frac{F^2 L}{2AE} \tag{5.12}$$

Comments: The first and third terms on the right side of this equation are the strain energy of the bar due to its own weight and the strain energy of the bar supporting only axial force F. The presence of the middle term shows that the strain energy produced by the two loads acting simultaneously is not equal to the sum of the strain energies associated with the loads acting separately.

CIRCULAR TORSION BARS

In the case of pure torsion of a bar, Eq. 3.11 for an arbitrary distance r from the centroid of the cross section gives $\tau = Tr/J$. The strain energy density, Eq. (5.3), becomes then $U_o = T^2 r^2 / 2J^2 G$. Inserting this into Eq. (5.7), the strain energy owing to torsion is

$$U_t = \int_0^L \frac{T^2}{2GJ^2} \left(\int r^2\,dA \right) dx \tag{5.13}$$

We have $dV = dA\,dx$; dA is an element of the cross-sectional area. By definition, the term in parentheses is the polar moment of inertia J of the cross-sectional area. Hence,

$$U_t = \frac{1}{2} \int_0^L \frac{T^2\,dx}{GJ} \tag{5.14}$$

For a prismatic bar subjected to end torques T (Figure 3.6), Eq. (5.14) appears as

$$U_t = \frac{T^2 L}{2GJ} \tag{5.15}$$

in which L is the length of the bar.

BEAMS

Consider a beam in pure bending. The flexure formula gives the axial normal stress $\sigma_x = My/I$. Using Eq. (5.1), the strain energy density is $U_o = M^2 y^2/2EI^2$. After carrying this into Eq. (5.7) and noting that $M^2/2EI^2$ is a function of x alone, we obtain

$$U_b = \int_0^L \frac{M^2}{2EI^2} \left(\int y^2 \, dA \right) dx \tag{5.16}$$

Since the integral in parentheses defines the moment of inertia I of the cross-sectional area about the neutral axis, the strain energy due to *bending* is

$$U_b = \frac{1}{2} \int_0^L \frac{M^2 \, dx}{EI} \tag{5.17}$$

This, integrating along beam length L, gives the required quantity.

For a beam of constant flexural rigidity EI, Eq. (5.17) may be written in terms of deflection by using Eq. (4.14) as follows:

$$U_b = \frac{EI}{2} \int_0^L \left(\frac{d^2 v}{dx^2} \right)^2 dx \tag{5.18}$$

The transverse shear force V produces shear stress τ_{xy} at every point in the beam. The strain energy density, inserting Eq. (3.20) into Eq. (5.3), is $U_o = V^2 Q^2/2GI^2 b^2$. Integration of this, over the volume of the beam of cross-sectional area A, results in the strain energy for beams in *shear*:

$$U_s = \int_0^L \frac{\alpha V^2 \, dx}{2AG} \tag{5.19}$$

In the foregoing, the *form factor for shear* is

$$\alpha = \frac{A}{I^2} \int_A \frac{Q^2}{b^2} \, dA \tag{5.20}$$

This represents a dimensionless quantity specific to a given cross-section geometry.

Example 5.3 illustrates the determination of the form factor for shear for a rectangular cross section. Other cross sections can be treated similarly. Table 5.1 furnishes several cases [4]. Subsequent to finding α, the strain energy due to shear is obtained using Eq. (5.19).

Table 5.1   Form factor for shear for various beam cross sections

Cross section	Form factor α
Rectangle	6/5
I section, box section, or channels*	A/A_{web}
Circle	10/9
Thin-walled circular	2

*A = area of the entire section, A_{web} = area of the web ht, where h is the beam depth and t is the web thickness.

EXAMPLE 5.3 | Determining the Total Strain Energy Stored in a Beam

A cantilevered beam with a rectangular cross section supports a concentrated load P as depicted in Figure 5.5. Find the total strain energy and compare the values of the bending and shear contributions.

Figure 5.5   Examples 5.3 and 5.4.

Solution: The first moment of the area, by Eq. (3.21), is $Q = (b/2)[(h/2)^2 - y_1^2]$. Inasmuch as $A/I^2 = 144/bh^5$, Eq. (5.20) gives

$$\alpha = \frac{144}{bh^5} \int_{-h/2}^{h/2} \frac{1}{4}\left(\frac{h^2}{4} - y_1^2\right)^2 b\,dy_1 = \frac{6}{5}$$

From the equilibrium requirements, the bending moment $M = -Px$ and the shear force $V = P$ at x (Figure 5.5). Carrying these and $\alpha = 6/5$ into Eqs. (5.17) and (5.19) then integrating, we obtain

$$U_b = \int_0^L \frac{P^2 x^2}{2EI}\,dx = \frac{P^2 L^3}{6EI} \tag{5.21}$$

$$U_s = \int_0^L \frac{6}{5}\frac{V^2}{2AG}\,dx = \frac{3P^2 L}{5AG} \tag{5.22}$$

Note that $I/A = h^2/12$. The total strain energy stored in the cantilever beam is

$$U = U_b + U_s = \frac{P^2 L^3}{6EI}\left[1 + \frac{3E}{10G}\left(\frac{h}{L}\right)^2\right] \tag{5.23}$$

Through the use of Eqs. (5.21) and (5.22), we find the ratio of the shear strain energy to the bending strain energy in the beam as follows:

$$\frac{U_s}{U_b} = \frac{3E}{10G}\left(\frac{h}{L}\right)^2 = \frac{3}{5}(1 + v)\left(\frac{h}{L}\right)^2 \tag{5.24}$$

Comments: When, for example, $L = 10h$ and $\nu = 1/3$, this quotient is only 1/125; the strain energy owing to the shear is less than 1 percent. For a slender beam, $h \ll L$, it is observed that the energy is due mainly to bending. Therefore, it is usual to neglect the shear in evaluating the strain energy in beams of ordinary proportions. Unless stated otherwise, we adhere to this practice.

PLATES

Based on the assumptions of Section 4.10, for thin plates σ_z, γ_{yz}, and γ_{xz} can be disregarded. Hence, introducing Eq. (5.4) together with Eq. (2.6) into Eq. (5.7); we have

$$U = \iiint_V \left[\frac{1}{2E}\left(\sigma_x^2 + \sigma_y^2 - 2\nu\sigma_x\sigma_y\right) + \frac{1}{2G}\tau_{xy}^2 \right] dx\, dy\, dz \qquad (5.25)$$

In the case of a plate of constant thickness, this equation may be written in terms of deflection w through the use of Eqs. (4.46) and (4.49). In so doing, we obtain

$$U = \frac{1}{2}\iint_A D\left[\left(\frac{\partial^2 w}{\partial x^2}\right) + \left(\frac{\partial^2 w}{\partial y^2}\right) + 2\nu\frac{\partial^2 w}{\partial x^2}\frac{\partial^2 w}{\partial y^2} + 2\nu(1-\nu)\left(\frac{\partial^2 w}{\partial x\partial y}\right)^2 \right] dx\, dy \qquad (5.26)$$

The preceding expression, the strain energy for plates in bending, is expressed in the following alternative form:

$$U = \frac{1}{2}\iint_A D\left\{ \left(\frac{\partial^2 w}{\partial x^2} + \frac{\partial^2 w}{\partial y^2}\right)^2 - 2(1-\nu)\left[\frac{\partial^2 w}{\partial x^2}\frac{\partial^2 w}{\partial y^2} - \left(\frac{\partial^2 w}{\partial x\partial y}\right)^2\right] \right\} dx\, dy \qquad (5.27)$$

Integrations over the area A of the plate surface yield the required quantity.

Applications of the equations developed in this section are illustrated in the formulation of various energy techniques and in the solution of sample problems.

5.5 THE WORK-ENERGY METHOD

The strain energy of a structure subjected to a set of forces and moments may be expressed in terms of the forces and resulting displacements. Suppose that all forces are applied gradually and the final values of the force and displacement are denoted by $P_k (k = 1, 2, \ldots, n)$ and δ_k. The total work W, $\frac{1}{2}\sum P_k\delta_k$, is equal to the strain energy gained by the structure, provided no energy is dissipated. That is,

$$U = W = \frac{1}{2}\sum_{k=1}^{n} P_k\delta_k \qquad (5.28)$$

In other words, the work done by the loads acting on the structure manifests as elastic strain energy.

Consider a member or structure subjected to a single concentrated load P. Equation (5.28) then becomes

$$U = \frac{1}{2}P\delta \qquad (5.29)$$

The quantity δ is the displacement through which the force P moves. In a like manner, it can be shown that

$$U = \frac{1}{2}M\theta \tag{5.30}$$

$$U = \frac{1}{2}T\phi \tag{5.31}$$

Note that M (or T) and θ (or ϕ) are, respectively, the moment (or torque) and the associated slope (or angle of twist) at a point of a structure.

The foregoing relationships provide a convenient approach for finding the displacement. This is known as the *method of work-energy*. In the next section, we present a more general approach that may be used to obtain the displacement at a given structure even when the structure carries combined loading.

EXAMPLE 5.4 | **Determination of Beam Deflection by the Work-Energy Method**

A cantilevered beam with a rectangular cross section is loaded as shown in Figure 5.5. Find the deflection v_A at the free end by considering the effects of both the internal bending moments and shear force.

Solution: The total strain energy U of the beam, given by Eq. (5.23), is equated to the work, $W = Pv_A/2$. Hence

$$v_A = \frac{PL^3}{3EI}\left[1 + \frac{3E}{10G}\left(\frac{h}{L}\right)^2\right] \tag{5.32}$$

Comment: If the effect of shear is disregarded, note that the relative error is identical to that found in the previous example. As already shown, this is less than 1 percent for a beam with ratio $L/h = 10$.

5.6   CASTIGLIANO'S THEOREM

Castigliano's theorems are in widespread use in the analysis of structural displacements and forces. They apply with ease to a variety of statically determinate as well as indeterminate problems. Two theorems were proposed in 1879 by A. Castigliano (1847–1884). The first theorem relies on a virtual (imaginary) variation in deformation and is discussed in Section 5.8. The second concerns the finite deformation experienced by a member under load. Both theorems are limited to small deformations of structures. The first theorem is pertinent to structures that behave nonlinearly as well as linearly. We deal mainly with Castigliano's second theorem, which is restricted to structures composed of linearly elastic materials. Unless specified otherwise we refer in this text to the second theorem as *Castigliano's theorem*.

Consider a linearly elastic structure subjected to a set of gradually applied external forces P_k ($k = 1, 2, \ldots, n$). Strain energy U of the structure is equal to the work done W by the applied forces, as given by Eq. (5.28). Let us permit a single load, say, P_i, to be increased a small amount dP_i, while the other applied forces P_k remain unchanged. The increase in strain energy is then $dU = (\partial U/\partial P_i)\, dP_i$, where $\partial U/\partial P_i$ represents the rate of change of the strain energy with respect to P_i. The total energy is

$$U' = U + \left(\frac{\partial U}{\partial P_i}\right) dP_i$$

Alternatively, an expression for U' may be written by reversing the order of loading. Suppose that dP_i is applied first, followed by the force P_k. Now the application of dP_i causes a small displacement $d\delta_i$. The work, $dP_i \cdot d\delta_i/2$, corresponding to this load increment, can be omitted because it is of the second order. The work done during the application of the forces P_k is unaffected by the presence of dP_i. But, the latter force dP_i performs work in moving an amount δ_i. Here δ_i is the displacement caused by the application of P_k. The total strain energy due to the work done by this sequence of loads is therefore

$$U' = U + dP_i \cdot \delta_i$$

Equating the preceding equations, we have the Castigliano's theorem:

$$\delta_i = \frac{\partial U}{\partial P_i} \tag{5.33}$$

The foregoing states that, for a linear structure, *the partial derivative of the strain energy with respect to an applied force is equal to component of displacement at the point of application and in the direction of that force.*

Castigliano's theorem can similarly be shown to be valid for applied moments M (or torques T) and the resulting slope θ (or angle of twist ϕ) of the structure. Therefore,

$$\theta_i = \frac{\partial U}{\partial M_i} \tag{5.34}$$

$$\phi = \frac{\partial U}{\partial T_i} \tag{5.35}$$

In using Castigliano's theorem, we must express the strain energy in terms of the external forces or moments. In the case of a prismatic beam, we have $U = \int M^2\, dx/2EI$. To obtain the deflection v_i corresponding to load P_i, it is often much simpler to differentiate under the integral sign. In so doing, we have

$$v_i = \frac{\partial U}{\partial P_i} = \frac{1}{EI} \int M \frac{\partial M}{\partial P_i}\, dx \tag{5.36}$$

Similarly, the slope may be expressed as

$$\theta_i = \frac{\partial U}{\partial M_i} = \frac{1}{EI} \int M \frac{\partial M}{\partial M_i}\, dx \tag{5.37}$$

Generally, the total strain energy in a straight or curved member subjected to a number of common loads (axial force F, bending moment M, shear force V, and torque T, equals the sum of the strain energies given by Eqs. (5.11), (5.17), (5.19), and (5.15). So, applying Eq. (5.33), the displacement at any point in the member is obtained in the following convenient form:

$$\delta_i = \frac{1}{AE} \int F \frac{\partial F}{\partial P_i}\, dx + \frac{1}{EI} \int M \frac{\partial M}{\partial P_i}\, dx + \frac{1}{AG} \int \alpha V \frac{\partial V}{\partial P_i}\, dx + \frac{1}{GJ} \int T \frac{\partial T}{\partial P_i}\, dx \quad \textbf{(5.38)}$$

Clearly, the last term of this equation applies only to circular bars. An expression may be written for the angle of rotation in a like manner. If it is necessary to obtain the displacement at a point where no corresponding load acts, the problem is treated as follows. We place a fictitious force Q (or couple C) at the point in question in the direction of the desired displacement δ (or θ). We then apply Castigliano's theorem and set the fictitious load $Q = 0$ (or $C = 0$) to obtain the desired displacements.

EXAMPLE 5.5 | Determining the Deflection of a Simple Beam by Castigliano's Theorem

A simple beam is subjected to a uniform load of intensity w, as shown in Figure 5.6a. What is the deflection v_C at an arbitrary distance a from the left support?

(a)   (b)

Figure 5.6 Example 5.5.

Solution: As the deflection is sought, a fictitious force Q is introduced at point C (Figure 5.6b). From the conditions of equilibrium, the reactions are found to be

$$R_A = \frac{wL}{2} + \frac{Qb}{L}, \qquad R_B = \frac{wL}{2} + \frac{Qa}{L}$$

The appropriate moment equations are then

$$M_{AC} = R_A x - \frac{wx^2}{2}, \qquad M_{BC} = R_B x' - \frac{wx'^2}{2}$$

Therefore,

$$\frac{\partial M_{AC}}{\partial Q} = \frac{bx}{L}, \qquad \frac{\partial M_{BC}}{\partial Q} = \frac{ax'}{L}$$

Introducing the foregoing into Eq. (5.38), we have

$$v_C = \frac{1}{EI}\left\{\int_0^a \left[\left(\frac{wL}{2}+\frac{Qb}{L}\right)x - \frac{wx^2}{2}\right]\left(\frac{bx}{L}\right)dx\right.$$

$$\left. + \int_0^b \left[\left(\frac{wL}{2}+\frac{Qa}{L}\right)x' - \frac{wx'^2}{2}\right]\left(\frac{ax'}{L}\right)dx'\right\}$$

Setting $Q = 0$ leads to

$$v_C = \frac{w}{2EIL}\left[\int_0^a (Lx - x^2)(bx)\,dx + \int_0^b (Lx' - x'^2)(ax')\,dx'\right]$$

Integration results in

$$v_C = \frac{wab}{24EIL}[4L(a^2 + b^2) - 3(a^3 + b^3)]$$

Note, as a check, that, for $a = b$, the preceding equation reduces to $5wL^4/384EI$ (see case 8 of Table A.9).

Determination of Deflection of a Curved Frame Using Castigliano's Theorem

EXAMPLE 5.6

A load of P is applied to a steel curved frame, as depicted in Figure 5.7a. Develop an expression for the vertical deflection δ_v of the free end by considering the effects of the internal normal and shear forces in addition to the bending moment. Calculate the value of δ_v for the following data:

$$a = 60 \text{ mm}, \qquad h = 30 \text{ mm}, \qquad b = 15 \text{ mm},$$

$$P = 10 \text{ kN}, \qquad E = 210 \text{ GPa}, \qquad G = 80 \text{ GPa}.$$

(a)   (b)

Figure 5.7 Example 5.6: (a) steel curved frame; (b) free-body diagram.

Solution: A free-body diagram of the part of the bar defined by angle θ is depicted in Figure 5.7b, where the internal forces (F and V) and moment (M) are positive as shown. Referring to the figure, we write

$$M = PR(1 - \cos\theta), \qquad V = P\sin\theta, \qquad F = P\cos\theta$$

Therefore,

$$\frac{\partial M}{\partial P} = R(1 - \cos\theta), \qquad \frac{\partial V}{\partial P} = \sin\theta, \qquad \frac{\partial F}{\partial P} = \cos\theta$$

The form factor for shear for the rectangular section is $\alpha = 6/5$ (Table 5.1). Substitution of the preceding expressions into Eq. (5.38) with $dx = R\,d\theta$ results in

$$\delta_v = \frac{PR^3}{EI} \int_0^\pi (1 - \cos\theta)^2\, d\theta + \frac{6PR}{5AG} \int_0^\pi \sin^2\theta\, d\theta + \frac{PR}{AE} \int_0^\pi \cos^2\theta\, d\theta$$

Using the trigonometric identities $\cos^2\theta = (1 + \cos 2\theta)/2$ and $\sin^2\theta = (1 - \cos 2\theta)/2$, we obtain, after integration,

$$\delta_v = \frac{3\pi PR^3}{2EI} + \frac{3\pi PR}{5AG} + \frac{\pi PR}{2AE} \tag{5.39}$$

The geometric properties of the cross section of the bar are

$$A = 0.015(0.03) = 4.5 \times 10^{-4}\ \text{m}^2, \qquad R = 0.075\ \text{m},$$

$$I = \frac{1}{12}[0.015(0.03)^3] = 337.5 \times 10^{-10}\ \text{m}^4$$

Carrying these values into Eq. (5.39) gives

$$\delta_v = (2.81 + 0.04 + 0.01) \times 10^{-3} = 2.86\ \text{mm}$$

Comments: If the effects of the normal and shear forces are neglected, we have $\delta_v = 2.81$ mm. Then, the error in deflection is approximately 1.7 percent. For this curved bar, where $R/c = 5$, the contribution of V and F to the displacement can therefore be disregarded. It is common practice to neglect the first and the third terms in Eq. (5.38) when $R/c > 4$ (see Section 16.8).

EXAMPLE 5.7 | **Determining Displacements of a Split Ring by Castigliano's Theorem**

A slender, cross-sectional, circular ring of radius R is cut open at $\theta = 0$, as shown in Figure 5.8a. The ring is fixed at one end and loaded at the free end by a z-directed force P. Find, at the free end A,

(a)   The z-directed displacement.

(b)   The rotation about the y axis.

Solution: Since the rotation is sought a fictitious couple C is applied at point A. The bending and twisting moments at any section are (Figure 5.8b)

$$M_\theta = -PR\sin\theta - C\sin\theta, \qquad T_\theta = PR(1 - \cos\theta) - C\cos\theta \tag{a}$$

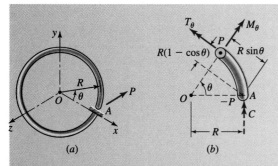

Figure 5.8 Example 5.7.

(a) Substitution of these quantities with $C = 0$ and $dx = R\,d\theta$ into Eq. (5.38) gives

$$\delta_z = \frac{\partial U}{\partial P} = \frac{1}{EI}\int_0^{2\pi}(-PR\sin\theta)(-R\sin\theta)R\,d\theta$$

$$+\frac{1}{GJ}\int_0^{2\pi}PR(1-\cos\theta)R(1-\cos\theta)R\,d\theta \qquad\qquad \textbf{(5.40a)}$$

$$=\frac{\pi PR^3}{EI}+\frac{3\pi PR^3}{GJ}$$

(b) Introducing Eqs. (a) into (5.37) with $dx = R\,d\theta$ and setting $C = 0$, we have

$$\theta_y = \frac{\partial U}{\partial C} = \frac{1}{EI}\int_0^{2\pi}(-PR\sin\theta)(-\sin\theta)R\,d\theta$$

$$+\frac{1}{GJ}\int_0^{2\pi}PR(1-\cos\theta)(-\cos\theta)R\,d\theta \qquad\qquad \textbf{(5.40b)}$$

$$=\frac{\pi PR^2}{EI}+\frac{\pi PR^2}{GJ}$$

Case Study 5-1 | WINCH CRANE FRAME DEFLECTION ANALYSIS BY THE ENERGY METHOD

A schematic representation of a winch crane frame is depicted in Figure 1.5. Find the deflection at the free end applying the energy method.

Given: The dimensions and loading of the frame is known from Case Study 3-1.

Data

$E = 200$ GPa, $\qquad G = 76.9$ GPa,

$I = 2.01(10^{-6})$ m^4

Assumption: The loading is static.

Requirement: Deflections owing to the bending and transverse shear are considered.

Solution: See Figure 3.31 (repeated on the following page) and Table 5.1.

(continued)

Case Study   (CONCLUDED)

$L_1 = 1.5$ m

Weight per length
$w = 130$ N/m

C

y

$b = 50$ mm

z $h = 100$ mm

D N.A.

$t = 6$ mm

$P = 3$ kN

Figure 3.31 (repeated)   Part CD of the crane frame shown in Figure 1.4.

The form factor for shear for the rectangular box section, from Table 5.1, is

$$\alpha = \frac{A}{A_{\text{web}}} = \frac{A}{ht}$$

Here h is the beam depth and t represents the wall thickness. The moment and shear force, at an arbitrary section x distance from the free end of the beam, are expressed as follows:

$$M = -Px - \frac{1}{2}wx^2, \qquad V = P + wx$$

Therefore,

$$\frac{\partial M}{\partial P} = -x, \qquad \frac{\partial V}{\partial P} = 1$$

After substitution of all the preceding equations, Eq. (5.38)

becomes

$$v_D = \frac{1}{EI} \int_0^{L_1} M \frac{\partial M}{\partial P} \, dx + \frac{1}{AG} \int_0^{L_1} \alpha V \frac{\partial V}{\partial P} \, dx$$

$$= \frac{1}{EI} \int_0^{L_1} \left(Px + \frac{1}{2}wx^2 \right)(x) \, dx$$

$$+ \frac{1}{htG} \int_0^{L_1} (P + wx)(1) \, dx$$

Integrating,

$$v_D = \frac{1}{EI} \left(\frac{PL_1^3}{3} + \frac{wL_1^4}{8} \right) + \frac{1}{htG} \left(PL_1 + \frac{wL_1^2}{2} \right)$$

(5.41)

Substituting the given numerical values into this equation, we obtain

$$v_D = \frac{1}{200 \times 10^3 (2.01)} \left[\frac{3000(1.5)^3}{3} + \frac{130(1.5)^4}{8} \right]$$

$$+ \frac{1}{100(6)(76.9 \times 10^3)} \left[3000(1.5) + \frac{130(1.5)^2}{2} \right]$$

$$= (8.6 + 0.1)10^{-3} = 8.7 \text{ mm}$$

Since P is vertical and directed downward, v_D represents a vertical displacement and is positive downward.

Comments: If the effect of shear force is omitted, $v_D = 8.6$ mm; the resultant error in deflection is about 1.2 percent. The contribution of shear force to the displacement of the frame can therefore be neglected.

APPLICATION TO TRUSSES

We now apply Castigliano's theorem to plane trusses. As pointed out in Section 1.9, it is assumed that the connection between the members is pinned and the only force in the member is an axial force, either tensile or compressive. Note that, in practice, members of a plane truss are usually riveted, welded, or bolted together by means of so-called gusset plates. However, due to the slenderness of the members, the internal forces can often be computed on the basis of frictionless joints that prevent translation in two directions, corresponding to reactions of two unknown force components. The methods of joints and sections are commonly used for the analysis of trusses. The *method of joints* consists of analyzing the truss, joint by joint, to determine the forces in the members by applying the conditions of equilibrium to the free-body diagram for each joint. This approach is most effective when the forces in all members of a truss are to be determined. If, however, the force in only a few members of a truss are desired, the method of sections applied to a portion (containing two or more joints) of the truss isolated by imagining that the members in which the forces are to be ascertained are cut.

The strain energy U for a truss is equal to the sum of the strain energies of its members. In the case of a truss consisting m members of length L_j, axial rigidity $A_j E_j$, and internal axial force F_j, the strain energy can be found from Eq. (5.11) as

$$U = \sum_{j=1}^{m} \frac{F_j^2 L_j}{2 A_j E_j} \tag{5.42}$$

The displacement δ_i of the joint of application of load P_i can be obtained by substituting Eq. (5.42) into Castigliano's theorem, Eq. (5.33). Therefore,

$$\delta_i = \frac{\partial U}{\partial P_i} = \sum_{j=1}^{m} \frac{F_j L_j}{A_j E_j} \frac{\partial F_j}{\partial P_i} \tag{5.43}$$

The preceding discussion applies to statically determinate and indeterminate linearly elastic trusses.

Determination of Displacements of a Truss by Castigliano's Theorem | **EXAMPLE 5.8**

A planar truss with pin and roller supports at A and B, respectively, is subjected to a vertical load P at joint E, as shown in Figure 5.9. Determine

(a) The vertical displacement of point E.

(b) The horizontal displacement of point E.

Figure 5.9 Example 5.8.

Assumption: All members are of equal axial rigidity AE.

Solution:

(a) The equilibrium conditions for the entire truss (Figure 5.9) results in $R_{Ax} = 2P$, $R_{Ay} = P$, and $R_B = 2P$. Applying the method of joints at A, E, C, and D, we obtain

$$F_{AD} = \sqrt{2}P, \qquad F_{AC} = F_{CE} = P, \qquad F_{DE} = -\sqrt{2}P, \qquad \text{(b)}$$

$$F_{BC} = F_{CD} = 0, \qquad F_{BD} = -2P$$

Through the use of Eq. (5.43),

$$\delta_v = \frac{1}{AE} \sum FL \frac{\partial F}{\partial P}$$

$$= \frac{1}{AE}[2(PL)(1) + 2(\sqrt{2}P)(\sqrt{2}L)(\sqrt{2}) + (-2P)(L)(-2)]$$

$$= 11.657 \frac{PL}{AE}$$

The positive sign for δ_v indicates that the displacement has the same sign as that assumed for P; it is downward.

(b) As the horizontal displacement is sought, a fictitious load Q is applied at point E (Figure 5.9). Now $F_{AC} = F_{CE} = P + Q$; all other forces remain the same as given by Eq. (b). From Eq. (5.43), we have

$$\delta_h = \frac{1}{AE} \sum FL \frac{\partial F}{\partial Q} = \frac{2L}{AE}(P + Q)$$

However, $Q = 0$, and the preceding reduces to

$$\delta_h = 2\frac{PL}{AE} \qquad \text{(5.44)}$$

5.7   STATICALLY INDETERMINATE PROBLEMS

Castigliano's theorem or the unit load method may be applied as a supplement to the equations of statics in the solutions of support reactions of a statically indeterminate structure. Consider, for instance, a structure indeterminate to the first degree. In this case, we select one of the reactions as the redundant (or unknown) load, say, R_1, by removing the corresponding support. All external forces, including both loads and redundant reactions, must generate displacements compatible with the original supports. We first express the strain energy in terms of R_1 and the given loads. Equation (5.36) may be applied at the removed support and equated to the given displacement:

$$\frac{\partial U}{\partial R_1} = \delta_1 = 0 \qquad \text{(5.45)}$$

This expression is solved for the redundant reaction R_1. Then, we can find from static equilibrium the other reactions.

In an analogous manner, the case of statically indeterminate structure with n redundant reactions, assuming no support movement, can be expressed in the form

$$\frac{\partial U}{\partial R_1} = 0, \ldots, \quad \frac{\partial U}{\partial R_n} = 0 \tag{5.46}$$

By solving these equations simultaneously, the magnitude of the redundant reactions are obtained. The remaining reactions are found from equations of equilibrium. Note that Eq. (5.46), Castigliano's second theorem, is also referred to as the *principle of least work* in some literature [10].

Analytical techniques are illustrated in solution of the following sample problems.

Determining the Reactions of a Propped Cantilevered Beam Using Castigliano's Theorem

EXAMPLE 5.9

A propped cantilevered beam carries a concentrated load P at its midspan (Figure 5.10a). Find the support reactions.

Figure 5.10 Example 5.9.

Solution: The free-body diagram of Figure 5.10b shows that the problem is statically indeterminate to the first degree. We select R_A as redundant, hence the expressions for the moments are

$$M_{AD} = R_A x, \qquad M_{DB} = R_A x - P\left(x - \frac{L}{2}\right)$$

It is important to note that the remaining unknowns, R_B and M_B, do not appear in preceding equations.

The deflection v_A at A must be 0. Equation (5.36) is thus

$$v_A = \frac{1}{EI}\left\{\int_0^{L/2} R_A x(x)\,\mathrm{d}x + \int_{L/2}^{L}\left[R_A x - P\left(x - \frac{L}{2}\right)\right]x\,\mathrm{d}x\right\} = 0$$

The preceding, after integrating, results in

$$R_A = \frac{5}{16}P$$

The other two reactions are $R_B = 11P/16$ and $M_B = 3PL/16$, as determined from the equations of statics.

EXAMPLE 5.10 | Determination of the Deflection of a Ring by Castigliano's Theorem

A ring of radius R is hinged and subjected to force P as shown in Figure 5.11a. Taking into account only the strain energy due to bending, determine the vertical displacement of point C.

Figure 5.11   Example 5.10: (a) ring; (b) quarter segment.

Solution: Owing to symmetry, it is necessary to analyze only a quarter segment (Figure 5.11b). Inasmuch as M_A and M_B are unknowns, the problem is statically indeterminate. The moment at any section a-a is

$$M_\theta = M_B - \frac{1}{2} P R (1 - \cos\theta)$$

Since the slope is 0 at B, substituting this expression into Eq. (5.37) with $dx = R\, d\theta$, we have

$$\theta_B = \frac{1}{EI} \int_0^{\pi/2} \left[M_B - \frac{1}{2} P R (1 - \cos\theta) \right] (1) R\, d\theta = 0$$

from which $M_B = P R [1 - (2/\pi)]/2$. The first equation then becomes

$$M_\theta = \frac{PR}{2} \left(\cos\theta - \frac{2}{\pi} \right)$$

The displacement of point C, by Eq. (5.36) is

$$\delta_C = \frac{4}{EI} \int_0^{\pi/2} M_\theta \frac{\partial M_\theta}{\partial P} R\, d\theta$$

$$= \frac{4}{EI} \int_0^{\pi/2} P \left[\frac{R}{2} \left(\cos\theta - \frac{2}{\pi} \right) \right]^2 R\, d\theta$$

Integration of the foregoing results in

$$\delta_C = 0.15 \frac{P R^3}{EI} \tag{5.47}$$

The positive sign of δ_C means that the displacement has the same sense as the force P, as shown.

Determining the Deflection and Reaction of a Frame Using Castigliano's Theorem | **EXAMPLE 5.11**

A frame of constant flexural rigidity EI supports a downward load P, as depicted in Figure 5.12a. Find

(a) The deflection at E.

(b) The horizontal reaction at E, if the point E is a fixed pin (Figure 5.12b).

(a)   (b)

Figure 5.12 Example 5.11.

Solution:

(a) Inasmuch as a displacement is sought, a fictitious force Q is introduced at point E (Figure 5.12a). Because of symmetry about a vertical axis through point C, we need to write expressions for moment associated with segments ED and DC, respectively:

$$M_1 = Qx, \qquad M_2 = Qa + \frac{Px}{2}$$

The horizontal displacement at E is found by substituting these equations into Eq. (5.38):

$$\delta_E = \frac{1}{EI} \int_0^a (Qx)x \, dx + \frac{2}{EI} \int_0^{b/2} \left(Qa + \frac{Px}{2} \right) a \, dx \qquad \text{(a)}$$

Setting $Q = 0$ and integrating,

$$\delta_E = \frac{Pab^2}{8EI} \qquad \text{(5.48a)}$$

(b) The problem is now statically indeterminate to the first degree; there is one unknown reaction after satisfying three equations of statics (Figure 5.12b). Since $\delta_E = 0$, setting $Q = -R$, we equate the deflection given by Eq. (a) to zero. In so doing, we have

$$\frac{1}{EI} \int_0^a (-Rx)x \, dx + \frac{2}{EI} \int_0^{b/2} \left(-Ra + \frac{Px}{2} \right) a \, dx = 0$$

The preceding gives

$$R = \frac{3}{8} \frac{Pb^2}{a^2 + 3ab} \qquad \text{(5.48b)}$$

5.8   VIRTUAL WORK AND POTENTIAL ENERGY

In connection with virtual work, we use the symbol δ to denote a virtual infinitesimally small quantity. An arbitrary or imaginary incremental displacement is termed a *virtual displacement*. A virtual displacement results in no geometric alterations of a member. It also must not violate the boundary or support conditions for a structure. In the brief development to follow, p_x, p_y, and p_z represent the x, y, and z components of the surface forces per unit area and the body forces are taken to be negligible.

THE PRINCIPLE OF VIRTUAL WORK

The virtual work, δW, done by surface forces on a member under a virtual displacement is given by

$$\delta W = \int (p_x \delta u + p_y \delta v + p_z \delta w)\, dA \tag{5.49}$$

The quantity A is the boundary surface area and δ_u, δ_v, and δ_w represent the x-, y-, and z-directed components of a virtual displacement.

In a like manner, the virtual strain energy, δU, acquired of a member of volume V caused by a virtual straining is expressed as follows:

$$\delta U = \int_V (\sigma_x \delta \varepsilon_x + \sigma_y \delta \varepsilon_y + \sigma_z \delta \varepsilon_z + \tau_{xy} \delta \gamma_{xy} + \tau_{yz} \delta \gamma_{yz} + \tau_{xz} \delta \gamma_{xz})\, dV \tag{5.50}$$

It can be shown that [1, 4] the total work done during the virtual displacement is 0: $\delta W - \delta U = 0$. Therefore,

$$\delta W = \delta U \tag{5.51}$$

This is called the *principle of virtual work*.

THE PRINCIPLE OF MINIMUM POTENTIAL ENERGY

Since the virtual displacements do not alter the shape of the member and the surface forces are considered constants, Eq. (5.51) can be written in the form

$$\delta \Pi = \delta(U - W) = 0 \tag{5.52}$$

In the foregoing, we have

$$\Pi = U - W \tag{5.53}$$

The quantity Π designates the *potential energy* of the member.

Equation (5.52) represents the condition of stationary potential energy of the member. An elastic member is in equilibrium if and only if the changes in potential energy $\delta \Pi$ are 0 for a virtual displacement. It can be verified that, for a stable equilibrium, the potential energy is a minimum. For all displacements fulfilling the prescribed boundary conditions and the equilibrium conditions, the potential energy has a minimum value. The preceding states the *principle of minimum potential energy*.

CASTIGLIANO'S FIRST THEOREM

Consider now a structure subjected to a set of external forces $P_k (k = 1, 2, \ldots, n)$. Suppose that the structure experiences a continuous virtual displacement in such a manner that it vanishes at all points of loading except under a single load, say, P_i. The virtual displacement in the direction of this force is denoted by $\delta(\delta_i)$. From Eq. (5.51), we have $\delta U = P_i \cdot \delta(\delta_i)$. In the limit, the principle of virtual work results in

$$P_i = \frac{\partial U}{\partial \delta_i} \qquad\qquad \textbf{(5.54)}$$

This is known as *Castigliano's first theorem:* For a linear or nonlinear structure, *the partial derivative of the strain energy with respect to an applied virtual displacement is equal to the load acting at the point in the direction of that displacement.* Similarly, it can be demonstrated that

$$M_i = \frac{\partial U}{\partial \theta_i} \qquad\qquad \textbf{(5.55)}$$

in which θ_i is the angular rotation and M_i represents the resulting moment.

Note that Castigliano's first theorem is also known as the theorem of virtual work. It is the basis for the derivation of the finite element stiffness equations. In applying the Castigliano's first theorem, the strain energy must be expressed in terms of the displacements.

*5.9   USE OF TRIGONOMETRIC SERIES IN ENERGY METHODS

Certain problems in the structural analysis and design are amenable to solutions by the use of trigonometric series. This technique offers a significant advantage because a single expression may apply to the entire length or surface of the member. A disadvantage in the trigonometric series is that arbitrary support conditions can make it impossible to write a series that is simple. The solution by trigonometric series is applied for variously loaded members in this and the sections to follow [1, 4, 5].

The method is now illustrated for the case of a simple beam loaded as depicted in Figure 5.13. The deflection curve can be represented by the following Fourier sine series:

$$v = \sum_{m=1}^{\infty} a_m \sin \frac{m \pi x}{L} \qquad\qquad \textbf{(5.56)}$$

that satisfies the boundary conditions ($v = 0$, $v'' = 0$ at $x = 0$ and $x = L$). The Fourier coefficients a_m are the maximum coordinates of the sine curves, and the values of m show the number of half-waves in the sine curves. The accuracy can be improved by increasing the number of terms in the series. We apply the principle of virtual work to determine the coefficients.

Figure 5.13   Simply supported beam under a force P at an arbitrary distance c from the left support.

The strain energy of the beam, substituting Eq. (5.56) into Eq. (5.18), is expressed in the form

$$U = \frac{EI}{2} \int_0^L \left(\frac{\mathrm{d}^2 v}{\mathrm{d}x^2} \right)^2 \mathrm{d}x = \frac{EI}{2} \int_0^L \left[\sum_{m=1}^{\infty} a_m \left(\frac{m\pi}{L} \right)^2 \sin \frac{m\pi x}{L} \right]^2 \mathrm{d}x \qquad \text{(a)}$$

The term in brackets, after expanding, can be expressed as

$$U = \sum_{m=1}^{\infty} \sum_{n=1}^{\infty} a_m a_n \left(\frac{m\pi}{L} \right)^2 \left(\frac{n\pi}{L} \right)^2 \sin \frac{m\pi x}{L} \sin \frac{n\pi x}{L}$$

For the orthogonal functions $\sin(m\pi x/L)$ and $\sin(n\pi/L)$, by direct integration it can be verified that

$$\int_0^L \sin \frac{m\pi x}{L} \sin \frac{n\pi x}{L} \, \mathrm{d}x = \begin{cases} 0, & \text{(for } m \neq n) \\ L/2, & \text{(for } m = n) \end{cases} \qquad \text{(5.57a)}$$

The strain energy given by Eq. (a) is therefore

$$U = \frac{\pi^4 EI}{4L^3} \sum_{m=1}^{\infty} m^4 a_m^2 \qquad \text{(5.58)}$$

The virtual work done by a force P acting through a virtual displacement at A increases the strain energy of the beam by δU. Applying Eq. (5.51),

$$-P \cdot \delta v_A = \delta U \qquad \text{(b)}$$

The minus sign means that P and δv_A are oppositely directed. So, by Eqs. (5.58) and (b),

$$-P \sum_{m=1}^{\infty} \sin \frac{m\pi c}{L} \delta a_m = \frac{\pi^4 EI}{4L^3} \sum_{m=1}^{\infty} m^4 \delta a_m^2$$

The foregoing gives

$$a_m = -\frac{2PL^3}{\pi^4 EI} \frac{1}{m^4} \sin \frac{m\pi c}{L}$$

Carrying this equation into Eq. (5.56), we have the equation of the deflection curve

$$v = -\frac{2PL^3}{\pi^4 EI} \sum_{m=1}^{\infty} \frac{1}{m^4} \sin \frac{m\pi c}{L} \sin \frac{m\pi x}{L} \qquad (5.59)$$

Using this infinite series, the deflection for any prescribed value of x can be readily obtained.

Determination of the Deflection of a Cantilevered Beam Using the Principle of Virtual Work

EXAMPLE 5.12

A cantilevered beam is subjected to a concentrated load P at its free end, as shown in Figure 5.14. Derive the equation of the deflection curve.

Assumptions: The origin of the coordinates is placed at the fixed end. Deflection is taken in the form

$$v = \sum_{m=1,3,5,\ldots}^{\infty} a_m \left(1 - \cos \frac{m\pi x}{2L} \right) \qquad (5.60)$$

Figure 5.14   Example 5.12.

Solution: The boundary conditions, $v(0) = 0$ and $v'(0) = 0$, are satisfied by the preceding equation. Substitution of this series into Eq. (5.18) gives

$$U = \frac{EI}{2} \int_0^L \left[\sum_{m=1,3,5,\ldots}^{\infty} a_m \left(\frac{m\pi}{2L} \right)^2 \cos \frac{m\pi x}{2L} \right]^2 dx$$

Squaring the term in brackets and noting the orthogonality relation,

$$\int_0^L \cos \frac{m\pi x}{2L} \cos \frac{n\pi x}{2L} \, dx = \begin{cases} 0, & (\text{for } m \neq n) \\ L/2, & (\text{for } m = n) \end{cases} \qquad (5.57b)$$

the strain energy becomes

$$U = \frac{\pi^4 EI}{64L^3} \sum_{m=1,3,5,\ldots}^{\infty} m^4 a_m^2 \qquad (5.61)$$

By the principle of virtual work, $-P \cdot \delta v_A = \delta U$, we have

$$-P \sum_{m=1,3,5,\ldots}^{\infty} \left(1 - \cos \frac{m\pi}{2} \right) \delta a_m = \frac{\pi^4 EI}{64L^3} \sum_{m=1,3,5,\ldots}^{\infty} m^4 \delta a_m^2$$

This results in $a_m = -32PL^3/m^4\pi^4 EI$. The beam deflection is found by inserting the value of a_m obtained into Eq. (5.60).

Comments: At the free end ($x = L$), retaining only the first three terms of the solution, we have the value of the maximum deflection $v_{\text{max}} = PL^3/3.001EI$. The exact solution owing to bending is $PL^3/3EI$.

*5.10   THE RAYLEIGH-RITZ METHOD

The Rayleigh-Ritz method is a convenient approach for determining solutions by the principle of minimum potential energy. This technique was suggested by Lord Rayleigh in 1877 and extended by W. Ritz in 1909. Here we discuss application of the method to beam and plate bending problems. The procedure also is used in determining beam and plate buckling loads in the next chapter. In a beam bending problem, the Rayleigh-Ritz method may be described as follows.

First choose an expression, such as polynomials or trigonometric series, for the deflection curve containing undetermined coefficients a_m ($m = 1, 2, \ldots$). This function must satisfy the geometric (deflection v and slope θ) boundary conditions. Static (moment M and shear V) boundary conditions need not be fulfilled. Obviously, a proper choice of the deflection expression is important to ensure good accuracy for the final solution. Next, using the selected displacement expression, obtain the potential energy function Π in terms of a_m. Note that the values of a_m govern the variation of the potential energy.

Inasmuch as the potential energy must be minimum at equilibrium, the Rayleigh-Ritz method is stated in the following form:

$$\frac{\partial \Pi}{\partial a_1} = 0, \ldots, \frac{\partial \Pi}{\partial a_m} = 0 \tag{5.62}$$

The preceding represents a set of algebraic equations solved to give the coefficients of a_m. Finally, carrying these values into the assumed function for deflection, we find the solution for a given problem. The advantages of the Rayleigh-Ritz technique lie in the relative ease with which mixed edge conditions can be treated. This method is among the simplest for solving plate and shell deflections by hand calculation.

EXAMPLE 5.13

Determining the Deflection of a Beam by the Rayleigh-Ritz Method

A simply supported beam is under a uniform load of intensity w, as shown in Figure 5.15. Develop an expression for the deflection curve, using the Fourier series.

Solution: Assume a solution for deflection v of the form given by Eq. (5.56):

$$v = \sum_{m=1}^{\infty} a_m \sin \frac{m\pi x}{L}$$

Figure 5.15 Example 5.13.

Note that this choice enables both v and v'' to be 0 at either end of the beam. Next, introduce v and its derivatives into $II = U - W$. Through the use of Eq. (5.57a), we have, after integration,

$$\Pi = \frac{EI}{2}\left(\frac{L}{2}\right)\sum_{m=1}^{\infty} a_m^2\left(\frac{m\pi}{L}\right)^4 - w\left(\frac{L}{\pi}\right)^2 \sum_{m=1}^{\infty}\frac{1}{m}a_m \cos\frac{m\pi x}{L}\Big|_0^L$$

We see that $\Pi = 0$ for even values of m, this integer assumes only odd values. Hence,

$$\Pi = \frac{\pi^4 EI}{4L^3}\sum_{m=1}^{\infty} a_m^2 m^4 + \frac{2wL}{\pi}\sum_{m=1,3,5,\ldots}^{\infty}\frac{a_m}{m}$$

From the minimizing condition, $\delta\Pi/\delta a_m = 0$, we have $a_m = -4wL^4/EI(m\pi)^5$, $m = 1$, $3, 5, \ldots$. The maximum deflection occurs at the midspan of the beam and is found by inserting this value of a_m into the first equation:

$$v_{\max} = -\frac{4wL^4}{EI\pi^5}\left(1 - \frac{1}{3^5} + \frac{1}{5^5} - \cdots\right) \tag{5.63}$$

The minus sign means a downward deflection.

Comment: Retaining only the first term, $v_{\max} = -wL^4/76.5EI$. The exact solution due to bending is $-wL^4/76.8EI$ (case 8 of Table A.9).

Determination of the Deflection of a Simply Supported Rectangular Plate Using the Rayleigh-Ritz Method   **EXAMPLE 5.14**

A simply supported rectangular plate is under a uniform load of intensity p_o (Figure 5.16). Derive an expression for the deflection surface.

Solution: The deflection surface can be represented by a double Fourier sine series:

$$w = \sum_{m=1}^{\infty}\sum_{n=1}^{\infty} a_{mn} \sin\frac{m\pi x}{a} \sin\frac{n\pi y}{b} \tag{5.64}$$

Figure 5.16
Example 5.14.

In this expression, a_{mn} are coefficients to be determined. The integers m and n represent half-sinusoidal curves in the x and y directions, respectively. Clearly, the boundary conditions

$$w = 0, \qquad \frac{\partial^2 w}{\partial x^2} = 0, \qquad (x = 0, x = a)$$

$$w = 0, \qquad \frac{\partial^2 w}{\partial y^2} = 0, \qquad (y = 0, y = b)$$

(5.65)

are satisfied by Eq. (5.64). The work done by the lateral surface loading p_o is

$$W = \iint_A w p_o \, dx \, dy$$

(5.66)

The quantity A represents the area of the plate surface.

It can be shown that [5] introducing Eq. (5.64) into (5.27) and (5.66), considering the orthogonality relations (5.57), and integrating, we obtain

$$U = \frac{\pi^4 ab}{8} D \sum_{m=1}^{\infty} \sum_{n=1}^{\infty} a_{mn} \left(\frac{m^2}{a^2} + \frac{n^2}{b^2} \right)^2$$

(5.67)

$$W = \frac{abp_o}{\pi^2} \sum_{m=1}^{\infty} \sum_{n=1}^{\infty} \frac{a_{mn}^2}{mn} (\cos m\pi - 1)(\cos n\pi - 1)$$

(5.68)

From the minimizing condition $\partial \Pi / \partial a_{mn} = 0$, it follows that

$$a_{mn} = \frac{16}{\pi^6 D} \frac{p_o}{mn[(m/a)^2 + (n/b)^2]^2} \qquad (m, n = 1, 3, \ldots)$$

(5.69)

Substituting Eq. (5.69) into Eq. (5.64) leads to the expression for the deflection. In so doing and carrying the resulting equation for the displacement into Eqs. (4.48), we obtain the components for the moments.

REFERENCES

1. Langhaar, H. L. *Energy Methods in Applied Mechanics*. Melbourne, FL: Krieger, 1989.
2. Oden, J. T., and E. A. Ripperger. *Mechanics of Elastic Structures*, 2nd ed. New York: McGraw-Hill, 1981.
3. Sokolnikoff, I. S. *Mathematical Theory of Elasticity*, 2nd ed. Melbourne, FL: Krieger, 1986.
4. Ugural, A. C., and S. K. Fenster. *Advanced Strength and Applied Elasticity*, 4th ed. Upper Saddle River, NJ: Prentice Hall, 2003.
5. Ugural, A. C. *Stresses in Plates and Shells*, 2nd ed. New York: McGraw-Hill, 1999.
6. Faupel, J. H., and F. E. Fisher. *Engineering Design*, 2nd ed. New York: Wiley, 1981.
7. Boresi, A. P., and R. J. Schmidt. *Advanced Mechanics of Materials*, 6th ed. New York: Wiley, 2003.
8. Burr, A. H., and J. B. Cheatham. *Mechanical Analysis and Design*, 2nd ed. Upper Saddle River, NJ: Prentice Hall, 1995.

9. Ugural, A. C. *Mechanics of Materials.* New York: McGraw-Hill, 1991.

10. Flügge, W., ed. *Handbook of Engineering Mechanics.* New York: McGraw-Hill, 1961.

11. Budynas, R. G. *Advanced Strength and Applied Stress Analysis,* 2nd ed. New York: McGraw-Hill, 1999.

12. Cook, R. D., and W. C. Young. *Advanced Mechanics of Materials,* 2nd ed. Upper Saddle River, NJ: Prentice Hall, 1999.

13. Chou, P. C., and N. J. Pagano. *Elasticity.* New York: Dover, 1992.

14. Timoshenko, S. P., and S. Woinowsky-Krieger. *Theory of Plates and Shells.* New York: McGraw-Hill, 1959.

15. Oden, J. T., and E. A. Ripperger. *Mechanics of Elastic Structures,* 2nd ed. New York: McGraw-Hill, 1981.

16. Wang, C. K., and C. G. Salmon. *Introductory Structural Analysis.* Upper Saddle River, NJ: Prentice Hall, 1984.

17. West, H. H. *Fundamentals of Structural Analysis,* 2nd ed. New York: Wiley, 2002.

18. Sack, R. L. *Structural Analysis.* New York: McGraw-Hill, 1984.

19. Peery, D. J., and J. J. Azar. *Aircraft Structures,* 2nd ed. New York: McGraw-Hill, 1982.

PROBLEMS

The beams, frames, and trusses described in the following problems have constant flexural rigidity EI, axial rigidity AE, and shear rigidity JG.

Sections 5.1 through 5.7

5.1 and **5.2** A beam of rectangular cross-sectional area A is supported and loaded as shown in Figures P5.1 and P5.2. Determine the strain energy of the beam caused by the shear deformation.

Figure P5.1   **Figure P5.2**

5.3 An overhanging beam ABC of modulus of elasticity E is loaded as shown in Figure P5.3. Taking into account only the effect of normal stresses, determine

(*a*)   The total strain energy.

(*b*)   The maximum strain energy density.

Assumption: The beam has a rectangular cross section of width b and depth h.

5.4 A simply supported rectangular beam of depth h, width b, and length L is under a uniform load w (Figure P5.2). Show that the maximum strain energy density due to bending is given by $U_{0,\text{max}} = 45U/8V$. The quantities U and V represent the strain energy and volume of the beam, respectively.

Figure P5.3

5.5 An overhang beam is loaded as shown in Figure P5.1. Determine the vertical deflection v_A at the free end A due to the effects of bending and shear. Apply the work-energy method.

5.6 A cantilever carries a concentrated load P as shown in Figure P5.6. Using Castigliano's theorem, determine the vertical deflection v_A at the free end A.

Figure P5.6

5.7 A cantilevered spring of constant flexural rigidity EI is loaded as depicted in Figure P5.7. Applying Castigliano's theorem, determine the vertical deflection at point B.

Assumption: The strain energy is attributable to bending alone.

Figure P5.7

5.8 Figure P5.8 shows a compound beam with a hinge at C. It is composed of two portions: a beam BC, simply supported at B, and a cantilever AC, fixed at A. Employing Castigliano's theorem, determine the deflection v_D at the point of application of the load P.

Figure P5.8

5.9 A continuous beam is subjected to a bending moment M_o at support C (Figure P5.9). Applying Castigliano's theorem, find the reaction at each support.

Figure P5.9

5.10 A steel I-beam is fixed at B and supported at C by an aluminum alloy tie rod CD of cross-sectional area A (Figure P5.10). Using Castigliano's theorem, determine the tension P in the rod caused by the distributed load depicted, in terms of w, L, A, E_a, E_s, and I, as needed.

Figure P5.10

5.11 and 5.12 A bent frame is supported and loaded as shown in Figures P5.11 and P5.12. Employing Castigliano's theorem, determine the horizontal deflection δ_A for point A.

Assumption: The effect of bending moment is considered only.

Figure P5.11   **Figure P5.12**

5.13 A semicircular arch is supported and loaded as shown in Figure P5.13. Using Castigliano's theorem, determine the horizontal displacement of the end B.

Assumption: The effect of bending moment is taken into account alone.

Figure P5.13

5.14 A frame is fixed at one end and loaded at the other end as depicted in Figure P5.14. Apply Castigliano's theorem to determine

(a) The horizontal deflection δ_A at point A.

(b) The slope θ_A at point A.

Assumption: The effects of axial force as well as shear are omitted.

Figure P5.14

5.15 A frame is fixed at one end and loaded as shown in Figure P5.15. Employing Castigliano's theorem, determine

(a) The vertical deflection δ_A at point A.

(b) The angle of twist at point B.

Assumption: The effect of bending moment is considered only.

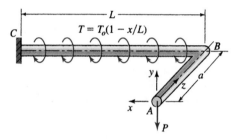

Figure P5.15

5.16 A simple truss carries a concentrated force P as indicated in Figure P5.16. Applying the work-energy method, determine the horizontal displacement of the joint D.

Given: $P = 60$ kN, $A = 250$ mm^2, $E = 210$ GPa.

Figure P5.16

5.17 The basic truss shown in Figure P5.17 carries a vertical load $2P$ and a horizontal load P at joint B. Apply Castigliano's theorem, to obtain horizontal displacement δ_C of point C.

Assumption: Each member has an axial rigidity AE.

Figure P5.17

5.18 A pin-connected truss supports the loads, as shown in Figure P5.18. Using Castigliano's theorem, find the vertical and horizontal displacements of the joint C.

Figure P5.18

5.19 The truss ABC shown in Figure P5.19 is subjected to a vertical load P. Employing Castigliano's theorem, determine the horizontal and vertical displacements of joint C.

Figure P5.19

5.20 A planar truss supports a horizontal load P, as shown in Figure P5.20. Apply Castigliano's theorem to obtain the vertical displacements of joint D.

5.21 A three-member truss carries load P, as shown in Figure P5.21. Applying Castigliano's theorem, determine the force in each member.

5.22 A curved frame of a structure is fixed at one end and simply supported at another, where a horizontal load P applies (Figure P5.22). Determine the roller reaction F at the end B, using Castigliano's theorem.

Assumption: The effect of bending moment is considered only.

Figure P5.20

Figure P5.21

Figure P5.22

5.23 Two-hinged frame ACB carries a concentrated load P at C, as shown in Figure P5.23. Determine, using Castigliano's theorem,

(a)   The horizontal displacement δ_B at B.

(b)   The horizontal reaction R at B, if the support B is a fixed pin.

Assumption: The strain energy is attributable to bending alone.

Figure P5.23

5.24 Figure P5.24 shows a structure that consists of a cantilever AB, fixed at A, and bars BC and CD, pin-connected at both ends. Find the vertical deflection of joint C, considering the effects of normal force and bending moment. Employ Castigliano's theorem.

Figure P5.24

5.25 A cantilevered beam with a rectangular cross section carries concentrated loads P at free end and at the center as shown in Figure P5.25. Determine, using Castigliano's theorem,

(a) The deflection of the free end, considering the effects of both the bending and shear.

(b) The error, if the effect of shear is neglected, for the case in which $L = 5h$ and the beam is made of ASTM-A36 structural steel.

Figure P5.25

5.26 A curved frame ABC is fixed at one end, hinged at another, and subjected to a concentrated load P, as shown in Figure P5.26. What are the horizontal H and vertical F reactions? Use Castigliano's theorem.

Assumption: The strain energy is attributable to bending only.

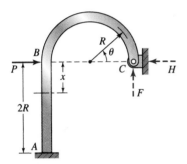

Figure P5.26

5.27 A pin-connected structure of three bars supports a load W at joint D (Figure P5.27). Apply Castigliano's theorem to determine the force in each bar.

Given: $a = 0.6L$, $h = 0.8L$

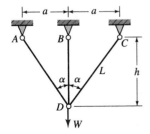

Figure P5.27

5.28 A fixed-ended beam supports a uniformly increasing load $w = kx$ (Figure P5.28), where k is a constant. Determine the reactions R_A and M_A at end A, using Castigliano's theorem.

Figure P5.28

Sections 5.8 through 5.10

5.29 Figure P5.29 shows a fixed-ended beam subjected to a concentrated load P at its midlength. Using the Rayleigh-Ritz method determine the expression for the deflection curve v.

Assumption: The deflection curve is of the form

$$v = \sum_{m=2,4,6,\ldots}^{\infty} a_m \left(1 - \cos \frac{m\pi x}{L} \right)$$

where a_m is a constant.

Figure P5.29

5.30 A simply supported beam carries a distributed load of intensity $w = w_o \sin \pi x/L$ as shown in Figure P5.30. Using the principle of virtual work, determine the expression for the deflection curve v.

Assumption: The deflection curve has the form $v = a \sin \pi x/L$, where a is to be found.

Figure P5.30

5.31 Applying Castigliano's first theorem, find the load W required to produce a vertical displacement of $\frac{1}{4}$ in. at joint D in the pin-connected structure depicted in Figure P5.27.

Given: $h = 10$ ft, $\alpha = 30°$, $E = 30 \times 10^6$ psi, $A = 1$ in.^{2}.

Assumption: Each member has the same cross-sectional area A.

5.32 A cantilevered beam is subjected to a concentrated load P at its free end (Figure 5.14). Apply the principle of virtual work to determine

(a) An expression for the deflection curve v.

(b) The maximum deflection and the maximum slope.

Assumption: Deflection curve of the beam has the form $v = ax^2(3L - x)/2L^3$, where a is a constant.

5.33 Resolve Problem 5.32, employing the Rayleigh-Ritz method.

5.34 A simply supported beam is loaded as shown in Figure 5.13. Using the Rayleigh-Ritz method, determine the deflection at point A.

Assumption: The deflection curve of the beam is of the form $v = ax(L - x)$, in which a is a constant.

5.35 A simply supported rectangular plate carries a concentrated load P at its center (Figure P5.35). Employing the Rayleigh-Ritz method, derive the equation of the deflection surface w.

Assumption: The deflection surface is of the form given by Eq. (5.64).

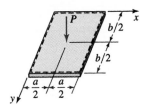

Figure P5.35

6

BUCKLING DESIGN OF MEMBERS

Outline

6.1 INTRODUCTION

Elastic stability relates to the ability of a member or structure to support a given load without experiencing a sudden change in configuration. A buckling response leads to instability and collapse of the member. Some designs may thus be governed by the possible instability of a system that commonly arises in buckling of components. Here, we are concerned primarily with the column buckling, which presents but one case of structural stability [1–15]. Critical stresses in rectangular plates are discussed briefly in Section 6.10. The problem of buckling in springs is examined in Section 14.6. Buckling of thin-walled cylinders under axial loading and pressure vessels are taken up in the last section of Chapter 16, after discussing the bending of shells.

Both equilibrium and energy methods are applied in determining the critical load. The choice depends on the particulars of the problem considered. Although the equilibrium approach gives exact solutions, the results obtained by the energy approach (sometimes approximate) usually is preferred due to the physical insight that may be more readily gained. A vast number of other situations involve structural stability, such as the buckling of pressure vessels under combined loading; twist-bend buckling of shafts in torsion; lateral buckling of deep, narrow beams; buckling of thin plates in the form of an angle or channel in compression. Analysis of such problems is mathematically complex and beyond the scope of this text.

6.2 BUCKLING OF COLUMNS

A prismatic bar loaded in compression is called *column*. Such bars are commonly used in trusses and the framework of buildings. They are also encountered in machine linkages, jack screws, coil springs in compression, and a wide variety of other elements. The buckling of a column is its sudden, large, lateral deflection due to a small increase in existing compressive load. A wooden household yardstick with a compressive load applied at its ends illustrates the basic buckling phenomenon. Failure from the viewpoint of instability may take place for a load that is 1% of the compressive load alone that would cause failure based on a strength criterion. That is, consideration of material strength (stress level) alone is insufficient to predict the behavior of such a member. Railroad rails, if subjected to an axial compression because of temperature rise, could fail similarly.

PIN-ENDED COLUMNS

Consider a slender pin-ended column centrically loaded by compressive forces P at each end (Figure 6.1a). In Figure 6.1b, load P has been increased sufficiently to cause a small lateral deflection. This is a condition between stability and instability or neutral equilibrium. The bending moment at any section is $M = -Pv$. So, Eq. (4.14) becomes

$$EI\frac{d^2v}{dx^2} = -Pv \tag{6.1}$$

or

$$\frac{d^2v}{dx^2} + k^2v = 0 \tag{6.2}$$

Figure 6.1 Column with pinned ends.

For simplification, the following notation is used:

$$k^2 = \frac{P}{EI} \tag{6.3}$$

The solution of Eq. (6.2) is

$$v = A \sin kx + B \cos kx \tag{a}$$

The constants A and B are obtained from the end conditions

$$v(0) = 0 \quad \text{and} \quad v(L) = 0$$

The first requirement gives $B = 0$ and the second leads to for $A = 0$: $\sin kL = 0$, which is satisfied by

$$\sqrt{\frac{P}{EI}} L = n\pi \qquad (n = 1, 2, \ldots) \tag{b}$$

The quantity n represents the number of half-waves in the buckled column shape. Note that, $n = 2, \ldots,$ are usually of no practical interest. The only way to obtain higher modes of buckling is to provide lateral support of the column at the points of 0 moments on the elastic curve, the so-called *inflection points*.

When $n = 1$, solution of Eq. (b) results in the value of the smallest critical load, the Euler buckling load:

$$P_{\text{cr}} = \frac{\pi^2 EI}{L^2} \tag{6.4}$$

This is also called the *Euler column formula*. The quantities I, L, and E are moment of inertia of the cross-sectional area, original length of the column, and modulus of elasticity, respectively. Note that the strength is not a factor in the buckling load. Introducing the foregoing results back in Eq. (a), we obtain the buckled shape of the column as

$$v = A \sin \frac{\pi x}{L}$$

The value of the maximum deflection, $v_{max} = A$, is undefined. Therefore, the critical load sustains only a small lateral deflection [1].

It is clear that EI represents the flexural rigidity for bending in the plane of buckling. If the column is free to deflect in any direction, it tends to bend about the axis having the smallest principal moment of inertia I. By definition, $I = Ar^2$, where A is the cross-sectional area and r is the *radius of gyration* about the axis of bending. We may consider the r of an area to be the distance from the axes at that entire area could be concentrated and yet have the same value for I. Substitution of the preceding relationship into Eq. (6.4) gives

$$P_{cr} = \frac{\pi^2 E A}{(L/r)^2} \qquad (6.5)$$

We seek the minimum value of P_{cr}; hence, the smallest radius of gyration should be used in this equation. The quotient L/r, called the *slenderness ratio,* is an important parameter in the classification of columns.

COLUMNS WITH OTHER END CONDITIONS

For columns with various combinations of fixed, free, and pinned supports, the Euler formula can be written in the form

$$P_{cr} = \frac{\pi^2 E I}{L_e^2} \qquad (6.6)$$

in which L_e is called the *effective length.* As shown in Figure 6.2, it develops that the effective length is the distance between the inflection points on the elastic curves. In a like manner, Eq. (6.5) can be expressed as

$$P_{cr} = \frac{\pi^2 E A}{(L_e/r)^2} \qquad (6.7)$$

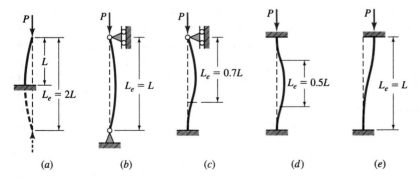

Figure 6.2 Effective lengths of columns for various end conditions: (a) fixed-free; (b) pinned-pinned; (c) fixed-pinned; (d) fixed-fixed; (e) fixed-nonrotating.

The quantity L_e/r is referred to as the *effective slenderness ratio* of the column. In the actual design of a column, the designer endeavors to configure the ends, using bolts, welds, or pins, to achieve the required ideal end condition.

Minimum AISI recommended *actual* effective lengths for *steel* columns [5] are as follows:

$$L_e = 2.1L \qquad \text{(fixed-free)}$$
$$L_e = L \qquad \text{(pinned-pinned)}$$
$$L_e = 0.80L \qquad \text{(fixed-pinned)} \qquad\qquad\text{(c)}$$
$$L_e = 0.65L \qquad \text{(fixed-fixed)}$$
$$L_e = 1.2L \qquad \text{(fixed-nonrotating)}$$

Note that only a steel column with pinned-pinned ends has the same actual length and the theoretical value noted in Figure 6.2b. Also observe that steel columns with one or two fixed ends always have actual lengths longer than the theoretical values. The foregoing apply to end construction, where ideal conditions are approximated. The distinction between the theoretical analyses and empirical approaches necessary in design is discussed in Sections 6.6 and 6.7.

6.3 • CRITICAL STRESS IN A COLUMN

As previously pointed out, a column failure is always sudden, total, and unexpected. There is no advance warning. The behavior of an ideal column is often represented on a plot of average critical stress P_{cr}/A versus the slenderness ratio L_e/r (Figure 6.3). Such a representation offers a clear rationale for the classification of compression bars. The range of L_e/r is a function of the material under consideration.

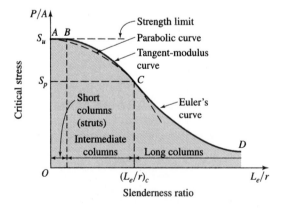

Figure 6.3   Average stress in columns versus the slenderness ratio.

LONG COLUMNS

For a long column, that is, a member with a sufficiently large slenderness ratio, buckling occurs elastically at stress that does not exceed the proportional limit of the material. Hence, the Euler's load of Eq. (6.7) is appropriate in this case, and the critical stress is

$$\sigma_{cr} = \frac{P_{cr}}{A} = \frac{\pi^2 E}{(L_e/r)^2} \tag{6.8}$$

The corresponding portion CD of the curve (Figure 6.3) is labeled as *Euler's curve*. The smallest value of the slenderness ratio for which Euler's formula applies is found by equating σ_{cr} to the proportional limit or yield strength of the specific material:

$$\left(\frac{L_e}{r}\right)_c = \pi\sqrt{\frac{E}{S_y}} \tag{6.9}$$

For instance, in the case of a structural steel with $E = 210$ GPa and $S_y = 250$ MPa, this equation gives $(L_e/r)_c = 91$.

We see from Figure 6.3 that very slender columns buckle at low levels of stress; they are much less stable than short columns. Equation (6.9) shows that the critical stress is increased by using a material of higher modulus of elasticity E or by increasing the radius of gyration r. A tubular column, for example, has a much larger value of r than a solid column of the same cross-sectional area. However, there is a limit beyond which the buckling strength cannot be increased. The wall thickness eventually becomes so thin as to cause the member to crumble due to a change in the shape of a cross section.

SHORT COLUMNS OR STRUTS

Compression members having low slenderness ratios (for instance, steel rods with $L_e/r < 30$) show essentially no instability and are called *short columns*. For these bars, failure occurs by yielding or crushing, without buckling, at stresses above the proportional limit of the material. Therefore, the maximum stress

$$\sigma_{max} = \frac{P}{A} \tag{6.10}$$

represents the strength limit of such a column, shown by horizontal line AB in Figure 6.3. This is equal to the yield strength or ultimate strength in compression.

INTERMEDIATE COLUMNS

Most structural columns lie in a region between the short and long classifications, represented by part BC in Figure 6.3. Such intermediate columns fail by inelastic buckling at stress levels above the proportional limit. Substitution of the tangent modulus E_t, slope of the stress-strain curve beyond the proportional or yield point, for the elastic modulus E is the only modification necessary to make Eq. (6.7) applicable in the inelastic range. Hence, the critical stress may be expressed by the generalized Euler buckling formula, the *tangent modulus formula:*

$$\sigma_{cr} = \frac{\pi^2 E_t}{(L_e/r)^2} \tag{6.11}$$

If the tangent moduli corresponding to the given stresses can be read from the compression stress-strain diagram, the value of L_e/r at which a column buckles can readily be calculated by applying Eq. (6.11). When, however, L_e/r is known and σ_{cr} is to be obtained, a trial-and-error approach is necessary.

Over the years, many other formulas have been proposed and employed for intermediate columns. A number of these formulas are based on the use of linear or parabolic relationships between the slenderness ratio and the critical stress. The parabolic J. B. *Johnson formula* has the form:

$$\sigma_{cr} = S_y - \frac{1}{E}\left(\frac{S_y}{2\pi}\frac{L_e}{r}\right)^2 \tag{6.12a}$$

or, in terms of the critical load,

$$P_{cr} = \sigma_{cr}A = S_y A\left[1 - \frac{S_y(L_e/r)^2}{4\pi^2 E}\right] \tag{6.12b}$$

where A represents the cross-sectional area of the column. The relation (6.12) seems to be the preferred one among designers in the machine, aircraft, and structural steel construction fields. Despite much scatter in the test results, the Johnson formula has been found to agree reasonably well with experimental data. However, the dimensionless form of tangent modulus curves have very distinct advantage when structures of new materials are analyzed [3].

Note that Eqs. (6.8), (6.9), and (6.11) or (6.12) determine the ultimate stresses, not the working stresses. It is therefore necessary to divide the right side of each formula by an appropriate factor of safety, often 2 to 3, depending on the material, to determine the allowable values. Some typical relationships for allowable stress are introduced in Section 6.6.

EXAMPLE 6.1	The Most Efficient Design of a Rectangular Column

A steel column of length L and an $a \times b$ rectangular cross section is fixed at the base and supported at the top, as shown in Figure 6.4. The column must resist a load P with a factor of safety n with respect to buckling.

(a) What is the ratio of a/b for the most efficient design against buckling?

(b) Design the most efficient cross section for the column, using $L = 400$ mm, $E = 200$ GPa, $P = 15$ kN, and $n = 2$.

Assumption: Support restrains end A from moving in the yz plane but allows it to move in the xz plane.

Solution: The radius of gyration r of the cross section is

$$r_z^2 = \frac{I_z}{A} = \frac{ba^3/12}{ab} = \frac{a^2}{12} \quad \text{or} \quad r_z = a/\sqrt{12}$$

Similarly, we obtain $r_y = b/\sqrt{12}$. The effective lengths of the column shown in Figure 6.4 with respect to buckling in the xy and xz planes, from Figures 6.2c and 6.2a, are $L_e = 0.7L$ and $L_e = 2L$, respectively. Thus

$$\frac{L_e}{r_z} = \frac{0.7L}{a/\sqrt{12}}$$

$$\frac{L_e}{r_y} = \frac{2L}{b/\sqrt{12}}$$

(a)

Figure 6.4
Example 6.1.

(a) For the most effective design, the critical stresses corresponding to the two possible modes of buckling are to be identical. Referring to Eq. (6.8), it is concluded therefore that

$$\frac{0.7L}{a/\sqrt{12}} = \frac{2L}{b/\sqrt{12}}$$

Solving,

$$\frac{a}{b} = 0.35 \tag{6.13}$$

(b) Designing for the given data, based on a safety factor of $n = 2$, we have $P_{cr} = 2(15) = 30$ kN and

$$\sigma_{cr} = \frac{P_{cr}}{A} = \frac{30(10^3)}{0.35b^2}$$

The second of Eqs. (a) gives $L_e/r = 2(0.4)\sqrt{12}/b = 2.771/b$. Through the use of Eq. (6.8), we write

$$\frac{30(10^3)}{0.35b^2} = \frac{\pi^2(200 \times 10^9)}{(2.771/b)^2}$$

from which $b = 24$ mm and hence $a = 8.4$ mm.

Development of Specific Johnson's Formula

EXAMPLE 6.2

Derive specific Johnson formula for the intermediate sizes of columns having

(a) Round cross sections.

(b) Rectangular cross sections.

Solution:

(a) For a solid circular section: $A = \pi d^2/4$, $I = \pi d^4/64$, and

$$r = \sqrt{\frac{I}{A}} = \frac{d}{4} \tag{b}$$

Using Eq. (6.12a),

$$\frac{4P_{cr}}{\pi d^2} = S_y - \frac{1}{E}\left(\frac{S_y}{2\pi}\frac{4L_e}{d}\right)^2$$

Solving,

$$d = 2\left(\frac{P_{cr}}{\pi S_y} + \frac{S_y^2 L_e^2}{\pi^2 E}\right)^{1/2} \tag{6.14}$$

(b) In the case of a rectangular section of height h and width b with $h \le b$: $A = bh$, $I = bh^3/12$; hence, $r^2 = h^2/12$. Introducing these into Eq. (6.12a),

$$\frac{P_{cr}}{bh} = S_y - \frac{1}{E}\left(\frac{S_y}{2\pi}\frac{\sqrt{12}L_e}{h}\right)^2$$

or

$$b = \frac{P_{cr}}{hS_y\left(1 - \dfrac{3L_e^2 S_y}{\pi^2 E h^2}\right)} \tag{6.15}$$

EXAMPLE 6.3

Analysis of the Load-Carrying Capacity of a Pin-Ended Braced Column

A steel column braced at midpoint C as shown in Figure 6.5. Determine the allowable load P_{all} on the basis of a factor of safety n.

Given: $L = 15$ in., $a = \frac{1}{4}$ in., $b = 1$ in., $S_y = 36$ ksi, $E = 30 \times 10^6$ psi

Assumptions: Bracing acts as simple support in the xy plane. Use $n = 3$.

Solution: Referring to Example 6.1, the cross sectional area properties are

$$r_z = \frac{a}{\sqrt{12}} = \frac{0.25}{\sqrt{12}} = 0.072 \text{ in.}$$

$$r_y = \frac{b}{\sqrt{12}} = \frac{1}{\sqrt{12}} = 0.289 \text{ in.}$$

and $A = ab = (0.25)(1) = 0.25$ in.^{2}

Figure 6.5   Example 6.3.

Buckling in the xz plane (unrestrained by the brace). The slenderness ratio is $L/r_y = 15/0.289 = 51.9 < 91$ and the Johnson equation is valid per Eq. (6.9). Relation (6.12b) results in

$$P_{cr} = (36,000)(0.25)\left[1 - \frac{36,000(51.9)^2}{4\pi^2(30 \times 10^6)}\right]$$

$$= 8.263 \text{ kips}$$

Buckling in the xy plane (braced). We have $L_e/r_z = 7.5/0.072 = 104.2$. Hence, applying the Euler's equation,

$$P_{cr} = \frac{\pi^2 EA}{(L_e/r)^2} = \frac{\pi^2(30 \times 10^6)(0.25)}{(104.2)^2}$$

$$= 6.818 \text{ kips}$$

Comment: The working load P_{all}, therefore, must be based on buckling in the xy plane:

$$P_{all} = \frac{P_{cr}}{n} = \frac{6.818}{3} = 2.273 \text{ kips}$$

Case Study 6-1 | BUCKLING STRESS ANALYSIS IN A BASIC TRUSS

A plane truss, consisting of members AB, BC, and AC, is subjected to vertical and horizontal loads P_1 and P_2 at point B and supported at A and C, as shown in Figure 6.6a. Although the members are actually joined together by riveted or welded connections, it is usual to consider them pinned together (Figure 6.6b). Further information on trusses may be found in Section 5.6. Determine the stresses in each member.

Given: The geometry of the assembly is known. Each member is 6061-T6 aluminum alloy pipe of outer diameter D and inner diameter d.

Data

$D = 80$ mm, $\quad d = 65$ mm, $\quad a = 4$ m,

$b = 2.25$ m, $\quad h = 3$ m,

$P_1 = 50$ kN, $\quad P_2 = 25$ kN,

$E = 70$ GPa, $\quad S_y = 260$ MPa, $\quad$ (from Table B.1)

Assumptions:

1. The weight of members is small compared to the applied loading and is neglected.

(a)

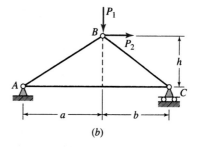

(b)

Figure 6.6 (a) Basic truss. (b) Schematic representation of basic truss.

(continued)

Case Study   (CONCLUDED)

2.   The loading is static.

3.   Each member is treated as a two-force member.

4.   Friction in the pin joints is ignored.

Solution: See Figure 6.6b.
      The axial forces in members, applying the method of joints, are found as

$$F_{AB} = 10 \text{ kN } (C), \qquad F_{BC} = 55 \text{ kN } (C),$$

$$F_{AC} = 33 \text{ kN } (T)$$

For a tube cross section, we have the following properties:

$$A = \frac{\pi}{4}(D^2 - d^2), \qquad I = Ar^2 = \frac{\pi}{64}(D^4 - d^4)$$

The radius of gyration is therefore

$$r = \frac{1}{4}\sqrt{D^2 + d^2} \qquad\qquad \textbf{(6.16)}$$

Substitution of the given numerical values results in

$$A = \frac{\pi}{4}(80^2 - 65^2) = 1708 \text{ mm}^2$$

$$r = \frac{1}{4}\sqrt{80^2 + 65^2} = 25.77 \text{ mm}$$

Hence, $L/r = 5000/25.77 = 194$ for member AB and $L/r = 3750/25.77 = 145.5$ for member BC.
      Using Eq. (3.1), the tensile stress in the member AC is

$$\sigma_{AC} = \frac{F_{AC}}{A} = \frac{33(10^3)}{1708(10^{-6})} = 19.32 \text{ MPa}$$

The critical stresses from Eq. (6.8), for members AB and BC, respectively, are

$$(\sigma_{cr})_{AB} = \frac{\pi^2(70 \times 10^9)}{(194)^2} = 18.36 \text{ MPa}$$

$$(\sigma_{cr})_{BC} = \frac{\pi^2(70 \times 10^9)}{(145.5)^2} = 32.63 \text{ MPa}$$

Comments: Interestingly, 18.36 and 32.63 MPa, compared with the yield strength of 260 MPa, demonstrate the significance of buckling analysis in predicting a safe working load. A somewhat detailed FEA analysis of the member forces, displacements, and design of a basic truss are discussed in Case Study 17-1.

6.4   INITIALLY CURVED COLUMNS

In an actual structure, it is not always possible for a column to be perfectly straight. As might be expected, the load-carrying capacity and deflection under load of a column are significantly affected by even a small initial curvature. To determine the extent of this influence, consider a pin-ended column with the unloaded form described by a half sine wave:

$$v_o = a_o \sin \frac{\pi x}{L} \qquad\qquad \textbf{(a)}$$

This is shown by the dashed lines in Figure 6.7, where a_o represents the maximum initial displacement.

TOTAL DEFLECTION
An additional displacement v_1 accompanies a subsequently applied load P. Therefore,

$$v = v_o + v_1$$

The differential equation of the column, using Eq. (4.14), is

$$EI\frac{d^2v_1}{dx^2} = -P(v_o + v_1) = -Pv \qquad \text{(b)}$$

Introducing Eq. (a) and setting $k^2 = P/EI$, we have

$$\frac{d^2v_1}{dx^2} + k^2v_1 = -\frac{P}{EI}a_o \sin\frac{\pi x}{L}$$

For simplicity, let b designate the ratio of the axial load to its critical value:

$$b = \frac{P}{P_{cr}} = \frac{PL^2}{\pi^2 EI} \qquad \text{(6.17)}$$

The trial particular solution of this equation, $v_{1p} = B\sin(\pi x/L)$, when inserted into Eq. (b) gives

$$B = \frac{Pa_o}{(\pi^2 EI/L^2) - P} = \frac{a_o}{(1/b) - 1}$$

The general solution of Eq. (b) is

$$v_1 = c_1 \sin kx + c_2 \cos kx + B\sin\frac{\pi x}{L}$$

The constants c_1 and c_2 are evaluated, from the end conditions $v_1(0) = v_1(L) = 0$, as $c_1 = c_2 = 0$. The column deflection is then

$$v = a_o \sin\frac{\pi x}{L} + B\sin\frac{\pi x}{L} = \frac{a_o}{1-b} \sin\frac{\pi x}{L} \qquad \text{(6.18)}$$

This equation indicates that the axial force P causes the initial deflection of the column to increase by the factor $1/(1-b)$. Since $b < 1$, this factor is always greater than unity. Clearly, if $b = 1$, deflection becomes infinitely large. Note that an initially curved column deflects with any applied load P in contrast to a perfectly straight column that does not bend until P_{cr} is reached.

Figure 6.7
Initially curved column with pinned ends.

CRITICAL STRESS

We begin with the formula applicable to combined axial loading and bending: $\sigma_x = (P/A) + (My/I)$, in which A and I are the cross-sectional area and the moment of inertia. Substituting $M = -Pv$ together with Eq. (6.18) into this expression, the maximum compressive stress at midspan is found as

$$\sigma_{max} = \frac{P}{A}\left(1 + \frac{a_o A}{S}\frac{1}{1-b}\right) \qquad \text{(6.19)}$$

The quantity S represents the section modulus I/c, in which c is the distance measured in the y direction from the centroid of the cross section to the outermost fibers.

By imposing the yield strength in compression (and tension) S_y as σ_{max}, we write Eq. (6.19) in the form

$$S_y = \frac{P_y}{A}\left(1 + \frac{a_o A}{S}\frac{1}{1-b}\right) \tag{6.20}$$

In the foregoing, P_y is the limit load that results in impending yielding and subsequent failure. Given S_y, a_o, E, and the column dimensions, Eq. (6.20) may be solved exactly by solving a quadratic or by trial and error for P. The allowable load P_{all} can then be obtained by dividing P_y by an appropriate factor of safety n.

6.5   ECCENTRIC LOADS AND THE SECANT FORMULA

In the preceding sections, we deal with the buckling of columns for which the load acts at the centroid of a cross section. We here treat columns under an eccentric load. This situation is obviously of great practical importance because, frequently, problems occur in which load eccentricities are unavoidable.

Let us consider a pinned-end column under compressive forces applied with a small eccentricity e from the column axis (Figure 6.8a). We assume the member is initially straight and that the material is linearly elastic. By increasing the load, the column deflects as depicted by the dashed lines in the figure. The bending moment at distance x from the midspan equals $M = -P(v + e)$. Then, the differential equation for the elastic curve appears in the form

$$EI\frac{d^2v}{dx^2} + P(v + e) = 0$$

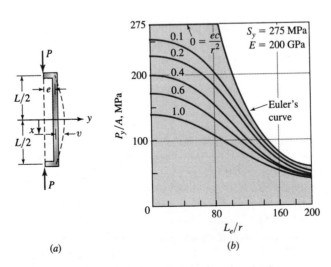

(a)   (b)

Figure 6.8   (a) Eccentrically loaded pinned-pinned column.
(b) Graph of the secant formula.

The boundary conditions are $v(L/2) = v(-L/2) = 0$. This equation is solved by following a procedure in a manner similar to that of Sections 6.2 and 6.3. In so doing, expressed in the terms of the critical load $P_{cr} = \pi^2 EI/L^2$, the midspan ($x = 0$) deflection is found as

$$v_{max} = e \left[\sec \frac{\pi}{2} \sqrt{\frac{P}{P_{cr}}} - 1 \right] \tag{6.21}$$

We observe from the foregoing expression that, as P approaches P_{cr}, the maximum deflection goes to infinity. It is therefore concluded that P should not be allowed to reach the critical value found in Section 6.2 for a column under a centric load.

The maximum compressive stress σ_{max} takes place at $x = 0$ on the concave side of the column. Hence,

$$\sigma_{max} = \frac{P}{A} + \frac{M_{max}c}{I}$$

The quantity r is the radius of gyration and c is the distance from the centroid of the cross section to the outermost fibers, both in the direction of eccentricity. Carrying $M_{max} = -P(v_{max} + e)$ and Eq. (6.21) into the foregoing expression,

$$\sigma_{max} = \frac{P}{A} \left[1 + \frac{ec}{r^2} \sec \left(\frac{\pi}{2} \sqrt{\frac{P}{P_{cr}}} \right) \right] \tag{6.22a}$$

Alternatively, we have

$$\sigma_{max} = \frac{P}{A} \left[1 + \frac{ec}{r^2} \sec \left(\frac{L}{2r} \sqrt{\frac{P}{AE}} \right) \right] \tag{6.22b}$$

This expression is referred to as the *secant formula*. The term ec/r^2 is called the *eccentricity ratio*.

We now impose the yield strength S_y as σ_{max} and hence $P = P_y$. Then, Eq. (6.22b) can be written as

$$\frac{P_y}{A} = \frac{S_y}{1 + (ec/r^2) \sec[(L_e/2r)\sqrt{P_y/AE}]} \tag{6.23}$$

Here the effective length is used; hence, the formula applies to columns with different end conditions. Figure 6.8b is a plot of the preceding expression for steel having a yield strength of 275 MPa and modulus of elasticity of 200 GPa. Note how the P/A contours asymtotically approach the Euler curve as L_e/r increases. Equation (6.23) may be solved for the load P_y by trial and error or a root-finding technique using numerical methods. Design charts in the fashion of Figure 6.8b can be used to good advantage (see Problem 6.9). Obviously, column design by the secant formula might best be programmed on a computer.

Formula (6.22) is an excellent description of column behavior, provided the eccentricity e of the load is known with a degree of accuracy. We point out that the foregoing derivation of the secant formula is on the assumption that buckling takes place in the xy plane.

It may also be necessary to analyze buckling in the yz plane; Eq. (6.22) does not apply in this plane. This possibility correlates especially to narrow columns.

EXAMPLE 6.4   Analysis of Buckling of a Column Using the Secant Formula

A 20-ft long pin-ended ASTM-A36 steel column of $S8 \times 23$ section (Figure 6.9a) is subjected to a centric load P_1 and an eccentrically applied load P_2, as shown in Figure 6.9b. Determine

(a) The maximum deflection.

(b) The factor of safety n against yielding.

Given: The geometry of the column and applied loading are known.

Data: $S_y = 36$ ksi, $E = 29 \times 10^6$ psi,   (from Table B.1).

(a)   (b)   (c)

Figure 6.9   Example 6.4.

Solution: See Figure 6.9 and Table A.7.

The loading may be replaced by a statically equivalent load $P = 100$ kips acting with an eccentricity $e = 1.2$ in. (Figure 6.9c). Using the properties of an $S8 \times 23$ section given in Table A.7, we obtain

$$\frac{P}{A} = \frac{100}{6.77} = 14.77 \text{ ksi}, \qquad \frac{L_e}{r} = \frac{20 \times 12}{3.10} = 77.42$$

$$\frac{ec}{r^2} = \frac{eA}{S} = \frac{1.2(6.77)}{16.2} = 0.501$$

$$P_{cr} = \frac{\pi^2 EI}{L^2} = \frac{\pi^2 (29 \times 10^6)(64.9)}{(20 \times 12)^2} = 322.5 \text{ kips}$$

(a) Carrying $P/P_{cr} = 1/3.225$ and $e = 1.2$ in. into Eq. (6.21) leads to the value of midspan deflection of the column as

$$v_{max} = e \left[\sec \left(\frac{\pi}{2} \sqrt{\frac{P}{P_{cr}}} \right) - 1 \right] = 1.2(1.559 - 1) = 0.671 \text{ in.}$$

(b) In a like manner, through the use of Eq. (6.22a), we have

$$\sigma_{max} = \frac{P}{A}\left[1 + \frac{ec}{r^2} \sec\left(\frac{\pi}{2}\sqrt{\frac{P}{P_{cr}}}\right)\right] = 26.31 \text{ ksi}$$

Hence,

$$n = \frac{S_y}{\sigma_{max}} = \frac{36}{26.31} = 1.37$$

Comments: Since the maximum stress and the slenderness ratio are within the elastic limit of 36 ksi and the slenderness ratio of about 200, the secant formula is applicable.

6.6 DESIGN OF COLUMNS UNDER A CENTRIC LOAD

In Sections 6.2 and 6.3, we obtain the critical load in a column by applying Euler's formula. This is followed by the investigation of the deformations and stresses in initially curved columns and eccentrically loaded columns by using the combined axial load and bending formula and the secant formula, respectively. In each case, we assume that all stresses remain below the proportional point or yield limit and that the column is a homogeneous prism.

The foregoing idealizations are important in understanding column behavior. However, the design of actual columns must be based on empirical formulas that consider the data obtained by laboratory tests. Care must be used in applying such "special purpose formulas." Specialized references should be consulted prior to design of a column for a particular application. Typical *design formulas* for centrically loaded columns made of three different materials follow. These represent specifications recommended by the American Institute of Steel Construction (AISC), the Aluminum Association, and the National Forest Products Association (NFPA). Various computer programs are readily available for the analysis and design of columns with any cross section (including variable) and any boundary conditions.

Column formulas for structural steel [5]

$$\sigma_{all} = \frac{S_y}{n}\left[1 - \frac{1}{2}\left(\frac{L_e/r}{C_c}\right)^2\right] \qquad \left(\frac{L_e}{r} < C_c\right) \tag{6.24a}$$

$$\sigma_{all} = \frac{\pi^2 E}{1.92(L_e/r)^2} \qquad \left(C_c \le \frac{L_e}{r} \le 200\right) \tag{6.24b}$$

where

$$C_c = \sqrt{2\pi^2 E/S_y} \tag{6.25a}$$

$$n = \frac{5}{3} + \frac{3}{8}\left(\frac{L_e/r}{C_c}\right) - \frac{1}{8}\left(\frac{L_e/r}{C_c}\right)^3 \tag{6.25b}$$

Column formulas for aluminum 6061-T6 alloy [6]

$$\sigma_{\text{all}} = 19 \text{ ksi} = 130 \text{ MPa} \qquad \left(\frac{L_e}{r} \leq 9.5\right) \tag{6.26a}$$

$$\sigma_{\text{all}} = \left[20.2 - 0.126\left(\frac{L_e}{r}\right)\right] \text{ksi}$$

$$= \left[140 - 0.87\left(\frac{L_e}{r}\right)\right] \text{MPa} \qquad \left(9.5 < \frac{L_e}{r} < 66\right) \tag{6.26b}$$

$$\sigma_{\text{all}} = \frac{51{,}000 \text{ ksi}}{(L_e/r)^2}$$

$$= \frac{350 \times 10^3}{(L_e/r)^2} \text{ MPa} \qquad \left(\frac{L_e}{r} \geq 66\right) \tag{6.26c}$$

Column formulas for timber of a rectangular cross section [7]

$$\sigma_{\text{all}} = \frac{0.3E}{(L_e/d)^2} \qquad \left(\frac{L_e}{d} \leq 50\right) \tag{6.27}$$

in which d is the smallest side dimension of the member. The allowable stress is not to exceed the value of stress for compression parallel to grain of the timber used.

Note that, for the structural steel columns, in Eqs. (6.24) and (6.25), C_c defines the limiting value of the slenderness ratio between intermediate and long bars. This is taken to correspond to one-half the yield strength S_y of the steel. By Eq. (6.8) we therefore have

$$C_c = \frac{L_e}{r} = \sqrt{\frac{2\pi^2 E}{S_y}} \tag{a}$$

Clearly, by applying a variable factor of safety, Eq. (6.25b) renders a consistent formula for intermediate and short columns. Also observe in Eq. (6.26) that for short and intermediate aluminum columns, σ_{all} is constant and linearly related to L_e/r. For long columns, a Euler-type formula is applied in both steel and aluminum columns [8]. Equation (6.27) for timber columns is also a Euler formula, adjusted by a suitable factor of safety.

EXAMPLE 6.5 | Design of a Wide-Flange Steel Column

Select the lightest wide-flange steel section to support an axial load of P on an effective length of L_e.

Given:　$S_y = 250$ MPa,　　$E = 200$ GPa,　　(from Table B.1),

　　　　　　$P = 408$ kN,　　$L_e = 4$ m

Solution: See Table A.6.

A suitable size for a prescribed shape may conveniently be obtained using tables in the AISC manual. However, we use a trial-and-error procedure here. Substituting the given data, Eq. (6.25a) results in

$$C_c = \sqrt{\frac{2\pi^2 E}{S_y}} = \sqrt{\frac{2\pi^2 (200 \times 10^3)}{250}} = 126$$

as the slenderness ratio.

First Try: Let $L_e/r = 0$. Equation (6.25b) yields $n = 5/3$, and by Eq. (6.24a), we have $\sigma_{all} = 250/n = 150$ MPa. The required area is then

$$A = \frac{P}{\sigma_{all}} = \frac{408 \times 10^3}{150 \times 10^6} = 2720 \, \text{mm}^2$$

Using Table A.6, we select a $W150 \times 24$ section having an area $A = 3060$ mm^2 (greater than 2720 mm^2) and with a minimum r of 24.6 mm. The value of $L_e/r = 4/0.0246 = 163$ is greater than $C_c = 126$, which was obtained in the foregoing. Therefore, applying Eq. (6.24b),

$$\sigma_{all} = \frac{\pi^2 (200 \times 10^9)}{1.92(163)^2} = 38.7 \, \text{MPa}$$

Hence, the permissible load, $38.7 \times 10^6 (3.06 \times 10^{-3}) = 118$ kN, is less than the design load of 408 kN, and a column with a larger A, a larger r, or both must be selected.

Second Try: Consider a $W150 \times 37$ section (see Table A.6) with $A = 4740$ mm^2 and minimum $r = 38.6$ mm. For this case, $L_e/r = 4/0.0386 = 104$ is less than 126. From Eq. (6.24a), we have

$$n = \frac{5}{3} + \frac{3}{8}\left(\frac{104}{126}\right) - \frac{1}{8}\left(\frac{104}{126}\right)^3 = 1.91$$

$$\sigma_{all} = \frac{250}{1.91}\left[1 - \frac{1}{2}\left(\frac{104}{126}\right)^2\right] = 86.3 \, \text{MPa}$$

The permissible load for this section, $86.3 \times 4740 = 409$ kN, is slightly larger than the design load.

Comment: A $W150 \times 37$ steel section is acceptable.

6.7 DESIGN OF COLUMNS UNDER AN ECCENTRIC LOAD

Recall from Section 6.5 that the secant formula is a rational equation and applies for all column lengths. On the other hand, this formula is quite difficult to employ in design, even using computers to facilitate the computation. Among various approaches employed in designing eccentrically loaded columns, the so-called interaction method seems the most simple. In this technique, the maximum stress in a member owing to an eccentric compressive load is given by the common formula

$$\sigma_{max} = \frac{P}{A} + \frac{Mc}{I}$$ (6.28)

It is obvious that, the first term is the axial stress developed by the eccentric load and the second term represents the magnitude of the superimposed bending stress. The moment equals $M = -Pe$, where e is the distance between the centroidal axis of the column and the axis through which the load is applied, the eccentricity.

SHORT COLUMNS

In the design of short columns or struts, the value of σ_{max} in Eq. (6.28) is not to exceed the allowable compressive strength of the material σ_{all}. Therefore,

$$\frac{P}{A} + \frac{Mc}{I} \leq \sigma_{all}$$

Division of the foregoing by σ_{all} gives

$$\frac{P/A}{\sigma_{all}} + \frac{Mc/I}{\sigma_{all}} \leq 1$$ (6.29)

INTERMEDIATE AND LONG COLUMNS

The design of intermediate and long columns is based on the assumption that the allowable stress generally is different for the two terms in the preceding expression. Accordingly, we introduce, for σ_{all} in the first and second terms, the values of corresponding allowable stress, respectively, to the centric loading and the pure bending. We have

$$\frac{P/A}{(\sigma_{all})_c} + \frac{Mc/I}{(\sigma_{all})_b} \leq 1$$ (6.30)

This is known as the *interaction formula,* a variety of which are used in application.

The value of $(\sigma_{all})_c$ is calculated from one of the axially loaded column formulas (6.24) through (6.27). It is to be mentioned that the *greatest value of the slenderness ratio* of the column should be used to compute $(\sigma_{all})_c$. Note also that allowable flexural stress of the material is denoted by $(\sigma_{all})_b$. The AISC specifications use Eq. (6.30) for $(P/A)/(\sigma_{all})_c < 0.15$. For other cases, the second term must be modified to include the lateral deflection of the column in the moment arm in determining the moment.

Column design by Eq. (6.29) is also a trial-and-error procedure, identical to that used in the preceding example. Obviously, any selected section that results in the largest sum (less than unity) of the terms on the left side of the formula is the most efficient section.

Determining the Allowable Load of an Eccentrically Loaded Column by the Interaction Formula

EXAMPLE 6.6

An $S\,310 \times 74$ steel column of effective length L_e is under an eccentric load, as shown in Figure 6.10. Calculate the maximum allowable value of the load P.

Figure 6.10   Example 6.6.

Given: $E = 200$ GPa and $S_y = 250$ MPa   (Table B.1), $L_e = 4$ m

From Table A.7, $A = 9.48 \times 10^3\,\text{mm}^2$, $r_y = 26.2$ mm,

$$r_z = 116 \text{ mm}, \qquad S = 833 \times 10^3\,\text{mm}^3$$

Assumption: Allowable stress in bending is 150 MPa.

Solution: The largest slenderness ratio of the column equals

$$\frac{L_e}{r_y} = \frac{4000}{26.2} = 153$$

Using Eq. (6.25a),

$$C_c = \sqrt{\frac{2\pi^2(200 \times 10^3)}{250}} = 126$$

which is smaller than 153. Hence, by Eq. (6.24b),

$$(\sigma_{\text{all}})_c = \frac{\pi^2(200 \times 10^9)}{1.92(153)^2} = 43.92 \text{ MPa}$$

The axial and bending stresses are, respectively,

$$\frac{P}{A} = \frac{P}{9.48 \times 10^{-3}}$$

$$\frac{Mc}{I} = \frac{Pe}{S} = \frac{P(0.2)}{0.833(10^{-3})}$$

Introducing the given numerical values and these equations into Eq. (6.30),

$$\frac{P/(9.48 \times 10^{-3})}{43.93 \times 10^6} + \frac{P(0.2)/(0.833 \times 10^{-3})}{150 \times 10^6} \leq 1$$

Solving, we have $P \leq 250$ kN. The maximum allowable load is therefore 250 kN.

6.8   BEAM-COLUMNS

Beams subjected simultaneously to axial compression and lateral loads are called *beam-columns*. Deflections in these members are not proportional to the magnitude of the axial load, like the previously discussed columns with initial curvature and eccentrically loaded columns. In this and the next sections, beam-columns of symmetrical cross section and with some common conditions of support and loading are analyzed.

Consider a beam subjected to an axial force P and a distributed lateral load w, as shown in Figure 6.11. The relationships between axial load P, shear force V, and bending moment M are obtained from the equilibrium of an isolated element of length dx between two cross sections taken normal to the original axis of the beam. On following a procedure similar to that used in Section 3.6, it can be shown that [1],

$$w = \frac{dV}{dx} \tag{6.31}$$

$$V = \frac{dM}{dx} + P\frac{dv}{dx} \tag{6.32}$$

We observe from Eq. (6.32) that, for beam-columns, shear force V, in addition to dM/dx as in beams, now also depend on the magnitude of axial force P and slope of the deflection curve.

For the analysis of beam-columns, it is sufficiently accurate to use the usual differential equation for deflection curve of beams. That is,

$$EI\frac{d^2v}{dx^2} = M \tag{6.33}$$

Figure 6.11
Beam-column with force P and lateral load w. Initial beam axis is shown by the dashed line.

where quantity EI is the flexural rigidity of the beam in the xy plane. However, in applying the foregoing expression, the bending moment due to the lateral loads as well as the axial forces must be written for the deflected member. Combining Eq. (6.33) with Eqs. (6.31) and (6.32), we can express the two alternative governing differential equations for beam columns:

$$\frac{d^2 M}{dx^2} + k^2 M = w \qquad (6.34)$$

$$\frac{d^4 v}{dx^4} + k^2 \frac{d^2 v}{dx^2} = \frac{w}{EI} \qquad (6.35)$$

In the preceding, as before, $k^2 = P/EI$. Clearly, if $P = 0$, the preceding equations reduce to the usual expressions for bending by lateral loads only.

Deflection and Critical Load Analysis of a Beam-Column by the Equilibrium Method | **EXAMPLE 6.7**

A pin-ended beam-column of length L is subjected to a concentrated transverse load F at its midspan as shown in Figure 6.12a. Determine

(a) An expression for the elastic curve.

(b) Maximum deflection and moment.

(c) The critical axial load.

Figure 6.12 Example 6.7: (a) a beam-column under lateral and axial forces; (b) free-body diagram for the deflected beam axis.

Assumption: The member has constant flexural rigidity EI.

Solution:

(a) The bending moment in the left segment of the beam (Figure 6.12b) is

$$M = -Pv - \frac{1}{2}Fx, \qquad \left(0 < x \le \frac{L}{2}\right) \qquad \text{(a)}$$

Equation (6.33) then becomes

$$EIv'' + Pv = -\frac{1}{2}Fx$$

The governing equation, setting $k^2 = P/EI$, is written as

$$\frac{d^2v}{dx^2} + k^2v = -\frac{k^2F}{2P}x, \qquad \left(0 < x \le \frac{L}{2}\right) \tag{6.36}$$

The general solution is

$$v = A\sin kx + B\cos kx - \frac{F}{2P}x \tag{b}$$

The boundary condition $v(0) = 0$ and the condition of symmetry $v'(L/2) = 0$ is applied to Eq. (b) to yield, respectively,

$$v(0) = 0: \qquad B = 0$$

$$v'\left(\frac{L}{2}\right) = 0: \qquad A = \frac{F}{2Fk\cos(kL/2)}$$

Substituting these constants into Eq. (b), we have

$$v = \frac{F}{2Pk}\frac{\sin kx}{\cos(kL/2)} - \frac{F}{2P}x \tag{6.37}$$

By differentiating this equation, we obtain the expressions for the slope (dv/dx) of the deflection curve and the bending moment $(EI\,d^2v/dx^2)$ at any section of the beam-column.

(b) The maximum deflection takes place at the center. Substituting $x = L/2$ into Eq. (6.37), after some simplifications, we obtain

$$v_{\max} = \frac{F}{2Pk}\left[\tan\frac{kL}{2} - \frac{kL}{2}\right] \tag{6.38}$$

The absolute maximum bending moment occurring at the midspan is readily obtained from Eq. (a) and (6.38):

$$M_{\max} = \frac{FL}{4} + Pv_{\max} = \frac{F}{2k}\tan\frac{kL}{2} \tag{6.39}$$

(c) Equations (6.37) through (6.39) become infinite if $kL/2$ approaches $\pi/2$, since then $\cos(kL/2) = 0$ and $\tan(kL/2) = \infty$. When $kL/2 = \pi/2$, we obtain

$$\frac{kL}{2} = \sqrt{\frac{P}{EI}}\frac{L}{2} = \frac{\pi}{2}$$

The critical load is therefore

$$P_{cr} = \frac{\pi^2 EI}{L^2} \tag{6.40}$$

Clearly, as $P \rightarrow P_{cr}$, even the smallest lateral load F produces considerable deflection and moment.

Comment: An alternate solution of this problem, obtained by the energy method (see Example 6.8), sheds further light on the deflection produced by the lateral load.

*6.9   ENERGY METHODS APPLIED TO BUCKLING

Energy approaches often more conveniently yield solution than equilibrium techniques in the analysis of elastic stability and buckling. The energy methods always result in buckling loads *higher* than the exact values if the assumed deflection of a slender member subject to compression differs from the true elastic curve. An efficient application of these approaches may be realized by selecting a series approximation for the deflection. Since a series involves a number of parameters, the approximation can be improved by increasing the number of terms in the series.

Reconsider the column hinged at both ends as depicted in Figure 6.1a. The configuration of this column in the first buckling mode is illustrated in Figure 6.1b. It can be shown [13] that the displacement of the column in the direction of load P is given by $\delta u \approx (1/2) \int (dv/dx)^2 \, dx$. Inasmuch as the load remains constant, the work done is

$$\delta W = \frac{1}{2} P \int_0^L \left(\frac{dv}{dx} \right)^2 dx \tag{6.41}$$

The strain energy associated with column bending is given by Eq. (5.18) in the form

$$U_1 = \int_0^L \frac{M^2}{EI} \, dx = \int_0^L \frac{EI}{2} \left(\frac{d^2v}{dx^2} \right)^2 dx$$

Likewise, the strain energy owing to a uniform compressive load P is, from Eq. (5.11),

$$U_2 = \frac{P^2 L}{2AE}$$

Because U_2 is constant, it does not enter to the analysis. Since the initial strain energy equals 0, the change in strain energy as the column proceeds from its initial to its buckled configuration is

$$\delta U = \int_0^L \frac{EI}{2} \left(\frac{d^2v}{dx^2} \right)^2 dx \tag{6.42}$$

From the principle of virtual work, $\delta W = \delta U$, it follows that

$$\frac{1}{2} \int_0^L P \left(\frac{dv}{dx}\right)^2 dx = \frac{1}{2} \int_0^L EI \left(\frac{d^2 v}{dx^2}\right)^2 dx \qquad \text{(6.43a)}$$

The preceding results in

$$P = \frac{\int_0^L EI (v'')^2}{\int_0^L (v')^2} \qquad \text{(6.43b)}$$

The end conditions are fulfilled by a deflection curve of the form

$$v = a \sin \frac{\pi x}{L}$$

in which a represents a constant. Carrying this deflection into Eq. (6.43b) and integrating, the critical load is found as

$$P_{cr} = \frac{\pi^2 EI}{L^2}$$

We observe from Eq. (6.43a) that, for $P > P_{cr}$, the work done by P exceeds the strain energy stored in the column; that is, a straight column is unstable if $P > P_{cr}$. This point, with regard to stability, corresponds to $A = 0$ in Section 6.2 and could not be found as readily by the equilibrium method. When $P = P_{cr}$, the column is in neutral equilibrium. If $P < P_{cr}$, a column is in stable equilibrium.

EXAMPLE 6.8   **Buckling Load and Deflection Analysis of a Beam-Column by the Principle of Virtual Work**

A simply supported beam-column is under a lateral force F at point A and axial loading P, as shown in Figure 6.13. Develop the equation of the elastic curve.

Figure 6.13   Example 6.8.

Solution: The total work done is obtained by addition of the work owing to the force F to Eq. (6.41). This problem has already been solved for $P = 0$ by employing the following series for deflection

$$v = \sum_{m=1}^{\infty} a_m \sin \frac{m\pi x}{L} \qquad \text{(a)}$$

On following a procedure similar to that used in Section 5.9, the condition $\delta U = \delta W$ now becomes

$$\frac{\pi^4 EI}{4L^3} \sum_{m=1}^{\infty} m^4 \delta\left(a_m^2\right) = \frac{\pi^2 P}{4L} \sum_{m=1}^{\infty} m^2 \delta\left(a_m^2\right) + F \sum_{m=1}^{\infty} \sin\frac{m\pi c}{L}\delta(a_m)$$

Solving,

$$a_m = \frac{2FL^3}{\pi^4 EI}\frac{1}{m^2}\frac{\sin(m\pi c/L)}{m^2 - b}$$

The quantity $b = P/P_{cr}$ is defined by Eq. (6.17). Carrying this equation into Eq. (a) results in

$$v = \frac{2FL^3}{\pi^4 EI} \sum_{m=1,3,...}^{\infty} \frac{\sin(m\pi c/L)}{m^2(m^2 - b)} \sin\frac{m\pi x}{L}, \qquad (0 \le x \le L) \tag{6.44}$$

Comments: When P approaches its critical value we have $b \to 1$ and the first term in Eq. (6.44),

$$v = \frac{2FL^3}{\pi^4 EI}\frac{1}{1 - b}\sin\frac{\pi c}{L}\sin\frac{\pi x}{L} \tag{6.45}$$

shows that the deflection becomes infinite, as expected. Comparing the preceding result with Eq. (5.59) we see that the axial force increases the deflection produced by the lateral force by a factor of $1/(1 - b)$.

---

*6.10   BUCKLING OF RECTANGULAR PLATES

The strength and deflection theory of plates, discussed in Sections 4.10 and 4.11, applies when such plates are in static equilibrium. Should a plate be compressed in its midplane, it becomes unstable and begins to buckle at a certain critical value of the in-plane force. The buckling of plates is qualitatively analogous to column buckling. However, a buckling analysis of the former case is not performed as readily as the latter. This is especially true in plates having other than simply supported edges. Often, in these cases, the energy method is used to good advantage to obtain the approximate buckling loads [1]. Thin plates or sheets, although quite capable of carrying tensile loadings, are poor in resisting compression. Usually, buckling or wrinkling phenomena observed in compressed plates (and shells) occur rather suddenly and are very dangerous. Fortunately, there is close correlation between theory and experimental data concerned with buckling of plates under a variety of loads and edge conditions.

In this section, we consider only simply supported rectangular plates subjected to uniaxial in-plane compressive forces per unit length N (Figure 6.14). It can be shown that the minimum value of N occurs when $n = 1$. That is, when the simply supported plate buckles, the buckling mode can be only one-half sine wave across the span, while several half sine waves can occur in the direction of compression.

The critical load N_{cr} per unit length of the plate is expressed as follows [9]:

$$N_{cr} = \frac{\pi^2 D}{b^2}\left(\frac{m}{r} + \frac{r}{n}\right)^2 = k^2 \frac{\pi^2 D}{b^2} \tag{6.46}$$

Figure 6.14
Compression of plate, simply supported on all four edges.

Here, the buckling load factor, aspect ratio, and flexural rigidity are, respectively,

$$k = \frac{m}{r} + \frac{r}{n}, \qquad r = \frac{a}{b}, \qquad D = \frac{Et^3}{12(1 - v^2)}$$

The quantity E represent the modulus of elasticity and v is the Poisson's ratio. The ratio $m/r = 1$ provides the following minimum value of the critical load

$$N_{cr} = \frac{4\pi^2 D}{b^2} \tag{6.47}$$

The corresponding critical stress, N_{cr}/t, is given by

$$\sigma_{cr} = \frac{\pi^2 E}{3(1 - v^2)} \left(\frac{t}{b}\right)^2 \tag{6.48}$$

where t is the thickness of the plate.

Variations in the buckling load factor k as functions of the aspect ratio r for $m = 1, 2, 3, 4$ are sketched in Figure 6.15. Obviously, for a specific m, the magnitude of k depends on r only. With reference to the figure, the magnitude of N_{cr} and the number of half-waves m for any value of the aspect ratio r can easily be found. When $r = 1.5$, for example, by Figure 6.14, $k = 4.34$ and $m = 2$. The corresponding critical load is equal to $N_{cr} = 4.34 \, \pi^2 D/b^2$, under which the plate buckles into two half sine waves in the direction of the loading, as depicted in Figure 6.14. We also see from Figure 6.15 that a plate m times as long as it is wide buckles in m half sine waves. Therefore, a long plate ($b \ll a$) with simply supported edges under a uniaxial compression tends to buckle into a number of square cells of side dimensions b and its critical load for all practical purposes is defined by Eq. (6.47).

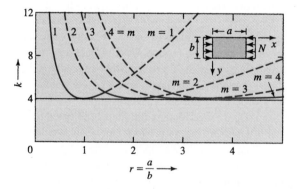

Figure 6.15   Variation of buckling load factor k with aspect ratio r for various number of half sine waves in the direction of compression.

REFERENCES

1. Timoshenko, S. P., and J. M. Gere. *Theory of Elastic Stability*. New York: McGraw-Hill, 1961.
2. Young, W. C., and R. C. Budynas. *Roark's Formulas for Stress and Strain,* 7th ed. New York: McGraw-Hill, 2001.
3. Peery, D. J., and J. J. Azar. *Aircraft Structures*. New York: McGraw-Hill, 1982.
4. Shigley, J. E., and C. R. Mischke, eds. *Standard Handbook of Machine Design*. New York: McGraw-Hill, 1986.
5. Manual of Steel Construction, 9th ed. New York: American Institute of Steel Construction, 1989.
6. *Specifications and Guidelines for Aluminum Structures*. Washington, DC: Aluminum Association, 2000.
7. *National Design Specifications for Wood Construction and Design Values for Wood Construction, NDS Supplements*. Washington, DC: National Forest Products Association, 1997.
8. Ugural, A. C. *Mechanics of Materials*. New York: McGraw-Hill, 1991.
9. Ugural, A. C. *Stresses in Plates and Shells,* 2nd ed. New York: McGraw-Hill, 1999.
10. Timoshenko, S. P., and S. Weinowsky-Krieger. *Theory of Plates and Shells,* 2nd ed. New York: McGraw-Hill, 1959.
11. Faupel, J. H., and F. E. Fisher. *Engineering Design,* 2nd ed. New York: Wiley, 1981.
12. Cook, R. D., and W. C. Young. *Advanced Mechanics of Materials,* 2nd ed. Upper Saddle River, NJ: Prentice Hall, 1999.
13. Ugural, A. C., and S. K. Fenster. *Advanced Strength and Applied Elasticity,* 4th ed. Upper Saddle River, NJ: Prentice Hall, 2003.
14. Brush, D. O., and B. O. Almroth. *Buckling of Bars, Plates, and Shells*. New York: McGraw-Hill, 1975.
15. Flügge, W., ed. *Stresses in Plates and Shells,* 2nd ed. Berlin: Springer, 1990.

PROBLEMS

Sections 6.1 through 6.5

6.1 An industrial machine requires a solid, round steel piston connecting rod of length L that carries a maximum compressive load P. Determine the required diameter.

Given: $L = 1.2$ m, $P = 50$ kN, $E = 210$ GPa, $S_y = 600$ MPa
Assumption: The ends are taken to be pinned.

6.2 Redo Problem 6.1 for a cold-rolled (510) bronze connecting rod.

Given: $L = 250$ mm, $P = 220$ kN, $E = 110$ GPa, $S_y = 520$ MPa (Table B.1)

6.3 Repeat Example 6.2 for a long column using the Euler formula.

6.4 A two-member pin-connected structure supports a concentrated load P at joint B as shown in Figure P6.4. Calculate the largest load P that may be applied with a factor of safety n.

Given: $n = 2.5$, $E = 210$ GPa.
Assumption: Buckling occurs in the plane of the structure.

Figure P6.4

6.5 A simple truss in which all members have the same axial rigidity AE is loaded as shown in Figure P6.5. Calculate the diameter d necessary for

(a) The bar AB.

(b) The bar BC.

Given: $E = 210$ GPa, $S_y = 250$ MPa.

Assumption: Buckling occurs in the plane of the truss. The Euler formula applies.

Figure P6.5

6.6 A structure that consists of the beam AB and the column CD is supported and loaded as shown in Figure P6.6. What is the largest load F that may be applied?

Given: The member CD is a round steel bar with $E = 200$ GPa.

Design Assumption: Failure is due to buckling only. Use the Euler formula with a factor of safety $n = 1.5$.

Figure P6.6

6.7 A compression member of 9-ft effective length is made by welding together two $3 \times 2 \times \frac{1}{4}$ in. steel angles ($E = 30 \times 10^6$ psi), as shown in Figure P6.7. Determine the allowable centric load for the member if a factor of safety of 2.5 is required.

Given: For one angle the area properties: $\bar{x} = 0.413$ in., $A = 1.19$ in.^{2}, $I_{\bar{x}} = 1.09$ in.^{4}, $I_{\bar{y}} = 0.392$ in.^{4}

Figure P6.7

6.8 A solid circular steel column of length L and diameter d is hinged at both ends. Calculate the load capacity for a safety factor of n.

Given: $S_y = 350$ MPa, $E = 210$ GPa, $L = 1$ m, $d = 60$ mm, $n = 3$
Assumption: The initial crookedness is 2 mm.

6.9 A steel pipe of outer diameter D and inner diameter d is employed as a 2-m column (Figure P6.9). Use Figure 6.8 to determine the largest allowable load P_y when the eccentricity e is

(a) 12 mm.

(b) 9 mm.

Given: $D = 100$ mm, $d = 88$ mm, $S_y = 275$ MPa, $E = 200$ GPa

Figure P6.9

Sections 6.6 and 6.7

6.10 Resolve Problem 6.7 using the AISC formulas.

Given: $S_y = 35$ ksi.

6.11 A steel rod of length L is required to support a concentric load P. Compute the minimum diameter required.

Given: $S_y = 350$ MPa, $E = 200$ GPa, $L = 3$ m, $P = 50$ kN
Assumption: Both ends of the rod are fixed.

6.12 A 4-m long fixed-ended timber column must safely carry a 100-kN centric load. Design the column using a square cross section.

Given: $E = 12.4$ GPa and $\sigma_{all} = 10$ MPa for compression parallel to the grain.

6.13 An aluminum alloy 6061-T6 pipe has an outer diameter D, inner diameter d, and length L. Determine the allowable axial load P if the pipe is used as a column fixed at one end and pinned at the other.

Given: $D = 14$ in., $d = 12$ in., $L = 20$ ft

6.14 A pin-ended steel column is subjected to a vertical load P. Determine the allowable stress.

Given: The cross section of the column is 80×120 mm and length is 3.5 m.

$$P = 600 \text{ kN}, \qquad E = 210 \text{ GPa}, \qquad S_y = 280 \text{ MPa}$$

6.15 A $W\,360 \times 216$ rolled-steel column (see Table A.6) with built-in ends is braced at midpoint C, as depicted in Figure P6.15. Calculate the allowable axial load P.

Given: $E = 200$ GPa, $S_y = 280$ MPa.
Assumption: Bracing acts as a simple support in the xy plane.

Figure P6.15

6.16 A 100×100 mm steel column of effective length L_e supports an eccentric load (Figure P6.16). Applying the interaction method, determine the maximum safe load P.

Given: $L_e = 4$ m, $E = 200$ GPa, $S_y = 280$ MPa, $(\sigma_{all})_b = 150$ MPa

Figure P6.16

6.17 A timber column 150 × 300 mm cross section is made of a grade of southern pine for which $E = 12$ GPa and $(\sigma_{all})_b = 10$ MPa for compression parallel to the grain. Employing the interaction method, determine the largest allowable effective length L_e for the eccentric loading shown in Figure P6.17.

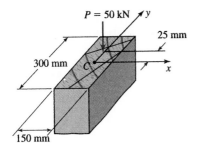

Figure P6.17

6.18 An aluminum alloy 6061-T6 rod of diameter d supports a compression load P with an eccentricity e. Applying the interaction method, with $(\sigma_{all})_b = 20$ ksi, calculate the largest effective length L_e that may be used.

Given: $d = 3$ in., $P = 10$ kips, $e = 0.75$ in.

Sections 6.8 through 6.10

6.19 A simply supported beam-column is subjected to end moments M_o and axial compression forces P, as depicted in Figure P6.19. Employ the equilibrium method to determine an expression for the deflection v.

Figure P6.19

6.20 A simply supported beam-column carries a uniformly distributed lateral load w and axial compression forces P (Figure P6.20). Using the equilibrium approach, determine an expression for the deflection v.

Figure P6.20

6.21 Redo Problem 6.20, employing the principle of virtual work. Let the deflection curve be expressed by Eq. (5.56).

6.22 Determine the critical load P that can be carried by a pin-ended column. Use the Rayleigh-Ritz method. Let the origin of the coordinates be placed at midspan and $v = a[1 - (2x/L)^2]$, where a is a constant.

6.23 The moment of inertia of a cantilever column varies as $I = I_o[1 - (x/2L)]$, in which I_o represents the constant moment of inertia at the fixed end $(x = 0)$. Determine the buckling load by the Rayleigh-Ritz method.

Assumption: The deflection curve is of the form $v = a(x/L)^2$, in which a is a constant.

FAILURE CRITERIA AND RELIABILITY

Outline

7.1   INTRODUCTION

A proper design includes the prediction of circumstances under which failure is likely to occur. In the most general terms, *failure* refers to any action that causes the member of the structure or machine to cease to function satisfactorily. The strength, stiffness, and stability of various load-carrying members are possible types or modes of failure. Failure may also be associated with poor appearance, poor adaptability to new demands, or other considerations not directly related to the ability of the structure to carry a load. Important variables associated with the failure include the type of material, configuration and rate of loading, the shape and surface peculiarities, and the operational environment.

This chapter is devoted to study of *static failure criteria* and the *reliability method* in design. We are concerned mainly with the failure of homogeneous and isotropic materials by yielding and fracture. The mechanical behavior of materials associated with failure also is discussed. In addition to possible failure by yielding or fracture, a member can fail at much lower stresses by crack propagation, should a crack of sufficient size be present. The *fracture mechanics theory* provides a means to predict a sudden failure on the basis of a computed stress intensity factor compared to a tested toughness criterion for the material (Sections 7.2 through 7.4). Other modes of failure include excessive elastic deflection of some element, rendering the machine or structure useless, or failure of a component by buckling. Various types of failure are considered in the problems presented as the subject unfolds [1–5].

Unless we are content to overdesign members, it is necessary to predict the most probable modes of failure and product reliability. Of necessity, the strength theories of failure are used in the majority of machine and structural designs. The actual failure mechanism in an element may be quite complicated; each failure theory is only an attempt to model the mechanism of failure for a given class of material. In each case, a factor of safety is employed to provide the required safety and reliability. Clearly, composite materials that do not exhibit uniform properties require more complex failure criteria [4].

7.2   INTRODUCTION TO FRACTURE MECHANICS

Fracture is defined as the separation or fragmentation of a member into two or more pieces. It normally constitutes a "pulling apart" associated with the tensile stress. A relatively brittle material fractures without yielding occurring throughout the fractured cross section. Thus, a brittle fracture occurs with little or no deformation or reduction in area and hence very little energy absorption. This type of failure usually takes place in some materials in an instant. The mechanisms of brittle fracture are the concern of fracture mechanics. It is based on a stress analysis in the vicinity of a crack, flaw, inclusion, or defect of unknown small radius in a part. A crack is a microscopic flaw that may exist under normal conditions on the surface or within the material.

As pointed out in Section 3.12, the stress concentration factors are limited to elastic structures for which all dimensions are precisely known, particularly the radius of the curvature in regions of high stress concentration. When there exists a crack, the stress concentration factor approaches infinity as the root radius approaches 0, thus rendering the concept of stress concentration factor useless. Furthermore, even if the radius of curvature of the flaw tip is known, the high local stresses there lead to some local plastic deformation. Elastic stress concentration factors become meaningless for this situation. On the basis of the foregoing, it may be concluded that analysis from the point of view of stress concentration factors is inadequate when cracks are present.

In 1920, A. A. Griffith postulated that an existing crack rapidly propagates (leading to rupture in an instant) when the strain energy released from the stressed body equals or exceeds that required to create the surfaces of the crack [6]. On this basis, it has been possible to calculate the average stress (if no crack were present) that causes crack growth in a part. For relatively brittle materials, the crack provides a mechanism by which energy is supplied continuously as the crack propagates.

Since major catastrophic failures of ships, bridges, and pressure vessels in the 1940s and 1950s, increasing attention has been given by design engineers to the conditions of the growth of a crack. Griffith's concept has been considerably expanded by G. R. Irwin [7]. Although Griffith's experiments dealt primarily with glass, his criterion has been widely applied to other materials, such as hard steels, strong aluminum alloys, and even low-carbon steel below the ductile-brittle transition temperatures. Inasmuch as failure does not occur then in an entirely brittle manner, application to materials that lie between relatively brittle and ductile requires modification of the theory and remains an active area of contemporary design and research in solid mechanics.

Adequate treatment of the subject of fracture mechanics is beyond the scope of this text. However, the basic principles and some important results simply are stated. Briefly, the fracture mechanics approach starts with an assumed initial minute crack (or cracks), for which the size, shape, and location can be defined. If brittle failure occurs, it is because the conditions of loading and environment are such that they cause an almost sudden propagation to failure of the original crack. Under fatigue loading, the initial crack may grow slowly until it reaches a critical size at which the rapid fracture occurs.

7.3 STRESS-INTENSITY FACTORS

In the fracture mechanics approach, a stress-intensity factor, K, is evaluated, as contrasted to stress-concentration factors. This can be thought of as a measure of the effective local stress at the crack root. The three modes of crack deformation of a plate are depicted in Figure 7.1. The most currently available values of K are for tensile loading normal to the crack, which is called *mode I* (Figure 7.1a). Accordingly, it is designated as K_I. Other types, modes II and III, essentially pertain to the in-plane and out-of-plane shear loads, respectively (Figures 7.1b and 7.1c). The treatment here concerns only mode I, and we eliminate the subscript and let $K = K_I$.

Acceptable solutions for many configurations, specific initial crack shapes, and orientations have been developed analytically and by computational techniques, including FEA [8]. For plates and beams, the *stress intensity factor* is defined in the form [9, 10]:

$$K = \lambda \sigma \sqrt{\pi a} \qquad (7.1)$$

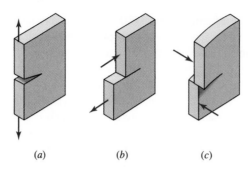

(a) (b) (c)

Figure 7.1 Crack deformation types:
(a) mode I, opening; (b) mode II, sliding;
(c) mode III, tearing.

where

σ = normal stress

λ = geometry factor, depends on a/w, listed in Table 7.1

a = crack length (or half crack length)

w = member width (or half-width of member)

We observe from Eq. (7.1) and Table 7.1 that the stress intensity factor depends on the applied load and geometry of the specimen as well as the size and shape of the crack. Clearly, the K may be increased by increasing either the stress or the crack size. The units of the stress-intensity factors are commonly MPa$\sqrt{m}$ in SI and ksi$\sqrt{in.}$ in the U.S. customary system.

Most cracks are not as simple as depicted in Table 7.1. They may be at an angle, embedded in a body, or sunken into a surface. A shallow surface crack in a member may be considered semi-elliptical. A circular or elliptical shape has proven to be adequate for various studies. Books on fracture mechanics provide methods of analysis, applications, and voluminous references [11–13].

Note that crack propagation occurring after an increase in load may be interrupted if a small inelastic zone forms ahead of the crack. But, stress intensity has risen with the increase in crack length and, in time, the crack may advance again a short amount. If stress continues to increase due to the reduced load-carrying area or otherwise, the crack may grow, leading to failure. The use of the stress intensity factors to predict the rate of growth of a fatigue crack is discussed in Section 8.14.

7.4 FRACTURE TOUGHNESS

In a toughness test of a given material, the stress-intensity factor at which a crack propagates is measured. This is the critical stress-intensity factor, known as the *fracture toughness* and designated by the symbol K_c. Usually, testing is done on an ASTM standard

Table 7.1 Geometry factors λ for some initial crack shapes

Case A. Tension of a long plate with a central crack

a/w	λ
0.1	1.01
0.2	1.03
0.3	1.06
0.4	1.11
0.5	1.19
0.6	1.30

Case B. Tension of a long plate with an edge crack

a/w	λ
0 ($w \to \infty$)	1.12
0.2	1.37
0.4	2.11
0.5	2.83

Case C. Tension of a long plate with double edge cracks

a/w	λ
0 ($w \to \infty$)	1.12
0.2	1.12
0.4	1.14
0.5	1.15
0.6	1.22

Case D. Pure bending of a beam with an edge crack

a/w	λ
0.1	1.02
0.2	1.06
0.3	1.16
0.4	1.32
0.5	1.62
0.6	2.10

specimen, either a beam or a tension member with an edge crack at the root of a notch. Loading is applied slowly and a record is made of load versus notch opening. The data are interpreted for the value of fracture toughness [14–17].

For a known applied stress acting on a part of known or assumed crack length, when the magnitude of stress intensity factor K reaches fracture toughness K_c, the crack propagates leading to rupture in an instant. The *factor of safety* for fracture mechanics, strength-to-stress ratio, is therefore

$$n = \frac{K_c}{K} \tag{7.2}$$

Substituting the stress intensity factor from Eq. (7.1), the foregoing becomes

$$n = \frac{K_c}{\lambda \sigma \sqrt{\pi a}} \tag{7.3}$$

Table 7.2 presents the values of the yield strength and fracture toughness for some metal alloys, measured at room temperature in a single edge-notch test specimen [18, 19]. For consistency of results, the ASTM specifications require a crack length a or member thickness t given by

$$a, t \geq 2.5 \left(\frac{K_c}{S_y} \right) \tag{7.4}$$

This ensures plane strain and flat crack surfaces. The values of a and t obtained by Eq. (7.4) are also included in the table.

Application of the preceding equations is illustrated in the solution of various numerical problems to follow.

Table 7.2   Yield strength S_y and fatigue toughness K_c for some engineering materials

Metals	S_y		K_c		Minimum values of a and t	
	MPa	(ksi)	MPa$\sqrt{m}$	(ksi$\sqrt{\text{in.}}$)	mm	(in.)
Steel						
AISI 4340	1503	(218)	59	(53.7)	3.9	(0.15)
Stainless steel						
AISI 403	690	(100)	77	(70.1)	31.1	(1.22)
Aluminum						
2024-T851	444	(64.4)	23	(20.9)	6.7	(0.26)
7075-T7351	392	(56.9)	31	(28.2)	15.6	(0.61)
Titanium						
Ti-6A1-4V	798	(116)	111	(101)	48.4	(1.91)
Ti-6A1-6V	1149	(167)	66	(60.1)	8.2	(0.32)

An Edge Crack on Aircraft Panel

EXAMPLE 7.1

An aircraft panel of width w and thickness t is loaded in tension, as shown in Figure 7.2. Estimate the maximum load P that can be applied without causing sudden fracture when an edge crack grows to length of a.

Given: $w = 100$ mm, $t = 16$ mm, $a = 20$ mm

Design Decision: The plate will be made of 7075-T7351 aluminum alloy. This decision should result in low weight.

Solution: From Table 7.2,

$$K_c = 31 \text{ MPa}\sqrt{\text{m}} = 31\sqrt{1000} \text{ MPa}\sqrt{\text{mm}}, \qquad S_y = 392 \text{ MPa}$$

for the aluminum alloy. Note that values of the length a and thickness t satisfy Table 7.2. Referring to Case B of Table 7.1, we have

$$\frac{a}{w} = \frac{20}{100} = 0.2, \qquad \lambda = 1.37$$

Equation (7.3) with $n = 1$,

$$\sigma_{\text{all}} = \frac{K_c}{\lambda\sqrt{\pi a}} = \frac{31\sqrt{1000}}{1.37\sqrt{\pi(20)}} = 90.27 \text{ MPa}$$

Hence,

$$P = \sigma_{\text{all}}(wt) = (90.27)(100 \times 16) = 144.4 \text{ kN}$$

Comment: The nominal stress at the fracture, $P/(16)(100 - 20) = 112.8$ MPa, is well below the yield strength of the material.

Figure 7.2 Example 7.1.

Design of a Wide Plate with a Central Crack

EXAMPLE 7.2

A large plate of width $2w$ carries a uniformly distributed tensile force P in a longitudinal direction with a safety factor of n. The plate has a central transverse crack that is $2a$ long. Calculate the thickness t required

    (a) To resist yielding.

    (b) To prevent sudden fracture.

Given: $w = 60$ mm, $P = 160$ kN, $n = 2.5$, $a = 9$ mm

Design Decision: The plate is made of Ti-6A1-6V alloy.

Solution: By Table 7.2,

$$K_c = 66\sqrt{1000}\ \text{MPa}\sqrt{\text{mm}}, \qquad S_y = 1149\ \text{MPa}$$

for the titanium alloy.

(a) The allowable tensile stress based on the net area is

$$\sigma_{\text{all}} = \frac{S_y}{n} = \frac{P}{2(w-a)t}$$

Hence,

$$t = \frac{Pn}{2(w-a)S_y} = \frac{160(10^3)(2.5)}{2(60-9)1149} = 3.4\ \text{mm}$$

(b) By Case A of Table 7.1,

$$\frac{a}{w} = \frac{9}{60} = 0.15, \qquad \lambda = 1.02$$

Through the use of Eq, (7.3) with $n = 2.5$, the stress at fracture is

$$\sigma = \frac{K_c}{\lambda n \sqrt{\pi a}} = \frac{66\sqrt{1000}}{1.02(2.5)\sqrt{\pi(9)}} = 153.9\ \text{MPa}$$

Since this stress is smaller than the yield strength, the *fracture governs* the design; $\sigma_{\text{all}} = 153.9$ MPa. Therefore,

$$t_{\text{red}} = \frac{P}{2w\sigma_{\text{all}}} = \frac{160(10^3)}{2(60)(153.9)} = 8.66\ \text{mm}$$

Comment: Use a thickness of 8.7 mm. Both values of a and t fulfill Table 7.2.

EXAMPLE 7.3 | The Load Capacity of a Bracket with an Edge Crack

A bracket having an edge crack and uniform thickness t carries a concentrated load, as shown in Figure 7.3a. What is the magnitude of the fracture load P with a safety factor of n for crack length of a?

(a)   (b)

Figure 7.3   Example 7.3: (a) bracket with edge crack; (b) free-body diagram of segment AB.

Given: $a = 0.16$ in., $\quad w = 2$ in., $\quad d = 4$ in., $\quad t = 1$ in., $\quad n = 1.2$

Assumptions: The bracket is made of AISI 4340 steel. A linear elastic stress analysis is acceptable.

Solution: Referring to Table 7.2,

$$K_c = 53.7 \text{ ksi}\sqrt{\text{in.}}, \qquad S_y = 218 \text{ ksi}$$

Observe that values of a and t both satisfy the table. At the section through the point B (Figure 7.3b), the bending moment is

$$M = P\left(d + \frac{w}{2}\right) = P(4 + 1) = 5P$$

Nominal stress for the *combined loading*, by superposition of two states of stress for axial force P and moment M, is expressed as

$$
\begin{aligned}
\lambda\sigma &= \lambda_a \sigma_a + \lambda_b \sigma_b \\
&= \lambda_a \frac{P}{wt} + \lambda_b \frac{6M}{tw^2}
\end{aligned}
\tag{7.5}
$$

Here w and t represent the width and thickness of the member, respectively.

The ratio of crack length to bracket width is $a/w = 0.08$. For Cases B and D of Table 7.1, we have $\lambda_a = 1.12$ and $\lambda_b = 1.02$, respectively. Equation (7.5) leads to

$$
\begin{aligned}
\lambda\sigma &= 1.12\,\frac{P}{2(1)} + 1.02\,\frac{6(5P)}{1(2)^2} \\
&= 0.56P + 7.65P = 8.21P
\end{aligned}
$$

Then, through the use of Eq. (7.3),

$$\lambda\sigma = \frac{K_c}{n\sqrt{\pi a}}; \qquad 8.21P = \frac{53.7(10^3)}{1.2\sqrt{\pi(0.16)}}$$

Solving, $P = 7.69$ kips.

Comments: The normal stress at fracture, $7.69/1(2 - 0.16) = 4.18$ ksi, is well below the yield strength and our assumption is valid.

7.5 YIELD AND FRACTURE CRITERIA

In many cases, a member fails when the material begins to yield or deform permanently. For metals, the shear stress plays an important role in yielding. Occasionally, a small dimensional change may be tolerated and a static load that exceeds the yield point permitted. The fracture at the ultimate strength of the material would then constitute failure. The fracture, or separation of the material under stress into two or more parts, usually is associated with a tensile stress. Particularly in a one-dimensional stress field, compression stress is generally considered less damaging than tensile stress. The brittle or ductile character of a

Table 7.3   Utilizable values of a material for states of stress in tension and torsion tests

Quantity	Tension test	Torsion test
Maximum shear stress	$S_{ys} = S_y/2$	S_{ys}
Maximum energy of distortion	$U_{od} = [(1 + \nu)/3E]S_y^2$	$U_{od} = S_{ys}^2(1 + \nu)/E$
Maximum octahedral shear stress	$\tau_{oct} = (\sqrt{2}/3)S_y$	$\tau_{oct} = (\sqrt{2}/\sqrt{3})S_{ys}$
Maximum principal stress	S_u	S_{us}

Notes: S_y = yield strength in tension, S_{ys} = yield strength in shear, S_u = ultimate strength in tension, S_{us} = ultimate strength in shear.

material is relevant to the mechanism of failure. The distinction between ductile and brittle material itself is not simple, however. The nature of the stress, the temperature, and material itself all play a role, as is discussed in Section 2.9, by defining the boundary between ductility and brittleness.

Let us consider an element subjected to a triaxial state of stress, where $\sigma_1 > \sigma_2 > \sigma_3$. Recall that subscripts 1, 2, and 3 denote the principal directions. The state of stress in a uniaxial loading is defined by σ_1, equal to the normal force divided by the cross-sectional area, and $\sigma_2 = \sigma_3 = 0$. Corresponding to the onset of yielding and fracture in a simple tension test, the stresses and strain energy shown in the second column of Table 7.3 are determined as follows. When specimen starts to yield, we have $\sigma_1 = S_y$. Therefore, the maximum shear stress is $\tau_{max} = \sigma_1/2 = S_y/2$ by Eq. (3.34), the maximum distortion energy density absorbed by the material is $U_{od} = S_y^2/6G$ by using Eq. (5.9) with $G = E/2(1 + \nu)$, and the maximum octahedral shear stress $\tau_{oct} = (\sqrt{2}/3)S_y$ from Eq. (3.58b). On the other hand, at an impending fracture, the maximum principal stress is $\sigma_1 = S_u$. Note that the forgoing quantities obtained in simple tension have special significance in predicting failure involving combined stress.

In the torsion test, the state of stress is specified by $\tau = \sigma_1 = -\sigma_2$ and $\sigma_3 = 0$. Here, the shear stress is calculated using the torsion formula. Corresponding to this case of pure shear, at the start of yielding and fracture, are the stresses and strain energy shown in the third column of the table. These quantities are readily obtained by a procedure similar to that described in the preceding for tension test.

The mechanical behavior of materials subjected to uniaxial normal stresses or pure shearing stresses is readily presented on stress-strain diagrams. The onset of failure by yielding or fracture in these cases is considerably more apparent than in situations involving combined stress. From the viewpoint of mechanical design, it is imperative that some practical guidelines be available to predict yielding or fracture under various conditions of stress, as they are likely to exist in service. To meet this need, a number of failure criteria or theories consistent with behavior and strength of material have been developed.

These strength theories are structured to apply to particular classes of materials. We discuss the two most widely accepted theories to predict the onset of inelastic behavior for ductile materials under combined stress in Sections 7.6 through 7.9. The three fracture theories pertaining to brittle materials under combined stress are presented in Sections 7.10 through 7.12. As we observe, the "theory" behind most static failure criteria is that *whatever is responsible for failure in the simple tensile test also is responsible for failure under combined loading.* It is important to note that yielding of ductile materials should be further qualified to yielding of ductile metals; many polymers are ductile but do not follow the standard yield theories.

7.6 MAXIMUM SHEAR STRESS THEORY

The maximum shear stress theory is developed on the basis of the experimental observation that a ductile material yields as a result of slip or shear along crystalline planes. Proposed by C. A. Coulomb (1736–1806), it is also known as the *Tresca yield criterion* in recognition of the contribution by H. E. Tresca (1814–1885) to its application. This theory states that yielding begins whenever the maximum shear stress at any point in the body becomes equal to the maximum shear stress at yielding in a simple tension test. Hence, according to Eq. (3.34) and Table 7.3,

$$\tau_{max} = \frac{1}{2}|\sigma_1 - \sigma_3| = S_{ys} = \frac{1}{2}S_y$$

The maximum shear stress theory is therefore given by

$$|\sigma_1 - \sigma_3| = \frac{S_y}{n} \tag{7.6}$$

for a factor of safety n.

In the case of *plane stress,* $\sigma_3 = 0$, two combinations of stresses are to be considered. When σ_1 and σ_2 have *opposite signs,* that is, one tensile and the other compressive, the maximum shear stress is $(\sigma_1 - \sigma_2)/2$. The yield condition then becomes

$$|\sigma_1 - \sigma_2| = \frac{S_y}{n} \tag{7.7}$$

The foregoing may be restated in the form, for $n = 1$,

$$\frac{\sigma_1}{S_y} - \frac{\sigma_2}{S_y} = \pm 1 \tag{7.8}$$

When σ_1 and σ_2 carry the *same sign,* the maximum shear stress is $(\sigma_1 - \sigma_3)/2 = \sigma_1/2$. Then, for $|\sigma_1| > |\sigma_2|$ and $|\sigma_2| > |\sigma_1|$, we have the yield conditions, respectively,

$$|\sigma_1| = \frac{S_y}{n} \quad \text{and} \quad |\sigma_2| = \frac{S_y}{n} \tag{7.9}$$

Equation (7.8) and Eq. (7.9) for $n = 1$ are represented graphically in Figure 7.4. Note that Eq. (7.8) applies to the second and fourth quadrants, while Eq. (7.9) applies to the first and third quadrants. The boundary of the hexagon thus marks the onset of yielding, with points outside the shaded region depicting a yielded state. We demonstrate in Section 7.9 that the maximum shear stress theory has reasonably good agreement with the experiment

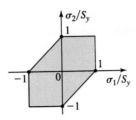

Figure 7.4   Yield criterion based on maximum shear stress.

for ductile materials. The theory offers an additional advantage in its ease of application. However, the maximum distortion criterion, discussed next, is recommended because it correlates better with the actual test data for yielding of ductile materials.

A TYPICAL CASE OF COMBINED LOADING

In a common case of combined plane bending, torsion, axial, and transverse shear loadings, such as in Figure 3.29, we have $\sigma_y = \sigma_z = \tau_{yz} = \tau_{xz} = 0$. Hence, the principal stresses reduce to

$$\sigma_{1,2} = \frac{\sigma_x}{2} \pm \tau_{max} \tag{7.10}$$

where

$$\tau_{max} = \left[\left(\frac{\sigma_x}{2} \right)^2 + \tau_{xy}^2 \right]^{1/2}$$

Substituting these into Eq. (7.7), the maximum shear stress criterion becomes

$$\frac{S_y}{n} = \left(\sigma_x^2 + 4\tau_{xy}^2 \right)^{1/2} \tag{7.11}$$

for the preceding special case.

EXAMPLE 7.4 | **Failure of Ductile Material under Combined Torsion and Axial Loading**

A circular rod, constructed of a ductile material of tensile yield strength S_y, is subjected to a torque T. Determine the axial tensile force P that can be applied simultaneously to the rod (Figure 7.5).

Given: $T = 500\pi$ N · m, $D = 50$ mm,   factor of safety $n = 1.2$

Design Decisions: The rod is made of steel of $S_y = 300$ MPa. Use the maximum shear stress failure criterion.

Figure 7.5   Example 7.4.

Solution: For the situation described, the critical stresses occur on the elements at the surface of the shaft. Based on the maximum shear stress theory, from Eq. (7.11),

$$\sigma_x = \left[\left(\frac{S_y}{n}\right)^2 - 4\tau_{xy}^2\right]^{1/2} \tag{a}$$

where

$$\sigma_x = \frac{P}{A} = \frac{4P}{\pi D^2}, \qquad \tau_{xy} = \frac{Tr}{J} = \frac{16T}{\pi D^3} \tag{b}$$

Substituting the given numerical values, Eq. (a) gives

$$\sigma_x = \left[\left(\frac{300 \times 10^6}{1.2}\right)^2 - 4\left(\frac{16 \times 500\pi}{\pi \times 0.05^3}\right)^2\right]^{1/2} = 214.75 \text{ MPa}$$

The first of Eqs. (b) is therefore

$$P = \frac{\pi(0.05)^2(214.75 \times 10^6)}{4} = 421.7 \text{ kN}$$

Comment: This is the maximum force that can be applied without causing permanent deformation.

7.7   MAXIMUM DISTORTION ENERGY THEORY

The maximum distortion energy theory or criterion was originally proposed by J. C. Maxwell in 1856, additional contributions were made in 1904 by M. T. Hueber, in 1913 by R. von Mises, and in 1925 by H. Hencky. Today it is mostly referred to as the *von Mises-Hencky theory* or simply *von Mises theory*. This theory predicts that failure by yielding occurs when, at any point in the body, the distortion energy per unit volume in a state of combined stress becomes equal to that associated with yielding in a simple tension test. Hence, in accordance with Eq. (5.9) and Table 7.3,

$$\frac{1}{12G}[(\sigma_1 - \sigma_2)^2 + (\sigma_2 - \sigma_3)^2 + (\sigma_3 - \sigma_1)^2] = \frac{1+\nu}{3E}S_y^2$$

where $G = E/2(1 + v)$. The maximum energy of distortion criterion for yielding is therefore

$$\frac{\sqrt{2}}{2}[(\sigma_1 - \sigma_2)^2 + (\sigma_2 - \sigma_3)^2 + (\sigma_3 - \sigma_1)^2]^{1/2} = \frac{S_y}{n} \qquad (7.12)$$

for a safety factor n.

It is often convenient to replace S_y/n by an *equivalent stress* σ_e in the preceding equation. In so doing, we have

$$\frac{\sqrt{2}}{2}[(\sigma_1 - \sigma_2)^2 + (\sigma_2 - \sigma_3)^2 + (\sigma_3 - \sigma_1)^2]^{1/2} = \sigma_e \qquad (7.13)$$

Commonly used names for the equivalent stress are the effective stress and the *von Mises stress*. Observe from Eqs. (7.12) and (7.13) that only the differences of the principal stresses are involved. Consequently, the addition of an equal amount to each stress does not affect the conclusion with respect to whether or not yielding occurs. In other words, inelastic action does not depend on hydrostatic tensile or compressive stress.

For *plane stress* $\sigma_3 = 0$, the maximum energy of distortion theory becomes

$$\left(\sigma_1^2 - \sigma_1\sigma_2 + \sigma_2^2\right)^{1/2} = \frac{S_y}{n} \qquad (7.14)$$

or

$$\left(\sigma_1^2 - \sigma_1\sigma_2 + \sigma_2^2\right)^{1/2} = \sigma_e \qquad (7.15)$$

Equation (7.14) may alternatively be represented in the following form for $n = 1$:

$$\left(\frac{\sigma_1}{S_y}\right)^2 - \left(\frac{\sigma_1}{S_y}\right)\left(\frac{\sigma_2}{S_y}\right) + \left(\frac{\sigma_2}{S_y}\right)^2 = 1$$

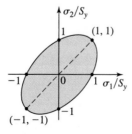

Figure 7.6 Yield criterion based on distortion energy.

This expression defines the ellipse shown in Figure 7.6. As in the case of the maximum shear stress theory, points within the shaded area represent nonyielding states. The boundary of the ellipse indicates the onset of yielding, with the points outside the shaded area representing a yielded state. The maximum energy of distortion theory of failure agrees quite well with test data for yielding of ductile materials and plane stress. It is commonly used in design and gives the same result as the octahedral shear stress theory, discussed in the next section.

A TYPICAL CASE OF COMBINED LOADING

Reconsider the particular case of combined loading, where $\sigma_y = \sigma_z = \tau_{yz} = \tau_{xz} = 0$ (see Section 7.6). Substitution of Eq. (7.10) into (7.14) leads to the expression

$$\frac{S_y}{n} = \left(\sigma_x^2 + 3\tau_{xy}^2\right)^{1/2} \tag{7.16}$$

Clearly, Eq. (7.16) is based on the maximum energy of distortion criterion for the foregoing special case.

7.8 OCTAHEDRAL SHEAR STRESS THEORY

The octahedral shear stress theory, also known as the *Mises-Hencky criterion* or simply the *Mises criterion,* predicts failure by yielding whenever the octahedral shear stress for any state of stress equals the octahedral shear stress for the simple tensile test. Accordingly, through the use of Eq. (3.58b) and Table 7.3, the octahedral shear stress theory is

$$\tau_{\text{oct}} = \frac{\sqrt{2}}{3} S_y \tag{7.17}$$

which gives Eq. (7.12). Eichinger (in 1926) and Nadai (in 1937) independently developed this theory.

The octahedral shear stress criterion may also be considered in terms of distortion energy. In a general state of stress, from Eq. (5.9), we have

$$U_{od} = \frac{3}{2} \frac{1+\nu}{E} \tau_{\text{oct}}^2$$

When the forgoing reaches the value given in Table 7.3, Eq. (7.17) again is found.

We see that the octahedral shear stress criterion is equivalent to the distortion energy theory; that is, the former criterion enables us to apply the latter theory while dealing with stress rather than energy. We use the procedure of the energy of distortion criterion in this text.

EXAMPLE 7.5

Finding the Safety Factor in the Yielding of Material under Combined Static Loading

The stresses shown in Figure 7.7 occur at a point in a ductile structural member of yield strength S_y. Determine the factor of safety n with respect to inelastic deformation.

Figure 7.7   Example 7.5.

Design Decisions: The member is made of steel of $S_y = 36$ ksi. Use the maximum shear stress and the energy of distortion failure criteria.

Solution: Following the usual sign convention, we write the stress components $\sigma_x = -10$ ksi, $\sigma_y = 2$ ksi, and $\tau_{xy} = 8$ ksi. Substituting into Eq. (3.33),

$$\sigma_{1,2} = \frac{-10+2}{2} \pm \sqrt{\left(\frac{-10-2}{2}\right)^2 + 8^2} = -4 \pm 10$$

or

$$\sigma_1 = 6 \text{ ksi}, \qquad \sigma_2 = -14 \text{ ksi}$$

Maximum shear stress theory.   Through the use of Eq. (7.7),

$$6 + 14 = \frac{36}{n}$$

from which $n = 1.8$.

Maximum energy of distortion theory.   Applying Eq. (7.14),

$$[(6)^2 - (6)(-14) + (-14)^2]^{1/2} = \frac{36}{n}$$

Solving, we obtain $n = 2.03$.

Comments: Inasmuch as the maximum distortion energy criterion is more accurate, it makes sense for a higher factor of safety to be obtained by this theory.

Case Study 7-1 | FAILURE ANALYSIS OF A CONICAL TANK

A thin-walled conical vessel, or tank, is supported on its edge and filled with a liquid, as depicted in Figure 7.8. Expressions for stress in this shell are developed in Case Study 16-2. Determine the vessel wall thickness on the basis of the maximum shear stress and the energy of distortion failure theories.

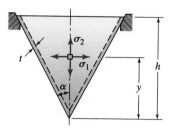

Figure 7.8   Conical tank filled with a liquid.

Given: The geometry and loading of the tank are known.

Assumptions:

1. The vessel is made of structural steel of yield strength S_y.
2. The factor of safety against yielding is n.
3. The vessel is taken to be simply supported from the top.

Solution: The tangential stress $\sigma_\theta = \sigma_1$ and meridional stress $\sigma_S = \sigma_2$ in the tank are expressed by Eqs. (16.76a) and (16.77a) as follows

$$\sigma_1 = \gamma(h - y)y\frac{\tan\alpha}{t\cos\alpha}$$

$$\sigma_2 = \gamma\left(h - \frac{2}{3}y\right)y\frac{\tan\alpha}{2t\cos\alpha} \qquad \textbf{(a)}$$

where

h = liquid height

t = vessel wall thickness

α = half angle at the apex of cone

γ = specific weight of liquid

The largest magnitudes of these principal stresses are given by

$$\sigma_{1,max} = \frac{\gamma h^2}{4t}\frac{\tan\alpha}{\cos\alpha}, \qquad \left(\text{at } y = \frac{h}{2}\right)$$

$$\sigma_{2,max} = \frac{3\gamma h^2}{16t}\frac{\tan\alpha}{\cos\alpha}, \qquad \left(\text{at } y = \frac{3h}{4}\right) \qquad \textbf{(b)}$$

Comment: Note that the maximum stresses occur at different locations.

Maximum shear stress criterion. Since σ_1 and σ_2 are of the same sign and $|\sigma_1| > |\sigma_2|$, the first of Eqs. (7.9) together with (b) results in

$$\frac{S_y}{n} = \frac{\gamma h^2}{4t}\frac{\tan\alpha}{\cos\alpha} \qquad \textbf{(7.18)}$$

The thickness of the vessel is obtained from the preceding equation in the form

$$t = 0.25\frac{\gamma h^2 n}{S_y}\frac{\tan\alpha}{\cos\alpha} \qquad \textbf{(7.19a)}$$

Maximum energy of distortion criterion. Inasmuch as the maximum magnitudes of σ_1 and σ_2 occur at different locations, we must first determine the section at which combined stresses are at a critical value. For this purpose, we substitute Eq. (a) into Eq. (7.14):

$$\frac{S_y^2}{n^2} = \left[\gamma(h-y)y\frac{\tan\alpha}{t\cos\alpha}\right]^2 + \left[\gamma\left(h - \frac{2}{3}y\right)y\frac{\tan\alpha}{2t\cos\alpha}\right]^2$$
$$- \left[\gamma(h-y)y\frac{\tan\alpha}{t\cos\alpha}\right]\left[\gamma\left(h - \frac{2}{3}y\right)y\frac{\tan\alpha}{2t\cos\alpha}\right] \qquad \textbf{(c)}$$

Differentiation of Eq. (c) with respect to the variable y and equating the result to 0 gives [20],

$$y = 0.52h$$

Introducing this value of y back into Eq. (c), the thickness of the vessel is found to be

$$t = 0.225\frac{\gamma h^2 n}{S_y}\frac{\tan\alpha}{\cos\alpha} \qquad \textbf{(7.19b)}$$

Comment: The thickness according to the maximum shear stress criterion is therefore 10% larger than that based on the maximum distortion energy criterion.

7.9   COMPARISON OF THE YIELDING THEORIES

Two approaches may be used to compare the theories of yielding heretofore discussed. The first comparison equates, for each theory, the critical values corresponding to uniaxial loading and torsion. Referring to Table 7.3,

Maximum shearing stress theory: $S_{ys} = 0.50 S_y$

Energy of distortion theory, or its
equivalent, the octahedral shear stress theory: $S_{ys} = 0.577 S_y$

(7.20)

We observe that the difference in strength predicted by these criteria is not substantial. A second comparison may be made by means of superposition of Figures 7.4 and 7.6. This is left as an exercise for the reader.

Experiment shows that, for ductile materials, the yield stress obtained from a torsion test is 0.5 to 0.6 times that determined from simple tension test. We conclude, therefore, that the energy of distortion criterion or octahedral shearing stress criterion is most suitable for ductile materials. However, the shear stress theory, which results in $S_{ys} = 0.50 S_y$, is simple to apply and offers a conservative result in design.

As a third comparison, consider a solid shaft of diameter D and tensile yield strength S_y subjected to combined loading consisting of tension P and torque T. The yield criteria based on the maximum shear stress and energy of distortion theories, for $n = 1$, are given by Eqs. (7.11) and (7.16):

$$S_y = \left(\sigma_x^2 + 4\tau_{xy}^2\right)^{1/2}, \qquad S_y = \left(\sigma_x^2 + 3\tau_{xy}^2\right)^{1/2} \tag{a}$$

In the preceding, σ_x and τ_{xy} represent axial tension and torsional stresses, respectively. Therefore,

$$\sigma_x = \frac{4P}{\pi D^2}, \qquad \tau_{xy} = \frac{16T}{\pi D^3}$$

Figure 7.9   Yield curves for torsion-tension shaft. The points shown in this figure are based on experimental data [21].

A dimensionless plot of Eqs. (a) and some experimental results are shown in Figure 7.9. We note again particularly good agreement between the maximum energy of distortion criterion and experimental data for ductile materials. The difference in results is not very great, however, and both theories are widely used in design of members.

7.10 MAXIMUM PRINCIPAL STRESS THEORY

In accordance with the maximum principal stress theory, credited to W. J. M. Rankine (1820–1872), a material fails by fracturing when the maximum principal stress reaches the ultimate strength S_u in a simple tension test. Thus, at the beginning of the fracture,

$$|\sigma_1| = \frac{S_u}{n} \quad \text{or} \quad |\sigma_3| = \frac{S_u}{n} \tag{7.21}$$

for safety factor n. That is, a crack starts at the most highly stressed point in a brittle material when the maximum principal stress at the point reaches S_u. This criterion is suggested by the observation that fracture surfaces in brittle materials under tension are planes that carry the maximum principal stress. Clearly, the maximum principal stress theory is based on the assumption that the ultimate strength of the material is the same in tension and compression: $S_u = |S_{uc}|$.

For the case of *plane stress* ($\sigma_3 = 0$), Eq. (7.21), the fracture condition, is given by

$$|\sigma_1| = \frac{S_u}{n} \quad \text{or} \quad |\sigma_2| = \frac{S_u}{n} \tag{7.22}$$

This may be restated in the form, for $n = 1$,

$$\frac{\sigma_1}{S_u} = \pm 1 \quad \text{or} \quad \frac{\sigma_2}{S_u} = \pm 1 \tag{7.23}$$

Figure 7.10 is a plot of Eq. (7.23). Note that points a, b, and c, d in the figure indicate the tensile and compressive principal stresses, respectively. As in other criteria, the boundary of the square indicates the onset of failure by fracture. The area within the boundary is therefore a region of no failure.

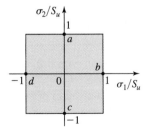

Figure 7.10 Fracture criterion based on maximum principal stress.

Note that, while a material may be weak in simple compression, it may nevertheless sustain very high hydrostatic pressure without fracturing. Furthermore, most brittle materials are much stronger in compression than in tension; that is, $S_{uc} \gg S_u$. These are inconsistent with the theory. Moreover, the theory makes no allowance for influences on the failure mechanism other than those of normal stresses. However, for brittle materials in all stress ranges, the maximum principal stress theory has good experimental verification, provided there is a tensile principal stress.

EXAMPLE 7.6

The Failure of a Pipe of Brittle Material under Static Torsion Loading

A cast pipe of outer diameter D and inner diameter d is made of an aluminum alloy having ultimate strengths in tension and compression S_u and S_{uc}, respectively. Determine the maximum torque that can be applied without causing rupture.

Given: $D = 100 \, \text{mm},$ $d = 60 \, \text{mm},$ $S_u = 200 \, \text{MPa},$ $S_{uc} = 600 \, \text{MPa}$

Design Decision: Use the maximum principal stress theory and a safety factor of $n = 2$.

Solution: The torque and the maximum shear stress are related by the torsion formula:

$$T = \frac{J\tau}{c} = \frac{\pi(0.05^4 - 0.03^4)\tau}{2(0.05)} = 170.9 \times 10^{-6}\tau \qquad \text{(a)}$$

The state of stress is described by

$$\sigma_1 = -\sigma_2 = \tau, \qquad \sigma_3 = 0$$

From Eqs. (7.22) and the preceding, we have $\tau = S_u/n$. Then, Eq. (a) results in

$$T = 170.9 \times 10^{-6} \left(\frac{200 \times 10^6}{2} \right) = 17.09 \, \text{kN} \cdot \text{m}$$

Comment: According to the maximum principal stress theory, the torque is limited to $17.09 \, \text{kN} \cdot \text{m}$ to avoid failure by fracture.

7.11   MOHR'S THEORY

The Mohr theory of failure is employed to predict the fracture of a material with different properties in tension and compression when the results of a variety of tests are available for that material. This criterion uses Mohr's circles of stress. Using the extreme values of principal stress enables one to apply the Mohr approach to either two- or three-dimensional cases.

Experiments are performed on a given material to determine the states of stress that result in failure. Each such state of stress defines a Mohr circle. When the data describing states limiting stress are derived from only simple tension, compression, and torsion tests,

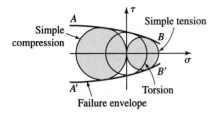

Figure 7.11 Mohr's fracture criterion.

the three resulting circles are sufficient to construct the envelope, labeled by lines AB and $A'B'$ in Figure 7.11.

Note that the Mohr envelope represents the locus of all possible failure states. Many solids, particularly those that are brittle, show greater resistance to compression than to tension. As a result, higher limiting shear stresses, for these materials, are found to the left of the origin as depicted in the figure.

7.12 THE COULOMB-MOHR THEORY

The Coulomb-Mohr theory, like the Mohr criterion, may be employed to predict the effect of a given state of stress on a brittle material having different properties in tension and in compression. The Mohr's circles for the uniaxial tension and compression tests are used to predict failure by Coulomb-Mohr theory as shown in Figure 7.12a. The points of contact of the straight-line envelopes (AB and $A'B'$) with the stress circles define the state of stress at a fracture. For example, if such points are C and C', the stresses and the planes on which they act can be obtained using the established procedure for Mohr's circle of stress.

In the case of *plane stress,* we have $\sigma_3 = 0$. When σ_1 and σ_2 have *opposite signs* (that is, one is tensile and the other is compressive), it can be verified that [22] the onset of fracture is expressed by

$$\frac{\sigma_1}{S_u} - \frac{\sigma_2}{|S_{uc}|} = \frac{1}{n} \tag{7.24}$$

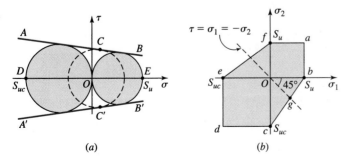

(a)   (b)

Figure 7.12 (a) Straight-line Mohr's envelopes; (b) Coulomb-Mohr fracture criterion.

for safety factor n. Here S_u and S_{uc} represent the ultimate strengths of the material in tension and compression, respectively. This equation may be rearranged into the form

$$n = \frac{S_u}{\sigma_1 - \sigma_2|S_u/S_{uc}|} \tag{7.25}$$

Relationships for the case where the principal stresses have the same sign may be deduced from Figure 7.12a. In the case of *biaxial tension,* the corresponding circle is represented by diameter OE. Hence, fracture occurs if either of the two tensile stresses achieves the value S_u; that is,

$$\sigma_1 = \frac{S_u}{n} \quad \text{or} \quad \sigma_2 = \frac{S_u}{n} \tag{7.26}$$

For *biaxial compression,* a Mohr's circle of diameter OD is obtained. Failure by fracture occurs if either of the compressive stresses attains the value S_{uc}; therefore,

$$\sigma_2 = -\frac{S_{uc}}{n} \quad \text{or} \quad \sigma_1 = -\frac{S_{uc}}{n} \tag{7.27}$$

The foregoing expressions are depicted in Figure 7.12b, for the case in which $n = 1$. Lines ab and af represent Eq. (7.26), and lines dc and de, Eq. (7.27). The boundary bc is obtained by applying Eq. (7.24). Line ef completes the hexagon in a way similar to Figure 7.4. Points within the shaded area represent states of nonfailure according to the Coulomb-Mohr theory. The boundary of the figure depicts the onset of failure due to fracture.

In the case of *pure shear,* the corresponding limiting point g represents the ultimate shear strength S_{us}. At point g, $\sigma_1 = S_{us}$ and $\sigma_2 = -\sigma_1 = -S_{us}$. Substituting for σ_1, σ_2, and $n = 1$ into Eq. (7.24), we have

$$S_{us} = \frac{S_u}{1 + |S_u/S_{uc}|} \tag{7.28}$$

When $|S_{uc}| = S_u$, $S_{uc} = 0.5S_u$. If $|S_{uc}| = 4S_u$, typical of ordinary gray cast iron, then $S_{us} = 0.8S_u$.

EXAMPLE 7.7 | Rework Example 7.6, Employing the Coulomb-Mohr Theory

Solution: We have the following results, from Example 7.6:

$$T = 170.9 \times 10^{-6}\tau$$

and $\sigma_1 = -\sigma_2 = \tau$. So, applying Eq. (7.24) with $n = 2$,

$$\frac{\tau}{200 \times 10^6} - \frac{-\tau}{600 \times 10^6} = \frac{1}{2}$$

Solving, $\tau = 75$ MPa. The first equation then gives $T = 12.82$ kN · m.

Comments: On the basis of the maximum principal stress theory, the torque that can be applied to the pipe 17.09 kN · m obtained in Example 7.6 is thus 25% larger than on the basis of the Coulomb-Mohr theory.

Determination of the Largest Load Supported by the Frame of a Punch Press

EXAMPLE 7.8

Figure 7.13 depicts a punch press frame made of ASTM A-48 gray cast iron having ultimate strengths in tension and compression S_u and S_{uc}, respectively. Calculate the allowable load P.

Figure 7.13   Example 7.8.

Given: $S_u = 170$ MPa, $S_{uc} = 650$ MPa

Design Decisions: Use the Coulomb-Mohr theory and a factor of safety of $n = 2.5$.

Solution: The centroid, total section area, and moment of inertia about the neutral axis (Figure 7.13) are

$$\bar{z} = \frac{(180 \times 80)(210) + (120 \times 240)60}{180 \times 80 + 120 \times 240} = 110 \text{ mm}$$

$$A = 180 \times 80 + 120 \times 240 = 43.2 \times 10^3 \text{ mm}^2$$

$$I = \frac{1}{12}(80)(180)^3 + (80 \times 180)(100)^2 + \frac{1}{12}(240)(120)^3 + (120 \times 240)(50)^2$$

$$= 289.44 \times 10^6 \text{ mm}^4$$

The internal force resultants in section A-B are equivalent to a centric force P and a bending moment $M = 0.51P$.

Note that cast iron has nonlinear stress-strain relationship. Therefore, bending stresses are not exactly given by the flexure formula, $\sigma = Mc/I$. For simplicity, however, it is generally used in the design of cast iron machine elements. Hence, the stress distribution across the section is taken to be linear.

The distances from the neutral axis to the extreme fibers are $c_A = 110$ mm and $c_B = 190$ mm. The greatest tensile and compressive stresses occur at points A and B, respectively:

$$\sigma_A = \frac{P}{A} + \frac{Mc_A}{I} = 23.148P + 193.823P$$

$$\sigma_B = \frac{P}{A} - \frac{Mc_B}{I} = 23.148P - 334.784P$$

We therefore have, on the tension and compression sides, respectively,

$$\sigma_1 = 216.971P, \qquad \sigma_2 = 0$$
$$\sigma_2 = -311.636P, \qquad \sigma_1 = 0$$

(a)

The maximum allowable load P is the smaller of the two loads calculated from Eqs. (a), (7.26), and (7.27):

$$|216.971P| = \frac{170(10^6)}{2.5} \quad \text{or} \quad P = 313.4 \text{ kN}$$

$$|-313.636P| = \frac{650(10^6)}{2.5} \quad \text{or} \quad P = 829 \text{ kN}$$

Comment: The tensile stress governs the allowable load $P = 313.4$ kN that the member can carry.

7.13   RELIABILITY

The concept of reliability is closely related to the factor of safety. Reliability is the probability that a member or structure will perform without failure a specific function under given conditions for a given period of time. It is very important for the designer and the manufacturer to know the reliability of the product. The reliability R can be expressed by a number that has the range

$$0 \leq R < 1$$

For instance, reliability of $R = 0.98$ means that there is 98% chance, under certain operating conditions, that the part will perform its proper function without failure; that is, if 100 parts are put into service and an average of 2 parts fail, then the parts proved to be 98% reliable.

Recall from Section 1.6 that, in the conventional design of members, the possibility of failure is reduced to acceptable levels by factor of safety based on judgment derived from past performances. In contrast, in the reliability method, the variability of material properties, fabrication-size tolerances, as well as uncertainties in loading and even design approximations can be appraised on a statistical basis. As far as possible, the proposed criteria are calibrated against well-established cases. The reliability method has the advantage of consistency in the safety factor, not only for individual members but also for complex structures and machines. Important risk analyses of complete engineering systems are based on the same premises. Clearly, the usefulness of the reliability approach depends on adequate information on the statistical distribution of loading applied to parts in service, from which can be calculated the significant stress and significant strength of production runs of manufactured parts.

Note that the reliability method to design is relatively new and analyses leading to an assessment of reliability address uncertainties. Of course, this approach is more expensive and time consuming than the factor of safety method of design, because a larger quantity of data must be obtained by testing. However, in certain industries, designing to a designated reliability is necessary. We use the factor of safety for most of the problems in this text. For an interactive statistics program from the engineering software database, see the website at www.mecheng.asme.org/database/STAT/MASTER.HTML.

7.14 NORMAL DISTRIBUTIONS

To obtain quantitative estimates of the percentages of anticipated failures from a study, we must look into the nature of the distribution curves for significant quantities involved. We consider only the case involving the normal or Gauss distribution, credited to K. F. Gauss (1777–1855). This is the most widely used model for approximating the dispersion of the observed data in applied probability [23].

Several other distributions might prove useful in situations where random variables have only positive values or asymmetrical distributions. A formula, introduced by W. Weibull, is often used in mechanical design. This formula does not arise from classical statistics and is flexible to apply. The Weibull distribution is used in work dealing with experimental data, particularly reliability. The distribution of bearing failures at a constant load can be best approximated by the Weibull distribution (see Section 10.14).

In analytical form, the Gaussian, or normal, distribution is given as follows:

$$p(x) = \frac{1}{\sqrt{2\pi}\,\sigma} e^{-(x-\mu)/2\sigma^2} \qquad (-\infty < x < \infty) \qquad (7.29)$$

where

$p(x) =$ probability or frequency function

$\sigma =$ standard deviation*

$\mu =$ mean value

$x =$ quantity.

The standard deviation is widely used and regarded as the usual index of dispersion or scatter of the particular quantity. The mean value and standard deviations are defined by

$$\mu = \frac{1}{n} \sum_{i=1}^{n} x_i \qquad (7.30a)$$

$$\sigma = \left[\frac{1}{n-1} \sum_{i=1}^{n} (x_i - \mu)^2 \right]^{1/2} \qquad (7.30b)$$

where n is the total number of elements, called the population.

A plot of Eq. (7.29), the standard normal distribution [24], is shown in Figure 7.14. This *bell-shaped curve* is symmetrical about the mean value μ. Since the probability that any value of x will fall between plus and minus infinity is 1, the area under the curve in the figure is unity. Note that about 68% of the population represented fall within the band $\mu + 1\sigma$, 95% fall within the band $\mu + 2\sigma$, and so on.

---

*The symbol σ used for standard deviation here should not be confused with the symbol of stress, although often the units are the same.

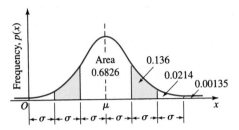

Figure 7.14   The standard normal distribution.

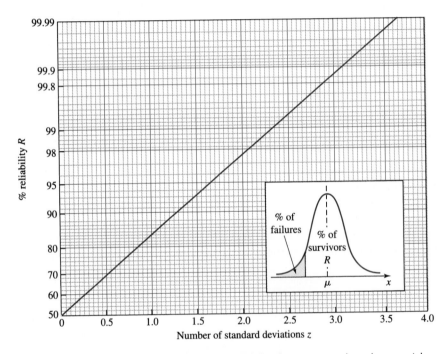

Figure 7.15   Reliability chart: Generalized normal distribution curve plotted on special probability paper.

The reliability, or rate of survival, R is a function of the *number of standard deviations z,* also referred to as the safety index:

$$z = \left| \frac{x - \mu}{\sigma} \right| \qquad (7.31)$$

The straight line of Figure 7.15, drawn on probability-chart paper, plots percentages of reliability R as an increasing function of number of standard deviations z. A larger z results in fewer failures; hence, a more conservative design. We note that the percentage of the population corresponding to any portion of the standard normal distribution (Figure 7.14) can be read from the reliability chart.

7.15 THE RELIABILITY METHOD AND THE MARGIN OF SAFETY

In the reliability method of design, for a given member, the distribution of loads and strengths are determined, then these two are related to achieve an acceptable success rate. The designer's task is to make a judicious selection of materials, processes, and geometry (size) to achieve a reliability goal. The approach of reliability finds considerably more application with members subjected to wear and fatigue loading. We here introduce it in the simpler context of static loading. Section 8.6 discusses the reliability factor for fatigue endurance strength of materials.

Consider the distribution curves for the two main random variables, load L and strength S, which is also called capacity or resistance (Figure 7.16). For a given member, the frequency functions $p(l)$ and $p(s)$ define the behavior of critical parameters load and strength. The mean value of strength is denoted by μ_s and the mean value of the load by μ_l. So, based on mean values, there would be a *margin of safety*

$$m = S - L \qquad (7.32)$$

However, the "interference" or shaded area of overlap in the figure indicates some possibility of a weak part in which failure could occur. The preceding margin of safety must not be confused with that used in aerospace industry (see Section 1.8).

Figure 7.17 shows a corresponding plot of the distribution of margin of safety. In this diagram, the probability of failure is given by the (shaded) area under the tail of the curve to the left of the origin. The member would survive in all instances to the right of the origin. By statistical theory, the difference between two variables with normal distributions has itself a normal distribution. Therefore, if the strength S and the load L are normally distributed, then the margin of safety m also has a normal distribution, as shown in Figures 7.16 and 7.17.

The margin of safety has a *mean value* μ_m and *standard deviation* σ_m expressed as follows [25, 26]:

$$\mu_m = \mu_s - \mu_l \qquad (7.33a)$$

$$\sigma_m = \sqrt{\sigma_s^2 + \sigma_l^2} \qquad (7.33b)$$

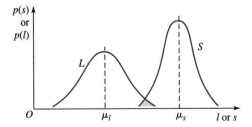

Figure 7.16 Normal distribution curves of load L and strength S.

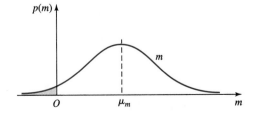

Figure 7.17 Normal distribution curve of margin of safety $m = S - L$.

Here σ_s and σ_l are, respectively, the standard deviations for the strength S and the load L. When $S > L$, m is positive. The designer is interested in the probability that $m > 0$; that is, the area to the right of 0 in Figure 7.17. At $x = m = 0$, Eq. (7.31) becomes $z = \mu/\sigma$. The number of standard deviations, on introducing Eq. (7.32) may now be written in the following form:

$$z = \frac{\mu_m}{\sigma_m} = \left| \frac{\mu_s - \mu_l}{\sqrt{\sigma_s^2 + \sigma_l^2}} \right| \tag{7.34}$$

For the prescribed mean and deviation values of the strength and load, Eq. (7.34) is solved to yield the number of standard deviations z. Then, the probability that a margin of safety exists may be read as the reliability R from the chart of Figure 7.15. Equation (7.34) is therefore called the coupling equation, because it relates the reliability, through z, to the statistical parameters of the normally distributed strength and load. For example, when the mean values of S and L are equal (that is, $\mu_s = \mu_l$) it follows that $z = 0$ and the reliability of a part is 50%. Reliability of an assembly or system of parts may be found from their individual reliability values [27].

Application of the reliability theory is illustrated in the solution of the following numerical problems.

EXAMPLE 7.9 | **A Shipment of Control Rods**

In a shipment of 600 control rods, the mean tensile strength is found to be 35 ksi and the standard deviation 5 ksi. How many rods can be expected to have

  (a) A strength of less than 29.5 ksi.

  (b) A strength of between 29.5 and 48.5 ksi.

Given: $\mu_s = 29.5$ and 48.5 ksi, $\mu_l = 35$ ksi, $\sigma_m = 5$ ksi

Assumption: Both loading and strength have normal distributions.

Solution:

  (a) Substituting given numerical values into Eq. (7.34) results in the number of standard deviations:

$$z = \left| \frac{29.5 - 35}{5} \right| = 1.10$$

The corresponding reliability, obtained from Figure 7.15 is 86.5%. Note that $1 - 0.865 = 0.135$ represents the proportion of the total rods having a strength less than 29.5 ksi. Hence, the number of rods with a strength less than 29.5 ksi is $600(0.135) = 81$.

(b) In this case, applying Eq. (7.31),

$$z = \left| \frac{48.5 - 35}{5} \right| = 2.7$$

From Figure 7.15, we then have $R = 99.65\%$. The number of rods expected to have strength between 29.5 and 48.5 ksi is therefore $600(0.9965 - 0.135) = 517$.

A Machine Part in Service

EXAMPLE 7.10

At the critical point of a machine part in service, the load-induced mean stress and standard deviation are 30 and 5 ksi, respectively. If the material has a yield strength of 50 ksi with a standard deviation of 4 ksi, determine the reliability against yielding. What percentage of failure is expected in service?

Given: $\mu_s = 50\,\text{ksi}, \qquad \mu_l = 30\,\text{ksi}, \qquad \sigma_s = 4\,\text{ksi}, \qquad \sigma_l = 5\,\text{ksi}$

Assumption: Both loading and strength have normal distribution.

Solution: Through the use of Eqs. (7.33), we have

$$\mu_m = 50 - 30 = 20\,\text{ksi}$$

$$\sigma_m = \sqrt{4^2 + 5^2} = 6.403\,\text{ksi}$$

Equation (7.34) then gives $z = 20/6.403 = 3.124$. Figure 7.15 shows that this corresponds to 99.91% reliability. So, the failure percentage expected in service would be $100 - 99.91 = 0.09\%$.

The Twisting-off Strength of Bolts

EXAMPLE 7.11

Bolts, each of which has a mean twisting-off strength of 25 N · m with a standard deviation of 1.5 N · m, are tightened with automatic wrenches on a production line (see Section 15.7). If the automatic wrenches have a standard deviation of 2 N · m, calculate mean value of wrench torque setting that results in an estimated 1 bolt in 400 twisting off during assembly.

Given: $\mu_s = 25\,\text{N} \cdot \text{m}$, $\sigma_s = 1.5\,\text{N} \cdot \text{m}$, $\sigma_l = 2\,\text{N} \cdot \text{m}$

Assumption: Both the wrench twist-off torque and the bolt twist-off strength have normal distributions.

Solution: Substitution of $\sigma_s = 1.5\,\text{N} \cdot \text{m}$ and $\sigma_l = 2\,\text{N} \cdot \text{m}$ in Eq. (7.33b) gives $\sigma_m = 2.5\,\text{N} \cdot \text{m}$. Figure 7.15 shows that a reliability of $399/400 = 0.9975$, or 99.75%, corresponds to 2.8 standard deviation. The mean value is then $\mu_m = z\sigma_m = 2.8(2.5) = 7\,\text{N} \cdot \text{m}$. Since $\mu_s = 25\,\text{N} \cdot \text{m}$, we have, from Eq. (7.33a), $\mu_l = 18\,\text{N} \cdot \text{m}$. This is the required value of wrench setting.

REFERENCES

1. Nadai, A. *Theory of Flow and Fracture of Solids.* New York: McGraw-Hill, 1950.
2. Marin, J. *Mechanical Behavior of Materials.* Upper Saddle River, NJ: Prentice Hall, 1962.
3. Van Vlack, L. H. *Elements of Material Science and Engineering,* 6th ed. Reading, MA: Addison-Wesley, 1989.
4. Dowling, N. E. *Mechanical Behavior of Materials.* Upper Saddle River, NJ: Prentice Hall, 1993.
5. American Society of Metals. *Metals Handbook.* Metals Park, OH: ASM, 1961.
6. Griffith, A. A. "The Phenomena on Rupture and Flow of Solids." Philosophical Transactions of the Royal Society (London) A221 (1921), pp. 163–98.
7. Irwin, G. R. "Fracture Mechanics." *Proceedings, First Symposium on Naval Structural Mechanics.* New York: Pergamon, 1958, p. 557.
8. Cook, R. D., et al. *Concepts and Applications of Finite Element Analysis,* 4th ed. New York: Wiley, 2001.
9. Boresi, A. P., and R. C. Schmidt. *Advanced Mechanics of Materials,* 6th ed. New York: Wiley, 2003.
10. Knott, J. F. *Fundamentals of Fracture Mechanics.* New York: Wiley, 1973.
11. Brock, D. *The Practical Use of Fracture Mechanics.* Dordrecht, the Netherlands: Kluwer, 1988.
12. Gdoutos, E. E. *Fracture Mechanics Criteria and Applications.* Dordrecht, the Netherlands: Kluwer, 1990.
13. Meguid, S. A. *Engineering Fracture Mechanics.* London: Elsevier, 1989.
14. Brown, W. F., ed. *Review of Developments in Plane-Strain Fracture Toughness Testing,* Tech. Pub. 463. Philadelphia: ATM, 1970.
15. Irwin, G. R. "Mechanical Testing." *Metals Handbook,* 9th ed., vol. 8. Metals Park, OH: ASM, 1989, pp. 437–493.
16. Dieter, G. E. *Mechanical Metallurgy,* 3rd ed. New York: McGraw-Hill, 1986.
17. Norton, R. L. *Machine Design: An Integrated Approach,* 2nd ed. Upper Saddle River, NJ: Prentice Hall, 2000.
18. *Damage Tolerant Design Handbook,* MCIC-HB-01. Kettering, OH: Air Force Materials Laboratory, Wright-Patterson Air Force Base, December 1972.
19. *ASM International Guide in Selecting Engineering Materials.* Metals Park, OH: American Society of Metals, 1989.

20. Fauppel, J. H., and F. E. Fisher. *Engineering Design,* 2nd ed. New York: Wiley, 1981.
21. Taylor, G. I., and H. Quinney. *The Plastic Distortion of Metals, Philosophical Transactions of the Royal Society,* Section A, No. 230, 1932, p. 323.
22. Ugural, A. C., and S. K. Fenster. *Advanced Strength and Applied Elasticity,* 4th ed. Upper Saddle River, NJ: Prentice Hall, 2003.
23. Shigley, J. E., and C. R. Mischke. *Mechanical Engineering Design,* 6th ed. New York: McGraw-Hill, 2001.
24. Juvinall, R. C., and K. M. Marshek. *Fundamentals of Machine Component Design,* 3rd ed. New York: Wiley, 2000.
25. Lipson, C., and J. Sheth. *Statistical Design and Analysis of Engineering Experiments.* New York: McGraw-Hill, 1973.
26. Kennedy, J. B., and A. M. Neville. *Basic Statistical Methods for Engineers and Scientists,* 3rd ed. New York: Harper and Row, 1986.
27. Burr, A. H., and J. B. Cheatham. *Mechanical Analysis and Design,* 2nd ed. Upper Saddle River, NJ: Prentice Hall, 1995.

Problems

Sections 7.1 through 7.4

7.1   An AISI-4340 steel ship deck of thickness t and width $2w$ is in tension. If a central transverse crack of length $2a$ is present (Case A of Table 7.1), estimate the maximum tensile load P that can be applied without causing sudden fracture. What is the nominal stress at fracture?

Given: $t = 25$ mm, $w = 250$ mm, $a = 25$ mm

7.2   Estimate the maximum load P that the plate shown in Case B of Table 7.1 can carry. What is the mode of failure?

Given: $S_y = 650$ MPa, $K_c = 100$ Pa$\sqrt{m}$, $w = 350$ mm, $a = 25$ mm, $t = 15$ mm, factor of safety $n = 1.2$

7.3   A 2024-T851 aluminum-alloy plate of width w and thickness t is subjected to a tensile loading. It contains a transverse crack of length a on one edge (Figure 7.2). There is concern that the plate will undergo sudden fracture. Calculate the maximum allowable axial load P. What is the nominal stress at fracture?

Given: $w = 125$ mm, $t = 25$ mm, $a = 20$ mm

7.4   An AISI-4340 steel pipe of diameter d and wall thickness t contains a crack of length $2a$. Estimate the pressure p that will cause fracture when

(*a*)   The crack is longitudinal as in Figure P7.4.

(*b*)   The crack is circumferential.

Given: $d = 50$ mm, $t = 4$ mm, $a = 5$ mm
Assumption: A factor of safety $n = 1.5$ and geometry factor $\lambda = 1.01$ are used (Table 7.1).

Figure P7.4

7.5   A 7075-T7351 aluminum alloy beam containing an edge crack of length a is in pure bending as shown in Case D of Table 7.1. Determine the maximum moment M that can be applied without causing sudden fracture.

Given: $a = 40$ mm, $w = 100$ mm, $t = 25$ mm

7.6   An AISI-4340 steel plate of width $w = 5$ in. and thickness $t = 1$ in. is under uniaxial tension. A crack of length a is present on the edge of the plate, as shown in Figure 7.2. Determine

(a)   The axial load possible P_{all} for the case in which $a = \frac{1}{2}$ in.

(b)   The critical crack length a, if the plate is made of Ti-6A1-6V titanium alloy and subjected to the P_{all} calculated in part a.

7.7   Rework Example 7.3, for the case in which the bracket is made of AISI-403 stainless steel and $a = 0.6$ in., $d = 6.25$ in., $w = 1.5$ in., $t = \frac{1}{2}$ in., and $n = 2$.

7.8   An AISI-4340 steel ship deck-panel of width w and thickness t is under tension. Calculate the maximum load P that can be applied without causing fracture when double-edge cracks grow to a length of a (Case C of Table 7.1).

Given: $t = 1$ in., $w = 2$ in., $a = 0.2$ in., $n = 1.4$

Sections 7.5 through 7.12

7.9   A solid steel bar having yield strength S_y and diameter D carries end loads P, M, and T. Calculate the factor of safety n, assuming that failure occurs according to the following criteria:

(a)   Maximum energy of distortion.

(b)   Maximum shear stress.

Given: $D = 100$ mm, $S_y = 260$ MPa, $P = 50$ kN, $M = 5$ kN $\cdot$ m, $T = 8$ kN $\cdot$ m

7.10   A steel bar AB of diameter D and yield strength S_y supports an axial load P and vertical load F acting at the end of the arm BC (Figure P7.10). Determine the largest value of F according to the maximum energy of distortion theory of failure.

Given: $D = 40$ mm, $S_y = 250$ MPa, $P = 20$ F
Assumptions: The effect of the direct shear is negligible and the factor of safety $n = 1.4$.

7.11   A cantilever WF aluminum alloy beam of yield strength S_y is loaded as shown in Figure P7.11. Using a factor of safety of n, determine whether failure occurs according to the maximum shear stress criterion.

Given: $S_y = 320$ MPa, $n = 2$, $I_z = 13.4 \times 10^6$ mm^4

Figure P7.10   **Figure P7.11**

7.12 Resolve Problem 7.11 applying the maximum energy of distortion theory.

7.13 A thin-walled cylindrical pressure vessel of diameter d and constructed of structural steel with yield strength S_y must withstand an internal pressure p. Calculate the wall thickness t required.

Given: $S_y = 36$ ksi, $d = 20$ in., $p = 500$ psi, $n = 1.5$
Design Decision: Use the following criteria:

(a) Maximum shear stress.
(b) Maximum energy of distortion.

7.14 Redo Problem 7.13, if the vessel is made of a material having $S_u = 50$ ksi and $S_{uc} = 90$ ksi.

Design Decision: Apply the following theories:

(a) Maximum principal stress.
(b) Coulomb-Mohr.

7.15 A cantilever WF cast iron beam of ultimate tensile strength S_u and ultimate compression strength S_{uc} is subjected to a concentrated load at its free end (Figure P7.11). What is the factor of safety n?

Given: $S_u = 280$ MPa, $S_{uc} = 620$ MPa
Assumption: Failure occurs in accordance with the following theories:

(a) Maximum principal stress.
(b) Coulomb-Mohr.

7.16 and 7.17 The state of stress shown (Figures P7.16 and P7.17) occurs at a critical point in an ASTM A-48 gray cast iron (Table B.1) component of a lawn mower. Calculate the factor of safety n with respect to fracture.

Figure P7.16   **Figure P7.17**

Design Decision: Apply the following criteria:

(a)   Maximum principal stress.

(b)   Coulomb-Mohr.

7.18   A thin-walled steel spherical storage tank is filled with liquid of density γ and supported on a ring, as shown in Figure P7.18. According to the maximum shear stress and maximum energy of distortion criteria, determine the factor of safety n, in terms of thickness t, mean radius a, yield strength S_y, and γ, as required.

Figure P7.18

Given: Tangential stress $\sigma_\theta = \sigma_1$ and meridian stress $\sigma_\phi = \sigma_2$ at $\phi = 150°$, are calculated from Eqs. (16.27) as follows:

$$\sigma_1 = 3.56 \frac{\gamma a^2}{t}, \qquad \sigma_2 = -1.70 \frac{\gamma a^2}{t}$$

where the negative sign indicates compression.

7.19   Resolve Problem 7.18 for the condition that the tank is made of cast iron having ultimate compression strength S_{uc} five times ultimate tensile strength S_u.

Design Decision: Maximum principal stress and Coulomb-Mohr criteria are used.

7.20   A cylindrical rod of diameter D is made of ASTM-A36 steel (Table B.1). Use the maximum shear stress criterion to determine the maximum end torque T that can be applied to the rod simultaneously with an axial load of $P = 100$ kips (Figure 7.5).

Given: $D = 2$ in.
Assumption: $n = 1.5$.

7.21   Redo Problem 7.20, applying the maximum energy of distortion criterion.

7.22   An ASTM-20 gray cast iron rod (Table B.2) is under pure torsion. Determine, with a factor of safety $n = 1.4$, the maximum shear stress τ that may be expected at impending rupture using

(a)   The Coulomb-Mohr criterion.

(b)   The principal stress criterion.

7.23   The state of stress shown in Figure P7.23 occurs at a critical point in a machine component made of ASTM-A47 malleable cast iron (Table B.1). Apply the Coulomb-Mohr theory to calculate the maximum value of the shear stress τ for a safety factor of $n = 2$.

Figure P7.23

7.24 Resolve Problem 7.23 for the condition that the machine component is made of an ASTM-A242 high-strength steel (Table B.1). Use

(a) The maximum energy of distortion criterion.

(b) The maximum shear stress criterion.

7.25 An ASTM-A36 steel shaft of length L carries a torque T and its own weight per unit length w (see Table B.1), as depicted in Figure P7.25. Determine the required shaft diameter D, using the maximum energy of distortion criterion with a safety factor of $n = 2.1$.

Given: $L = 6$ m, $T = 400$ N $\cdot$ m
Assumption: The bearings at the ends act as simple supports.

Figure P7.25

Sections 7.13 through 7.15

7.26 At a critical location in a component in tension, the load induced stresses are $\mu_l = 250$ MPa and $\sigma_l = 35$ MPa. What is the reliability R against yielding?

Given: The material yield strengths are $\mu_s = 400$ MPa and $\sigma_s = 30$ MPa.

7.27 Calculate the diameter d of a bar subjected to an axial tensile load P for a desired reliability of $R = 99.7\%$.

Given: The material yield strengths of $\mu_s = 50$ ksi and $\sigma_s = 5$ ksi. The loads of $\mu_l = 40$ kips and $\sigma_l = 6$ kips.

7.28 Determine the mean μ and the standard deviation σ for the grades of a sample of 12 students shown in the accompanying table.

n	1	2	3	4	5	6	7	8	9	10	11	12
x	77	85	48	94	80	60	65	96	70	86	69	82
y	78.2	82.1	60.3	91.5	84.6	70.7	68.4	90.8	75.0	92.5	61.8	80.1

| Notes: $n =$ number of students, $x =$ final examination grade, $y =$ course grade.

7.W Search and download the statistics shareware program on the website at www.mecheng.asme. org/database/STAT/MASTER.HTML for computing the mean and standard deviations for a normal distribution. Resolve Problem 7.28 using this program.

7.29 A total of 68 cold-drawn steel bars have been tested to obtain the 0.2% offset yield strength S_y in ksi. The results are as follows:

S_y	74	66	62	78	81	82	85	86	89	94
n	7	2	5	5	10	18	8	3	6	4

Based on normal distribution, determine

(a) The mean μ and standard deviation σ of the population.

(b) The reliability for a yield strength of $S_y = 75$ ksi.

7.30 A bar under a maximum load of 5 kips was designed to carry a load of 6 kips. The maximum load is applied with standard deviation of 600 lb and shaft strength standard deviation of 400 lb, both are normally distributed. Calculate the expected reliability.

7.31 A structural member is subjected to a maximum load of 20 kN. Assume that the load and strength have normal distributions with standard variations of 3 and 2.5 kN, respectively. If the member is designed to withstand a load of 25 kN, determine failure percentage that would be expected.

FATIGUE

Outline

8.1   INTRODUCTION

A member may fail at stress levels substantially below the yield strength of the material if it is subjected to time-varying loads rather than static loading. The phenomenon of progressive fracture due to repeated loading is called *fatigue*. Its occurrence is a function of the magnitude of stress and a number of repetitions, so it is called *fatigue failure*. We observe throughout this chapter that the fatigue strength of a component is significantly affected by a variety of factors. A fatigue crack most often is initiated at a point of high stress concentration, such as at the edge of a notch, or by minute flows in the material. Fatigue failure is of a brittle nature even for materials that normally behave in a ductile manner. The usual fracture occurs under tensile stress and with no warning. For combined fluctuating loading conditions, it is common practice to modify the static failure theories and material strength for the purposes of design.

The fatigue failure phenomenon was first recognized in the 1800s when railroad axles fractured after only a limited time in service. Until about the middle of the 19th century, repeated and static loadings were treated alike, with the exception of the use of safety factors. Poncelet's book in 1839 used the term *fatigue* owing to the fluctuating stress. At the present time, the development of modern high-speed transportation and machinery has increased the importance of the fatigue properties of materials. In spite of periodic inspection of parts for cracks and other flaws, numerous major railroad and aircraft accidents have been caused by fatigue failures. The basic mechanism associated with fatigue failure is now reasonably well understood, although research continues on its many details [1–31]. The complexity of the problem is such that rational design procedures for fatigue are difficult to develop. The great variation in properties makes it necessary to apply statistical methods in the evaluation of the fatigue strength.

8.2   THE NATURE OF FATIGUE FAILURES

The type of fracture produced in ductile metals subjected to fatigue loading differs greatly from that of fracture under static loading, considered in Section 2.3. In fatigue fractures, two regions of failure can be detected: the beachmarks (so termed because they resemble ripples left on sand by retracting waves) zone produced by the gradual development of the crack and the sudden fracture zone. As the name suggests, the latter region is the portion that fails suddenly when a crack reaches its size limit. Figure 8.1 depicts fatigue fracture surfaces of two common cross sections under high nominal stress conditions [1]. Note that the curvature of the beachmarks serves to indicate where the failure originates. The beachmarked area, also referred to as the *fatigue zone,* has a smooth, velvety texture. This contrasts with the sudden-fracture region, which is relatively dull and rough, looking like a static "brittle" type.

Microscopic examinations of ductile metal specimen subjected to fatigue stressing reveal that little if any distortion occurs, whereas failure due to static overload causes excessive distortion. The appearance of the surfaces of fracture greatly aids in identifying the cause of crack initiation to be corrected in redesign. For this purpose, numerous photographs and schematic representations of failed surfaces have been published in technical

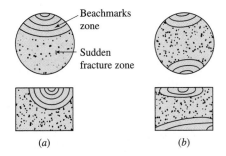

Figure 8.1 Schematic representation of fatigue fracture surfaces of circular and rectangular cross sections subjected to (a) tension-tension or tension-compression; (b) reversed bending [1].

Figure 8.2 Effect of state of stress on fatigue fracture of circular smooth and notched cross sections under various loading conditions [2].

literature [1, 5]. Figure 8.2 is a simplified sketch of the effect of state of stress on the origin, appearance, and location of fatigue fracture for variously loaded sections. For all axial and bending stress conditions, as well as the high-torsion smooth stress condition, the crack growth in the beachmarks region is indicated by curved vectors, starting from the point of crack initiation.

Note that the sudden fracture region can be a small portion of the original cross section, particularly under bending and torsion fatigue stressing. Unless interrupted by notches, the fatigue crack under bending is normal to the tensile stresses; that is perpendicular to the axis of a shaft. In torsional fatigue failures, the crack is at a 45° angle to the axis of a notch-free shaft (or spring wire) under high nominal stress conditions (Figure 8.2). Finally, note that, if cracks initiate at several circumferential points, the sudden fracture zone is more centered.

8.3   FATIGUE TESTS

To determine the strength of materials under the action of fatigue loads, four types of tests are performed: tension, torsion, bending, and combinations of these. In each test, specimens are subjected to repeated forces at specified magnitudes while the cycles or stress reversals to rupture are counted. A widely used fatigue testing device is the R. R. Moore high-speed rotating-beam machine (Figure 8.3a). To perform a test, the specimen is loaded with a selected weight W. Note that turning on the motor rotates the specimen, however, not the weight. There are various other types of fatigue testing machines [3]. A typical rotating-beam fatigue testing machine has an adjustable-speed spindle, operating

Figure 8.3   Bending fatigue: (a) schematic of the R. R. Moore rotating-beam fatigue testing machine; (b) standard round specimen; (c) completely reversed (sinusoidal) stress.

at speeds in ranges of 500–10,000 rpm. The device can apply a moment up to 200 lb·in. to the specimen.

REVERSED BENDING TEST

In the rotating-beam test, the machine applies a pure bending moment to the highly polished, so-called mirror finish, specimen of circular cross section (Figure 8.3b). As the specimen rotates at a point on its outer surface, the bending stress varies continuously from maximum tension to maximum compression. This fully or completely reversed bending stress can be represented on the stress S-cycles N axes by the curves of Figure 8.3c. It is obvious that the highest level of stress is at the center, where the smallest diameter is about 0.3 in. The large radius of curvature avoids stress concentration. Various standard types of fatigue specimens used, including those for axial, torsion, and bending stresses described in the ASTM manual on fatigue testing.

In some fatigue testing machines, constant-speed (usually 1750 rpm) motors are used, which give the sinusoidal type or fully reversed cyclic stress variation shown in the figure. It takes about one-half of a day to reach 10^6 cycles and about 40 days to reach 10^8 cycles on one specimen. A series of tests performed with various weights and using multiple specimens, carefully made to be nearly the same as possible, gives results or the fatigue data.

8.4 THE S-N DIAGRAMS

Fatigue test data are frequently represented in the form of a plot of *fatigue strength S* or completely reversed stress versus the number of cycles to failure or *fatigue life N* with a semi-logarithmic scale; that is, S against $\log N$. Sometimes data are represented by plotting S versus N or $\log S$ versus $\log N$. Inasmuch as fatigue failures originate at local points of relative weakness, usually the data contain a large amount of scatter. In any case, an average curve, tending to conform to certain generalized pattern, is drawn to represent the test results.

Figure 8.4 shows two typical S-N diagrams corresponding to rotating beam tests on a series of identical round steel and aluminum specimens subjected to reversed flexural loads

Figure 8.4   Fully reversed rotating-beam S-N curves for two typical materials.

of different magnitude. As may be seen from the figure, when the applied maximum stress is high, a relatively small number of cycles cause fracture. Note that, most often, fatigue data represent the mean values based on a 50% survival rate (50% reliability) of specimens.

THE ENDURANCE LIMIT AND FATIGUE STRENGTH

The endurance limit and fatigue strength are two important cyclic properties of the materials. The *fatigue strength* (S'_n), sometimes also termed *endurance strength,* is the completely reversed stress under which a material fails after a specified number of cycles. Therefore, when a value for the fatigue strength of a material is stated, it must be accompanied by the number of stress cycles. The *endurance limit* (S'_e) or fatigue limit is usually defined as the maximum completely reversed stress a material can withstand "indefinitely" without fracture. The endurance limit is therefore stated with no associated number of cycles to failure.

Bending Fatigue Strength

For *ferrous materials,* such as steels, the stress where the curve levels off is the endurance limit S'_e (Figure 8.4). Note that the curve for steel displays a decided break or "knee" occurring before or near 1×10^6 cycles. This value is often used as the basis of the endurance limit for steel. Beyond the point (S'_e, N_e), failure does not occur, even for an infinitely large number of loading cycles. At $N = N_f$ cycles, rupture occurs at approximately static fracture stress $S_f \approx 0.9 S_u$, where S_u is the ultimate strength in tension.

On the other hand, for *nonferrous metals,* notably aluminum alloys, the typical *S-N* curve indicates that the stress at failure continue to decrease as the number of cycles increases (Figure 8.4). That is, nonferrous materials do *not* show a break in their *S-N* curves, and as a result, a distinct endurance limit cannot truly be specified. For such materials, the stress corresponding to some arbitrary number of $5(10^8)$ cycles is commonly assigned as the endurance strength, S'_n.

It is necessary to make the assumption that most ferrous materials must not be stressed above the endurance limit S'_e if about 10^6 or more cycles to failure is required. This is illustrated in Figure 8.5, presenting test results for wrought steels having ultimate strength $S_u < 200$ ksi. Note the large scatter in fatigue life N corresponding to a given stress level and the small scatter in fatigue stress corresponding to a prescribed life. The preceding is

Figure 8.5   Fully reversed rotating beam *S-N* curve for wrought steels of $S_u < 200$ ksi with superimposed data points [20].

typical of fatigue strength tests. The figure also depicts that samples run at higher reversed stress levels break after fewer cycles, some (labeled in the dotted circle) do not fail at all prior to their tests being stopped (here at 10^7 cycles). The data are bracketed by solid lines. Interestingly, at the lower bound of the scatter band, the endurance limit can be conservatively estimated as $0.5S_u$ for design purposes. We mention that, for most wrought steels, the endurance limit varies between 0.45 and 0.60 of the ultimate strength.

Axial Fatigue Strength

Various types of fatigue servohydraulic testing machines have been developed for applying fluctuating axial compression. A specimen similar to that used in static tensile tests (see Section 2.3) is used. The most common types apply an axial reversed sinusoidal stress as shown in Figure 8.3c. A comparison of the strengths obtained for uniaxial fatigue stresses and bending fatigue stresses indicate that, in some cases, the former strengths are about 10 to 30% lower than the latter strengths for the same material [10–14]. Data for completely reversed axial loading test on AISI steel ($S_u = 125$ ksi) are shown in Figure 8.6. Observe the slope at around 10^3 cycles and the change to basically no slope at about 10^6 cycles corresponding to the endurance limit S'_e.

Torsional Fatigue Strength

A limited number of investigations have been made to determine the torsional fatigue strengths of materials using circular or cylindrical specimens subjected to complete stress reversal. For ductile metals and alloys, it was found that the torsional fatigue strength

Figure 8.6   Fully reversed axial S-N curve for AISI 4310 steel, showing breaks at about the low-cycle/high-cycle transition and an endurance limit [13].

(or torsional endurance limit) for complete stress reversal is about equal to 0.577 times the fatigue strength (or endurance limit) for complete bending stress reversal [19].

For brittle materials, the ratio of the fatigue strength in reversed bending to reversed torsion is higher and may approach the value of 1. The failure points for reversed bending and reversed torsion in biaxial-stress tests are similar to that for static loading failure (e.g., Figure 7.8). We conclude therefore that the relationship between torsional strength and bending strength in cyclic loading is the same as in the static loading case.

FATIGUE REGIMES

The stress-life (*S-N*) diagrams indicate different types of behavior as the cycles to failure increase. Two essential regimes are the low-cycle fatigue ($1 < N < 10^3$) and the high-cycle fatigue ($10^3 \leq N$). Note that there is no sharp dividing line between the two regions. In this text, we assume high-cycle fatigue starting at around $N = 10^3$ cycles. The infinite life begins at about 10^6 cycles, where failures may occur with only negligibly small plastic strains. The finite life portion of the curve is below about 10^7 cycles. The boundary between the infinite life and finite life lies somewhere between 10^6 and 10^7 cycles for steels, as shown in Figure 8.6. For low-cycle fatigue, the stresses are high enough to cause local yielding.

A fracture mechanics approach is applied to finite life problems in Section 8.14. We use the stress-life data in treating the high-cycle fatigue of components under any type of loading. For further details, see [5–7], which also provide discussion of the strain-life approach to fatigue analysis.

8.5   ESTIMATING THE ENDURANCE LIMIT AND FATIGUE STRENGTH

Many criteria have been suggested for interpreting fatigue data. No correlation exists between the endurance limit and such mechanical properties as yield strength, ductility, and so on. However, experiments show that the endurance limit, endurance strength, endurance limit in shear, and ultimate strength in shear can be related to the ultimate strength in tension.

Experimental values of ultimate strengths in tension S_u and shear should be used if they are available (see Appendix B). Recall from Section 2.10 that S_u can be estimated from a nondestructive hardness test. For reference purposes, Table 8.1 presents the relationships among the preceding quantities for a number of commonly encountered loadings and materials [12, 14]. Steel product manufacturers customarily present stress data of this kind in terms of the ultimate tensile strength, because S_u is the easiest to obtain and most reliable experimental measure of part strength.

If necessary, the ultimate strength of a steel can be estimated by Eq. (2.22) as $S_u = 3500H_B$ in MPa or $S_u = 500H_B$ in ksi. Here H_B denotes the Brinell hardness number (Bhn). It should be noted, however, that Eq. (8.1) can be relied on only up to Bhn values of about 400. Test data show that the endurance limit S'_e may or may not continue to increase for greater hardness, contingent on the composition of the steel [14].

In the absence of test data, the values given in the table can be used for preliminary design calculations. The relations are based on testing a polished laboratory specimen of a fixed size and geometric shape and on a 50% survival rate. Therefore, these data must

Table 8.1 Approximate fatigue strength of the specimens for fully reversed loads

Reversed Bending

Steels	$S'_e = 0.5S_u$ $S'_e = 700$ MPa (200 ksi)	$[S_u < 1400$ MPa (200 ksi)$]$ $[S_u \geq 1400$ MPa (200 ksi)$]$	**(8.1)**
Irons	$S'_e = 0.4S_u$ $S'_e = 160$ MPa (24 ksi)	$[S_u < 400$ MPa (60 ksi)$]$ $[S_u \geq 400$ MPa (60 ksi)$]$	**(8.2a)**
Aluminums	$S'_n = 0.4S_u$ $S'_n = 130$ MPa (19 ksi)	$[S_u < 330$ MPa (48 ksi)$]$ $[S_u \geq 330$ MPa (48 ksi)$]$	**(8.2b)**
Copper alloys	$S'_n = 0.4S_u$ $S'_n = 100$ MPa (14 ksi)	$[S_u < 280$ MPa (40 ksi)$]$ $[S_u \geq 280$ MPa (40 ksi)$]$	**(8.2c)**

Axial Loading

Steels	$S'_e = 0.45S_u$	**(8.3)**

Torsional Loading

Steels	$S'_{es} = 0.29S_u$	
Irons	$S'_{es} = 0.32S_u$	**(8.4)**
Copper alloys	$S'_{es} = 0.22S_u$	

Also

Steels	$S_{us} = 0.67S_u$	**(8.5)**

Note: S'_e = endurance limit, S'_{es} = endurance limit in shear, S_u = ultimate tensile strength, S_{us} = ultimate strength in shear, S'_n = endurance strength.

be modified by those factors adversely affecting results determined under laboratory conditions, discussed in the next section.

8.6 MODIFIED ENDURANCE LIMIT

The specimen used in the laboratory to determine the endurance limit is prepared very carefully and tested under closely controlled conditions. However, it is unrealistic to expect the endurance limit of a machine or structural member to match the values obtained in the laboratory. Material, manufacturing, environmental, and design conditions influence fatigue. Typical effects include the size, shape, and composition of the material; heat treatment; mechanical treatment; stress concentration; residual stresses; corrosion; temperature; speed; type of stress; and life of the member [3].

To account for the most important of these effects, various endurance limit modifying factors are used. These empirical factors, when applied to steel parts, lead to results of good accuracy, because most of the data on which they are based are obtained from testing steel specimens. The corrected or modified endurance limit, also referred to as the *endurance limit,* representing the endurance limit of the mechanical element, is defined as follows:

$$S_e = C_f C_r C_s C_t (1/K_f) S_e' \tag{8.6}$$

where

S_e = modified endurance limit

S_e' = endurance limit of the test specimen

C_f = surface finish factor

C_r = reliability factor

C_s = size factor

C_t = temperature factor

K_f = fatigue stress-concentration factor.

This working equation for the endurance limit is extremely important in fatigue problems. It should be used when actual fatigue test data that pertain closely to the particular application are not available. Equation (8.6) can be applied with great confidence to steel components, since the data on which correction factors rely usually come from testing steel specimens.

Recall from Section 8.4 that nonferrous materials show no break in their *S-N* curves; hence, a definite endurance limit of test specimen cannot be specified. For these materials, the fatigue strength S_n' replaces S_e' in Eq. (8.6). Likewise, for the case of reversed torsion loading, the modified endurance limit in shear S_{es} and endurance limit in shearing test specimen S_{es}', supersede S_e and S_e', respectively, in the equation.

8.7   ENDURANCE LIMIT REDUCTION FACTORS

The endurance limit modifying or reduction factors must be used in design application with great care, since the available information is related to specific specimens and tests. Only limited data are available for material strength in severe environments. Manufacturing processes can have significant effects on fatigue life characteristics. Most of the miscellaneous factors affecting the endurance limit, such as heat treatment, corrosion, mechanical surface treatment, and welding, have no quantitative values. Random variations occur in these factors, which are experimentally determined. The values assigned depend on the designer's experience and judgment.

The endurance limit modifying factors are nearly 1.0 for bending loads under 10^3 cycles. They increase progressively in some manner with the increase in the number of cycles. In the following brief discussion, some representative or approximate values for reduction factors are presented. These values are abstracted from [12, 13, 19].

SURFACE FINISH FACTOR

Fatigue strength is sensitive to the condition of the surface, because the maximum stresses occur here in bending and torsion. As already noted, the rotating-beam specimen is polished to a mirror finish to preclude surface imperfections serving as stress raisers; rougher finishes lower the fatigue strength.

The surface finish factor C_f, which depends on the quality of the finish and tensile strength, may be expressed in the form

$$C_f = AS_u^b \tag{8.7}$$

where the ultimate strength S_u is in either MPa or ksi. Table 8.2 presents the values of the factor A and exponent b for a variety of finishes applied to steels. We observe from Eq. (8.7) that (since $b < 0$) the values of C_f decrease with increases in tensile strength S_u. The surface factor for "mirror-polish" finish steels equals approximately 1, $C_f = 1$.

Equation (8.7) has the advantage of being computer programmable and eliminating the need to refer to charts such as Figure 8.7. Note that the surface conditions in this figure are

Table 8.2 Surface finish factors C_f

Surface finish	A		b
	MPa	(ksi)	
Ground	1.58	(1.34)	−0.085
Machine or cold drawn	4.51	(2.7)	−0.265
Hot rolled	57.7	(14.4)	−0.718
Forged	272.0	(39.9)	−0.995

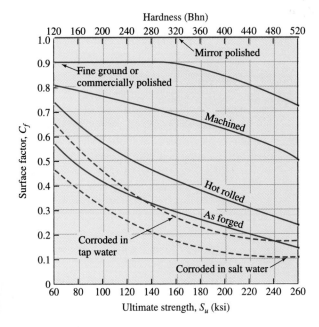

Figure 8.7 Surface factors for various finishes on steel [20].

poorly defined (e.g., the machined surface texture or degree of roughness). Interestingly, Figure 8.7 shows that corrosive environments drastically reduce the endurance limit.

Table 8.2 may also be applied to aluminum alloys and other ductile metals with caution. It is important to mention that testing of actual parts under service loading conditions must be done in critical applications. The surface factor for ordinary cast irons is also taken as approximately 1, $C_f = 1$, since their internal discontinuities dwarf the effects of a rough surface.

RELIABILITY FACTOR

The factor of reliability C_r accounts for material variation in fatigue data and depends on survival rate. It is defined by the following commonly used formula:

$$C_r = 1 - 0.08z \tag{8.8}$$

The quantity z is the number of standard deviations, discussed in Section 7.14.

For a required survival rate or percent reliability, Figure 7.15 gives the corresponding z, then using Eq. (8.8) we calculate a reliability factor. Table 8.3 presents a number of values of the C_r. Observe that a 50% reliability has a factor of 1 and the factor reduces with increasing survival rate.

SIZE FACTOR

The influence of size on fatigue strength can be a significant factor. The endurance limit decreases with increasing member size. This is owing to the probability that a larger part is more likely to have a weaker metallurgical defect at which a fatigue crack will start. There is notable scatter in reported values of the size factor C_s. Various researchers have suggested different formulas for estimating it.

The approximate results, for *bending* and *torsion* of a part of diameter D may be stated as follows:

$$C_s = \begin{cases} 0.85 & (13 \text{ mm} < D \leq 50 \text{ mm}) \qquad (\frac{1}{2} < D \leq 2 \text{ in.}) \\ 0.70 & (D > 50 \text{ mm}) \qquad\quad (D > 2 \text{ in.}) \end{cases} \tag{8.9}$$

Table 8.3   Reliability factors

Survival rate (%)	C_r
50	1.00
90	0.89
95	0.87
98	0.84
99	0.81
99.9	0.75
99.99	0.70

This applies to cylindrical parts, and for members of other shapes, few consistent data are available. For a rotating part of rectangular cross section of width b and depth h, use in the foregoing equation [13]:

$$D = 0.8(bh)^{1/2} \tag{8.10}$$

Prudent design would suggest employing a factor $C_s = 0.7$, lacking other information. Note that, for *axial loading,* there is no size effect: $C_s = 1$.

TEMPERATURE FACTOR

Temperature effects vary with the material in most cases, and values of ultimate strength should be modified before determining the endurance limit S'_e in Eq. (8.6). Alternatively, for steels, a temperature factor C_t can be approximated at moderately high temperatures by the formula [12]:

$$C_t = \begin{cases} 1 & T \leq 450°C \quad (840°F) \\ 1 - 0.0058(T - 450) & 450°C < T \leq 550°C \\ 1 - 0.0032(T - 840) & 840°F < T \leq 1020°F \end{cases} \tag{8.11}$$

A more accurate estimation of C_t is presented in [13, 17]. Unless otherwise specified, we assume throughout the text the operating temperature is normal or room temperature and take $C_t = 1$.

FATIGUE STRESS CONCENTRATION FACTOR

As pointed out earlier, the stress concentration is a very significant factor in failure by fatigue. For dynamic loading, the theoretical stress-concentration factor K_t (see Section 3.14) needs to be modified on the basis of the notch sensitivity of the material. *Notch* is a generic term in this context and can be a hole, a groove, or a fillet.

The fatigue stress concentration factor may be defined as

$$K_f = \frac{\text{endurance limit of notch-free specimen}}{\text{endurance limit of notched specimen}} \tag{8.12}$$

The tests show that K_f is often equal to or less than the K_t, owing to internal irregularities in the material structure. Therefore, even unnotched samples may suffer from these internal notches. An extreme case in point is gray cast iron.

The foregoing situation is dealt with by using a notch factor. The two stress concentration factors are related by the ratio the *notch sensitivity q:*

$$q = \frac{K_f - 1}{K_t - 1} \tag{8.13a}$$

This expression can be written in the form

$$K_f = 1 + q(K_t - 1) \tag{8.13b}$$

We observe from Eq. (8.13) that q varies between 0 (giving $K_f = 1$) and 1 (giving $K_f = K_t$). Generally, the more ductile the material response, the less notch sensitive it is. Materials showing brittle behavior are more notch sensitive. Obviously, notch sensitivity also depends on the notch radius. Contrary to K_t, as notch radii approach 0, the q decreases and approaches 0.

Figures 8.8 and 8.9 provide approximate data for steels and 2024 aluminum alloys subjected to reversed bending, reversed axial loads, and reversed torsion [5, 19]. Note

Figure 8.8 Fatigue notch sensitivity curves for bending and axial loads [19].

Figure 8.9 Fatigue notch sensitivity curves for torsion [19].

that the actual test data from which the curves were plotted exhibit a large amount of scatter. So, it is always safe to use $K_f = K_t$ when there is doubt about the true value of q. The curves show that q is not far from unity for large notch radii. For larger notch radii, use the values of q corresponding to $r = 0.16$ in. (4 mm). In concluding this discussion, we note that the notch sensitivity of cast iron is very low, $0 < q \leq 0.20$, depending on tensile strength. If one is uncertain, it would be conservative to use a value $q = 0.20$ for all grades of cast iron.

Determining the Endurance Limit of a Torsion Bar | **EXAMPLE 8.1**

A round torsion bar machined from steel is under reversed torsional loading. Because of the design of the ends, a fatigue stress concentration factor K_f exists. Estimate the modified endurance limit.

Given: The diameter of the bar $d = 1\frac{5}{8}$ in. and $K_f = 1.2$. The operating temperature is 500°C maximum.

Assumption: Reliability is 98%.

Design Decision: The bar is made of AISI 1050 cold-drawn steel.

Solution: From Table B.3, we find the ultimate strength in tension as $S_u = 100$ ksi. Then, applying Eq. (8.4), the endurance limit of the test specimen is

$$S'_{es} = 0.29S_u = 0.29(100) = 29 \text{ ksi}$$

By Eq. (8.7) and Table 8.2, the surface finish factor is

$$C_f = AS_u^b = 2.7(100)^{-0.265} = 0.80$$

The reliability factor corresponding to 98% is $C_r = 0.84$ (Table 8.3). Using Eq. (8.9), the size factor $C_s = 0.85$. Applying Eq. (8.11),

$$C_t = 1 - 0.0058(500 - 450) = 0.71$$

Hence, the endurance limit for design is found to be

$$S_{es} = C_f C_r C_s C_t (1/K_f) S'_{es}$$
$$= (0.80)(0.84)(0.85)(0.71)(1/1.2)(29)$$
$$= 9.8 \text{ ksi}$$

(a)

EXAMPLE 8.2   Finding the Endurance Limit for a Stepped Shaft in Reversed Bending

Rework Example 8.1 for the condition that the critical point on the shaft is at a diameter change from d to D with a full fillet where there is reversed bending and no torsion, as shown in Figure 8.10.

Figure 8.10   Example 8.2.

Given: $d = 1\frac{5}{8}$ in., $D = 1\frac{7}{8}$ in., $S_u = 100$ ksi

Solution: We now have, by Eq. (8.1): $S'_e = 0.5(100) = 50$ ksi. From the given dimensions, the full fillet radius $r = (1\frac{7}{8} - 1\frac{5}{8})/2 = 0.125$ in. Therefore,

$$\frac{r}{d} = \frac{0.125}{1.625} = 0.08, \qquad \frac{D}{d} = \frac{1.875}{1.625} = 1.15$$

Referring to Figure C.9, $K_t = 1.7$. For $r = 0.125$ in. and $S_u = 100$ ksi, by Figure 8.8, $q = 0.82$. Hence, through the use of Eq. (8.13b),

$$K_f = 1 + q(K_t - 1)$$

$$= 1 + 0.82(1.7 - 1) = 1.57$$

The endurance limit, given by Eq. (a) of Example 8.1, becomes

$$S_e = C_f C_r C_s C_t (1/K_t) S'_e = (0.80)(0.84)(0.85)(0.71)(1/1.57)(50) = 12.92 \text{ ksi}$$

8.8   FLUCTUATING STRESSES

Any loads varying with time can actually cause fatigue failure. The type of these loads may vary amply from one application to another. Hence, it is necessary to determine the fatigue resistance of parts corresponding to stress situations other than complete reversals discussed so far.

A common fluctuating stress pattern consists of an alternating (usually sinusoidal) stress superimposed on a uniform mean stress (Figure 8.11a). This loading is typical for an engine valve spring that is preloaded at installation, then further compressed when the valve is opened. The case in which the mean stress is 0 is called *fully* or *completely reversed stress* (Figure 8.11b), discussed in Section 8.3. Figure 8.11c shows repeated stress, where the minimum value equals 0. Figure 8.11d shows pulsating stress, varying between 0 and the maximum value with each application of load, as on the teeth of gears. Note that the shape of the wave of the stress-time relation has no important effect on the fatigue failure, so usually the relation is schematically depicted as a sinusoidal or sawtooth wave.

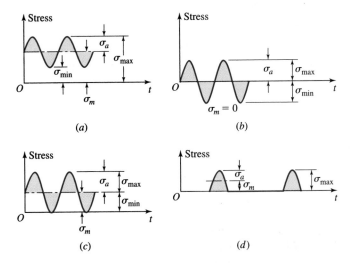

Figure 8.11 Some cyclic stress-time relations: (a) fluctuating; (b) completely reversed; (c) repeated; (d) pulsating.

Irrespective of the form of the stress-time relation, the stress varies from a maximum stress σ_{max} to a minimum stress σ_{min}. Therefore, the definitions of *mean stress* and range or *alternating stress* are

$$\sigma_m = \frac{1}{2}(\sigma_{max} + \sigma_{min})$$

$$\sigma_a = \frac{1}{2}(\sigma_{max} - \sigma_{min})$$

(8.14)

Clearly, these components of the fluctuating stress are also independent of the shape of the stress-time curve. Two ratios can be formed:

$$R = \frac{\sigma_{min}}{\sigma_{max}}, \qquad A = \frac{\sigma_a}{\sigma_m}$$

Here R is the stress ratio and A is the amplitude ratio. When the stress is fully reversed ($\sigma_m = 0$), we have $R = 1$ and $A = \infty$.

The mean stress is analogous to a static stress, which may have any value between σ_{max} and σ_{min}. We see that the presence of a mean stress component can have a significant effect on the fatigue life. The alternating stress represents the amplitude of the fluctuating stress. We have occasion to apply the subscripts (a and m) of these components to shear stresses as well as normal stresses.

8.9 THEORIES OF FATIGUE FAILURE

To predict whether the state of stress at a critical point in an element would result in failure, a fatigue criterion based on the mean and alternating stresses is used. Such a theory utilizes static and cyclic material characteristics. Table 8.4 presents frequently employed

Table 8.4   Failure criteria for fatigue

Fracture theory	Goodman	Gerber	SAE
Equation	$\dfrac{\sigma_a}{S_e} + \dfrac{\sigma_m}{S_u} = 1$	$\dfrac{\sigma_a}{S_e} + \left(\dfrac{\sigma_m}{S_u}\right)^2 = 1$	$\dfrac{\sigma_a}{S_e} + \dfrac{\sigma_m}{S_f} = 1$

Yield theory	Soderberg	Modified Goodman	
Equation	$\dfrac{\sigma_a}{S_e} + \dfrac{\sigma_m}{S_y} = 1$	$\dfrac{\sigma_a}{S_e} + \dfrac{\sigma_m}{S_u} = 1 \quad \left(\text{for} \quad \dfrac{\sigma_a}{\sigma_m} \geq \beta\right)$	
		$\dfrac{\sigma_a + \sigma_m}{S_y} = 1 \quad \left(\text{for} \quad \dfrac{\sigma_a}{\sigma_m} \leq \beta\right)$	

Subscript notation: a = alternating, y = static tensile yield, m = mean, u = static tensile ultimate, e = modified endurance limit, f = fracture. Material constant, $\beta = S_e(S_u - S_y)/S_u(S_y - S_e)$.

fatigue failure theories or *criteria,* also called the mean stress-alternating stress relations [20–26]. Equations as written apply only to materials with an endurance limit; either S_e or S_e' can be used in these relationships. For a finite life (a given number of cycles) the corresponding fatigue strength S_n may be substituted for the endurance limit.

The criteria in the table together with given material properties form the basis for practical fatigue calculations for members subjected to a simple fluctuating loading. Note that the modified Goodman criterion is algebraically more involved, as there are two inequalities to check rather than one in the other relations. In the case of combined fluctuating loading, the static failure theories are modified according to a mean stress-alternating stress relation listed in Table 8.4, as will be shown in Section 8.12.

Figure 8.12 shows the foregoing relationships, plotted on mean stress (σ_m) versus alternating stress (σ_a) axes. A *fatigue failure diagram* of this type is usually constructed for

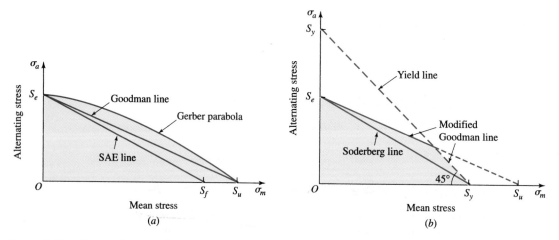

Figure 8.12   Fatigue diagrams showing various theories of failure: (a) fracture criteria; (b) yield criteria.

analysis and design purposes; it is easy to use, and the results can be scaled off directly. For each criterion, points on or inside the respective line guards against failure. The mean stress axes of the diagrams have the fracture strength S_f, ultimate strength S_u, and yield strength S_y. Clearly, the yield strength plotted on the ordinate as well as endurance limit in Figure 8.12b indicate that yielding rather than fatigue might be the criterion of failure. The yield line connecting S_y on both axes shown in the figure serves as a limit on the first cycle of stress.

8.10 COMPARISON OF THE FATIGUE CRITERIA

A comparison of the failure theories for fatigue may be made referring to Figure 8.12. We see from Figure 8.12a that the Gerber criterion leads to least conservative results for fracture. The Gerber (parabolic) line is a good fit to experimental data, making it useful for the analysis of failed parts. The Goodman criterion is more conservative than the SAE criterion. For hard steels, both theories give identical solutions, since for brittle materials $S_u = S_f$.

Figure 8.12b shows that the modified Goodman criterion resembles the Soderberg criterion, except that the former is slightly less conservative. Two line segments form the modified Goodman failure line, as shown in the figure. The 45° line segment implies failure when the maximum mean stress exceeds the yield strength. The modified Goodman line is particularly better for highly localized yielding occurring in many machine parts. Note that the Soderberg theory eliminates the need to involve the yield line.

Recall from Section 8.1 that, fatigue failures appear brittle, even if the material shows some ductility in a static tension test, as high stresses and yielding are localized near the crack. Thus, the *Goodman criterion,* which gives reasonably good results for brittle materials while conservative values for ductile materials, is a realistic scheme for *most materials.* For most metals, the Soderberg relation also leads to conservative estimates. Both theories are in widespread use for mild steel. In this text, the Goodman criterion is used to derive readily the basic equations for the design and analysis of common components. The easy and quick graphical approach is applied for the modified Goodman criterion. The Soderberg line is employed less often.

EXAMPLE 8.3

Determining the Allowable Fully Reversed Load of an Actuating Rod

A round rod is subjected to an axial tensile force F_m. Calculate the limiting value of the completely reversed load F_a that can be applied with a $\frac{1}{2}$ in. eccentricity without causing fatigue failure at 10^6 cycles.

Given: The diameter of the rod $D = 2$ in., and $F_m = 40$ kips.

Assumptions: The rod is made of ASTM-A36 steel having $S_y = 36$ ksi, $S_u = 58$ ksi (Table B.1), $S_n = 30$ ksi at 10^6 cycles.

Design Decisions: Apply the Soderberg and Goodman criteria.

Solution: The mean and maximum alternating stresses are

$$\sigma_m = \frac{F_m}{A} = \frac{40}{\pi (1)^2} = 12.73 \text{ ksi}$$

$$\sigma_a = \frac{M_a c}{I} = \frac{0.5 F_a (1)}{\pi (1)^4/4} = 0.637 F_a$$

Soderberg criterion.   Substituting these and the given numerical values into Table 8.4,

$$\frac{0.637 F_a}{30 \times 10^3} + \frac{12.73 \times 10^3}{36 \times 10^3} = 1 \quad \text{or} \quad F_a = 30.4 \text{ kips}$$

Goodman criterion.   Similar to the preceding, we now have

$$\frac{0.637 F_a}{30 \times 10^3} + \frac{12.73 \times 10^3}{58 \times 10^3} = 1 \quad \text{or} \quad F_a = 36.8 \text{ kips}$$

Comments: According to the Goodman theory, the eccentric load that can be carried by the rod is thus about 17% larger than on the basis of the Soderberg theory.

8.11   DESIGN FOR SIMPLE FLUCTUATING LOADS

When tensile stress at a point occurs by an alternating stress σ_a and a mean stress σ_m as shown in Figure 8.11a, both these components contribute to failure. The failure line (Figure 8.12) is an approximate depiction of this effect. Usually, a fatigue theory of failure is not applied to problems where mean stress is negative.

As noted previously, the Goodman criterion may be used safely with almost any material for which the endurance limit S_e and ultimate strength S_u are known. For design purposes, these quantities are replaced by S_e/n and S_u/n, respectively, where n represents the factor of safety. In so doing, the Goodman criterion, given in Table 8.4, becomes

$$\frac{\sigma_a}{S_e} + \frac{\sigma_m}{S_u} = \frac{1}{n} \tag{8.15}$$

where σ_a and σ_m are defined by Eq. (8.14). We note that, if a stress concentration exists at the cross section for which the stresses are computed, for ductile materials, it is commonly neglected as far as the mean stress is concerned (see Section 3.13). However, stress concentration must be taken into account for calculating the modified endurance limit S_e from Eq. (8.6).

The Goodman criterion, Eq. (8.15), may be rearranged in the following convenient form:

$$\frac{S_u}{n} = \sigma_m + \frac{S_u}{S_e}\sigma_a \qquad (8.16)$$

The right-hand side of this equation can then be considered the static equivalent of the fluctuating state of stress. Hence, we define the *equivalent normal stress* as

$$\sigma_e = \sigma_m + \frac{S_u}{S_e}\sigma_a \qquad (8.17)$$

Although Eq. (8.17) refers to normal stress, the development could have been made equally well for shear stress by replacing σ by τ [11]. The equation for *equivalent shear stress* is then

$$\tau_e = \tau_m + \frac{S_u}{S_e}\tau_a \qquad (8.18)$$

In this expression, it is assumed that

$$\frac{S_{us}}{S_{es}} \approx \frac{S_u}{S_e} \qquad (8.19)$$

because data for the ultimate strength in shear S_{us} and modified endurance limit in shear S_{es} are ordinarily not available. However, recall from Sections 8.4 through 8.6 that there are methods for estimating these quantities.

In some situations, a member is to withstand a given ratio of the alternating load to mean load. Then, a solution may readily be obtained if the ratio of alternating stress to mean stress can be determined from some known stress-load relationship. In such cases, it is suitable to recast Eq. (8.16) into the form

$$\sigma_m = \frac{S_u/n}{\dfrac{\sigma_a}{\sigma_m}\dfrac{S_u}{S_e} + 1} \qquad (8.20)$$

Once mean stress σ_m is obtained, the stress-load relationship is used to determine the required dimension of the element. For shear stress, the foregoing equation may be expressed as

$$\tau_m = \frac{S_{us}/n}{\dfrac{\tau_a}{\tau_m}\dfrac{S_u}{S_e} + 1} \qquad (8.21)$$

A reasonable design procedure ensures a significant safety factor against fatigue failure in the material. For a fluctuating stress, by inversion of Eq. (8.16) we express the factor of safety n as follows:

$$n = \frac{S_u}{\sigma_m + \dfrac{S_u}{S_e}\sigma_a} \qquad (8.22)$$

The equations for the safety factor become, with simple steady stress,

$$n = \frac{S_u}{\sigma_m} \qquad (8.23)$$

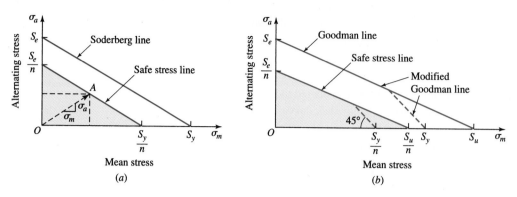

Figure 8.13   (a) Soderberg diagram; (b) Goodman diagrams.

and with simple alternating stress,

$$n = \frac{S_e}{\sigma_a} \tag{8.24}$$

It should be pointed out that, Eqs. (8.15) through (8.23) could also be written on the basis of the Soderberg criterion by substituting S_y for S_u.

DESIGN GRAPHS OF FAILURE CRITERIA

The graphical representation of the Soderberg, Goodman, and modified Goodman theories are shown in Figure 8.13. Note that other criteria listed in Table 8.4 may be plotted similarly. The Soderberg failure line is drawn between the yield point and the endurance limit on mean stress-range stress coordinates (Figure 8.13a). It is an approximate representation of the fatigue.

The *safe stress* line through any point $A(\sigma_m, \sigma_a)$ is constructed parallel to the Soderberg line. This line is the locus of all sets of σ_m and σ_a stresses having a factor of safety n. Any point on or below the safe line represents safe loading. The Goodman criteria are interpreted in a like manner (Figure 8.13b). The graphical approach permits rapid solution of the mean or range stress and provides an overview of the failure theory. Graphical solutions serve as a check to analytically obtained results.

The Examples that follow illustrate the application of the Goodman criteria to design of members under a simple fluctuating loading.

EXAMPLE 8.4	Design of a Cylindrical Pressure Vessel for Fluctuating Loading

A thin-walled cylindrical pressure vessel of diameter d is subjected to an internal pressure p varying continuously from p_{min} to p_{max}. Determine the thickness t for an ultimate strength S_u, a modified endurance limit S_e, and a safety factor of n.

Given: $D = 1.5$ m, $p_{min} = 0.8$ MPa, $p_{max} = 4$ MPa,
$S_y = 300$ MPa, $S_u = 400$ MPa, $S_e = 150$ MPa

Design Decision: The Goodman theories, based on maximum normal stress and a safety factor of $n = 2$, are used.

Solution: The state of stress on the cylinder wall is considered to be biaxial (see Section 3.4). Maximum principal stress, that is, tangential stress, in the cylinder has the mean and range values

$$\sigma_m = \frac{p_m r}{t}, \qquad \sigma_a = \frac{p_a r}{t} \tag{a}$$

where

$$p_m = \frac{1}{2}(p_{max} + p_{min}) = \frac{1}{2}(4 + 0.8) = 2.4 \text{ MPa}$$

$$p_a = \frac{1}{2}(p_{max} - p_{min}) = \frac{1}{2}(4 - 0.8) = 1.6 \text{ MPa}$$

Since stresses are proportional to pressures, we have

$$\frac{\sigma_a}{\sigma_m} = \frac{p_a}{p_m} = \frac{1.6}{2.4} = \frac{2}{3}$$

Substitution of the given data into Eq. (8.20) gives

$$\sigma_m = \frac{400/2}{\frac{2}{3}\frac{400}{150} + 1} = 72 \text{ MPa}$$

Using Eq. (a), we have

$$t = \frac{p_m r}{\sigma_m} = \frac{2.4(750)}{72} = 25 \text{ mm}$$

This is the minimum safe thickness for the pressure vessel.

Alternatively, a graphical solution of σ_m by the modified Goodman criterion is obtained by plotting the given data to scale, as shown in Figure 8.14. We observe from the figure that the locus of points representing σ_m and σ_a for any thickness is a line through the origin with a slope of 2/3. Its intersection with the safe stress line gives the state of stress σ_m, σ_a for the minimum safe value of the thickness t. The corresponding value of the mean stresses is $\sigma_m = 72$ MPa.

Figure 8.14   Example 8.4. Goodman criteria applied to design of pressure vessel.

EXAMPLE 8.5   Determining the Safety Factor against Fatigue Failure of a Tensile Link

A tensile link of thickness t with two fillets is subjected to a load fluctuating between P_{min} and P_{max} (Figure 8.15). Calculate the factor of safety n for unlimited life.

Figure 8.15   Example 8.5.

Given: $D = 120$ mm, $d = 80$ mm, $r = 16$ mm, $t = 15$ mm,
$P_{min} = 90$ kN, $P_{max} = 210$ kN

Design Decisions: The link is made of steel with $S_u = 700$ MPa. The fillets and adjacent surfaces are ground. A reliability of 99.9% is desired.

Solution: The minimum cross-sectional area equals $A = 80 \times 15 = 1200$ mm^2. The maximum mean and range stresses are

$$\sigma_m = \frac{(210 + 90)(10^3)}{2(1200 \times 10^{-6})} = 125 \text{ MPa}$$

$$\sigma_a = \frac{(210 - 90)(10^3)}{2(1200 \times 10^{-6})} = 50 \text{ MPa}$$

Referring to Figure C.1, for $D/d = 1.5$ and $r/d = 0.2$, we obtain $K_t = 1.72$. Inasmuch as the given 16-mm fillet radius is large, we use the value of the notch sensitivity for the steel having $S_u = 700$ MPa, corresponding to $r = 4$ mm in Figure 8.8; that is $q = 0.85$. Hence, $K_f = 1 + 0.85(1.72 - 1) = 1.61$.

For axial loading there is no size factor, $C_s = 1.0$. Corresponding to a round surface finish, from Eq. (8.7) and Table 8.2,

$$C_f = A(S_u)^b$$

$$= 1.58(700)^{-0.085} = 0.91$$

By Table 8.3, for 99.9% material reliability, $C_r = 0.75$. The temperature is not elevated, $C_t = 1.0$. The modified endurance limit, using Eqs. (8.3) and (8.6),

$$S_e = C_f C_r C_s C_t (1/K_f)(0.45 S_u)$$

$$= (0.91)(0.75)(1.0)(1.0)(1/1.61)(0.45 \times 700) = 133.5 \text{ MPa}$$

The factor of safety, applying Eq. (8.22), is therefore

$$n = \frac{700}{125 + \frac{700}{133.5}(50)} = 1.81$$

Comments: If the load is well controlled and there is no impact, this factor guards the link against the fatigue failure.

Case Study 8-1 | CAMSHAFT FATIGUE DESIGN OF INTERMITTENT-MOTION MECHANISM

Figure 3.40 illustrates a rotating camshaft of an intermittent-motion mechanism in its peak lift position. The cam exerts a force P on the follower, because of a stop mechanism (not shown), only during less than half a shaft revolution. Calculate the factor of safety for the camshaft according to the Goodman criterion.

Given: The geometry is known and the shaft supports a pulsating force with P_{max} and P_{min}. The material of all parts is AISI 1095 steel, carburized on the cam surface oil quenched and tempered (OQ&T) at 650°C. The fillet and adjacent surfaces are fine ground.

Data

$P_{max} = 1.6$ kips, $\qquad P_{min} = 0$

$S_u = 130$ ksi, $\qquad S_y = 80$ ksi $\qquad$ (from Table B.4)

$L_1 = 2.8$ in., $\qquad L_2 = 3.2$ in.,

$L_3 = 0.5$ in., $\qquad L_4 = 1.5$ in.

$L_5 = L_1 + \dfrac{1}{2}(L_3 + L_4) = 3.8$ in.,

$L_6 = L_2 + \dfrac{1}{2}(L_3 + L_4) = 4.2$ in.

$D_s = 1$ in., $\qquad D_c = 1.6$ in.,

$r_c = 1.5$ in., $\qquad r = 0.1$ in.

$I/c = \pi D_s^3/32 = 98.175(10^{-3})$ in.^{3}

Assumptions:

1. Bearings act as simple supports.
2. The operating temperature is normal.
3. The torque can be regarded negligible.
4. A material reliability of 99.9% is required.

Solution: See Figures 3.40 (repeated here) and 8.16.

Alternating and mean stresses. The reactions at the supports A and B are determined by the conditions of equilibrium as

$$R_A = \frac{L_6}{L_5 + L_6} P_{max}$$

$$= \frac{4.2}{8}(1600) = 840 \text{ lb}$$

$$R_B = P_{max} - R_A = 760 \text{ lb}$$

and noted in Figure 8.16a.

Figure 3.40 (repeated) Layout of camshaft and follower of an intermittent-motion mechanism.

(continued)

Case Study   (CONTINUED)

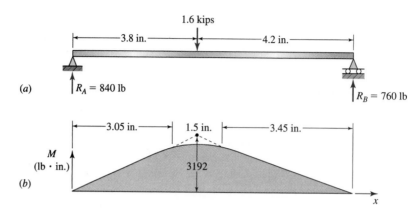

Figure 8.16   Diagrams of camshaft shown in Figure 3.40: (a) load; (b) bending moment.

The plot of the moment diagram, from a "maximum" moment of $760 \times 4.2 = 3192$ lb·in., is shown in Figure 8.16b. We observe that the moment on the right side,

$$M = R_B \left(L_6 - \frac{1}{2} L_4 \right)$$

$$= 760(3.45) = 2622 \text{ lb·in.}$$

is larger than (2562 lb·in.) at the left side. We have

$$\sigma_{\text{max}} = \frac{M}{I/c} = \frac{2622}{98.175(10^{-3})} = 26.71 \text{ ksi}$$

$$\sigma_{\text{min}} = 0$$

Equation (8.14) results in

$$\sigma_a = \sigma_m = \frac{26.71}{2} = 13.36 \text{ ksi}$$

Stress concentration factors.   The step in the shaft is asymmetrical. Stress at point E is influenced by the radius $r = 1.5$ in. (equivalent to a diameter of 3.0 in.) and at point F by the 0.8 in. cam radius (equivalent to $D_c = 1.6$ in. diameter). Hence, we obtain the following values:
    At point E,

$$\frac{r}{d} = \frac{0.1}{1.0} = 0.1, \qquad \frac{D}{d} = \frac{3.0}{1.0} = 3.0,$$

$$K_t = 1.8 \quad \text{(from Figure C.9)}$$

At point F,

$$\frac{r}{d} = \frac{0.1}{1.0} = 0.1, \qquad \frac{D}{d} = \frac{1.6}{1.0} = 1.6,$$

$$K_t = 1.7 \quad \text{(from Figure C.9)}$$

For $r = 0.10$ in. and $S_u = 130$ ksi, by Figure 8.8, $q = 0.86$. It follows, from Eq. (8.13b), that

$$(K_f)_E = 1 + 0.86(1.8 - 1) = 1.69$$

$$(K_f)_F = 1 + 0.86(1.7 - 1) = 1.60$$

Comments: Note that the maximum stress in the shaft is well under the material yield strength. The stress concentration at E is only 5% larger than that at F. Therefore, fatigue failure is expected to begin at point F, where the stress pulses are tensile, and compressive at E.
    Modified endurance limit.   Through the use of Eq. (8.6), we have

$$S_e = C_f C_r C_s C_t \left(\frac{1}{K_f} \right) S'_e$$

where

$$C_f = 1.34(130)^{-0.085} = 0.886 \quad \text{(from Table 8.2)}$$

$$C_r = 0.75 \quad \text{(by Table 8.3)}$$

$$C_s = 0.85 \quad \text{(from Eq. 8.9)}$$

$$C_t = 1.0 \quad \text{(room temperature)}$$

$$K_f = 1.6$$

$$S'_e = 0.5(130) = 65 \text{ ksi} \quad \text{(by Eq. 8.1)}$$

Case Study (CONCLUDED)

Hence,

$$S_e = (0.886)(0.75)(0.85)(1.0)\left(\frac{1}{1.6}\right)(65)$$

$$= 22.95 \text{ ksi}$$

Factor of safety. The safety factor guarding against fatigue failure at point F is determined using Eq. (8.22):

$$n = \frac{S_u}{\sigma_m + \dfrac{S_u}{S_e}\sigma_a} = \frac{130}{13.36 + \frac{130}{22.95}(13.36)} = 1.46$$

Comments: If the load is properly controlled so that there is no impact, the foregoing factor seems well sufficient. Inasmuch as lift motion is involved, the deflection needs to be checked accurately by FEA.

8.12 DESIGN FOR COMBINED FLUCTUATING LOADS

In numerous practical situations, structural and machine components are subject to combined fluctuating bending, torsion, and axial loading; for example, propeller shafts, crankshafts, airplane wings, and so on. Often, under conditions of a general cyclic state of stress, static failure theories are modified for analysis and design. In this section, we consider the maximum shear stress and maximum distortion energy theories associated with the Goodman criterion. Note that the expressions to follow can also be written based on the Soderberg criterion, substituting the yield strength S_y for the ultimate strength S_u, as required.

For combined stresses, we treat the fatigue effect first by defining equivalent values of each principal stress. We designate the mean component of σ_1 owing to a steady loading by σ_{1m} and the alternating component due to the reversed load by σ_{1a}. Based on the Goodman relation, the equivalent principal stresses are then

$$\sigma_{ie} = \sigma_{im} + \frac{S_u}{S_e}\sigma_{ia} \qquad (i = 1, 2, 3) \tag{8.25}$$

These equivalent values are then used in the expressions of static failure criteria applied to fatigue loading.

Equation (7.6), with S_u replacing the quantity S_y used thus far, and Eq. (8.25) lead to the maximum shear stress theory applied to fatigue loading. Therefore,

$$\frac{S_u}{n} = (\sigma_1 - \sigma_3)_e = \sigma_{1m} - \sigma_{3m} + \frac{S_u}{S_e}(\sigma_{1a} - \sigma_{3a}) \tag{8.26}$$

Clearly, it is assumed that this modified static yield failure theory applies to brittle behavior as well. For the special case, where $\sigma_y = \sigma_z = \tau_{yz} = \tau_{xz} = 0$, Eq. (7.11) results in

$$\frac{S_u}{n} = \left(\sigma_x^2 + 4\tau_{xy}^2\right)_e^{1/2} \tag{8.27}$$

Introducing the equivalent stresses σ_e and τ_e from Eqs. (8.17) and (8.18) for σ_x and τ_{xy}, respectively, into Eq. (8.27), we have the maximum shear stress theory incorporated with

Goodman criterion:

$$\frac{S_u}{n} = \left[\left(\sigma_{xm} + \frac{S_u}{S_e}\sigma_{xa} \right)^2 + 4\left(\tau_{xym} + \frac{S_u}{S_e}\tau_{xya} \right)^2 \right]^{1/2} \tag{8.28}$$

Similarly, Eqs. (8.13a) and (8.25) give the maximum distortion energy theory applied to fatigue loading as

$$\frac{S_u}{n} = \left\{ \frac{1}{2}[(\sigma_1 - \sigma_2)^2 + (\sigma_2 - \sigma_3)^2 + (\sigma_3 - \sigma_1)^2] \right\}^{1/2}_e \tag{8.29}$$

The two-dimensional equivalent ($\sigma_3 = 0$) is

$$\frac{S_u}{n} = \left(\sigma_1^2 - \sigma_1\sigma_2 + \sigma_2^2 \right)^{1/2}_e \tag{8.30}$$

For the special case in which $\sigma_y = \sigma_z = \tau_{yz} = \tau_{xz} = 0$, using Eq. (7.16),

$$\frac{S_u}{n} = \left(\sigma_x^2 + 3\tau_{xy}^2 \right)^{1/2}_e \tag{8.31}$$

Substitution of the equivalent stresses σ_e for σ_x and τ_e for τ_{xy} into this expression gives, the *maximum energy of distortion theory* combined with *Goodman criterion:*

$$\frac{S_u}{n} = \left[\left(\sigma_{xm} + \frac{S_u}{S_e}\sigma_{xa} \right)^2 + 3\left(\tau_{xym} + \frac{S_u}{S_e}\tau_{xya} \right)^2 \right]^{1/2} \tag{8.32}$$

ALTERNATIVE DERIVATION

Equivalent alternating stress/equivalent mean stress fatigue criteria are represented in Table 8.4, replacing σ_a and σ_m with σ_{ea} and σ_{em}. In so doing, the Goodman criterion, for example, becomes

$$\frac{1}{n} = \frac{\sigma_{ea}}{S_e} + \frac{\sigma_{em}}{S_u} \tag{8.33}$$

The static failure theories may also be modified, substituting a and m in the expressions given in Sections 7.6 through 7.12. So, the von Mises stresses for the alternating and mean components for a triaxial and biaxial states of stress are obtained by applying Eqs. (7.13) and (7.15), respectively. In a like manner, relations for the special case in which $\sigma_y = \sigma_z = \tau_{yz} = \tau_{xy} = 0$, through the use of Eq. (7.16), may be written as

$$\sigma_{ea} = \left(\sigma_{xa}^2 + 3\tau_{xya}^2 \right)^{1/2}, \qquad \sigma_{em} = \left(\sigma_{xm}^2 + 3\tau_{xym}^2 \right)^{1/2} \tag{8.34}$$

Carrying Eqs. (8.34) into Eqs. (8.33) yields the energy of distortion theory associated with the Goodman relation in the following alternate form:

$$\frac{1}{n} = \frac{1}{S_e} \left(\sigma_{xa}^2 + 3\tau_{xya}^2 \right)^{1/2} + \frac{1}{S_u} \left(\sigma_{xm}^2 + 3\tau_{xym}^2 \right)^{1/2} \tag{8.35}$$

in which n represents the factor of safety.

In conclusion, we note that Eqs. (8.28) and (8.32) or (8.35) can be employed to develop a series of design formulas. Their application to the design of transmission shafts is illustrated in the next chapter. Obviously, fatigue analysis should be considered wherever a simple or combined fluctuating load is present. Springs, for example, frequently fail in fatigue. Chapter 14 treats spring design by using the Soderberg and Goodman criteria. We discuss the preloaded threaded fasteners in fatigue in Section 15.12.

8.13 PREDICTION OF CUMULATIVE FATIGUE DAMAGE

Machine and structural members are not always subjected to the constant stress cycles, as shown in Figure 8.11. Many parts may be under different severe levels of reversed stress cycles or randomly varying stress levels. Examples include automotive suspension and aircraft structural components operating at stress levels between the fracture strength S_f and endurance limit S'_e, say, S (Figure 8.4). If the reversed stress is higher than the endurance limit, S replaces S'_e in Eq. (8.6) and the design may again be based on the formulas developed in the preceding section. However, when a machine part is to operate for a finite time at higher stress, the cumulative damage must be examined.

It is important to note that predicting the cumulative damage of parts stressed above the endurance limit is at best a rough procedure. This point is demonstrated by the typical scatter band depicted in Figure 8.5 for completely reversed loads. Clearly, for parts subjected to randomly varying loads, the damage prognosis is further complicated.

Miner's Cumulative Rule

The simplest, most widely accepted criterion used to explain cumulative fatigue damage is called the *Miner's rule*. The procedure, also known as the linear cumulative damage rule, is expressed in the form

$$\frac{n_1}{N_1} + \frac{n_2}{N_2} + \cdots = 1 \tag{8.36}$$

Here n is the number of cycles of higher stress S applied to the specimen and N is the life (in cycles) corresponding to S, as taken from the appropriate S-N curve.

The Miner's equation assumes that the damage to the material is directly proportional to the number of cycles at a given stress. The rule also presupposes that the stress sequence does not matter and the rate of damage accumulation at a stress level is independent of the stress history. These have not been completely verified in tests. Sometimes specifications are used in which the right side of Eq. (8.36) is taken to be between 0.7 and 2.2.

A typical set of plots of S versus N, for different types of surfaces, is shown in Figure 8.17 [5, 20, 21]. The values of N_1, N_2, and so on, may be obtained from such curves. Employing these values, Eq. (8.36) becomes the design criterion. The use of Miner's rule is illustrated in the solution of the following numerical problem.

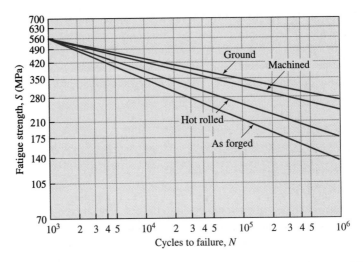

Figure 8.17   Allowable stress-cycle diagram for steel parts with 187 to 207 Bhn.

EXAMPLE 8.6	Cumulative Fatigue Damage of a Machine Bracket

A steel bracket of a machine is subjected to a reversed bending stress of S_1 for n_1 cycles, S_2 for n_2 cycles, and S_3 for n_3 cycles. Determine whether failure will occur.

Given: $S_1 = 420$ MPa, $S_2 = 350$ MPa, $S_3 = 280$ MPa
  $n_1 = 5,000$ cycles, $n_2 = 20,000$ cycles, $n_3 = 30,000$ cycles.

Design Decisions: The bracket has a machined surface and Bhn = 200. Miner's cumulative damage rule is used.

Solution: The appropriate limiting number of cycles corresponding, respectively, to the preceding stress values are, from Figure 8.17,

   $N_1 = 12,000$ cycles, $N_2 = 50,000$ cycles, $N_3 = 280,000$ cycles

Applying Eq. (8.36),

$$\frac{5000}{12,000} + \frac{20,000}{50,000} + \frac{30,000}{280,000} = 0.924$$

Comment: Since $0.924 < 1$, the member is safe.

8.14   FRACTURE MECHANICS APPROACH TO FATIGUE

As discussed previously, the fatigue failures are progressive, starting with a very small crack at or near a surface, followed by their gradual increase in width and depth, then sudden fracture through the remaining zone. We now present a procedure for estimating the life remaining in a part after the discovery of a crack. The method, known as the *fracture-mechanics approach to fatigue,* applies to elastic isotropic materials.

To develop fatigue strength data in terms of a fracture mechanics approach, numerous specimens of the same material are tested to failure at certain levels of cyclical stress range $\Delta\sigma$. The test is usually done in an axial fatigue testing machine. The crack growth rate da/dN is continuously measured as the applied stress varies from σ_{min} to σ_{max} during the test. Here a represents the initial crack length and N is the number of stress cycles.

For each loading cycle, the stress intensity range factor ΔK is defined as

$$\Delta K = K_{max} - K_{min} \tag{8.37a}$$

where K_{max} and K_{min} are the maximum and minimum stress intensity factors, respectively, around a crack. From Eq. (7.1), we have

$$K_{max} = \lambda\sqrt{\pi a}\,\sigma_{max}, \qquad K_{min} = \lambda\sqrt{\pi a}\,\sigma_{min} \tag{8.38}$$

in which λ is a geometry factor. Substitution of this into Eq. (8.37a) gives

$$\Delta K = \lambda\sqrt{\pi a}(\sigma_{max} - \sigma_{min}) \tag{8.37b}$$

The quantities σ_{max} and σ_{min} are the maximum and minimum nominal stresses, respectively. The critical or final crack length a_f at fracture, from Eq. (7.3) taking factor of safety $n = 1$, may be expressed as

$$a_f = \frac{1}{\pi}\left(\frac{K_c}{\lambda\sigma_{max}}\right)^2 \tag{8.39}$$

where K_c is the fracture toughness.

Crack growth rate da/dN is often plotted on log-log paper against the stress intensity range factor ΔK. The major central portion of the curve plots as a straight line and is of interest in predicting fatigue life. The relationship in this region is defined in the form [16, 27]

$$\frac{da}{dN} = A(\Delta K)^n \tag{8.40}$$

This is known as the *Paris equation* after P. C. Paris. The empirical values of the factor A and exponent n for a number of steels are listed in Table 8.5. Details of the experimental basis for this expression are given in the technical literature [30].

Equation (8.40) is integrated to give the number of cycles N to increase the crack length from an initial value a to the critical length a_f at fracture, the remaining fatigue life, based on a particular load, geometry, and material parameters for a particular application.

Table 8.5 Paris equation parameters for various steels [28]

Steel	A		n
	SI units	(U.S. units)	
Ferritic-pearlitic	6.90×10^{-12}	(3.60×10^{-10})	3.00
Martensitic	1.35×10^{-10}	(6.60×10^{-9})	2.25
Austenitic stainless	5.60×10^{-12}	(3.00×10^{-10})	3.25

Considering λ independent of the initial crack length, it can be shown that [11, 31]

$$N = \frac{a_f^{1-n/2} - a^{1-n/2}}{A\left(1 - \frac{n}{2}\right)(1.77\lambda\Delta\sigma)^n} \tag{8.41}$$

where

N = fatigue life in cycles
a = initial crack length
a_f = crack length at fracture
λ = geometry factor (see Table 7.1)
$\Delta\sigma$ = stress range (in MPa or ksi)

The application of Eq. (8.41), the fatigue life determination procedure, is illustrated in the simple example that follows.

EXAMPLE 8.7   Fatigue Life Analysis of an Instrument Panel with a Crack

A long plate of an instrument is of width $2w$ and thickness t. The panel is subjected to an axial tensile load that varies from P_{min} to P_{max} with a complete cycle every 15 seconds. Before loading, on inspection a central transverse crack of length $2a$ is detected on the plate. Estimate the expected life.

Given: $a = 0.3$ in., $t = 0.8$ in., $w = 2$ in., $P_{max} = 2P_{min} = 144$ kips

Assumption: The plate is made of an AISI 4340 tempered steel.

Solution: See Tables 8.5, 7.2, and 7.1.
    The material and geometric properties of the panel are

$A = 3.6 \times 10^{-10}$, $n = 3$, $K_c = 53.7$ ksi$\sqrt{\text{in.}}$, $S_y = 218$ ksi,

$\lambda = 1.02$,   for $a/w = 0.15$ (Case A of Table 7.1)

Note that the values of a and t satisfy Table 7.2. The largest and smallest normal stresses are

$$\sigma_{max} = \frac{P_{max}}{2wt} = \frac{144}{2(2)(0.8)} = 45 \text{ ksi}, \qquad \sigma_{min} = 22.5 \text{ ksi}$$

The cyclical stress range is then $\Delta\sigma = 45 - 22.5 = 22.5$ ksi.
    The final crack length at fracture, from Eq. (8.39) is found to be

$$a_f = \frac{1}{\pi}\left(\frac{K_c}{\lambda\sigma_{max}}\right)^2 = \frac{1}{\pi}\left(\frac{53.7}{1.02 \times 45}\right)^2 = 0.436 \text{ in.}$$

Substituting the numerical values, Eq. (8.41) results in

$$N = \frac{0.436^{-0.5} - 0.3^{-0.5}}{3.6(10^{-10})(-0.5)[(1.77)(1.02)(22.5)]^3} = 25,800 \text{ cycles}$$

With a period of 15 sec, approximate fatigue life L is

$$L = \frac{25,800(15)}{60(60)} = 107.5 \text{ hr}$$

8.15   SURFACE FATIGUE FAILURE: WEAR

So far, we have dealt with the modes of failure of components by yielding, fracture, and fatigue. A variety of types of failure can also occur to the surface of a component generally called *wear*. Surface failure or damage may render the member unfit for use. The severity of wear can be reduced by using a lubricant (i.e., oil, grease, or solid film) between the mating surfaces. *Scoring* or *scuffing* is surface failure in the absence of adequate lubricant. *Abrasion* is wear owing to the presence of foreign materials between the rubbing surfaces. Abrasive wear arises when two interacting surfaces are in direct physical contact and one is significantly harder than the other.

The discussion of Section 3.14 shows that, when two solid members are pressed together, contact stresses are produced. *Pitting* is a surface fatigue failure or fatigue wear due to many repetitions of high contact stress: Small pieces of material are lost from the surface, leaving behind pits. Pits grow into larger areas of flaked-off surface material, which is then termed *spalling*. An audible warning is often noticeable when the pitting process commences. In machine components such as rolling-element bearings, gears, friction drives, cams, and tappets, a prevalent form of failure is fatigue wear. In these situations, the removal of material results from a cyclic load variation.

Surface fatigue occurs in pure-rolling or roll-sliding contact, owing to many thousands of cycles of repeated contact stress. A typical stress-life diagram on the basis of computed maximum contact pressure p_o (see: Sections 3.14 and 3.15) is shown in Figure 8.18. Note that the degree of sliding usually increases from the parallel rollers (top line) to spur gear teeth (bottom line). Observe from the figure that the tendency of surface fatigue failure can be reduced by decreasing the sliding and decreasing loads.

High-strength smooth materials are required in contact stress applications. No material has an endurance limit against surface fatigue. Therefore, a contact stress or surface fatigue

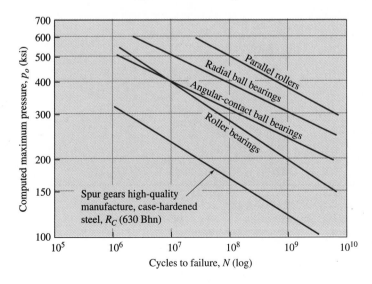

Figure 8.18   Average *S-N* curves for contact stresses, 10% failure probability [14].

strength value for only a particular number of cycles is given for the materials. Usually, increased surface *hardness* increases resistance to surface fatigue. Also *compressive residual stresses* in the contacting surfaces increase resistance to surface fatigue failure. These contact stresses can be introduced by methods such as surface treatments, thermal treatments, and mechanical treatments. Thermal stressing occurs whenever a part is heated and cooled, as in heat treatment. The most common methods for introducing surface compressive stresses are shot peening and cold forming (see Section 2.11). Mechanical prestressing refers to the prearranged overloading of the part in the same direction as its service loading, before its being placed in service.

STRESSES AFFECTING SURFACE FATIGUE

When two surfaces are in *pure-rolling* contact, shear stress τ (existing at any point below the surface and a distance from the load axis) reverses while going through the contact zone from A to A', as shown in Figure 8.19. This fully reversed shear stress, as well as the subsurface maximum shear stress occurring along the load axis and maximum contact pressure p_o, may be the cause of pits that begin at the subsurface.

  If some *sliding accompanies rolling,* as shown in Figure 8.20, both fully reversed tangential surface shear and normal stresses are produced [14] as any point on the surface rolls through the contact region: Pitting begins at the surface. The resulting surface tensile stress leads to the propagation of surface fatigue cracks. In addition, significant factors that influence stresses in contact zone include highly localized heating and thermal expansion produced by sliding friction and the increase in viscosity of the oil due to high pressure in elastohydrodynamic lubrication (see Section 10.15).

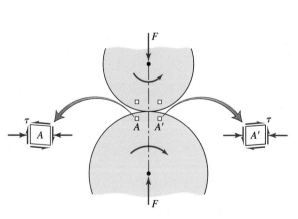

Figure 8.19   Two rotating cylinders compressed by force *F.* Note the subsurface shear stress that reverses when rolling through the contact zone.

Figure 8.20   Two rotating and sliding thick-walled cylinders compressed by force *F.* Tangential normal and shear stresses due to the sliding friction between the members have the largest values at the surface in the locations shown.

REFERENCES

1. "Failure Analysis and Prevention." *Metals Handbook,* 9th ed. Metals Park, OH: ASM, 1986.

2. Engel, L., and H. Klingele. *An Atlas of Metal Damage.* Munich: Hanser Verlag, 1981.

3. Marin, J. *Mechanical Behavior of Materials.* Upper Saddle River, NJ: Prentice Hall, 1962.

4. Dowling, N. E. *Mechanical Behavior of Materials.* Upper Saddle River, NJ: Prentice Hall, 1993.

5. Rice, R. C., ed. *Fatigue Design Handbook,* 2nd ed. Warrendale, PA: Society of Automotive Engineers, 1988.

6. Fuchs, W. O., and R. I. Stevens. *Metal Fatigue Engineering.* New York: Wiley, 1980.

7. Bannantine, J. A., J. J. Comer, and J. L. Handrock. *Fundamentals of Metal Fatigue.* Upper Saddle River, NJ: Prentice Hall, 1990.

8. Mischke, C. R. "Prediction of Stochastic Endurance Strength." *Transactions of the ASME, Journal of Vibration Acoustics; Stress, and Reliability in Design* 109, no.1 (January 1987), pp. 113–22.

9. Madayag, J. A. *Metal Fatigue: Theory and Design.* New York: Wiley, 1969.

10. Haywood, R. B. *Designing Against Fatigue of Metals.* New York: Reinhold, 1962.

11. Burr, A. H., and J. B. Cheatham. *Mechanical Analysis and Design,* 2nd ed. Upper Saddle River, NJ: Prentice Hall, 1995.

12. Norton, R. L. *Machine Design: An Integrated Approach,* 2nd ed. Upper Saddle River, NJ: Prentice Hall, 2000.

13. Shigley, J. E., and C. R. Mischke. *Mechanical Engineering Design,* 6th ed. New York: McGraw-Hill, 2001.

14. Juvinall, R. C., and K. M. Marshek. *Fundamentals of Machine Component Design,* 3rd ed. New York: Wiley, 2000.

15. *Damage Tolerant Design Handbook,* MCIC-HB-01. Kettering, OH: Air Force Materials Laboratory, Wright-Patterson Air Force Base, December 1972.

16. Paris, P. C., M. P. Gomez, and W. P. Anderson. "A Rational Analytic Theory of Fatigue." *Trend Engineering* [University of Washington] 13, no. 1 (1961), pp. 9–14.

17. Shigley, J. E., and L. D. Mitchel. *Mechanical Engineering Design,* 4th ed. New York: McGraw-Hill, 1983.

18. Dimarogonas, A. D. *Machine Design: A CAD Approach.* New York: Wiley, 2001.

19. Sines, G., and J. L. Waisman, eds. *Metal Fatigue.* New York: McGraw-Hill, 1959, pp. 296–98.

20. Juvinall, R. C. *Stress, Strain, and Strength.* New York: McGraw-Hill, 1967.

21. Spotts, M. F., and T. E. Shoup. *Design of Machine Elements,* 7th ed. Upper Saddle River, NJ: Prentice Hall, 1998.

22. Boresi, A. P., and R. J. Schmidt. *Advanced Mechanics of Materials,* 6th ed. New York: Wiley, 2003.

23. Budynas, R. G. *Advanced Strength and Applied Stress Analysis,* 2nd ed. New York: McGraw-Hill, 1999.

24. Hamrock, B. J., B. Jacobson, and S. R. Schmid. *Fundamentals of Machine Elements.* New York: McGraw-Hill, 1999.

25. Ugural, A. C., and S. K. Fenster. *Advanced Strength and Applied Elasticity,* 4th ed. Upper Saddle River, NJ: Prentice Hall, 2003.

26. Sullivan, J. L. "Fatigue Life under Combined Stress." *Machine Design,* January 25, 1979.

27. Paris, P. C., and F. Erdogan. "A Critical Analysis of Crack Propagation Laws." *Transactions of the ASME, Journal of Basic Engineering,* 85(4): p. 528, 1963.

28. Barsom, J. M. "Fatigue-Crack Propagation in Steels at Various Yield Strengths." *Transactions of the ASME, Journal of Engineering Industry,* Series B (4), (1971), p. 1190.

29. Barsom, J. M., and S. T. Rolfe. *Fracture and Fatigue Control in Structures,* 2nd ed. Upper Saddle River, NJ: Prentice Hall, 1987.

30. Stulen, F. B., H. N. Cummings, and W. C. Schulte. "Preventing Fatigue Failures—Part 5."
    Machine Design 33, November 13, 1961.

31. Suresh, S. *Fatigue of Materials.* Cambridge: Cambridge University Press, 1991.

PROBLEMS

Sections 8.1 through 8.7

8.1 A structural steel bar of thickness t with full-fillets is loaded by reversed axial force P
(Figure P8.1). Calculate

(*a*) The maximum stress.

(*b*) The maximum fatigue stress concentration factor.

Given: $P = 15$ kN, $t = 10$ mm

30 mm r_c 38 mm r_d 34 mm

Figure P8.1

8.2 A bar with full-fillets is forged from a structural steel (Figure P8.1). Determine the value of the
endurance limit S_e.

Assumptions: A survival rate of 95% is used. The operating temperature is 475°C maximum.

8.3 A machined and full-filleted AISI 4140 annealed steel bar carries a fluctuating axial loading, as
shown in Figure P8.3. What is the value of endurance limit S_e?

Given: $b = 20$ mm, $D = 30$ mm, $r = 2$ mm
Assumptions: A reliability of 90% is used.

Figure P8.3

8.4 A stepped cantilever beam of diameters d and D, machined from an AISI 1060 annealed steel
bar, is subjected to a fluctuating moment M, as depicted in Figure P8.4. Determine the modified endurance limit S_e.

Given: $d = 25$ mm, $D = 35$ mm, $r = 4$ mm
Design Assumption: Reliability is 90%.

Figure P8.4

8.5 A notched beam, machined from AISI 1030 hot-rolled steel, is subjected to reversed bending. Determine the endurance limit S_e.

Assumptions: A survival rate of 98% and $C_s = 0.7$ are used. The fatigue stress concentration factor is $K_f = 2.5$.

Sections 8.8 through 8.15

8.6 A stepped cantilevered beam, machined from steel having ultimate tensile strength S_u, is under reversed bending (Figure P8.4). Determine the maximum value of the bending moment M, using the Goodman criterion.

Given: $d = 1$ in., $\quad D = 1.5$ in., $\quad r = 0.05$ in., $\quad S_u = 100$ ksi
Design Assumptions: A survival rate of 95% is used. The factor of safety $n = 1.5$.

8.7 A cold drawn AISI 1020 annealed steel link is subjected to axial loading (that fluctuates from 0 to F) by pins that go through holes (Figure P8.7). What is the maximum value of F with a factor of safety of n, according to the Goodman criterion?

Given: $R = 10$ mm, $\quad r = 4$ mm, $\quad t = 2.5$ mm, $\quad n = 1.4$
Assumption: A reliability of 99.99% is used.

Figure P8.7

8.8 A cold drawn AISI 1050 steel plate with a central hole is under a tension load P that varies from 5 kN to 25 kN (Figure P8.8). Based on the Goodman criterion, determine the factor of safety n

(a) Against yielding.

(b) Against fatigue failure.

Given: $D = 25$ mm, $\quad d = 5$ mm, $\quad t = 10$ mm
Assumption: A reliability of 98% and $C_r = 0.7$ are used.

Figure P8.8

8.9 Resolve Problem 8.8 for the condition that the load varies from –5 kN to 25 kN.

Assumption: Buckling does not occur.

8.10 A machined AISI 4130 normalized steel bar of diameter D carries an axial load P, as shown in Figure P8.10. Calculate the value of

(a) The static force P to produce fracture.

(b) The completely reversed force P to produce fatigue failure.

Given: $D = 2\frac{1}{8}$ in.
Assumptions: The survival rate is 95%. The operating temperature is 900°F maximum.

Figure P8.10

8.11 Redo Problem 8.10 for the case of a grooved shaft shown in Figure P8.11.

Given: $D = 2\frac{1}{8}$ in., $d = 2$ in., $r = 0.05$ in.

Figure P8.11

8.12 A stepped shaft ground from AISI 1040 annealed steel is subjected to torsion, as shown in Figure P8.12. Determine the value of

(a) The torque T to produce static yielding.

(b) The torque T to produce fatigue failure.

Given: $D = 50$ mm, $d = 25$ mm, $r = 1.25$ mm
Assumption: Reliability is 98%.

Figure P8.12

8.13 Repeat Problem 8.12 for the condition that the shaft is subjected to axial loading and no torsion.

8.14 Redo Problem 8.12 for the case in which the shaft is machined from an AISI 1095 hot-rolled steel.

8.15 A shaft with a transverse hole ground from AISI 1095 annealed steel is under bending moment M that varies from 0.5 to 1.4 kip in. (Figure P8.15). Determine the factor of safety n against fatigue failure, using the Goodman criterion.

Given: $D = 1$ in., $\quad d = \frac{1}{8}$ in.
Assumption: A reliability of 99% is used.

Figure P8.15

8.16 Resolve Problem 8.15 for the condition that shaft is ground from AISI 1060 HR steel and is under axial loading varying from 5 to 15 kips.

8.17 A thin-walled cylindrical vessel of diameter d is subjected to an internal pressure varying from 60 to 300 psi continuously. Apply the maximum energy of distortion theory incorporated with the Soderberg relation to design the vessel.

Given: $d = 80$ in., $\quad S_y = 40$ ksi, $\quad S_e = 30$ ksi, $\quad n = 2.5$

8.18 A thin-walled cylindrical vessel of diameter d and thickness t is under internal pressure varying from 0.6 to 2.8 MPa continuously.

Given: $d = 1.5$ m, $\quad t = 25$ mm, $\quad S_u = 350$ MPa, $\quad S_e = 150$ MPa
Design Decision: Use the maximum energy of distortion theory incorporated with the Goodman relation. Determine the factor of safety n.

8.19 A small leaf spring, 10 mm wide × 100 mm long and h mm deep, is subjected to a concentrated center load P varying continuously from 0 to 20 N. The spring may be approximated to be a simply supported beam (Figure P8.19). Calculate the required depth for a factor of safety of 4.

Given: $S_u = 980$ MPa, $\quad S_e = 400$ MPa
Design Decision: Apply the Goodman theory, based on the maximum normal stress.

Figure P8.19

8.20 Redo Problem 8.19 using the Soderberg criterion and yield strength of $S_y = 620$ MPa.

8.21 An electrical contact includes a flat spring in the form of a cantilever, $\frac{1}{8}$ in. wide × 1.5 in. long and h in. deep, is subjected at its free end to a load P that varies continuously from 0 to 0.5 lb (Figure P8.21). Compute the value of h for a factor of safety $n = 1.2$.

Given: $S_u = 150$ ksi, $S_e = 72$ ksi
Design Decision: Employ the Goodman criterion, based on the maximum normal stress.

Figure P8.21

8.22 A cantilever spring is subjected to a concentrated load P varying continuously from 0 to P_o (Figure P8.22). What is the greatest allowable load P_o for $n = 4$?

Given: $S_y = 850$ MPa, $S_e = 175$ MPa, $b = 5$ mm, $h = 10$ mm, $K_f = 2$
Assumption: Failure occurs due to bending stress at the fillet.
Design Decision: Use the Soderberg criterion.

Figure P8.22

8.23 Resolve Problem 8.22 for the load varying from $P_o/2$ upward to P_o downward, $n = 2$.

8.24 A 24-mm wide, 4-mm thick, and 3-mm long leaf spring, made of AISI 1050CD steel, is straight and unstressed when the cam and shaft are removed (Figure P8.24). Use the Goodman theory to calculate the factor of safety n for the spring.

Given: $S_e = 250$ MPa, $E = 200$ GPa, $v = 0.3$
Assumption: The cam rotates continuously. Leaf spring is considered as a *wide* cantilever beam.

Figure P8.24

8.25 Repeat Problem 8.24 for the case in which the cantilevered spring is made of normalized AISI 1095 steel and employing the Soderberg criterion.

8.26 Figure P8.26 shows a circular aluminum bar having two shoulder fillets supporting a concentrated load P at its midspan. Determine the allowable value for diameter D if stress conditions at the fillets are to be satisfactory for conditions of operation. Dimensions shown are in millimeters.

Given: $S_u = 600$ MPa, $\qquad S_y = 280$ MPa, $\qquad n = 2.5$, $\qquad K_f S_e = 150$ MPa

Assumptions: The load P varies from 2 kN to 6 kN. The Soderberg relation is employed.

Figure P8.26

8.27 The filleted flat bar shown in Figure P8.27 is made from 1040 steel OQ&T at 650°C. What is the factor of safety n, if the bending moment M varies from 0.6 to 3 kN · m?

Given: $K_f S_e = 400$ MPa, $\qquad D = 120$ mm, $\qquad d = 60$ mm, $\qquad r = 4$ mm, $\qquad t = 20$ mm

Design Assumption: The Goodman criterion of fatigue failure is applied.

Figure P8.27

PART

II

APPLICATIONS

SHAFTS AND ASSOCIATED PARTS

Outline

9.1   INTRODUCTION

Shafts are used in a variety of ways in all types of mechanical equipment. A *shaft,* usually a slender member of round cross section, rotates and transmits power or motion. However, a shaft can have a noncircular cross section and need not be rotating. An *axle,* a nonrotating member that carries no torque, is used to support rotating members. A *spindle* designates a short shaft. A *flexible shaft* transmits motion between two points (e.g., motor and machine), where the rotational axes are at an angle with respect to one another. The customary shaft types are straight shafts of constant or stepped cross section and crankshafts (Figure 9.1). The former two carry rotating members such as gears, pulleys, grooved pulleys (sheaves), or other wheels. The latter are used to convert reciprocating motion into rotary motion or vice versa.

Most shafts are under fluctuating loads of combined bending and torsion with various degrees of stress concentration. Many shafts are not subjected to shock or impact loading; however, some applications arise where such load takes place (Section 9.5). Thus, the associated considerations of static strength, fatigue strength, and reliability play a significant role in shaft design. A shaft designed from the preceding viewpoint satisfies strength requirements. Of equal importance in design is the consideration of shaft deflection and rigidity requirements. Excessive lateral shaft deflection can cause bearing wear or failure and objectionable noise. The operating speed of a shaft should not be close to a critical speed (Section 9.7), or large vibrations are likely to develop.

In addition to the shaft itself, the design usually must include calculation of the necessary keys and couplings. Keys, pins, snap rings, and clamp collars are used on shafts to secure rotating elements. The use of shaft shoulder is an excellent means of axially positioning the shaft elements. Figure 9.2 shows a stepped shaft supporting a gear, a crowned pulley, and a sheave. The mounting parts, discussed in Section 9.8, as well as shaft shoulders, are a source of stress raisers, and they must be properly selected and located to minimize the resulting stress concentrations. Press and shrink fits (Section 9.6) are also used for mounting. Shafts are carried in bearings, in a simply supported form, cantilevered or overhang, depending on the machine configuration. Couplings connect a shaft to a shaft

(a)

(b)

(c)

Figure 9.1   Common shaft types: (a) constant diameter; (b) stepped; (c) crank shaft.

Figure 9.2 A stepped shaft with various elements attached.

of power source or load. Parameters that must be considered in the selection of a coupling to connect two shafts include the angle between the shafts, transmitted power, vibrations, and shock loads. The websites www.pddnet.com, www.powertransmission.com, and www.grainger.com present general information on shaft couplings.

9.2 MATERIALS USED FOR SHAFTING

To minimize deflections, *shaft materials* are generally cold-drawn or machined from hot-rolled, plain-carbon steel. The shaft ends should be made with chamfers to facilitate forcing on the mounted parts and to avoid denting the surfaces. Cold drawing improves the physical properties. It raises considerably the values of ultimate tensile and yield strengths of steel. Where toughness, shock resistance, and greater strength are needed, alloy steels are used. The foregoing materials can be heat treated to produce the desired properties. If the service requirements demand resistance to wear rather than extreme strength, it is customary to harden only the surface of the shaft, and a carburizing grade steel can be used. Note that the hardening treatment is applied to those surfaces requiring it; the remainder of the shaft is left in its original condition.

Thick-walled seamless tubing is available for simpler, smaller shafts. Large-diameter members (> about 75-mm diameter), such as railroad axles, press cranks, and so on, are usually forged and machined to the required size. In addition to steels, high strength nodular cast iron is used to make shaped shafts; for example, automotive engine crankshafts. Bronze or stainless steel is sometimes used for marine or other corrosive environments. Because *keys* and *pins* are loaded in shear, they are made of ductile materials. Soft, low-carbon steel is in widespread usage. Most keys and pins are usually made from cold-rolled bar stock, cut to length and tapered if needed.

9.3 DESIGN OF SHAFTS IN STEADY TORSION

In the design of circular slender shafts that transmit power at a specified speed, the material and the dimensions of the cross section are selected to not exceed the allowable shearing stress or a limiting angle of twist when rotating. Therefore, a designer needs to know the torque acting on the power-transmitting shaft (see Section 1.12). Equations (1.15) through (1.17) may be used to convert the power supplied to the shaft into a constant torque exerted on it during rotation. After having determined the torque to be transmitted, the design of circular shafts to meet strength requirements can be accomplished by using the process outlined in Section 3.2:

1.   Assume that, as is often the case, shear stress is closely associated with failure. Note, however, that in some materials, the maximum tensile and compression stresses occurring on planes at 45° (see Figure 3.27) to the shaft axis may cause the failure.

2.   An important value of the shear stress is defined by $\tau_{max} = Tc/J$.

3.   The maximum usable value of τ_{max} without failure is the yield shear strength S_{ys} or ultimate shear strength S_{us}.

4.   A factor of safety n is applied to τ_{max} to determine the allowable stress $\tau_{all} = S_{ys}/n$ or $\tau_{all} = S_{us}/n$. The required parameter J/c of the shaft based on the strength specification of is

$$\frac{J}{c} = \frac{T}{\tau_{all}}$$

(9.1)

For a given allowable stress, Eq. (9.1) can be used to design both solid and hollow circular shafts carrying torque only.

EXAMPLE 9.1 | **Design of a Shaft of Round Cross Section for Steady Torsion Loading**

A solid shaft is to transmit 500 kW at $n = 1200$ rpm without exceeding the yield strength in shear of S_{ys} or a twisting through more than 4° in a length of 2 m. Calculate the required diameter of the shaft.

Design Decisions: The shaft is made of steel having $S_{ys} = 300$ MPa and $G = 80$ GPa. A safety factor of 1.5 is used.

Solution: The torque, applying Eq. (1.15),

$$T = \frac{9549 \text{ kW}}{n} = \frac{9549(500)}{1200} = 3979 \text{ N} \cdot \text{m}$$

Strength Specification.   Through the use of Eq. (9.1), we have

$$\frac{\pi}{2}c^3 = \frac{3979(1.5)}{300(10^6)}$$

The foregoing gives $c = 23.3$ mm.

Distortion Specification.   The size of the shaft is now obtained from Eq. (4.9):

$$\frac{\phi_{all}}{L} = \frac{T}{GJ}$$

(9.2)

Substituting the given numerical values,

$$\frac{4°(\pi/180)}{2} = \frac{3979}{(80 \times 10^9)\pi c^4/2}$$

This yields $c = 30.9$ mm.

Comments: The minimum allowable diameter of the shaft must be 61.8 mm. A 62-mm shaft should be used.

9.4 COMBINED STATIC LOADINGS ON SHAFTS

The shaft design process is far simpler when only static loads are present than when the loading fluctuates. However, even with the fatigue loading, a preliminary estimate of shaft diameter may be needed many times, as is observed in the next section. Hence, the results of the rational design procedure of Section 3.2, presented here, is useful in getting the first estimate of shaft diameter for any type of combined static loading conditions.

Bending, Torsion, and Axial Loads

Consider a solid circular shaft of diameter D, acted on by bending moment M, torque T, and axial load P. To begin with, we determine the maximum normal and shear stresses occurring in the outer fibers at a critical section:

$$\sigma_x = \frac{32M}{\pi D^3} + \frac{4P}{\pi D^2}$$

$$\tau_{xy} = \frac{16T}{\pi D^3} \tag{9.3}$$

in which the axial component of σ_x may be either additive or subtractive. The foregoing equations are used with a selected design criterion. Note that, for a hollow shaft, the preceding expressions become

$$\sigma_x = \frac{32M}{\pi D^3[1 - (d/D)^4]} + \frac{4P}{\pi(D^2 - d^2)} \tag{9.4}$$

$$\tau_{xy} = \frac{16T}{\pi D^3[1 - (d/D)^4]} \tag{9.5}$$

The quantities D and d represent the outer and inner diameters of the shaft, respectively.

Substituting Eqs. (9.3) into Eq. (7.11), a shaft design formula based on the maximum shear theory of failure is

$$\frac{S_y}{n} = \frac{4}{\pi D^3}[(8M + PD)^2 + (8T)^2]^{1/2} \tag{9.6}$$

Similarly, carrying Eqs. (9.3) into Eq. (7.16), the maximum energy of distortion theory of failure results in

$$\frac{S_y}{n} = \frac{4}{\pi D^3}[(8M + PD)^2 + 48T^2]^{1/2} \tag{9.7}$$

BENDING AND TORSION

Under many conditions, the axial force P in the preceding expressions is either 0 or so small that it can be neglected. Substituting $P = 0$ into Eqs. (9.6) and (9.7), we have the following shaft design equations based on the maximum shear stress theory of failure:

$$\frac{S_y}{n} = \frac{32}{\pi D^3}[M^2 + T^2]^{1/2} \tag{9.8}$$

And the *maximum energy of distortion theory* of failure is

$$\frac{S_y}{n} = \frac{32}{\pi D^3}\left[M^2 + \frac{3}{4}T^2\right]^{1/2} \tag{9.9}$$

Equations (9.6) through (9.9) can be used to determine the factor of safety n if the diameter D is given or to find the diameter if a safety factor is selected.

9.5　DESIGN OF SHAFTS FOR FLUCTUATING AND SHOCK LOADS

Shafts are used in a wide variety of machine applications. The design process for circular torsion members is described in Section 9.3; we are now concerned with the members carrying fluctuating and shock loads of combined bending and torsion, which is the case for most transmission shafts [1–10]. Referring to Section 8.8, the definitions of the mean and alternating moments and torques, are

$$
\begin{aligned}
M_m &= \tfrac{1}{2}(M_{max} + M_{min}) \\
T_m &= \tfrac{1}{2}(T_{max} + T_{min})
\end{aligned}
\quad \text{and} \quad
\begin{aligned}
M_a &= \tfrac{1}{2}(M_{max} - M_{min}) \\
T_a &= \tfrac{1}{2}(T_{max} - T_{min})
\end{aligned}
\tag{9.10}
$$

Although in practice design usually includes considerations for associated keys and couplings, these are neglected in the ensuing procedure. We note that all the shaft design formulas to be presented assume an infinite life design of a material with an endurance limit.

For a solid round shaft of diameter D subjected to bending moment M and torsion T, we have on an outermost element

$$\sigma_x = \frac{32M}{\pi D^3} \quad \text{and} \quad \tau_{xy} = \frac{16T}{\pi D^3}$$

We can replace σ_{xm}, σ_{xa}, τ_{xym}, and τ_{xya} by these formulas (using the appropriate subscripts on σ, τ, M, and T) to express the equations developed in Section 8.12 in terms of the bending moment and torque.

The maximum shear stress theory combined with Goodman fatigue criterion, applying Eq. (8.28), is thus obtained as

$$\frac{S_u}{n} = \frac{32}{\pi D^3}\left[\left(M_m + \frac{S_u}{S_e}M_a\right)^2 + \left(T_m + \frac{S_u}{S_e}T_a\right)^2\right]^{1/2} \tag{9.11}$$

In a similar manner, the *maximum energy of distortion theory* incorporated with *Goodman fatigue criterion,* from Eq. (8.32):

$$\frac{S_u}{n} = \frac{32}{\pi D^3}\left[\left(M_m + \frac{S_u}{S_e}M_a\right)^2 + \frac{3}{4}\left(T_m + \frac{S_u}{S_e}T_a\right)^2\right]^{1/2} \tag{9.12}$$

The quantities S_u and S_e represent the ultimate strength and endurance limit, respectively.

Note that an alternate form of the maximum energy of distortion theory associated with Goodman fatigue relation, through the use of Eq. (8.35), may be expressed in the form

$$\frac{S_u}{n} = \frac{32}{\pi D^3}\left\{\left[\left(\frac{S_u}{S_e}M_a\right)^2 + \frac{3}{4}\left(\frac{S_u}{S_e}T_a\right)^2\right]^{1/2} + \left[(M_m)^2 + \frac{3}{4}(T_m)^2\right]^{1/2}\right\} \tag{9.12'}$$

Alternatively, the maximum shear stress criterion, Eq. (9.11), may also be readily written.

SHOCK FACTORS

Effect of shock load on shaft has been neglected in the preceding derivation. For some equipment, where operation is jerky, this condition requires special consideration. To account of shock condition, multiplying coefficients (that is, correction factors K_{sb} in bending and K_{st} in torsion) may be used in the foregoing equations. Thus, the maximum shear stress theory associated with Goodman fatigue relation, Eq. (9.11) becomes

$$\frac{S_u}{n} = \frac{32}{\pi D^3}\left[K_{sb}\left(M_m + \frac{S_u}{S_e}M_a\right)^2 + K_{st}\left(T_m + \frac{S_u}{S_e}T_a\right)^2\right]^{1/2} \tag{9.13}$$

Likewise, the *maximum energy of distortion theory* combined with *Goodman criterion,* through the use of Eq. (9.12):

$$\frac{S_u}{n} = \frac{32}{\pi D^3}\left[K_{sb}\left(M_m + \frac{S_u}{S_e}M_a\right)^2 + \frac{3}{4}K_{st}\left(T_m + \frac{S_u}{S_e}T_a\right)^2\right]^{1/2} \tag{9.14}$$

The values for K_{sb} and K_{st} are listed in Table 9.1 [1]. Equations (9.13) and (9.14) represent the general form of design formulas of solid transmission shafts. As shown previously, for hollow shafts of outer diameter D and inner diameter d, D^3 is replaced by $D^3[1 - (d/D)^4]$ in these equations.

STEADY-STATE OPERATION

Operation of shafts under steady loads involves a completely reversed alternating bending stress (σ_a) and an approximate torsional mean stress (τ_m). This is the case of a rotating shaft with constant moment $M = M_{\max} = -M_{\min}$ and torque $T = T_{\max} = T_{\min}$. Therefore,

Table 9.1   Shock factors in bending and torsion

Nature of loading	K_{sb}, K_{st}
Gradually applied or steady	1.0
Minor shocks	1.5
Heavy shocks	2.0

Eqs. (9.10) give

$$M_m = \frac{1}{2}[M + (-M)] = 0 \qquad M_a = \frac{1}{2}[M - (-M)] = M$$

$$T_m = \frac{1}{2}(T + T) = T \qquad T_a = \frac{1}{2}(T - T) = 0$$

(9.15)

and Eqs. (9.11) through (9.14) are simplified considerably. Then, Eq. (9.12) results in

$$D^3 = \frac{32n}{\pi S_u}\left[\left(\frac{S_u}{S_e}M_a\right)^2 + \frac{3}{4}T_m^2\right]^{1/2}$$

(9.16a)

The preceding expressions could also be written on the basis of the *Soderberg criterion*, replacing the quantity S_u by the yield strength S_y, as needed. In so doing, for instance, Eq. (9.16a) becomes

$$D^3 = \frac{32n}{\pi S_y}\left[\left(\frac{S_y}{S_e}M_a\right)^2 + \frac{3}{4}T_m^2\right]^{1/2}$$

(9.16b)

This is essentially the *ASME shaft design equation* [2]. Note that, for a shaft with varying diameters or other causes of stress concentration, the section of worst combination of moment and torque may not be obvious. It might therefore be necessary to apply design equations at several locations. Clearly, unsteady operation produces fluctuations on the shaft torque (Example 9.2); hence, $T_a \neq 0$ in Eqs. (9.15).

The foregoing discussion shows that the design of shafts subjected to fluctuating and shock loads cannot be carried out in a routine manner, as in the case of static loads. Usually, the diameter of a shaft must be assumed and a complete analysis performed at a critical section where the maximum stress occurs. A design of this type may require several revisions. The FEA is in widespread use for such cases for final design. Alternatively, experimental methods are used, since formulas of solid mechanics may not be sufficiently accurate.

DISPLACEMENTS

Shaft deflections frequently can be a critical factor, since excessive displacements cause rapid wear of shaft bearings, misalignments of gears driven from the shaft, and shaft vibrations (see Section 9.6). Deflection calculations require that the entire shaft geometry be defined. Hence, a shaft typically is first designed for strength, then the displacements are calculated once the geometry is completely prescribed. Both transverse and twisting displacements must be analyzed. Approaches used in obtaining the deflections of a shaft include the methods of Chapters 4, 5, and 17.

Shaft Design for Repeated Torsion and Bending

EXAMPLE 9.2

Power is transmitted from a motor through a gear at E to pulleys at D and C of a revolving solid shaft AB with ground surface. Figure 9.3a shows the corresponding load diagram of the shaft. The shaft is mounted on bearings at the ends A and B. Determine the required diameter of the shaft by employing the maximum energy of distortion theory of failure incorporating the Soderberg fatigue relation.

Figure 9.3   Example 9.2.

Given: The shaft is made of steel with an ultimate strength of 810 MPa and a yield strength of 605 MPa. Torque fluctuates 10% each way from the mean value. The fatigue stress-concentration factor for bending and torsion is equal to 1.4. The operating temperature is 500°C maximum.

Design Assumptions: Bearings act as simple supports. A factor of safety of $n = 2$ is used. The survival rate is taken to be 50%.

Solution: The reactions at A and B, as obtained from the equations of statics, are noted in Figure 9.3a. The determination of the resultant bending moment of $(M_y^2 + M_z^2)^{1/2}$ is facilitated by using the moment diagrams (Figures 9.3b and 9.3c). At point C, we have

$$M_C = [(0.1)^2 + (1.5)^2]^{1/2} = 1.503 \text{ kN} \cdot \text{m}$$

Similarly, at D and E,

$$M_D = 2.121 \text{ kN} \cdot \text{m} \qquad M_E = 1.304 \text{ kN} \cdot \text{m}$$

The maximum bending moment is at D. Note from Figure 9.3d that the torque is also maximum at D, $T_D = 1$ kN·m. The exact location along the shaft where the maximum stress occurs, the critical section, is therefore at D. Hence, at point D,

$$M_m = 0 \qquad\qquad M_a = 2.121 \text{ kN·m}$$

$$T_m = 1 \text{ kN·m} \qquad T_a = 0.1(1) = 0.1 \text{ kN·m}$$

Using Eq. (8.1), the endurance limit of the material is

$$S_e' = 0.5(S_u) = 0.5(810) = 405 \text{ MPa}$$

By Eq. (8.7) and Table 8.2, we determine that, for a ground surface,

$$C_f = A S_u^b = 1.58(810^{-0.085}) = 0.894$$

For reliability of 50%, we have $C_r = 1$ from Table 8.3. *Assuming* that the shaft diameter will be larger than 51 mm, $C_s = 0.70$ by Eq. (8.9). The temperature factor is found applying Eq. (8.11):

$$C_t = 1 - 0.0058(T - 450) = 1 - 0.0058(500 - 450) = 0.71$$

We can now determine the modified endurance limit by Eq. (8.6):

$$S_e = C_f C_r C_s C_t (1/K_f) S_e' = (0.894)(1)(0.70)(0.71)(1/1.4)(405 \times 10^6) = 128.5 \text{ MPa}$$

Because the loading is smooth, $K_{sb} = K_{st} = 1$ from Table 9.1.

Substituting the $S_y = 605$ MPa for S_u and the numerical values obtained into Eq. (9.12), we have

$$\frac{605(10^6)}{2} = \frac{32}{\pi D^3} \left[(1)\left(0 + \frac{605 \times 2121}{128.5} \right)^2 + (1)\left(\frac{3}{4} \right)\left(1000 + \frac{605 \times 100}{128.5} \right)^2 \right]^{1/2}$$

Solving,

$$D = 0.0697 \text{ m} = 69.7 \text{ mm}$$

Comments: Since this is larger than 51 mm, our assumption is *correct*. A diameter of 70 mm is therefore quite satisfactory.

EXAMPLE 9.3

Determining Factor of Safety for a Stepped Shaft under Torsional Shock Loading

A stepped shaft of diameters D and d with a shoulder fillet radius r has been machined from AISI 1095 annealed steel and fixed at end A (Figure 9.4). Determine the factor of safety n, using the maximum shear stress theory incorporated with the Goodman fatigue relation.

Given: Free end C of the shaft is made to rotate back and forth between $1.0°$ and $1.5°$ under torsional minor shock loading. The shaft is at room temperature.

Figure 9.4   Example 9.3.

Data:

$$L = 300 \text{ mm}, \qquad d = 30 \text{ mm}, \qquad D = 60 \text{ mm}, \qquad r = 2 \text{ mm},$$

$$K_{st} = 1.5 \quad \text{(by Table 9.1)}, \qquad G = 79 \text{ GPa} \quad \text{(from Table B.1)}$$

$$S_u = 658 \quad \text{and} \quad H_B = 192 \quad \text{(by Table B.4)}$$

Design Assumption: A reliability of 95% is used.

Solution: From the geometry of the shaft, $D = 2d$ and $L_{AB} = L_{BC} = L$. The polar moment of inertia of the shaft segments are

$$J_{BC} = \frac{\pi d^4}{32} \qquad J_{AB} = \frac{\pi D^4}{32} = 16 J_{BC}$$

in which

$$J_{BC} = \frac{\pi}{32}(0.030)^4 = 79.52(10^{-9}) \text{ m}^4$$

The total angle of twist is

$$\phi = \frac{TL}{G}\left(\frac{1}{16 J_{BC}} + \frac{1}{J_{BC}}\right)$$

or

$$T = \frac{16\, G J_{BC} \phi}{17\, L}$$

Substituting the numerical data, this becomes $T = 19{,}708.5\,\phi$. Accordingly, for $\phi_{\max} = 0.0262$ rad and $\phi_{\min} = 0.0175$ rad, it follows that $T_{\max} = 516.4$ N·m and $T_{\min} = 344.9$ N·m. Hence,

$$T_m = 430.7 \text{ N·m} \qquad T_a = 85.8 \text{ N·m}$$

The modified endurance limit, using Eq. (8.6), is

$$S_e = C_f C_r C_s C_t \left(\frac{1}{K_f}\right) S_e'$$

where

$$C_f = AS_u^b = 4.51(658^{-0.265}) = 0.808 \quad \text{(by Eq. (8.7) and Table 8.2)}$$

$$C_r = 0.87 \quad \text{(from Table 8.3)}$$

$$C_s = 0.85 \quad \text{(by Eq. (8.9))}$$

$$C_t = 1 \quad \text{(for normal temperature)}$$

$$S_e' = 0.29S_u = 190.8 \text{ MPa} \quad \text{(applying Eq. (8.4))}$$

and

$$K_t = 1.6 \quad \text{(from Figure C.8, for } D/d = 2 \text{ and } r/d = 0.067)$$

$$q = 0.92 \quad \text{(from Figure 8.9, for } r = 2 \text{ mm and } H_B = 192 \text{ annealed steel)}$$

$$K_f = 1 + 0.92(1.6 - 1) = 1.55 \quad \text{(using Eq. (8.13b))}$$

Therefore

$$S_e = (0.808)(0.87)(0.85)(1)(1/1.55)(190.8) = 73.55 \text{ MPa}$$

We now use Eq. (9.13) with $M_m = M_a = 0$ to estimate the factor of safety:

$$\frac{S_u}{n} = \frac{32}{\pi d_{BC}^3}\left[K_{st}\left(T_m + \frac{S_u}{S_e}T_a \right)^2 \right]^{1/2} \tag{9.17}$$

Introducing the numerical values,

$$\frac{658(10^6)}{n} = \frac{32}{\pi(0.03)^3}\left[1.5\left(430.2 + \frac{658}{73.55}86.2 \right)^2 \right]^{1/2}$$

This gives $n = 1.19$.

9.6   INTERFERENCE FITS

Fits between parts, such as a shaft fitting in a hub, affect the accuracy of relative positioning of the components. *Press* or *shrink fit,* also termed *interference fit,* can sustain a load without relative motion between the two mating parts. A clearance fit provides ease with which the members can slide with respect to one another. The tolerance, or permissible variation in size of a part, affects both function and fabrication cost.

The preferred limits and fits for cylindrical parts are given by the American National Standards Institute (ANSI) Standard B4.1-1967. This is widely used for establishing tolerances for various fits. The American Gear Manufacturers Association (AGMA) Standard 9003-A91, Flexible Couplings-Keyless Fits, contains formulas for the calculation of interference fits.

The interference fits are usually characterized by maintenance of constant pressures between two mating parts through the range of sizes. The amount of interference needed to create a tight joint varies directly with the diameter of the shaft. A simple rule of thumb is to use 0.001 in. of interference for diameters up to 1 in. and 0.002 in. for diameters between 1 through 4 in. A detailed discussion of the state of stresses in shrink fits is found in Chapter 16, where we consider applications to various members.*

9.7 CRITICAL SPEED OF SHAFTS

A rotating shaft becomes dynamically unstable at certain speeds, and large amplitudes of lateral vibration develop stresses to such a value that rupture may occur. The speed at which this phenomenon occurs is called a *critical speed*. Texts on vibration theory show that the frequency for free vibration when the shaft is *not* rotating is the same as its critical speed. That is, the critical speed of rotation numerically corresponds to the lateral natural frequency of vibration, which is induced when rotation is stopped and the shaft center is displaced laterally, then suddenly released. Hence, a natural frequency is also called a *critical frequency* or *critical speed* [11].

Equating the kinetic energy due to the rotation of the mounted shaft masses to the potential energy of the deflected shaft results in an expression, called the *Rayleigh equation*. This expression defines the critical speed of the shaft. A shaft has as many critical speeds as there are rotating masses. Unless otherwise specified, the term *critical speed* is used to refer to the lowest or fundamental critical speed. The critical speed n_{cr} (in cycles per second, cps) for a shaft on two supports and carrying multiple masses is defined as follows:

$$n_{cr} = \frac{1}{2\pi} \left[\frac{g(W_1\delta_1 + W_2\delta_2 + \cdots + W_m\delta_m)}{W_1\delta_1^2 + W_2\delta_2^2 + \cdots + W_m\delta_m^2} \right]^{1/2} = \frac{1}{2\pi} \sqrt{\frac{g \sum W\delta}{\sum W\delta^2}} \qquad (9.18)$$

The quantity W_m represents the concentrated weight (including load) of a rotating mass and δ_m is the respective static deflection of the weight, as shown in Figure 9.5. The acceleration of gravity is represented by g. In SI units, $g = 9.81$ m/s^2.

Note that the Rayleigh equation only estimates the critical speed. It ignores the effects of the weight of the shaft, self-damping of the material, the flexibility of the bearings or supports, and assumes that all weights are concentrated. Tests have shown that the foregoing factors tend to lower the calculated critical speed. Thus, the approximate values of n_{cr} calculated from Eq. (9.18) are always higher than the true fundamental frequency. A good rule of thumb in practice is to keep the actual operating speed about 25% lower than the calculated critical speed. More accurate approaches for determining the critical frequency, such as a modified Rayleigh's method (Rayleigh-Ritz) and Holzer's method, exist but are somewhat more complicated to implement [12, 13].

Only rotations sufficiently below or above the critical speed result in dynamic stability of the shafts. In unusual situations, in very high-speed turbines, sometimes satisfactory operation is provided by quickly going through the critical speed, then running well above the critical speed. This practice is to be avoided if possible, as vibration may develop from

*Some readers may prefer to study Sections 16.3 and 16.4 as a potential assembly method.

Figure 9.5   Simply supported shaft with concentrated loads (deflection greatly exaggerated).

other causes in the operation above critical speeds, even though the operation is stable. Interestingly, the critical speed of a shaft on three supports is also equal to the natural frequency of the shaft in lateral vibration [12].

Shaft critical speeds may readily be estimated by calculating static deflections at one or several points. The maximum allowable deflection of a shaft is usually determined by the critical speed and gear or bearing requirements. Critical speed requirements vary greatly with the specific application.

EXAMPLE 9.4 | Determining Critical Speed of a Hollow Shaft

A shaft with inner and outer diameters of d and D, respectively, is mounted between bearings and supporting two wheels as shown in Figure 9.6. Calculate the critical speed in rpm.

Figure 9.6   Example 9.4.

Given: $d = 30$ mm, $D = 50$ mm

Assumptions: The shaft is made of steel having $E = 210$ GPa. The weight of the shaft is ignored. Bearings act as simple supports.

Solution: The moment of inertia of the cross section is $I = \frac{\pi}{4}(25^4 - 15^4) = 267 \times 10^3$ mm^4. The concentrated forces are $W_C = 20 \times 9.81 = 196.2$ N and $W_D = 30 \times 9.81 = 294.3$ N. Static deflections at C and D can be obtained by the equations for Case 6 of Table A.9.

Deflection at C.   Due to the load at C,

$$\delta'_C = \frac{196.2(1)(0.5)(1.5^2 - 1^2 - 0.5^2)}{6(1.5)(267 \times 10^{-9})(210 \times 10^9)} = 0.194 \text{ mm}$$

Owing to load at D,

$$\delta''_C = \frac{294.3(0.4)(0.5)(1.5^2 - 0.4^2 - 0.5^2)}{6(1.5)(267 \times 210)} = 0.215 \text{ mm}$$

The total deflection is then

$$\delta_C = 0.194 + 0.215 = 0.409 \text{ mm}$$

Deflection at D.   Similarly, we obtain

$$\delta_D' = \frac{196.2(0.4)(0.5)(1.5^2 - 0.4^2 - 0.5^2)}{6(1.5)(267 \times 210)} = 0.143 \text{ mm}$$

$$\delta_D'' = \frac{294.3(0.4)(1.1)(1.5^2 - 0.4^2 - 1.1^2)}{6(1.5)(267 \times 210)} = 0.226 \text{ mm}$$

and hence

$$\delta_D = 0.143 + 0.226 = 0.369 \text{ mm}$$

Using Eq. (9.18) with $m = 2$, we have

$$n_{cr} = \frac{1}{2\pi} \left[\frac{9.81 \left(196.2 \times 0.409 \times 10^{-3} + 294.3 \times 0.369 \times 10^{-3}\right)}{196.2(0.409 \times 10^{-3})^2 + 294.3(0.369 \times 10^{-3})^2} \right]^{1/2}$$

$$= 25.37 \text{ cps} = 1522 \text{ rpm}$$

Determination of Critical Speed of a Stepped Shaft | **EXAMPLE 9.5**

Figure 9.7a shows a stepped round shaft supported by two bearings and carrying the flywheel weight *W*. Calculate the critical speed in rpm.

Given: The moment of inertia *I* of the shaft in its central region is twice that the moment of inertia in the end parts:

$$W = 400 \text{ N}, \qquad L = 1 \text{ m}, \qquad I = 0.3 \times 10^{-6} \text{ m}^4.$$

Assumptions: The shaft is made of steel with $E = 200$ GPa. The shaft weight is ignored. Bearings act as simple supports.

Solution: The application of the moment-area method to obtain the static deflection at the midspan *C* is illustrated in Figure 9.7. The bending moment diagram is given in Figure 9.7b and the *M/EI* diagram in Figure 9.7c. Note that, in the latter figure, C_1 and C_2 denote the centroids of the triangular and trapezoidal areas, respectively.

The first moment of the various parts of the *M/EI* diagram are used to find the deflection. From the symmetry of the beam, the tangent to the deflection curve at *C* is horizontal. Hence, according to the second-moment area theorem defined by Eq. (4.23), the deflection δ_C is obtained by taking the

Figure 9.7 Example 9.5. Calculation of deflections of a stepped shaft by the moment-area method.

moment of the M/EI area diagram between A and C about point A. That is,

$$\delta_C = (\text{first moment of triangle}) + (\text{first moment of trapezoid})$$

$$= \left(\frac{L}{6}\right)\left(\frac{WL^2}{64EI}\right) + \left(\frac{L}{4} + \frac{5L}{36}\right)\left(\frac{3WL^2}{128EI}\right) = \frac{3WL^3}{256EI} \tag{9.19}$$

Substituting this, $m = 1$, and $\delta_1 = \delta_C$ into Eq. (9.18), we have

$$n_{cr} = \frac{1}{2\pi}\sqrt{\frac{g}{\delta_C}}$$

$$= \frac{1}{2\pi}\left[\frac{9.81(256)(200 \times 10^9)(0.3 \times 10^{-6})}{3(400)(1)^3}\right]^{1/2}$$

$$= 56.40 \text{ cps} = 3384 \text{ rpm}$$

9.8 MOUNTING PARTS

Mounting parts, such as keys, pins, screws, ring, collars, and splines, are usually used on shafts to attach the hub of rotating members such as gears, pulleys, sprockets, cams, and flywheels. Note that the portion of the mounted members in contact with the shaft is the *hub*. The hub is attached to the shaft in variety of ways, using one of the foregoing mounting elements. Each mounting configuration has its own advantages and disadvantages.

Tables of dimensions for the mounting parts may be found in engineering handbooks and manufacturer's catalogs.

KEYS

A *key* enables the transmission of torque from the shaft to the hub. Numerous kinds of keys are used to meet various design requirements. They are standardized as to size and shape in several styles. Figure 9.8 illustrates a variety of keys. The grooves in the shaft and hub into which the key fits form the *keyways* or key seats. The square, flat type of keys are most common in machine construction.

The *gib-head key* is tapered so that, when firmly driven, it prevents relative axial motion. Another advantage is that the hub position can be adjusted for the best location. A tapered key may have no head or a gib head (as in Figure 9.8d) to facilitate removal.

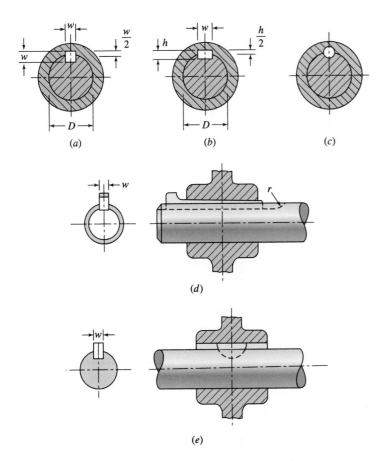

Figure 9.8 Common types of shaft keys: (a) square key ($w \approx D/4$); (b) flat key ($w \approx D/4$, $h \approx 3w/4$); (c) round key (often tapered); (d) gib-head key; (e) Woodruff key.

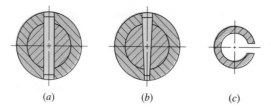

Figure 9.9 Some types of pins: (a) straight round pin; (b) tapered round pin; (c) cross section of a split tubular pin or so-called roll pin.

The *Woodruff key* is semicircular in plan and of constant width (w). It is utilized widely in the automotive and machine tool industries. Woodruff keys yield better concentricity after assembly of the hub and shafts. They are self-aligning and accordingly preferred for tapered shafts.

PINS

A pin is employed for axial positioning and the transfer of relatively light torque or axial load (or both) to the hub. Some types of shaft pins are the straight round pin, tapered round pin, and roll pin (Figure 9.9). The so-called roll pin is a split-tubular spring pin. It has sufficient flexibility to accommodate itself to small amounts of misalignment and variations in hole diameters, so it does not come loose under vibrating loads.

SCREWS

Very wide keys can be held in place with countersunk flat head or *cap screws* if the shaft is not weakened. In addition to a key or pin, *setscrews* are often employed to keep the hub from shifting axially on the shaft. For light service, the rotation between shaft and hub also may be prevented by setscrews alone. Setscrews are sometimes used in combination with keys. Various types of screws and standardized screw threads are discussed in Chapter 15.

RINGS AND COLLARS

Retaining rings, commonly referred to as *snap rings,* are available in numerous varieties and require that a small groove of specific dimensions be machined in the shaft. Keys, pins, and snap rings can be avoided by the use of *clamp collars* that squeeze the outside diameter of the shaft with high pressure to clamp something to it. The hub bore and clamp collar have a matching slight taper. The clamp collar with axial slits is forced into the space between hub and shaft by tightening the bolts.

METHODS OF AXIALLY POSITIONING OF HUBS

Figure 9.10 shows common methods of axially positioning and retaining hubs into shafts [14]. Axial loads acting on shafts or members mounted on the shaft are transmitted as follows: light loads by clamp joints, setscrews, snap rings, and tapered keys (Figure 9.8);

Figure 9.10 Various means of securing hubs for axial motion: (a) clamp collar; (b) setscrew; (c) snap rings; (d) nut; (e) tapered pin; (e) interference fit.

medium loads by nuts, pins, and clamp joints; heavy loads by press or shrink fits. Interference fits are also used to position and retain bearings into hubs.

9.9 STRESSES IN KEYS

The distribution of the force on the surfaces of a key is very complicated. Obviously, it depends on the fit of the key in the grooves of the shaft and hub. The stress varies nonuniformly along the key length; it is highest near the ends.

Owing to many uncertainties, an exact stress analysis cannot be made. However, it is commonly assumed in practice that a key is fitted as depicted in Figure 9.11. This implies that the entire torque T is carried by a tangential force F located at the shaft surface and uniformly distributed along the full length of the key:

$$T = Fr \tag{9.20}$$

where r is the shaft radius.

Shear and compressive or bearing stresses are calculated for the keys from force F, using a sufficiently large factor of safety. For steady loads, a *factor of safety* of 2 is commonly applied. On the other hand, for minor to high shock loads, a factor of safety of 2.5 to 4.5 should be used.

For keyways, the concentration of stress depends on the values of the fillet radius at the ends and along the bottom of the keyways. For end-milled key seats in shafts under either bending or torsion loading, the theoretical *stress concentration factors* range from 2 to about 4, contingent on the ratio r of r/D [15]. The quantity r represents the fillet radius (see Figure 9.8d) and D is the shaft diameter. The approximate values of the fatigue stress concentration factor range between 1.3 and 2 [4].

Figure 9.11 Forces on a key tightly fitted at top and bottom.

EXAMPLE 9.6 | Design of a Shaft Key

A shaft of diameter D rotates at 600 rpm and transmits 100 hp through a gear. A square key of width w is to be used (Figure 9.8a) as a mounting part. Determine the required length of the key.

Given: $D = 50$ mm, $w = 12$ mm

Design Decisions: The shaft and key will be made of AISI 1035 cold-drawn steel having the same tensile and compressive yield strength and that yield strength in shear is $S_{ys} = S_y/2$. The transmitted power produces intermittent minor shocks and a factor of safety of $n = 2.5$ is used.

Solution: From Table B.3, for AISI 1035 CD steel, we find $S_y = 460$ MPa. Through the use of Eq. (1.16),

$$T = \frac{7121(100)}{600} = 1.187 \text{ kN} \cdot \text{m}$$

The force F at the surface of the shaft (Figure 9.11) is

$$F = \frac{T}{r} = \frac{1.187}{0.025} = 47.48 \text{ kN}$$

On the basis of shear stress in the key,

$$\frac{S_y}{2n} = \frac{F}{wL} \quad \text{or} \quad L = \frac{2Fn}{S_y w} \tag{9.21}$$

Substitution of the given numerical values yields

$$L = \frac{2(47,480)(2.5)}{460(10^6)(0.012)} = 43 \text{ mm}$$

Based on compression or bearing on the key or shaft (Figure 9.8a),

$$\frac{S_{yc}}{n} = \frac{F}{(w/2)L} \quad \text{or} \quad L = \frac{2Fn}{S_{yc} w} \tag{9.22}$$

Inasmuch as $S_y = S_{yc}$, this also results in $L = 43$ mm.

9.10   SPLINES

When axial movement between the shaft and hub is required, relative rotation is prevented by means of splines machined on the shaft and into the hub. For example, splines are used to connect the transmission output shaft to the drive shaft in automobiles, where the suspension movement causes axial motion between the components. Splines are essentially "built-in keys." They can transform more torque than can be handled by keys. There are two forms of splines (Figure 9.12): straight or square tooth splines and involute tooth splines. The former is relatively simple and employed in some machine tools, automatic equipment, and so on. The latter has an involute curve in its outline, which is in widespread use on gears. The involute tooth has less stress concentration than the square tooth and,

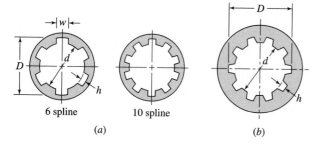

Figure 9.12 Some common types of splines: (a) straight-sided; (b) involute.

hence, is stronger. Also easier to cut and fit, the involute splines are becoming the prominent spline form.

Formulas for the dimensions of splines are based on the nominal shaft diameter [6, 16]. Figure 9.12a shows the standard SAE 6 and 10 straight spline fittings. Note that the values of root diameter d, width w, and depth h of the internal spline are based on the nominal shaft diameter D or about the root diameter of the external spline. According to the SAE, the torque capacity (in lb·in.) of *straight-sided splines* with sliding is

$$T = pnr_m h L_c \qquad (9.23)$$

where

T = theoretical torque capacity

n = number of splines

$r_m = (D + d)/4$, mean or pitch radius (see Figure 9.12)

h = depth of the spline

L_c = length of the spline contact

p = spline pressure

The SAE states that, in actual practice, owing to the inaccuracies in spacing and tooth form, the contact length L_c is about 25% of the spline length.

Involute splines (Figure 9.12b) have a general form of internal and external involute gear teeth, discussed in detail in Chapter 11, with modified dimensions. The length L_c of spline contact required to transmit a torque, as suggested by the SAE, is

$$L_c = \frac{D^2 \left(1 - d_i^4/D^4\right)}{d_m^3} \qquad (9.24)$$

The quantities D = nominal shaft diameter, d_m = the mean or pitch diameter, and d_i = internal diameter (if any) of a hollow shaft.

The shear area at the mean diameter of the spline is $A_s = \pi d_m L_c/2$. By the SAE assumption, only one-quarter of the shear area is to be stressed. The shear stress is estimated as

$$\tau = \frac{T}{(d_m/2)(A_s/4)} = \frac{8T}{d_m A_s}$$

or

$$\tau = \frac{16T}{\pi d_m^2 L_c}$$

(9.25)

Here, T represents the torque on the shaft and L_c is given by Eq. (9.23) or (9.24). If bending is present, the flexure stress in the spline must also be calculated.

9.11   COUPLINGS

Couplings are used to connect two shafts. A wide variety of commercial shaft couplings are available. They may be grouped into two broad classes: rigid and flexible. A rigid coupling locks the two shafts together, allowing no relative motion between them, although some axial adjustment is possible at assembly. No provision is made for misalignment between the two shafts connected, nor does it reduce shock or vibration across it from one shaft to the other. However, shafts are often subject to some radial, angular, and axial misalignment. In these situations, flexible couplings must be used. Severe misalignment must be corrected; slight misalignment can be absorbed by flexible couplings. This prevents fatigue failure or destruction of bearings.

CLAMPED RIGID COUPLINGS

Collinear shafts can be connected by clamp couplings that are made in several designs. The most-common two-piece split coupling clamps around both shafts by means of bolts and transmits torque. It is necessary to key the shafts to the coupling. The torque is transmitted mainly by friction due to the clamping action and partially by the key. Clamp couplings are widely used in heavy-duty service.

FLANGED RIGID COUPLINGS

Collinear shafts can also be connected by flanged couplings, similar to those shown in Figure 9.13. The flanged portion at the outside diameter serves a safety function by shielding the bolt heads and nuts. The load is taken to be divided equally among the bolts. Rigid couplings are simple in design. They are generally restricted to relatively low-speed applications where good shaft alignment or shaft flexibility can be expected.

Keyed couplings are the most widely used rigid couplings (Figure 9.13a). They can transmit substantial torques. The coupling halves are attached to the shaft ends by keys. As can be seen in the figure, flange alignment is obtained by fitting a shallow machined projection on one flange face to a female recess cut in the face of the other flange. Another common way to obtain flange alignment is to permit one shaft to act as a pilot and enter the mating flange. Keyed couplings employ standard keys as discussed in Section 9.8.

Compression couplings have a split double cone that does not move axially but is squeezed against the shaft by the wedging of the flanges, as shown in Figure 9.13b. This kind of coupling transmits torque only by the frictional force between the shaft and the split double cone, eliminating the need for a key and keyway in the coupling.

In specifying a rigid coupling using ground and fitted flange bolts, the designer should check the strength of various parts. These include direct shear failure of the bolts, bearing

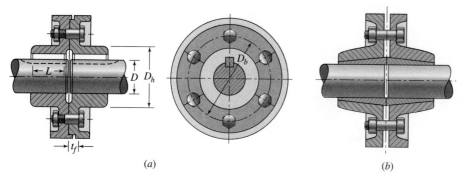

Figure 9.13   Flanged rigid couplings: (a) keyed type; (b) compression type.

of the projected area of the bolt in contact with the side of the hole, shear of the flange at the hub, and shear or crushing of the key. Note that, in contrast to fitted bolts, a flange coupling designed on the basis of friction-torque capacity requires a somewhat different analysis than that just described [17].

For flanged rigid couplings, it is usually assumed that shear stress in any one bolt is uniform and governed by the distance from its center to the center of the coupling. Friction between the flanges is disregarded. Then, if the shear stress in a bolt is multiplied by its cross-sectional area, the force in the bolt is ascertained. The moment of the forces developed by the bolts around the axis of a shaft estimates the torque capacity of a coupling.

Determining the Torque Capacity of a Rigid Coupling   **EXAMPLE 9.7**

A flanged keyed coupling is keyed to a shaft (Figure 9.13a). Calculate the torque that can be transmitted.

Given: There are 6 bolts of 25-mm diameter. The bolt circle diameter is $D_b = 150$ mm.

Assumptions: The torque capacity is controlled by an allowable shear strength of 210 MPa in the bolts.

Solution: Area in shear for one bolt is

$$A = \left(\tfrac{1}{4}\right)\pi(25)^2 = 491 \text{ mm}^2$$

Allowable force for one bolt:

$$P_{all} = AS_{ys} = 491(210) = 103.1 \text{ kN}$$

Inasmuch as six bolts are available at a 75-mm distance from the central axis, we have

$$T_{all} = 103.1 \times 10^3 \times 0.075 \times 6 = 46.4 \text{ kN·m}$$

Figure 9.14   A jaw coupling showing jaws and an elastomer insert. (Courtesy of Magnaloy Coupling Co., Alpena, MI.)

FLEXIBLE COUPLINGS

Flexible couplings are employed to connect shafts subject to some small amount of misalignment. One class of flexible couplings contains a flexing insert such as rubber or spring. The insert cushions the effect of shock and impact loads that could be transferred between shafts. A shear type of rubber-inserted coupling can be used for higher speeds and horsepowers. A chain coupling type consists of two identical sprockets coupled by a roller chain.

Figure 9.14 shows the two identical hubs of a *square-jawed coupling* with an elastomer (i.e., rubber) insert. In operation, the halves slide along the shafts on which they are mounted until they engage with the elastomer. The clearances permit some axial, angular, and parallel misalignment. Clearly, the jaws are subjected to bearing and shear stresses. The force acting on the jaw producing these stresses depends on the horsepower and speed that the coupling is to transmit.

Many other types of flexible couplings are available. Details, dimensions, and load ratings may be found in the catalogs of various manufacturers or mechanical engineering handbooks.

9.12   UNIVERSAL JOINTS

A universal joint is a kinematics linkage used to connect two shafts that have permanent intersecting axes. Universal joints permit substantial misalignment of shafts. They come in two common types: the *Hooke* or Cardan *coupling*, which does not have constant velocity across a single joint, and the Rzeppa coupling, which does. Both types can deal with very large angular misalignment. Shaft angles up to 30° may be used [18].

Hooke's coupling is the simplest kind of universal joint. It consists of a yoke on each shaft connected by a central cross-link. Figure 9.15 depicts a double–Hooke joint, where plain bearings are used at the yoke-to-cross connections. These joints are employed mostly with equal yoke alignment angles (α) in the two joints, as shown in the figure. The use of equal angles provides uniform angular velocity in the driven shaft. A pair of Hooke's couplings is often used in a rear-drive automobile drive shaft. Note that a familiar application of the Rzeppa coupling is in front-wheel-drive automobiles, where the drive shaft is short and shaft angles can be large. For further information, see texts on mechanics of machinery [19, 11] and manufacturers' product literature on universal joints.

Figure 9.15 Two arrangements of a pair of Hooke's couplings for achieving constant velocity ratio.

Case Study 9-1 | SHAFTING DESIGN OF A WINCH CRANE GEAR BOX

Figure 9.16 shows the input shaft of the crane gear box, supported in the gear box by bearings A and B and driven by electric motor (see Case Study 1-1). Determine

(a) The factor of safety n for the shaft using the maximum energy of distortion theory incorporated with the Goodman criterion.

(b) The rotational displacements or slopes at the bearings.

(c) The stresses in the shaft key.

Given: The geometry and dimensions of the hollow shaft and square shaft key are known.

Data: Refer to Figure 9.16a and Case Study 11-1.

$F_t = 206$ N, $\qquad F_r = 75$ N, $\qquad T = 2.06$ N·m,

$a = 66$ mm, $\qquad b = 84$ mm, $\qquad L = 150$ N·m,

$d = 6$ mm, $\qquad d_p = 20$ mm, $\qquad D = 12$ mm,

$w = 2.4$ mm, $\qquad L_k = 25$ mm,

$I = \dfrac{\pi}{64}(D^4 - d^4) = 954.3$ mm^4

The operating environment is room air at a maximum temperature of 50°C.

Assumption: Bearings act as simple supports.

Design Decisions:

1. The shaft and shaft key are made of 1030 CD steel with machined surfaces:

   $S_u = 520$ MPa, $\quad S_y = 440$ MPa (from Table B.3),

   $E = 210$ GPa

2. At the keyway, $K_f = 2$.

3. The shaft rotates and carries steady loading at normal temperature.

4. The factor of safety is $n = 3$ against shear of shaft key.

5. A survival rate of 99.9% is used.

Solution: See Figure 9.16 and Table A.9.

(a) The reactions at A and B, as determined by the conditions of equilibrium, are indicated in Figure 9.16b. The moment and torque diagrams are obtained in the usual manner and drawn in Figure 9.16c. Observe that the critical section is at point C. We have

$$M_C = [(2.77)^2 + (7.62)^2]^{1/2} = 8.11 \text{ N·m}$$

$$T_C = 2.06 \text{ N·m}$$

(continued)

Case Study (CONTINUED)

Figure 9.16 Drive shaft supported in the gear box of the winch crane shown in Figure 1.4: (a) shaft layout; (b) loading diagram; (c) moment and torque diagrams.

The mean and alternating moments and torques, by Eq. (9.10), are then

$$M_m = 0 \qquad M_a = 8.11 \text{ N·m}$$

$$T_m = 2.06 \text{ N·m} \qquad T_a = 0$$

The modified endurance limit, through the use of Eqs. (8.1) and (8.6) and referring to Section 8.7, is

$$S_e = C_f C_r C_s C_t \left(\frac{1}{K_f}\right) S'_e$$

where

$$C_f = 4.51(520^{-0.265}) = 0.86$$

$$C_r = 0.75 \quad \text{(by Table 8.3)}$$

$$C_s \approx 0.85 \quad \text{(using Eq. (8.9))}$$

$$C_t = 1$$

$$K_f = 2$$

$$S'_e = 0.5(520) = 260 \text{ MPa} \quad \text{(from Eq. (8.1))}$$

(continued)

Case Study (CONCLUDED)

Hence

$$S_e = (0.86)(0.75)(0.85)(1)\left(\frac{1}{2}\right)(260) = 71.27 \text{ MPa}$$

Since the loading is steady, the shock factors $K_{sb} = K_{st} = 1$ by Table 9.1.

Substituting the numerical values into Eq. (9.12) and replacing D^3 with $D^3[1 - (d/D)^4]$, we obtain

$$\frac{520(10^6)}{n} = \frac{32}{\pi(0.012)^3[1 - (6/12)^4]}$$

$$\times \left[\left(0 + \frac{520 \times 8.11}{71.27}\right)^2 + \frac{3}{4}(2.06)^2\right]^{1/2}$$

from which

$$n = 1.4$$

(b) The slopes at the ends A and B (Figure 9.16b) are given by Case 6 of Table A.9. Note that $L^2 - b^2 = (L + b)(L - b) = (L + b)a$ and similarly $L^2 - a^2 = (L + a)b$. Introducing the given data, the results are

$$\theta_A = -\frac{F_t ab(L + b)}{6EIL}$$

$$= -\frac{206(66)(84)(150 + 84)}{6(210 \times 10^3)(954.3)(150)}$$

$$= -1.482(10^{-3}) \text{ rad} = -0.085°$$

$$\theta_B = \frac{F_t ab(L + a)}{6EIL}$$

$$= \frac{206(66)(84)(150 + 66)}{6(210 \times 10^3)(954.3)(150)}$$

$$= 1.368(10^{-3}) \text{ rad} = 0.078°$$

where a minus sign means a clockwise rotation.

Comments: Inasmuch as the bearing and gear stiffnesses are ignored, the negligibly small values of θ_A and θ_B estimated by the preceding equations represent higher angles than the true slopes. Therefore, self-aligning bearings are not necessary.

(c) The compressive forces acting on the sides of the shaft key equal $F_t = 206$ N (Figure 9.16). The shear stress in the shaft key is

$$\tau = \frac{F_t}{wL_k} = \frac{206}{(0.0024)(0.025)} = 3.433 \text{ MPa}$$

We have, from Eq. (7.20),

$$S_{ys} = 0.577S_y = (0.577)(440) = 253.9 \text{ MPa}$$

The allowable shear stress in the shaft key is

$$\tau_{all} = \frac{S_{ys}}{n} = \frac{253.9}{3} = 84.63 \text{ MPa}$$

Since $\tau_{all} \gg \tau$, shear should not occur at shaft key. We obtain the same result on the basis of compression or bearing on key (see Section 9.9).

Comments: On following a procedure similar to that in the preceding solution, the design of the remaining three shafts and the associated keys in the gear box of the winch crane can be analyzed in a like manner.

REFERENCES

1. Berchardt, H. A. "A Comprehensive Method for Designing Shafts to Insure Adequate Fatigue Life." *Machine Design* (April 25, 1963).
2. Design of Transmission Shafting, ANSI/ASME B106.1M-1985.
3. Loewenthal, S. H. "Proposed Design Procedure for Transmission Shafting Under Fatigue Loading." Technical Note TM-78927, NASA, 1978.

4. Juvinall, R. C., and K. M. Marshek. *Fundamentals of Machine Component Design,* 3rd ed. New York: Wiley, 2000.

5. Norton, R. L. *Machine Design—An Integrated Approach,* 2nd ed. Upper Saddle River, NJ: Prentice Hall, 2000.

6. Oberg, E., et al. *Machinery's Handbook,* 25th ed. New York: Industrial Press, 1996.

7. Burr, A. H., and J. B. Cheatham. *Mechanical Analysis and Design,* 2nd ed. Upper Saddle River, NJ: Prentice Hall, 1995.

8. Spotts, M. F., and T. E. Shoup. *Design of Machine Elements,* 7th ed. Upper Saddle River, NJ: Prentice Hall, 1998.

9. Hamrock, B. J., B. Jacobson, and S. R. Schmid. *Fundamentals of Machine Elements.* New York: McGraw-Hill, 1999.

10. Wilson, C. E. *Computer Integrated Machine Design.* Upper Saddle River, NJ: Prentice Hall, 1997.

11. Norton, R. L. *Design of Machinery,* 2nd ed. New York: McGraw-Hill, 1999.

12. Timoshenko, S., D. H. Young, and W. Weaver, Jr. *Vibration Problems in Engineering,* 4th ed. New York: Wiley, 1974.

13. Mischke, C. R. *Engineering Analysis.* Reading, MA: Addison-Wesley, 1963.

14. Dimarogonas, A. D. *Machine Design—A CAD Approach.* New York: Wiley, 2001.

15. Peterson, R. E. *Stress Concentration Factors.* New York: Wiley, 1974.

16. Society of Automotive Engineers, Inc. *SAE Handbook.* New York: SAE, 1954.

17. Deutschman, A. D., W. J. Michels, and C. E. Wilson. *Machine Design.* New York: Macmillan, 1975.

18. Avollone, E. A., and T. Baumeister III. *Mark's Standard Handbook for Mechanical Engineers,* 10th ed. New York: McGraw-Hill, 1996.

19. Hinkle, R. T. *Kinematics of Machinery.* Upper Saddle River, NJ: Prentice Hall, 1964.

Problems

The bearings of the shafts described in the following problems act as simple supports.

Sections 9.1 through 9.4

9.1 Design a solid shaft for a 15-hp motor operating at a speed n.

Given: $G = 80$ GPa, $S_{ys} = 150$ MPa, $n = 2500$ rpm

The angle of twist is limited to $2°$ per meter length.

Design Assumptions: The shaft is made of steel. A factor of safety of 3 is used.

9.2 A 40-hp motor, through a set of gears, drives a shaft at a speed n, as shown in Figure P9.2.

Figure P9.2

(a) Based on a safety factor of 2, design solid shafts AC and BC.

(b) Determine the total angle of twist between A and B.

Given: $G = 82$ GPa, $\quad S_{ys} = 210$ MPa, $\quad n = 1200$ rpm

9.3 Because of atmospheric corrosion an ASTM-A36 steel shaft of diameter D_s is to be replaced by an aluminum alloy 2014-T6 shaft. Determine the diameter of aluminum shaft D_a in terms of D_s. What is the weight ratio of the shafts?

Assumption: Both shafts have the same angular stiffness.

9.4 A solid steel shaft of diameter D carries end loads P, M, and T. Determine the factor of safety n, assuming that failure occurs according to the following criteria:

(a) Maximum shear stress.

(b) Maximum energy of distortion.

Given: $S_y = 260$ MPa, $\quad D = 100$ mm, $\quad P = 50$ kN, $\quad M = 5$ kN·m, $\quad T = 8$ kN·m

9.5 A solid steel shaft carries belt tensions (at an angle α from the y axis in the yz plane) at pulley C, as shown in Figure P9.5. For $\alpha = 0$ and a factor of safety of n, design the shaft according to the following failure criteria:

(a) Maximum shear stress.

(b) Maximum energy of distortion.

Given: $S_y = 250$ MPa, $\quad n = 1.5$

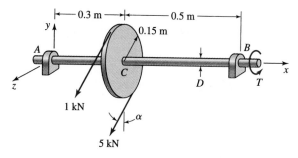

Figure P9.5

9.6 A solid shaft is used to transmit 90 hp to a series of chemical mixing vats at a speed of n. Calculate the shaft diameter according to the following failure criteria:

(a) Maximum shear stress.

(b) Maximum energy of distortion.

Given: $n = 110$ rpm

Design Decisions: The shaft is made of type 302 cold-rolled stainless steel. Since the atmosphere may be corrosive, a safety factor of 4 is used.

Section 9.5

9.7 A revolving solid steel shaft AB with a machined surface carries minor shock belt tensions (in the yz plane) for the case in which $\alpha = 30°$ (Figure P9.5). Design the shaft segment BC by using the maximum shear stress theory incorporated with the Goodman failure relation.

Given: $S_u = 520$ MPa, $n = 1.5$, and $K_f = 1.2$

The operating temperature is $500°$C maximum.

Assumptions: The survival rate is 90%.

9.8 A solid shaft of diameter D rotates and supports the loading depicted in Figure P9.8. Determine the factor of safety n for the shaft on the basis of maximum energy of distortion theory of failure combined with the Goodman criterion.

Given: $D = 2.5$ in. and $S_u = 180$ ksi. The torque fluctuates 10% each way from mean value, and the survival rate is 99%.
Design Decision: The shaft is grounded from unnotched steel.

Figure P9.8

9.9 A solid shaft of diameter D rotates and supports the loading shown in Figure P9.9. Calculate the factor of safety n for the shaft using the maximum shear stress theory of failure incorporated with the Soderberg criterion.

Given: $D = 3.5$ in., $S_y = 130$ ksi, and $S_u = 200$ ksi. The torque involves heavy shocks and fluctuates 20% each way from the mean value, and the survival rate is 90%. The maximum operating temperature is $950°$F.
Design Decision: The shaft is to be hot-rolled from an unnotched steel.

Figure P9.9

9.10 A rotating solid shaft is acted on by repetitive steady moments M and heavy shock torques T at its ends. Calculate, on the basis of the maximum energy of distortion failure criterion associated with the Goodman theory, the required shaft diameter D.

Given: $M = 200$ N $\cdot$ m, $\quad T = 500$ N $\cdot$ m, $\quad S_u = 455$ MPa, $\quad n = 1.5$, $\quad K_f = 2.2$
Design Decisions: The shaft is machined from 1020 HR steel. A survival rate of 95% is used.

9.11 Figure P9.11 shows a rotating stepped shaft supported in (frictionless) ball bearings at A and B and loaded by nonrotating force P and torque T. All dimensions are in millimeters. Determine the factor of safety n for the shaft, based on the maximum shear stress theory of failure incorporated with the Soderberg fatigue relation.

Given: $P = 5$ kN, $\quad T = 600$ kN$\cdot$m, $\quad S_y = 600$ MPa, $\quad S_u = 1000$ MPa, $\quad K_t = 1.8$

The torque fluctuates 15% each way from mean value, and the survival rate is 98%.

Assumption: The shaft is to be hot rolled from steel.

Figure P9.11

9.12 A solid shaft of diameter D rotates and carries the minor shock loading as shown in Figure P9.12. Calculate the factor of safety n for the shaft using the maximum energy of distortion theory of failure combined with the Goodman criterion.

Given: $D = 75$ mm, $S_y = 550$ MPa, $S_u = 660$ MPa. The torque fluctuates 5% each way from the mean value, and the shaft is to be machined from unnotched steel.

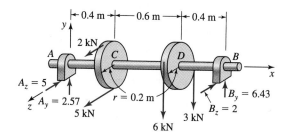

Figure P9.12

9.13 Redo Problem 9.12 using the maximum shear stress theory of failure incorporated with the Soderberg fatigue relation and a survival rate of 90%.

9.14 A revolving shaft, made of solid AISI 1040 cold-drawn steel, supports the loading depicted in Figure P9.14. The pulley weighs 300 lb and the gear weighs 100 lb. Design the shaft by using the maximum energy of distortion theory of failure incorporated with the Goodman fatigue criterion.

Given: $K_f = 1.8$, $\quad n = 1.6$
Assumption: A survival rate of 95% is used.

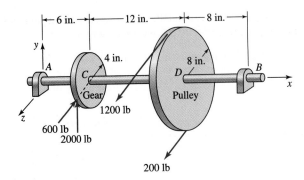

Figure P9.14

9.15 Redo Problem 9.14 using the maximum shear stress theory of failure incorporated with the Soderberg criterion, a survival rate of 99.9%, $K_f = 1.2, n = 2$, and neglecting the weights of pulley and gear.

Sections 9.6 through 9.12

9.16 A solid steel shaft of diameter D is supported and loaded as shown in Figure P9.16. Determine the critical speed n_{cr} in rpm.

Given: $D = 25$ mm, $E = 210$ GPa

Figure P9.16

9.17 Calculate the critical speed n_{cr} in rpm of the steel shaft of Figure P9.16, if the maximum allowable static deflection is 0.5 mm.

Given: $E = 200$ GPa

9.18 A uniform steel shaft with an overhang is loaded as shown in Figure P9.18. Determine the critical speed n_{cr} in rpm.

Given: $E = 210$ GPa

Figure P9.18

9.19 A solid shaft of diameter D has a $\frac{3}{4} \times \frac{3}{8}$ in. flat key. Determine the required length L of key based on the maximum steady torque that can be transmitted by the shaft.

Given: $D = 3$ in., $S_{ys} = 0.58S_y$, $n = 2$

Design Decision: The shaft and key are made of cold-drawn steels of AISI 1030 and AISI 1020, respectively.

9.20 A $\frac{3}{8} \times \frac{3}{8} \times 3$ in. key is used to hold a 3 in. long hub in a $1\frac{1}{2}$ in. diameter shaft. What is the factor of safety against shear failure of the key if the torque transmitted is 3.5 kip·in.

Assumption: Key and shaft are of the same material with an allowable stress of 10 ksi.

9.W Through the use of the website at www.grainger.com conduct a search for flexible couplings, both rated for $\frac{1}{2}$ hp at 1725 rpm:

(a) $\frac{1}{2}$ in. bore, $2\frac{1}{2}$ in. long

(b) $\frac{1}{2}$ in. bore, $3\frac{1}{2}$ in. long

List the manufacturer and description in each case.

9.21 For the coupling shown in Figure 9.13a, the key is $\frac{9}{16} \times \frac{9}{16} \times 3\frac{1}{2}$ in., bolt diameter $D_b = 6$ in., hub diameters $D_h = 4$ in. and $D = 2$ in. Six $\frac{3}{4}$-in. bolts are used and flange thickness is $t_f = \frac{7}{8}$ in. Determine

(a)   The shear and bearing stresses in the key.

(b)   The shear stress in the bolts.

(c)   The bearing stress on bolts in the flange.

(d)   The shear stress in the flange at the hub or web.

Given: The shaft carries a steady load of 60 hp at 200 rpm.

CHAPTER

10

BEARINGS AND LUBRICATION

Outline

10.1 INTRODUCTION

The goal of a bearing is to provide relative positioning and rotational freedom while transmitting a load between two parts, commonly a shaft and its housing. The object of lubrication is to reduce the friction, wear, and heating between two surfaces moving relative to each other. This is done by inserting a substance, called *lubricant,* between the moving surfaces. The study of lubrication and the design of bearings are concerned mainly with phenomena related to the oil film between the moving parts. Note that *tribology* may be defined as the study of the lubrication, friction, and wear of moving or stationary parts. The literature on this complex subject is voluminous. Much is collected in the *CRC Handbook of Lubrication,* sponsored by the American Society of Lubrication Engineers [1]. Also see [2–6]. The website www.machinedesign.com includes general information on bearings and lubrication.

There are two parts in this chapter. In Part A, the fundamentals of lubrication with particular emphasis on the design of journal (so-called sleeve or sliding) bearings is discussed. The basic forms of journal bearings are simple. In Part B, concern is with rolling bearings, also known as *rolling-element bearings,* and antifriction bearings. We describe the most common types of rolling bearings, bearing dimensions, bearing load, and bearing life. There is also a brief discussion on materials, mounting, and lubricants of rolling bearings. Rolling-element bearings are employed to transfer the main load through elements in rolling contact, and they have been brought to their present state of perfection only after a long period of development. Either ball bearings or roller bearings, they are made by all major bearing manufacturers worldwide.

Part A   Lubrication and Journal Bearings

Journal bearings support loads perpendicular to the shaft axis by pressure developed in the liquid. A journal bearing is a typical sliding bearing requiring sliding of the load-carrying member on its support. Sleeve thrust bearings support loads in the direction of the shaft axis. We begin with a description of the lubrications and journal bearings. The general relationship between film velocity rate, viscosity, coefficient of friction, and load are then developed. This is followed by discussions of the hydrodynamic lubrication theory, design, and heat balance of bearings. Techniques for supplying oil to bearings and bearing materials are also considered.

10.2 LUBRICANTS

As noted previously, the introduction of a lubricant to a sliding surface reduces the coefficient of friction. In addition, lubricants can act as contaminants to the metal surfaces and coat them with monolayers of molecules that inhibit adhesion between metals. Although usually in the liquid state, solids, and gases are also used as lubricants. A brief description of the classification and characteristics of lubricants are as follows. Lubricant manufacturers should be consulted for particular application.

LIQUID LUBRICANTS

Liquid lubricants are largely petroleum-based or synthetic *oils*. They are characterized by their viscosity, but other properties are also important. Characteristics such as acidity, resistance to oxidation, antifoaming, pour, flash, and fire deterioration are related to the quality of oil needed for a particular operation. Many oils are marketed under the name of application, such as compressor or turbine oils. Oils for vehicle engines are classified by their viscosity as well as by the presence of additives for various service conditions (see Section 10.4).

Synthetic lubricants are mainly silicones. They have high-temperature stability, low–temperature fluidity, and high-internal resistance. Because of their higher cost, synthetic lubricants are used only when their special properties are needed; for instance, in the hydraulic control systems of aircraft. *Water* and *air* are used as lubricants where contamination by oil is prohibitive. In addition, often the lubricant is the water or air in which the machine is immersed. Air or an inert gas has very low internal resistance. It operates well from low to high temperatures. Gas lubricants are necessary at extremely high speeds involving low loading conditions.

Greases are liquid lubricants that have been thickened (by mixing with soaps) to provide properties not available in the liquid lubricant alone. Mineral oils are the most commonly used liquid for this purpose. Greases are often used where the lubricant is required to stay in position. Unlike oils, greases cannot circulate and thereby serve a cooling and cleaning function; however, they are expected to accomplish all functions of fluid lubricants. The many types of greases have properties suitable for a wide variety of operating conditions. Typical uses of greases include vehicle suspension and steering; for gears and bearings in lightly loaded and intermittent service, with infrequent lubrication by hand or grease gun.

SOLID LUBRICANTS

Solid lubricants are of two types: graphite and powdered metal. They are used for bearing operating at high temperatures (e.g., in electric motors). Other kinds include teflon and some chemical coatings. Solid lubricants may be brushed or sprayed directly into the bearing surfaces. To improve retention, they are mixed with adhesives. Determination of composite bearing materials with low wear rates as well as frictional coefficients is an active area of contemporary design and research.

10.3   TYPES OF JOURNAL BEARINGS AND LUBRICATION

The journal bearing or sleeve bearing supports a load in the radial direction. It has two main parts: a shaft called the *journal* and a hollow cylinder or sleeve that carries the shaft, called the *bearing* (Figure 10.1). When assembly operations do not require that a bearing be of two pieces, the bearing insert can be made as a one-piece cylindrical shell pressed into a hole in the housing. This insert is also called a *bushing*.

Figure 10.1   (a) Full journal bearing. (b) Partial journal bearing.

A *full-journal bearing,* or so-called 360° journal bearing, is made with the full bearing thickness around the whole circumference, as depicted in Figure 10.1a. Circumferential or any (usually axial or diagonal) grooving may be cut in the two-piece and one-piece bearings, respectively. Preferably, the oil is brought at the center of the bearing so that it will flow out both ends, thus increasing the flow and cooling action. In most applications, the journal rotates (with a speed *n*) within a stationary bearing and the relative motion is sliding. However, the journal may remain stationary and the bearing rotates or both the journal and bearing rotate. All situations require an oil film between the moving parts to minimize friction and wear. The pressure distribution around a bearing varies greatly. The coefficient of friction, *f*, is the ratio of the tangential friction force, discussed in Section 10.5, to the load carried by the bearing.

Sleeve bearings are employed in numerous fields. Two typical services a bearing is to perform are as follows. The crankshaft and connecting rod bearings of an automobile engine must operate for thousands of miles at high temperatures and under variable loading. The journal bearings used in the steam turbines and power generator sets must have very high reliability. *Gas bearings* using air or more inert gases as the lubricant and film find applications for lightly loaded, high-speed shafts, such as in gas-cycle machinery, gyros, and high-speed dental drills. Also, when the loads are light and the service relatively unimportant, a *nylon bearing* that requires no lubrication must be used.

A *partial bearing* is used when the radial load on a bearing always acts in one direction; hence, the bearing surface needs to extend only part way around the periphery. Often, an oil cap is placed around the remainder of the circumference. An angle (e.g., $\theta = 60°$) describes the angular length of a partial bearing (Figure 10.1b). Rail freight car axle bearings are an example. A partial bearing having *zero* clearance is known as a *fitted bearing.* Zero clearance means that the radii of the journal and bearing are equal. We consider only the more common full bearing.

FORMS OF LUBRICATION

Lubrications commonly are classified according to the degree with which the lubricant separates the sliding surfaces. Five distinct forms or types of lubrication occur in bearings: hydrodynamic, mixed, boundary, elastohydrodynamic, and hydrostatic. The bearings are often designated according to the form of lubrication used.

Figure 10.2   The change in the coefficient of friction f with shaft speed n in a journal bearing.

To gain insight into the possible lubrication states, consider the experimentally determined curve between the shaft speed n and coefficient of friction f in a journal bearing (Figure 10.2). Clearly, the numerical values for the curve in the figure depend on features of the particular bearing design. Note that bearings operate under boundary conditions at startup or shut down. At slow speeds, the coefficient of friction remains about the same in region of boundary lubrication. As n is increased, a mixed lubrication situation is initiated (point A), and f drops rapidly until hydrodynamic lubrication is established (point B). At higher speeds, f rises slowly. For extremely large velocities (beyond point C), instability and turbulance may be established in the lubricant. Note that regions to the left and right of point B represent thin-film and thick-film lubrications, respectively. We now briefly discuss the conditions that induce the foregoing lubrication states.

Hydrodynamic lubrication means that load-carrying surfaces of the bearing are separated by a (relatively thick) layer of fluid, called *fluid film*. For this condition to occur, a relative motion must exist between the two surfaces and a pressure must be developed. The pressure is created internally by the relative velocity, viscosity of the fluid, and the wedging action that results when the two surfaces are not parallel. This technique does not depend on the introduction of the lubricant under pressure. It does require, however, the existence of an adequate fluid supply at all times.

In a journal bearing at rest, the shaft sits in contact with the bottom of the bearing. As soon as the shaft rotates, its centerline shifts eccentrically within the bearing. Thus, a flow is set up within the small thickness of the oil film. When the rotating speed increases sufficiently, the shaft moves up on a wedge of pumped oil, and ends its metal contact with bearing; hydrodynamic lubrication is established. In a hydrodynamically lubricated sleeve-bearing surface, wear does not occur. Friction losses originate only within the lubricant film. Typical minimum film thickness (denoted h_o) ranges from 0.008 to 0.020 mm. Coefficients of friction f commonly range from 0.002 to 0.010. Hydrodynamic lubrication is also known as *fluid film* or *fluid lubrication*. The design of journal bearings is based on this most desirable type of lubrication.

Mixed lubrication describes a combination of partial lubricant film plus intermittent contact between the surfaces. Under this condition, the wear between the surfaces depends on the properties of the surfaces and the lubricant viscosity. Typical values of the coefficient of friction are 0.004 to 0.10.

Boundary lubrication refers to the situations in which the fluid film gets thinner and partial metal-to-metal contact can occur. This depends on such factors as surface finish and wear-in and surface chemical reaction. Boundary lubrication occurs in journal bearings at low speeds and high loads, as when starting or stopping a rotating machinery. The properties of the sliding metallic surfaces and the lubricant are significant factors in limiting wear. The coefficient of friction is about 0.10. Boundary lubrication is less desirable than the other types, inasmuch as it allows the surface asperities to contact and wear rapidly. Design for this type of lubrication is largely emprical.

Note that the initial boundary lubrication can be avoided by the introduction of pressurized oil on the loaded side of the journal; thereby hydraulically lifting it at startup and again at shut down. This is a common practice on large machines (e.g., power turbines), to provide sleeves and shaft a wear-free long life. The foregoing, called hydrostatic lubrication, is discussed later.

Elastohyrodynamic lubrication is concerned with the interrelation between the hydrodynamic action of full-fluid films and the elastic deformation of the supporting materials. It occurs when the lubricant is introduced between surfaces in rolling contact, such as mating gears and rolling bearings. Under loaded contact, balls and rollers, as well as cams and gear teeth, develop a small area of contact because of local elastic deformation owing to high stress (e.g., 700–3500 MPa). Factors that have major effect on creating elastohyrodynamic lubrication are increased relative velocity, increased oil viscosity, and increased radius of curvature at the contact. The mathematical explanation requires the Hertz contact stress analysis, as discussed in Chapter 3 and fluid mechanics [1].

Hydrostatic lubrication refers to the continuous supply of flow of lubricant to the sliding interface at some elevated hydrostatic pressure. It does not require motion of the surfaces relative to another. This mechanism creates full-film lubrication. Some special applications involving hydrostatic lifts, thrust bearings, and oil lifts needed during the startup of heavily loaded bearings are of the hydrostatic forms. Obviously, in hydrostatic lubrication, the pressure is developed externally by a pump, and the fluid (typically oil) enters the bearing opposite the load. The advantages of this technique include notably low friction and high load-carrying capacity at low speeds at all times. Disadvantages are the cost and the need for an external source of fluid pressurization.

Consider a simplified sketch of a vertical shaft *hydrostatic thrust bearing* shown in Figure 10.3. The rotating shaft supports a vertical load W. High-pressure oil at p is supplied into the recess of radius r_o at the center of the bearing from an external pump. Oil flows radially outward the annulus of depth h, finally escapes at the periphery of the shaft, and then finally returns through a system of piping to the reservoir at about atmospheric pressure. The oil film is present whether the shaft rotates or not. It can be shown that [7] the load-carrying capacity is given in the form

$$W = \frac{p\pi}{2}\left[\frac{r^2 - r_o^2}{\ln(r/r_o)}\right] \tag{10.1}$$

The preceding is applicable even if the recess is eliminated. In this case, r_o becomes the radius of the inlet oil-supply pipe.

Figure 10.3   Schematic representation of a hydrostatic thrust bearing.

Hydrostatic bearings are used in various special applications. Some examples are telescopes and radar tracking units subjected to heavy loads at very low speeds as well as the machine tools and gyroscopes under high speed but light loads. Further details may be found in [6, 7].

10.4   LUBRICANT VISCOSITY

When two plates having relative motions are separated by a lubricant (e.g., oil) a flow takes place. In most lubrication problems, conditions are such that the flow is laminar. In *laminar flow,* the fluid is in layers that are maintained as the flow progresses. When this condition is not met, the flow is called *turbulant.* The laminar flow and internal resistance to shear of the fluid can be demonstrated by referring to the system depicted in Figure 10.4a. The figure shows that the lower plate is stationary, while the upper plate moves to the right with a velocity U under the action of the force F. Inasmuch as most fluids tend to "wet" and adhere to solid surfaces, it can be taken that, when the plate moves, it does not slide along on top of the film (Figure 10.4b). The plot of fluid velocity u against y across the film (shown in the figure) is known as the *velocity profile*.

Figure 10.4   Laminar flow: (a) flat plate moving on fluid film; (b) a fluid element.

Newton's law of viscous flow states that the shear stress in the fluid is proportional to the rate of change of velocity with respect to y (Figure 10.4a). That is,

$$\tau = \eta \frac{du}{dy} \tag{10.2}$$

The factor of proportionality η is called the *absolute viscosity* or simply the *viscosity*. The viscosity is a measure of the ability of the fluid *to resist* shear stress. Newtonian fluids include air, water, and most oils. Those fluids to which Eq. (10.2) does not apply are called non-Newtonian. Examples are lubricating greases and some oils with additives. Let the distance between the two plates, the film thickness, be denoted by h, as shown in Figure 10.4. Because the velocity varies linearly across the film, we have $du/dy = U/h$ and $\tau = F/A$. Substitution of these relations into Eq. (10.2) results in

$$F = \eta \frac{AU}{h} \tag{10.3}$$

In the foregoing, A represents the area of the upper plate.

UNITS OF VISCOSITY

In SI, viscosity is measured in newton-second per square meter ($N \cdot s/m^2$) or pascals-seconds. The U.S. customary unit of viscosity is the pound-force-second per square inch ($lb \cdot s/in.^2$), called the *reyn*. The conversion between the two units is the same as stress:

$$1 \text{ reyn} = 6890 \text{ Pa} \cdot s$$

The reyn and pascal-second are such large units that microreyn (μreyn) and millipascal second (mPa $\cdot$ s) are more commonly used.

In the former metric system, centimeter-gram-second (cgs), the unit of viscosity, is poise (p), having dimensions of dyne-second per square centimeter (dyne $\cdot$ s/cm^2). Note that 1 centipoise is equal to 1 millipascal-second (1 cp = 1 mPa $\cdot$ s). It has been customary to use the centipoise (cp), which is $1/100$ of a poise. The conversion from cgs units to U.S. customary units is as follows: 1 reyn = 6.89 (10^6) cp. To obtain viscosity in μreyn, multiply the cp value by 0.145.

VISCOSITY IN TERMS OF SAYBOLT UNIVERSAL SECONDS

The American Society for Testing and Materials (ASTM) standard method for determining viscosity employs an instrument known as the Saybolt universal viscometer. The approach consists of measuring the time in seconds needed for 60 cm^3 of oil at a specified temperature to flow through a capillary tube 17.6-mm diameter and 12.25-mm long (Figure 10.5). The time, measured in seconds, is known as *Saybolt universal seconds*, S.

Kinematic viscosity, also called *Saybolt universal viscosity* (SUV) in seconds, is defined by

$$\nu = \frac{\text{absolute viscosity}}{\text{mass density}} = \frac{\eta}{\rho} \tag{10.4}$$

Figure 10.5   Saybolt universal viscosimeter.

Figure 10.6   Variation in viscosity with temperature of several fluids.

The mass density ρ is in g/cm^3 of oil (which is numerically equal to the specific gravity). In SI, the kinematic viscosity ν has the unit of m^2/s. In the former metric system, a unit of cm^2/s was named a *stoke*, abbreviated *St.*

Absolute viscosity is needed for calculation of oil pressure and flows within a bearing. It can be found from Saybolt viscosimeter measurements by the formulas:

$$\eta = \left(0.22S - \frac{180}{S} \right)\rho \qquad (\text{mPa} \cdot \text{s, or cp}) \tag{10.5a}$$

$$\eta = 0.145\left(0.22S - \frac{180}{S} \right)\rho \qquad (\mu\text{reyn}) \tag{10.5b}$$

Here Saybolt time S is in seconds. Interestingly, for petroleum oils, the mass density at 60°F (15.6°C) is approximately 0.89 g/cm^3. The mass density, at any other temperature, is given by

$$\rho = 0.89 - 0.00063(°\text{C} - 15.6) \tag{10.6a}$$

$$\rho = 0.89 - 0.00035(°\text{F} - 60) \tag{10.6b}$$

both in g/cm^3.

EFFECTS OF TEMPERATURE AND PRESSURE

The viscosity of a liquid varies inversely with temperature and directly with pressure, both nonlinearly. In contrast, gases such as air have an increased viscosity with increased temperature. Figure 10.6 shows the absolute viscosity of various fluids and how they vary. The Society of Automotive Engineers (SAE) and the International Standards Organization

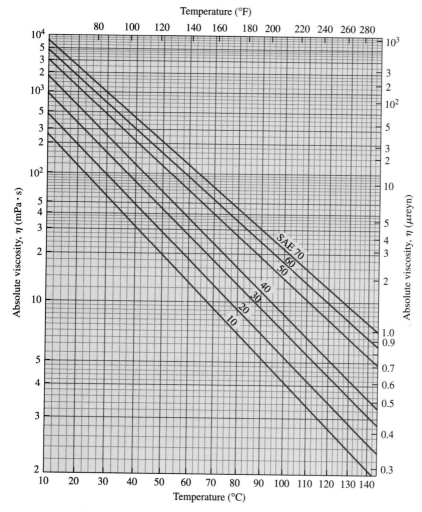

Figure 10.7 Viscosity vs. temperature curve for typical SAE graded oils.

(ISO) classify oils according to viscosity. Viscosity-temperature curves for typical SAE numbered oils are given in Figure 10.7. These oil types must exhibit particular viscosity behavior at 100°C. In addition, the SAE classifies identifications such as 10W, 20W, 30W, and 40W. Accordingly, for instance, a 20W-40 *multigrade,* also called *multiviscosity,* oil must satisfy the 20W behavior at −18°C and the SAE 40 viscosity behavior at 100°C. The viscosity of multigrade oils vary less with temperature than that of single grade oils.

A widely used means for specifying the rate of change of viscosity with temperature is known as the *viscosity index,* abbreviated *VI.* It compares an oil to oils with very small and very large rates of change in viscosity. A contemporary basis for viscosity index rating is given in the American National Standards Institute (ANSI) /ASTM Specification D2270.

EXAMPLE 10.1   Determination of the Viscosity and SAE Number of an Oil

An engine oil has a kinematic viscosity at 80°C corresponding to 62 seconds as found from a Saybolt viscosimeter. Calculate the absolute viscosity in millipascal-second and microreyns. What is the corresponding SAE number?

Solution: Through the use of Eq. (10.6a), we have

$$\rho = 0.89 - 0.00063(80 - 15.6) = 0.849 \ \text{g/cm}^3$$

Then, applying Eq. (10.5a),

$$\eta = 0.849 \left[0.22(62) - \frac{180}{62} \right] = 9.116 \ \text{mPa} \cdot \text{s}$$

Equation (10.5b) gives

$$\eta = 0.145(9.116) = 1.322 \ \mu\text{reyn}$$

Referring to Figure 10.7, the viscosity at 80°C is near to that of an SAE 20 oil.

10.5   PETROFF'S BEARING EQUATION

The phenomenon of bearing friction was first explained by N. Petroff in 1883. He analyzed a journal bearing based on the assumption that the shaft is concentric in bearing. Obviously, this operation condition could not occur in an actual journal bearing. However, by Petroff's approach, if the load applied is very low and the speed and viscosity are fairly high, approximate results are obtained. Usually, Petroff's equation is applied in preliminary design calculations.

FRICTION TORQUE

Let us assume that the moving flat plate shown in Figure 10.4a is wrapped into a cylindrical shaft (Figure 10.8). Now the thickness h becomes the radial clearance c that is taken to be completely filled with lubricant and from which leakage is negligible. Note that the radial clearance represents the difference in radii of the bearing and journal.

The developed journal area A is $2\pi r L$. Carrying A and h into Eq. (10.3) yields the tangential friction force $F = 2\pi \eta U r L/c$, in which the tangential velocity U of the journal is $2\pi rn$. The frictional torque owing to resistance of fluid equals $T_f = Fr$. Equation for no-load torque is then

Figure 10.8
Journal-centered bearing.

$$T_f = \frac{4\pi^2 \eta L r^3 n}{c} \tag{10.7}$$

where

T_f = frictional torque

η = absolute viscosity

L = length of bearing

r = journal radius

n = journal speed, revolutions per second, rps

c = radial clearance or film thickness

Figure 10.9
Lightly loaded journal bearing.

When a small load W is supported by the bearing, the pressure P of the projected area equals $P = W/2rL$ (Figure 10.9). The frictional force is fW, where f represents the coefficient of friction. Hence, the friction torque due to load is

$$T_f = fWr = 2r^2 fLP \qquad (10.8)$$

Clearly, load W will cause the shaft to become somewhat eccentric in its bearing.

According to Petroff's approach, the effect of load W in Eq. (10.7) can be considered negligible. Therefore, Eq. (10.7) can be equated to Eq. (10.8). In so doing, we obtain the coefficient of friction in the form

$$f = 2\pi^2 \frac{\eta n}{P} \frac{r}{c} \qquad (10.9)$$

This is known as *Petroff's equation* or *Petroff's law*. Through the use of Eq. (10.9), reasonable estimates of the coefficent of friction in lightly loaded bearings can be obtained. The two dimensionless quantities n/P and r/c are significant parameters in lubrication, as is observed in Section 10.7.

FRICTION POWER

Having the expression for the friction torque available, friction power for the bearing may be obtained from the general relations given in Section 1.12. In the SI units, by Eqs. (1.15) and (1.16),

$$kW = \frac{T_f n}{159} \qquad (T_f \text{ in } N \cdot m) \qquad (10.10)$$

$$hp = \frac{T_f n}{119} \qquad (T_f \text{ in } N \cdot m) \qquad (10.11)$$

where n is in rps.

In U.S. customary units, using Eq. (1.17), we have

$$hp = \frac{T_f n}{1050} \qquad (T_f \text{ in } lb \cdot in.) \qquad (10.12)$$

As before, journal speed n is in rps.

EXAMPLE 10.2 | Determining Friction Power Using Petroff's Approach

A 80-mm diameter shaft is supported by a full journal bearing of 120-mm length with a radial clearance of 0.05 mm. It is lubricated by SAE 10 oil at 70°C. The shaft rotates 1200 rpm and is under a radial load of 500 N. Apply Petroff's equation to determine (a) the bearing coefficient of friction and (b) the friction torque and power loss.

Solution: From Figure 10.7, $\eta = 9.2$ mPa · s. We have

$$P = \frac{500}{(0.08)(0.12)} = 52.08 \text{ kPa}$$

$$n = \frac{1200}{60} = 20 \text{ rps}$$

(a)   Substitution of the given data into Eq. (10.9) gives

$$f = 2\pi^2 \frac{(0.0092)(20)}{52,080} \frac{40}{0.05} = 0.0558$$

(b)   Equations (10.8) and (10.10) are therefore

$$T_f = fWD/2 = (0.0558)(500)(0.04) = 1.116 \text{ N} \cdot \text{m}$$

$$kW = \frac{1.116(20)}{159} = 0.14$$

10.6   HYDRODYNAMIC LUBRICATION THEORY

Recall from Section 10.3 that, in hydrodynamic lubrication, oil is drawn into the wedge-shaped opening produced by two nonparallel surfaces having relative motion. The velocity profile of the lubricant is different at the wider and narrower sections. As a result, sufficient pressure is built-up in the oil film to support the applied vertical load without causing metal-to-metal contact. This technique is utilized in the thrust bearings for hydraulic turbines and propeller shafts of ships as well as in the conventional journal bearings for piston engines and compressors.

REYNOLDS'S EQUATION OF HYDRODYNAMIC LUBRICATION

Hydrodynamic lubrication theory is based on Osborne Reynolds's study of the laboratory investigation of railroad bearings by Beauchamp Tower in the early 1880s in England [8]. The initial Reynolds's differential equation for hydrodynamic lubrication was used by him to explain Tower's results. A simplifying assumption of Reynolds's analysis was that the oil films were so thin in comparison with the bearing radius that the curvature could be disregarded. This enabled him to replace the curved partial bearing with a flat bearing. Other presuppositions include those discussed in Section 10.4. The following is a brief outline of the development of Reynolds's fluid flow equation for two typical bearings.

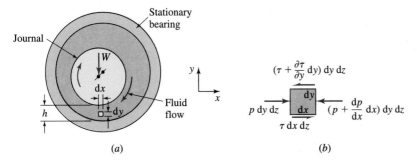

Figure 10.10   (a) An eccentric journal. (b) Pressure and viscous forces acting on an oil fluid element of sides dx, dy, and dz, isolated from part a.

Long Bearings

Consider a journal rotating in the clockwise direction supported by a lubricant film of variable thickness h on a fixed sleeve (Figure 10.10a). Assume that the lubricant velocity u and shear stress τ vary in both the x and y directions, while pressure p depends on the x direction alone and bearing *side leakage* is *neglected*. The summation of the x-directed forces on the fluid film (Figure 10.10b) gives

$$p\,dydz + \tau\,dxdz - \left(p + \frac{dp}{dx}dx\right)dydz - \left(\tau + \frac{\partial \tau}{\partial y}dy\right)dxdz = 0$$

This reduces to

$$\frac{dp}{dx} = \frac{\partial \tau}{\partial y} \tag{a}$$

From Newton's law of flow,

$$\tau = -\eta\frac{\partial u}{\partial y} \tag{b}$$

in which the minus sign indicates a negative velocity ingredient.

Carrying Eq. (b) into Eq. (a) and rearranging, we obtain

$$\frac{\partial^2 u}{\partial y^2} = \frac{1}{\eta}\frac{dp}{dx}$$

Integrating twice with respect to y results in

$$u = \frac{1}{\eta}\left(\frac{dp}{dx}\frac{y^2}{2} + c_1 y + c_2\right) \tag{c}$$

Assuming that no slip occurs between the lubricant and the boundary surfaces (Figure 10.11) leads to

$$u = 0 \quad \text{(at } y = 0), \quad u = U \quad \text{(at } y = h)$$

Figure 10.11   Velocity profile of the oil.

The quantity U represents the journal surface velocity. The constants c_1 and c_2 are evaluated by introducing these conditions into Eq. (c):

$$c_1 = \frac{U\eta}{h} - \frac{h}{2}\frac{dp}{dx}, \qquad c_2 = 0$$

Hence,

$$u = \frac{1}{2\eta}\frac{dp}{dx}(y^2 - hy) + \frac{U}{h}y \tag{10.13}$$

It is interesting to observe from this equation that the velocity distribution across the film is obtained by superimposing a parabolic distribution (the first term) onto a linear distribution (the second term). The former and the latter are indicated by the solid and dashed lines in Figure 10.11, respectively.

Let the volume of the lubricant per unit time flowing (in the x direction) across the section containing the element in Figure 10.10 be denoted by Q. For a width of unity in the z direction, using Eq. (10.13), we have

$$Q = \int_0^h u\,dy = \frac{Uh}{2} - \frac{h^3}{12\eta}\frac{dp}{dx} \tag{d}$$

Based on the assumptions of lubricant incompressibility and no side leakage, the flow rate must be identical for all sections: $dQ/dx = 0$. So, differentiating Eq. (d) and setting the result equal to 0 yield

$$\frac{d}{dx}\left(\frac{h^3}{\eta}\frac{dp}{dx}\right) = 6U\frac{dh}{dx} \tag{10.14}$$

This is the *Reynolds's equation for one-dimensional flow.*

For the case in which the axial (z-directed) fluid flow *includes leakage,* the preceding expression may be generalized to obtain the *two-dimensional Reynolds's equation:*

$$\frac{\partial}{\partial x}\left(\frac{h^3}{\eta}\frac{\partial p}{\partial x}\right) + \frac{\partial}{\partial z}\left(\frac{h^3}{\eta}\frac{\partial p}{\partial z}\right) = 6U\frac{\partial h}{\partial x} \tag{10.15}$$

The solutions of Eqs. (10.14) and (10.15) provide reasonable approximations for bearings of $L/D > 1.5$. Here L and D represent the length and diameter of the bearing, respectively.

Long bearings are sometimes used to restrain a shaft from vibration and position the shaft accurately in transmission shafts and machine tools, respectively [7].

Short Bearings

The circumferential flow of oil around the bearing may be taken to be negligible in comparison to the flow in the z direction for a short bearing. On the basis of this premise, F. W. Ocvirk and G. B. Dubois [9] proposed that the x term in Eq. (10.15) may be omitted. In so doing, we obtain

$$\frac{\partial}{\partial z}\left(\frac{h^3}{\eta}\frac{\partial p}{\partial z}\right) = 6U\frac{\partial h}{\partial x} \qquad (10.16)$$

The foregoing equation can readily be integrated to give an expression for pressure in the oil film. Often, this procedure is referred to as *Ocvirk's short bearing approximation*. The solution of Eq. (10.16) has moderate accuracy for bearings of L/D ratios up to about 0.75. In modern power machines, the trend is toward the use of short bearings.

We should mention that the exact solution of the Reynolds's equation is a challenging problem that has interested many investigators ever since then, and it is still the starting point for lubrication studies. A mathematical treatment of the hydrodynamic lubrication is beyond the scope of this volume. Fortunately, it is possible to make design calculations from the graphs obtained by mathematical analysis, as will be observed in the next section.

10.7 THE DESIGN OF JOURNAL BEARINGS

In actual bearings, a full continuous fluid film does not exist. The film ruptures, and bearing load W is supported by a partial film located beneath the journal. Petroff's law may be applied only to estimate the values of coefficient of friction. As noted previously, mathematical solutions to Reynolds's equations give reasonably good results for hydrodynamic or journal bearings of some commonly encountered proportions.

The design of journal bearings usually involves two suitable combinations of variables: *variables under control* (viscosity, load, radius and length of bearing, and clearance) and *dependent variables* or performance factors (coefficients of friction, temperature rise, oil flow, and minimum oil-film thickness). Essentially, in bearing design, limits for the latter group of variables are defined. Then, the former group is decided on so that these limitations are not exceeded. The following is a brief discussion of the quantities under control.

Lubricants

Recall that lubricants are characterized by their viscosity (η). Their choice is based on such factors as type of machine, method of lubrication, and load features.

Bearing Load

Usually, the load acting on a bearing is particularized. The value of the load per projected area, P, depends on the length and diameter of the bearing. Obviously, the smaller is P, the greater the bearing life.

Length/Diameter Ratio

Various factors are considered in choosing proper length-to-diameter ratios, or L/D values. Bearings with a length-to-diameter ratio less than 1 (short bearings) accommodate the shaft deflections and misalignments that are expected to be severe. Long bearings ($L/D > 1$) must be used in applications where shaft alignment is important.

Clearance

The effects of varying dimensions and clearance ratios are very significant in a bearing design. The radial clearance c (Figure 10.9) is contingent to some extent on the desired quality. Suitable values to be used for radial bearing clearance rely on factors that include materials, manufacturing accuracy, load-carrying capacity, minimum film thickness, and oil flow. Furthermore, the clearance may increase because of wear. The clearance ratios (r/c) typically vary from 0.001 to 0.002 and occasionally as high as 0.003. It would seem that large clearances increase the flow that reduces film temperature and hence increase bearing life. However, very large clearances result in a decrease in minimum film thickness. Therefore, some iteration ordinarily is needed to obtain a proper value for the clearance.

DESIGN CHARTS

A. A. Raimondi and J. Boyd applied digital computer techniques toward the solution of Reynolds's equation and present the results in the form of design charts and tables [10]. These provide accurate results for bearings of all proportions. Most charts utilize the bearing characteristic number, or the *Sommerfeld number:*

$$S = \left(\frac{r}{c}\right)^2 \frac{\eta n}{P} \tag{10.17}$$

where

> S = bearing characteristic number, dimensionless
>
> r = journal radius
>
> c = radial clearance
>
> η = viscosity, reyns
>
> n = relative speed between journal and bearing, rps
>
> P = load per projected area

Notations used in the charts are illustrated in Figure 10.12. The center of the journal is shown at O and the center of the bearing is at O'. The minimum oil-film thickness h_0 occurs at the line of centers. The distance between these centers represents the eccentricity, denoted by e. The eccentricity ratio ϵ is defined by

$$\epsilon = \frac{e}{c} \tag{10.18}$$

Figure 10.12 Radial pressure distribution in a journal bearing.

The minimum film thickness is then

$$h_0 = c - e = c(1 - \epsilon) \tag{10.19}$$

The foregoing gives

$$\epsilon = 1 - \frac{h_0}{c} \tag{10.20}$$

As depicted in the figure, the angular location of the minimum oil-film thickness is designated by Φ. The terminating position and position of maximum film pressure p_{max} of the lubricant are denoted by ϕ_0 and ϕ_{max}, respectively.

Load per projected area, the average pressure or so-called unit loading, is

$$P = \frac{W}{DL} \tag{10.21}$$

Here $W = $ load, $D = $ journal diameter, and $L = $ journal length. Note that, L and D are also referred to as the *bearing length* and *diameter*, respectively. Table 10.1 furnishes some representative values of P in common use.

Design charts by Raimondi and Boyd provide solutions for journal bearings having various length-diameter ratios (L/D). Only portions of three selected charts are reproduced in Figures 10.13 through 10.15, for *full bearings*. All charts give the plots of *dimensionless* bearing parameters as functions of the dimensionless Sommerfeld variable, S. Note that the S scale on the charts is logarithmic except for a linear portion between 0 and 0.01.

Table 10.1   Average sleeve bearing pressures in current practice

Application	Average pressure $P = W/DL$	
	MPa	(psi)
Relatively Steady Loads		
Centrifugal pumps	0.7–1.3	(100–180)
Gear reducer	0.8–1.7	(120–250)
Steam turbines	1.0–2.1	(150–300)
Electric motors	0.8–1.7	(120–250)
Rapidly Fluctuating Loads		
Automotive gasoline engines		
Main bearings	4–5	(600–750)
Connecting rod bearings	12–16	(1700–2300)
Diesel Engines		
Main bearings	6–12	(900–1700)
Connecting rod bearings	8–16	(1150–2300)

| SOURCE: [11].

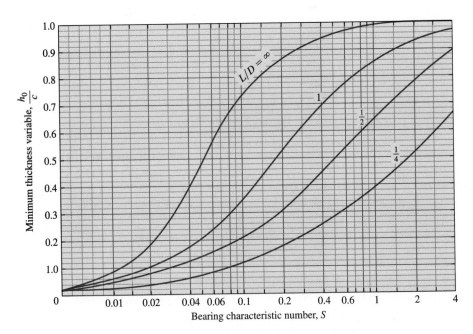

Figure 10.13   Chart for minimum film-thickness variable [10].

Figure 10.14 Chart for coefficient of friction variable [10].

Figure 10.15 Chart for film maximum pressure [10].

Space does not permit the inclusion of charts for partial bearings and thrust bearings. Those seeking more complete information can find it in the references cited.

The use of the design charts are illustrated in the solution of the following numerical problem.

EXAMPLE 10.3 | Determining the Performance Factors of Journal Bearings Using the Design Charts

A full-journal bearing of diameter D, length L, with a radial clearance c, carries a load of W at a speed of n. It is lubricated by SAE 30 oil, supplied at atmospheric pressure, and the average temperature of the oil film is t. Using the design charts, analyze the bearing.

Given: The numerical values are (see Figure 10.12)

$$D = 60 \text{ mm}, \quad r = 30 \text{ mm}, \quad L = 30 \text{ mm}, \quad c = 0.05 \text{ mm},$$

$$n = 30 \text{ rps}, \quad W = 3.6 \text{ kN}, \quad t = 60°\text{C}$$

Solution: The variables under control of the designer are

$$P = \frac{W}{DL} = \frac{3600}{(0.03)(0.06)} = 2 \text{ MPa}$$

$$\eta = 27 \text{ mPa} \cdot \text{s} \qquad \text{(from Figure 10.7)}$$

$$S = \left(\frac{r}{c}\right)^2 \left(\frac{\eta n}{P}\right) = \left(\frac{30}{0.05}\right)^2 \frac{(0.027)(30)}{2 \times 10^6} = 0.146$$

The determination of the dependent variables proceeds as described in detail in the following paragraphs. Note that the procedure can also be carried out conveniently in tabular form.

Minimum film thickness (Figure 10.13).   Use $S = 0.146$ with $L/D = 1/2$ to enter the chart in this figure:

$$\frac{h_0}{c} = 0.25 \quad \text{or} \quad h_0 = 0.0125 \text{ mm}$$

Then, by Eq. (10.19),

$$e = c - h = 0.05 - 0.0125 = 0.0375 \text{ mm}$$

The eccentricity ratio is then $\epsilon = e/c = 0.0375/0.05 = 0.75$.

Comments: The permissible oil-film thickness depends largely on the surface roughness of the journal and bearing [9]. The surface finish should therefore be specified and closely controlled if the design calculations indicate that the bearing operates with a very thin oil film.

Coefficient of friction (Figure 10.14).   Use $S = 0.146$ with $L/D = 1/2$. Hence, from the chart in this figure,

$$\frac{r}{c} f = 4.8 \quad \text{or} \quad f = 0.008$$

The friction torque is then, applying Eq. (10.8),

$$T_f = fWr = 0.008(3600)(0.03) = 0.864 \text{ N} \cdot \text{m}$$

The frictional power lost in the bearing, from Eq. (10.10), is

$$\text{kW} = \frac{T_f n}{159} = \frac{(0.864)(30)}{159} = 0.163$$

Film pressure (Figure 10.15). Use $S = 0.146$ with $L/D = 1/2$ to enter the chart in this figure:

$$\frac{P}{p_{max}} = 0.32$$

The foregoing gives $p_{max} = 2/0.32 = 6.25$ MPa.

Comments: A temperature rise Δt in the oil film owing to the fluid friction can be determined based on the assumption that the oil film carries away all the heat generated [12]. The Raimondi-Boyd papers also contain charts to obtain oil flow Q, side leakage Q_s, and conservative estimates of the Δt. In addition, they include charts to find the angular locations of the minimum film thickness, maximum pressure, and the terminating position of the oil. These charts are not presented in this book.

10.8 LUBRICANT SUPPLY TO JOURNAL BEARINGS

The hydrodynamic analysis assumes that oil is available to flow into the journal bearing at least as fast as it leaks out at the ends. A variety of methods of lubrication are used for journal bearings. The system chosen for a specific problem depends to a large extent on the type of service the bearing is to perform. Some typical techniques for supplying oil to the bearing are briefly described as follows.

SPLASH METHOD

The splash system of lubrication is used effectively when a machine has a rotating part, such as crank or gear enclosed in a housing. The moving part runs through a reservoir of oil in the enclosed casing. This causes a spray of oil to soak the casing, lubricating the bearing. The term *oil bath* refers to a system where oil is supplied by partially submerging the journal into the oil reservoir, as in the railroad partial bearings.

MISCELLANEOUS METHODS

A number of simple methods of lubrication also are used. Bearings that are used in low-speed, light-load applications can be lubricated by hand oiling. A wick-feed oiler, as the name implies, depends on an absorbent material serving as a wick to supply oil to the bearing. A drop-feed oiler permits oil from a reservoir to flow through a needle valve to the bearing. A ring-oiled bearing uses a ring that is often located over the journal at the center of the bearing.

Self-contained bearings contain the lubricant in the bearing housing, which is sealed to prevent oil loss. Oil may be gravity fed from a reservoir or cup above the bearing. Obviously, a bearing of this type is economically more desirable because it requires no expensive cooling or lubricant-circulating system. Self-contained bearings are known as *pillow-block* or *pedestal bearings*.

PRESSURE-FED SYSTEMS

In the pressure-fed lubrication systems a continuous supply of oil is furnished to the bearing by a small pump. The oil is returned to a reservoir after circulating through the bearing. The pump may be located within the machine housing above the sump and driven by one of the shafts. This complete system is the commonly used method. An example is the pressure-fed lubrication system of a piston-type engine or compressor. Here, oil supplied by the pump fills grooves in the main bearings. Holes drilled in the crankshaft transfer oil from these grooves to the connection rod sleeve bearings. Note that, in most automotive engines, the piston pins are splash lubricated.

METHODS FOR OIL DISTRIBUTION

Figure 10.16a illustrates a bearing with a *circumferential groove* used to distribute oil in a tangential direction. The oil flows either by gravity or under pressure into the groove through an oil supply hole placed in the groove opposite the portion of the oil film supporting the load. The effect of the groove is to create two half bearings, each having a

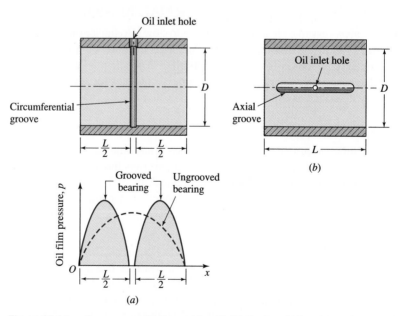

Figure 10.16   Common methods used for oil distribution: (a) bearing with circumferential groove and comparison of the axial pressure distribution with or without a groove; (b) bearing with axial groove.

smaller L/D ratio than the original. As a result, the pressure distribution does not vary, as the smooth curve shown by the dashed line indicates; hydrodynamic pressure drops to nearly 0 at the groove. Although the oil film is broken in half, the efficient cooling obtained allows these bearings to carry larger loads without overheating.

An *axial groove* fed by the oil hole (Figure 10.16b) generally gives sufficient flow at low or ambient oil pressure. A wide variety of groove types give even better oil distribution. In all flow problems, it is assumed that provision has been made to keep the entrance full.

10.9 HEAT BALANCE OF JOURNAL BEARINGS

The frictional loss of energy in a bearing is transferred into heat, raising the temperature of the lubricant and the adjacent parts in a bearing. The *heat balance of a bearing* refers to the balance between the heat developed and dissipated in a bearing. The usual desired value for the average oil temperature is about 70°C for a satisfactory balance. If the average temperature rises above 105°C, deterioration of the lubricant as well as the bearing material can occur [13].

In a pressure-fed system, as the oil flows through the bearing, it absorbs heat from the bearing. The oil is then returned to a sump, where it is cooled before being recirculated. Based on this method, the lubricant carries most of the generated heat, and hence design charts give a reasonably accurate value of a temperature rise in the oil.

HEAT DISSIPATED

Here, we consider heat balance in self-contained bearings, where the lubricant is stored in the bearing housing itself. Bearings of this type dissipate heat to the surrounding atmosphere by conduction, convection, and radiation heat transfer. Practically, a precise value of the rate of heat flow cannot be calculated with any accuracy. The heat dissipated from the bearing housing may only be approximated by

$$H = CA(t_o - t_a) \qquad (10.22)$$

The foregoing gives

$$t_o = t_a + \frac{H}{AC} \qquad (10.23)$$

where

$H =$ time rate of heat lost, watts

$C =$ overall heat transfer coefficient, watts/m$^2 \cdot$ °C

$A =$ surface area of housing, m^2

$t_o =$ average oil film temperature, °C

$t_a =$ temperature of surrounding air, °C.

Rough estimates of values for coefficient C are given in Table 10.2. For simple ring-oiled bearings, the bearing housing area may be estimated as 12.5 times the bearing

Table 10.2   Heat transfer coefficient C for self-contained bearings

Lubrication system	Conditions	C
Oil ring	Still air	7.4
	Average air circulation	8.5
Oil bath	Still air	9.6
	Average air circulation	11.3

projected area (i.e., $12.5DL$). It is to be emphasized that Eq. (10.23) should be used only when "ballpark" results are sufficient.

HEAT DEVELOPED

Under equilibrium conditions, the rate at which heat develops within a bearing is equal to the rate at which heat dissipates:

$$fW(2\pi rn) = H \tag{10.24}$$

In this expression, f = coefficient of friction, W = load, r = journal radius, n = journal speed (as defined in Section 10.5), and H is given by Eq. (10.22). A heat-balance computation, involving finding average film temperature at the equilibrium, is a trial and error procedure [14, 15].

10.10 MATERIALS FOR JOURNAL BEARINGS

The operating conditions for journal bearing materials are such that rather strict requirements must be placed on the material to be used. For instance, in thick-film lubrication, any material with sufficient compressive strength and a smooth surface is an adequate bearing material. Small bushings and thrust bearings are often expected to run with thin-film lubrication. Any foreign particles larger than the minimum film thickness present in the oil damage the shaft surface unless they can become imbedded in a relatively soft bearing material. In this section, we discuss some of the types of bearing materials in widespread usage. Special uses are for many other materials, such as hard wood, glass, silver, ceramics, and sapphires [16–19].

ALLOYS

Babbitt alloys are the most commonly used materials, usually having a tin or lead base. They posses low melting points, moduli of elasticity, yield strength, and good plastic flow. In a bearing, the foregoing give good conformability and embeddability characteristics. *Comformability* measures the capability of the bearing to adapt to shaft misalignment and deflection. *Embeddability* is the bearing's capability to ingest harder, foreign particles. Shafts for babbitt bearings should have a minimum hardness of 150–200 Bhn and a ground surface finish.

Compressive and fatigue strengths of babbitts are low, particularly above about 77°C. Babbitts can rarely be used above about 121°C. However, these shortcomings are improved by using a thin internal babbitt surface on a steel (or aluminum) backing. For small and

Figure 10.17 Babbitt metal bearing cast into a steel shell.

medium bearings under higher pressure (as in internal combustion engines), babbitt layers 0.025–2.5-mm thick are used; while in medium and large bearings under low pressure, the babbitt is often cast in thicknesses of 3–13 mm into a thicker steel shell (Figure 10.17).

Copper alloys are principally bronze and aluminum alloys. They are generally stronger and harder, have greater load capacity and fatigue strength, but less compatible (i.e., antiweld and antiscoring) than babbitt bearings. Owing to their thermal conductivity, corrosion resistance, and low cost, aluminum alloys are in widespread usage for bearings in internal combustion engines. A thin layer of babbitt is placed inside an aluminum bearing to improve its comformability and embeddability.

SINTERED MATERIALS

Sintered materials, porous metal bearings or insertable powder-metallurgy bushings, have found wide acceptance. These self-lubricated bearings have interconnected pores in which oil is stored in the factory. The pores act as a reservoir for oil, expelling it when heated by shaft rubbing, reabsorbing it when inactive. The low cost and lifetime use in a machine, without further lubrication, are their prime advantages.

NONMETALLIC MATERIALS

A variety of *plastics* are used as bearing materials. No corrosion, quiet operation, moldability, and excellent compatibility are their advantages. The last characteristic often implies that no lubrication is required. *Carbon-graphite* bearings can be used at high temperatures. They are chemically inert. These bearings are useful in ovens and in pumps for acids and fuel oils. *Rubber* and other elastomers are excellent bearing material for water pumps and propellers. They are generally placed inside a noncorrodible metal shell and can provide vibration isolation, compensate for misalignment, and have good conformability.

Part B   Rolling-Element Bearings

Recall from Section 10.1 that rolling-element bearings are also known as *rolling bearings* or *antifriction bearings*. The Anti-friction Bearing Manufacturing Association (AFBMA) and the ISO standardized bearing dimensions and the basis for their selection. The load, speed, and operating viscosity of the lubricant affect the friction characteristics of a rolling bearing. These bearings provide coefficients of friction between 0.001 and 0.002. The designer must deal with such matters as fatigue, friction, heat, lubrication, kinematic problems, material properties, machining tolerances, assembly, use, and cost. A complete history of the rolling element bearings is given in [20].

COMPARISON OF ROLLING AND SLIDING BEARINGS

Some *advantages* of rolling-element bearings over the sliding or journal bearings are as follows:

1. Low starting and good operating friction torque.
2. Ease of lubrication.
3. Requiring less axial space.
4. Generally, taking both radial and axial loads.
5. Rapid replacement.
6. Warning of impending failure by increasing noisiness.
7. Good low-temperature starting.

The *disadvantages* of rolling-element bearings compared to sliding bearings include

1. Greater diametral space.
2. More severe alignment requirements.
3. Higher initial cost.
4. Noisier normal operation.
5. Finite life due to eventual failure by fatigue.
6. Ease of damage by foreign matter.
7. Poor damping ability.

10.11   TYPES AND DIMENSIONS OF ROLLING BEARINGS

Rolling bearings can carry radial, thrust or combinations of the two loads, depending on their design. Accordingly, most rolling bearings are categorized in one of the three groups: radial for carrying loads that are primarily radial, *thrust* or axial contact for supporting loads that are primarily axial, and *angular contact* for carrying combined axial and radial loads. As noted earlier, the rolling-element bearings are of two types: ball bearings and roller bearings. The latter are capable of higher speeds, and the former can take greater loads. The rolling bearings are precise, yet simple machine elements. They are made in a wide variety of types and sizes (Figure 10.18). Most bearing manufacturers provide engineering manuals and brochures containing descriptions of the various kinds available. Only some common types are considered here.

BALL BEARINGS

A ball bearing is employed in almost every type of machine or mechanism with rotating parts. Figure 10.19 illustrates the various parts, surfaces, and edges of a ball bearing.

Figure 10.18   Various rolling-element bearings
(Courtesy of SKF).

Figure 10.19   Ball bearing geometry and nomenclature
(Courtesy of New Departure-Hyatt Division, General Motors
Corporation).

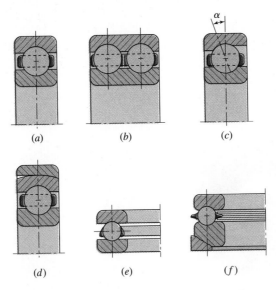

Figure 10.20   Some types of ball bearings: (a) deep groove (Conrad); (b) double-row; (c) angular contact; (d) external self aligning; (e) thrust; (f) self-aligning thrust. (Courtesy of the Timken Company.)

Observe that the basic bearing consists of an inner ring, an outer ring, the balls, and the separator (also known as cage or retainer). To increase the contact area and hence permit larger loads to be carried, the balls run in curvilinear grooves in the rings called *raceways*. The radius of the raceway is very little larger than the radius of the ball.

The *deep-groove* (Conrad-type) *bearing* (Figure 10.20a) can stand a radial load as well as some thrust load. The balls are inserted into grooves by moving the inner ring to an eccentric position. They are separated after loading, and then the retainers are inserted. Obviously, an increase in radial load capacity may be obtained by using rings with deep grooves or by employing a double-row radial bearing (Figure 10.20b).

The *angular-contact bearing* (Figure 10.20c) has a two-shouldered ball groove in one ring and a single-shouldered ball groove in the other ring. It can support greater thrust capacity in one direction as well as radial loads. The cutaway shoulder allows bearing assembly and use of a one-piece machined cage. The contact angle α is defined in the figure. Typical values of α for angular ball bearings vary from 15° to 40°.

The *self-aligning bearing* has an outer raceway ball path ground in a spherical shape so it can accommodate large amounts of angular misalignments or shaft deflections. These bearings can support both radial and axial loads and are available in two types: self-aligning external (Figure 10.20d) and self-aligning internal. *Thrust bearings* are designed to carry a pure axial load only, as shown in Figures 10.20e and 10.20f. They are made exclusively for machinery with vertically oriented shafts and have modest speed capacity.

Figure 10.21   Some types of roller bearings:
(a) straight cylindrical; (b) spherical; (c) tapered
thrust; (d) needle; (e) tapered. (Courtesy of the
Timken Company.)

ROLLER BEARINGS

A roller bearing uses straight, tapered, or contoured cylindrical rollers. When shock and
impact loads are present or when a large bearing is needed, these bearings usually are
employed. Roller bearings can support much higher static and dynamic (shock) loads than
comparably sized ball bearings, since they have line contact instead of point contact. A
roller bearing generally consists of the same elements as a ball bearing. These bearings can
be grouped into five basic types: cylindrical roller bearings, spherical roller bearings,
tapered thrust roller bearings, needle roller bearings, and tapered roller bearings (Figure 10.21). Straight roller bearings provide purely radial load support in most applications;
they cannot resist thrust loads. The spherical roller bearings have the advantage of accom-
modating some shaft misalignments in heavy-duty rolling mill and industrial gear drives.
Needle bearings are in widespread usage where radial space is limited.

   Tapered roller bearings combine the advantages of ball and straight roller bearings, as
they can stand either radial or thrust loads or any combination of the two. The centerlines
of the conical roller intersect at a common apex on the centerline of rotation. Tapered roller
bearings have numerous features that make them complicated [12], and space does not
permit their discussion in this text. Note that pairs of single-row roller bearings are usually
employed for wheel bearings and some other applications. Double-row and four-row roller
types are used to support heavier loads. Selection and analysis of most bearing types are
identical to that presented in the following sections.

(a)   (b)

Figure 10.22   Special bearings: (a) pillow block; (b) flange. (Courtesy of Emerson Power Transmission, Sealmaster Bearings, Aurora, IL.)

SPECIAL BEARINGS

Rolling element bearings are available in many other types and arrangements. Detailed information is available in the literature published by the several manufacturers and other references. Two common samples are shown in Figure 10.22. Note that these bearings package standard ball or roller bearings in cast-iron housings. They can be readily attached to horizontal or vertical surfaces.

STANDARD DIMENSIONS FOR BEARINGS

The AFBMA established standard boundary dimensions for the rolling-element bearings, shafts, and housing shoulders. These dimensions are illustrated in Figure 10.23: D = bearing bore, D_o = outside diameter (OD), w = width, d_s = shaft shoulder diameter, d_h = housing diameter, and r = fillet radius. For a given bore, there are various widths and outside diameters. Similarly, for a particular outside diameter, we can find many bearings with different bores and widths.

In basic AFBMA plan, the bearings are identified by a two-digit number, called the *dimension series code*. The first and second digits represent the width series and the diameter series, respectively. This code does not disclose the dimensions directly, however; and it is required to resort to tabulations. Tables 10.3 and 10.4 furnish dimensions of some 02 and 03 series of ball and cylindrical roller bearings. The load ratings of these bearings, discussed in the next section, are also included in the table. More detailed information is readily available in the latest AFBMA Standards [21], engineering handbooks, and manufacturers' catalogs and journals [22, 23].

Figure 10.23 Dimensions of ball bearing, shaft, and housing.

Table 10.3 Dimensions and basic load ratings for 02-series ball bearings

Bore D (mm)	OD D_o (mm)	Width w (mm)	Fillet radius r (mm)	Load ratings (kN)			
				Deep groove		Angular contact	
				C	C_s	C	C_s
10	30	9	0.6	5.07	2.24	4.94	2.12
12	32	10	0.6	6.89	3.10	7.02	3.05
15	35	11	0.6	7.80	3.55	8.06	3.65
17	40	12	0.6	9.56	4.50	9.95	4.75
20	47	14	1.0	12.7	6.20	13.3	6.55
25	52	15	1.0	14.0	6.95	14.8	7.65
30	62	16	1.0	19.5	10.0	20.3	11.0
35	72	17	1.0	25.5	13.7	27.0	15.0
40	80	18	1.0	30.7	16.6	31.9	18.6
45	85	19	1.0	33.2	18.6	35.8	21.2
50	90	20	1.0	35.1	19.6	37.7	22.8
55	100	21	1.5	43.6	25.0	46.2	28.5
60	110	22	1.5	47.5	28.0	55.9	35.5
65	120	23	1.5	55.5	34.0	63.7	41.5
70	125	24	1.5	61.8	37.5	68.9	45.5
75	130	25	1.5	66.3	40.5	71.5	49.0
80	140	26	2.0	70.2	45.0	80.6	55.0
85	150	28	2.0	83.2	53.0	90.4	63.0
90	160	30	2.0	95.6	62.0	106	73.5
95	170	32	2.0	108	69.5	121	85.0

SOURCE: [24].

Note: Bearing life capacities, C, for 10^6 revolution life with 90% reliability.

Table 10.4   Dimensions and basic load ratings for straight cylindrical bearings

Bore D (mm)	02 series			03 series		
	OD D_o (mm)	Width w (mm)	Load rating C (kN)	OD D_o (mm)	Width w (mm)	Load rating C (kN)
25	52	15	16.8	62	17	28.6
30	62	16	22.4	72	19	36.9
35	72	17	31.9	80	21	44.6
40	80	18	41.8	90	23	56.1
45	85	19	44.0	100	25	72.1
50	90	20	45.7	110	27	88.0
55	100	21	56.1	120	29	102
60	110	22	64.4	130	31	123
65	120	23	76.5	140	33	138
70	125	24	79.2	150	35	151
75	130	25	91.3	160	37	183
80	140	26	106	170	39	190
85	150	28	119	180	41	212
90	160	30	142	190	43	242
95	170	32	165	200	45	264

SOURCE: [24].

Note: Bearing life capacities, C, for 10^6 revolution life with 90% reliability.

10.12   ROLLING BEARING LIFE

When the ball or roller of an antifriction bearing rolls into a loading region, contact (i.e., Hertzian) stresses occur on the raceways and on the rolling element. Owing to these stresses, which are higher than the endurance limit of the material, the bearing has a limited life. If a bearing is well maintained and operating at moderate temperatures, metal fatigue is cause of failure alone. Failure consists of pitting, spalling, or chipping load carrying surfaces, as discussed in Section 8.15.

Practically, the life of an individual bearing or any one group of identical bearings cannot be accurately predicted. Hence, the AFBMA established the following definitions associated with the life of a bearing. We note that *bearing life* is defined as the number of revolutions or hours at some uniform speed at which the bearing operates until fatigue failure.

Rating life L_{10} refers to the number of revolutions (or hours at a uniform speed) that 90% of a group of identical roller bearings will complete or exceed before the first evidence of fatigue develops. The term *minimum life* is also used to denote the rating life. *Median life* refers to the life that 50% of the group of bearings would complete or exceed. Test results show that the median life is about five times the L_{10} life.

Basic dynamic load rating C is the constant radial load that a group of apparently identical bearings can take for a rating life of 1 million (i.e., 10^6) revolutions of the inner ring in a stationary load (outer ring does not rotate).

Basic static load rating C_s refers to the maximum allowable static load that does not impair the running characteristics of the bearing.

The basic load ratings for different types of bearings are listed in Tables 10.3 and 10.4. The value of C_s depends on the bearing material, number of rolling elements per row, the bearing contact angle, and the ball or roller diameter. Except for an additional parameter relating the load pattern, the value of C is based on the same factors that determine C_s.

10.13   EQUIVALENT RADIAL LOAD

Catalog ratings are based on only the radial load. However, with the exception of thrust bearings, bearings are usually operated with some combined radial and axial loads. It is then necessary to define an equivalent radial load that has the same effect on bearing life as the applied loading. The AFBMA recommends, for rolling bearings, the maximum of the values of these two equations:

$$P = XVF_r + YF_a \qquad\qquad \textbf{(10.25)}$$

$$P = VF_r \qquad\qquad \textbf{(10.26)}$$

where

P = equivalent radial load

F_r = applied radial load

F_a = applied axial load (thrust)

V = a rotation factor

$\quad = \begin{cases} 1.0 & \text{(for inner-ring rotation)} \\ 1.2 & \text{(for outer-ring rotation)} \end{cases}$

X = a radial factor

Y = a thrust factor

The equivalent load factors X and Y depend on the geometry of the bearing, including the number of balls and the ball diameter. The AFBMA recommendations are based on the ratio of the axial load F_a to the basic static load rating C_s and a variable reference value e. For deep-groove (single-row and double-row) and angular-contact ball bearings, the values of X and Y are given in Tables 10.5 and 10.6. Straight cylindrical roller bearings are very limited in their thrust capacity because axial loads produce sliding friction at the roller ends. So, the equivalent load for these bearings can also be estimated using Eq. (10.26).

Table 10.5   Factors for deep-groove ball bearings

F_a/C_s	e	$F_a/VF_r \leq e$		$F_a/VF_r > e$	
		X	Y	X	Y
0.014*	0.19				2.30
0.21	0.21				2.15
0.028	0.22				1.99
0.042	0.24				1.85
0.056	0.26				1.71
0.070	0.27	1.0	0	0.56	1.63
0.084	0.28				1.55
0.110	0.30				1.45
0.17	0.34				1.31
0.28	0.38				1.15
0.42	0.42				1.04
0.56	0.44				1.00

SOURCE: [24].
*Use 0.014 if $F_a/C_s < 0.014$.

Table 10.6   Factors for commonly used angular-contact ball bearings

Contact angle (α)	e	$\frac{iF_a}{C_s}$*	Single-row bearing $F_a/VF_r > e$		Double-row bearing			
					$F_a/VF_r \leq e$		$F_a/VF_r > e$	
			X	Y	X	Y	X	Y
	0.38	0.015		1.47		1.65		2.39
	0.40	0.029		1.40		1.57		2.28
	0.43	0.058		1.30		1.46		2.11
	0.46	0.087		1.23		1.38		2.00
15°	0.47	0.12	0.44	1.19	1.0	1.34	0.72	1.93
	0.50	0.17		1.12		1.26		1.82
	0.55	0.29		1.02		1.14		1.66
	0.56	0.44		1.00		1.12		1.63
	0.56	0.58		1.00		1.12		1.63
25°	0.68		0.41	0.87	1.0	0.92	0.67	1.41
35°	0.95		0.37	0.66	1.0	0.66	0.60	1.07

SOURCE: [24].
*i is the number of rows of balls.

Table 10.7 Shock or service factors K_s

Type of load	Ball bearing	Roller bearing
Constant or steady	1.0	1.0
Light shocks	1.5	1.0
Moderate shocks	2.0	1.3
Heavy shocks	2.5	1.7
Extreme shocks	3.0	2.0

EQUIVALENT SHOCK LOADING

Some applications have various degrees of shock loading, which has the effect of increasing the equivalent radial load. Therefore, a shock or service factor, K_s can be substituted into Eqs. (10.25) and (10.26) to account for any shock and impact conditions to which the bearing may be subjected. In so doing, the equivalent radial load becomes the larger of the values given by the two equations:

$$P = K_s(XVF_r + YF_a) \tag{10.27}$$

$$P = K_s VF_r \tag{10.28}$$

Values to be used for K_s depend on the judgment and experience of the designer, but Table 10.7 may serve as a guide.

10.14 SELECTION OF ROLLING BEARINGS

Each group of seemingly identical bearings may differ slightly metallurgically, in surface finish, in roundness of rolling elements, and so on. Consequently, no two bearings within the same family may have the exact number of operating hours to fatigue failure after having been subjected to the identical speed and load condition. Therefore, the selection of rolling bearings is often made from tables of standard types and sizes containing data on their load and life ratings.

Usually, the basic static load rating C_s has little effect in the ball or roller bearing selection. However, if a bearing in a machine is stationary over an extended period of time with a load higher than C_s, local permanent deformation can occur. In general, the bearings cannot operate at very low speeds under loading that exceeds the basic static load rating.

The basic dynamic load rating C enters directly into the process of selecting a bearing, as is observed in the following formulation for the bearing life. Extensive testing of rolling bearings and subsequent statistical analysis has shown that load and life of a bearing are related statistically. This relationship can be expressed as

$$L_{10} = \left(\frac{C}{P}\right)^a \tag{10.29}$$

where

L_{10} = rating life, in 10^6 revolutions

C = basic load rating (from Tables 10.3 and 10.4)

P = equivalent radial load (from Section 10.13)

$a = \begin{cases} 3 & \text{(for ball bearings)} \\ 10/3 & \text{(for roller bearings)} \end{cases}$

We note that the load C is simply a reference value (see Section 10.12) that permits bearing life to be predicted at any level of actual load applied. Alternatively, the foregoing equation may be written in the form

$$L_{10} = \frac{10^6}{60n}\left(\frac{C}{P}\right)^a \qquad (10.30)$$

in which

L_{10} = rating life, in hours

n = rotational speed, in rpm

When *two groups* of identical bearings are run with different loads P_1 and P_2, the ratio of their rating lives L'_{10} and L''_{10}, by Eq. (10.29), is

$$\frac{L'_{10}}{L''_{10}} = \left(\frac{P_2}{P_1}\right)^a \qquad (10.31)$$

Good agreement between this relation and experimental data has been realized. Rearranging the foregoing, we have

$$L'_{10}P_1^a = L''_{10}P_2^a = 10^6 C^a \qquad (10.32)$$

Clearly, the terms in Eq. (10.32) are constant $10^6 C^a$, as previously defined.

RELIABILITY REQUIREMENT

Recall from Section 10.12 that the definition of rating life L_{10} is based on a 90% reliability (or 10% failure). In some applications, the foregoing survival rate cannot be tolerated (e.g., nuclear power plant controls, medical and hospital equipment). As mentioned in Section 7.14, the distribution of bearing failures at a constant load can be best approximated by the Weibull distribution.

Using the general Weibull equation [11, 24, 25] together with extensive experimental data, the AFBMA formulated recommended *life adjustment factors, K_r,* plotted in Figure 10.24. This curve can be applied to both ball and roller bearings but is restricted to reliabilities no greater than 99%. The expected bearing life is the product of the rating life and the adjustment factor. Combining this factor with Eq. (10.29), we have

$$L_5 = K_r\left(\frac{C}{P}\right)^a \qquad (10.33)$$

The quantity L_5 represents the rating life for any given reliability *greater* than 90%.

Figure 10.24 Reliability factor K_r.

Table 10.8 Representative rolling bearing design lives

Type of application	Life (kilohours)
Instruments and apparatus for infrequent use	Up to 0.5
Aircraft engines	0.5–2
Machines used intermittently	
Service interruption is minor importance	4–8
Reliability is of great importance	8–14
Machines used in an 8-hour service working day	
Not always fully utilized	14–20
Fully utilized	20–30
Machines for continuous 24-hour service	50–60
Reliability is of extreme importance	100–200

SOURCE: [12].

Most manufacturers' handbooks contain specific data on bearing design lives for many classes of machinery. For reference, Table 10.8 may be used when such information is unavailable.

Determination of the Median Life of a Deep-Groove Ball Bearing　　　　　**EXAMPLE 10.4**

A 50-mm bore (02-series) deep-groove ball bearing carries a combined load of 9 kN radially and 6 kN axially at 1200 rpm. Calculate

    (a)   the equivalent radial load

    (b)   the median life in hours.

Assumptions: The inner ring rotates and the load is steady.

Solution: Referring to Table 10.3, we find that, for a 50-mm bore bearing, $C = 35.1$ kN and $C_s = 19.6$ kN.

(a) To obtain the values of the radial load factors X and Y, it is necessary to obtain

$$\frac{F_a}{C_s} = \frac{6}{19.6} = 0.306, \qquad \frac{F_a}{VF_r} = \frac{6}{1(9)} = 0.667$$

We find from Table 10.5 that $F_a/VF_r > e$; $X = 0.56$ and $Y = 1.13$ by interpolation. Applying Eq. (10.25),

$$P = XVF_r + YF_a = (0.56)(1)(9) + (1.13)(6) = 11.82 \text{ kN}$$

Through the use of Eq. (10.26), $P = VF_r = 1(9) = 9$ kN.

(b) Since $11.95 > 9$ kN, the larger value is used for life calculation. The rating life, from Eq. (10.29), is

$$L_{10} = \left(\frac{C}{P}\right)^a = \left(\frac{35.1}{11.82}\right)^3 = 26.19(10^6) \text{ rev}$$

By Eq. (10.30),

$$L_{10} = \frac{10^6}{60n}\left(\frac{C}{P}\right)^a = \frac{10^6(26.19)}{60(1200)} = 364 \text{ hr}$$

The median life is therefore $5L_{10} = 1820$ hr.

EXAMPLE 10.5 | The Median Life of a Deep-Groove Ball Bearing under Moderate Shock

Redo Example 10.4, but the outer ring rotates, and the bearing is subjected to a moderate shock load.

Solution: We now have $V = 1.2$; hence,

$$\frac{F_a}{VF_r} = \frac{6}{1.2(9)} = 0.556$$

Table 10.5 shows that still $F_a/VF_r > e$; therefore, $X = 0.56$ and $Y = 1.13$, as before.

(a) Applying Eq. (10.27),

$$P = K_s(XVF_r + YF_a) = 2(0.56 \times 1.2 \times 9 + 1.13 \times 6) = 25.66 \text{ kN}$$

From Eq. (10.28), $P = K_sVF_r = 2(1.2 \times 9) = 21.6$ kN.

(b) Inasmuch as $25.66 > 21.6$ kN, we use the larger value for calculating the rating life. Through the use of Eq. (10.30),

$$L_{10} = \frac{10^6}{60n}\left(\frac{C}{P}\right)^a = \frac{10^6}{60(1200)}\left(\frac{35.1}{25.66}\right)^3 = 35.5 \text{ hr}$$

and the median life is $5L_{10} = 177.5$ hr.

Extending a Ball Bearing's Expected Life

EXAMPLE 10.6

What change in the loading of a ball bearing will increase the expected life by 25%?

Solution: Let L'_{10} and P_1 be the initial life and load L''_{10} and P_2 be the new life and load. Then, Eq. (10.32) with $a = 3$ and $L''_{10} = 1.25L'_{10}$ gives

$$P_2^3 = \frac{L'_{10}P_1^3}{1.25L'_{10}} = 0.8P_1^3$$

from which $P_2 = 0.928P_1$.

Comment: A reduction of the load to about 93% of its initial value causes a 25% increase in the expected life of a ball bearing.

The Expected Life of a Ball Bearing with a Low Rate of Failure

EXAMPLE 10.7

Determine the expected life of the bearing in Example 10.4, if only a 6% probability of failure can be permitted.

Solution: From Figure 10.24, for a reliability of 94%, $K_r = 0.7$. Using Eq. (10.33), the expected rating life is

$$L_5 = K_r \left(\frac{C}{P}\right)^a = 0.7(364) = 254.8 \text{ hr}$$

Comment: To improve the reliability of the bearing in Example 10.4 from 90 to 94%, a reduction of median life from 1820 hr to $5L_{10} = 1274$ hr is required.

Case Study 10-1 | BALL BEARING SELECTION FOR A SHAFT IN THE GEAR BOX
OF A WINCH CRANE

Figure 10.25 shows a flanged ball bearing of the input shaft in the gear box (see Case Study 11-1) of a winch crane. Analyze the load-carrying capacity of the bearing.

Given: The shaft has 12-mm diameter and operates at 1725 rpm. Rating life is 30 khr.

Assumptions: Thrust loads are negligible. Bearings at both ends of the shaft are taken to be identical 02-series

Figure 10.25 Flanged ball bearing at left end of input shaft in gear box of winch crane shown in Figure 1.4.

(continued)

Case Study (CONCLUDED)

deep groove and subjected to light-shock loading. The inner ring rotates.

Solution: See Figures 9.16, 10.25, and 11.20; Tables 10.3, 10.5, and 10.7.

Referring to Figure 9.16b, the forces acting on bearings at the shaft end are

$$R_A = [(42)^2 + (115.4)^2]^{1/2} = 122.8 \text{ N}$$

$$R_B = [(33)^2 + (90.6)^2]^{1/2} = 96.42 \text{ N}$$

Since $R_A > R_B$, we analyze the bearing at the left end A of the shaft.

Through the use of Eq. (10.27) with axial thrust $F_a = 0$, the equivalent radial load is

$$P = K_s X V F_r$$

Here

$$F_r = 122.8 \text{ N}$$

$$K_s = 1.5 \quad \text{(from Table 10.7)}$$

$$X = 1.0, \quad Y = 0 \quad \text{(by Table 10.5)}$$

$$V = 1 \quad \text{(from Section 10.13)}$$

Therefore, we have

$$P = 1.5(1.0)(1)(122.8) = 184.2 \text{ N}$$

The basic dynamic load rating C, applying Eq. (10.30), is given by

$$C = P \left(\frac{60 n L_{10}}{10^6} \right)^{1/a}$$

where $a = 3$ for ball bearings. Introducing the given data into this equation,

$$C = 184.2 \left[\frac{60(1725)(30,000)}{10^6} \right]^{1/3} = 2.69 \text{ kN}$$

We see from Table 10.3 that a 02-series deep-groove ball bearing with a bore of 12-mm has a load rating of $C = 6.89$ kN. This is well above the estimated value of 2.69 kN, and the bearing is quite satisfactory. Following this procedure, other bearings for the shafts in the gear box may be analyzed in a like manner.

Comment: The final selection of the bearings would be made on the basis of standard shaft and housing dimensions.

10.15   MATERIALS AND LUBRICANTS OF ROLLING BEARINGS

Most *balls* and *rings* are made from high-carbon chromium steel (SAE 52100), heat treated to high strength and hardness, and the surfaces smoothly ground and polished. *Separators* are usually made of low-carbon steel and copper alloy, such as bronze. Unlike ball bearings, roller bearings are often fabricated of case-hardened steel alloys. Modern steel manufacturing processes have resulted in bearing steels with reduced level of impurities.

As pointed out in Section 10.3, elastohydrodynamic lubrication occurs in rolling bearings in which deformation of the parts must be taken into account as well as increased viscosity of the oil owing to the high pressure. This small elastic flattening of parts, together with the increase in viscosity, provides a film, although very thin, that is much thicker than would prevail with complete rigid parts [26–28]. In addition to providing a film between the sliding and rolling parts, a lubricant may help distribute and dissipate heat, prevent corrosion of the bearing surfaces, and protect the parts from the entrance of foreign particles.

Depending on the load, speed, and temperature requirements, bearing lubricants are either *greases* or *oils*. Where bearing speeds are higher or loading is severe, oil is preferred.

Synthetic and *dry lubricants* are also widely used for special applications. Greases are suitable for low-speed operation and permit bearings to be prepacked.

10.16   MOUNTING AND CLOSURE OF ROLLING BEARINGS

Rolling-element bearings are generally mounted with the rotating inner or outer ring with a press fit. Then the stationary ring is mounted with a push fit. Bearing manufacturers' literature contain extensive information and illustrations on mountings. Here, we discuss only the basic principle of mounting ball bearings properly.

Figure 10.26 shows a common method of mounting, where the inner rings are backed up against the shaft shoulders and held in position by round nuts threaded into the shaft. As noted, the outer ring of the left-hand bearing is backed up against a housing shoulder and retained in position, but the outer ring of the right-hand bearing floats in the housing. This allows the outer ring to slide both ways in its mounting to avoid thermal-expansion-induced axial forces on the bearings, which would seriously shorten their life. An alternative bearing mounting is illustrated in Figure 10.27. Here the inner ring is backed up against the shaft shoulder, as before, however, no retaining device is needed; threads are eliminated. With this assembly, the outer rings of both bearings are completely retained. As a result, accurate dimensions in axial direction or the use of adjusting devices is required.

Duplexing of angular contact ball bearings arises when maximum stiffness and resistance to shaft misalignment is required, such as in machine tools and instruments. Bearings for *duplex mounting* have their rings ground with an offset, so that, when a pair of bearings is rigidly assembled, a controlled axial preload is automatically achieved [14, 15]. Figures 10.28a and 10.28b show a face-to-face (DF) and back-to-back (DB) mounting arrangements, respectively, that take heavy radial and thrust loads

Figure 10.26   A common bearing mounting. (Courtesy of the Timken Company.) Note: The outer ring of the left-hand bearing is held in position by a device (not shown).

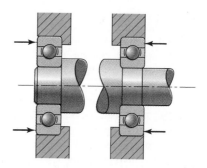

Figure 10.27   An alternative bearing mounting. (Courtesy of the Timken Company.) Note: The outer rings of both bearings are held in position by devices (not shown).

Figure 10.28   Mounting arrangements of angular ball bearings: (a) face to face; (b) back to back; (c) tandem. (Courtesy of the Timken Company.)

from either direction. The latter has greater mounting stiffness. Clearly, a tandem (DT) mounting arrangement is employed when the thrust is in the same direction (Figure 10.28c). Single-row ball bearings are often loaded by the axial load built in during assembly, as shown in Figure 10.26. Preloading helps to remove the internal clearance often found in bearings to increase the fatigue life and decrease the shaft slope at the bearings.

Note that the majority of bearings may be supplied with side shields. The shields are not complete closures, but they offer a measure of protection against dust or dirt. A sealed bearing is generally to be lubricated for life. The roller bearings are not often supplied in a sealed and self-lubricated form, as are most ball bearing types.

REFERENCES

1. Booser, E. R., ed. *CRC Handbook of Lubrication,* vols. I and II. Boca Raton, FL: CRC Press, 1983 and 1984.
2. Lansdown, A. R. *Lubrication.* Oxford: Pergamon Press, 1982.
3. Hamrock, B. J. *Fundamentals of Fluid Film Lubrication.* New York: McGraw-Hill, 1993.
4. Cameron, A. *Basic Lubrication Theory.* New York: Wiley, 1976.
5. Szeri, A. Z. *Tribology.* New York: McGraw-Hill, 1980.
6. Wills, J. G. *Lubrication Fundamentals.* New York: Marcel Decker, 1980.
7. Avallone, A. E., and T. Beaumeister III. *Mark's Standard Handbook for Mechanical Engineers,* 10th ed., New York: McGraw-Hill, 1996.
8. Reynolds, O. "On the Theory of Lubrication and Its Application to Mr. Beauchamp Tower's Experiments." *Philosophical Transactions of the Royal Society* (London) 177(1886), pp. 157–234 .
9. Ocvirk, F. W. "Short Bearing Approximation for Full Journal Bearings." TN 2208, NACA, 1952. Also see Dubois, G. B., and F .W. Ocvirk. "The Short Bearing Approximation for Plain Journal Bearings." *Transactions of the ASME* 77 (1955), pp. 1173–178; Ocvirk, F. W., and G. B. Dubois. "Surface Finish and Clearance Effects on Journal-Bearing Load Capacity and Friction." *Transactions of the ASME* 81 (1959), *Journal of Basic Engineering,* p. 245.
10. Raimondi, A. A., and J. Boyd. "A Solution for Finite Journal Bearings and its Application to Analysis and Design." Parts I, II, and III. *Transactions of the ASLE* 1, no. 1, pp. 159–209; reprinted in *Lubrication Science and Technology.* New York: Pergamon Press, 1958.
11. Juvinall, R. E., and M. Marshek. *Fundamentals of Machine Component Design,* 3rd ed. New York: Wiley, 2000.

12. Shigley, J. E., and C. R. Mischke. *Mechanical Engineering Design,* 6th ed. New York: McGraw-Hill, 2001.

13. Rothbart, H. A., ed. *Mechanical Design and Systems Handbook,* 2nd ed. New York: McGraw-Hill, 1985.

14. Hamrock, B. J., B. Jacobson, and S. R. Schmid. *Fundamentals of Machine Elements.* New York: McGraw-Hill, 1999.

15. Deutschman, A. D., and C. E. Wilson. *Machine Design.* New York: Macmillan, 1975.

16. Burr, A. H., and J. B. Cheatham. *Mechanical Analysis and Design,* 2nd ed. Upper Saddle River, NJ: Prentice Hall, 1995.

17. Fuller, D. D. *Theory and Practice of Lubrication for Engineers,* 2nd ed. New York: Wiley, 1984.

18. O'Conner, J. J., and J. Boyd. *Standard Handbook of Lubrication Engineering.* New York: McGraw-Hill, 1968.

19. Norton, R. E. *Machine Design—An Integrated Approach,* 2nd ed. Upper Saddle River, NJ: Prentice Hall, 2000.

20. Hamrock, J., and D. Dowson. *Ball Bearing Lubrication—The Elastohydrodynamics of Elliptical Contacts.* New York: Wiley, 1981.

21. *Standards of the Anti-friction Bearing Manufacturing Association.* New York: 1990.

22. Ball Bearing General Catalog. Sandusky, OH: New Departure-Hyatt Bearing Division, General Motors Corporation.

23. Timken Engineering Journal. Canton, OH: The Timken Roller Bearing Company.

24. Bamberger, E. N., et al. "Life Adjustment Factors for Ball and Roller Bearings." *ASME Engineering Design Guide,* 1971.

25. Harris, T. A. *Rolling Bearing Analysis.* 3rd ed. New York: Wiley, 1991.

26. Cheng, H. S. "Elastohydrodynamic Lubrication." In E. R. Booser, ed., *Handbook of Lubrication.* Boca Raton, FL: CRC Press, 1983, pp. 155–60.

27. Dowson, D., and G. Higginson. "A Numerical Solution to the Elastohdrodynamic Problems." *Journal of Mechanical Engineering Science* 1, no. 1 (1959), p. 6.

28. Hamrock, B., and D. Dowson. "Isothermal Elastohydrodynamic Lubrication of Point Contacts. Part III, Fully Flooded Results." *ASME Journal of Lubrication Technology* 99 (1977), pp. 264–76.

PROBLEMS

Sections 10.1 through 10.5

10.1 A lightly loaded journal bearing 220 mm in length and 160 mm in diameter consumes 2 hp in friction when running at 1200 rpm. Diametral clearance is 0.18 mm and SAE 30 oil is used. Find the temperature of the oil film.

10.2 A journal bearing has a 4-in. length, a 3-in. diameter, a c/r ratio of 0.002, carries a 500-lb radial load at 24,000 rpm, and is supplied with an oil having a viscosity of 0.6 μreyns. Using the Petroff approach, estimate

(a) The frictional torque developed.

(b) The frictional horsepower.

(c) The coefficient of friction.

10.3 A Petroff bearing has a 120-mm length, a 120-mm diameter, a 0.05-mm radial clearance, a speed of 600 rpm, and a radial load of 8 kN. Assume that the coefficient of friction is 0.01 and the average oil-film temperature is 70°C. Determine

(a) The viscosity of the oil.

(b) The approximate SAE grade of the oil.

10.4 A journal bearing having a 125-mm diameter, a 125-mm length, and c/r ratio of 0.0004 carries a radial load of 12 kN. A frictional force of 80 N is developed at a speed of 240 rpm. What is the viscosity of the oil according to the Petroff approach?

10.5 A journal bearing 6 in. in diameter and 1.5 in. long carries a radial load of 500 lb at 1500 rpm; $c/r = 0.001$. It is lubricated by SAE 30 oil at 180°F. Estimate, using the Petroff approach,

(a) The bearing coefficient of friction.

(b) The friction power loss.

Sections 10.6 through 10.10

10.6 A 4-in. diameter × 2-in. long bearing turns at 1800 rpm; $c/r = 0.001$; $h_0 = 0.001$ in. SAE 30 oil is used at 200°F. Through the use of the design charts, find the load W.

10.7 Redo Problem 10.4 employing the design charts.

10.8 Resolve Problem 10.5 using the design charts.

10.9 A 4-in. diameter shaft is supported by a bearing 4 in. long with a minimum oil-film thickness of 0.001 in. and radial clearance of 0.0025 in. It is lubricated by SAE 20 oil. The bearing carries a load of 100 psi of projected area at 900 rpm. Employing the design charts, determine

(a) The temperature of the oil film.

(b) The coefficient of friction.

(c) The friction power.

10.10 A 25-mm diameter × 25-mm long bearing carries a radial load of 1.5 kN at 1000 rpm; $c/r = 0.0008$; $\eta = 50$ mPa · s. Using the design charts, determine

(a) The minimum oil-film thickness.

(b) The friction power loss.

10.11 A 80-mm diameter × 40-mm long bearing supports a radial load of 4 kN at 600 rpm; $c/r = 0.002$. SAE 40 oil is used at 65°C. Employing the design charts, determine

(a) The minimum oil-film thickness.

(b) The maximum oil pressure.

10.12 A 50-mm diameter × 50-mm long bearing having a c/r ratio of 0.001 consumes 0.16 hp in friction at an operating speed of 1630 rpm. It is lubricated by SAE 20 oil at 83°C. (Hint: Try $S = 0.03$.) Using the design charts, determine

    (a) The radial load for the bearing.

    (b) The minimum oil-film thickness.

    (c) The eccentricity ratio.

10.13 A journal bearing having an L/D ratio of 1/2, a 100-mm diameter, a c/r ratio of 0.0015, and an operating speed of 900 rpm carries a radial load of 8 kN. The minimum oil-film thickness is to be 0.025 mm. Using the design charts, determine

    (a) The viscosity of the oil.

    (b) The friction force and power developed.

10.14 A 100-mm diameter × 50-mm long ring-oiled bearing supports a radial load 6 kN at 300 rpm in still air; $c/r = 0.001$; and $\eta = 20$ mPa · s. If the temperature of the surrounding air of the housing is 20°C, estimate the average film temperature.

10.15 Redo Problem 10.14 for an oil-bath lubrication system in an average air circulation condition when the temperature of the air surrounding air of the housing is $t_a = 30$°C.

Sections 10.11 through 10.16

10.W Use the website at www.grainger.com to conduct a search for roller bearings. Locate a thrust ball bearing $\frac{1}{4}$ in. bore, $\frac{9}{16}$ in. OD, and $\frac{7}{32}$ in. width. List the manufacturer and description.

10.16 A 25-mm (02-series) deep-groove ball bearing carries a combined load of 2 kN radially and 3 kN axially at 1500 rpm. The outer ring rotates and the load is steady. Determine the rating life in hours.

10.17 Resolve Problem 10.16, for a single-row, angular-contact ball bearing having 35° contact angle.

10.18 Redo Problem 10.16, if the inner ring rotates and the bearing is subjected to a light shock load.

10.19 What percentage change in loading of a ball bearing causes the expected life be doubled?

10.20 Resolve Problem 10.19 for a roller bearing.

10.21 A 60-mm bore (02-series) double-row, angular-contact ball bearing has a 15° contact angle. The outer ring rotates, and the bearing carries a combined steady load of 5 kN radially and 1.5 kN axially at 1000 rpm. Calculate the median life in hours.

10.22 Determine the expected rating lives in hours of a 35-mm bore (02- and 03-series) straight cylindrical bearings operating at 2400 rpm. Radial load is 5 kN, with heavy shock, and the outer rings rotate.

10.23 Calculate the median lives in hours of a 75-mm bore (02- and 03-series) straight cylindrical bearings operating at 2000 rpm. Radial load is 25 kN, with light shock, and inner rings rotate.

10.24 Select two (02- and 03-series) straight cylindrical bearings for an industrial machine intended for a rating life of 24-hour operation at 2400 rpm. The radial load is 12.5 kN, with extreme shock, and the inner rings rotate.

10.25 Select a (02-series) deep-groove ball bearing for a machine intended for a median life of 40 hours operation at 900 rpm. The bearing is subjected to a radial load of 8 kN, with heavy shock, and the outer ring rotates.

10.26 Determine the expected rating life of the deep-groove ball bearing in Problem 10.16, if only a 5% probability of failure can be permitted at 1200 rpm.

10.27 Calculate the expected median life of the straight cylindrical bearing in Problem 10.22, if only a 2% probability of failure can be permitted.

SPUR GEARS

Outline

11.1   INTRODUCTION

Gears are used to transmit torque, rotary motion, and power from one shaft to another. They have a long history. In about 2600 B.C., the Chinese used primitive gearsets, most likely made of wood, their teeth merely pegs inserted in wheels. In the 15th century A.D., Leonardo da Vinci showed many gear arrangements in his drawings. Presently, a wide variety of gear types have been developed that operate quietly and with very low friction losses. Smooth, vibrationless action is secured by giving the proper geometric form to the outline of the teeth.

Compared to various other means of power transmission (e.g., belts and chains), gears are the most rugged and durable. They have transmission *efficiency* as high as 98%. However, gears are generally more costly than belts and chains. As we shall see, two modes of failure affect gear teeth: fatigue fracture owing to fluctuating bending stress at the root of the tooth and fatigue (wear) of the tooth surface. Both must be checked when designing the gears. The shapes and sizes of the teeth are standardized by the American Gear Manufacturers Association (AGMA). The methods of AGMA are widely employed in design and analysis of gearing. Selection of the proper materials to obtain satisfactory strength, fatigue, and wear properties is important. The AGMA approach requires extensive use of charts and graphs accompanied by equations that facilitate application of computer-aided design. Gear design strength and life rating equations have been computer modeled and programmed by most gear suppliers. It is not necessary for designers to create their own computer programs [1–4].

There are four principal *types of gearing:* spur, helical, bevel, and worm gears (Figure 11.1). Note that spur and helical gears have teeth parallel and inclined to the axis of

Figure 11.1   A variety of gears, including spur gears, rack and pinion, helical gears, bevel gears, worm, and worm gear. (Courtesy of Quality Transmission Components.)

rotation, respectively. Bevel gears have teeth on conical surfaces. The geometry of a worm is similar to that of a screw. Of all types, the spur gear is the simplest. Here, we introduce the general gearing terminology, develop fundamental geometric relationships of the tooth form, and deal mainly with spur gears. A review of the nomenclature and kinematics is followed by a detailed discussion of the stresses and a number of factors influencing gear design. The basis of the AGMA method and its use are illustrated. Other gear types are dealt with in the next chapter. For general information on gear types, gear drives, and gearboxes, see the website at www.machinedesign.com. The site at www.powertransmission.com lists websites for numerous manufacturers of gears and gear drives.

11.2 GEOMETRY AND NOMENCLATURE

Consider two virtual friction cylinders (or disks) having no slip at the point of contact, represented by the circles in Figure 11.2a. A friction cylinder can be transformed into *spur gear* by placing teeth on it that run parallel to the axis of the cylinder. The surfaces of the rolling cylinders, shown by the dashed lines in the figures then become the *pitch circles*. The diameters are the pitch diameters, and the cylinders represent the pitch cylinders. The teeth, which lie in axial paths on the cylinder, are arranged to extend both outside and inside the pitch circles (Figure 11.2b). All calculations are based on the pitch circle. Note that spur gears are used to transmit rotary motion between parallel shafts.

A *pinion* is the smaller of the two mating gears, which is also referred to as a *pair of gears* or *gearset*. The larger is often called the *gear*. In most applications, the pinion is the driving element whereas the gear is the driven element. This reduces speed, but increases torque, from the power source (engine, motor, turbine): Machinery being driven runs slower. In some cases, gears with teeth cut on the inside of the rim are needed. Such a gear is known as an *internal gear* or an *annulus* (Figure 11.3a). A *rack* (Figure 11.3b) can be thought of as a segment of an internal gear of infinite diameter.

PROPERTIES OF GEAR TOOTH

The face and flank portion of the tooth surface are divided by the pitch cylinder. The *circular pitch p* is the distance, on the pitch circle, from a point on one tooth to a

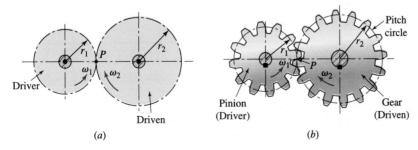

Figure 11.2 Spur gears are used to connect parallel shafts: (a) friction cylinders; (b) an external gearset.

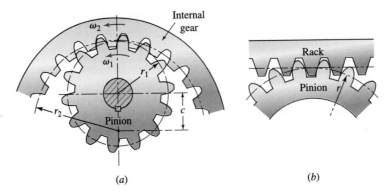

Figure 11.3   Gearsets: (a) internal gear and pinion; (b) rack and pinion.

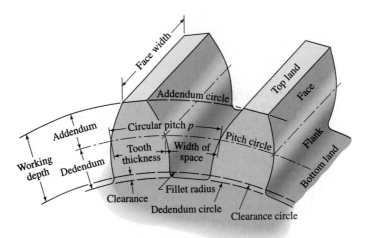

Figure 11.4   Nomenclature of the spur gear teeth.

corresponding point on the next. This leads to the definition

$$p = \frac{\pi d}{N}$$ (11.1)

where

     p = circular pitch, in.

     d = pitch diameter, in.

     N = number of teeth

The *diametral pitch P* is defined as the number of teeth in the gear per inch of pitch diameter. Therefore,

$$P = \frac{N}{d}$$ (11.2)

This measure is used in the U.S. specification of gears. The units of P are teeth/in. or in.$^{-1}$. Both circular and diametral pitches prescribe the tooth size. The latter is a more convenient definition. Combining Eqs. (11.1) and (11.2) yields the useful relationship

$$pP = \pi \qquad \text{(11.3)}$$

For two gears to mesh, they must have the same pitch.

In SI units, the size of teeth is specified by the *module* (denoted by m) measured in *millimeters*. We have

$$m = \frac{d}{N} \qquad \text{(11.4)}$$

where pitch diameter d and pitch radius r must be in millimeters and N is the number of teeth. Carrying the foregoing expression into Eq. (11.1) results in the circular pitch in millimeters:

$$p = \pi m \qquad \text{(11.5a)}$$

The diametral pitch, using Eq. (11.3), is then

$$P = \frac{1}{m} \qquad \text{(11.5b)}$$

It is measured in teeth/mm, or mm^{-1}. Note that metric gears are not interchangeable with U.S. gears, as the standards for tooth size are different.

The addendum a is radial distance between the top land and the pitch circle as shown in Fig. 11.4. The dedendum b_d represents the radial distance from the bottom land to the pitch circle. The *face width b* of the tooth is measured along the axis of the gear. The whole depth h is the sum of the addendum and dedendum. The clearance circle represents a circle tangent to the addendum circle of the mating gear. The clearance f represents the amount by which the dedendum in a given gear exceed the addendum of the mating gear. Clearance is required to prevent the end of the tooth of one gear from riding on the bottom of the mating gear. The difference between the whole depth and clearance represents the working depth h_k. The distance between the centers of the two gears in mesh is called the *center distance c*. Using Eq. (11.2) with $d = 2r$,

$$c = r_1 + r_2 = \frac{N_1 + N_2}{2P} \qquad \text{(11.6)}$$

Here, subscripts 1 and 2 refer to driver and driven gears, respectively.

The width of space between teeth must be made slightly larger than the gear tooth thickness t, both measured on the pitch circle. Otherwise, the gears cannot mesh without jamming. The difference between the foregoing dimensions is known as *backlash*. That is, the backlash is the gap between mating teeth measured along the circumference of the pitch circle. Manufacturing tolerances preclude a 0 backlash, since all teeth cannot be exactly the same dimensions and all must mesh without jamming. The amount of backlash must be limited to the minimum amount necessary to ensure satisfactory meshing of gears. Excessive backlash increases noise and impact loading whenever torque reversals occur.

11.3   FUNDAMENTALS

The main requirement of gear-tooth geometry is the provision that angular ratios are exactly constant. We assume that the teeth are perfectly formed, perfectly smooth, and absolutely rigid. Although manufacturing inaccuracies and tooth deflections induce slight deviations in velocity ratio, acceptable tooth profiles are based on theoretical curves that meet this criterion.

BASIC LAW OF GEARING

For quiet, vibrationless operation, the velocities of two mating gears must be the same at all times. This condition is satisfied when the pitch circle of the driver is moving with constant velocity, the velocity of the pitch circle of the driven gear neither increases nor decreases at any instant while the two teeth are touching. The basic law of gearing states that *as the gears rotate, the common normal at the point of contact between the teeth must always pass through a fixed point on the line of centers*. The fixed point is called the *pitch point P* (Figure 11.2). If two gears in mesh satisfy the basic law, the gears are said to produce *conjugate action*.

According to the fundamental law, when two gears are in mesh, their pitch circles roll on one another without slipping. Denoting the pitch radii by r_1 and r_2 and angular velocities as ω_1 and ω_2, respectively, the *pitch line velocity* is then

$$V = r_1\omega_1 = r_2\omega_2 \tag{11.7}$$

Several useful relations for determining the *speed ratio* may be written as follows:

$$r_s = \frac{\omega_2}{\omega_1} = \frac{n_2}{n_1} = \frac{N_1}{N_2} = \frac{d_1}{d_2} \tag{11.8}$$

where

r_s = speed or velocity ratio

ω = angular velocity, rad/sec

n = speed, rpm

N = number of teeth

d = pitch circle diameter

Subscripts 1 and 2 refer to the driver and driven gears, respectively.

INVOLUTE TOOTH FORM

To obtain conjugate action, most gear profiles are cut to conform to an involute curve [5]. Our discussions are limited to toothed wheel gearing of the involute form. The involute curve may be generated graphically by wrapping a string around a fixed cylinder, then tracing the path a point on the string (kept taut) makes as the string is unwrapped from the cylinder. When the involute is applied to gearing, the cylinder around which the string is wrapped is defined as the *base circle* (Figure 11.5). Gear teeth are cut in the shape of an involute curve between the base and the addendum circles, while that part of the tooth between the base and dedendum circles is generally a radial line. Figure 11.6 shows two involutes, on separate cylinders in mesh, representing the gear teeth. Note especially that conjugate involute action can occur only outside both base circles.

Figure 11.5 Development of the involute curve.

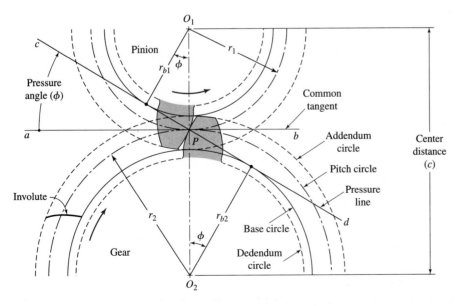

Figure 11.6 Involute gear teeth contact form and pressure angle.

11.4   GEAR TOOTH ACTION AND SYSTEMS OF GEARING

To illustrate the action occurring when two gears are in mesh, consider Figure 11.6. The pitch radii r_1 and r_2 are mutually tangent along the line of centers O_1O_2, at the pitch point P. Line ab is the common tangent through the pitch point. Note that line cd is normal to the teeth that are in contact and always passes through P at an angle ϕ to ab. Line cd is also tangent to both base circles. This line, called *line of action* or *pressure line,* represents (neglecting the sliding friction) the direction in which the resultant force acts between the gears. The angle ϕ is known as the *pressure angle,* which is measured in a direction opposite to the direction of rotation of the driver. The involute is the only geometric profile satisfying the basic law of gearing that maintains a constant-pressure angle as the gears rotate. Gears to be run together must be cut to the same nominal pressure angle.

As pointed out, the base circle is tangent to the pressure line. Referring to Figure 11.6, the radius of the base circle is then

$$r_b = r \cos \phi \qquad \text{(a)}$$

where r represents the pitch circle radius. Note that changing the center distance has no effect on the base circle, because this is used to generate the tooth profiles. That is, the base circle is basic to a gear. Increasing the center c distance increases the pressure angle ϕ, but the teeth are still conjugate; the requirement for uniform motion transmission is still satisfied. Therefore, with an involute tooth form, center distance errors do not affect the velocity ratio.

STANDARD GEAR TEETH

Most gears are cut to operate with standard pressure angles of 20° or 25°. The tooth proportions for some involute, spur-gear teeth are given in Table 11.1 in terms of the diameteral pitch P. Full-depth involute is a commonly used system of gearing. The table shows that the 20° stub-tooth involute system has shorter addenda and dedenda than the full-depth systems. The short addendum reduces the duration of contact. Because of insufficient overlap of contact, vibration may occur, especially in gears with few stub teeth. As a general rule, spur gears should be designed with a *face width b* greater than $9/P$ and less than $13/P$. Unless otherwise specified, we use the term pressure angle to refer to a pressure angle of *full-depth* teeth.

Table 11.1   Commonly used standard tooth systems for spur gears

Item	20° full depth	20° stub	25° full depth
Addendum a	$1/P$	$0.8/P$	$1/P$
Dedendum b_d	$1.25/P$	$1/P$	$1.25/P$
Clearance f	$0.25/P$	$0.2/P$	$0.25/P$
Working depth h_k	$2/P$	$1.6/P$	$2/P$
Whole depth h	$2.25/P$	$1.8/P$	$2.25/P$
Tooth thickness t	$1.571/P$	$1.571/P$	$1.571/P$

Figure 11.7 Actual size gear teeth of various diametral pitches. (Courtesy of Bourn & Koch Machine Tool Co., Rockford, IL.)

Figure 11.7 depicts the actual sizes of $20°$ pressure angle, standard, full-depth teeth, for several standard pitches from $P = 4$ to $P = 80$. Note the inverse relationship between P and tooth size. With SI units, the *standard* values of *metric module mm* are listed in the following:

0.3	0.4	0.5	0.8	1	1.25
1.5	2	3	4	5	6
8	10	12	16	20	25

The conversion from one standard to the other is $m = 24.4/P$. The most widely used pressure angle ϕ, in both U.S. customary and SI units, is $20°$.

Determining Gear Tooth and Gear Mesh Parameters

EXAMPLE 11.1

Two parallel shafts A and B with center distance c are to be connected by 2-teeth/in. diametral pitch, $20°$ pressure angle, spur gears 1 and 2 providing a velocity ratio of r_s (Figure 11.8). Determine, for each gear,

(a) The number of teeth N.

(b) The radius of the base circle r_b and outside diameter d_o.

(c) Clearance f.

(d) The pitch-line velocity V, if gear 2 rotates at speed n_2.

Figure 11.8   Example 11.1.

Given: $n_2 = 500$ rpm, $r_s = 1/3$, $c = 14$ in., $P = 2$ in.$^{-1}$, $\phi = 20°$.

Design Decision: Common stock gear sizes are considered.

Solution:

(a) Using Eqs. (11.6) and (11.7), we have $r_1 + r_2 = c = 14$ in., $r_1/r_2 = 1/3$. Hence, $r_1 = 3.5$ in., $r_2 = 10.5$ in., or $d_1 = 7$ in., $d_2 = 21$ in. Equation (11.2) leads to

$$N_1 = 7(2) = 14, \qquad N_2 = 21(2) = 42$$

(b) Base circle radii, applying Eq. (a),

$$r_{b1} = 3.5 \cos 20° = 3.289 \text{ in.}$$

$$r_{b2} = 10.5 \cos 20° = 9.867 \text{ in.}$$

From Table 11.1, the addendum $a = 1/2 = 0.5$ in. Then,

$$d_{o1} = 7 + 2(0.5) = 8 \text{ in.}$$

$$d_{o2} = 21 + 2(0.5) = 22 \text{ in.}$$

(c) We have $f = b_d - a$. Table 11.1 gives the dedendum $b_d = 1.25/2 = 0.625$ in. and hence

$$f = 0.625 - 0.5 = 0.125 \text{ in.}$$

for the pinion and gear. Note as a check that, from Table 11.1, $f = 0.25/2 = 0.125$ in.

(d) Substituting the given data, Eq. (11.7) results in

$$V = r_2 \omega_2 = \frac{10.5}{12} \left(500 \times \frac{2\pi}{60} \right) = 45.81 \text{ fps}$$

11.5 CONTACT RATIO AND INTERFERENCE

Inasmuch as the tips of gear teeth lie on the addendum circle, *contact* between two gears *starts* when the addendum circle of the driven gear intersects the pressure line and *ends* when the addendum circle of the driver intersects the pressure line. The length of action or *length of contact Z* can be derived from the mating gear and pinion geometry [5–7] in the form

$$Z = [(r_p + a_p)^2 - (r_p \cos \phi)^2]^{1/2} + [(r_g + a_g)^2 - (r_g \cos \phi)^2]^{1/2} - c \sin \phi \quad \text{(11.9)}$$

Here r = pitch radius, a = addendum, c = center distance, and ϕ = pressure angle. The subscripts p and g present pinion and gear, respectively. When two gears are in mesh, it is desirable to have at least one pair of teeth in contact at all times.

The method often used to show how many teeth are in contact is the *contact ratio* m_p, which is defined as the length of contact divided by the base pitch:

$$m_p = \frac{Z}{p_b} \quad \text{(11.10)}$$

The base pitch p_b refers to the distance measured on the base circle between corresponding points of adjacent teeth:

$$p_b = p \cos \phi$$

where p is the circular pitch. Obviously, the length of contact must be somewhat greater than a base pitch, so that a new pair of teeth come into contact before the pair that had been carrying the load separate. If the contact ratio is 1, then one tooth is leaving contact just as the next is beginning contact. Most gears are designed with contact ratios between 1.2 and 1.6. For instance, a ratio of 1.5 indicates that one pair of teeth is always in contact and a second in contact 50% of the time. Generally, the greater is the contact ratio or considerable overlap of gear actions, the smoother and quieter the operation of gears.

Since the part of a gear tooth below the base line is cut as a radial line and not an involute curve, if contact should take place below the base circle, nonconjugate action would result. Hence, the basic law of gearing would not hold. The contact of these portions of tooth profiles that are not conjugate is called *interference*. When interference occurs, the gears do not operate without modification. Removal of the portion of tooth below the base circle and cutting away the interfering material results in an *undercut tooth*. Undercutting causes early tooth failure. Interference and its attendant undercutting can be prevented as follows: remove a portion of the tips of the tooth, increase the pressure angle, or use minimum required tooth numbers [8, 9]. The method to be used depends largely on the application and the designer's experience.

11.6 GEAR TRAINS

Up to now in our discussion of gears, we have been concerned with no more than a pair of gears in mesh. Various applications exist where many pairs of gears are in mesh. Such a system is generally called a *gear train*. Typical examples include the gear trains in

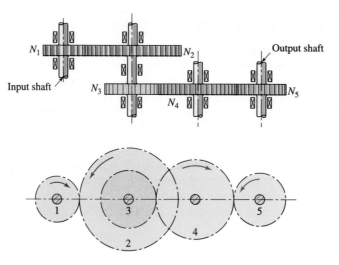

Figure 11.9   Gear train.

odometers and mechanical watches or clocks. A gearset, the simplest form of gear train, is often limited to a ratio of 10:1. Gear trains are used to obtain a desired velocity or speed of an output shaft while the input shaft runs at different speed. The velocity ratio between the input and output gears is constant. Detailed kinematic relationships for gear trains may be found in [3]. AGMA suggests equations that can be used to determine thermal capacity for gear trains (see also Section 12.11).

The speed ratio of a conventional gear train can be readily obtained from an expanded version of Eq. (11.8), if the number of teeth in each driver and driven gear is known. Consider, for example, a gear train made of five gears, with gears 2 and 3 mounted on the same shaft (Figure 11.9). The speed ratio between gear 5 and 1 is given by

$$\frac{n_5}{n_1} = \left(-\frac{N_1}{N_2}\right)\left(-\frac{N_3}{N_4}\right)\left(-\frac{N_4}{N_5}\right) \tag{a}$$

In the foregoing expression, the *minus* signs indicate that the pinion and gear rotate in *opposite* directions, as depicted in the figure. The intermediate gears, called *idler gears,* do not influence the overall speed ratio. In this case, gear 4 is an idler (its tooth numbers cancel in the preceding equation), hence it affects only the direction of rotation of gear 5.

Additional ratios can be inserted into Eq. (a) if the train consists of a larger number of gears. This equation can be generalized for any number of gearsets in the train to obtain the so-called *gear value:*

$$e = \pm \frac{\text{product of number of teeth on driver gears}}{\text{product of number of teeth on driven gears}} \tag{11.11}$$

Clearly, to ascertain the correct algebraic sign for the overall train ratio, the signs of the ratios of the individual pairs must be indicated in this expression. Note also that pitch

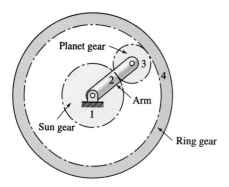

Figure 11.10 A planetary gear train.

diameters can be used in Eq. (11.11) as well. For spur gears, e is positive when the last gear rotates in the identical sense as the first; it is negative when the last rotates in the opposite sense. If the gear has internal teeth, its diameter is negative and the members rotate in the same direction.

PLANETARY GEAR TRAINS

Also referred to as the *epicylic* trains, planetary gear trains permit some of the gear axes to rotate about one another. Such trains always include a *sun gear,* an *arm,* and one or more *planet gears* (Figure 11.10). It is obvious that the maximum number of planets is limited by the space available and the teeth of each planet must align simultaneously with the teeth of the sun and the ring.

A planetary train must have two inputs: the motion of any two elements of the train; for example, the sun gear rotates at a speed of n_s (CW) and that the ring rotates at n_r (CCW) in Figure 11.10. The output would then be the motion of the arm. While power flow through a conventional gear train and the sense of motion for its members may be seen readily, it is often difficult to ascertain the behavior of a planetary train by observation. Planetary gear trains are thus more complicated to analyze than ordinary gear trains.

However, planetary gear trains have several advantages over conventional trains. These include higher train ratios obtainable in smaller packages and bidirectional outputs available from a single unidirectional input. The foregoing features make planetary trains popular as automatic transmissions and drives in motor vehicles [4], where they provide desired forward gear reductions and a reverse motion. Manufacturing precision and the use of the helical gears contribute greatly to the quietness of planetary systems.

It can be shown that [5, 10], the *gear value* of any *planetary train* is given in following convenient form:

$$e = \frac{\omega_L - \omega_A}{\omega_F - \omega_A} \qquad (11.12)$$

Here, we have

e = gear value, defined by Eq. (11.11)

ω_F = angular velocity of the first gear in the train

ω_L = angular velocity of the last gear in the train

ω_A = angular velocity of the arm

Note that both the first and last gears chosen must not be orbiting. When two of the velocities are specified, Eq. (11.12) can be used to compute the unknown velocity. That is, either the velocities of the arm and one gear or the velocities of the first and last gears must be known.

EXAMPLE 11.2

Analysis of a Planetary Gear Train

In the epicyclic gear train illustrated in Figure 11.10, the sun gear is driven clockwise at 120 rad/m and has N_1 teeth, the planet gear N_3 teeth, and the ring gear N_4 teeth. The sun gear is the input and the arm is the output. The ring gear is held stationary. What is the velocity of the arm?

Given: $N_1 = 20$, $N_3 = 30$, $N_4 = 80$

Assumption: The sun gear is the first gear in the train and the ring gear is the last.

Solution: Refer to Figure 11.10.
The gear value, through the use of Eq. (11.11), is

$$e = \left(-\frac{N_1}{N_3}\right)\left(+\frac{N_3}{N_4}\right) = \left(-\frac{20}{30}\right)\left(+\frac{30}{80}\right) = -0.25$$

Observe the signs on the gearset ratios: One is an external set $(-)$ and one an internal set $(+)$. Substitution of this equation together with $\omega_F = \omega_1 = 120$ rad/m and $\omega_L = \omega_4 = 0$ into Eq. (11.12) give

$$-0.25 = \frac{\omega_4 - \omega_2}{\omega_1 - \omega_2} = \frac{0 - \omega_2}{120 - \omega_2}$$

from which $\omega_2 = 24$ rad/m.

Comment: The arm rotates five times as slow and in the same direction as the sun gear.

11.7   TRANSMITTED LOAD

With a pair of gears or gearset, power is transmitted by the load that the tooth of one gear exerts on the tooth of the other. As pointed out in Section 11.4, the transmitted load F_n is normal to the tooth surface; therefore, it acts along the pressure line or the line of action

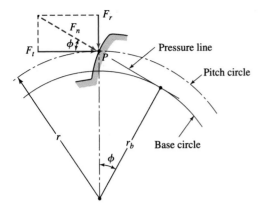

Figure 11.11 Gear-tooth force F_n, shown resolved at pitch point P.

(Figure 11.11). This force between teeth can be resolved into *tangential force* and radial force components, respectively:

$$F_t = F_n \cos \phi$$
$$F_r = F_n \sin \phi = F_t \tan \phi$$

(11.13)

The quantity ϕ is the pressure angle in degrees. Tangential component F_t, when multiplied by the pitch line velocity, accounts for the power transmitted, as is shown in Section 1.12. However, radial component F_r does no work but tends to push the gears apart.

The velocity along the pressure line is equal to the tangential velocity of the base circles. The tangential *velocity* of the *pitch circle* (in feet per minute, fpm) is given by

$$V = \frac{\pi dn}{12}$$

(11.14)

where d represents the pitch diameter in inches and n is the speed in rpm.

In design, we assume that the tangential force remains *constant* as the contact between two teeth moves from the top of the tooth to the bottom of the tooth. The applied torque and the transmitted load are related by

$$T = \frac{d}{2} F_n \cos \phi = \frac{d}{2} F_t$$

(11.15)

The horsepower is defined by

$$\text{hp} = \frac{Tn}{63,000}$$

(1.17)

in which the torque T is in pounds-inch and n is in rpm. Carrying Eqs. (11.14) and (11.15) into the preceding expression, we obtain the *tangential load transmitted:*

$$F_t = \frac{33,000 \text{ hp}}{V} \qquad \text{(11.16)}$$

where V is given by Eq. (11.14). Recall from Section 1.12 that, 1 hp equals 0.7457 kW. In SI units, the forgoing equations are given by the relationships

$$kW = \frac{F_t V}{1000} = \frac{Tn}{9549} \qquad \text{(1.15)}$$

$$hp = \frac{F_t V}{745.7} = \frac{Tn}{7121} \qquad \text{(1.16)}$$

$$F_t = \frac{1000 \text{ kW}}{V} = \frac{745.7 \text{ hp}}{V} \qquad \text{(11.17)}$$

Here,

F_t = transmitted tangential load (N)

d = gear diameter

n = speed (rpm)

T = torque (N·m)

$V = \pi dn/60$ = pitch line velocity (in meters per second, m/s)

DYNAMIC EFFECTS

The tangential force F_t is readily obtained by Eq. (11.16). However, this is not the entire force that acts between the gear and teeth. Tooth inaccuracies and deflections, misalignments, and the like produce dynamic effects that also act on the teeth. The *dynamic load F_d* or total gear-tooth load, in U.S. customary units, is estimated using one of the following formulas:

$$F_d = \frac{600 + V}{600} F_t \qquad \text{(for } 0 < V \leq 2000 \text{ fpm)} \qquad \text{(11.18a)}$$

$$F_d = \frac{1200 + V}{1200} F_t \qquad \text{(for } 2000 < V \leq 4000 \text{ fpm)} \qquad \text{(11.18b)}$$

$$F_d = \frac{78 + \sqrt{V}}{78} F_t \qquad \text{(for } V > 4000 \text{ fpm)} \qquad \text{(11.18c)}$$

where V is the pitch line velocity in fpm. To convert to m/s *divide* the given values in these equations by 196.8. Clearly, the dynamic load occurs in the time while a tooth goes through mesh. Note that the preceding relations form the basis of the AGMA dynamic factors, discussed in Section 11.9.

Gear Force Analysis

EXAMPLE 11.3

The three meshing gears shown in Figure 11.12a have a module of 5 mm and a 20° pressure angle. Driving gear 1 transmits 40 kW at 2000 rpm to idler gear 2 on shaft B. Output gear 3 is mounted to shaft C, which drives a machine. Determine and show, on a free-body diagram,

(a) The tangential and radial forces acting on gear 2.

(b) The reaction on shaft B.

Figure 11.12   Example 11.2: (a) a gearset; and (b) free-body diagram of the forces acting on gear 2 and reaction on shaft B.

Assumptions: The idler gear and shaft transmit power from the input gear to the output gear. No idler shaft torque is applied to the idler gear. Friction losses in the bearings and gears are omitted.

Solution: The pitch diameters of gears 1 and 3, from Eq. (11.4), are $d_1 = N_1 m = 20(5) = 100$ mm and $d_3 = N_3 m = 30(5) = 150$ mm.

(a)   Through the use of Eq. (1.15),

$$T = \frac{9549\ kW}{n} = \frac{9549(40)}{2000} = 191\ N \cdot m$$

By Eqs. (11.15) and (11.13), the tangential and radial forces of gear 1 on gear 2 are then

$$F_{t,12} = \frac{T_1}{r_1} = \frac{191}{0.05} = 3.82\ kN, \qquad F_{r,12} = 3.82 \tan 20° = 1.39\ kN$$

Inasmuch as gear 2 is an idler, it carries no torque; so the tangential reaction of gear 3 on 2 is also equal to $F_{t,12}$. Accordingly, we have

$$F_{t,32} = 3.82\ kN, \qquad F_{r,32} = 1.39\ kN$$

The forces are shown in proper directions in Figure 11.12b.

(b)   Equilibrium of x- and y-directed forces acting on the idler gear gives $R_{Bx} = R_{By} = 3.82 + 1.39 = 5.21$ kN. The reaction on the shaft B is then

$$R_B = \sqrt{5.21^2 + 5.21^2} = 7.37 \text{ kN}$$

acting as depicted in Figure 11.12b.

Comments: When a combination of numerous gears is used as in a gear train, usually the shafts supporting the gears lie in different planes and the problem becomes a little more involved. For this case, the tangential and radial force components of one gear must be further resolved into components in the same plane as the components of the meshing gear. Hence, forces along two mutually perpendicular directions may be added algebraically [11].

11.8   THE BENDING STRENGTH OF A GEAR TOOTH: THE LEWIS FORMULA

Wilfred Lewis was the first to present the application of the bending equation to a gear tooth. The formula was announced in 1892, and it still serves the basis for gear tooth-bending stress analysis. Simplifying *assumptions* in the Lewis approach are as follows [12]:

1. A full load is applied to the tip of a single tooth.
2. The radial load component is negligible.
3. The load is distributed uniformly across the full face width.
4. The forces owing to tooth sliding friction are negligible.
5. The stress concentration in the tooth fillet is negligible.

To develop the basic Lewis equation, consider a cantilever subjected to a load F_t, uniformly distributed across its width b (Figure 11.13a). We have the section modulus

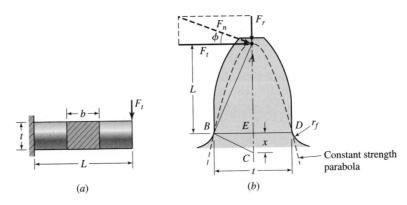

Figure 11.13   Beam strength of a gear tooth.

$I/c = bt^2/6$. So, the maximum bending stress is

$$\sigma = \frac{Mc}{I} = \frac{6F_t L}{bt^2} \qquad \text{(a)}$$

This flexure formula yields results of acceptable accuracy at cross sections away from the point of load application (see Section 3.1).

We now treat the tooth as a cantilever fixed at BD (Figure 11.13b). It was noted already that the normal force F_n is considered as acting through the corner tip of the tooth along the pressure line. The radial component F_r causes a uniform compressive stress over the cross section. This compressive stress is small enough compared to the bending stress, due to the tangential load F_t, to be ignored in determining the strength of the tooth. Clearly, the compressive stress increases the bending stress on the compressive side of the tooth and decreases the resultant stress on the tensile side. Therefore, for many materials that are stronger in compression than in tension, the assumption made results in a stronger tooth design. Also note that, because gear teeth are subjected to *fatigue failures* that start on the tension side of the tooth, the compressive stress reduces the resultant tensile stress and thus strengthens the tooth.

UNIFORM STRENGTH GEAR TOOTH

In a gear tooth of *constant* strength, the stress is uniform, hence $b/6F_t = \text{constant} = C$ Equation (a) then leads to $L = Ct^2$. The foregoing expression represents a parabola inscribed through point A, as shown by the dashed lines in Figure 11.13b. This parabola is tangent to the tooth profile at points B and D, where the maximum compressive and tensile stresses occur, respectively. The tensile stress is the cause of fatigue failure in a gear tooth. Referring to the figure, by similar triangles ABE and BCE, we write $(t/2)/x = L/(t/2)$ or $L = t^2/4x$. Carrying this into Eq. (a) and multiplying the numerator and denominator by the circular pitch p, we have

$$\sigma = \frac{F_t p}{b(2/3)xp} \qquad \text{(b)}$$

The Lewis form factor is defined as

$$y = \frac{2x}{3p} \qquad \text{(11.19)}$$

Finally, substitution of the preceding into Eq. (b) results in the original Lewis formula:

$$\sigma = \frac{F_t}{bpy} \qquad \text{(11.20)}$$

Because the diametral pitch rather than circular pitch is often used to designate gears, the following substitution may be made: $p = \pi/P$ and $Y = \pi y$. Then, the *Lewis form factor* is expressed as

$$Y = \frac{2xP}{3} \qquad \text{(11.21)}$$

Similarly, the *Lewis formula* becomes

$$\sigma = \frac{F_t P}{bY} \qquad \text{(11.22)}$$

Table 11.2   Values of the Lewis form factor for some common full-depth teeth

No. of teeth	20° Y	25° Y	No. of teeth	20° Y	25° Y
12	0.245	0.277	26	0.344	0.407
13	0.264	0.293	28	0.352	0.417
14	0.276	0.307	30	0.358	0.425
15	0.289	0.320	35	0.373	0.443
16	0.295	0.332	40	0.389	0.457
17	0.302	0.342	50	0.408	0.477
18	0.308	0.352	60	0.421	0.491
19	0.314	0.361	75	0.433	0.506
20	0.320	0.369	100	0.446	0.521
21	0.326	0.377	150	0.458	0.537
22	0.330	0.384	200	0.463	0.545
24	0.337	0.396	300	0.471	0.554
25	0.340	0.402	Rack	0.484	0.566

When using SI units, in terms of module $m = 1/P$,

$$\sigma = \frac{F_t}{mbY} \tag{11.23}$$

Both Y and y are functions of tooth shape (but not size) and thus vary with the number of teeth in the gear. Some values of Y determined from Eq. (11.21) are listed in Table 11.2. For nonstandard gears, the factor Y (or y) can be obtained by a graphical layout of the gear or digital computation.

Let bending stress σ in Eq. (11.22) be designated by the allowable static bending stress σ_o and so tangential load F_t by the *allowable bending load* F_b. Then, this equation becomes

$$F_b = \sigma_o b \frac{Y}{P} \tag{11.24}$$

or, in SI units,

$$F_b = \sigma_o bYm \tag{11.25}$$

The values of σ_o for some materials of different hardness are listed in Table 11.3. Note that the tip-load condition assumed in the preceding derivation occurs when another pair of the teeth is still in contact. Actually, the heaviest loads occur near the middle of the tooth while a single pair of teeth is in contact. For this case, the derivation of the Lewis equation would follow exactly as in the previous case.

EFFECT OF STRESS CONCENTRATION

The stress in a gear tooth is greatly influenced by size of the fillet radius r_f (Figure 11.13b). It is very difficult to obtain the theoretical values of stress concentration factor K_t for the

Table 11.3 Allowable static bending stresses for use in the Lewis equation

Material	Treatment	σ_o ksi	σ_o (MPa)	Average Bhn
Cast iron				
ASTM 35		12	(82.7)	210
ASTM 50		15	(103)	220
Cast steel				
0.20% C		20	(138)	180
0.20% C	WQ&T	25	(172)	250
Forged steel				
SAE 1020	WQ&T	18	(124)	155
SAE 1030		20	(138)	180
SAE 1040		25	(172)	200
SAE 1045	WQ&T	32	(221)	205
SAE 1050	OQ&T	35	(241)	220
Alloy steels				
SAE 2345	OQ&T	50	(345)	475
SAE 4340	OQ&T	65	(448)	475
SAE 6145	OQ&T	67	(462)	475
SAE 65 (phosphor bronze)		12	(82.7)	100

I Note: WQ&T = water-quenched and tempered, OQ&T = oil-quenched and tempered.

rather complex shape of gear tooth. Experimental techniques and the finite-element method are used for this purpose [12–14]. Since the gear tooth is subjected to fatigue loading, the factor K_t should be modified by the notch sensitivity factor q to obtain the fatigue stress concentration factor K_f. The Lewis formula can be modified to include the effect of the stress concentration. In so doing, Eqs. (11.22) and (11.24) become, respectively,

$$\sigma = \frac{K_f F_t P}{bY} \qquad (11.26)$$

$$F_b = \frac{\sigma_o b}{K_f} \frac{Y}{P} \qquad (11.27)$$

As a reasonable approximation, $K_f = 1.5$ may be used in these equations.

REQUIREMENT FOR SATISFACTORY GEAR PERFORMANCE

The load capacity of a pair of gears is based on either the bending or wear (Section 11.10) capacity, whichever is smaller. For *satisfactory gear performance,* it is necessary that the dynamic load should not exceed the allowable load capacity. That is,

$$F_b \geq F_d \qquad (11.28)$$

in which the dynamic load F_d is given by Eq. (11.18). Note that, this "dynamic-load" approach can be used for all gear types [6].

The Lewis equation is important, since it serves as the basis for the AGMA approach to the bending strength of the gear tooth, discussed in the next section. Equations (11.27) and (11.28) are quite useful in estimating the capacity of gear drives when the life and reliability are not significant considerations. They are quite useful in preliminary gear design for a variety of applications. When a gearset is to be designed to transmit a load F_b, the gear material should be chosen so that the values of the product $\sigma_o Y$ are approximately the same for both gears.

| EXAMPLE 11.4 | Power Transmitted by a Gear Based on Bending Strength and Using the Lewis Formula |

A 25° pressure angle, 25-tooth spur gear having a module of 2 mm and a 45-mm face width is to operate at 900 rpm. Determine

(a) The allowable bending load applying the Lewis formula.

(b) The maximum tangential load and power that the gear can transmit.

Design Decisions: The gear is made of SAE 1040 steel. A fatigue stress concentration factor of 1.5 is used.

Solution: We have $Y = 0.402$ for 25 teeth (Table 11.2) and $\sigma_o = 172$ MPa (Table 11.3). The pitch diameter is $d = mN = 2(25) = 50$ mm and $V = \pi dn = \pi(0.05)(15) = 2.356$ m/s $= 463.7$ fpm.

(a) Using Eq. (11.27) with $1/P = m$, we have

$$F_b = \frac{\sigma_o bYm}{K_f} = \frac{1}{1.5}(172 \times 45 \times 0.402 \times 2) = 4.149 \text{ kN}$$

(b) From Eq. (11.18a), the dynamic load is

$$F_d = \frac{600 + 463.7}{600} F_t = 1.77 F_t$$

The limiting value of the transmitted load, applying Eq. (11.28), is

$$4.149 = 1.77 F_t \quad \text{or} \quad F_t = 2.344 \text{ kN}$$

The corresponding gear power, by Eq. (1.15), is

$$kW = \frac{F_t \pi dn}{60}$$

$$= \frac{(2.344)\pi(0.05)900}{60} = 5.52$$

11.9 DESIGN FOR THE BENDING STRENGTH OF A GEAR TOOTH: THE AGMA METHOD

The fundamental formula for the bending stress of a gear tooth is the AGMA modification of the Lewis equation. This formula applies to the original Lewis equation correction factors that compensate for some of the simplifying presuppositions made in the derivation as well as for important factors not initially considered. In the AGMA method to the design and analysis of gearing, the bending strength of a gear tooth is also modified by various factors to obtain the allowable bending stress.

In this section, we present selective AGMA bending strength equations for a gear tooth. They are based on certain assumptions about the tooth and gear-tooth geometry. It should be mentioned that some definitions and symbols used are different than that given by the AGMA. Nevertheless, procedures introduced here and in Section 11.11 are representative of current practice [15–18]. For further information, see the latest AGMA Standards [1] and the relevant literature.

Bending stress is defined by the formula

$$\sigma = F_t K_o K_v \frac{P}{b} \frac{K_s K_m}{J} \qquad \text{(U.S. customary units)} \qquad \text{(11.29)}$$

$$\sigma = F_t K_o K_v \frac{1.0}{bm} \frac{K_s K_m}{J} \qquad \text{(SI units)} \qquad \text{(11.29')}$$

where

σ = calculated bending stress at the root of the tooth

F_t = transmitted tangential load

K_o = overload factor

K_v = velocity or dynamic factor

P = diametral pitch

b = face width

m = metric module

K_s = size factor

K_m = mounting factor

J = geometry factor

Allowable bending stress, or the design stress value, is

$$\sigma_{\text{all}} = \frac{S_t K_L}{K_T K_R} \qquad \text{(11.30)}$$

in which

σ_{all} = allowable bending stress

S_t = bending strength

K_L = life factor

K_T = temperature factor

K_R = reliability factor

As a *design specification,* the bending stress must not exceed the design stress value:

$$\sigma \leq \sigma_{all} \tag{11.31}$$

Note that there are three groups of terms in Eq. (11.29): The first refers to the loading characteristics, the second to the gear geometry, and the third to the tooth form. Obviously, the essence of this equation is the Lewis formula with the updated geometry factor J introduced for the form factor Y. The K factors are modifiers to account for various conditions. Equation (11.30) defines the allowable bending stress. The specification in the AGMA approach for designing for strength is given by Eq. (11.31). That is, the calculated stress σ of Eq. (11.29) must always be less than or equal to the allowable stress σ_{all} as determined by Eq. (11.30). To facilitate the use of the Eqs. (11.29) through (11.31), the following concise description of the correction factors is given.

The *overload factor* K_o is used to compensate for situations in which the actual load exceeds the transmitted load F_t. Table 11.4 gives some suggested values for K_o. The velocity or dynamic factor K_v shows the severity of impact as successive pairs of teeth engage. This depends on pitch velocity and manufacturing accuracy. Figure 11.14 depicts some commonly employed approximate factors pertaining to representative gear manufacturing processes. It is seen from the figure that dynamic factors become higher, when hobs or milling cutters are used to form the teeth or inaccurate teeth are generated. For more detailed information, consult the appropriate AGMA standard.

The *size factor* K_s attempts to account for any nonuniformity of the material properties. It depends on the tooth size, diameter of parts, and other tooth and gear dimensions. For most standard steel gears, the size factor is usually taken as unity. A value of 1.25 to 1.5 would be a conservative assumption in cases of very large teeth. The *mounting factor* K_m reflects the accuracy of mating gear alignment. Table 11.5 is used as a basis for rough estimates. The *geometry factor* J relies on the tooth shape, the position at which the highest

Table 11.4   Overload correction factor K_o

Source of power	Load on driven machine		
	Uniform	**Moderate shock**	**Heavy shock**
Uniform	1.00	1.25	1.75
Light shock	1.25	1.50	2.00
Medium shock	1.50	1.75	2.25

Figure 11.14 Dynamic factor K_v.
I SOURCE: The AGMA and [7].

Notes:

Curve A: $K_v = \sqrt{\dfrac{78 + \sqrt{V}}{78}}$,   Curve B: $K_v = \dfrac{78 + \sqrt{V}}{78}$

Curve C: $K_v = \dfrac{50 + \sqrt{V}}{50}$,   Curve D: $K_v = \dfrac{1200 + V}{1200}$

Curve E: $K_v = \dfrac{600 + V}{600}$,

where V is in feet per minute, fpm. To covert to meters per second (m/s), *divide* the given values by 196.8.

Table 11.5 Mounting correction factor K_m

	Face width (in.)			
Condition of support	**0 to 2**	**6**	**9**	**16 up**
Accurate mounting, low bearing clearances, maximum deflection, precision gears	1.3	1.4	1.5	1.8
Less rigid mountings, less accurate gears, contact across the full face	1.6	1.7	1.8	2.2
Accuracy and mounting such that less than full-face contact exists		Over 2.2		

(a)

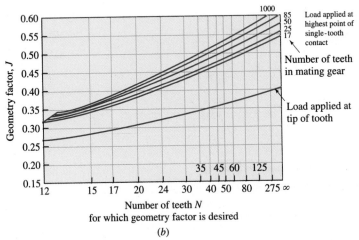

(b)

Figure 11.15   Geometry factors for spur gears (based on tooth fillet radius of 0.35/P): (a) 20° full-depth teeth; (b) 25° full-depth teeth.
| SOURCE: AGMA 218.01.

load is applied, and the contact ratio. The equation for J includes a modified value of the Lewis factor Y and a fatigue stress concentration factor K_f. Figure 11.15 may be used to estimate the geometry factor for only 20° and 25° standard spur gears.

The bending strength S_t for standard gear materials varies with such factors as material quality, heat treatment, mechanical treatment, and material composition. Some selected values for AGMA fatigue-strength for bending are found in Table 11.6. These values are based on a reliability of 99%, corresponding to 10^7 tooth load cycles. Note that, in the table, Bhn and R_C denote the Brinell and Rockwell hardness numbers, respectively.

Table 11.6 Bending strength S_t of spur, helical, and bevel gear teeth

Material	Heat treatment	Minimum hardness or tensile strength	S_t ksi	(MPa)
Steel	Normalized	140 Bhn	19–25	(131–172)
	Q&T	180 Bhn	25–33	(172–223)
	Q&T	300 Bhn	36–47	(248–324)
	Q&T	400 Bhn	42–56	(290–386)
	Case carburized	55 R_C	55–65	(380–448)
		60 R_C	55–70	(379–483)
	Nitrided AISI-4140	48 R_C case	34–45	(234–310)
		300 Bhn core		
Cast iron				
AGMA Grade 30		175 Bhn	8.5	(58.6)
AGMA Grade 40		200 Bhn	13	(89.6)
Nodular iron ASTM Grade:				
60-40-18			15	(103)
80-55-06	Annealed		20	(138)
100-70-18	Normalized		26	(179)
120-90-02	Q&T		30	(207)
Bronze, AGMA 2C	Sand cast	40 ksi (276 MPa)	5.7	(39.3)

SOURCE: AGMA 218.01.
Q&T = Quenched and tempered.

Table 11.7 Life factor K_L for spur and helical steel gears

Number of cycles	160 Bhn	250 Bhn	450 Bhn	Case carburized (55–63 R_C)
10^3	1.6	2.4	3.4	2.7–4.6
10^4	1.4	1.9	2.4	2.0–3.1
10^5	1.2	1.4	1.7	1.5–2.1
10^6	1.1	1.1	1.2	1.1–1.4
10^7	1.0	1.0	1.0	1.0

SOURCE: AGMA 218.01.

The *life factor* K_L rectifies the allowable stress for the required number of stress cycles other than 10^7. Values of this factor are furnished in Table 11.7. The *temperature factor* K_T is applied to adjust the allowable stress for the effect of operating temperature. Usually, for gear lubricant temperatures up to $T < 160°F$, $K_T = 1$ is used. For $T > 160°F$, use $K_T = (460 + T)/620$. The *reliability factor* K_R corrects the allowable stress for the

Table 11.8   Reliability factor K_R

Reliability (%)	90	99	99.9	99.99
Factor K_R	0.85	1.00	1.25	1.50

| SOURCE: The AGMA.

reliabilities other than 99%. Table 11.8 lists some K_R values applied to the fatigue strength for bending of the material.

The use of the AGMA formulas and graphs is illustrated in the solution of the following numerical problem.

EXAMPLE 11.5 | Design of a Speed Reducer for Bending Strength by the AGMA Method

A conveyor drive involving heavy-shock torsional loading is to be operated by an electric motor turning at a speed of n, as shown schematically in Figure 11.16. The speed ratio of the spur gears connecting the motor and conveyor or speed reducer is to be $r_s = 1 : 2$. Determine the maximum horsepower that the gearset can transmit, based on bending strength and applying the AGMA formulas.

Figure 11.16   Example 11.4, schematic arrangement of motor, gear, and conveyor drive.

Given: Both gears are of the same 300 Bhn steel and have a face width of $b = 1.5$ in. Pinion rotates at $n = 1600$ rpm. $P = 10$ in.$^{-1}$ and $N_p = 18$.

Design Decisions: Rational values of the factors are chosen, as indicated in the parentheses in the solution.

Solution: The pinion pitch diameter and number of teeth of the gear are

$$d_p = \frac{N_p}{P} = \frac{18}{10} = 1.8 \text{ in.,} \qquad N_g = N_p \left(\frac{1}{r_s}\right) = 18(2) = 36$$

The pitch-line velocity, using Eq. (11.14), is

$$V = \frac{\pi d_p n_p}{12} = \frac{\pi (1.8)(1600)}{12} = 754 \text{ fpm}$$

The allowable bending stress is estimated from Eq. (11.30):

$$\sigma_{all} = \frac{S_t K_L}{K_T K_R} \tag{a}$$

where

$S_t = 41.5$ ksi (from Table 11.6, for average strength)

$K_L = 1.0$ (from Table 11.7, for indefinite life)

$K_T = 1$ (oil temperature should be $< 160°F$)

$K_R = 1.25$ (by Table 11.8, for 99.9% reliability)

Carrying the foregoing values into Eq. (a) results in

$$\sigma_{all} = \frac{41.5(1)}{1(1.25)} = 33.2 \text{ ksi}$$

The maximum allowable transmitted load is now obtained, from Eq. (11.29) by setting $\sigma_{all} = \sigma$, as

$$F_t = \frac{33,200}{K_o K_v} \frac{b}{P} \frac{J}{K_s K_m} \tag{b}$$

In the foregoing, we have

$P = 10$ and $b = 1.5$ in.

$K_v = 1.55$ (from curve C of Figure 11.14)

$J = 0.235$ (from Figure 11.15a, the load acts at the tip of the tooth, $N_p = 18$)

$K_o = 1.75$ (by Table 11.4)

$K_s = 1.0$ (for standard gears)

$K_m = 1.6$ (from Table 11.5)

Equation (b) yields

$$F_t = \frac{33,200(1.5)(0.235)}{(1.75)(1.55)(10)(1.0)(1.6)} = 270 \text{ lb}$$

The allowable power is then, by Eq. (11.16),

$$\text{hp} = \frac{F_t V}{33,000} = \frac{270(754)}{33,000} = 6.2$$

11.10   THE WEAR STRENGTH OF A GEAR TOOTH: THE BUCKINGHAM FORMULA

The failure of the surfaces of gear teeth is called *wear*. As noted in Section 8.15, wear is a broad term, which encompasses a number of kinds of surface failures. So, it is evident that gear-tooth surface durability is a more complex matter than the capacity to withstand gear-tooth bending failure. Tests have shown that pitting, a surface fatigue failure due to repeated high-contact stress, occurs on those portions of a gear tooth that have relatively little sliding compared with rolling. Clearly, spur gears and helical gears have pitting near the pitch line, where the motion is almost pure rolling. We therefore are concerned with only gear-tooth-surface fatigue failure or pitting and designate it by the term wear [19, 20].

As was the case with the rolling bearings, gear teeth are subjected to Hertz contact stresses, and the lubrication is often elastohyrodynamic (Section 10.15). The gear teeth must be sufficiently strong to carry the wear load. The surface stresses in gear teeth were first investigated in a systematic way by E. Buckingham [6], who considered the teeth as two parallel cylinders in contact. The radii of the teeth as two parallel cylinders are taken as the radii of curvature for the involutes when the teeth make contact at pitch point P (Figure 11.17). Hence,

$$R_1 = \frac{d_p}{2} \sin \phi, \qquad R_2 = \frac{d_g}{2} \sin \phi = \frac{N_g d_p}{2N_p} \sin \phi \qquad \textbf{(a)}$$

where ϕ is the pressure angle, d_p and d_g are the pitch diameters of the pinion and gear, respectively. The last form in the second of these equations is from the relationship $d_p/N_p = d_g/N_g$.

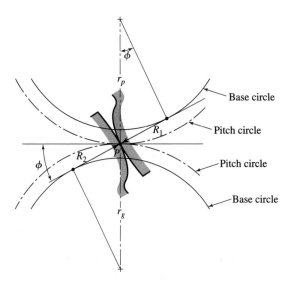

Figure 11.17   Radii of curvature R_1 and R_2 for tooth surfaces at pitch point P.

Generally, good correlation has been observed between spur gear surface fatigue failure and the computed elastic contact stress (see Section 3.16). The maximum contact pressure p_o between the two cylinders may be computed from the equation given in Table 3.2, for $\nu = 0.3$:

$$p_o = 0.592 \left[\frac{F_a E_p E_g}{E_p + E_g} \left(\frac{1}{R_1} + \frac{1}{R_2} \right) \right]^{1/2} \qquad \textbf{(b)}$$

Here F_a is the load per axial length pressing the cylinders together, E_p and E_g are the modulii of elasticity for the materials composing the gears. We observe from this equation that stress increases only as the square root of the load F_a. Likewise, stress decreases with decreased moduli of elasticity E_p and E_g. Moreover, larger gears have greater radii of curvature, hence lower stress.

We now let the maximum contact pressure p_o represent the *surface endurance limit* in compression, S_e, for the pinion and gear material. Substituting Eq. (a) into Eq. (b), we have

$$S_e^2 = 0.35 \frac{F_a E_p E_g}{E_p + E_g} \frac{2}{d_p \sin \phi} \left(1 + \frac{N_p}{N_g} \right) \qquad \textbf{(c)}$$

Usually, both sides of this expression are multiplied by gear width b, and the total load $F_a b$ denoted by F_w. In so doing and solving Eq. (c), we obtain the *allowable wear load:*

$$F_w = d_p b Q K \qquad \textbf{(11.32)}$$

where

$$K = \frac{S_e^2 \sin \phi}{1.4} \left(\frac{1}{E_p} + \frac{1}{E_g} \right) \qquad \textbf{(11.33)}$$

$$Q = \frac{2N_g}{N_p + N_g} \qquad \textbf{(11.34)}$$

Equation (11.32) is known as the *Buckingham formula.* The typical values of S_e and *wear load factor K* for materials of different Brinell surface hardnesses Bhn, recommended by Buckingham, are shown in Table 11.9. Note that the values of K are obtained from Eq. (11.33) for pressure angles of 20° and 25°, full-depth teeth only. The allowable wear load serves as the basis for analyzing gear-tooth surface durability.

For *satisfactory gear performance,* the usual requirement is that

$$F_w \geq F_d \qquad \textbf{(11.35)}$$

Here the dynamic load F_d is defined by Eqs. (11.18). To prevent too much pinion wear, particularly for high-speed gearing, a medium-hard pinion with low hardness gear is often used. This has the advantage of giving some increase in load capacity and slightly lower coefficient of friction on the teeth.

Table 11.9   Surface endurance limit S_e and wear load factor K for use in the Buckingham equation

| Materials in pinion and gear | S_e | | K | | | |
| | ksi | (MPa) | $\phi = 20°$ | | $\phi = 25°$ | |
			psi	(MPa)	psi	(MPa)
Both gears steel, with average						
Bhn of pinion and gear						
150	50	(345)	41	(0.283)	51	(0.352)
200	70	(483)	79	(0.545)	98	(0.676)
250	90	(621)	131	(0.903)	162	(1.117)
300	110	(758)	196	(1.352)	242	(1.669)
350	130	(896)	270	(1.862)	333	(2.297)
400	150	(1034)	366	(2.524)	453	(3.124)
Steel (150 Bhn) & cast iron	50	(354)	60	(0.414)	74	(0.510)
Steel (200 Bhn) & cast iron	70	(483)	119	(0.821)	147	(1.014)
Steel (250 Bhn) & cast iron	90	(621)	196	(1.352)	242	(1.669)
Steel (150 Bhn) & phosphor bronze	59	(407)	62	(0.428)	77	(0.531)
Steel (200 Bhn) & phosphor bronze	65	(448)	100	(0.690)	123	(0.848)
Steel (250 Bhn) & phosphor bronze	85	(586)	184	(1.269)	228	(1.572)
Cast iron & cast iron	90	(621)	264	(1.821)	327	(2.555)
Cast iron & phosphor bronze	83	(572)	234	(1.614)	288	(1.986)

EXAMPLE 11.6   Maximum Load Transmitted by a Gear Based on Wear Strength and Using the Buckingham Formula

The pinion of Example 11.4 is to be mated with a gear. Determine

(a) The allowable load for wear for the gearset using the Buckingham formula.

(b) The maximum load that can be transmitted.

Given: $N_p = 25$, $d_p = 50$ mm, $b = 45$ mm, $N_g = 60$

Design Decision: The mating gear is made of 60-tooth cast iron.

Solution:

(a)   Using Eq. (11.34),

$$Q = \frac{2N_g}{N_p + N_g} = \frac{120}{85} = \frac{24}{17}$$

By Table 11.9,

$$K = 1.014 \text{ MPa} \quad \text{for} \quad 200 \text{ Bhn}$$

Applying Eq. (11.32),

$$F_w = d_p b Q K = (50)(45)\left(\frac{24}{17}\right)(1.014) = 3.221 \text{ kN}$$

(b) The dynamic load, from Example 11.4, is $1.77 F_t$. The wear-limiting value of the transmitted load, using Eq. (11.35), is then

$$3.221 = 1.77 F_t \quad \text{or} \quad F_t = 1.82 \text{ kN}$$

11.11 DESIGN FOR THE WEAR STRENGTH OF A GEAR TOOTH: THE AGMA METHOD

The Buckingham equation serves as the basis for analyzing only the contact stress on the gear tooth. The AGMA formula applies several factors, influencing the actual state of stress at the point of contact, not considered in the previous section. Similarly, in the AGMA method, the surface fatigue strength of the gear tooth is modified by a variety of factors to determine the allowable contact stress. The selective AGMA wear formulas for gears are to follow [1].

Contact stress is defined by the formula

$$\sigma_c = C_p\left(F_t K_o K_v \frac{K_s}{bd} \frac{K_m C_f}{I}\right)^{1/2} \tag{11.36}$$

where

$$C_p = 0.564 \left[\frac{1}{\dfrac{1 - v_p^2}{E_p} + \dfrac{1 - v_g^2}{E_g}}\right]^{1/2} \tag{11.37a}$$

$$I = \frac{\sin\phi\cos\phi}{2m_N}\frac{m_G}{m_G + 1} \tag{11.37b}$$

In the foregoing,

σ_c = calculated contact stress

C_p = elastic coefficient

K_v = velocity or dynamic factor

K_s = size factor

b = face width

d = pitch diameter

$$K_m = \text{load distribution factor}$$

$$C_f = \text{surface condition factor}$$

$$I = \text{geometry factor}$$

$$m_G = \text{gear ratio} = d_g/d_p = N_g/N_p \text{ (for internal gears } m_G \text{ is negative)}$$

$$m_N = \text{load sharing ratio}$$

$$= 1 \text{ (for spur gears)}$$

$$E = \text{modulus of elasticity}$$

$$\nu = \text{Poisson's ratio}$$

$$\phi = \text{pressure angle}$$

Allowable contact stress or the design stress value is

$$\sigma_{c,\text{all}} = \frac{S_c C_L C_H}{K_T K_R} \qquad \qquad \textbf{(11.38)}$$

in which

$$\sigma_{c,\text{all}} = \text{allowable contact stress}$$

$$S_c = \text{surface fatigue strength}$$

$$C_L = \text{life factor}$$

$$C_H = \text{hardness ratio factor}$$

$$K_T = \text{temperature factor}$$

$$K_R = \text{reliability factor}$$

Design specification, the contact stress must not exceed the design stress value:

$$\sigma_c \leq \sigma_{c,\text{all}} \qquad \qquad \textbf{(11.39)}$$

A comparison of the foregoing fundamental equations with those given in Section 11.9 shows that some bending factors and wear factors are equal and so indicated by the same symbols. A brief description of the new wear factors follows.

The elastic coefficient C_p, defined by Eq. (11.37a), accounts for differences in tooth material. In this expression E_p and E_g are the modulii of elasticity for, respectively, pinion and gear, and ν_p and ν_g are their respective Poisson's ratios. The units of C_p are $\sqrt{\text{psi}}$ or $\sqrt{\text{MPa}}$, depending on the system of units used. For convenience, rounded values of C_p are given in Table 11.10, where $\nu = 0.3$ in all cases.

The *surface condition factor* C_f is used to account for such considerations as surface finish, residual stress, and plasticity effects. The C_f is usually taken as unity for a smooth surface finish. When rough finishes are present or the possibility of high residual stress exists, a value of 1.25 is reasonable. If both rough finish and residual stress exist, 1.5 is the suggested value.

The surface fatigue strength S_c represents a function of such factors as the material of the pinion and gear, number of cycles of load application, size of the gears, type of heat treatment, mechanical treatment, and the presence of residual stresses. Table 11.11 may be

Table 11.10 AGMA elastic coefficients C_p for spur gears, in $\sqrt{psi}$ and ($\sqrt{MPa}$)

Pinion material	E ksi (GPa)	Gear material			
		Steel	Cast iron	Aluminum bronze	Tin bronze
Steel	30,000 (207)	2300 (191)	2000 (166)	1950 (162)	1900 (158)
Cast iron	19,000 (131)	2000 (166)	1800 (149)	1800 (149)	1750 (145)
Aluminum bronze	17,500 (121)	1950 (162)	1800 (149)	1750 (145)	1700 (141)
Tin bronze	16,000 (110)	1900 (158)	1750 (145)	1700 (141)	1650 (137)

Table 11.11 Surface fatigue strength or allowable contact stress S_c

Material	Minimum hardness or tensile strength		S_c	
			ksi	(MPa)
Steel	Through hardened			
	180 Bhn		85–95	(586–655)
	240 Bhn		105–115	(724–793)
	300 Bhn		120–135	(827–931)
	360 Bhn		145–160	(1000–1103)
	400 Bhn		155–170	(1069–1172)
	Case carburized			
	55 R_C		180–200	(1241–1379)
	60 R_C		200–225	(1379–1551)
	Flame or induction hardened			
	50 R_C		170–190	(1172–1310)
Cast iron				
AGMA grade 20			50–60	(345–414)
AGMA grade 30	175 Bhn		65–75	(448–517)
AGMA grade 40	200 Bhn		75–85	(517–586)
Nodular (ductile) iron			90–100% of the S_c	
Annealed	165 Bhn		value of steel with the	
Normalized	210 Bhn		same hardness	
OQ&T	255 Bhn			
Tin bronze				
AGMA 2C(10–12% tin)	40 ksi	(276 MPa)	30	(207)
Aluminum bronze				
ASTM B 148-52 (alloy 9C-H.T.)	90 ksi	(621 MPa)	65	(448)

SOURCE: AGMA 218.01

OQ&T = Oil quenched and tempered; H.T. = heat treated.

Figure 11.18   Life factor for steel gears.
| SOURCE: AGMA 218.01.

used to estimate the values for S_c. The life factor C_L accounts for the expected life of the gear. Figure 11.18 can be used to obtain approximate values for C_L.

The hardness-ratio factor C_H is used *only* for the gear. Its intent is to adjust the surface strengths for the effect of the hardness. The values of C_H are calculated from the expression

$$C_H = 1.0 + A\left(\frac{N_g}{N_p} - 1.0\right) \qquad \left(\text{for } \frac{H_{Bp}}{H_{Bg}} < 1.70\right) \tag{11.40}$$

where

$$A = 8.98(10^{-3})\left(\frac{H_{Bp}}{H_{Bg}}\right) - 8.29(10^{-3})$$

The quantities H_{Bp} and H_{Bg} represent the Brinell hardness of the pinion and gear, respectively.

EXAMPLE 11.7 | Design of a Speed Reducer for Wear by the AGMA Method

Determine the maximum horsepower that the speed reducer gearset in Example 11.5 can transmit, based on wear strength and applying the AGMA method.

Given: Both gears are made of the same 300 Bhn steel of $E = 30 \times 10^6$ psi, $v = 0.3$, and have a face width of $b = 1.5$ in. $P = 10$ in.$^{-1}$ and $N_p = 18$.

Design Decisions: Rational values of the factors are chosen, as indicated in the parentheses in the solution.

Solution: Allowable contact stress is estimated from Eq. (11.38) as

$$\sigma_{c,\text{all}} = \frac{S_c C_L C_H}{K_T K_R} \tag{a}$$

In the preceding,

$S_c = 127.5$ ksi (from Table 11.11, for average strength)

$C_L = 1.0$ (from Figure 11.18, for indefinite life)

$$C_H = 1.0 + A\left(\frac{N_g}{N_p} - 1.0\right) = 1.0 \quad \text{(by Eq. (11.40))}$$

$K_T = 1.0, \quad K_R = 1.25$ (both from Example 11.5)

Hence,

$$\sigma_{c,\text{all}} = \frac{127{,}500(1.0)(1.0)}{(1.0)(1.25)} = 102 \text{ ksi}$$

The maximum allowable transmitted load is now determined, from Eq. (a) setting $\sigma_{c,\text{all}} = \sigma_c$, as

$$F_t = \left(\frac{102{,}000}{C_p}\right)^2 \frac{1.0}{K_o K_v} \frac{bd}{K_s} \frac{I}{K_m C_f}$$

where

$C_p = 2300 \sqrt{\text{psi}}$ (by Table 11.10)

$b = 1.5$ in., $\quad d_p = 1.8$ in.

$K_v = 1.55, \quad K_o = 1.75, \quad K_s = 1, \quad K_m = 1.6$ (all from Example 11.5)

$C_f = 1.0$ (for smooth surface finish)

$$I = \frac{\sin\phi \cos\phi}{2} \frac{m_G}{m_G + 1} = 0.107 \quad \text{(using Eq. 11.37b)}$$

Therefore,

$$F_t = \left(\frac{102{,}000}{2300}\right)^2 \frac{1.0}{(1.75)(1.55)} \frac{1.5(1.8)}{1.0} \frac{0.107}{1.6(1.0)} = 131 \text{ lb}$$

This value applies to both mating gear-tooth surfaces. The corresponding power, using Eq. 11.16 with $V = 754$ fpm (from Example 11.5), is

$$\text{hp} = \frac{F_t V}{33{,}000} = \frac{131(754)}{33{,}000} = 2.99$$

11.12 MATERIALS FOR GEARS

Gears are made from a wide variety of materials, both metallic and nonmetallic, covered in Chapter 2. The material used depends on which of several criteria is most important to the problem at hand. When high strength is the prime consideration, steel should be chosen rather than cast iron or other materials. Test data for fatigue strengths of most materials can be found in contemporary technical literature and current publications of the AGMA (see Tables 11.6 and 11.11). For situations involving noise abatement, nonmetallic materials perform better than metallic ones. The characteristics of some common gear materials follow.

Cast irons have low cost, ease of casting, good machinability, high wear resistance, and good noise abatement, which make them one of the most commonly used gear materials. Cast-iron gears typically have greater surface-fatigue strength than bending fatigue strength. Nodular cast iron gears, containing a material such as magnesium or cerium, have higher bending strength and good surface durability. The combination of a steel pinion and cast iron gear represents a well-balanced design.

Steels usually require heat treatment to produce a high surface endurance capacity. Heat-treated steel gears must be designed to resist distortion; hence, alloy steels and oil quenching are often preferred. Through-hardened gears usually have 0.35 to 0.6% carbon. Case-hardened gears are generally processed by flame hardening, induction hardening, carburizing, or nitriding. When gear accuracy is required, the gear must be ground.

Nonferrous metals such as the copper alloys (known as *bronzes*) are most often used for gears. They are useful in situations where corrosion resistance is important. Owing to their ability to reduce friction and wear, bronzes are generally employed for making worm wheel in a worm gearset. Aluminum, zinc, and titanium are also used to obtain alloys that are serviceable for gear materials.

Plastics such as acetal, polpropylene, nylon, and other nonmetallic materials have often been used to make gears. Teflon is sometimes added to these materials to lower the coefficient of friction. Plastic gears are generally quiet, durable, reasonably priced, and can operate under light loads without lubrication. However, they are limited in torque capacity by their low strength. Furthermore, plastics have low heat conductivity, resulting in heat distortion and weakening of the gear teeth.

Reinforced thermoplastics, formulated with fillers such as glass fiber, are desirable gear materials owing to their versatility. For best wear resistance, nonmetallic gears are often mated with cast iron or steel pinions having a hardness of at least 300 Bhn. The *composite gears* of thermosetting phenolic have been used in applications such as the camshaft-drive gear driven by a steel pinion in some gasoline engines. Design procedures of gears made of plastic are identical to that of metal gears, but not yet as reliable. Prototype testing is therefore more significant than for gears made of metals.

11.13   GEAR MANUFACTURING

Various methods are employed to manufacture gears. These can be divided into two classes: forming and finishing. Gear teeth are formed numerous ways by milling and generating processes. For high speed and heavy loads, a finishing operation may be required to bring the tooth outline to a sufficient degree of accuracy and surface finish. Finishing operations typically remove little or no material. Gear errors may be diminished somewhat by finishing the tooth profiles. A general discussion of some manufacturing processes is found in Chapter 2. This section can provide only a brief description of gear forming and finishing methods [10, 15].

FORMING GEAR TEETH

Milling refers to removal of the material between the teeth from a blank on a milling machine that uses a formed circular cutter. The cutter must be made to the shape of the gear tooth for

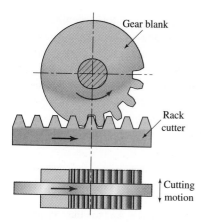

Figure 11.19 The generating action of a cutting tool, a rack cutter reciprocating in the direction of the axis of a gear blank.

the tooth geometry and number of teeth of each particular gear. Gears having large-size teeth are often made by formed cutters. *Shaping* is the generation of teeth with a rack cutter, or with a pinion cutter called a *shaper* (i.e., a small circular gear). The cutter works in a rapid reciprocating motion parallel to the axis of the gear blank while slowly translating with and into the blank, as depicted in Figure 11.19. When the blank and cutter have rolled a distance equal to the circular pitch, the cutter is returned to the starting point and the process is continued until all the teeth have been cut. The process requires a special cutting machine that translates the tool and rotates the gear blank with the same velocity as though a rack were driving a gear. Internal gears can be cut with this method as well. *Hobbing* refers to a process that accounts for the major portion of gears made in high-quantity production. A hob is a cutting tool shaped like a worm or screw. Both the hob and blank must be rotated at the proper angular velocity ratio. The hob is then fed slowly across the face of the blank until all the teeth have been cut.

Other gear forming methods include die casting, drawing, extruding, sintering, stamping, and injection molding. These processes produce large volumes of gears that are low in cost but poor in quality. Gears are *die cast* by forcing molten metal under pressure into a form die. In a cold *drawing* process, the metal is drawn through several dies and emerges as a long piece of gear from which gears of smaller widths can be sliced. On the other hand, in an *extruding* process, the metal is pushed rather than pulled through the dies. Nonferrous materials such as copper and aluminum alloys are usually extruded. A *sintering* method consists of applying pressure and heat to a powdered metal (PM) to form the gear. A *stamping* process uses a press and a die to cut out the gear shape. An *injection molding* method is applied to produce nonmetallic gears in a variety of thermoplastics such as nylon and acetal.

FINISHING PROCESSES

Grinding is accomplished by the use of some form of abrasive grinding wheel. It can be used to give the final form to teeth after heat treatment. *Shaving* refers to a machine operation that removes small amounts of material. It is done prior to hardening and is widely used for gears made in large quantities. *Burnishing* runs the gear to be smoothed against a specially hardened gear, which flattens and spreads minute surface irregularities only. *Honing* employs a tool, known as a *hone,* to drive the gear to be finished. It makes minor tooth-form corrections and improves the smoothness of the surface of the hardened gear surface. *Lapping* runs a gear with another that has some abrasive material embedded in it. Sometimes two mating gears are similarly run.

Case Study 11-1 | DESIGN OF THE SPUR GEAR TRAIN FOR A WINCH CRANE

The spur gearbox of the winch crane (Figure 1.4) is illustrated in Figure 11.20. Analyze the design of each gearset using the AGMA method.

Given: The geometry and properties of each element are known. A 0.5 hp, 1725 rpm electric motor at 95% efficiency delivers 0.475 hp to the 85-mm diameter drum (see Case Study 1-1). The maximum capacity of the crane is $P = 3$ kN. All gears have $\phi = 20°$ pressure angle. Shafts 1 or 2, 3, and 4 are supported by 12, 19, and 25-mm bore flanged bearings, respectively.

Figure 11.20   Gearbox of the winch crane shown in Figure 1.4.

Case Study (CONTINUED)

Data:

	Module m (mm)	Number of teeth N	Pitch diameter d (mm)	Face width b (mm)
Gear 1 (pinion)	1.3	15	20	14
Gear 2	1.3	60	80	14
Gear 3 (pinion)	1.6	18	28.8	20
Gear 4	1.6	72	115.2	20
Gear 5 (pinion)	2.5	15	37.5	32
Gear 6	2.5	60	150	32

Assumptions:

1. The pinions are made of carburized 55 R_c steel. Gears are Q&T, 180 Bhn steel. Hence, by Tables 11.6 and 11.11, we have

   Pinions: $S_t = 414$ MPa, $S_c = 1310$ MPa
   Gears: $S_t = 198$ MPa, $S_c = 620$ MPa

2. All gears and pinions are high precision shaved and ground; manufacturing quality corresponds to curve A in Figure 11.14.

3. Loads are applied at highest point of single-tooth contact.

Design Decisions: The following reasonable values of the bending and wear strength factors for pinions and gears are chosen (from Tables 11.4, 11.5, 11.7, 11.8, 11.10):

$K_o = 1.5$, $K_s = 1.0$ (from Section 11.9)

$K_m = 1.6$, $K_L = 1.1$

$K_T = 1.0$ (from Section 11.9), $K_R = 1.25$,

$C_H = 1.0$ (by Equation (11.40), $C_p = 191\sqrt{\text{MPa}}$

$C_f = 1.25$ (from Section 11.11),

$C_L = 1.1$ (from Figure 11.18)

Solution: See Figures 1.4 and 11.20, Sections 11.9 and 11.11.

The operating line velocity of the hoist at the maximum load is, by Eq. (1.16),

$$V = \frac{745.7 \text{ hp}}{P} = \frac{745.7(0.475)}{3{,}000} = 0.12 \text{ m/s}$$

The operating speed of the drum shaft is

$$n = \frac{V}{\pi d} = \frac{(0.12)(60)}{\pi(0.085)} \approx 27 \text{ rpm}$$

This agrees with the values suggested by several catalogs for light lifting. The gear train in Figure 11.20 fits all the parameters in the lifting system.

Gearset I: (Pair of gears 1 and 2, Figure 11.20)
The input torque and the transmitted load on gear 1 are

$$T_1 = \frac{7121 \text{ hp}}{n_1} = \frac{7121(0.5)}{1725} = 2.06 \text{ N} \cdot \text{m}$$

$$F_{t1} = \frac{T_1(2)}{d_1} = \frac{2.06(2)}{0.02} = 206 \text{ N}$$

The radial load is then

$$F_{r1} = F_{t1} \tan\phi = 206 \tan 20° = 75 \text{ N}$$

The pitch-line velocity is determined as

$$V_1 = \pi d_1 n_1 = \pi(0.02)\left(\frac{1725}{60}\right) = 1.81 \text{ m/s} = 356 \text{ fpm}$$

Then, from curve A of Figure 11.14, the dynamic factor is

$$K_v = \sqrt{\frac{78 + \sqrt{356}}{78}} = 1.11$$

Equation (11.37b) with $m_g = 4$ gives

$$I = \frac{\sin 20° \cos 20°}{2} \frac{4}{4+1} = 0.129$$

By Figure 11.15a, we have

$J = 0.25$ (for pinion, $N_p = 15$ and $N_g = 60$)

$J = 0.42$ (for gear, $N_g = 60$ and $N_p = 15$)

(continued)

Case Study   (CONTINUED)

Gear 1 (pinion).   Substituting the numerical values into Eqs. (11.29) and (11.30):

$$\sigma = F_{t1}K_oK_v\frac{1.0}{bm}\frac{K_sK_m}{J}$$

$$= 206(1.5)(1.11)\frac{1.0}{14(1.3)10^{-6}}\frac{1.0(1.6)}{0.25}$$

$$= 120.6 \text{ MPa}$$

$$\sigma_{\text{all}} = \frac{S_tK_L}{K_TK_R} = \frac{414(1.1)}{(1.0)(1.25)} = 364.3 \text{ MPa}$$

Similarly, Eqs. (11.36) and (11.38) lead to

$$\sigma_c = C_p\left(F_{t1}K_oK_v\frac{K_s}{bd}\frac{K_mC_f}{I}\right)^{1/2}$$

$$= 191\left[206(1.5)(1.11)\frac{1.0}{14(20)10^{-6}}\frac{1.6(1.25)}{0.129}\right]^{1/2}$$

$$= 832.4 \text{ MPa}$$

$$\sigma_{c,\text{all}} = \frac{S_cC_LC_H}{K_TK_R} = \frac{1310(1.1)(1.0)}{(1.0)(1.25)} = 1153 \text{ MPa}$$

Gear 2.   We have $F_{t2} = F_{t1} = 206$ N. Substitution of the data into Eqs. (11.29′) and (11.30) gives

$$\sigma = F_{t2}K_oK_v\frac{1.0}{bm}\frac{K_sK_m}{J}$$

$$= 206(1.5)(1.11)\frac{1.0}{14(1.3)10^{-6}}\frac{1.0(1.6)}{0.42}$$

$$= 71.79 \text{ MPa}$$

$$\sigma_{\text{all}} = \frac{S_tK_L}{K_TK_R} = \frac{198(1.1)}{(1.0)(1.25)} = 174.2 \text{ MPa}$$

In a like manner, through the use of Eqs. (11.36) and (11.38),

$$\sigma_c = C_p\left(F_{t2}K_oK_v\frac{K_s}{bd}\frac{K_mC_f}{I}\right)^{1/2}$$

$$= 191\left[206(1.5)(1.11)\frac{1.0}{14(80)10^{-6}}\frac{1.6(1.25)}{0.129}\right]^{1/2}$$

$$= 416.2 \text{ MPa}$$

$$\sigma_{c,\text{all}} = \frac{S_cC_LC_H}{K_TK_R}$$

$$= \frac{620(1.1)(1.0)}{(1.0)(1.25)} = 545.6 \text{ Pa}$$

Comments: Inasmuch as $\sigma < \sigma_{\text{all}}$ and $\sigma_c < \sigma_{c,\text{all}}$, the pair of gears 1 and 2 is *safe* with regard to the AGMA bending and wear strengths, respectively.

Gearsets II and III: (Pairs of gears 3–4 and 5–6, Figure 11.20)

Shaft 2 rotates at the speed

$$n_2 = n_1\frac{N_1}{N_2} = 1725\left(\frac{15}{60}\right) \approx 431 \text{ rpm}$$

Hence, for gear 3 (pinion), we have

$$T_3 = \frac{7121 \text{ hp}}{n_2} = \frac{7121(0.5)}{431} = 8.261 \text{ N} \cdot \text{m}$$

$$F_{t3} = \frac{T_3(2)}{d_3} = \frac{8.261(2)}{0.0288} = 573.7 \text{ N}$$

$$F_{r3} = 573.7 \tan 20° = 208.8 \text{ N}$$

$$V_3 = \pi d_3n_3 = \pi(0.0288)\left(\frac{431}{60}\right) = 0.65 \text{ m/s}$$

Shaft 3 runs at

$$n_3 = n_2\frac{N_3}{N_4} = 431\left(\frac{18}{72}\right) \approx 108$$

Case Study (CONCLUDED)

It follows for gears 5 (pinion) and 6 that

$$T_5 = \frac{7121(0.5)}{108} = 32.97 \text{ N} \cdot \text{m}$$

$$F_{t5} = \frac{32.97(2)}{0.0375} = 1758 \text{ N} = F_{t6}$$

$$F_{r5} = 1758 \tan 20° = 639.9 \text{ N} = F_{r6}$$

$$V_5 = \pi d_5 n_3 = \pi(0.0375)\left(\frac{108}{60}\right) = 0.21 \text{ m/s} = V_6$$

The output shaft rotates at

$$n_4 = n_3 \frac{N_5}{N_6} = 108\left(\frac{15}{60}\right) \approx 27 \text{ rpm}$$

The speed ratio between the output and input shafts (or gears 6 and 1) of the spur gear train can now be

obtained as

$$r_s = \frac{n_4}{n_1} = \left(-\frac{N_1}{N_2}\right)\left(-\frac{N_3}{N_4}\right)\left(-\frac{N_5}{N_6}\right)$$

$$= \left(-\frac{15}{60}\right)\left(-\frac{18}{72}\right)\left(-\frac{15}{60}\right) = -\frac{1}{64}$$

Here, the minus sign means that the pinion and gear rotate in opposite directions.

Comments: Having the tangential forces and pitch-line velocities available, the design analysis for gearsets 2 and 3 can readily be made by following a procedure identical to that described for gearset 1.

REFERENCES

1. *Standards of the American Gear Manufacturers Association.* Arlington, VA: AGMA, 1990.
2. Townsend, D. P., ed. *Dudley's Gear Handbook,* 2nd ed. New York: McGraw-Hill, 1992.
3. Drago, R. J. *Fundamentals of Gear Design.* Boston: Butterworth Publishers, 1988.
4. Avallone, E. A., and T. Baumeister III, eds. *Marks' Standard Handbook for Mechanical Engineers,* 10th ed. New York: McGraw-Hill, 1996.
5. Shigley, E. J., and C. R. Mischke. *Mechanical Engineering Design,* 6th ed. New York: McGraw-Hill, 2001.
6. Buckingham, E. *Analytical Mechanics of Gears.* New York: McGraw-Hill, 1949.
7. Juvinall, R. C., and K. M. Marshek. *Fundamentals of Machine Component Design,* 3rd ed. New York: Wiley, 2000.
8. Lipp, R. "Avoiding Tooth Interference in Gears." *Machine Design* 54, no. 1 (1982), p. 122.
9. Mabie, H. H., and C. F. Reinholtz. *Mechanisms and Dynamics of Machinery,* 4th ed. New York: Wiley, 1987.
10. Norton, R. L. *Machine Design: An Integrated Approach,* 2nd ed. Upper Saddle River, NJ: Prentice Hall, 2000.
11. Deutchman, A. D., W. J. Michels, and C. E. Wilson. *Machine Design: Theory and Practice.* New York: Macmillan, 1975.
12. Lewis, W. "Investigation of the Strength of Gear Teeth." Philadelphia: Proceedings of the Engineers Club, 1893, pp. 16–23; reprinted in *Gear Technology* 9, no. 6 (Nov.–Dec. 1992), p. 19.
13. Dolan, T. J., and E. L. Broghamer. "A Photoelastic Study of Stresses in Gear Tooth Profiles." Bulletin No. 335, Engineering Experimental Station, University of Illinois, Urbana, 1942.

14. Peterson, R. E. *Stress Concentration Factors.* New York: Wiley, 1974.

15. Shigley, E. J., and C. E. Mischke, eds. *Standard Handbook of Machine Design.* New York: McGraw-Hill, 1986.

16. Hamrock, B. J., B. Jacobson, and S. R. Schmid. *Fundamentals of Machine Elements.* New York: McGraw-Hill, 1999.

17. Burr, A. H., and J. B. Cheatham. *Mechanical Analysis and Design,* 2nd ed. Upper Saddle River, NJ: Prentice Hall, 1995.

18. Mott, R. I. *Machine Elements in Mechanical Design,* 2nd ed. New York: Merrill, 1992.

19. Dudley, D. W. "Gear Wear." In M. B. Peterson and W. O. Winer, eds., *Wear Control Handbook.* New York: ASME, 1980, p. 764.

20. Dudley, D. W. *Handbook of Practical Gear Design.* New York: McGraw-Hill, 1984.

PROBLEMS

Sections 11.1 through 11.7

11.W1 Use the website at www.grainger.com to conduct a search for spur gears. Select and list the manufacturer and description of spur gears with

    (*a*)   A 16 pitch, 14.5° pressure angle, and 48 teeth.

    (*b*)   A 24 pitch, 14.5° pressure angle, and 12 teeth.

11.W2 Search the website at www.powertransmission.com. List 10 websites for manufacturers of gears and gear drives.

11.1 A 20° pressure angle gear has 32 teeth and a diametral pitch of 4 teeth/in. Determine the whole depth, working depth, base circle radius, and outside radius.

11.2 Two modules of 3-mm gears are to be mounted on a center distance of 360 mm. The speed ratio is 1/3. Determine the number of teeth in each gear.

11.3 Determine the approximate center distance for an external 25° pressure angle gear having a circular pitch of 0.5234 in. that drives an internal gear having 84 teeth, if the speed ratio is to be 1/4.

11.4 A gearset has a module of 4 mm and a speed ratio of 1/4. The pinion has 22 teeth. Determine the number of teeth of the driven gear, gear and pinion diameters, and the center distance.

11.5 The gears shown in Figure P11.5 have a diametral pitch of 3 teeth/in. and a 25° pressure angle. Determine, and show on a free-body diagram,

    (*a*)   The tangential and radial forces acting on each gear.

    (*b*)   The reactions on shaft *C*.

    Given: Driving gear 1 transmits 30 hp at 4000 rpm through the idler to gear 3 on shaft *C*.

11.6 Redo Problem 11.5 for gears having a diametral pitch of 6 teeth/in., a 20° pressure angle, and a clockwise rotation of the driving gear 1.

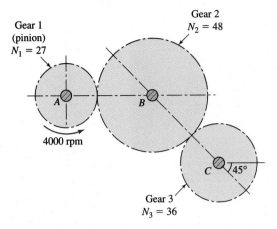

Figure P11.5

11.7 The gears shown in Figure P11.7 have a diametral pitch of 4 teeth/in. and a 20° pressure angle. Determine, and indicate on a free-body diagram,

(a) The tangential and radial forces on each gear.

(b) The reaction on shaft B.

Design Decision: Driving gear 1 transmits 50 hp at 1200 rpm through the idler pair mounted on shaft B to gear 4.

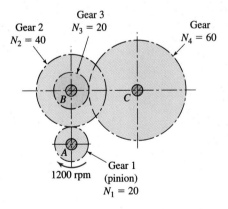

Figure P11.7

11.8 The gears shown in Figure P11.8 have a module of 6 mm and a 20° pressure angle. Determine, and show on a free-body diagram,

(a) The tangential and radial loads on each gear.

(b) The reactions on shaft C.

Given: Driving gear 1 transmits 80 kW at 1600 rpm through the idler to gear 3 mounted on shaft C.

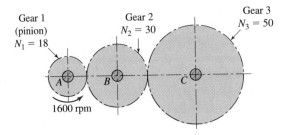

Gear 1
(pinion)
$N_1 = 18$

Gear 2
$N_2 = 30$

Gear 3
$N_3 = 50$

1600 rpm

Figure P11.8

11.9 Resolve Problem 11.8, for gears having a module of 8 mm and a 25° pressure angle.

11.10 The gears shown in Figure P11.10 have a diametral pitch of 5 teeth/in. and 25° pressure angle. Determine, and indicate on free-body diagrams,

(a)   The tangential and radial forces on gears 2 and 3.

(b)   The reactions on shaft C.

Design Decision: Driving gear 1 transmits 10 hp at 1500 rpm through idler pair mounted on shaft B to gear 4.

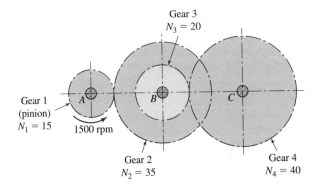

Gear 3
$N_3 = 20$

Gear 1
(pinion)
$N_1 = 15$

1500 rpm

Gear 2
$N_2 = 35$

Gear 4
$N_4 = 40$

Figure P11.10

11.11 The gears shown in Figure P11.11 have a diametral pitch of 6 teeth/in. and a 25° pressure angle. Determine, and show on a free-body diagram,

(a)   The tangential and radial forces on gears 2 and 3.

(b)   The reactions on shaft B.

Given: Driving gear 1 transmits 20 hp at 1800 rpm through an idler pair mounted on shaft B to gear 4.

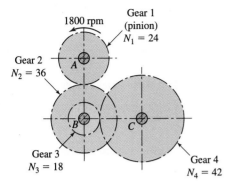

Figure P11.11

Sections 11.8 through 11.13

11.12 The gears shown in Figure P11.10 have a diametral pitch of 5 teeth/in., 25° pressure angle, and a tooth width of $\frac{1}{2}$ in. Determine

(a) The allowable bending load, using the Lewis equation and $K_f = 1.6$, for the tooth of gear 2.

(b) The allowable load for wear, applying the Buckingham equation, for gears 1 and 2.

(c) The maximum tangential load that gear 2 can transmit.

Design Decisions: All gears are made of cast steel (0.20% C WQ&T, Table 11.3), gears 2 and 3 are mounted on shaft B.

11.13 The gears shown in Figure P11.10 have a module of 5 mm, tooth width of 15 mm, and a 20° pressure angle. Determine

(a) The allowable bending load, applying the Lewis equation and $K_f = 1.5$, for the tooth of gear 3.

(b) The allowable load for wear, using the Buckingham equation, for gears 3 and 4.

(c) The maximum tangential load that gear 3 can transmit.

Design Decisions: All gears are made of cast steel (0.20% C WQ&T, Table 11.3); gears 2 and 3 are mounted on shaft B.

11.14 The gears shown in Figure P11.11 have a diametral pitch of 6 teeth/in., a 25° pressure angle, and a tooth width of $\frac{1}{2}$ in. Determine

(a) The allowable bending load, applying the Lewis equation and $K_f = 1.4$, for the tooth of gear 2.

(b) The allowable load for wear, using the Buckingham equation, for gears 3 and 4.

(c) The maximum tangential load that gear 2 can transmit, based on bending strength.

Design Decisions: All gears are made of steel hardened to 200 Bhn; gears 2 and 3 are mounted on shaft B.

11.15 The gears shown in Figure P11.11 have a module of 10 mm, a 20° pressure angle, and a tooth width of 15 mm. Determine

(a) The allowable bending load, using the Lewis equation and $K_f = 1.5$, for the tooth of gear 4.

(b) The allowable load for wear, applying the Buckingham equation, for gears 1 and 2.

Design Decisions: All gears are made of hardened steel (200 Bhn), and gears 2 and 3 are mounted on shaft *B*.

11.16 The 20° pressure angle, and tooth width of $\frac{5}{8}$ in. gears in a speed reducer are specified as follows:

Pinion: 1600 rpm,   24 teeth,   180-Bhn steel,   hp = 1.2, $P = 12$ in.$^{-1}$,

$C_L = C_f = 1$, $K_m = 1.6$, $K_o = K_s = K_T = K_L = 1$, $K_R = 1.25$

Gear: 60 teeth, AGMA 30 cast iron.

Are the gears safe with regard to the AGMA bending strength?

Assumption: The manufacturing quality of the pinion and gear corresponds to curve *D* of Figure 11.14.

11.17 Determine whether the gears in Problem 11.16 are safe with regard to the AGMA wear strength.

11.18 A pair of cast iron (AGMA grade 40) gears have a diametral pitch of 5 teeth/in., a 20° pressure angle, and a width of 2 in. A 20-tooth pinion rotating at 90 rpm and drives a 40-tooth gear. Determine the maximum horsepower that can be transmitted, based on wear strength and using the Buckingham equation.

11.19 Resolve Problem 11.18 using the AGMA method, if the life is to be no more than 10^6 cycles corresponding to a reliability of 99%.

Given: $E = 19 \times 10^6$ psi and $\nu = 0.3$.
Assumption: The gears are manufactured with precision.

11.20 A pair of gears have a 20° pressure angle, and a diametral pitch of 6 teeth/in. Determine the maximum horsepower that can be transmitted, based on bending strength and applying the Lewis equation and $K_f = 1.4$.

Design Decisions: The gear is made of phosphor bronze, has 60 teeth, rotates at 240 rpm. The pinion is made of SAE 1040 steel and rotates at 600 rpm.

Given: Both gears have a width of 3.5 in.

11.21 Redo Problem 11.20 for two meshing gears that have widths of 80 mm and a module of 4 mm.

11.22 Resolve Problem 11.20, based on a reliability of 90% and moderate shock on the driven machine. Apply the AGMA method for $K_L = K_T = K_s = 1$.

Assumption: The manufacturing quality of the gearset corresponds to curve *C* of Figure 11.14.

11.23 Two meshing gears have face widths of 2 in. and diametral pitches of 5 teeth per inches. The pinion is rotating at 600 rpm, has 28 teeth, and the velocity ratio is to be 4/5. Determine the horsepower that can be transmitted, based on wear strength and using the Buckingham equation.

Design Decisions: The gears are made both of steel hardened to a 350 Bhn and have a 20° pressure angle.

11.24 Resolve Problem 11.23 for a gearset that has a width of 60 mm and a module of 6 mm.

11.25 Redo Problem 11.23 based on a reliability of 99.9% and applying the AGMA method.

Given: $C_L = C_f = 1$, $K_o = K_s = K_T = 1$, $E = 30 \times 10^6$ psi, $\nu = 0.3$
Assumption: The gearset is manufactured with high precision, shaved and ground.

12

HELICAL, BEVEL, AND WORM GEARS

Outline

12.1   INTRODUCTION

In Chapter 11, the kinematic relationships and factors that must be considered in designing straight-toothed or spur gears were presented. Several specialized forms of gears exist. We now deal with the three principal types of nonspur gearing: helical, worm, and bevel gears.* The geometry of these different types of gearing is considerably more complicated than for spur gears (see Figure 11.1). However, much of the discussion of the previous chapter applies equally well to the present chapter. Therefore, the treatment here of nonspur gears is relatively brief. As noted in Section11.9, the reader should consult the American Gear Manufacturing Association (AGMA) standards for more information when faced with a real design problem involving gearing.

Helical, bevel, and worm gears can meet specific geometric or strength requirements that cannot be obtained from spur gears. Helical gears are very similar to spur gears. They have teeth that lie in helical paths on the cylinders instead of teeth parallel to the shaft axis. Bevel gears, with straight or spiral teeth cut on cones, can be employed to transmit motion between intersecting shafts. A worm gearset, consisting of screw meshing with a gear, can be used to obtain a large reduction in speed. The analysis of power screw force components, to be discussed in Section 15.3, also applies to worm gears. By noting that the thread angle of a screw corresponds to the pressure angle of the worm, expressions of the basic definitions and efficiency of power screws are directly used for a wormset in Sections 12.9 and 12.11.

12.2   HELICAL GEARS

Like spur gears, helical gears are cut from a cylindrical gear blank and have involute teeth. The difference is that their teeth are at some helix angle to the shaft axis. These gears are used for transmitting power between parallel or nonparallel shafts. The former case is shown in Figure 12.1a. The helix can slope in either the upward or downward direction. The terms *right-hand* (R.H.) and *left-hand* (L.H.) *helical gears* are used to distinguish between the two types, as indicated in the figure. Note that the rule for determining whether a helical gear is right- or left-handed is the same as that used for determining right- and left-hand screws.

Herringbone gear refers to a helical gear having half its face cut with teeth of one hand the other half with the teeth of opposite hand (Figure 12.2). In nonparallel, nonintersecting shaft applications, gears with helical teeth are known as *crossed helical gears,* as shown in Figure 12.1b. Such gears have point contact, rather than the line contact of regular helical gears. This severely reduces their load-carrying capacity. Nevertheless, crossed helical gears are frequently used for the transmission of relatively small loads, such as distributor and speedometer drives of automobiles. We consider only helical gears on parallel shafts.

*For variations of the foregoing basic gear types and further details: See, for example, [1–15] and the websites at www.machinedesign.com and www.powertransmission.com.

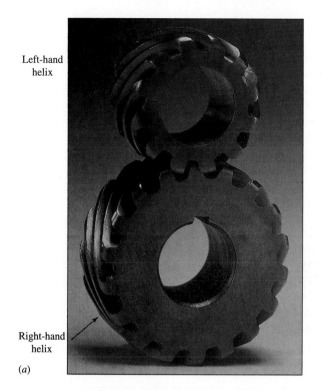

Left-hand
helix

Right-hand
helix

(a)

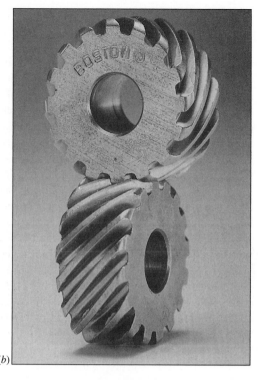

(b)

Figure 12.1   Helical gears (courtesy of Boston Gear): (a) opposite-hand pair meshed on parallel axes (most common type); (b) same-hand pair meshed on crossed axes.

c

Figure 12.2   A typical herringbone gearset.

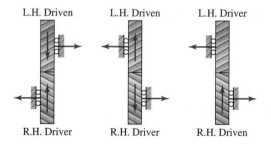

L.H. Driven   L.H. Driven   L.H. Driver

R.H. Driver   R.H. Driver   R.H. Driven

Figure 12.3   Direction, rotation, and thrust load for three helical gearsets with parallel shafts.

Figure 12.3 illustrates the thrust, rotation, and hand relations for some helical gearsets with parallel shafts. Note that the direction in which the thrust load acts is determined by applying the right- or left-hand rule to the driver. That is, for the right-hand driver, if the fingers of the right hand are pointed in the direction of rotation of the gears, the thumb points in the direction of the thrust. The driven gear then has a thrust load acting in the direction opposite to that of the driver, as shown in the figure.

The *advantages* of helical gears over other basic gear types include more teeth in contact simultaneously and the load transferred gradually and uniformly as successive teeth come into engagement. Gears with helical teeth operate more smoothly and carry larger loads at higher speeds than spur gears. The line of contact extends diagonally across the face of mating gears. When employed for the same applications as spur gears, these gears have quiet operation. The *disadvantages* of helical gears are greater cost than spur gears and the presence of an axial force component that requires thrust bearings on the shaft.

12.3   HELICAL GEAR GEOMETRY

Helical gear tooth proportions follow the same standards as those for spur gears. The teeth form the helix angle ψ with the gear axis, measured on an imaginary cylinder of pitch diameter d. The usual range of values of the helix angle is between 15° and 30°. Various relations may readily be developed from geometry of a basic rack, Figure 12.4, illustrated with dimensions in transverse plane (At), normal plane (An), and axial plane (Ax). The distances between similar pitch lines from tooth to tooth are the *circular pitch p*, the *normal circular pitch p_n*, and the *axial (circular) pitch p_a*.

Observe from the figure that the p and the pressure angle ϕ are measured in the transverse plane or the plane of rotation, as with spur gears. Hence, the p and the ϕ are also referred to as the *transverse circular pitch* and *transverse pressure angle*, respectively. The quantities p_n and normal pressure angle ϕ_n are measured in a normal plane. Referring to the triangles *ABC* and *ADC*, we write

$$p_n = p \cos \psi, \qquad p_a = p \cot \psi = p_n / \sin \psi \qquad (12.1)$$

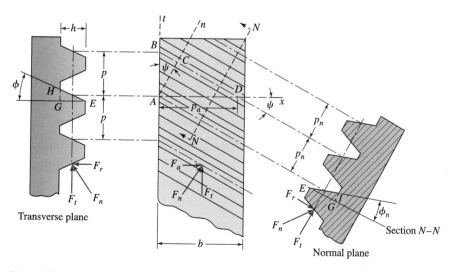

Figure 12.4   Portion of a helical rack.

Diametral pitch is more commonly employed than circular pitch to define tooth size. The product of circular and diametral pitch equals π for normal as well as transverse plane. Therefore,

$$Pp = \pi, \qquad P_n p_n = \pi, \qquad P_n = P/\cos\psi, \qquad P = N/d \qquad \text{(12.2)}$$

where $P =$ diametral pitch, $P_n =$ normal dimateral pitch, and $N =$ number of teeth. In the triangles EGH and EGI (Figure 12.4): $EG = h$; however, $GH/GI = p/p_n = 1/\cos\psi$ from the first of Eqs. (12.1). Here, h is the whole depth of the gear teeth. We also have $\tan\phi = GH/h$, $\tan\phi_n = GI/h$, or $\tan\phi/\tan\phi_n = GH/GI$. Hence, pressure angles are related by

$$\tan\phi_n = \tan\phi\,\cos\psi \qquad \text{(12.3)}$$

Other geometric quantities are expressed similarly to those for spur gears:

$$d = \frac{Np}{\pi} = \frac{Np_n}{\pi\cos\psi} = \frac{N}{P_n\cos\psi} \qquad \text{(12.4)}$$

$$c = \frac{d_1 + d_2}{2} = \frac{p}{2\pi}(N_1 + N_2) = \frac{N_1 + N_2}{2P_n\cos\psi} \qquad \text{(12.5)}$$

in which c represents the center distance of mating gears (1 and 2).

Virtual Number of Teeth

Intersection of the normal plane N-N and the pitch cylinder of diameter d is an *ellipse* (Figure 12.4). The shape of the gear teeth generated in this plane, using the radius of curvature of the ellipse, would be a nearly virtual spur gear having the same properties as the actual helical gear. From analytic geometry, the radius of curvature r_c at the end of a semi-minor axis of the ellipse is

$$r_c = \frac{d/2}{\cos^2\psi} \qquad \text{(12.6)}$$

The number of teeth of the equivalent spur gear in the normal plane, known as either the virtual or equivalent number of teeth, is then

$$N' = \frac{2\pi r_c}{p_n} = \frac{\pi d}{p_n\cos^2\psi} \qquad \text{(12.7a)}$$

Carrying the values for $\pi d/p_n$ from Eq. (12.4) into Eq. (12.7a), the *virtual number of teeth* N' may be expressed in the convenient form

$$N' = \frac{N}{\cos^3\psi} \qquad \text{(12.7b)}$$

As noted previously, N is the number of actual teeth. It is necessary to know the virtual number of teeth in design, as discussed in Section 12.5.

CONTACT RATIOS

The transverse contact ratio m_p is the same as defined for spur gears by Eq. (11.10). Owing to the helix angle, the axial contact ratio m_F is given by

$$m_F = \frac{b}{p_a} \qquad (12.8)$$

The quantities b and p_a represent the face width and axial pitch, respectively. The preceding ratio, indicating the degree of helical overlap in the mesh, should be higher than 1.15.

Determination of the Geometric Quantities for Helical Gears

EXAMPLE 12.1

Two helical gears have a center distance of 10 in., a pressure angle of $\phi = 25°$, a helix angle of $\psi = 30°$, and a diametral pitch of $P = 6$ in.$^{-1}$. If the speed ratio is to be 1/3, calculate

   (a) The (transverse) circular, normal circular, and axial pitches.

   (b) The number of teeth of each gear.

   (c) The normal diametral pitch and normal pressure angle.

Solution:

   (a) Applying Eqs. (12.1) and (12.2),

$$p = \frac{\pi}{6} = 0.5236 \text{ in.}$$

$$p_n = 0.5236 \cos 30° = 0.453 \text{ in.}$$

$$p_a = 0.5236 \cot 30° = 0.907 \text{ in.}$$

   (b) Through the use of Eq. (11.8),

$$r_s = \frac{1}{3} = \frac{N_1}{N_2} \quad \text{or} \quad N_2 = 3N_1$$

Equation (12.5) gives then

$$10 = \frac{0.5236}{2\pi}(N_1 + 3N_1) \quad \text{or} \quad N_1 = 30$$

and hence $N_2 = 90$.

   (c) From Eq. (12.2), we have

$$P_n = \frac{6}{\cos 30°} = 6.928$$

By Eq. (12.3),

$$\tan \phi_n = \tan 25° \cos 30°$$

from which

$$\phi_n = 22°$$

12.4   HELICAL GEAR TOOTH LOADS

This section is concerned with the applied forces or loads acting on the tooth of a helical gear. As in the case of spur gears, the points of application of the force are in the pitch plane and in the center of the gear face. Again, the load is normal to the tooth surface and indicated by F_n. The transmitted load F_t is the same for spur or helical gears.

Figure 12.4 schematically shows the three components of the force acting against a helical gear tooth. Obviously, the inclined tooth develops the axial component, which is not present with spur gearing. We see from Figure 12.5 that the normal load F_n is at a compound angle defined by the normal pressure angle ϕ_n and the helix angle ψ in combination. The projection of F_n on the plane of rotation is inclined at an angle ϕ to the radial force component. Hence, the values of components of the tooth force are

$$F_r = F_n \sin \phi$$

$$F_t = F_n \cos \phi_n \cos \psi$$ (12.9)

$$F_a = F_n \cos \phi_n \sin \psi$$

in which

F_n = normal load or applied force

F_r = radial component

F_t = tangential component, also called the *transmitted load*

F_a = axial component, also called the thrust load

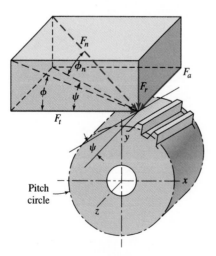

Figure 12.5   Components of tooth force acting on a helical gear.

Usually, the transmitted load F_t is obtained from Eq. (11.16) and the other forces are desired. It is therefore convenient to rewrite Eqs. (12.9) as

$$F_r = F_t \tan \phi$$

$$F_a = F_t \tan \psi \qquad \text{(12.10)}$$

$$F_n = \frac{F_t}{\cos \phi_n \cos \psi}$$

In the foregoing, the pressure angle ϕ is related to the helix angle ψ and normal pressure angle ϕ_n by Eq. (12.3).

The thrust load requires the use of bearings that can resist axial forces as well as radial loads. Sometimes, the need for a thrust-resistant bearing can be eliminated by using a herringbone gear. Obviously, the axial thrust forces for each set of teeth in a herringbone gear cancel each other.

12.5 HELICAL GEAR-TOOTH BENDING AND WEAR STRENGTHS

The equations for the bending and wear strengths of helical gear teeth are similar to those of spur gears. However, slight modifications must be made to take care of the effects of the helix angle ψ. The discussions of the factors presented in Chapter 11 on spur gears also apply to the helical gearsets. Adjusted forms of the fundamental equations are introduced in the following discussion.

THE LEWIS EQUATION

The allowable bending load of the helical gear teeth is given by the expression:

$$F_b = \frac{\sigma_o b}{K_f} \frac{Y}{P_n} \qquad \text{(11.27')}$$

The value of Y is obtained from Table 11.2 using the virtual teeth number N'.

THE BUCKINGHAM EQUATION

The limit load for wear for helical gears on parallel shafts may be written in the form:

$$F_w = \frac{d_p b Q K}{\cos^2 \psi} \qquad \text{(11.32')}$$

where

$$Q = \frac{2N_g}{N_p + N_g} \qquad \text{(11.34)}$$

The wear load factor K can be taken from Table 11.9 for the normal pressure angle ϕ_n.

For satisfactory helical gear performance, the usual *requirement* is that $F_b \geq F_d$ and $F_w \geq F_d$. The dynamic load F_d acting on helical gears can be estimated by the formula:

$$F_d = \frac{78 + \sqrt{V}}{78} F_t \qquad \text{(for } 0 < V > 4000 \text{ fpm)} \qquad \textbf{(11.18c')}$$

in which the pitch line velocity, V in fpm, is defined by Eq. (11.14). To convert to m/s *divide* the given values in this expression by 196.8.

THE AGMA EQUATIONS

The formulas used for spur gears also apply to the helical gears. They were presented in Sections 11.9 and 11.11 with explanation of the terms. So, the equation for *bending stress* is

$$\sigma = F_t K_o K_v \frac{P}{b} \frac{K_s K_m}{J} \qquad \textbf{(11.29)}$$

$$\sigma = F_t K_o K_v \frac{1.0}{bm} \frac{K_s K_m}{J} \qquad \textbf{(11.29')}$$

Similarly, for *wear strength*, we have

$$\sigma_c = C_p \left(F_t K_o K_v \frac{K_s}{bd} \frac{K_m C_f}{I} \right)^{1/2} \qquad \textbf{(11.36)}$$

where

$$I = \frac{\sin\phi \cos\phi}{2m_N} \frac{m_G}{m_G + 1} \qquad \textbf{(11.37b)}$$

The charts and graphs previously given in Chapter 11 are valid equally well, with the exception of the geometry factor J. The values of this factor for helical gears is obtained from Figure 12.6. The size factor $K_s = 1$ for helical gears.

The calculation of the geometry factor I through the use of Eq. (11.37b), for helical gears, requires the values of the load-sharing factor:

$$m_N = \frac{p_{nb}}{0.95Z} \qquad \textbf{(12.11)}$$

Here $p_{nb} =$ normal base pitch $= p_n \cos\phi_n$, and $Z =$ length of action in transverse plane. Equation (11.9) may be used to compute the value for Z, in which the addendum equals $1/P_n$. Consult the appropriate AGMA standard for further details.

Allowable bending and *surface stresses* are calculated from equations given in Sections 11.9 and 11.11, repeated here, exactly as with spur gears,

$$\sigma_{\text{all}} = \frac{S_t K_L}{K_T K_R} \qquad \textbf{(11.30)}$$

$$\sigma_{c,\text{all}} = \frac{S_c C_L C_H}{K_T K_R} \qquad \textbf{(11.38)}$$

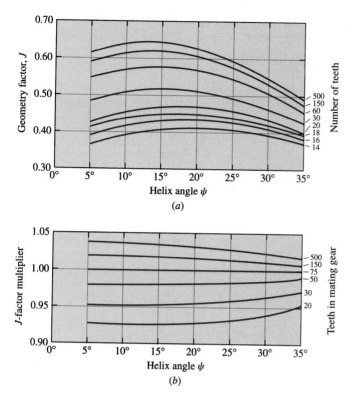

Figure 12.6 Helical gears with normal pressure angle $\phi_n = 20°$: (a) AGMA helical gear geometry factor J; (b) J-factor multipliers for use when the mating gear has other than 75 teeth (ANSI/AGMA 218.01).

The design specifications and applications are the same as discussed in Chapter 11 for spur gears. Example 12.2 and Case Study 12-1 further illustrate the analysis and design of helical gears.

Electric Motor Geared to Drive a Machine | **EXAMPLE 12.2**

A motor at about $n = 2400$ rpm drives a machine by means of a helical gearset as shown in Figure 12.7. Calculate

(a) The value of the helix angle.

(b) The allowable bending and wear loads using the Lewis and Buckingham formulas.

(c) The horsepower that can be transmitted by the gearset.

Given: The gears have the following geometric quantities:

$P_n = 5$ in.$^{-1}$, $\phi = 20°$, $c = 9$ in., $N_1 = 30$, $N_2 = 42$, $b = 2$ in.

Figure 12.7   Example 12.2, schematic arrangement of motor, gear, and driven machine.

Design Assumptions: The gears are made of SAE 1045 steel, water-quenched and tempered (WQ&T), and hardened to 200 Bhn.

Solution:

(a)   From Eqs. (12.1) through (12.5), we have

$$P = \frac{1}{2c}(N_1 + N_2) = \frac{72}{18}$$

$$d_1 = \frac{N_1}{P} = \frac{30}{72}(18) = 7.5 \text{ in.}$$

$$d_2 = \frac{N_2}{P} = \frac{42}{72}(18) = 10.5 \text{ in.}$$

$$\cos \psi_1 = \frac{N_1}{P_n d_1} = \frac{30}{5(7.5)} = 0.8 \quad \text{or} \quad \psi_1 = \psi_2 = 36.9°$$

(b)   The virtual number of teeth, using Eq. (12.7), is

$$N' = \frac{N}{\cos^3 \psi} = \frac{30}{(0.8)^3} = 58.6$$

Hence, interpolating in Table 11.2, $Y = 0.419$. By Table 11.3, $\sigma_o = 32$ ksi. Applying the Lewis equation, (11.27′) with $K_f = 1$,

$$F_b = \sigma_o b \frac{Y}{P_n} = 32(2)\frac{0.419}{5} = 5.363 \text{ kips}$$

By Table 11.9, $K = 79$ ksi. From Eq. (11.34),

$$Q = \frac{2N_g}{N_p + N_g} = \frac{2(42)}{72} = \frac{7}{6}$$

The Buckingham formula, Eq. (11.32′), yields

$$F_w = \frac{d_1 b Q K}{\cos^2 \psi} = \frac{7.5(2)(7)(79)}{6(0.8)^2} = 2.16 \text{ kips}$$

(c) The horsepower capacity is based on F_w since it is smaller than F_b. The pitch-line velocity equals

$$V = \frac{\pi d_1 n_1}{12} = \frac{\pi(7.5)(2400)}{12} = 4712 \text{ fpm}$$

The dynamic load, using Eq. (11.18c′), is

$$F_d = \frac{78 + \sqrt{4712}}{78} F_t = 1.88 F_t$$

Equation (11.35), $F_w \geq F_d$, results in

$$2.16 = 1.88 F_t \quad \text{or} \quad F_t = 1.15 \text{ kips}$$

The corresponding gear power is therefore

$$\text{hp} = \frac{F_t V}{33,000} = \frac{1150(4712)}{33,000} = 164$$

Comments: Observe that the dynamic load is about twice the transmitted load, as expected for reliable operation.

Case Study 12-1 | HIGH-SPEED TURBINE GEARED TO DRIVE A GENERATOR

A turbine rotates at about $n = 8000$ rpm and drives, by means of a helical gearset, a 250-kW (335-hp) generator at 1000 rpm, as depicted in Figure 12.8. Determine

(a) The gear dimensions and the gear-tooth forces.

(b) The load capacity based on the bending strength and surface wear using the Lewis and Buckingham equations.

(c) The AGMA load capacity on the basis of strength only.

Given: Gearset helix angle $\psi = 30°$

Pinion: $N_p = 35$, $\phi_n = 20°$, $P_n = 10$ in.$^{-1}$

Design Assumptions:

1. Moderate shock load on the generator and a light shock on the turbine.

2. Mounting is accurate.

3. Reliability is 99.9%.

4. Both pinion and gear are through hardened, precision shaped, and ground to permit to run at high speeds.

5. The pinion is made of steel with 150 Bhn and gear is cast iron.

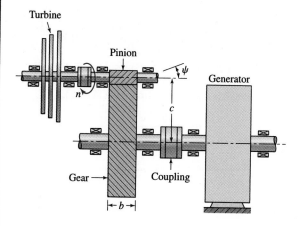

Figure 12.8 Schematic arrangement of turbine, gear, and generator.

(continued)

Case Study (CONTINUED)

6. The gearset goes on a maximum of $c = 18\frac{1}{2}$-in. center distance. However, to keep pitch-line velocity down, gears are designed with as small a center distance as possible.

7. To keep stresses down, a wide face width $b = 8$-in. large gearset is used.

8. The generator efficiency is 95%.

Solution: See Figures 12.6, 12.8, Tables 11.5 through 11.9.

(a) The geometric quantities for the gearset are obtained by using Eqs. (12.1) through (12.5). Therefore,

$$\phi = \tan^{-1}\frac{\tan\phi_n}{\cos\psi} = \tan^{-1}\frac{\tan 20°}{\cos 30°} = 22.8°$$

$$P = P_n \cos\psi = 10 \cos 30° = 8.66$$

$$d_p = \frac{N_p}{P} = \frac{35}{8.66} = 4.04 \text{ in.}$$

$$N_g = N_p\left(\frac{n_p}{n_g}\right) = 35\left(\frac{8000}{1000}\right) = 280$$

$$d_g = \frac{N_g}{P} = \frac{280}{8.66} = 32.3 \text{ in.}$$

It follows that

$$c = \frac{1}{2}(d_p + d_g)$$

$$= \frac{1}{2}(4.04 + 32.3) = 18.17 \text{ in.}$$

Comment: The condition that the center distance is not to exceed 18.5 is satisfied.

The pitch-line velocity equals

$$V = \frac{\pi dn}{12} = \frac{\pi(4.04)8000}{12} = 8461 \text{ fpm}$$

The power that the gear must transmit is about $335/0.95 = 353$ hp. The transmitted load is then

$$F_t = \frac{33{,}000 \text{ hp}}{V} = \frac{33{,}000(353)}{8461} = 1.38 \text{ kips}$$

As a result, the radial, axial, and normal components of tooth force are, using Eq. (12.10),

$$F_r = F_t \tan\phi = 1380 \tan 22.8° = 580 \text{ lb}$$

$$F_a = F_t \tan\psi = 1380 \tan 30° = 797 \text{ lb}$$

$$F_n = \frac{F_t}{\cos\phi_n \cos\psi} = \frac{1380}{\cos 20° \cos 30°} = 1696 \text{ lb}$$

(b) From Eq. (12.7), we have

$$N' = \frac{N}{\cos^3\psi} = \frac{35}{\cos^3 30°} = 53.9$$

Then, for 53.9 teeth and $\phi = 22.8°$ by interpolation from Table 11.2, $Y = 0.452$. Using Table 11.3, $\sigma_o \approx 18$ ksi. Applying Eq. (11.27′) with $K_f = 1$,

$$F_b = \sigma_o b\frac{Y}{P_n} = 18(8)\frac{0.452}{10} = 6.51 \text{ kips}$$

Corresponding to $\phi = 22.8°$, interpolating in Table 11.9, $K \approx 68$ psi. Through the use of Eq. (11.34),

$$Q = \frac{2N_g}{N_p + N_g} = \frac{2(280)}{35 + 280} = \frac{112}{63}$$

The limit load for wear, by Eq. (11.32′), is

$$F_w = \frac{d_p b Q K}{\cos^2\psi}$$

$$= \frac{4.04(8)(68)(112)}{63(\cos^2 30°)} = 5.21 \text{ kips}$$

Because the permissible load in wear is less than that allowable in bending, it is used as the dynamic load F_d. So, from Eq. (11.18c′),

$$F_d = \frac{78 + \sqrt{V}}{78}F_t$$

$$5.21 = \frac{78 + \sqrt{8461}}{78}F_t = 2.18F_t$$

or

$$F_t = 2.39 \text{ kips}$$

(c) Application of Eq. (11.30) leads to

$$\sigma_{all} = \frac{S_t K_L}{K_T K_R}$$

Case Study (CONCLUDED)

where

$S_t = 20.5$ ksi   (interpolating, Table 11.6 for a Bhn of 150)

$K_L = 1.0$ (indefinite life, Table 11.7)

$K_T = 1.0$ (from Section 11.9)

$K_R = 1.25$ (by Table 11.8)

The preceding equation is then

$$\sigma_{all} = \frac{20,500(1.0)}{(1.0)(1.25)} = 16.4 \text{ ksi}$$

The tangential force, by Eq. (11.29) with $\sigma = \sigma_{all}$, is

$$F_t = \frac{\sigma_{all}}{K_o K_v} \frac{b}{P} \frac{J}{K_s K_m} \qquad \text{(a)}$$

Here, we have

$K_o = 1.5$ (by Table 11.4);

$K_v = 2.18$ (from curve B of Figure 11.14);

$K_s = 1.0$ (from Section 12.5);

$K_m = 1.5$ (by Table 11.5);

$J = 0.47$ (for $N_p = 35$ and $\psi = 30°$, Figure 12.6a);

J-multiplier $= 1.02$ (for $N_g = 280$ and $\psi = 30°$, Figure 12.6b);

$$J = 1.02 \times 0.47 = 0.48$$

Equation (a) is then

$$F_t = \frac{16,400(8)(0.48)}{(1.5)(2.18)(8.66)(1.0)(1.5)} = 1.48 \text{ kips}$$

compared to the approximate result 2.39 kips for part b.

Comments: The tangential load capacity of the gearset, 1.48 kips, is larger than the force to be transmitted, 1.38 kips (allowance is made for 95% generator efficiency); the gears are safe. Since a wide face width is used, the design should be checked for combined bending and torsion at the pinion [1].

12.6   BEVEL GEARS

Bevel gears are cut on conical blanks to be used to transmit motion between intersecting shafts. The simplest bevel gear type is the straight-tooth bevel gear or *straight bevel gear* (Figure 12.9). As the name implies the teeth are cut straight, parallel to the cone axis, like spur gears. Clearly, the teeth have a taper and, if extended inward, would intersect each other at the axis. Although bevel gears are usually made for a shaft angle of 90°, the only type we deal with here, they may be produced for almost any angle as well as with teeth lying in spiral paths on the pitch cones. When two straight bevel gears intersect at right angles and the gears have the same number of teeth, they are called *miter gears.*

Spiral teeth bevel gears, called just *spiral bevel gears* or *hypoid gears,* have shafts that do not intersect, as shown in Figure 11.1. We see that the teeth are cut at a spiral angle to the cone axis, analogous to helical gears. It is therefore possible to connect continuous non-intersecting shafts by such gears. Often spiral gears are most desirable for those applications involving large speed-reduction ratios and those requiring great smoothness and quiet operation. These gears are in widespread usage for automotive applications. *Zerol bevel gears* have curved teeth like spiral bevels, however, they have a 0 spiral angle similar to

Figure 12.9   Bevel gears. (Courtesy of Boston Gear.)

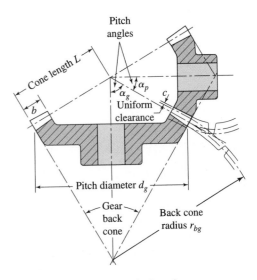

Figure 12.10   Notation for bevel gears.

straight bevel gears. Bevel gears are noninterchangeable. Usually, they are made and replaced as matched pinion-gearsets.

STRAIGHT BEVEL GEARS

Straight bevel gears are the most economical of the various bevel gear types. These gears are used primarily for relatively low-speed applications with pitch line velocities up to 1000 fpm, where smoothness and quiet are not significant consideration. However, with the use of a finishing operation (e.g., grinding), higher speeds have been successfully handled by straight bevel gears.

Geometry

The geometry of bevel gears is shown in Figure 12.10. The size and shape of the teeth are defined at the large end, on the back cones. They are similar to those of spur gear teeth. Note that the pitch cone and (developed) back cone elements are perpendicular. The *pitch angles* (also called pitch cone angles) are defined by the pitch cones joining at the apex. Standard straight bevel gears are cut by using a 20° pressure angle and full-depth teeth, which increases the contact ratio and the strength of the pinion.

The diametral pitch refers to the back cone of the gear. Therefore, the relationships between the geometric quantities and the speed for bevel gears are given as follows:

$$d_p = \frac{N_p}{P}, \qquad d_g = \frac{N_g}{P} \qquad\qquad (12.12\text{a})$$

$$\tan \alpha_p = \frac{N_p}{N_g}, \qquad \tan \alpha_g = \frac{N_g}{N_p} \qquad\qquad (12.12\text{b})$$

$$r_s = \frac{\omega_g}{\omega_p} = \frac{N_p}{N_g} = \frac{d_p}{d_g} = \tan \alpha_p = \cot \alpha_g \qquad\qquad (12.13)$$

where

d = pitch diameter

P = diametral pitch

N = number of tooth

α = pitch angle

ω = angular speed

r_s = speed ratio

In the preceding equations the subscripts p and g refer to the pinion and gear, respectively.

It is to be noted that, for 20° pressure angle straight-bevel gear teeth, the face width b should be made equal to

$$b = \frac{L}{3} \quad \text{or} \quad b = \frac{10}{P}$$

whichever is smaller. The uniform clearance is given by the formula

$$c = \frac{0.188}{P} + 0.002 \text{ in.}$$

The quantities L and c represent the pitch cone length and clearance, respectively (Figure 12.10).

VIRTUAL NUMBER OF TEETH

In the discussion of helical gears, it was pointed out that the tooth profile in the normal plane is a spur gear having an ellipse as its radius of curvature. The result is that the form factors for spur gears apply, provided the equivalent or a virtual number of teeth N' are used in finding the tabular values. The identical situation exists with regard to bevel gears.

Figure 12.10 depicts the gear teeth profiles at the back cones. They relate to those of spur gears having radii of r_{bg} (gear) and r_{bp} (pinion). The preceding is referred to as the *Tredgold's approximation*. Accordingly, the *virtual number of teeth N'* in these imaginary spur gears:

$$N'_p = 2r_{bp}P, \qquad N'_g = 2r_{bg}P \qquad\qquad (12.14)$$

This may be written in the convenient form

$$N'_p = \frac{N_p}{\cos \alpha_p}, \qquad N'_g = \frac{N_g}{\cos \alpha_g} \qquad\qquad \text{(12.15)}$$

in which r_b is the back cone radius and N represents the actual number of teeth of bevel gear.

12.7   TOOTH LOADS OF STRAIGHT BEVEL GEARS

In practice, the resultant tooth load is taken to be acting at the midpoint of the tooth face (Figure 12.11a). While the actual resultant occurs somewhere between the midpoint and the large end of the tooth, there is only a small error in making this simplifying assumption. The transmitted *tangential load* or tangential component of the applied force, acting at the pitch point P, is then

$$F_t = \frac{T}{r_{\text{avg}}} \qquad\qquad \text{(12.16)}$$

Here T represents the torque applied and r_{avg} is the average pitch radius of the gear under consideration.

The resultant force normal to the tooth surface at point P of the gear has value $F_n = F_t \sec \phi$ (Figure 12.11b). The projection of this force in the axial plane, $F_t \tan \phi$, is divided into the axial and radial components:

$$F_a = F_t \tan \phi \sin \alpha$$
$$F_r = F_t \tan \phi \cos \alpha \qquad\qquad \text{(12.17)}$$

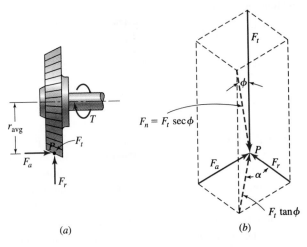

(a)   (b)

Figure 12.11   Forces at midpoint of bevel-gear tooth:
(a) three mutually perpendicular components; (b) the total
(normal) force and its projections.

where

F_t = tangential force

F_a = axial force

F_r = radial force

ϕ = pressure angle

α = pitch angle

It is obvious that the three components F_t, F_a, and F_r are at right angles to each other. These forces can be used to ascertain the bearing reactions, by applying the equations of statics.

Determining the Tooth Loads of a Bevel Gearset

EXAMPLE 12.3

A set of 20° pressure angle straight bevel gears is to be used to transmit 20 hp from a pinion operating at 500 rpm to a gear mounted on a shaft that intersects the shaft at an angle of 90° (Figure 12.12a). Calculate

(a) The pitch angles and average radii for the gears.

(b) The forces on the gears.

(c) The torque produced about the gear shaft axis.

Figure 12.12 Example 12.3, pitch cones of bevel gearset: (a) data; (b) axial and radial tooth forces.

Solution:

(a) Equation (12.13) gives

$$r_s = \frac{200}{500} = \frac{1}{2.5} \quad \text{or} \quad d_g = 2.5d_p = 25 \text{ in.}$$

$$\alpha_p = \tan^{-1}\left(\frac{1}{2.5}\right) = 21.8° \quad \text{and} \quad \alpha_g = 90° - \alpha_p = 68.2°$$

Hence,

$$r_{g,\text{avg}} = r_g - \frac{b}{2} \sin \alpha_g = 12.5 - (1) \sin 68.2° = 11.6 \text{ in.}$$

$$r_{p,\text{avg}} = r_p - \frac{b}{2} \sin \alpha_p = 5 - (1) \sin 21.8° = 4.6 \text{ in.}$$

(b) Through the use of Eq. (11.16),

$$F_t = \frac{33,000 \text{ hp}}{\pi d_{p,\text{avg}} n_p / 12} = \frac{33,000(20)(12)}{\pi (9.2)500} = 548 \text{ lb}$$

From Eqs. (12.17), the pinion forces are

$$F_a = F_t \tan \phi \sin \alpha_p = 548 (\tan 20°)(\sin 21.8°) = 74 \text{ lb}$$

$$F_r = F_t \tan \phi \cos \alpha_p = 548 (\tan 20°)(\cos 21.8°) = 185 \text{ lb}$$

As shown in Figure 12.12b, the pinion thrust force equals the gear radial force and the pinion radial force equals the gear thrust force.

(c) Torque, $T = F_t(d_g/2) = 548(12.5) = 6.85 \text{ kip} \cdot \text{in.}$

12.8 BEVEL GEAR-TOOTH BENDING AND WEAR STRENGTHS

The expressions for bending and wear strengths are analogous to those for spur gears. However, slight modifications must be made to take care of the effects of the cone angle α. The adjusted forms of the basic formulas are introduced in the following paragraphs.

THE LEWIS EQUATION

It is assumed that the bevel gear tooth is equivalent to a spur gear tooth whose cross section is the same as the cross section of the bevel tooth at the midpoint of the face b. The allowable *bending load* is given by

$$F_b = \frac{\sigma_o b}{K_f} \frac{Y}{P} \tag{11.27}$$

The factor Y is read from Table 11.2, for a gear of N' virtual number of teeth.

THE BUCKINGHAM EQUATION

Due to the difficulty in achieving a bearing along the entire face width b, about three-quarters of b alone is considered as effective. So, the allowable *wear load* can be expressed as

$$F_w = \frac{0.75 d_p b K Q'}{\cos \alpha_p} \tag{11.32'}$$

where

$$Q' = \frac{2N'_g}{N'_p + N'_g} \qquad (11.34')$$

In the preceding, we have

d_p = pitch diameter measured at the back of the tooth

N' = virtual tooth number

α = pitch angle

K = wear load factor (from Table 11.9)

For the satisfactory operation of the bevel gearsets, the usual *requirement* is that $F_b \geq F_d$ and $F_w \geq F_d$, where the dynamic load F_d is given by Eq. (11.18).

THE AGMA EQUATIONS

The formulas are the same as those presented in the discussions of the spur gears. But only some of the values of the correction factors are applicable to bevel gears [5]. For a complete treatment, consult the appropriate AGMA publications and the references listed [2]. We present only a brief summary of the method to bevel gear design as an introduction to the subject.

The equation for the *bending stress* at the root of bevel gear tooth is the same as for spur or helical gears. Therefore,

$$\sigma = F_t K_o K_v \frac{P}{b} \frac{K_s K_m}{J} \qquad (11.29)$$

$$\sigma = F_t K_o K_v \frac{1.0}{bm} \frac{K_s K_m}{J} \qquad (11.29')$$

The tangential load F_t is obtained from Eq. (12.16). Figure 12.13 gives the values for the foregoing geometry factor J for straight bevel gears. The AGMA standard also provides charts of the factors for zerol and spiral bevel gears. The remaining factors in Eq. (11.29) can be taken to be the same as defined in Section 11.9. The allowable bending stress of bevel gear tooth is calculated from Eq. (11.30), exactly as for spur and helical gears.

Surface stress for the wear of a bevel gear tooth is computed in a manner like that of spur or helical gears. Hence,

$$\sigma_c = C_p \left(F_t K_o K_v \frac{K_s}{bd} \frac{K_m C_f}{I} \right)^{1/2} \qquad (11.36)$$

Only two modifications are required: Values of C_p are 1.23 times the values listed in Table 11.10 and values of I are taken from Figure 12.14. The *allowable* surface stress of a bevel gear tooth is obtained by Eq. (11.38), exactly as for spur or helical gears.

Figure 12.13   Geometry factors *J* for straight bevel gears. Pressure angle 20° and shaft angle 90° (AGMA Information Sheet 226.01).

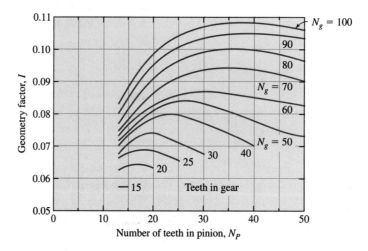

Figure 12.14   Geometry factors *I* for straight bevel gears. Pressure angle 20° and shaft angle 90° (AGMA Information Sheet 215.91).

12.9   WORM GEARSETS

Worm gearing can be employed to transmit motion between nonparallel nonintersecting shafts, as shown in Figures 11.1, 12.15, and 12.18. A worm gearset, or simply wormset, consists of a *worm* (resembles a screw) and the *worm gear* (a special helical gear). The

Worm

Gear

Figure 12.15 A single-enveloping wormset. (Courtesy of Martin Sprocket and Gear Co.)

shafts on which the worm and gear are mounted are usually 90° apart. The meshing of two teeth takes place with a sliding action, without shock relevant to spur, helical, or bevel gears. This action, while occurring in quiet operation, may generate overheating, however. It is possible to obtain a large *speed reduction* (up to 360:1) and a high increase of torque by means of wormsets. Typical applications of worm gearsets include positioning devices that take advantage of their locking property (see Section 15.3).

Only a few materials are available for wormsets. The worms are highly stressed and usually made of case-hardened alloy steel. The gear is customarily made of one of the bronzes. The gear is hobbed, while the worm is ordinarily cut and ground or polished. The teeth of the worm must be properly shaped to provide conjugate surfaces. Tooth forms of worm gears are not involutes, and there are large sliding-velocity components in the mesh.

WORM GEAR GEOMETRY

Worms can be made with single, double, or more threads. Worm gearing may be either single enveloping (commonly used) or double enveloping. In a *single-enveloping* set (Figures 12.15 and 12.16b), the helical gear has its width cut into a concave surface, partially enclosing the worm when in mesh. To provide more contact, the worm may have an hourglass shape, in which case the set is referred to as *double enveloping*. That is, with the helical gear cut concavely, the double-enveloping type also has the worm length cut concavely: Both the worm and the gear partially enclose each other. The geometry of the worm is very complicated, and reference should be made to the literature for details.

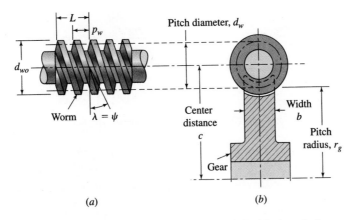

Figure 12.16 Notation for a worm gearset: (a) double-threaded worm; (b) worm gear (shown in a half-sectional view).

The *terminology* used to describe the worm (Figure 12.16a) and power screws (see Section 15.3) is very similar. In general, the worm is analogous to a screw thread and the worm gear is similar to its nut. The *axial pitch* of the worm gear p_w is the distance between corresponding points on adjacent teeth. The *lead L* is the axial distance the worm gear (nut) advances during one revolution of the worm. In a multiple thread worm, the lead is found by multiplying the number of threads (or teeth) by the axial pitch.

The pitch diameter of a worm d_w is not a function of its number of threads N_w. The speed ratio of a wormset is obtained by the ratio of gear teeth to worm threads:

$$r_s = \frac{\omega_g}{\omega_w} = \frac{N_w}{N_g} = \frac{L}{\pi d_g} \qquad (12.18)$$

As in the case of a spur or helical gear, the pitch diameter of a worm gear is related to its circular pitch and number of teeth using Eq. (12.1):

$$d_g = \frac{N_g p}{\pi} \qquad (12.19)$$

The center distance between the two shafts equals $c = (d_w + d_g)/2$, as shown in the figure.

The *lead angle* of the worm (which corresponds to the screw lead angle) is the angle between a tangent to the helix angle and the plane of rotation. The lead and the lead angle of the worm have the following relationships:

$$L = p_w N_w \qquad (12.20)$$

$$\tan \lambda = \frac{L}{\pi d_w} = \frac{V_g}{V_w} \qquad (12.21)$$

Table 12.1 Various normal pressure angles for wormsets

Pressure angle, ϕ_n (degrees)	Maximum lead angle, λ (degrees)	Lewis form factor, Y
$14\frac{1}{2}$	15	0.314
20	25	0.392
25	35	0.470
30	45	0.550

in which

$$L = \text{lead}$$
$$p_w = \text{axial pitch}$$
$$N_w = \text{number of threads}$$
$$\lambda = \text{lead angle}$$
$$d_w = \text{pitch diameter}$$
$$V = \text{pitch line velocity}$$

For a 90° *shaft angle* (Figure 12.16), the lead angle of the worm and helix angle of the gear are equal:

$$\lambda = \psi$$

Note that λ and ψ are measured on the pitch surfaces.

We show in Section 15.3 that the normal pressure angle ϕ_n of the worm corresponds to the thread angle α_n of a screw. Normal pressure angles are related to the lead angle and the Lewis form factor Y, as shown in Table 12.1.

In conclusion, we point out that worm gears usually contain no less than 24 teeth, and the number of gear teeth plus worm threads should be more than 40. The face width b of the gear should not exceed half of the worm outside diameter d_{wo}. AGMA recommends that the magnitude of the pitch diameter as $d_w \approx 3p_g$, where p_g represents the circular pitch of the gear.

Determination of the Geometric Quantities of a Worm | **EXAMPLE 12.4**

A triple threaded worm has a lead L of 75 mm. The gear has 48 teeth and is cut with a hob of modulus $m_n = 7$ mm perpendicular to the teeth. Calculate

(a) The speed ratio r_s.

(b) The center distance c between the shafts if they are 90° apart.

Solution: For a 90° shaft angle, we have $\lambda = \psi$.

(a)   The velocity ratio of the worm gearset is

$$r_s = \frac{N_w}{N_g} = \frac{3}{48} = \frac{1}{16}$$

(b)   Using Eq. (12.20),

$$p_w = \frac{L}{N_w} = \frac{75}{3} = 25 \text{ mm}$$

From Eq. (12.2) with $m_n = 1/P_n$, we obtain

$$p_n = \pi m_n = 7\pi = 21.99 \text{ mm}$$

Equation (12.1) results in

$$\cos \lambda = \frac{p_n}{p_w} = \frac{21.99}{25} = 0.88 \quad \text{or} \quad \lambda = 28.4°$$

Application of Eq. (12.21) gives

$$d_w = \frac{L}{\pi \tan \lambda} = \frac{75}{\pi \tan 28.4°} = 44.15 \text{ mm}$$

Through the use of Eq. (12.18),

$$d_g = \frac{L}{\pi r_s} = \frac{75}{\pi/16} = 381.97 \text{ mm}$$

We then have

$$c = \frac{1}{2}(d_w + d_g) = \frac{1}{2}(44.15 + 381.97) = 213.1 \text{ mm}$$

12.10   WORM GEAR BENDING AND WEAR STRENGTHS

The approximate bending and wear strengths of worm gearsets can be obtained by equations analogous to those used for spur gears. Nevertheless, adjustments are made to account for the effects of the normal pressure angle ϕ_n and lead angle λ of the worm. The fundamental formulas for the allowable bending and wear loads for the gear teeth are as follows.

THE LEWIS EQUATION

The bending stresses are much higher in the gear than in the worm. The following slightly *modified Lewis equation* is therefore applied to worm gear:

$$F_b = \frac{\sigma_o b}{K_f} \frac{Y}{P_n} \tag{11.27'}$$

The value of Y can be taken from Table 12.1. It is to be noted that the normal diametral pitch P_n is defined by Eq. (12.2).

THE LIMIT LOAD FOR WEAR

The *wear equation* by Buckingham, frequently used for rough estimates, has the form

$$F_w = d_g b K_w \qquad \text{(12.22)}$$

Here,

F_w = allowable wear load

d_g = pitch diameter

b = face width of the gear

K_w = a material and geometry factor, obtained from Table 12.2

For a satisfactory worm gearset, the usual *requirement* is that $F_b \geq F_d$ and $F_w \geq F_d$. The dynamic load F_d acting on worm gears can be approximated by

$$F_d = \frac{1200 + V}{1200} F_t \qquad \text{(for } 0 < V > 4000 \text{ fpm)} \qquad \text{(11.18b')}$$

As before, the pitch line velocity V is in fpm in this formula.

THE AGMA EQUATIONS

The design of worm gearsets is more complicated and dissimilar to that of other gearing. The AGMA prescribes an input power rating formula for wormsets. This permits the worm gear dimensions to be obtained for a given power or torque-speed combination [5]. A wider variation in procedures is employed for estimating bending and surface strengths. Moreover, worm gear capacity is frequently limited not by fatigue strength but heat dissipation or cooling capacity. The latter is discussed in the next section.

Table 12.2 Worm gear factors K_w

Material		K_w (psi)		
Worm	Gear	$\lambda < 10°$	$\lambda < 25°$	$\lambda > 25°$
Steel, 250 Bhn	Bronze*	60	75	90
Steel, 500 Bhn	Bronze*	80	100	120
Hardened steel	Chilled bronze	120	150	180
Cast iron	Bronze*	150	185	225

| *Sand cast.

12.11   THERMAL CAPACITY OF WORM GEARSETS

The power capacity of a wormset in continuous operation is often limited by the *heat-dissipation capacity* of the housing or casing. Lubricant temperature commonly should not exceed about 93°C (200°F). The basic relationship between temperature rise and heat dissipation can be expressed as follows:

$$H = CA\Delta t \tag{12.23}$$

where

H = time rate of heat dissipation, lb · ft/minute

C = heat transfer, or cooling rate, coefficient (lb · ft per minute per square foot of housing surface area per °F)

A = housing external surface area, ft^2

Δt = temperature difference between oil and ambient air, °F

The values of A for a conventional housing, as recommended by AGMA, may be estimated by the formula:

$$A = 0.3c^{1.7} \tag{12.24}$$

Here A is in square feet and c represents the distance between shafts (in inches). The approximate values of heat transfer rate C can be obtained from Figure 12.17. Note from the figure that C is greater at high velocities of the worm shaft, which causes a better circulation of the oil within the housing.

Figure 12.17   Heat transfer coefficient C for worm gear housing [10].

Figure 12.18 Worm gear speed reducer. (Courtesy of Cleveland Gear Company.)

The manufacturer usually provides the means for cooling, such as external fins on housing and a fan installed on worm shaft, as shown in Figure 12.18. It is observed from the figure that, the worm gear has spiral teeth and shaft at right angle to the worm shaft. Clearly, an extensive sump and corresponding large quantity of oil helps increase heat transfer, particularly during overloads. In some warm gear reduction units, oil in the sump may be externally circulated for cooling.

The heat dissipation capacity H of the housing, as determined by Eq. (12.23), in terms of horsepower is given in the form

$$(hp)_d = \frac{CA\Delta t}{33,000} \tag{12.25}$$

This loss of horsepower equals the difference between the input horse power $(hp)_i$ and output horsepower $(hp)_o$. Inasmuch as $e = (hp)_o/(hp)_i$, we have $(hp)_d = (hp)_i - e(hp)_o$.

The input horse power capacity is therefore

$$(hp)_i = \frac{(hp)_d}{1 - e}$$

(12.26)

The quantity e represents the efficiency.

WORM GEAR EFFICIENCY

The expression of the *efficiency e* for a worm gear reduction is the same as that used for a power screw and nut, developed in Section 15.4. In the notation of this chapter, Eq. (15.13) is written as follows:

$$e = \frac{\cos \phi_n - f \tan \lambda}{\cos \phi_n + f \cot \lambda}$$

(12.27)

where f is the coefficient of friction and ϕ_n represents the normal pressure angle. The value of f depends on the velocity of sliding V_s between the teeth:

$$V_s = \frac{V_w}{\cos \lambda}$$

(12.28)

The quantity V_s is the pitch-line velocity of the worm. Table 12.3 furnishes the values of the coefficient of friction.

Table 12.3   Worm gear coefficient of friction f for various sliding velocities V_s

V_s, fpm	f	V_s, fpm	f	V_s, fpm	f
0	0.150	120	0.0519	1200	0.0200
1	0.115	140	0.0498	1400	0.0186
2	0.110	160	0.0477	1600	0.0175
5	0.099	180	0.0456	1800	0.0167
10	0.090	200	0.0435	2000	0.0160
20	0.080	250	0.0400	2200	0.0154
30	0.073	300	0.0365	2400	0.0149
40	0.0691	400	0.0327	2600	0.0146
50	0.0654	500	0.0295	2800	0.0143
60	0.0620	600	0.0274	3000	0.0140
70	0.0600	700	0.0255	4000	0.0131
80	0.0580	800	0.0240	5000	0.0126
90	0.0560	900	0.0227	6000	0.0122
100	0.0540	1,000	0.0217		

| SOURCE: From ANSI/AGMA Standard 6034-A87.

Design Analysis of a Worm Gear Speed Reducer

EXAMPLE 12.5

A worm gearset and its associated geometric quantities are schematically shown in Figure 12.19. Estimate

$n = 1000$ rpm

Worm

$c = 8$ in.

Gear

$N_w = 4$
$d_w = 3$ in.
$\lambda = 15°$
$\phi_n = 25°$

$N_g = 60$

Figure 12.19   Example 12.5.

(a)   The heat dissipation capacity.

(b)   The efficiency.

(c)   The input and output horsepower.

Assumptions: The gearset is designed for continuous operation based on a limiting 100°F temperature rise of the housing without fan.

Solution: The speed ratio of the worm gearset is

$$r_s = \frac{N_w}{N_g} = \frac{4}{60} = \frac{1}{15}$$

(a)   Through the use of Eq. (12.24),

$$A = 0.3c^{1.7} = 0.3(8)^{1.7} = 10.29 \text{ ft}^2$$

From Figure 12.17, we have

$$C = 42 \text{ lb ft/(min ft}^2 \cdot {}°\text{F)}$$

Carrying the data into Eq. (12.25),

$$(\text{hp})_d = \frac{CA\Delta t}{33,000} = \frac{42(10.29)(100)}{33,000} = 1.31$$

(b)   The pitch-line velocity of the worm is

$$V_w = \frac{\pi d_w n_w}{12} = \frac{\pi(3)(1000)}{12} = 785.4 \text{ fpm}$$

Applying Eq. (12.28),

$$V_s = \frac{V_w}{\cos \lambda} = \frac{785.4}{\cos 15°} = 813 \text{ fpm}$$

By Table 12.3, $f = 0.0238$. Introducing the numerical values into Eq. (12.27),

$$e = \frac{\cos 25° - 0.0238(\tan 15°)}{\cos 25° + 0.0238(\cot 15°)} = 0.904 \quad \text{or} \quad 90.4\%$$

(c) Using Eq. (12.26), the input horsepower is equal to

$$(\text{hp})_i = \frac{(\text{hp})_d}{1 - e} = \frac{1.31}{1 - 0.904} = 13.65$$

The output horsepower is then

$$(\text{hp})_o = 13.65 - 1.31 = 12.3$$

Comments: Because of the sliding friction inherent in the tooth action, usually worm gearsets have significantly lower efficiencies than those of spur gear drives. The latter can have efficiencies as high as 98% (Section 11.1).

REFERENCES

1. Dudley, D. W. *Handbook of Practical Gear Design.* New York: McGraw-Hill, 1984.
2. Townsend, D. P., ed. *Dudley's Gear Handbook,* 2nd ed. New York: McGraw-Hill, 1992.
3. Buckingham, E., and H. H. Ryffel. *Design of Worm and Spiral Gears.* New York: Industrial Press, 1960.
4. "1979 Mechanical Drives Reference Issue." *Machine Design,* June 29, 1979.
5. American Manufacturers Association. *Standards of the American Manufacturers Association.* Arlington, VA: AMA, 1993.
6. Avallone, E. A., and T. B. Baumeister III. *Mark's Standard Handbook for Mechanical Engineers,* 10th ed. New York: McGraw-Hill, 1996.
7. Shigley, J. E., and C. R. Mischke. *Mechanical Engineering Design,* 6th ed. New York: McGraw-Hill, 2001.
8. Norton, R. L. *Machine Design—An Integrated Approach,* 2nd ed. Upper Saddle River, NJ: Prentice Hall, 2000.
9. Burr, A. H., and J. B. Cheatham. *Mechanical Analysis and Design,* 2nd ed. Upper Saddle River, NJ: Prentice Hall, 1995.
10. Juvinall, R. C., and K. M. Marshek. *Fundamentals of Machine Component Design,* 3rd ed. New York: Wiley, 2000.
11. Mott, R. L. *Machine Elements in Mechanical Design,* 2nd ed. New York: Macmillan, 1992.
12. Deutchman, A. D., W. J. Michels, and C. E. Wilson. *Machine Design: Theory and Practice.* New York: Macmillan, 1975.
13. Dimarogonas, A. D. *Machine Design: A CAD Approach.* New York: Wiley, 2001.
14. Rothbart H. A., ed. *Mechanical Design and Systems Handbook,* 2nd ed. New York: McGraw-Hill, 1985.
15. Hamrock, B. J., B. O. Jacobson, and S. R. Schmid. *Fundamentals of Machine Elements.* New York: McGraw-Hill, 1999.

PROBLEMS

Sections 12.1 through 12.5

12.1 A helical gearset consists of a 20-tooth pinion rotating in a counterclockwise direction and driving a 40-tooth gear. Determine

(a) The normal, transverse, and axial circular pitches.

(b) The diametral pitch and pressure angle.

(c) The pitch diameter of each gear.

(d) The directions of the thrusts, and show these on a sketch of the gearset.

Given: The pinion has a right-handed helix angle of 30°, a normal pressure angle of 25°, and a normal diametral pitch of 6 teeth/in.

12.2 Redo Problem 12.1 for a helical gearset that consists of an 18-tooth pinion having a ψ of 20° L. H., a ϕ_n of $14\frac{1}{2}°$, a P_n of 8 in.$^{-1}$, and driving a 55-tooth gear.

12.3 A 32-tooth gear has a pitch diameter of 260 mm, a normal module of $m_n = 6$ mm, a normal pressure angle of 20°. Calculate the kW transmitted at 800 rpm.

Given: The force normal to the tooth surface is 10 N.

12.4 A left-handed helical pinion has a ϕ_n of 20°, a P_n of 10 in.$^{-1}$, a ψ of 45°, a N_p of 32, an F_n of 90 lb, runs at 2400 rpm in the counterclockwise direction. The driven gear has 60 teeth. Determine and show on a sketch

(a) The tangential, axial, and radial forces acting on each gear.

(b) The torque acting on the shaft of each gear.

12.5 The helical gears depicted in Figure P12.5 have a normal diametral pitch of 4 teeth/in., a 25° pressure angle, and a helix angle of 20°. Calculate and show on a sketch

(a) The tangential, radial, and axial forces acting on each gear.

(b) The torque acting on each shaft.

Given: Gear 1 transmits 20 hp at 1500 rpm through the idler to gear 3 on shaft C; the speed ratio for gears 3 to 1 is to be 1/2.

Gear 2
$N_2 = 36$

Gear 3

B

C

Gear 1 (pinion)
$N_1 = 24$, L.H.

A

1500 rpm

Figure P12.5

12.6 A 22-tooth helical pinion has a normal pressure angle of 20°, a normal diametral pitch of 8 teeth/in., a face width of 2 in., and a helix angle of 25°; it rotates at 1800 rpm and transmits 30 hp to a 40-tooth gear. Determine the factor of safety based on bending strength, employing the Lewis equation.

Given: Fatigue stress concentration factor $K_f = 1.5$.
Assumption: The pinion and gear are both steel, hardened to 250 Bhn.

12.7 Resolve Problem 12.6 based on wear strength and using the Buckingham equation.

12.8 Two meshing helical gears are both made of SAE 1020 (WQ&T) steel, hardened to 150 Bhn and are mounted on (parallel) shafts. Calculate the horsepower capacity of the gearset.

Requirement: Use the Lewis equation for bending strength and the Buckingham equation for wear strength.
Given: The number of teeth are 30 and 65, $\phi_n = 25°$, $\psi = 35°$, $P_n = 6$ in.$^{-1}$, and $b = 1.5$ in.; the pinion rotates at 2400 rpm.

12.9 Two meshing helical gears are made of SAE 1040 steel, hardened about to 200 Bhn, and are mounted on parallel shafts 6 in. apart. Determine the horsepower capacity of the gearset

(a) Applying the Lewis equation and $K_f = 1.4$ for bending strength and the Buckingham equation for wear strength.

(b) Applying the AGMA method on the basis of strength only.

Given: The gears are to have a speed ratio of 1/3. A ϕ_n of 20°, a ψ of 30°, a P of 15 in.$^{-1}$, and a b of 2.5 in.; the pinion rotates at 900 rpm.
Design Assumptions:

1. The mounting is accurate.
2. Reliability is 90%.
3. The gearset has indefinite life.
4. Light shock loading acts on the pinion and uniform shock loading on the driven gear.
5. The pinion and gear are both high precision ground.

12.W Review the website at www.bisonger.com and select a $\frac{1}{4}$-horsepower motor for

(a) A single-speed reduction unit.

(b) A double-speed reduction unit.

Sections 12.6 through 12.8

12.10 A 20° pressure angle straight bevel pinion having 20 teeth and a diametral pitch of 8 teeth/in. drives a 42-tooth gear. Determine

(a) The pitch diameters.

(b) The pitch angles.

(c) The face width.

(d) The clearance.

12.11 A pair of bevel gears is to transmit 15 hp at 500 rpm with a speed ratio of 1/2. The 20° pressure angle pinion has an 8-in. back cone pitch diameter, 2.5-in. face width, and a diametral pitch of 7 teeth/in. Calculate and show on a sketch the axial and radial forces acting on each gear.

12.12 If the gears in Problem 12.11 are made of SAE 1020 steel (WQ&T), will they be satisfactory from a bending viewpoint? Employ the Lewis equation and $K_f = 1.4$.

12.13 If the pinion and gear in Problem 12.11 are made of steel (200 Bhn) and phospor bronze, respectively, will they be satisfactory from wear strength viewpoint? Use the Buckingham equation.

12.14 A pair of 20° pressure angle bevel gears of $N_1 = 30$ and $N_2 = 60$ have a diameteral pitch P of 3 teeth/in. at the outside diameter. Calculate the horsepower capacity of the pair, based on the Lewis and Buckingham equations.

Given: Width of face b is 2.8 in.; $K_f = 1.5$; the pinion runs at 720 rpm.
Design Assumption: The gears are made of steel and hardened to about 200 Bhn.

12.15 A pair of 20° pressure angle bevel gears of $N_1 = 30$ and $N_2 = 60$ have a module m of 8.5 mm at the outside diameter. Determine the power capacity of the pair, using the Lewis and Buckingham equations.

Given: Face width b is 70 mm; $K_f = 1.5$; the pinion rotates at 720 rpm.
Design Assumption: The gears are made of steel and hardened to about 200 Bhn.

Sections 12.9 through 12.11

12.16 Two shafts at right angles, with center distance of 6 in., are to be connected by worm gearset. Determine the pitch diameter, lead, and number of teeth of the worm.

Given: a speed ratio of 0.025, lead angle of worm of 35°, and a normal pitch of $\frac{3}{8}$ in. for the worm gear.

12.17 A double-threaded, 3-in. diameter worm has an input of 40 hp at 1800 rpm. The worm gear has 80 teeth and is 10 in. in diameter. Calculate the tangential force on the gear teeth if the efficiency is 90%.

12.18 A quadruple-threaded worm having 2.5 in. diameter meshes with a worm gear with a diametral pitch of 6 teeth/in. and 90 teeth. Find

(a) The lead.

(b) The lead angle.

(c) The center distance.

12.19 A 10-hp, 1000-rpm electric motor drives a 50-rpm machine through a worm gear reducer with a center distance of 7 in. (Figure P12.19). Determine

(a) The value of the helix angle.

(b) The transmitted load.

(c) The power delivered to the driven machine.

Figure P12.19

12.20 If the gears of Figure P12.19 are made of cast steel (WQ&T), will they be satisfactory from the bending strength viewpoint? Use the modified Lewis equation and $K_f = 1.4$.

12.21 If the worm and gear of Figure P12.19 are made of cast iron and bronze, respectively, will they be satisfactory from the wear viewpoint? Employ Eq. (12.22) by Buckingham.

12.22 The worm gear reducer of Figure P12.19 is to be designed for continuous operation and a limiting 100°F temperature rise of the housing with a fan. Estimate the heat dissipation capacity. Will overheating be a problem?

Belts, Chains, Clutches, and Brakes

Outline

13.1   INTRODUCTION

In contrast with bearings, friction is a useful and essential agent in belts, clutches, and brakes. Frictional forces are commonly developed on flat or cylindrical surfaces in contact with shorter pads or linkages or longer bands or belts. A number of these combinations are employed for brakes and clutches, and the band (chain) and wheel pair is used in belt (chain) drives as well. Hence, only a few different analyses are required, with surface forms affecting the equations more than the functions of the elements. Also common operating problems relate to pressure distribution and wear, temperature rise and heat dissipation, and so on. The foregoing devices are thus effectively analyzed and studied together.

A *belt* or *chain drive* provides a convenient means for transferring motion from one shaft to another by means of a belt or chain connecting pulleys on the shafts. Part A of this chapter is devoted to the discussion of the flexible elements: belts and chains. In many cases, their use reduces vibration and shock transmission, simplifies the design of a machine substantially, and reduces the cost. Power is to be transferred between parallel or nonparallel shafts separated by a considerable distance. Thus, the designer is provided considerable flexibility in location of driver and driven machinery [1–16]. The websites www.machinedesign.com and www.powertransmission.com on mechanical systems include information on belts and chains as well as on clutches and brakes.

Brakes and clutches are essentially the same devices. Each is usually associated with rotation. The *brake* absorbs the kinetic energy of moving bodies and thus controls the speed. The function of the brake is to turn mechanical energy into heat. The *clutch* transmits power between two shafts or elements that must be often connected and disconnected. A break acts likewise, with exception that one element is fixed. Clutches and brakes, treated in Part B of the chapter, are all of the friction type that relies on sliding between solid surfaces. Other kinds provide a magnetic, hydraulic, or mechanical connection between the two parts. The clutch is in common use to maintain constant torque on a shaft and serve an emergency disconnection device to decouple the shaft from the motor in the event of a machine jam. In such cases, a brake also is fitted to bring the shaft (and machine) to a rapid stop in urgency. Brakes and clutches are used extensively in production machines of all types as well as in vehicle applications. They are *classified* as follows: disk or axial types, cone types, drum types with external shoes, drum types with internal shoes, and miscellaneous types [17–23].

Part A   Flexible Elements

In addition to gears (Chapters 11 and 12), belts and chains are in widespread use. Belts are frequently necessary to reduce the higher relative speeds of electric motors to the lower values required by mechanical equipment. Chains also can be employed for large reduction in speed if required. Belt drives are relatively quieter than chain drives, while chain drives have greater life expectancy. However, neither belts nor chains have an infinite life and should be replaced at the first sign of wear or lack of elasticity.

13.2 BELTS

There are four *main belt types:* flat, round, V, and timing. Flat and round belts may be used for long center distances between the pulleys in a belt drive. On the other hand, V and timing belts are employed for limited shorter center distances. Excluding timing belts, there is some slip and creep between the belt and the pulley, which is usually made of cast iron or formed steel. Characteristics of the principal belt types are furnished in Table 13.1. Catalogues of various manufacturers of the belts contain much practical information.

Flat and Round Belts

Flat belts and round belts are made of urethane or rubber-impregnated fabric reinforced with steel or nylon cords to take the tension load. One or both surfaces may have friction surface coating. Flat belts find considerable use in applications requiring small pulley diameters. Most often, both driver and driven pulleys lie in the same vertical plane. Flat belts are quiet and efficient at high speeds, and they can transmit large amounts of power. However, a flat belt must operate with higher tension to transmit the same torque as a V belt. *Crowned pulleys* are used for flat belts. Round belts run in grooved pulleys or *sheaves*. Deep-groove pulleys are employed for the drives that transmit power between horizontal and vertical shafts, or so-called quarter-turn drives, and for relatively long center distances.

V Belts

A V belt is rubber covered with impregnated fabric and reinforced with nylon, dacron, rayon, glass, or steel tensile cords. V belts are the most common means of transmitting power between electric motors and driven machinery. They are also used in other household,

Table 13.1 Characteristics of some common belts [1]

Belt type	Geometric form	Size range	Center distance between pulleys
Flat		$t = 0.75$ to 5 mm	No upper limit
Round		$d = 3$ to 19 mm	No upper limit
V		$a = 13$ to 38 mm $b = 8$ to 23 mm $2\beta = 34°$ to $40°$	Limited
Timing		$p = 2$ mm and up	Limited

Figure 13.1 Standard cross sections of V belts. All dimensions are in inches.

Figure 13.2 Multiple V-belt drive. (Courtesy of T. B. Wood's Incorporated.)

automotive, and industrial applications. Usually, V-belt speed should be in the range of about 4000 fpm. These belts are produced in two series: the *standard* V belt, as shown in Figure 13.1, and the high-capacity V-belt. Note that each standard section is designated by a letter for sizes in inch dimensions. Metric sizes are identified by numbers. V belts are slightly less efficient than flat belts. The *included angle* 2β for V belts, defined in the table, is usually from 34° to 40°.

Crowned pulleys and sheaves are also employed for V belts. The "wedging action" of the belt in the groove leads to a large increase in the tractive force produced by the belt, as discussed in Section 13.4. *Variable-pitch pulleys* permit an adjustment in the width of the groove. The effective pitch diameter of the pulley is thus varied. These pulleys are employed to change the input to output speed ratio of a V-belt drive. Some variable-pitch drives can change speed ratios when the belt is transmitting power. As for the number of V belts, as many as 12 or more can be used on a single sheave, making it a *multiple drive.* All belts in such a drive should stretch at the same rate to keep the load equally divided among them. A multiple V-belt drive (Figure 13.2) is used to satisfy high-power transmission requirements.

TIMING BELTS

A timing belt is made of rubberized fabric and steel wire and has evenly spaced teeth on the inside circumference. Also known as a *toothed* or *synchronous belt,* a timing belt

Figure 13.3　Toothed or timing-belt drive for precise speed ratio. (Courtesy of T. B. Wood's Incorporated.)

does not stretch or slip and hence transmits power at a constant angular velocity ratio. This permits timing belts to be employed for many applications requiring precise speed ratio, such as driving an engine camshaft from the crankshaft. Toothed belts also allow the use of small pulleys and small arcs of contact. They are relatively lightweight and can efficiently operate at speeds up to at least 16,000 fpm. A timing belt fits into the grooves cut on the periphery of the wheels or *sprockets,* as shown in Figure 13.3. The sprockets come in sizes from 0.60-in. diameter to 35.8 in. and with teeth numbers ranging from 10 to 120. The *efficiency* of a toothed belt drive ranges from about 97 to 99%.

Figure 13.4 illustrates a portion of the timing-belt drive. The teeth are coated with nylon fabric. The tension member, usually steel wire, of a timing belt is positioned at the belt pitch line. The pitch length therefore is the same regardless of the backing thickness. Note that, as in the case of gears, the circular pitch p is the distance, measured on the pitch circle, from a point on the tooth to a corresponding point on an adjacent tooth. Since timing belts are toothed, they provide *advantages* over ordinary belts; for example, for a timing belt, no initial tension is necessary and a fixed center drive may be used at any slow or fast speed. The *disadvantages* are the cost of the belt, the necessity of grooving the sprocket, and the dynamic fluctuations generated at the belt-tooth meshing frequency.

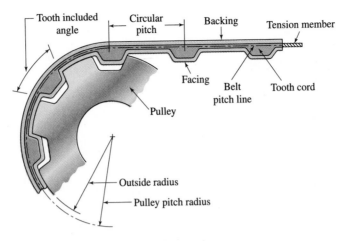

Figure 13.4   Timing-belt drive nomenclature.

13.3   BELT DRIVES

As noted previously, a belt drive transfers power from one shaft to another by using a belt and connecting pulleys on the shafts. Flat-belt drives produce very little noise and absorb more torsional vibration from the system than either V-belt or other drives. A flat-belt drive has an efficiency of around 98%, which is nearly the same as for a gear drive. A V-belt drive can transmit more power than a flat belt drive, as will be shown in the next section. However, the efficiency of a V-belt drive varies between 70 and 96% [3].

We present a conventional analysis that has long been used for the belt drives. Note that a number of theories describe the mechanics of the belt drives in more detailed mathematical form [1–5]. Figure 13.5a illustrates the usual belt drive, where the belt tension is such that the sag is visible when the belt is running. The friction force on the belt is taken to be uniform throughout the entire arc of contact. Due to friction of the rotation of the driver pulley, the *tight-side* tension is greater than the *slack-side* tension. Referring to the figure, the following basic relationships may be developed.

TRANSMITTED POWER

For belt drives, the torque on a pulley is given as follows

$$T = (F_1 - F_2)r \tag{13.1}$$

in which

r = pitch radius

F_1 = tension on the tight side

F_2 = tension on the slack side

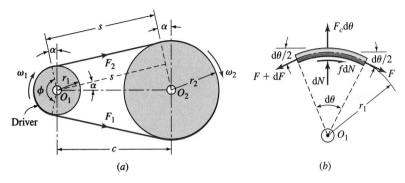

Figure 13.5 Belt drive: (a) forces in moving belt; (b) belt element on the verge of slipping on the small pulley.

We have $F_1 > F_2$. Note that the *pitch radius* is approximately measured from the center of the pulley to the neutral axis of the belt. The required *initial tension* F_i depends on the elastic characteristics of the belt. However, it is usually satisfactory to take

$$F_i = \frac{1}{2}(F_1 + F_2) \tag{13.2}$$

The transmitted power, through the use of Eq. (1.17),

$$\text{hp} = \frac{(F_1 - F_2)V}{33,000} = \frac{Tn}{63,000} \tag{13.3}$$

where

$$V = \frac{\pi dn}{12} \tag{13.4}$$

In the foregoing, we have

T = torque, lb · in.

V = belt velocity, fpm

n = speed of the pulley, rpm

d = pitch diameter of the pulley, in.

The speed ratio of the belt drive is given by

$$\frac{n_1}{n_2} = \frac{r_2}{r_1} \tag{13.5}$$

The numbers 1 and 2 refer to input and output, or small and large, pulleys, respectively.

CONTACT ANGLE

From the geometry of the drive (Figure 3.5a), the angle α is found to be

$$\sin \alpha = \frac{r_2 - r_1}{c} \tag{13.6}$$

The *contact angle* on the small pulley ϕ, or so-called *angle of wrap,* is therefore

$$\phi = \pi - 2\alpha \tag{13.7}$$

where

r_1 = pitch radius of the small pulley

r_2 = pitch radius of the large pulley

c = center distance

The capacity of the belt drive is determined by the value of ϕ. This angle is particularly critical with pulleys of greatly differing size and shorter center distances.

BELT LENGTH AND CENTER DISTANCE

The wrap angles on the small and large pulleys are $\pi - 2\alpha$ and $\pi + 2\alpha$, respectively. The distance between the beginning and end of contact, or span, $s = [c^2 - (r_2 - r_1)^2]^{1/2}$. The pitch length of the belt is obtained by the summation of the two arc lengths, $r_1(\pi - 2\alpha) + r_2(\pi + 2\alpha)$, with twice the span, $2s$. In so doing, we have

$$L = 2[c^2 - (r_2 - r_1)^2]^{1/2} + r_1(\pi - 2\alpha) + r_2(\pi + 2\alpha) \tag{13.8}$$

The span, the term in brackets on the right-hand side of this expression, can be approximated by two terms of a binomial expression and $\sin\alpha$ from Eq. (13.6) substituted for α. Then, we have the *belt pitch length* estimated by [6, 7]

$$L = 2c + \pi(r_1 + r_2) + \frac{1}{c}(r_2 - r_1)^2 \tag{13.9}$$

This gives the approximate *center distance:*

$$c = \frac{1}{4}\left[b + \sqrt{b^2 - 8(r_2 - r_1)^2}\right] \tag{13.10}$$

in which

$$b = L - \pi(r_2 + r_1) \tag{13.11}$$

The values of *actual pitch lengths* of some standard V belts are listed in Table 13.2. These values are substituted into Eq. (13.11) to obtain the actual center distances. Observe from the table that long center distances are not recommended for the V belt, since the excessive vibration of the slack side shortens the material life. In the case of flat belts, there is virtually no limit to the center distance, as noted in Table 13.1. Table 13.3 shows standard pitches with their letter identifications for toothed belts.

For nonstandard V belts, sometimes the center distance is given by the larger of

$$c = 3r_1 + r_2 \quad \text{or} \quad c = 2r_2 \tag{13.12}$$

This value of c may be used in Eq. (13.9) to estimate the belt pitch length L. It is important to note that belt drives should be designed with provision for center distance adjustments, unless an idler pulley is employed, since belts tend to stretch in application.

Table 13.2   Pitch lengths (in inches) of standard V belts

				Cross section			
A	B	C	A	B	C	D	E
27.3			113.3	113.8	114.9		
36.3	36.8		121.3	121.8	122.9	113.3	
43.3	43.8			145.8	146.9	147.3	
52.3	52.8	53.9		159.8	160.9	161.3	
61.3	61.8	62.9		174.8	175.9	173.3	
69.3	69.8	70.9		181.8	182.9	183.3	184.4
76.3	76.8	77.9		211.8	212.9	213.3	214.5
86.3	86.8	87.9		240.3	240.9	240.8	241.0
91.3	91.8	92.9		270.3	270.9	270.8	271.0
106.3	106.8	107.9		300.3	300.9	300.8	301.0

Table 13.3   Standard pitches of typical timing belts

Service	Designation	Pitch p (in.)
Extra light	XL	$\frac{1}{5}$
Light	L	$\frac{3}{8}$
Heavy	H	$\frac{1}{2}$
Extra heavy	XH	$\frac{7}{8}$
Double extra heavy	XXH	$1\frac{1}{4}$

Determination of the Geometric Quantities of a V-Belt Drive | **EXAMPLE 13.1**

A V belt is to operate on sheaves of 8-in. and 12-in. pitch diameters (Figure 13.5a). Calculate

   (a)   The center distance.

   (b)   The contact angle.

Assumption: B-section V belt is used, having the actual pitch length of 69.8 in. (Table 13.2).

Solution:

   (a)   Through the use of Eq. (13.11),

$$b = L - \pi(r_1 + r_2) = 69.8 - \pi(6 + 4) = 38.38 \text{ in.}$$

       Equation (13.10) is therefore

$$c = \frac{1}{4}\left[38.38 + \sqrt{38.38^2 - 8(6-4)^2}\right] = 19.09 \text{ in.}$$

   (b)   Applying Eq. (13.7),

$$\phi = \pi - 2\alpha = 180° - 2\sin^{-1}\left(\frac{150 - 100}{480}\right) = 168°$$

Figure 13.6   Weighed idler used to maintain slack-side tension.

MAINTAINING THE INITIAL TENSION OF THE BELT

A flat belt stretches over a period of time, and some initial tension is lost. Of course, the simplest solution is to have excessive initial tension. However, this would overload the shafts and bearings and shorten the belt life. A *self-tightening derive* that automatically maintains the desired tension is illustrated in Figure 13.6. Note that a third pulley is forced against the slack side of the belt on top by weights (as in the figure) or by a spring. The extra pulley that rotates freely is termed an *idler pulley* or simply *idler*. The idler is positioned so that it increases the contact angle ϕ and thus the capacity of the drive.

There are various other approaches of maintaining necessary belt tension. These include using a pivoted-overhung motor drive, changing the belt and pulley materials to increase the coefficient of friction, and increasing the center distance during operation by employing a drive with an adjustable center distance. We note that, because of the resistance to stretch of their interior tension cords, timing and V belts do not require frequent adjustment of initial tension.

13.4   BELT TENSION RELATIONSHIPS

The discussion of the preceding section pertains to belts that run slowly enough that centrifugal loading can be disregarded. We now develop a relationship between the tight- and slack-side tensions for the belt operating at maximum capacity. For this purpose, we first define the *centrifugal force* F_c, representing the inertia effect of the belt, in the following form

$$F_c = \frac{w}{g}V^2 \qquad\qquad \text{(13.13)}$$

where

$\quad w =$ belt weight per unit length

$\quad V =$ belt velocity

$\quad g =$ acceleration of gravity

In SI units, F_c is expressed in N, w in N/m, V in m/s, and g in 9.81 m/s^2; in U.S. customary units, F_c is measured in lb, w in lb/ft, V in fps, g in 32.2 ft/s^2.

FLAT OR ROUND BELT DRIVES

Reconsider the belt drive of Figure 13.5a, running at its largest capacity. The free-body diagram of a belt element on the verge of slipping on small pulley is depicted in Figure 13.5b. The element is under normal force dN, tension F, centrifugal force $F_c\, d\theta$, and friction force $f\, dN$, where f represents the coefficient of friction. Equilibrium of the forces in the horizontal direction is satisfied by

$$(F + dF)\cos\frac{d\theta}{2} - f\, dN - F\cos\frac{d\theta}{2} = 0$$

Simplifying, and noting for small angles $\cos(d\theta/2) = 1$, we have

$$dF = f\, dN \tag{13.14}$$

Likewise, equilibrium of the vertical forces gives

$$dN + F_c\, d\theta - (F + dF + F)\sin\frac{d\theta}{2} = 0$$

We can take $\sin(d\theta/2) = d\theta/2$, since $d\theta$ is a small angle, and neglect the higher-order term $dF \cdot d\theta$. In so doing and introducing the value of dN from Eq. (13.14), the preceding equation becomes

$$\frac{dF}{F - F_c} = f\, d\theta$$

The solution of this expression is obtained by integrating from minimum tension F_2 to maximum tension F_1 through the angle of contact ϕ of the belt (Figure 13.5a). Hence,

$$\int_{F_2}^{F_1} \frac{dF}{F_1 F_c} = \ln\left(\frac{F_1 - F_c}{F_2 - F_c}\right) = f\phi \tag{13.15}$$

This may be written in the following convenient form:

$$\frac{F_1 - F_c}{F_2 - F_c} = e^{f\phi} \tag{13.16}$$

We see from this relation that centrifugal force tends to reduce the angles of contact ϕ.

V- BELT DRIVES

Figure 13.7 illustrates how a V belt rides in the sheave groove: with contact on the sides and clearance at the bottom. Obviously, this "wedging action" increases the normal force on the belt element from dN (Figure 13.5b) to $dN/\sin\beta$. Following a procedure similar to that used in the preceding discussion, for a V belt drive, we therefore obtain

$$\frac{F_1 - F_c}{F_2 - F_c} = e^{f\phi/\sin\beta} \tag{13.17}$$

Figure 13.7   V belt in a sheave groove.

The quantity β is half the included angle of the V belt. It is interesting to observe that the (smaller) contact angle ϕ of the *driver pulley* leads to larger belt tension and hence is *critical*. Hence, design of belt drive is on the basis of small pulley geometry.

13.5   DESIGN OF V-BELT DRIVES

Of special concern is the design of belt drives for maximum tension and expected life or durability. In this section, attention is directed mainly to the former. Inasmuch as V-belt cross sections vary considerably, a rational design of belt drive usually relies on tables, charts, and guidelines given by the manufacturers Design data are based on theory as well as results of extreme testing [8–11].

During circuit around sheaves, the force on the belt varies considerably, as depicted in Figure 13.8. Note the additional equivalent *tension* forces F_{b1} and F_{b2} in the cord due to *bending* around the pulleys. The peak or total force F_B in the belt at point B is the sum of the tight-side tension and the equivalent tension force owing to the bending around the

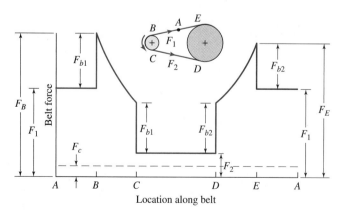

Figure 13.8   Forces in moving V belt.

small pulley. Therefore,

$$F_B = F_1 + F_{b1}$$

Likewise, at point E, the total force may be expressed as $F_E = F_2 + F_{b2}$. The tensions F_1 and F_2 are obtained from Eq. (13.3). For standard V belts, the bending and centrifugal forces are given by emprical formulas. The peak forces F_B and F_E are key to the design of V-belt drives. Durability design is somewhat complicated by the induced flexure stresses in the belt. Expected V-belt life refers to a certain number of peak forces a belt can sustain before failure by fatigue. The fatigue performance of a V-belt drive is best obtained by experimental tests.

We now develop an *approximate design equation* for maximum tension in the belt. Let the speed and power of the belt drive be prescribed. Then, from Eq. (13.3), the torque at the smaller pulley is

$$T = \frac{63,000 \, (\text{hp})}{n_1} \qquad (13.18)$$

in which n_1 represents the speed of the smaller pulley in rpm. The slack-side tension, as obtained from Eq. (13.1), is

$$F_2 = F_1 - \frac{T_1}{r_1} \qquad (13.19)$$

The quantity r_1 is the pitch radius of the smaller pulley. Substituting Eq. (13.19) into Eq. (13.17), after rearrangement, the tight-side tension may be expressed in the convenient form

$$F_1 = F_c + \left(\frac{\gamma}{\gamma - 1}\right)\frac{T_1}{r_1} \qquad (13.20)$$

where

$$\gamma = e^{f\phi/\sin\beta} \qquad (13.21)$$

Here the coefficient of friction f between rubber and dry steel is *usually* taken to be about 0.3.

In the case of a *flat belt* ($\beta = 90°$), $\sin\beta = 1$. We therefore see from Eq. (13.20) that, for a given maximum tension F_1, a V belt can transmit more torque (and power). Consequently, V belts are usually preferred over flat belts. Table 13.4 may be used to estimate

Table 13.4 Ratios of V-belt tensions for various contact angles

Contact angle, ϕ	F_1/F_2	Contact angle, ϕ	F_1/F_2	Contact angle, ϕ	F_1/F_2
180°	5.00	150°	3.82	120°	2.92
175°	4.78	145°	3.66	115°	2.80
170°	4.57	140°	3.50	110°	2.67
165°	4.37	135°	3.34	105°	2.56
160°	4.18	130°	3.20	100°	2.44
155°	4.00	125°	3.06	90°	2.24

V-belt tensions. Note that, if the sheave diameters are equal, the contact angle is 180° and hence tight-side tension F_1 is five times as great as the slack-side tension F_2. Practically, the minimum contact angle is 90°, for which $F_1 = 2.24F_2$.

Since V belts are usually made from reinforced rubber, the required belt strength is governed mainly by the tension; that is, the effect of the additional force in the belt due to bending around the pulley may be neglected. However, the tight-side tension should be multiplied by a service factor K_s. The *maximum tension* is then

$$F_{max} = K_s F_1 \qquad\qquad (13.22)$$

Service factors are listed by the manufacturers in great detail, usually based on the number of hours per day of overload, variations in loading the driving and driven shafts, starting overload, and variations in environmental conditions. Examples of the driven equipment in V-belt drives are blowers, pumps, compressors, fans, light-duty conveyors, dough mixers, generators, laundry machinery, machine tools, punches, presses, shears, printing machinery, bucket elevators, textile machinery, mills, and hoists. Typical service factors, relying on the characteristics of the driving and driven machinery, are given in Table 13.5. The horsepower should be multiplied by a service factor when selecting belt sizes. The design of a V-belt drive should use the largest possible pulleys. As the sheave sizes become smaller, the belt tension increases for a given horsepower output. Recommended pulley diameters for use with three electric motor sizes are given in Table 13.6.

Table 13.5   Service factors K_s for V-belt drives

	Driver (motor or engine)	
Driven machine	**Normal torque characteristic**	**High or nonuniform torque**
Uniform	1.0 to 1.2	1.1 to 1.3
Light shock	1.1 to 1.3	1.2 to 1.4
Medium shock	1.2 to 1.2	1.4 to 1.6
Heavy shock	1.3 to 1.5	1.5 to 1.8

Table 13.6   Recommended sheave pitch diameters (in.) for V-belt drives

	Motor speed, rpm				
Motor hp	**575**	**695**	**870**	**1160**	**1750**
$\frac{1}{2}$	2.5	2.5	2.5	—	—
$\frac{3}{4}$	3.0	2.5	2.5	2.5	—
1.00	3.0	3.0	2.5	2.5	2.25

Finally, we note that the shaft load at the pulley consists of torque T and force F_s. The latter is the vector sum of tensions F_1 and F_2. Referring to Figure 13.5a, the *shaft force* may therefore be expressed as

$$F_s = [(F_1 + F_2 \cos 2\alpha)^2 + (F_2 \sin 2\alpha)^2]^{1/2} \qquad \textbf{(13.23)}$$

The angle α is defined by Eq. (13.6). Equation (13.23) yields results approximately equal to the scalar sum $F_1 + F_2$ in most cases. The designer will find it useful to have the shaft load for determining the reactions at the shaft bearings.

Design Analysis of a V-Belt Drive | **EXAMPLE 13.2**

The capacity of a V-belt drive is to be 10 kW, based on a coefficient of friction of 0.3. Determine the required belt tensions and the maximum tension.

Given: A driver sheave has a radius of $r_1 = 100$ mm, a speed of $n_1 = 1800$ rpm, and a contact angle of $\phi = 153°$. The belt weighs 2.25 N/m and the included angle is 36°.

Assumptions: The driver is a normal torque motor and driven machine involves light shock load.

Solution: We have $\phi = 153° = 2.76$ rad and $\beta = 18°$. The tight-side tension is estimated from Eq. (13.20) as

$$F_1 = F_c + \left(\frac{\gamma}{\gamma - 1}\right)\frac{T_1}{r_1} \qquad \text{(a)}$$

where

$$F_c = \frac{w}{g}V^2 = \frac{2.25}{9.81}\left(\frac{\pi \times 0.2 \times 1800}{60}\right)^2 = 81.5 \text{ N}$$

$$\gamma = e^{f\phi/\sin\beta} = e^{0.3(2.67)/\sin 18°} = 13.36$$

$$T_1 = \frac{9549 \text{ kW}}{n} = \frac{9549(10)}{1800} = 53.05 \text{ N} \cdot \text{m}$$

Carrying the preceding values into Eq. (a), we have

$$F_1 = 81.5 + \left(\frac{13.36}{13.36 - 1}\right)\frac{53.05}{0.1} = 655 \text{ N}$$

Then, by Eq. (13.19), the slack-side tension is

$$F_2 = 655 - \frac{53.05}{0.1} = 124.5 \text{ N}$$

Based upon a service factor of 1.2 (Table 13.5) to F_1, Eq. (13.22) gives a maximum tensile force

$$F_{max} = 1.2(655) = 786 \text{ N}$$

applied to the belt.

The design of timing-belt drives is the same as that of flat- or V-belt drives. The manufacturers provide detailed information on sizes and strengths. The following case study illustrates an application.

Case Study 13-1 | BELT DESIGN OF A HIGH-SPEED CUTTING MACHINE

A high-speed knife cutter assembly for flexible materials (including PVC and other plastics) is shown in Figure 13.9. The unit is compact and designed for bench-top installation. The drive wheels are part of the automatic feeding mechanism (not shown in the figure). These wheels drive the material through the feed tube to the cutting wheel. The feed mechanism includes a compression spring for smooth operation. The variable cut lengths and rates (as high as 1000 per minute) are accomplished by changing the number of blades in the rotary cutting wheel and changing the reduction drive ratio between the motor and shaft on which cutter wheel is keyed. Determine

(a)   The belt length.

(b)   The maximum center distance.

(c)   The maximum belt tension.

Design Requirements: The center distance between the motor (driver) pulley and driven pulley should not exceed $c = 17$ in. A belt coefficient of friction of $f = 1.0$ is used.

Given: A 2-hp, $n_1 = 1800$ rpm, AC motor is used. The belt weighs $w = 0.007$ lb/in. The driver pulley radius, $r_1 = 1\frac{1}{4}$ in. Driven pulley radius, $r_2 = 2\frac{1}{4}$ in.

Assumptions: The driver is a normal torque motor. The cutter, and hence the driven shaft, resists heavy shock loads. The machine cuts uniform lengths of flexible materials of cross sections up to 2 in. in diameter. Operation is fully automatic, requiring minimal operator involvement.

Figure 13.9   A compact high-speed cutting machine.

Case Study (CONCLUDED)

Solution: See Figure 13.9 and Table 13.5.

(a) The appropriate belt pitch length is determined using Eq. (13.9),

$$L = 2c + \pi(r_1 + r_2) + \frac{1}{c}(r_2 - r_1)^2$$

Substitution of the given data yields

$$L = 2(17) + \pi(1.25 + 2.25) + \frac{1}{17}(2.25 - 1.25)^2$$

$$= 45.05 \text{ in.}$$

Comment: A standard toothed or timing belt with maximum length of 45 in. is selected.

(b) An estimate of the center distance is given by Eq. (13.10):

$$c = \frac{1}{4}\left[b + \sqrt{b^2 - 8(r_2 - r_1)^2}\right] \qquad \textbf{(b)}$$

where

$$b = L - \pi(r_2 + r_1) \qquad \textbf{(c)}$$

Carrying the given numerical values into Eq. (c), we have

$$b = 45 - \pi(2.25 + 1.25) = 34 \text{ in.}$$

Equation (b) is then

$$c = \frac{1}{4}\left[34 + \sqrt{(34)^2 - 8(2.25 - 1.25)^2}\right] = 16.97 \text{ in.}$$

Comment: The requirement that $c < 17$ in. is satisfied.

(c) The contact angle ϕ, from Eqs. (13.7) and (13.6), equals

$$\phi = \pi - 2\sin^{-1}\left(\frac{r_2 - r_1}{c}\right)$$

$$= \pi - 2\sin^{-1}\left(\frac{2.25 - 1.25}{16.97}\right) = 3.024 \text{ rad}$$

The tight-side tension in the belt is calculated, through the use of Eq. (3.20) in which, for a toothed (or flat) belt, $\sin\beta = 1$ and

$$F_c = \frac{w}{g}V^2$$

$$= \frac{0.007}{386.4}\left[\frac{\pi(2.5)1800}{60}\right]^2 = 1.006 \text{ lb}$$

$$\gamma = e^{f\phi} = e^{(1)(3.024)} = 20.57$$

$$T_1 = \frac{33,000 \text{ hp}}{n_1}$$

$$= \frac{33,000(2)}{1800} = 36.67 \text{ lb} \cdot \text{in.}$$

Therefore

$$F_1 = 1.006 + \left(\frac{20.57}{20.57 - 1}\right)\frac{36.67}{1.25} = 31.84 \text{ lb}$$

It follows that

$$F_2 = F_1 - \frac{T_1}{r_1}$$

$$= 31.34 - \frac{36.67}{1.25} = 2.50 \text{ lb}$$

The service factor, from Table 13.5, $K_s = 1.4$. The maximum belt tensile load is then obtained by Eq. (13.22) as

$$F_{max} = K_s F_1$$

$$= 1.4(31.84) = 44.58 \text{ lb}$$

Comment: Recall from Section 13.2 that toothed belts can provide safe operation at speeds up to at least 16,000 fpm. This is well above the belt velocity, $V = \pi(2.5)(1800)/12 = 1178$ fpm, of the cutting machine.

13.6   CHAIN DRIVES

As pointed out in Section 13.1, chains are used for power transmission between parallel shafts. They can be employed for high loads and where precise speed ratios must be sustained. While precise location and alignment tolerances are not required, as with gear drives, the best performance can be expected when input and output sprockets lie in the same vertical plane [12, 13]. Chain drives have shorter service lives than typical gear drives. They present no fire hazard and are unaffected by relatively high temperatures. Sometimes, an adjustable *idler* sprocket (toothed wheel) is placed on the outside of the chain near the driving sprocket to remove sag on loose side and increase the number of teeth in contact. The only maintenance required after careful alignment of the elements is proper lubrication. Usually, a chain should have a sheet of metal casing for protection from atmospheric dust and to facilitate lubrication. Chains should be washed regularly in kerosene and then soaked in oil.

The *speed ratio* of a chain drive is expressed by the equation

$$\frac{n_2}{n_1} = \frac{N_1}{N_2}$$

In the preceding, we have

n_2 = output speed

n_1 = input speed

N_2 = number of teeth in the output sprocket

N_1 = number of teeth in the input sprocket

An odd number of teeth on the driving sprocket (17, 19, 21, ...) is recommended, typically 17 and 25. Usually, an odd number of sprocket teeth causes each small sprocket tooth to contact many or all chain links, minimizing wear. The larger sprocket is ordinarily limited to about 120 teeth.

Center distance c should be between a value that just allows the sprockets to clear. It is $c = 2(r_1 + r_2)$; for smaller speed ratios, $n_1/n_2 < 3$. Here, r_1 and r_2 refer to pitch radii of the input and output sprockets, respectively. When longer chains are used, idlers may be required on the slack side of the chain. For the cases in which speed ratios $n_1/n_2 \geq 3$, the center distance should be $c = 2(r_2 - r_1)$. Having a tentative center distance c between shafts selected, *chain length L* may be estimated applying Eq. (13.7). Finally, the center distance is recalculated through the use of Eq. (13.8). The angle of contact for the chain drive is given by Eq. (13.6). Note that, for a small sprocket, the angle of contact should not be less than 120°.

Chain pitch p represents the length of an individual link from pin center to pin center. The pitch radius of a sprocket with N teeth may be defined as

$$r = \frac{Np}{2\pi}$$

An even number of pitches in the chain is preferred to avoid a special link. *Chain velocity V* is defined as the number of feet coming off the sprocket in unit time. Therefore,

$$V = \frac{Npn}{12} \tag{13.24}$$

The *tensile force* that a chain transmits F_1 may be obtained from Eq. (1.17) in the form

$$F_1 = \frac{33,000\,(hp)}{V} = \frac{396,000\,(hp)}{pNn} \qquad \textbf{(13.25)}$$

In Eqs. (13.24) and (13.25), the chain pitch p is measured in in. and the sprocket speed n is in revolutions per minute, rpm. The total force or tension in the chain includes transmitted force F_1, a centrifugal force F_c, a small catenary tension, and a force from link action. There are also impact forces when link plates engage sprocket teeth, discussed in the next section.

13.7   COMMON CHAIN TYPES

There are various types of power transmission chains, however, roller chains are the most widely employed. Types of driven equipment with roller chain drives include bakery machinery, blowers and fans, boat propellers, compressors, conveyors, clay-working machinery, crushers, elevators, feeders, food processors, dryers, machine tools, mills, pumps, pulp grinders, printing presses, carding machinery, and wood working machinery. In some of these applications as well as in cranes, hoists, generators, ice machines, and a variety of laundry machinery, inverted chains are also used. Both common chain types are used on sprockets, well suited for heavy loads, and have high efficiency.

ROLLER CHAINS

Of its diverse applications, the most familiar is the roller chain drive on a bicycle. A roller chain is generally made of hardened steel and sprockets of steel or cast iron. Nevertheless, stainless steel and bronze chains are obtainable where corrosion resistance is needed. The geometry of a roller chain is shown in Figure 13.10. The rollers rotate in bushings that are press fitted to the inner link plates. The pins are prevented from turning by the outer links' press-fit assembly. Roller chains have been standardized according to size by the American National Standards Institute [14]. The characteristics of representative standard sizes are

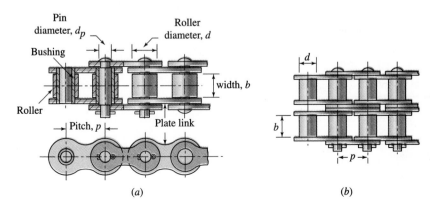

Figure 13.10   Portion of a roller chain: (a) single strand; (b) double strand.

Table 13.7   Sizes and strengths of standard roller chains

| Chain no. | Pitch, p, in. | Roller | | Pin diameter, d_p, in. | Link plate thickness, t | Minimum ultimate strength lb |
		Diameter d, in.	Width, b, in.			
25	$\frac{1}{4}$	0.130	$\frac{1}{8}$	0.0905	0.030	780
35	$\frac{3}{8}$	0.200	$\frac{3}{16}$	0.141	0.050	1760
41	$\frac{1}{2}$	0.306	$\frac{1}{4}$	0.141	0.050	1500
40	$\frac{1}{2}$	$\frac{5}{16}$	$\frac{5}{16}$	0.156	0.060	3125
50	$\frac{5}{8}$	0.400	$\frac{3}{8}$	0.200	0.080	4480
60	$\frac{3}{4}$	$\frac{15}{32}$	$\frac{1}{2}$	0.234	0.094	7030
80	1	$\frac{5}{8}$	$\frac{5}{8}$	0.312	0.125	12,500
100	$1\frac{1}{4}$	$\frac{3}{4}$	$\frac{3}{4}$	0.375	0.156	19,530
120	$1\frac{1}{2}$	$\frac{7}{8}$	1	0.437	0.187	28,125
140	$1\frac{3}{4}$	1	1	0.500	0.219	38,280
160	2	$1\frac{1}{8}$	$1\frac{1}{4}$	0.562	0.250	50,000
180	$2\frac{1}{4}$	$1\frac{13}{32}$	$1\frac{13}{32}$	0.687	0.2811	63,280
200	$2\frac{1}{2}$	$1\frac{9}{16}$	$1\frac{1}{2}$	0.781	0.312	78,125
240	3	$1\frac{7}{8}$	$1\frac{7}{8}$	0.937	0.375	112,500

| SOURCE: ANSI/ASME Standard B29.1M-1993.

listed in Table 13.7. These chains are manufactured in single (Figure 13.10a), double (Figure 13.10b), triple, and quadruple strands. Clearly, the use of multistrands increases the load capacity of a chain and sprocket system.

Chordal Action

The instantaneous chain velocity varies from the average velocity given by Eq. (13.24). Consider a sprocket running at constant speed n and driving a roller chain in a counterclockwise direction, as illustrated in Figure 13.11. A chord representing the link between

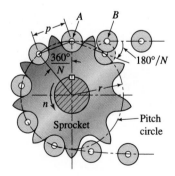

Figure 13.11   A roller chain and sprocket engagement.

centers has the pitch length of p. The chord subtends the pitch angle of the sprocket, $360°/N$. From the geometry, $\sin(180°/2) = (p/2)/r$. We therefore have

$$r = \frac{p}{2\sin(180°/N)}$$

where

r = radius of the sprocket

p = chain pitch

N = number of teeth in the sprocket

The angle $180°/N$ through which the link (AB) swings until the roller (B) seats on the sprocket is referred to as the *angle of articulation.* When the sprocket is in the position shown in the figure, the chain velocity is $2\pi rn$. This velocity changes to $2\pi r_1 n$, where $r_1 = r\cos(180°/N)$, after the sprocket is turned the angle of articulation. The change of velocity ΔV is called the *chordal action:*

$$\Delta V = 2\pi rn[1 - \cos(180°/N)]$$

Rotation of the link through the angle of articulation causes impact among the rollers and the sprocket teeth as well as wear in the chain joint. The movement of the link up and down with rotation through the articulation angle develops an uneven chain exit velocity. Consequently, the driven shaft of a roller chain drive may be given a pulsating motion, particularly at high-speed operations. The angle of articulation (hence the impact and chordal action) should be reduced as much as practicable, by increasing the number of sprocket teeth [1, 15].

THE POWER CAPACITY OF ROLLER CHAINS

Equations and tables for roller-chain power capacity and selection were developed through the American Chain Association (ACA), as the result of many years of laboratory testing and field observation. Rated horsepower capacities are usually given in tabular form for each type of single-strand chain corresponding to a life expectancy of 15 khr for a variety of sprocket speeds. Table 13.8 is an example for ANSI No. 60 roller chains. Listed in Table 13.9 are the service factors that account for the abruptness associated with load application. Table 13.10 shows multiple-strand factors.

At lower speeds, the power capacity of roller chains is based upon the fatigue strength of the link plate. On the other hand, at higher speeds, the power relies on roller and bushing impact life. At extremely high speeds, the power capacity is on the basis of the galling or welding between pin and bushings. The *design power capacity* may be expressed as follows:

$$H_d = H_r K_1 K_2 \qquad (13.26)$$

where

H_r = horsepower rating (from Table 13.8)

K_1 = service factor (from Table 13.9)

K_2 = multiple-strand factor (from Table 13.10)

Usually a medium or light mineral oil is used as the lubricant. We observe from Table 13.8 that the proper lubrication of roller chains is essential to their performance. As

Table 13.8 Rated horsepower capacity (H_r) for a standard single strand roller chain—ANSI No. 60, $\frac{3}{4}$-in. pitch

No. of teeth in small sprocket	Revolutions per minute—small sprocket																								
	10	25	50	100	150	200	300	400	500	600	700	800	900	1000	1100	1200	1400	1600	1800	2000	2500	3000	3500	4000	4500
11	0.18	0.41	0.77	1.44	2.07	2.69	3.87	5.02	6.13	7.23	8.30	9.36	10.4	11.4	12.5	11.9	9.41	7.70	6.45	5.51	3.94	3.00	2.38	1.95	1.63
12	0.20	0.45	0.85	1.58	2.28	2.95	4.25	5.51	6.74	7.94	9.12	10.3	11.4	12.6	13.7	13.5	10.7	8.77	7.35	6.28	4.49	3.42	2.71	2.22	1.86
13	0.22	0.50	0.92	1.73	2.49	3.22	4.64	6.01	7.34	8.65	9.94	11.2	12.5	13.7	14.9	15.2	12.1	9.89	8.29	7.08	5.06	3.85	3.06	2.50	0
14	0.24	0.54	1.00	1.87	2.69	3.49	5.02	6.51	7.96	9.37	10.8	12.1	13.5	14.8	16.2	17.0	13.5	11.1	9.26	7.91	5.66	4.31	3.42	2.80	0
15	0.25	0.58	1.08	2.01	2.90	3.76	5.41	7.01	8.57	10.1	11.6	13.1	14.5	16.0	17.4	18.8	15.0	12.3	10.3	8.77	6.28	4.77	3.79	3.10	0
16	0.27	0.62	1.16	2.16	3.11	4.03	5.80	7.52	9.19	10.8	12.4	14.0	15.6	17.1	18.7	20.2	16.5	13.5	11.3	9.66	6.91	5.26	4.17	3.42	0
17	0.29	0.66	1.24	2.31	3.32	4.30	6.20	8.03	9.81	11.6	13.3	15.0	16.7	18.3	19.9	21.6	18.1	14.8	12.4	10.6	7.57	5.76	4.57	3.74	0
18	0.31	0.70	1.31	2.45	3.53	4.58	6.59	8.54	10.4	12.3	14.1	15.9	17.7	19.5	21.2	22.9	19.7	16.1	13.5	11.5	8.25	6.28	4.98	4.08	0
19	0.33	0.75	1.39	2.60	3.74	4.85	6.99	9.05	11.1	13.0	15.0	16.9	18.8	20.6	22.5	24.3	21.4	17.5	14.6	12.5	8.95	6.81	5.40	4.42	0
20	0.35	0.79	1.47	2.75	3.96	5.13	7.38	9.57	11.7	13.8	15.8	17.9	19.8	21.8	23.8	25.7	23.1	18.9	15.8	13.5	9.66	7.35	5.83	0	
21	0.36	0.83	1.55	2.90	4.17	5.40	7.78	10.1	12.3	14.5	16.7	18.8	20.9	23.0	25.1	27.1	24.8	20.3	17.0	14.5	10.4	7.91	6.28	0	
22	0.38	0.87	1.63	3.05	4.39	5.68	8.19	10.6	13.0	15.3	17.5	19.8	22.0	24.2	26.4	28.5	26.6	21.8	18.2	15.6	11.1	8.48	6.73	0	
23	0.40	0.92	1.71	3.19	4.60	5.96	8.59	11.1	13.6	16.0	18.4	20.8	23.1	25.4	27.7	29.9	28.4	23.3	19.5	16.7	11.9	9.07	7.19	0	
24	0.42	0.96	1.79	3.35	4.82	6.24	8.99	11.6	14.2	16.8	19.3	21.7	24.2	26.6	29.0	31.3	30.3	24.8	20.8	17.8	12.7	9.66	7.67	0	
25	0.44	1.00	1.87	3.50	5.04	6.52	9.40	12.2	14.9	17.5	20.1	22.7	25.3	27.8	30.3	32.7	32.2	26.4	22.1	18.9	13.5	10.3	8.15	0	
26	0.46	1.05	1.95	3.65	5.25	6.81	9.80	12.7	15.5	18.3	21.0	23.7	26.4	29.0	31.6	34.1	34.2	28.0	23.4	20.0	14.3	10.9	8.65	0	
28	0.50	1.13	2.12	3.95	5.69	7.37	10.6	13.8	16.8	19.8	22.8	25.7	28.5	31.4	34.2	37.0	38.2	31.3	26.2	22.4	16.0	12.2	0		
30	0.54	1.22	2.28	4.26	6.13	7.94	11.4	14.8	18.1	21.4	24.5	27.7	30.8	33.8	36.8	39.8	42.4	34.7	29.1	24.8	17.8	13.5	0		
32	0.57	1.31	2.45	4.56	6.57	8.52	12.3	15.9	19.4	22.9	26.3	29.7	33.0	36.3	39.5	42.7	46.7	38.2	32.0	27.3	19.6	14.9	0		
35	0.63	1.44	2.69	5.03	7.24	9.38	13.5	17.5	21.4	25.2	29.0	32.7	36.3	39.9	43.5	47.1	53.4	43.7	36.6	31.3	22.4	17.0	0		
40	0.73	1.67	3.11	5.81	8.37	10.8	15.6	20.2	24.7	29.1	33.5	37.7	42.0	46.1	50.3	54.4	62.5	53.4	44.7	38.2	27.3	0			
45	0.83	1.89	3.53	6.60	9.50	12.3	17.7	23.0	28.1	33.1	38.0	42.9	47.7	52.4	57.1	61.7	70.9	63.7	53.4	45.6	32.6	0			

Type A   Type B   Type C

1   SOURCE: ANSI/ASME Standard B29.1M-1993.

Table 13.9 Service factors (K_1) for single strand roller chains

Type of driven load	Type of input power		
	I.C. engine hydraulic drive	Electric motor or turbine	I.C. engine mechanical drive
Smooth	1.0	1.0	1.2
Moderate shock	1.2	1.3	1.4
Heavy shock	1.4	1.5	1.7

| SOURCE: ANSI/ASME Standard B29.1M-1993.

Table 13.10 Multiple-strand factors (K_2) for roller chains

Number of strands	Multiple-strand factor
2	1.7
3	2.5
4	3.3

| SOURCE: ANSI/ASME Standard B29.1M-1993.

speed increases, this requirement becomes more rigorous. The following *types of lubrication* systems are satisfactory:

Type A. Manual or drip lubrication, oil applied periodically with brush or spout can.

Type B. Bath or disk lubrication, oil level is maintained in the casing at predetermined height.

Type C. Oil stream lubrication, oil supplied by circulating pump inside chain loop or lower span.

Note that the limiting revolution per minute (rpm) for each lubrication type is read from the column to the left of the boundary line shown in the table. The chain manufacturer should be consulted for drives that exceed the speed and power requirements for the preceding lubrication kinds.

Analysis of a Roller Chain Drive | **EXAMPLE 13.3**

A three-strand ANSI No. 60, $\frac{3}{4}$-in. pitch, rolling chain transmits power from a N_1-tooth driver sprocket operating at n_1 rpm. Determine

(a) The design power capacity.

(b) The tension in the chain.

(c) The factor of safety n of the chain on the basis of ultimate strength.

Given: $N_1 = 19$, $p = \frac{3}{4}$ in., $n_1 = 1000$ rpm

Assumptions: The input power type is an internal combustion (I.C.) engine, mechanical drive. The type of driven load is moderate shock. With the exception of the tensile force, all forces are taken to be negligible.

Solution: See Tables 13.7 through 13.10.

(a) For driver sprocket $H_r = 20.6$ hp, type B lubrication is required (Table 13.8). Service factor $K_1 = 1.4$ (Table 13.9). From Table 13.10 for three strands, $K_2 = 2.5$. Applying Eq. (13.26), we have

$$H_d = 20.6(1.4)(2.5) = 72.1 \text{ hp}$$

(b) The average chain velocity, by Eq. (13.24), is

$$V_1 = \frac{19(0.75)(1000)}{12} = 1187.5 \text{ fpm}$$

Equation (13.25) results in

$$F_1 = \frac{33,000(72.1)}{1187.5} = 2.0 \text{ kips}$$

(c) The ultimate strength, for a single-strand chain, is 7.03 kips (Table 13.7). The allowable load for a three-strand chain is then $F_{all} = 7.03(3) = 21.09$ kips. Hence, the factor of safety is

$$n = \frac{F_{all}}{F_1} = \frac{21.09}{2.0} = 10.5$$

Comment: The analysis is based on 15 khr of chain life, since other estimates are not available.

INVERTED-TOOTH CHAINS

The inverted-tooth chain, also referred to as the *silent chain,* is composed of a series of toothed link plates that are pin connected to allow articulation. The chain pitch is defined in Figure 13.12. An inverted-tooth chain ordinarily has guide links on the sides or in the center to keep it on the sprocket. To increase the chain life, different details of joint

Figure 13.12 Portion of an inverted tooth or "silent" chain.

construction are used. Enclosures for the chain are customarily needed. Therefore, silent chains are more expensive than roller chains. Usually, when properly lubricated, at full load, drive efficiency is as high as 99%.

As the name suggests, these chains are quieter than roller chains. They may be run at higher speeds, because there is minor impact force when the chain link engages the sprocket. The inverted-tooth chain has a smooth flat surface, which can be conveniently used for conveying. Power capacities of silent chains are listed in tables analogous to those for roller chains. However, these chains reach maximum power at maximum speed, while roller chains reach highest power far below their maximum speed. Most of the remarks in the foregoing paragraphs relate as well to inverted-tooth chains and sprockets, which are also standardized by ANSI [16]. Regular pitches vary between $\frac{3}{8}$ and 2 in. Sprockets may have 21 to 150 teeth. Center distance adjustment is periodically needed to compensate for wear.

Part B   High-Friction Devices

Our concern was with two flexible elements in the preceding sections. In Part B, we turn to clutches and brakes, which are high-friction devices. We consider the most commonly used types, having two or more surfaces pressed together with a normal force to produce a friction torque. *Performance analysis* of clutches and brakes involves the determination of the actuating force, torque transmitted, energy absorption, and temperature rise. The transmitted torque is associated with the actuating force, the coefficient of friction, and the geometry of the device. The temperature rise is related to energy absorbed in the form of friction heat during braking or clutching.

13.8   MATERIALS FOR BRAKES AND CLUTCHES

The materials used for clutches and brakes are of two types, those used for the disk or drums and those used for friction materials or linings. In the design of these devices, the selection of the friction materials is critical. Most linings are attached to the disks or drums by either riveting or bonding. The former has the advantage of low cost and relative ease in installation. The latter affords more friction area and greater effective thickness but is more expensive.

Drums are ordinarily made of *cast iron* with some alloying material added. Materials like stainless steel, Monel, and so on are used when good heat conduction is important. Many railroad brakes employ cast iron shoes, which are bearing on cast iron wheels or drums. Friction, thermal conductivity, resistance to wear, and thermal fatigue characteristics of drums are very important. They must have a sufficiently smooth surface finish to minimize wear of the lining.

Linings are often made of molded, woven, sintered, or solid materials. They are composed mainly of reinforcing fibers (to render strength and ability to resist high temperatures), metal particles (to obtain wear resistance and higher coefficient of friction), and bonding materials. The binder is ordinarily a thermosetting resin or rubber. In addition, a friction material should have good heat conductivity and impenetrability to moisture.

Because of health hazards associated with asbestos, it is banned on all current production, and alternative reinforcing materials are now in use.

A *woven* (cotton) lining is made as a fabric belt impregnated with metal particles and polymerized. These belts have flexibility, as required by band brakes. *Molded* linings typically use polymeric resin to bind a variety of powdered fillers or fibrous materials. Brass or zinc chips are sometimes added to improve heat conduction and wear resistance and reduce scoring of the drum and disks. These materials are the most commonly used in drum brakes and the least costly. *Sintered metal* pads are made of a mixture of copper or iron particles having friction modifiers molded then heated to blend the material. They are the most costly but also the best suited for heavy-duty applications. Sintered metal-ceramic friction pads are similar, except that ceramic particles are added prior to sintering.

For sufficient performance of the brake or clutch, the requirements imposed on friction materials include the following: a high coefficient of friction having small variation on changes in pressure, velocity, and temperature; resistance to wear, seizing, and the tendency to grab; heat and thermal fatigue resistance. Tables 13.11 and 13.12 list

Table 13.11   Properties of common brake and clutch friction materials, operating dry

Material*	Dynamic coefficient of friction, f	Maximum pressure p_{max}		Maximum drum temperature	
		MPa	psi	°C	°F
Molded	0.25–0.45	1.03–2.07	150–300	204–260	400–500
Woven	0.25–0.45	0.35–0.69	50–100	204–260	400–500
Sintered metal	0.15–0.45	1.03–2.07	150–300	232–677	450–1250
Cork	0.30–0.50	0.06–0.10	8–14	82	180
Wood	0.20–0.25	0.35–0.63	50–90	93	200
Cast iron, hard steel	0.15–0.25	0.70–0.17	100–250	260	500

SOURCE: [1, 13].
*When rubbing against smooth cast iron or steel.

Table 13.12   Values of friction coefficients of common brake/clutch friction materials, operating in oil

Material*	Dynamic coefficient of friction, f
Molded	0.06–0.09
Woven	0.08–0.10
Sintered metal	0.05–0.08
Cork	0.15–0.25
Wood	0.12–0.16
Cast iron, hard steel	0.03–0.06

SOURCE: [1, 13].
*When rubbing against smooth cast iron or steel.

approximate data related to allowable pressures and the coefficient of friction for a few linings. For longer life, the lower values of the maximum pressure given should be used. As seen from the tables, the coefficients of friction are much smaller in oil than under dry friction, as expected. For more accurate information, consult the manufacturer or obtain test data [17–21].

13.9 INTERNAL EXPANDING DRUM CLUTCHES AND BRAKES

Drum or rim clutches and brakes consist of three parts: the mating frictional surfaces, the means of transmitting the torque to and from the surfaces, and the actuating mechanism. Often, they are classified according to the operating mechanism. Figure 13.13 shows an internal expanding drum centrifugal clutch that engages automatically when the shaft speed exceeds a certain magnitude. The friction material is placed around the outer surface of the drum to engage the clutch.

The centrifugal clutch is in widespread use for automatic operation, such as to couple an engine to the drive train. When the engine speed increases, it automatically engages the clutch. This is particularly practical for electric motor drives, where during starting, the driven machine comes up to speed without shock. Used in chain saws for the same purpose, centrifugal clutches serve as an overload release that slips to allow the motor to continue running when the chain jams in the wood.

Magnetic, hydraulic, and pneumatic drum clutches are also useful in drives with complex loading cycles and in automatic machinery or robots. The expanding drum clutch is frequently used in textile machinery, excavators, and machine tools. Inasmuch as the analysis of drum clutch is similar to that for drum brakes, to be taken up in Sections 13.13 and 13.14, we do not discuss them at this time.

Figure 13.13 An internal expanding centrifugal-acting drum clutch. (Courtesy of Hillard Corporation.)

13.10   DISK CLUTCHES AND BRAKES

Basic disk clutches and brakes are considered in this section. The former transmits torque from the input to the output shaft by the frictional force developed between the two disks or plates when they are pressed together. The latter is basically the same device, but one of the shafts is fixed. One of the friction surfaces of the clutch or brake is typically metal (cast iron or steel) and the other is usually a friction material or lining. Magnetic, hydraulic, and pneumatic operating mechanisms are also available in disk, cone, and multiple disk clutches and brakes.

Uniform pressure and *uniform wear* are two basic conditions or *assumptions* that may occur at interface of the friction surfaces. The designer must decide on which assumption more closely approximates the particular clutch or brake being analyzed. The uniform-wear assumption leads to lower calculated clutch or brake capacity than the assumption of uniform pressure, as observed in Example 13.4. Hence, disk clutches and brakes are ordinarily designed on the basis of uniform wear that gives conservative results. The preceding analysis can be used as a guide. Design considerations also include the characteristics of the machine whose brakes or clutches are to be a part and the environment in which the machine operates.

DISK CLUTCHES

Friction clutches reduce shock by slipping during the engagement period. The single-plate or *disk clutch,* shown schematically in Figure 13.14, is employed in both automotive and industrial service. These devices are larger in diameter to give adequate torque capacity. Note that, in an automotive-type disk clutch, the input disk (flywheel) rotates with the crankshaft. The hub of the clutch output disk is spline-connected to the transmission shaft. Clearly, the device is disengaged by depressing the clutch pedal. The torque that can be transferred depends on the frictional force developed between the disks, coefficient of friction, and the geometry of the clutch. The axial force typically is quite large and can be applied mechanically (by spring, as in the figure), hydraulically, or electromagnetically. An advantage of the disk clutch over the drum clutch is the absence of centrifugal effects and efficient heat dissipation surfaces.

Figure 13.14   Basic disk clutch, shown in a disengaged position (for a brake, the "output" member is stationary).

Figure 13.15   Half-section view of a multiple-disk clutch, hydraulically operated.

Multiple-disk clutches can have the friction lining on facing sides of a number (as many as 24) of alternative driving and driven disks or plates. The disks are usually thin (about 1.5 mm) with small diameters. Thus, additional torque capacity with only a small increase in axial length is obtained. When the clutch is disengaged, the alternate disks are free to slide axially to disjoin. After the clutch is engaged, the disks are clamped tightly together to provide a *number of active friction surfaces N*. Disk clutches can be designed to operate either "dry" or "wet" with oil. The advantages of the latter are reduced wear, smoother action, and lower operating temperatures. As a result, most multiple-disk clutches operate either immersed in oil or in a spray. Multiple disk clutches are compact and suitable for high-speed operations in various machinery. They are often operated automatically by either air or hydraulic cylinders (e.g., in automotive automatic transmissions).

Figure 13.15 shows a hydraulically operated clutch. In this device, the axial piston motion and force are produced by oil in an annular chamber, which is connected by an oil passage to an external pressure source. We see from the figure that, with the housing keyed to the input shaft, two disks and the piston are internally splined and an end plate is fastened. These are the driving disks. The six driven disks are externally splined to the housing keyed to the output shaft.

We develop the torque capacity equations for a single pair of friction surfaces, as in Figure 13.14. However, they can be modified for multiple disks by merely multiplying the values obtained by the number of active surfaces N. For example, $N = 6$ in the device depicted in Figure 13.15.

Uniform Wear

When the clutch disks are sufficiently rigid, it can be assumed that wear over the lining is uniform. This condition applies after an initial wearing-in has occurred. The uniform wear rate, which is taken to be proportional to the product of pressure and velocity pV is constant. Note that the velocity at any point on the clutch surface varies with the radius and the angular velocity. Therefore, assuming a uniform angular velocity,

$$pr = C$$

where p = pressure, r = radius, and C = constant. This equation indicates that the *maximum pressure* p_{max} takes place at the inside radius $r = d/2$ (Figure 13.14). Hence,

$$pr = C = p_{max}\frac{d}{2} \qquad (13.27)$$

The total normal force that must be exerted by the actuating spring in Figure 13.14 is found by multiplying the area $2\pi r\, dr$ by the pressure $p = p_{max}d/2r$ and integrating over the friction surface. Hence, the *actuating force* F_a required equals

$$F_a = \int_{d/2}^{D/2} \pi p_{max}\, d\, dr = \frac{1}{2}\pi p_{max}\, d(D-d) \qquad (13.28)$$

The friction torque or *torque capacity* is obtained by multiplying the force on the element by the coefficient of friction f and the radius and integrating over the area. It follows that

$$T = \int_{d/2}^{D/2} (\pi p_{max}\, d) f r\, dr = \frac{1}{8}\pi f p_{max}\, d(D^2 - d^2) \qquad (13.29)$$

An expression relating the torque capacity to actuating force is obtained by solving Eq. (13.28) for p_{max} and inserting its value into Eq. (13.29). In so doing, we have

$$T = \frac{1}{4}F_a f(D+d) = F_a f r_{avg} \qquad (13.30)$$

where r_{avg} is the average disk radius. Note the simple physical interpretation of this equation. As previously pointed out, for a multiple-disk clutch, Eqs. (13.29) and (13.30) must be multiplied by the number of active surfaces N.

In the design of clutches, the ratio of inside to outside diameters is an important parameter. It can be verified, applying Eq. (13.29), that the maximum torque capacity for a prescribed outside diameter is attained when

$$d = \frac{D}{\sqrt{3}} = 0.577D \qquad (13.31)$$

Usually employed proportions vary between $d = 0.45D$ and $d = 0.80D$.

Uniform Pressure

If the clutch disks are relatively flexible, the pressure p_{max} can approach a uniform distribution over the entire lining surface. For this condition, the wear is not constant. Referring to Figure 13.14, we readily obtain the actuating force and the torque capacity as follows:

$$F_a = \int_{d/2}^{D/2} 2\pi p_{max} r\, dr = \frac{1}{4}\pi p_{max}(D^2 - d^2) \qquad (13.32)$$

$$T = \int_{d/2}^{D/2} (2\pi p_{max} r) f r\, dr = \frac{1}{12}\pi f p_{max}(D^3 - d^3) \qquad (13.33)$$

These can be combined to yield the torque as a function of actuating force:

$$T = \frac{1}{3}F_a f \frac{D^3 - d^3}{D^2 - d^2}$$

(13.34)

The torque capacity for a multiple-disk clutch is obtained on multiplying Eqs. (13.33) and (13.34) by the number of active surfaces N.

Design of a Disk Clutch

EXAMPLE 13.4

A disk clutch with a single friction surface has an outer diameter D and inner diameter d (Figure 13.14). Determine the torque that can be transmitted and the actuating force required of the spring, on the basis of

   (a)   Uniform wear.

   (b)   Uniform pressure.

Given: $D = 500$ mm, $d = 200$ mm

Design Decisions: Molded friction material and a steel disk are used, having $f = 0.35$ and $p_{max} = 1.5$ MPa (see Table 13.11).

Solution:

   (a)   Through the use of Eq. (13.29), we have

$$T = \frac{1}{8}\pi(0.35)(1500)(0.2)(0.5^2 - 0.2^2) = 8.659 \text{ kN} \cdot \text{m}$$

      From Eq. (13.28),

$$F_a = \frac{1}{2}\pi(1500)(0.2)(0.5 - 0.2) = 141.4 \text{ kN}$$

   (b)   Applying Eq. (13.33),

$$T = \frac{1}{12}\pi(0.35)(1500)(0.5^3 - 0.2^3) = 16.08 \text{ kN} \cdot \text{m}$$

      By Eq. (13.32),

$$F_a = \frac{1}{4}\pi(1500)(0.5^2 - 0.2^2) = 247.4 \text{ kN}$$

Comment: The preceding results indicate that the uniform wear condition yielded a smaller torque and actuating force; it is therefore the more conservative of the two assumptions in terms of clutch capacity.

Figure 13.16   Caliper-disk brake, hydraulically operated. (Courtesy of Ausco Products, Inc., Benton Harbor, MI.)

DISK BRAKES

A disk brake is very similar to the disk clutch shown in Figure 13.14, with the exception that one of the shafts is replaced by a fixed member. Loads are balanced by locating friction linings or pads on both sides of the disk. Servo action can be obtained by addition of several machine parts. The torque capacity and actuating force requirements of disk brakes may readily be ascertained through the use of the foregoing procedures. The equations for the disk clutch can be adapted to the disk brake, if the brake pad is shaped like a sector of a circle and calculations are made accordingly, as illustrated in the next example.

The *caliper disk brake* includes a disk-shaped rotor attached to the machine to be controlled and friction pads. The latter cover only a small portion of the disk surface, allowing the remainder exposed to dissipate heat. The linings are contained in a fixed-caliper assembly and forced against the disk by air pressure or hydraulically (Figure 13.16). Disk brakes have been employed in automotive applications, due to their equal braking torque for either direction of rotation as well as greater cooling capacity than drum brakes (see Section 13.14). They are also often preferred in heavy-duty industrial applications. Caliper disk brakes are also widely used on the front wheel of most motorcycles. The common bicycle is another example, where the wheel rim forms the disk.

EXAMPLE 13.5   Design of a Disk Brake

A disk brake has two pads of included angle $\gamma = 60°$ each, $D = 10$ in., $d = 5$ in. (Figure 13.14). Determine

(a)   The actuating force required to apply one shoe.

(b)   The torque capacity for both shoes.

Design Decision: Sintered-metal pads and cast iron disk are used with $f = 0.2$ and $p_{max} = 200$ psi.

Solution:

(a) Equation (13.28) may be written in the form

$$F_a = \frac{\gamma}{360}\left[\frac{1}{2}\pi p_{max}\, d(D - d)\right]$$

(13.35)

Introducing the given numerical values,

$$F_a = \frac{60}{360}\left[\frac{1}{2}\pi(200)(5)(10 - 5)\right] = 1309 \text{ lb}$$

(b) From Eq. (13.30), we obtain

$$T = 2\left[\frac{1}{4}F_a f(D + d)\right]$$

$$= 2\left[\frac{1}{4}(1309)(0.2)(10 + 5)\right] = 1964 \text{ lb} \cdot \text{in.}$$

13.11   CONE CLUTCHES AND BRAKES

The cone clutch, Figure 13.17, can be considered as the general case of a disk clutch having a cone angle of 90°. Due to the increased frictional area and the wedging action of the parts, cone clutches convey a larger torque than disk clutches with the identical outside diameters and actuating forces. Practically, a cone clutch can have no more than one friction interface; hence, $N = 1$. Cone clutches are often used in low-speed applications. They could also be employed as cone brakes with some slight modifications.

UNIFORM WEAR

The presupposition is made that the normal wear is proportional to the product of the normal pressure p and the radius. Let the radius r in Figure 13.17 locate the ring element running around the cone. The differential area is then equal to $dA = 2\pi r\, dr/\sin \alpha$. The normal force in the element equals $dF_n = p\, dA$, in which $p = p_{max}d/2r$. As before, p_{max} represents the maximum pressure. Hence, the total normal force is

$$F_n = \int_{d/2}^{D/2} \frac{p_{max}d}{2r}\frac{2\pi r\, dr}{\sin \alpha} = \frac{\pi d p_{max}}{2 \sin \alpha}(D - d)$$

(13.36)

The corresponding axial force is $F_n \sin \alpha$. The actuating force is then

$$F_a = \frac{1}{2}\pi p_{max}d(D - d)$$

(13.37)

which is the same as for a disk clutch, Eq. (13.28).

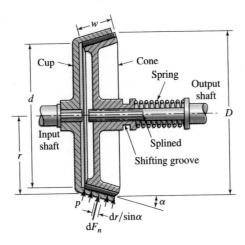

Figure 13.17   Cone clutch.

The torque that can be transmitted by the ring element is equal to $dT = dF_n f r = 2\pi p f r^2 \, dr/\sin\alpha$. The torque capacity of the clutch T is obtained by integrating the forgoing expression over the conical surface. In so doing, we obtain

$$T = \frac{\pi f p_{\max} d}{\sin\alpha} \int_{d/2}^{D/2} r \, dr = \frac{\pi f p_{\max} d}{8 \sin\alpha}(D^2 - d^2) \tag{13.38}$$

In terms of the actuating force, we have

$$T = \frac{F_a f}{4 \sin\alpha}(D + d) \tag{13.39}$$

UNIFORM PRESSURE

An analysis analogous to the uniform pressure made for disk clutches in Section 13.10 results in the following equations (Problem 13.16):

$$F_a = \frac{1}{4}\pi p_{\max}(D^2 - d^2) \tag{13.40}$$

$$T = \frac{\pi f p_{\max}}{12 \sin\alpha}(D^3 - d^3) = \frac{F_a f}{3 \sin\alpha} \frac{D^3 - d^3}{D^2 - d^2} \tag{13.41}$$

The cone angle α, cone diameter, and cone face width w are essential design parameters. The smaller the cone angle, the less actuating force is needed. This angle has a minimum value of $8°$. An angle of $12°$ is ordinarily regarded about optimum. The generally used values of α are in the range of $8°$ to $15°$.

13.12 BAND BRAKES

The *band brake,* the simplest of many braking devices, is employed in power excavators and hoisting and other machinery. Usually, the band is made of steel and lined with a woven friction material for flexibility. The braking action is secured by tightening the band wrapped around the drum that is to be slowed or halted. The difference in tensions at each end of the band ascertains the torque capacity.

Figure 13.18 shows a band brake with the drum rotating clockwise. For this case, friction forces acting on the band increase the tight-side tension F_1 and decrease the slack-side tension F_2. Consider the drum and band portion *above* the sectioning plane as a free body. Then, summation of the moments about the center of rotation of the drum gives the *torque capacity,* which is the same as for a belt drive:

$$T = (F_1 - F_2)r \qquad (13.42)$$

The quantity r is the radius of the drum. Likewise, considering the lever and hand portion *below* the sectioning plane as free body, the *actuating force* is

$$F_a = F_2 \frac{c}{a} \qquad (13.43)$$

The brake is actuated by the application of force F_a at the free end of the lever. It is obvious that a smaller force F_a is needed for operation when the tight side of the band is connected to the fixed support and the slack side attached to the lever, as shown in the figure.

An expression relating band tensions F_1 and F_2 is derived on following the same procedure used for flexible belts, with the exception that the centrifugal force acting on belts does not exist. Hence, referring to Section 13.3, the band tension relationship has the form

$$\frac{F_1}{F_2} = e^{f\phi} \qquad (13.44)$$

Figure 13.18 Simple band brake.

Figure 13.19   Differential brake.

Here

$$F_1 = \text{the larger tensile force}$$
$$F_2 = \text{the smaller tensile force}$$
$$f = \text{coefficient of friction}$$
$$\phi = \text{angle of contact between band and drum or the angle of wrap}$$

Let the analysis of belt shown in Figure 13.5b be applied to the band at the point of tangency for F_1. We now have $F_c = 0$ and $dN = p_{max} wr\, d\theta$. The inward components of the band forces are equal to $dN = F_1\, d\theta$. These two forces are set equal to each other to yield

$$F_1 = wr p_{max} \tag{13.45}$$

The quantity p_{max} is the maximum pressure between the drum and lining, and w represents the width of band. An expression similar to this can also be written for the slack side.

The ***differential band brake*** is analogous to the simple band brake, with the exception that the tight-side tension helps the actuating force (Figure 13.19). A brake of this type is termed *self-energized,* since the friction force assists in applying the band. For a differential brake, Eq. (13.43) becomes

$$F_a = \frac{1}{a}(cF_2 - sF_1) \tag{13.46}$$

In the case of a *self-locking* brake, the product sF_1 is greater than cF_2. Note that, when a brake is designed to be self-locking for one direction of rotation, it can be free to rotate in the opposite direction. A self-locking brake can then be employed when rotation is in one direction only.

EXAMPLE 13.6   Design of a Differential Band Brake

A differential band brake similar to that in Figure 13.19 uses a woven lining having design values of $f = 0.3$ and $p_{max} = 375$ kPa. Determine

(a)   The torque capacity.

(b)   The actuating force.

(c) The power capacity.

(d) The value of dimension s that would cause the brake to be self-locking.

Given: The speed is 250 rpm, $a = 500$ mm, $c = 150$ mm, $w = 60$ mm, $r = 200$ mm, $s = 25$ mm, and $\phi = 270°$.

Solution:

(a) Through the use of Eq. (13.45), we obtain

$$F_1 = wrp_{max} = (0.06)(0.2)(375) = 4.5 \text{ kN}$$

Applying Eq. (13.44),

$$F_2 = \frac{F_1}{e^{f\phi}} = \frac{4.5}{e^{0.3(1.5\pi)}} = 1.095 \text{ kN}$$

Then, Eq. (13.42) gives $T = (4.5 - 1.095)(0.2) = 0.681 \text{ kN} \cdot \text{m}$.

(b) By Eq. (13.46),

$$F_a = \frac{150(1.095) - 25(4.5)}{0.5} = 103.5 \text{ N}$$

(c) From Eq. (1.15),

$$kW = \frac{Tn}{9549} = \frac{681(250)}{9549} = 17.8$$

(d) Using Eq. (13.46), we have $F_a = 0$ for $s = 150(1.095)/4.5 = 36.5$ mm.

Comment: The brake is self-locking if $s \geq 36.5$ mm.

13.13 SHORT-SHOE DRUM BRAKES

A short-shoe drum brake consists of a short shoe pressed on the revolving drum by a lever. The schematic representation of a brake of this type is depicted in Figure 13.20. Inasmuch as the shoe is relatively short (i.e., the angle of contact is small, $\phi < 45°$), a *uniform pressure distribution* may be taken between drum and shoe. Accordingly, the resultant normal force and the friction force act at the center of contact.

The projected area A of the shoe is the width multiplied by the chord length subtended by a $\phi°$ arc of the radius of the drum. From the geometry of the figure, $A = 2[r \sin(\phi/2)]w$. Hence, the normal force on the shoe is

$$F_n = p_{max} \left[2 \left(r \sin \frac{\phi}{2} \right) \right] w \tag{13.47}$$

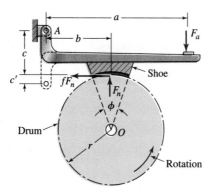

Figure 13.20   Short-shoe drum brake.

In the foregoing, we have

F_n = normal force

p_{max} = maximum pressure between the drum and shoe

r = radius of the drum

ϕ = angle of contact

w = width of the shoe

The value of the friction force is fF_n. The sum of moments about point O for the free-body diagram of the drum yields the *torque capacity* of the brake as

$$T = fF_n r \tag{13.48}$$

The quantity f represents the coefficient of friction.

We now consider the lever as the free body. Then taking moments about the pivot A, we have

$$F_a a + fF_n c - bF_n = 0$$

The preceding leads to the *actuating force:*

$$F_a = \frac{F_n}{a}(b - fc) \tag{13.49}$$

in which a, b, and c represent the distances shown in Figure 13.20.

SELF-ENERGIZING AND SELF-LOCKING BRAKES

For the brake with the *direction* of the rotation shown in the figure, the moment of the friction force assists in applying the shoe to the drum; this makes the brake *self-energizing*. If $b = fc$ or $b < fc$, the force F_a required to actuate the brake becomes 0 or negative, respectively. The brake is then said to be *self-locking* when

$$b \leq fc \tag{13.50}$$

A self-locking brake requires only that the shoe be brought in contact with the drum (with $F_a = 0$) for the drum to be "loaded" against rotation in one direction. The self-energizing feature is useful, but the self-locking effect is generally undesirable. To secure proper utilization of the self-energizing effect while avoiding self-lock, the value of b must be at least 25–50% greater than fc.

Note that, if the brake drum *rotation* is *reversed* from that indicated in Figure 13.20, the sign of fc in Eq. (13.49) becomes negative and the brake is then *self-deenergized*. Also, if the pivot is located on the other side of the line of action of fF_n, as depicted by the dashed lines in the figure, the friction force tends to unseat the shoe. Then, the brake would not be self-energizing. Clearly, both pivot situations discussed are reversed if the direction of rotation is reversed.

Design of a Short-Shoe Drum Brake

EXAMPLE 13.7

The brake shown in Figure 13.20 uses a cork lining having design values of $f = 0.4$ and $p_{max} = 150$ psi. Determine

  (a) The torque capacity and actuating force.

  (b) The reaction at pivot A.

Given: $a = 12$ in., $w = 3$ in., $b = 5$ in., $c = 2$ in., $r = 4$ in., $\phi = 30°$

Solution:

  (a) From Eq. (13.47), we have

$$F_n = 150 \left[2 \left(4 \sin \frac{30°}{2} \right) \right] 3 = 931.7 \text{ lb}$$

  Equation (13.48) yields

$$T = (0.4)(931.7)(4) = 1.491 \text{ kip} \cdot \text{in.}$$

  Applying Eq. (13.49),

$$F_a = \frac{931.7(5 - 0.4 \times 2)}{12} = 326.1 \text{ lb}$$

  (b) The conditions of equilibrium of the horizontal (x) and vertical (y) forces give

$$R_{Ax} = 931.7(0.4) = 372.7 \text{ lb}, \qquad R_{Ay} = 605.6 \text{ lb}$$

  The resultant radial reactional force is

$$R_A = \sqrt{372.7^2 + 605.6^2} = 711.1 \text{ lb}$$

Case Study 13-2 | The Brake Design of a High-Speed Cutting Machine

A short-shoe brake is used on the drum, which is keyed to the center shaft of the high-speed cutter as shown in Figure 13.9. The driven pulley is also keyed to that shaft. For details see Case Study 13-1. Determine the actuating force F_a.

Assumptions: The brake shoe material is molded asbestos. The drum is made of iron. The lining rubs against the smooth drum surface, operating dry.

Given: The drum radius $r = 3$ in., torque $T = 270$ lb·in. (CW), $a = 12$ in., $b = 1.2$ in., $d = 2.5$ in., the width of shoe $w = 1.5$ in. (Figure 3.9). By Table 13.11, $p_{max} = 200$ psi and $f = 0.35$.

Requirement: Shoe must be self-actuating.

Solution: The normal force, through the use of Eq. (13.48), is

$$F_n = \frac{T}{fr}$$

$$= \frac{270}{(0.35)3} = 257.1 \text{ lb}$$

The angle of contact, applying Eq. (13.47), is then

$$\phi = 2\sin^{-1}\frac{F_n}{2p_{max}rw}$$

$$= 2\sin^{-1}\frac{257.1}{2(200)(3)(1.5)} = 16.4°$$

The actuating force is obtained from Eq. (13.49) with $d = c$ as follows:

$$F_a = \frac{F_n}{a}(b - fc)$$

$$= \frac{257.1}{12}(1.2 - 0.5 \times 2.5)$$

$$= -1.07 \text{ lb}$$

Comments: Since $\phi < 45°$, the short-shoe drum brake approximations apply. A negative value of F_a means that the brake is self-energizing, as required.

13.14 LONG-SHOE DRUM BRAKES

When the angle of contact between the shoe and the drum is about 45° or more, the short-shoe equations can lead to appreciable errors. Most shoe brakes have contact angles of 90° or greater, so a more accurate analysis is needed. The obvious problem relates the determination of the pressure distribution. The analysis that includes the effects of deflection is complicated and not warranted here. In the following development, we make the usual simplifying assumption: the pressure varies directly with the distance from the shoe pivot point. This is equivalent to the presupposition made earlier, that the wear is proportional to the product of pressure and velocity.

External Long-Shoe Drum Brakes

Figure 13.21 illustrates an external long-shoe drum brake. The pressure p at some arbitrary angle θ is proportional to $c\sin\theta$. However, since c is a constant, p varies directly with $\sin\theta$.

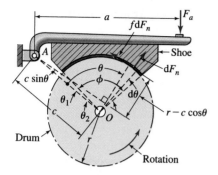

Figure 13.21 External long-shoe drum brake.

As a result,

$$p = p_{max} \frac{\sin \theta}{(\sin \theta)_m} \tag{13.51}$$

Here p_{max} = maximum pressure between the lining and the drum, and $(\sin \theta)_m$ = maximum value of $\sin \theta$. Based on the geometry,

$$(\sin \theta)_m = \begin{cases} 1.0 & \text{(if } \theta_2 > 90°) \\ \sin \theta_2 & \text{(if } \theta_2 \leq 90°) \end{cases} \tag{13.52}$$

Note from Eq. (13.52) that the maximum pressure takes place at the location having the value of $(\sin \theta)_m$. External long-shoe brakes are customarily designed for $\theta_1 \geq 5°$, $\theta_2 < 120°$, and $\phi = 90°$, where ϕ is the angle of contact.

Let w represent the width of the lining. Then, the area of a small element, cut by two radii an angle $d\theta$ apart, is equal to $wr \, d\theta$. Multiplying by the pressure p and the arm $c \sin \theta$ and integrating over the entire shoe, the moment of normal forces, M_n, about pivot A results:

$$M_n = \int_{\theta_1}^{\theta_2} p(wr \, d\theta)c \sin \theta$$

$$= \frac{wrcp_{max}}{(\sin \theta)_m} \int_{\theta_1}^{\theta_2} \sin^2 \theta \, d\theta$$

from which

$$M_n = \frac{wrcp_{max}}{4(\sin \theta)_m}[2\phi - \sin 2\theta_2 + \sin 2\theta_1] \tag{13.53}$$

In a like manner, the moment of friction forces, M_f, about A is written in the form

$$M_f = \int_{\theta_1}^{\theta_2} fpwr \, d\theta(r - c \cos\theta)$$

$$= \frac{fwrp_{max}}{(\sin\theta)_m} \int_{\theta_1}^{\theta_2} \left(r \sin\theta - \frac{c}{2} \sin 2\theta\right) d\theta$$

or

$$M_f = \frac{fwrp_{max}}{4(\sin\theta)_m} [c(\cos 2\theta_2 - \cos 2\theta_1) - 4r(\cos\theta_2 - \cos\theta_1)] \qquad \textbf{(13.54)}$$

Now, summation of the moments about the pivot point A results in the actuating force:

$$F_a = \frac{1}{a}(M_n \mp M_f) \qquad \textbf{(13.55)}$$

In this equation, the upper sign is for a self-energizing brake and the lower one for a self-deenergizing brake. Self-locking occurs when

$$M_f \geq M_n \qquad \textbf{(13.56)}$$

As noted previously, it is often desirable to make a brake shoe self-energizing but not self-locking. This can be accomplished by designing the brake so that the ratio M_f/M_n is no greater than about 0.7.

The torque capacity of the brake is found by taking moments of the friction forces about the center of the drum O. In so doing, we have

$$T = \int_{\theta_1}^{\theta_2} fpwr \, d\theta r$$

$$= \frac{fwrp_{max}}{(\sin\theta)_m} \int_{\theta_1}^{\theta_2} \sin\theta \, d\theta$$

from which

$$T = \frac{fwr^2 p_{max}}{(\sin\theta)_m}(\cos\theta_1 - \cos\theta_2) \qquad \textbf{(13.57)}$$

Finally, pin reactions at A and O can readily be obtained from horizontal and vertical force equilibrium equations. Note that reversing the direction of the rotation changes the sign of the terms containing the coefficient of friction in the preceding equations.

Design of a Long-Shoe Drum Brake

EXAMPLE 13.8

The long-shoe drum brake is actuated by a mechanism that exerts a force of $F_a = 4$ kN (Figure 13.22). Determine

(a) The maximum pressure.

(b) The torque and power capacities.

Figure 13.22   Example 13.8.

Design Decision: The lining is a molded asbestos having a coefficient of friction $f = 0.35$ and a width $w = 75$ mm.

Solution: The angle of contact is $\phi = 90°$ or $\pi/2$ rad. From the geometry, $\alpha = \tan^{-1}(200/150) = 53.13°$. Hence,

$$\theta_1 = 8.13°, \qquad \theta_2 = 98.13°$$

$$c = \sqrt{200^2 + 150^2} = 250 \text{ mm}$$

Inasmuch as $\theta_2 > 90°$, $(\sin\theta)_m = 1$.

(a) Through the use of Eq. (13.53),

$$M_n = \frac{(0.075)(0.15)(0.25)p_{max}}{4(1)} \left[2\left(\frac{\pi}{2}\right) - \sin 196.26° + \sin 16.26° \right]$$

$$= 2.6(10^{-3})p_{max}$$

From Eq. (13.54),

$$M_f = \frac{0.35(0.075)(0.15)p_{max}}{4}[(0.25)(\cos 196.26° - \cos 16.26°)$$

$$- 4(0.15)(\cos 98.13° - \cos 8.13°)]$$

$$= 0.196(10^{-3})p_{max}$$

Applying Eq. (13.55), we then have

$$4,000(0.45) = (2.6 + 0.196)(10^{-3})p_{max}$$

or

$$p_{max} = 644 \text{ kPa}$$

(b) Using Eq. (13.57),

$$T = \frac{(0.35)(0.075)(0.15^2)(0.644 \times 10^6)}{1}(\cos 8.13° - \cos 98.13°)$$

$$= 429.7 \text{ N} \cdot \text{m}$$

By Eq. (1.15), the corresponding power is

$$kW = \frac{Tn}{9549} = \frac{429.7(250)}{9549} = 11.25$$

INTERNAL LONG-SHOE DRUM BRAKES

Figure 13.23 shows an internal long-shoe drum brake. A brake of this type is widely used in automotive services. We see from the figure that both shoes pivot about anchor pins (A and B) and are forced against the inner surface of the drum by a piston in each end of the hydraulic wheel cylinder. The actuating forces are thus exerted hydraulically by pistons. The light return spring applies only enough force to take in the shoe against the adjusting

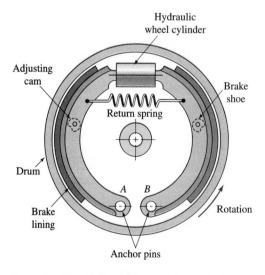

Figure 13.23   Brake with internal long-shoe; automotive type brake.

cams. Each adjusting cam functions as a "stop" and is utilized to minimize the clearance between the shoe and drum.

The method of analysis and the resulting expressions for the internal brakes are identical with those of long-shoe external drum brakes just discussed. That is, Eqs. (13.51) through (13.54) apply as well to internal-shoe drum brakes. Note that, now, a positive result for M_n indicates clockwise moment about A of the left shoe or counterclockwise moment about B of the right-side shoe. A positive or negative result for friction moment M_f should be interpreted in the same manner as for a brake with external shoe.

Typically, in Figure 13.23, the *left shoe* is *self-energizing* and the right shoe is deenergizing. Should the direction of the rotation be reversed, the right shoe would be self-energizing and the left shoe would not. For a prescribed actuating force, the braking capacity with both shoes self-energizing is clearly higher than if only one were. Interestingly, automotive brakes are also made using two hydraulic wheel cylinders with both shoes self-energizing. Of course, this results in braking ability in reverse that is much less than for forward motion. Recently, the caliper-type disk brakes discussed in Section 13.10 have replaced front drum brakes on most passenger cars due to their greater cooling capacity and other good qualities.

13.15 ENERGY ABSORPTION AND COOLING

The primary role of a brake is to absorb energy and dissipate the resulting heat without developing high temperatures. Clutches also absorb energy and dissipate heat but at a lower rate, since they connect two moving elements. The quality of heat dissipation depends on factors such as the size, shape, and condition of the surface of the various parts. Obviously, by increasing exposed surface areas (such as by fins and ribs) and the flow of the surrounding air, these devices can be cooled more conveniently. In addition, the length of time and the interval of brake application affect the temperature. With an increase of the temperature of the brake (or clutch), its coefficient of friction decreases. The result is *fading;* that is, the effectiveness of the device may be sharply deteriorated. The torque and power capacity of a brake or clutch is thus limited by the characteristics of the material and the ability of the device to dissipate heat. A satisfactory braking or clutching performance requires that the heat generation should not exceed the heat dissipation.

ENERGY SOURCES

The energy equation depends on the type of motion a body is going under. Let us consider a body of the weight W, mass m, and mass moment of inertia about its axis of rotation I. The sources of energy to be absorbed from the body by the clutch or brake are mainly as follows. Kinetic energy of translation is

$$E_k = \frac{1}{2}mv^2 \qquad (13.58)$$

Kinetic energy of rotation:

$$E_k = \frac{1}{2}I\omega^2 \qquad (13.59)$$

Potential energy is

$$E_p = Wh \tag{13.60}$$

In the foregoing, v = velocity, ω = angular velocity, and h = vertical distance.

To clarify the relevance to brakes of the kinetic and potential energies, refer to winch crane of Figure 1.4. Suppose that the crane lowers a mass m of weight W translating at time t_1 with a velocity v_1 at elevation h_1 and the gear shafts with mass moments of inertia I rotating at angular velocities ω_1. Shafts may be rotating at different speeds. If at time t_1 the internal brake (in the motor) is applied, then, at a time t_2 quantities will have reduced to v_2, ω_2, and h_2. Therefore, during the time interval $t_2 - t_1$, we have [6]: W_b = work done by the brake; W_r = work done by rolling friction, bearing friction, and air resistance; W_m = work done by drive motor. The conservation of the energy requires that the total work equals the change in energy:

$$W_b + W_r + W_m = \frac{1}{2}m\left(v_1^2 - v_2^2\right) + \sum \frac{1}{2}I\left(\omega_1^2 - \omega_2^2\right) + W(h_1 - h_2)$$

Here, the summation consists of multiplications made for different mass moments of inertia at their corresponding angular velocities.

The work required of the brake to stop, slow, or maintain speed is obtained by the solution of the preceding equation for W_b. This presents the mechanical energy transformed into heat at the brake and can be used to predict the temperature rise. Note that, in many machines such as slow speed hoists and winch cranes, W_r and W_m are negligible. Clearly, omitting these quantities results in a safer brake design.

TEMPERATURE RISE

When the motion of the body is halted or transmitted by a braking or clutching operation, the frictional energy E developed appears in the form of heat in the device. The temperature rise may be related to the energy absorbed in the form of friction heat by the familiar formula:

$$\Delta t = \frac{E}{Cm} \tag{13.61}$$

where

Δt = temperature rise, °C

E = frictional energy the brake or clutch must absorb, J

C = specific heat (use 500 J/kg°C for steel or cast iron)

m = mass of brake or clutch parts, kg

The frictional energy E is ascertained as briefly discussed in the preceding. Then, through the use of Eq. (13.61), the temperature rise of the brake or clutch assembly is obtained. The limiting temperatures for some commonly used brake and clutch linings are furnished in Table 13.11. These temperatures represent the largest values for steady operation.

Table 13.13   Representative pV values for shoe brakes

Operation	Heat dissipation	pV	
		(MPa)(m/s)	(ksi)(ft/min)
Continuous	Poor	1.05	30
Occasional	Poor	2.10	60
Continuous	Good	3.00	85

Equations (13.58) through (13.61) illustrate what happens when a brake or clutch is operated. Many variables are involved, however; and such an analysis may only estimate experimental results. In practice, the rate of energy absorption and heat dissipated by a brake or clutch is of utmost importance. Brake and lining manufacturers include the effect of the rate of energy dissipation by assigning the appropriate limiting values of pV, a product of pressure and velocity, for specific kinds of brake design and service conditions [20–23]. Typical values of pV used in industry are given in Table 13.13.

REFERENCES

1. Shigley, J. E., and C. R. Mischke. *Mechnaical Engineering Design,* 6th ed. New York: McGraw-Hill, 2001.
2. Firbank, T. C. "Mechanics of the Belt Drive." *International Journal of Mechanical Science* 12 (1970), pp. 1053–63.
3. Wallin, A. W. "Efficiency of Synchronous Belts and V-Belts." Proceedings of the National Conference Power Transmission, vol. 5. Illinois Institute of Technology, Chicago, November 7–9, 1978, pp. 265–71.
4. Gerbert, B. G. "Pressure Distribution and Belt Deformation in V-Belt Drives." *Journal of Engineering for Industry, Transactions of the ASME* 97 (1975), pp. 976–82.
5. Alciatore, D. G., and A. E. Traver. "Multiple Belt Drive Mechanics: Creep Theory vs. Shear Theory." *Journal of Mechanisms, Transmission, and Automation in Design, Transactions of the ASME* 112 (1990), pp. 65–70.
6. Burr, A. H., and J. B. Cheatham. *Mechanical Analysis and Design,* 2nd ed. Upper Saddle River, NJ: Prentice Hall, 1995.
7. Deutcshman, A. D., W. J. Michels, and C. E. Wilson. *Machine Design: Theory and Practice.* New York: Macmillan, 1975.
8. Gates Rubber Co. *V-Belt Drive Design Manual.* Denver, CO: Gates Rubber Co., 1995.
9. Spotts, M. F., and T. E. Shoup. *Design of Machine Elements,* 7th ed. Upper Saddle River, NJ: Prentice Hall, 1998.
10. Rubber Manufacturers Association. *Specifications for Drives Using Classical Multiple V Belts.* American National Standard, IP-20. Washington, DC: Rubber Manufacturers Association, 1988.
11. Erickson, W. D. *Belt Selection and Application for Engineers.* New York: Marcel Dekker, 1987.
12. American Chain Association (ACA). *Chains for Power Transmission and Material Handling.* New York: Marcel Dekker, 1982.
13. Binder, R. C. *Mechanics of Roller Chain Drive.* Upper Saddle River, NJ: Prentice Hall, 1956.
14. ANSI/ASME. *Precision Power Transmission Roller Chains, Attachments, and Sprockets.* ANSI/ASME Standard B29.1M. New York: ASME, 1993.

15. Juvinall, R. C., and R. M. Marshek. *Fundamentals of Machine Component Design,* 3rd ed. New York: Wiley, 2000.

16. ANSI/ASME. *Inverted-Tooth (Silent) Chains and Sprockets.* ANSI/ASME Standard B29,2M-82. New York: American Society of Mechanical Engineers, 1993.

17. Rothbart, H. A., ed. *Mechanical Design and Systems Handbook,* 2nd ed. New York: McGraw-Hill, 1985.

18. Orthwein, W. C. *Clutches and Brakes: Design and Selection.* New York: Marcel Dekker, 1986.

19. Proctor, J. "Selecting Clutches for Mechanical Devices." *Product Engineering* (June 19, 1961), pp. 43–58.

20. Remling, J. *Brakes.* New York: Wiley, 1978.

21. Neale, M. J., ed. *Tribology Handbook.* New York: Wiley, 1973.

22. Baker, A. K. *Vehicle Braking.* London: Pentech Press Limited, 1987.

23. Crouse, W. H. "Automotive Brakes." In *Automotive Chasis and Body,* 4th ed. New York: McGraw-Hill, 1971.

PROBLEMS

Sections 13.1 through 13.7

13.1 A flat belt 4 in. wide and $\frac{3}{16}$ in. thick operates on pulleys of diameters 5 in. and 15 in. and transmits 10 hp. Determine

   (*a*) The required belt tensions.

   (*b*) The belt length.

   Given: Speed of the small pulley is 1500 rpm, the pulleys are 5 ft apart, the coefficient of friction is 0.30, and the weight of the belt material is 0.04 lb/in.^{3}.

13.2 A plastic flat belt 60 mm wide and 0.5 mm thick transmits 10 kW. Calculate

   (*a*) The torque at the small pulley.

   (*b*) The contact angle.

   (*c*) The maximum tension and stress in the belt.

   Given: The input pulley has a diameter of 300 mm, it rotates at 2800 rpm, and the output pulley speed is 1600 rpm; the pulleys are 700 mm apart, the coefficient of friction is 0.2, and belt weight is 25 kN/m^3.
   Assumptions: The driver is a high torque motor and the driven machine is under a medium shock load.

13.3 A V-belt drive has a 200-mm diameter small sheave with a 170° contact angle, 38° included angle, coefficient of friction of 0.15, 1600-rpm driver speed, belt weight of 8 N/m, and a tight-side tension of 3 kN. Determine the power capacity of the drive.

13.4 A V-belt drive has an included angle of 38°, belt weight of 3 N/m, belt cross-sectional area of 145 mm^2, coefficient of friction of 0.25, $r_1 = 150$ mm, $\phi = 160°$, $n_1 = 3000$ rpm, and $F_2 = 800$ kN. Calculate

   (*a*) The maximum power transmitted.

   (*b*) The maximum stress in the belt.

13.5 A V-belt drive with an included angle of 34° is to have a capacity of 15 kW based on a coefficient of friction of 0.2 and a belt weight of 2.5 N/m. Determine the required maximum belt tension at full load.

Assumptions: The driver is a normal torque motor and the driven machine involves heavy shocks.
Design Decision: Speed is to be reduced from 2700 rpm to 1800 rpm using a 200-mm diameter small sheave; shafts are 500 mm apart.

13.6 A $\frac{5}{8}$-in. pitch roller chain operates on 22-tooth drive sprocket rotating at 4000 rpm and a driven sprocket rotating at 1000 rpm. Calculate the minimum center distance.

13.7 A $\frac{9}{16}$-in. pitch inverted chain operating on a 14-tooth drive sprocket rotating at 4600 rpm and a driven sprocket at 2100 rpm. Determine the minimum center distance.

Sections 13.8 through 13.11

13.W Search the website at www.sepac.com. List selection (application procedure and application) factors to consider prior to choosing a brake or clutch.

13.8 A disk clutch has a single pair of friction surfaces of 250-mm outside diameter $\times$ 150-mm inside diameter. Determine the maximum pressure and the torque capacity, using the assumption of

(a) Uniform wear.

(b) Uniform pressure.

Given: The coefficient of friction is 0.3 and the actuating force equals 6 kN.

13.9 A disk clutch that has both sides effective, an outside diameter four times the inside diameter, used in an application where 40 hp is to be developed at 500 rpm. Determine, based on uniform pressure condition,

(a) The inside and outside diameters.

(b) The actuating force required.

Design Decisions: A friction material with $f = 0.25$ and $p_{max} = 20$ psi is used.

13.10 Resolve Problem 13.9 based on the assumption of uniform wear.

13.11 A multiple disk clutch having four active faces, 12-in. outer diameter, 6-in. inner diameter, and $f = 0.2$ is to carry 50 hp at 400 rpm. Determine, using the condition of uniform wear,

(a) The actuating force required.

(b) The average pressure between the disks.

13.12 A 10-in. outside diameter cone clutch with 8° cone angle is to transmit 50 hp at 800 rpm. Calculate the face width w of the cone, on the basis of the uniform pressure assumption.

Design Decision: The maximum lining pressure will be 60 psi and the coefficient of friction $f = 0.3$.

13.13 Redo Problem 13.12 using the condition of uniform wear.

13.14 A cone clutch has a mean diameter of 500 mm, a cone angle of 10°, and a cone face width of $w = 80$ mm. Determine, using the uniform wear assumption,

    (*a*) The actuating force and torque capacity.

    (*b*) The power capacity for a speed of 500 rpm.

    Design Decision: The lining has $f = 0.2$ and $p_{max} = 0.5$ MPa.

13.15 A cone clutch has an average diameter of 250 mm, a cone angle of 12°, and $f = 0.2$. Calculate the torque that the brake can transmit.

    Assumptions: A uniform pressure of 400 kPa. Actuating force equals 5 kN.

13.16 Verify, based on the assumption of uniform pressure, that the actuating force and torque capacity for a cone clutch (Figure 13.17) are given by Eqs. (13.40) and (13.41).

Section 13.12

13.17 A band brake uses a 100 mm wide woven lining having design values of $f = 0.3$ and $p_{max} = 0.7$ MPa (Figure P13.17). Determine band tensions and power capacity at 150 rpm.

    Given: $\phi = 240°$ and $r = 200$ mm.

13.18 The drum of the band brake depicted in Figure P13.18 has a moment of inertia of $I = 20$ lb · in. · s² about point O. Calculate the actuating force F_a necessary to decelerate the drum at a rate of $\alpha = 200$ rad/s². Note that the torque is expressed by $T = I\alpha$.

    Given: $\phi = 210°$, $a = 12$ in., $r = 5$ in., and $f = 0.3$.

Figure P13.17

Figure P13.18

13.19 The band brake shown in Figure P13.18 has a power capacity of 40 kW at 600 rpm. Determine the belt tensions.

    Given: $\phi = 250°$, $r = 250$ mm, $a = 500$ mm, and $f = 0.4$.

13.20 The band brake depicted in Figure P13.18 uses a woven lining having design values of $p_{max} = 0.6$ MPa and $f = 0.4$. Calculate

    (*a*) The band tensions and the actuating force.

    (*b*) The power capacity at 200 rpm.

    Given: The band width $w = 75$ mm, $\phi = 240°$, $r = 150$ mm, and $a = 400$ mm.

13.21 The differential brake depicted in Figure P13.21 is to absorb 10 kW at 220 rpm. Determine

(a) The angle of wrap.

(b) The length of arm s from the geometry of the brake.

Given: The maximum pressure between the lining and the drum is 0.8 MPa, $f = 0.14$, and $w = 60$ mm.

Figure P13.21

13.22 The differential brake depicted in Figure 13.19 has $a = 12$ in., $c = 2$ in., $s = 3.2$ in., $r = 4$ in., $n = 300$ rpm, $\phi = 210°$, $f = 0.12$, and $F_a = 300$ lb. If F_a is acting upward, determine the horsepower capacity.

13.23 The differential band brake shown in Figure 13.19 has $a = 250$ mm, $c = 100$ mm, $s = 50$ mm, $r = 200$ mm, $\phi = 210°$, and a woven lining material with $f = 0.4$. Determine the actuating force F_a required. Will the brake be self-locking?

Requirement: A power of 15 kW is to be developed at 900 rpm.

13.24 Redo Problem 13.23 for counterclockwise rotation of the drum.

Sections 13.13 through 13.15

13.25 A short-shoe drum brake having $f = 0.25$, $a = 1$ m, $b = 0.4$ m, $c = 50$ mm, $r = 0.3$ m is to absorb 25 kW at 800 rpm (Figure 13.20). Determine

(a) The actuating force and whether the brake is self-locking.

(b) The pin reaction at A.

13.26 Resolve Problem 13.25 for clockwise rotation of the drum.

13.27 Redo Example 13.8 using short-shoe analysis, that is, assuming that the total normal and friction forces are concentrated at point B. Compare the results with the more exact results of Example 13.8.

13.28 Figure P13.28 depicts a long-shoe drum brake. Determine the value of dimension b in terms of the radius r so that the friction forces neither assist nor resist in applying the shoe to the drum.

Figure P13.28

13.29 The long-shoe brake shown in Figure P13.29 has $p_{max} = 900$ kPa, $f = 0.3$, and $w = 50$ mm. Calculate

(a) The actuating force.

(b) The power capacity at 600 rpm.

Figure P13.29

13.30 A long-shoe brake is shown in Figure P13.30. Determine

(a) The actuating force.

(b) The power capacity at 500 rpm.

Given: $b = 150$ mm, $d = 250$ mm, $r = 200$ mm, $w = 60$ mm, $f = 0.3$, and $p_{max} = 800$ kPa.

Figure P13.30

SPRINGS

Outline

14.1   INTRODUCTION

Springs are used to exert forces or torques in a mechanism or primarily to store the energy of impact loads. These flexible members often operate with high values for the ultimate stresses and with varying loads. Helical springs are round or rectangular wire, and flat springs (cantilever or simply supported beams) are in widespread usage. Springs come in a number of other kinds, such as disk, ring, spiral, and torsion bar springs. Numerous standard spring configurations are available as stock catalog items from spring manufacturers. Figure 14.1 shows various compression, tension, and torsion springs. The designer must understand and appropriately apply spring theory to specify or design a component.

Pneumatic springs of diverse types take advantage of the elastic compressibility of gases, as compressed air in automotive air shock absorbers. For applications involving very large forces with small displacements, hydraulic springs have proven very effective. Our concern in this text is only with springs of common geometric form made of solid metal or rubber. For more information on others, see [1–5]. As discussed in Section 1.4, mechanical components are usually designed on the basis of strength. Generally, displacement is of minor significance. Often deflection is checked whether it is reasonable. However, in the

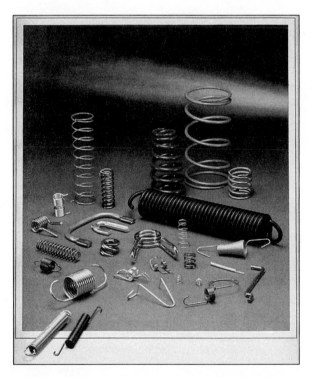

Figure 14.1   A collection of wire springs. (Courtesy of Rockford Spring Co.)

design of springs, displacement is as important as strength. A notable deflection is essential to most spring applications.

14.2 TORSION BARS

A *torsion bar* is a straight hollow or solid bar fixed at one end and twisted at the other, where it is supported. This is the simplest of all spring forms, as shown by the portion *AB* in Figure 14.2a. Typical applications include counterbalancing for automobile hoods and trunk lids. A torsion bar with splined ends (Figure 14.2b) is used for a vehicle suspension spring or sway bar. Usually, one end fits into a socket on the chassis, the other into the pivoted end of an arm. The arm is part of a linkage, permitting the wheel to rise and fall in approximately parallel motion. Note that, in a passenger car, the bar may have about $\frac{3}{4}$-m length, 25-mm diameter, and twist 30° to 45°.

The stress in a torsion bar is mainly one of torsional shear. Hence, the equations for stress, angular displacement, and stiffness are given in Sections 3.5 and 4.3. Referring to Figure 14.2a, we can write

$$T = PR, \qquad \delta = \phi R, \qquad k = \frac{T}{\phi}$$

in which, the angle of twist $\phi = TL/GJ$. For the solid *round torsion bar*, the moment of inertia is $J = \pi d^4/32$. We therefore have the formulas

$$\tau = \frac{16PR}{\pi d^3} \tag{14.1}$$

$$\delta = \frac{TLR}{GJ} = \frac{32PLR^2}{\pi d^4 G} \tag{14.2}$$

$$k = \frac{\pi d^4 G}{32L} \tag{14.3}$$

Figure 14.2 Torsion bar springs: (a) rod with bent ends; (b) rod with splined ends.

Here,

τ = torsional shear stress

P = load

δ = relative displacement between ends

G = modulus of rigidity

d = bar diameter

R = moment arm

L = bar length

k = spring rate

The foregoing basic equations are supplemented, in the case of torsional springs with noncircular cross sections, by Table 3.1. Note that, at the end parts of the spring that do not lie between supports A and B, there is a shear load P and an associated direct shear stress acting on cross sectional areas. Usually, the effects of the curvature and the effect of bending are neglected at these portions of the bar. When designing a torsion bar, the required diameter d is obtained by Eq. (14.1). Then, based on the allowable shear strength, Eq. (14.2) gives the bar length L necessary to provide the required deflection δ.

14.3   HELICAL TENSION AND COMPRESSION SPRINGS

In this section, attention is directed to closely coiled standard helical tension and compression springs. They provide a push or pull force and are capable of large deflection. The standard form has constant coil diameter, pitch (axial distance between coils), and spring rate (slope of its deflection curve). It is the most common spring configuration. Variable-pitch, barrel, and hourglass springs are employed to minimize resonant surging and vibration.

A helical spring of circular cross section is composed of a slender *wire of diameter d* wound into a helix of *mean coil diameter D* and *pitch angle* λ. The top portion, isolated from the compression spring of Figure 14.3a, is shown in Figure 14.3b. A section taken perpendicular to the axis of the spring's wire can be considered nearly vertical. Hence, centric load P applied to the spring is resisted by a transverse shear force P and a torque $T = PD/2$ acting on the cross section of the coil, as depicted in the figure. Figure 14.3c shows a helical tension spring.

For a helical spring, the ratio of the mean coil diameter to wire diameter is termed the *spring index C*:

$$C = \frac{D}{d}$$

(14.4)

The springs of ordinary geometry have $C > 3$ and $\lambda < 12°$. In the majority of springs, C varies from about 6 to 12. At $C > 12$, the spring is likely to buckle and also tangles readily when handled in bulk. The outside diameter $D_o = D + d$ and the inside diameter $D_i = D - d$ are of interest primarily to define the smallest hole in which the spring will fit or the largest pin over which the spring can be placed. Usually, the minimum diametral

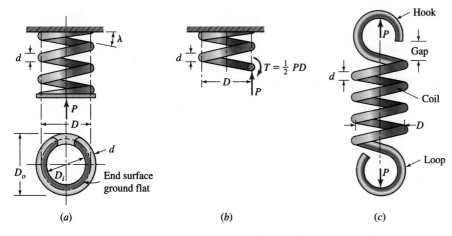

Figure 14.3 Helical springs: (a) compression spring; (b) free-body of top portion of compression spring; (c) tension spring.

clearance between the D_o and the hole or between D_i and a pin are about $0.1D$ for $D < 13$ mm or $0.05D$ for $D > 13$ mm.

STRESSES

An exact analysis by the theory of elasticity shows that the transverse or direct shear stress acting on an element at the inside coil diameter has the value

$$\tau_d = 1.23\frac{P}{A} = 1.23\frac{P}{\pi d^2/4}$$

(14.5)

This expression may be rewritten in the form

$$\tau_d = \frac{8PD}{\pi d^3} \times \frac{0.615}{C}$$

The torsional shear stress, neglecting the initial curvature of the wire, is

$$\tau_t = \frac{16T}{\pi d^3} = \frac{8PD}{\pi d^3}$$

The superposition of the preceding stresses gives the maximum or *total shear stress* in the wire on the inside of the coil:

$$\tau = K_s\frac{8PD}{\pi d^3} = K_s\frac{8PC}{\pi d^2}$$

(14.6)

In the foregoing,

$$K_s = 1 + \frac{0.615}{C}$$

(14.7)

is called the *direct shear factor*.

For a slender wire, the C has large values, and clearly, the maximum shear stress is caused primarily by torsion. In this case, a helical compression or tension spring can be thought of as a torsion bar wound into a helix. On the other hand, in a heavy spring, where C has small values, the effect of direct shear stress cannot be disregarded.

The intensity of the torsional stress increases on the inside of the spring because of the curvature. The following more accurate relationship, known as the *Wahl formula,* includes the curvature effect [2]:

$$\tau = K_w \frac{8PD}{\pi d^3} = K_w \frac{8PC}{\pi d^2} \tag{14.8}$$

The *Wahl factor* K_w is defined by

$$K_w = \frac{4C - 1}{4C - 4} + \frac{0.615}{C} \tag{14.9}$$

The first term in Eq. (14.9), which accounts for the effect of curvature, is basically a stress concentration factor. The second term gives a correction factor for direct shear only. The Wahl factor may be used for most calculations. A more exact theory shows that it is accurate within 2% for $C \geq 3$. Figure 14.4 illustrates the variation of the K_w as a function of C.

After some minor local yielding under static loading, typical spring materials relieve the local stress concentration due to the curvature (see Section 3.15). Therefore, we use Eqs. (14.8) and (14.6) for *alternating* loading and *static* or *mean* loading, respectively. We note that, occasionally,

$$K_s = 1 + \frac{0.5}{C} \tag{14.7'}$$

is used in static applications for compression springs, instead of Eq. (14.7). This is based on the assumption that the transverse shear stresses are uniformly distributed subsequent to some yielding under static loading.

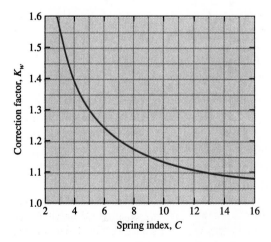

Figure 14.4   Stress correction factors for curvature and direct shear for helical springs.

Interestingly, the free-body diagram of Figure 14.3b contains no bending loading for closely coiled springs. However, for springs with a pitch angle of λ greater than 15° and deflection of each coil greater than $D/4$, bending stresses should be taken into account [2]. In addition, it is rarely possible to have exactly centric axial loading P, and any eccentricity introduces bending and changes the torsional moment arm. This gives rise to stresses on one side of the spring higher than indicated by the foregoing equations. Also observe from Figure 14.3b that, in addition to creating a transverse shear stress, a small component of force P produces axial compression of the spring wire. In critical spring designs involving relatively large values of λ, this factor should be considered.

DEFLECTION

In determining the deflection of a closely coiled spring, it is common practice to ignore the effect of direct shear. Therefore, the twist causes one end of the wire segment to rotate an angle $d\phi$ relative to the other, where $\phi = TL/GJ$. This corresponds to a deflection $d\delta$, at the axis of the spring,

$$d\delta = \frac{D}{2}d\phi = \frac{D}{2}d\left(\frac{TL}{GJ}\right)$$

The total deflection δ, of spring of length $L = \pi D N_a$, is then

$$\delta = \frac{8PD^3N_a}{Gd^4} = \frac{8PC^3N_a}{Gd} \qquad (14.10)$$

in which $N_a = $ *the number of active coils* and $G = $ modulus of rigidity. An alternative derivation of this equation may readily be accomplished using Castigliano's theorem.

SPRING RATE

The elastic behavior of a spring may be expressed conveniently by the slope of its force-deflection curve or *spring rate k*. Through the use of Eq. (14.10), we have

$$k = \frac{P}{\delta} = \frac{Gd^4}{8D^3N_a} = \frac{dG}{8C^3N_a} \qquad (14.11)$$

Also referred to as the spring constant, or spring scale, the spring rate has units of N/m in SI and lb/in. in the U.S. customary system. The standard helical spring has a spring rate k that is basically linear over most of its operating range. The first and last few percent of its deflection have a nonlinear rate. Often, in spring design, the spring rate is defined between about 15 and 85% of its total and working deflections [1].

Occasionally helical compression springs are wound in the form of a cone (Figure 14.5), where the coil radius and hence the torsional stresses vary throughout the length. The maximum stress in a *conical spring* is given by Eq. (14.8). The deflection and spring rate can be estimated from Eqs. (14.10) and (14.11), using the average value of mean coil diameter for D.

Figure 14.5
Conical-helical
compression spring.

EXAMPLE 14.1

Determination of the Spring Rate

A helical compression spring of an average coil diameter D, wire diameter d, and number of active coils N_a, supports an axial load P. Calculate the value of P that will cause a shear stress of τ_{all}, the corresponding deflection, and rate of the spring.

Given: $D = 48$ mm, $d = 6$ mm, $N_a = 5$, $\tau_{\text{all}} = 360$ MPa

Design Decisions: A steel wire of $G = 79$ GPa is used.

Solution: The mean diameter of the spring is $D = 48 - 6 = 42$ mm. The spring index is equal to $C = 42/6 = 7$. Applying Eq. (14.6), we have

$$360 = \frac{8P(42)}{\pi(6)^3}\left(1 + \frac{0.615}{7}\right) = 0.539P$$

Solving,

$$P = 668 \text{ N}$$

From Eq. (14.10),

$$\delta = \frac{8(668)(7)^3(5)}{(79 \times 10^3)(6)} = 19.34 \text{ mm}$$

The spring rate is therefore $k = 668/0.01934 = 34.54$ kN/m.

14.4 SPRING MATERIALS

Springs are manufactured either by hot- or cold-working processes, depending on the size and strength properties needed. Ordinarily, preheated wire should not be used if spring index $C < 4$ in. or if diameter $d > \frac{1}{4}$ in.; that is, small sizes should be wound cold. Heavy-duty springs (e.g., vehicle suspension parts) are usually hot worked. Winding of the springs causes residual stresses owing to bending. Customarily, in spring forming, such stresses are relieved by heat treatment.

A limited number of materials is suitable for usage as springs. These include carbon steels, alloy steels, phosphor bronze, brass, berylium copper, and a variety of nickel alloys. Plastics are used when loads are light. Blocks of rubber often form springs, as in bumpers and vibration isolation mountings of various machines such as electric motors and internal combustion engines. The UNS steels (see Section 2.12) listed in Table B.3 should be used in designing hot-rolled or forged heavy-duty coil springs, as well as flat springs, leaf springs, and torsion bars. The typical spring material has a high ultimate and yield strengths, to provide maximum energy storage. For springs under dynamic loading, the fatigue strength properties of the material are of main significance. The website www.acxesspring.com includes information on commonly used spring materials.

Experiment shows that, for common spring materials, the ultimate strength obtained from a torsion test can be estimated in terms of the ultimate strength determined from a

Table 14.1 Some common spring wire materials

Material	ASTM no.	Description
Hard-drawn wire 0.60–0.70C	A227	Least-expensive general-purpose spring steel. Suitable for static loading only and in temperature range 0°C to 120°C. Available in diameters 0.7 to 16 mm.
Music wire 0.80–0.95C	A228	Toughest high-carbon steel wire widely used in the smaller coil diameters. It has the highest tensile and fatigue strengths of any spring material. The temperature restrictions are the same as for hard-drawn wire. Available from 0.1 to 6.5 mm in diameter.
Oil-tempered wire 0.60–0.70C	A229	Used for many types of coil springs and less expensive than music wire. Suitable for static loading only and in the temperature range 0°C to 180°C. Available in diameters 0.5 to 16 mm.
Chrome vanadium	A232	Suitable for severe service conditions and shock loads. Widely used for aircraft engine valve springs, where fatigue resistance and long endurance needed, and for temperatures to 220°C. Available in diameters from 0.8 to11 mm.
Chrome silicon	A401	Suitable for fatigue loading and in temperatures up to 250°C. Second highest in strength to music wire. Available from 1.6 to 9.5 mm in diameter.

| SOURCE: [1, 4].

simple tension test, as noted in Section 8.5. The ultimate strength in shear is then

$$S_{us} = 0.67 S_u \qquad (8.5)$$

The quantity S_u is the ultimate strength or tensile strength.

SPRING WIRE

Round wire is the most-often utilized spring material. It is ready available in a selection of alloys and wide range of sizes. Rectangular wire is also attainable but only in limited sizes. A brief description of commonly used high-carbon (C) and alloy spring steels is given in Table 14.1.

Ultimate Strength in Tension

Spring materials may be compared by examining their tensile strengths varying with the wire size. The material and its processing also have an effect on tensile strength. The strength properties for some common spring steels may be estimated by the formula

$$S_u = Ad^b \qquad (14.12)$$

Here, S_u = ultimate tensile strength, A = a coefficient, b = an exponent, d = wire diameter (in. or mm). Values of coefficient A and exponent b pertaining to the materials presented in Table 14.1, are furnished in SI and U.S. customary units in Table 14.2. Likewise, the strengths of stainless steel wire and hard phosphor bronze wire are given in the form of

Table 14.2   Coefficients and exponents for Eq. (14.12)

Material	ASTM no.	b	A	
			MPa	ksi
Hard-drawn wire	A227	−0.201	1510	137
Music wire	A228	−0.163	2060	186
Oil-tempered wire	A229	−0.193	1610	146
Chrome vanadium wire	A232	−0.155	1790	173
Chrome silicon wire	A401	−0.091	1960	218

| SOURCE: [1].

Figure 14.6   Yield strength in shear of spring wire [4–6].

graphs, tables, and formulas. The preceding equation provides a convenient means to calculate steel wire tensile strength within a spring-design *computer program* and allows fast iterating to a proper design solution.

Yield Strength in Shear and Endurance Limit in Shear

Data of extensive testing [4–6] indicate that a semilogarithmic plot of torsional yield strength S_{ys} (and hence S_u) versus wire diameter is almost a straight line for some materials (Figure 14.6). Note from the figure that the strength increases with a reduction in diameter. There is also ample experimental evidence that the relationships between the ultimate strength in tension, the yield strength in shear, and the endurance limit in shear S'_{es} are as given in Table 14.3. Observe that the test data values of S'_{es} were developed with actual conditions of surface and size factors of the wire materials, to be discussed in Section 14.7. We use these values, assuming 50% reliability.

Table 14.3 Approximate strength ratios of some common spring materials

Material	S_{ys}/S_u	S'_{es}/S_u
Hard-drawn wire	0.42	0.21
Music wire	0.40	0.23
Oil-tempered wire	0.45	0.22
Chrome vanadium wire	0.52	0.20
Chrome silicon wire	0.52	0.20

SOURCE: [1].

Notes: S_{ys} = yield strength in shear, S_u = ultimate strength in tension, S'_{es} = endurance limit (or strength) in shear.

Determining the Allowable Load of a Helical Compression Spring

EXAMPLE 14.2

A helical compression spring for mechanical device is subjected to an axial load P. Determine

(a) The yield strength in the shear of the wire.

(b) The allowable load P corresponding to yielding.

Design Decisions: Use a 0.0625-in. music wire. The mean diameter of the helix is $D = 0.5$ in. A safety factor of 1.5 is applied due to uncertainty about the yielding.

Solution: The spring index is $C = D/d = 0.5/0.0625 = 8$.

(a) Through the use of Eq. (14.12) and Table 14.2, we have

$$S_u = Ad^b = 186(0.0625^{-0.163}) = 292 \text{ ksi}$$

Then, by Table 14.3, $S_{ys} = 0.4(292) = 117$ ksi.

(b) The allowable load is obtained by applying Eq. (14.6) as

$$P_{all} = \frac{\tau_{all}\pi d^2}{8K_s C}$$

where

$$\tau_{all} = \frac{S_{ys}}{n} = \frac{117}{1.5} = 78 \text{ ksi}$$

$$K_s = 1 + \frac{0.615}{8} = 1.077 \qquad \text{(from Eq. 14.7)}$$

Hence,

$$P_{all} = \frac{\pi(78,000)(0.0625)^2}{8(1.077)(8)} = 13.9 \text{ lb}$$

14.5   HELICAL COMPRESSION SPRINGS

End details are four "standard" types on helical compressive springs. They are plain, plain-ground, squared, and squared-ground, as shown in Figure 14.7. A spring with plain ends has ends that are the same as if a long spring had been cut into sections. A spring with plain ends that are squared, or closed, is obtained by deforming the ends to 0° helix angle. Springs should always be both *squared and ground* for significant applications, because a better transfer of load is obtained. A spring with squared and ground ends compressed between rigid plates can be considered to have fixed ends. This represents the most-common end condition.

Figure 14.7 shows how the type of end used affects the number of active coils N_a and the solid height of the spring. Square ends effectively decrease the *number of total coils N_t* by approximately two; that is,

$$N_t = N_a + 2 \tag{14.13}$$

Grinding by itself removes one active coil. To obtain basically uniform contact pressure over the full end turns, special end members must be used (such as countered end plates) for all end conditions except squared and ground.

Working deflection corresponds to the working load P_w on a compression spring. Referring to Figure 14.8, the *solid deflection δ_s* is defined as follows:

$$\delta_s = h_f - h_s \tag{14.14}$$

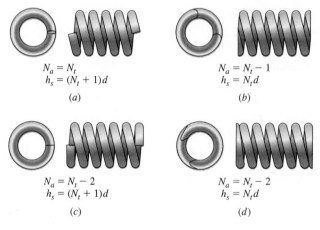

$$N_a = N_t$$
$$h_s = (N_t + 1)d$$
(a)

$$N_a = N_t - 1$$
$$h_s = N_t d$$
(b)

$$N_a = N_t - 2$$
$$h_s = (N_t + 1)d$$
(c)

$$N_a = N_t - 2$$
$$h_s = N_t d$$
(d)

Figure 14.7   Common types of ends for helical compression springs and corresponding spring solid-height equations: (a) plain ends; (b) plain-ground ends; (c) squared ends; (d) squared-ground ends.

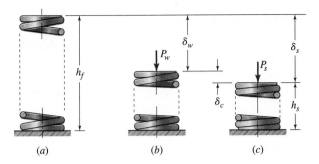

Figure 14.8 Deflections of a helical compression spring:
(a) free height; (b) working deflection; (c) solid deflection.

Here, h_f = free (no load) height and h_s = solid height or shut height under solid load P_s. For *special* applications where space is limited, solid height of ground springs can be obtained by the expression

$$h_s = (N_t - 0.5)(1.01d) \qquad (14.15)$$

Clash allowance (r_c) refers to difference in spring length between maximum load and spring solid position. It is defined as a ratio of margin of extra deflection or clash deflection δ_c to the working deflection δ_w:

$$r_c = \frac{\delta_c}{\delta_w} \qquad (14.16)$$

Usually, a minimum clash allowance of 10–15% is used to avoid reaching the solid height in service. A maximum clash allowance of 20% is satisfactory for most applications. Based on this value, an overload P_s of 20% deflects the spring to its maximum deflection δ_s and higher overload has no effect on deflection or stress. Hence, with a sufficient safety factor, a compression spring is protected against failure after it reaches its solid deflection.

DESIGN PROCEDURE FOR STATIC LOADING

The two *basic requirements* of a helical spring design are an allowable stress level and the desired spring rate. The stress requirement can be fulfilled by many combinations of D and d. Having D and d selected, N_a is determined on the basis of the required spring rate. Finally, the free height can be obtained for a prescribed clash allowance. Note that, in some situations, the outside diameter, inside diameter, or working deflection may be limited. Clearly, when the spring comes out too large or too heavy, a stronger material must be used.

If the resulting design is likely to fail by buckling (Section 14.6), the process would be repeated with another combination of D and d. In any case, spring design is essentially an iterative problem. Some assumptions must be made to prescribe the values of enough variables to calculate the stresses and deflections. Usually, charts, nomographs, and computer programs have been used to simplify the spring design problem [7–11].

14.6   BUCKLING OF HELICAL COMPRESSION SPRINGS

A compression spring is loaded as a column and can buckle if it is too slender. In this section, we examine the problem of buckling of springs by their resistance to bending. For this purpose, consider a spring of length L and coil radius $D/2$ subjected to bending moment M (Figure 14.9a). The effect is an angular rotation θ. The bending and twisting moments at any section are (Figure 14.9b)

$$M_\alpha = M \sin\alpha, \qquad T_\alpha = M \cos\alpha \tag{a}$$

Derivation of the equation for helical spring deflection is readily accomplished using Castigliano's theorem as follows. Application of Eq. (5.37) gives

$$\theta = \frac{1}{EI} \int_0^L M_\alpha \frac{\partial M_\alpha}{\partial C}\, dx + \frac{1}{GJ} \int_0^L T_\alpha \frac{\partial T_\alpha}{\partial C}\, dx \tag{b}$$

in which $C = M$. Introducing Eqs. (a) into (b) together with $dx = ds = (D/2)\, d\alpha$, we obtain

$$\theta = M \int_0^{2\pi N_a} \left(\frac{\sin^2\alpha}{EI} + \frac{\cos^2\alpha}{GJ} \right) \left(\frac{D}{2}\, d\alpha \right)$$

Here $G = E/2(1+v)$ and for a round wire $J = 2I = \pi d^4/32$. Hence, the angular rotation of the entire spring is, after integrating,

$$\theta = \frac{64MDN_a}{Ed^4} \left(1 + \frac{v}{2} \right) \tag{14.17}$$

By analogy to a simple beam in pure bending, we may write using Eqs. (4.15) and (4.14)

$$\theta = \frac{ML}{EI_e} \tag{c}$$

The *equivalent* moment of inertia of the spring coil I_e is obtained by eliminating θ from Eqs. (14.17) and (c). In so doing, we have

$$I_e = \frac{Ld^4}{64DN_a(1+v/2)} \tag{14.18}$$

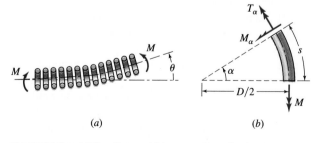

Figure 14.9   (a) Bending and (b) moment resultants at a cut section of a helical compression spring.

The preceding result may be used directly in Eq. (6.6) to ascertain the Euler buckling load of the spring in the form

$$P_{cr} = \frac{\pi^2 E I_e}{L_e^2} \qquad (14.19)$$

The quantity L_e denotes the effective column length (see Figure 6.2). The allowable value of compressive load is then found from $P_{all} = P_{cr}/n$, in which n represents a factor of safety.

ASPECT RATIO

It is important to point out that the measure of slenderness ratio for solid columns is not directly applicable to springs due to their much different form. An identical slenderness ratio is established as the *aspect ratio* of free length to mean coil diameter: h_f/D. In compression springs, it is important that the aspect ratios be not so great that buckling occurs. Figure 14.10 shows the results for the two end conditions given in Figures 6. 2c and 6.2d [1, 12–19].

Curve A in Figure 14.10 represents the springs supported between flat surfaces, a commonly used case. Observe from the figure that buckling occurs for conditions above and to the right of each curve. Clearly, as in the case of solid columns, the end conditions of the spring affect its tendency to buckle. *Curve B* renders the springs having one end free to tip. In these cases, the springs will buckle with smaller aspect ratios, as depicted in the figure.

We use Figure 14.10, rather than Eq. (14.19), to check readily for possible buckling of the spring. Note that, if buckling is indicated, the preferred solution is to redesign the spring. Otherwise, the buckling can be avoided by placing the spring either inside or outside a tube that provides a small clearance.

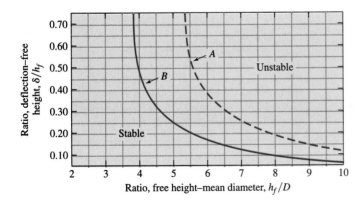

Figure 14.10 Buckling conditions for helical springs: (A) with parallel end plates (depicts the case of Figure 6.2d); (B) one end plate is free to tip (depicts the case of Figure 6.2c).

Case Study 14-1 | SPRING DESIGN OF A FEED MECHANISM FOR A HIGH-SPEED
CUTTING MACHINE

A helical compression spring of feed mechanism for a high-speed cutter (refer to Case Study 13-1) is shown in Figure 14.11. The spring is to support a load P without exceeding a deflection δ. Determine a satisfactory design. Will the spring buckle in service?

Figure 14.11   Compression spring for feed mechanism of high-speed cutter shown in Figure 13.9.

Given: $P = 60$ N, $\delta = 15$ mm

Assumptions: Clash allowance $r_c = 20\%$, spring index $C = 6$, safety factor $n = 2.2$. Loading is applied steadily. Ends are squared-ground and supported between flat surfaces.

Design Decision: Hard-drawn ASTM A227 wire of $G = 79$ GPa is used.

Solution: See Figures 13.9 and 14.11.

Arbitrarily *select* a 2-mm diameter wire. Then, using Eq. (14.12) and Table 14.2,

$$S_u = Ad^b = 1510(2^{-0.201}) = 1314 \text{ MPa}$$

Corresponding yield strength in shear is $S_{ys} = 0.42(1314) = 552$ MPa (from Table 14.3).

Stress Requirement. Rearranging Eq. (14.6) and setting $\tau = S_{ys}/n$,

$$d^2 = \frac{8PCn}{\pi S_{ys}}\left(1 + \frac{0.615}{C}\right)$$
$$= \frac{8(60)(6)(2.2)}{\pi(552 \times 10^6)}\left(1 + \frac{0.615}{6}\right)$$
$$d = 2.01 \text{ mm}$$

We then have $D = 6(2.01) = 12.06$ mm. Since $S_u = 1510(2.01^{-0.201}) = 1312 < 1314$ MPa, $d = 2.01$ mm is satisfactory.

Spring Rate Requirement. By Eq. (14.11),

$$\frac{P}{\delta} = \frac{dG}{8C^3 N_a}, \qquad \frac{60}{15} = \frac{2.01(79,000)}{8(6)^3 N_a}$$

Solving, $N_a = 22.97$. From Figure 14.7c, $h_s = (N_a + 2)d = 50.2$ mm. With a 20% clash allowance, the solid deflection is 120% of the working deflection. Hence,

$$\delta_s = 1.2\frac{P}{k} = 1.2\frac{60}{4} = 18 \text{ mm}$$

The free height is given by $h_f = h_s + \delta_s = 68.2$ mm.

Check for Buckling. For the extreme case of deflection ($\delta = \delta_s$),

$$\frac{\delta_s}{h_f} = \frac{18}{68.2} = 0.26, \qquad \frac{h_f}{D} = \frac{68.2}{12.06} = 5.66$$

Curve A in Figure 14.10 shows that the spring is far outside the buckling region and clearly safe.

Comment: Having the foregoing values of D, N_a, and h_f available, a technician can draw or make the compression spring for the high-speed cutter.

14.7   FATIGUE OF SPRINGS

Spring failures under fatigue loads are typical of that in torsional shear. A crack initiates at the surface on the inside of the coil and acts at 45° to the radial shear plane in the direction perpendicular to the tensile stress (see Figure 3.27). We note that helical springs are never used as both compression and extension springs: They do not normally experience a stress reversal. Moreover, these springs are assembled with a preload in addition to the working stress. The stress is thus prevented from being 0. The extreme case occurs if the preload drops to 0; that is, minimum shear stress τ_{min} equals 0.

Inasmuch as most failures are caused by fatigue, a poor surface is the worst disadvantage of hot-formed springs. *Presetting* refers to a process used in the manufacture of compression springs to produce residual stresses (see Section 3.13). This is done by making the spring longer than required and then compressing it to its solid height. Shot peening, discussed in Section 2.11, and presetting are two operations that add to the strength and durability of steel springs. The former, done after cooling, introduces a layer of compressive residual stresses. In a like manner, the latter always initiates surface residual stresses opposite to those caused by subsequent load application in the same direction as the presetting load. Maximum fatigue strengthening can be acquired using both the foregoing operations. The "set" spring loses some free length but gains the benefits described in the preceding. On the other hand, shot peening is most effective against cyclic loading in fatigue, while it has little benefit for statically loaded springs.

Data on fatigue strengths of round-wire helical springs are voluminous. Note that Eq. (8.1) defines the uncorrected endurance limit for fully reversed bending of steels as $S'_e = 700$ MPa for $S_u \geq 1400$ MPa. It can readily be verified by Eq. (14.12) and Table 14.2 that most spring wires smaller than 10-mm diameter are in this strength category. We conclude therefore that the torsional endurance limit of these spring-wire materials may be regarded as independent of size or their particular alloy composition. On this basis, the best data for the torsional endurance limit of spring steel wire of $d < 10$ mm is by Zimmerli [20]:

$$S'_{es} = 45 \text{ ksi} \quad (310 \text{ MPa}) \qquad \text{for unpeened springs}$$

$$S'_{es} = 67.5 \text{ ksi} \quad (465 \text{ MPa}) \qquad \text{for peened springs} \tag{14.20}$$

Equations (14.20) apply for infinite life with $\tau_{min} = 0$. As in the case of Table 14.2, the foregoing values were corrected for surface condition and size.

It should be mentioned that *corrosion*, even in a mild form, greatly reduces the fatigue strength. Also, if the spring operates under conditions of elevated temperature, there is a danger of creep or permanent deformation unless very low fluctuating stress values are used. Such effects become noticeable above 350°C and the ordinary spring steels cannot be used, as noted in Table 14.1.

14.8 DESIGN OF HELICAL COMPRESSION SPRINGS FOR FATIGUE LOADING

Springs are almost always subject to fluctuating or fatigue loads. The design process for dynamic loading is analogous to that for static loading with some significant variations. It is still an iterative problem. The design of helical springs for both static and fatigue loading can be readily computerized [11].

As pointed out in the preceding section, the stress-time diagram of Figure 14.12a expresses the worst condition that could occur for helical springs for pulsating shear when

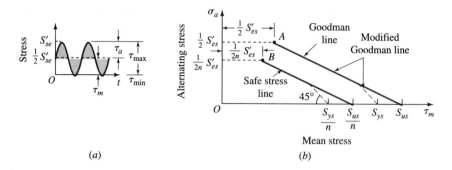

Figure 14.12   Fatigue loading: (a) endurance limit in pulsating shear test; (b) modified Goodman criteria for spring.

there is no preload; that is, when $\tau_{\min} = 0$. We assume that the endurance limit in shear S'_{es} is the value of shear (see Table 14.3) for which a part on the verge of failure after an "infinite" number of cycles. In many cases, S'_{es} may be based on 1 million or 10 million cycles per shear loading.

A dynamically loaded spring operates between two force levels, $P_{\max}$ and $P_{\min}$. Therefore, referring to Section 8.8, we define the *mean* and *alternating* axial *spring forces* as

$$P_m = \frac{1}{2}(P_{\max} + P_{\min}), \qquad P_a = \frac{1}{2}(P_{\max} - P_{\min})$$

The most common spring loading situation may involve both positive $P_{\max}$ and $P_{\min}$. The direct shear factor K_s is used for the mean stress τ_m only (see Section 14.3). We apply the Wahl factor K_w to the alternating stress τ_a. Equations (14.6) and (14.8) become then

$$\tau_m = K_s \frac{8 P_m C}{\pi d^2} \tag{14.21}$$

$$\tau_a = K_w \frac{8 P_a C}{\pi d^2} \tag{14.22}$$

The factors K_s and K_w are given by Eqs. (14.7) and (14.9), respectively. The notch sensitivity of high-hardness steels is near unity, $q \approx 1$. Hence, for analyzing fatigue loading, there is no need to correct K_w to the fatigue stress concentration factor K_f. Note that the *clash allowance* in for fatigue design should be based on a *maximum load*.

GOODMAN CRITERIA HELICAL SPRINGS

When the shear endurance limit S'_{es} of a spring wire is given, the Goodman or other fatigue failure criteria listed in Table 8.4 may be used. A torsional Goodman diagram can be constructed for any spring-loading situation. For $\tau_{\min} = 0$, the alternating stress is equal to the mean stress or $S'_{es}/2$. Hence, the line of failure can be drawn from point A to ultimate strength in shear S_{us} on the τ_m axis shown in Figure 14.12b. The line representing the safe stress is parallel to the line of failure and can be drawn (from point B) after dividing the endurance limit in shear S'_{es} by the factor of safety n. Recall from Chapter 8 that the stress points (τ_m, τ_a) falling on or below the safe stress line constitute a satisfactory spring design. The modified Goodman criterion includes another line (shown dashed in the figure), drawn from the yield strength in shear S_{ys} on the τ_m axis with a slope upward and to the left at $45°$.

The equation of the safe stress line is found by substituting the two stress points in the general equation of a line. In so doing, for the Goodman criterion,

$$\tau_a = \frac{\frac{1}{2}S'_{es}(S_{us}/2n - \tau_m)}{S_{us} - \frac{1}{2}S_{es}} \tag{14.23}$$

From Section 14.4, the ultimate strength in shear is given by $S_{us} = 0.67S_u$, in which S_u represents the ultimate tensile strength. When the ratio of range to mean stress is known, it may be convenient to rewrite Eq. (14.23) as follows:

$$\tau_m = \frac{S_{us}/n}{\dfrac{(\tau_a/\tau_m)(2S_{us} - S'_{es})}{S'_{es}} + 1} \tag{14.24}$$

An alternative form of Eq. (14.23) gives the factor of safety guarding against a failure:

$$n = \frac{S_{us} S'_{es}}{\tau_a(2S_{us} - S'_{es}) + \tau_m S'_{es}} \tag{14.25}$$

We note that Eqs. (14.23), (14.24), and, (14.25) could also be written based on the Soderberg criterion by replacing S_{us} by S_{ys}. Having found the mean shear stress τ_m, we may use it and the mean load P_m to obtain wire diameter d. Hence, through the use of Eq. (14.21),

$$d^3 = K_s \frac{8 P_m D}{\pi \tau_m} \quad \text{or} \quad d^2 = K_s \frac{8 P_m C}{\pi \tau_m} \tag{14.26}$$

When the safety factor is too low, the wire diameter, spring index, or the material can be altered to improve the result. The complete design includes consideration of the buckling discussed in Section 14.6 and surging of the springs, as is illustrated in the next example. Subsequent to several iterations, a reasonable combination of parameters can often be obtained.

COMPRESSION SPRING SURGE

A sudden compression of the end of a helical spring may form a compression wave that travels along the spring and is reproduced at the far end. This vibration wave, when it approaches resonance, is termed *surging*. It causes the coils to impact one another. The large forces due to both the excessive coil deflection and impacts fail the spring. To prevent this condition, the spring should not be cycled at a frequency close to its natural frequency.

Typically, the natural frequency of the spring should be greater than about 13 times that of any applied forcing frequency.

The *natural frequency* f_n of a helical compression spring depends on its end conditions. It can be shown that [2], for a spring with *fixed-fixed ends*,

$$f_n = \frac{d}{2\pi D^2 N_a} \sqrt{\frac{Gg}{2\gamma}} \text{ Hz} \tag{14.27}$$

This is twice that of a spring with *fixed-free ends*. Here, g is the acceleration of gravity and γ represents the weight per unit volume of the spring material. When d and D in inches, we have $g \approx 386$ in./s^2. For steel spring, $G = 11.5 \times 10^6$ psi, $\gamma = 0.285$ lb/in^3. Equation (14.27) then becomes

$$f_n = \frac{14,040d}{D^2 N_a} \text{ Hz} \tag{14.28}$$

In SI units, when d and D are in millimeters,

$$f_n = \frac{356,620d}{D^2 N_a} \text{ Hz} \tag{14.29}$$

The surge of a spring decreases the ability of the spring to control the motion of the machine part involved. For example, the engine valve (shown in a closed position in Figure P14.15) might tend to oscillate rather than to operate properly. In addition, the spring material under a compression wave is subjected to higher stresses, which may cause early fatigue failure. It is obvious therefore that springs used in high-speed machinery must have natural frequencies of vibration considerably in excess of the natural frequency of the motion they control.

EXAMPLE 14.3   A Helical Compression Spring: Design for Cyclic Loading

A helical compression spring for a cam follower is subjected to the load that varies between P_{min} and P_{max}. Apply the Goodman criterion to determine

    (a)   The wire diameter.

    (b)   The free height.

    (c)   The surge frequency.

    (d)   Whether the spring will buckle in service.

Given: $P_{min} = 300$ N, $P_{max} = 600$ N

Design Decisions: We use a chrome-vanadium ASTM A232 wire of $G = 79$ GPa; $r_c = 20\%$, $N_a = 10$, and $C = 7$. Both ends of spring are squared and ground. A safety factor of 1.3 is used due to uncertainty about the load.

Solution: The mean and alternating loads are

$$P_m = \frac{1}{2}(600 + 300) = 450 \text{ N}, \qquad P_a = \frac{1}{2}(600 - 300) = 150 \text{ N}$$

Equations (14.7) and (14.9) give

$$K_s = 1 + \frac{0.615}{7} = 1.088, \qquad K_w = \frac{28 - 1}{28 - 4} + \frac{0.615}{7} = 1.213$$

So we have, using Eqs. (14.22) and (14.23), $\tau_a/\tau_m = K_w P_a/K_s P_m = 0.372$.

(a)   Tentatively *select* a 6-mm wire diameter. Then, from Eq. (14.12) and Table 14.2, we have

$$S_u = Ad^b = 1790(6^{-0.155}) = 1356 \text{ MPa}$$

By Eq. (8.5) and Table 14.3, $S_{us} = 0.67(1356) = 908.5$ MPa and $S'_{es} = 0.2(1356) = 271$ MPa. Substitution of the numerical values into Eq. (14.24) results in

$$\tau_m = \frac{908.5/1.3}{\dfrac{(0.372)(2 \times 908.5 - 271)}{271} + 1} = 224 \text{ MPa}$$

Applying Eq. (14.26),

$$d^2 = K_s \frac{8P_m C}{\pi \tau_m} = 1.088 \frac{8(450)(7)}{\pi(224 \times 10^6)}$$

$$d = 6.24 \text{ mm}$$

Hence, $D = 7(6.24) = 43.68$ mm. Inasmuch as $S_u = 1790(6.24^{-0.155}) = 1348 < 1356$ MPa, $d = 6.24$ mm is satisfactory.

(b)   From Figure 14.7b, $h_s = (N_a + 2)d = 74.88$ mm. Using Eq. (14.11),

$$k = \frac{dG}{8C^3 N_a} = \frac{(6.24)(79,000)}{8(7)^3(10)} = 17.96 \text{ N/mm}$$

With a 20% clash allowance,

$$\delta_s = 1.2\frac{P_{\max}}{k} = 1.2(33.41) = 40.09 \text{ mm}$$

Thus,

$$h_f = 74.88 + 40.09 = 115 \text{ mm}$$

(c)   Through the use of Eq. (14.29),

$$f_n = \frac{356,620d}{D^2 N_a} = \frac{356,620(6.24)}{(43.68)^2(10)}$$

$$= 116.6 \text{ cps} = 6996 \text{ cpm}$$

Comment: If this corresponds to operating speeds (for equipment mounted on this spring), it may be necessary to redesign the spring.

(d) Check for buckling for extreme case of deflection ($\delta = \delta_s$):

$$\frac{\delta_s}{h_f} = \frac{40.09}{115} = 0.35, \qquad \frac{h_f}{D} = \frac{115}{43.68} = 2.63$$

Since (2.63, 0.35) is inside of the stable region of curve A in Figure 14.10, the spring will not buckle.

14.9 HELICAL EXTENSION SPRINGS

Figure 14.3c illustrates a round-wire helical extension spring. Observe that a hook and loop are provided to permit a pull force to be applied. The significant dimensions of a standard end hook or loop are shown in Figure 14.13. Most of the preceding discussion of compression springs applies equally to helical extension springs. The natural frequency of a helical extension spring with both ends fixed against axial deflection is the same as that for a helical spring in compression.

In extension springs, however, the coils are usually close wound so that there is an *initial tension* or so-termed preload P. No deflection therefore occurs until the initial tension built into the spring is overcome; that is, the applied load P becomes larger than initial tension ($P > P_i$). It is recommended that [1] the preload be built so that the resulting stress $\tau_{initial}$ calculated from Eq. (14.6) equals about $0.6S_u/C$. Here, S_u and C present ultimate strength and spring index, respectively.

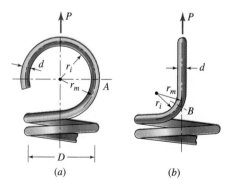

Figure 14.13 Points of maximum stress in hook in conventional extension springs: (a) stress at the cross section through A is due to axial force and bending; (b) stress at the cross section through B is due primarily to torsion.

Coil deflection of helical extension springs, through the use of Eq. (14.10) with $P = P - P_i$, is given as follows:

$$\delta = \frac{8N_a C^3 (P - P_i)}{dG} \tag{14.30}$$

The reduced coil diameter results in a lower stress because of the shorter moment arm. Hence, hook stresses can be reduced by winding the last few coils with a decreasing diameter D. No stress concentration factor is needed for the axial component of the load.

Active coils refers to all coils in the spring, not counting the end coils, which are bent to form a hook (Figure 14.3c). Depending on the details of the design, each end hook adds the equivalent of 0.1 to 0.5 helical coil. For an extension spring with two end hooks, the total number of coils is then

$$N_t = N_a + 2(0.1 \text{ to } 0.5) \tag{14.31}$$

As before, N_a represents the number of active coils.

The *spring rate* is expressed, by the application of Eq. (14.30), in the form

$$k = \frac{P - P_i}{\delta} = \frac{dG}{8N_a C^3} \tag{14.32}$$

The spring load is therefore

$$P = P_i + k\delta \tag{14.33}$$

The quantities k and δ are given by Eqs. (14.32) and (14.30), respectively.

Critical stresses occur in the end hooks or end loops of extension springs. The hooks must be designed so that the stress concentration effects produced by the presence of bends are decreased as much as possible. It is obvious that sharp bends should be avoided, since the stress concentration factor is higher for sharp bends. *Maximum bending stress* at section A (Figure 14.13a) and *maximum torsional stress* at section B (Figure 14.13b) in the bend of the end coil may be approximated respectively, by the formulas,

$$\sigma_A = K \frac{16PD}{\pi d^3} \tag{14.34}$$

$$\tau_B = K \frac{PD}{\pi d^3}$$

In each case, the stress concentration factor K is given by

$$K = \frac{r_m}{r_i} \tag{14.35}$$

where r_m is the mean radius and r_i represents the inside radius.

The *stresses in coils* are obtained from the same formulas as used in compression springs. In extension springs, a mechanical stop is desirable to limit deflection to an allowable value; while in compression springs, deflection is restricted by the solid deflection. Maximum stress values may be 70% of those used for extension springs of the identical compression springs.

14.10   TORSION SPRINGS

Torsion springs are of two general types: helical and spiral. The primary stress in a torsion spring is bending, with a moment being applied to each end of the wire. The analysis of curved beams discussed in Sections 3.7 and 16.8 is applicable. Springs of this kind are employed in door hinges, automotive starters, and so on, where torque is needed.

The yield strength S_y for torsion springs can be estimated from Table 14.3. Based on the energy of distortion criterion, we divide the S_{ys} in each part in Table 14.3 by the quantity 0.577. The endurance limit S_e for torsion springs can be found in a like manner: the S'_{es} in each part in Table 14.3 is divided by 0.577. The process of designing of torsion springs is very similar to that of the helical compression springs.

HELICAL TORSION SPRINGS

As depicted in Figure 14.14, helical torsion springs are wound in a way similar to extension or compression springs but with the ends shaped to transmit torque. These coil ends can have a variety of forms to suit the application. The coils are usually close wound like an extension spring but have no initial tension. We note that forces (P) should always be applied to arms of helical torsion springs to close the coil, as shown in the figure, rather than open it. The spring is usually placed over a supporting rod. The rod diameter is about 90% smaller than the inside diameter of the spring. Square or rectangular wire is in widespread use in torsion springs. However, round wire is often used in ordinary applications, since it costs less. The torque about the axis of the helix acts as a bending moment on each section of the wire. The material is therefore stressed in flexure. The bending stress can be obtained from curved beam theory. It is convenient to write the flexure formula in the form

$$\sigma = K\frac{Mc}{I} \tag{a}$$

Here,

$\sigma =$ maximum bending stress

$M =$ bending moment

$c =$ distance from the neutral axis to the extreme fiber

$I =$ moment of inertia about the neutral axis

$K =$ stress concentration factor

Figure 14.14   Helical torsion spring.

Wahl analytically determined the values for the stress concentration factors [2]. For *round wire*,

$$K_i = \frac{4C^2 - C - 1}{4C(C - 1)}$$

$$K_o = \frac{4C^2 + C - 1}{4C(C + 1)}$$

(14.36)

In the foregoing, the spring index $C = D/d$; the subscripts i and o refer to the inner and outer fibers, respectively. For *rectangular wire*,

$$K_i = \frac{3C^2 - C - 0.8}{3C(C - 1)}$$

$$K_o = \frac{3C^2 + C - 0.8}{3C(C + 1)}$$

(14.37)

where $C = D/h$. The quantity h represents the depth of the rectangular cross section. We see from these expressions that $K_i > K_o$, as expected.

The *maximum compressive bending stress* at the inner fiber of the helical torsion spring is therefore

$$\sigma_i = K_i \frac{Mc}{I}$$

(14.38)

Carrying the bending moment $M = Pa$ and the section modulus I/c of round and rectangular wires into Eq. (14.38) gives the bending stress. In so doing, stress on the inner fiber of the coil is:

$$\sigma_i = \frac{32Pa}{\pi d^3} K_i \qquad \text{(round wire)}$$

(14.39)

$$\sigma_i = \frac{6Pa}{bh^2} K_i \qquad \text{(rectangular wire)}$$

(14.40)

The quantity b is the width of rectangular cross section.

For commonly employed values of the spring index, $k = M/\theta_{\text{rev}}$, the curvature has no effect on the *angular deflection*. Through the use of Eqs. (4.15) and (4.14), we have

$$\theta_{\text{rev}} = \frac{1}{2\pi} \theta_{\text{rad}} = \frac{1}{2\pi} \frac{ML_w}{EI}$$

(14.41)

where

θ = angular deflection

L_w = length of wire = πDN_a

EI = flexural rigidity

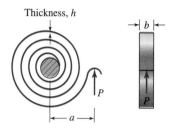

Figure 14.15   Spiral torsion spring.

For springs of round wire, to account for the friction between coils, based on experience [1], Eq. (4.41) is multiplied by the factor of 1.06. Interestingly, angle θ in some cases can be many complete turns, as in Example 14.4.

FATIGUE LOADING

A dynamically loaded torsion spring operates between two moment levels M_{max} and M_{min}. The tensile stress components occurring at the outside coil diameter of a round wire helical torsion is then

$$\sigma_{o,max} = K_o \frac{32M_{max}}{\pi d^3}, \qquad \sigma_{o,min} = K_o \frac{32M_{min}}{\pi d^3}$$

Hence, the mean and alternating stresses are

$$\sigma_{o,m} = \frac{\sigma_{o,max} + \sigma_{o,min}}{2}, \qquad \sigma_{o,a} = \frac{\sigma_{o,max} - \sigma_{o,min}}{2}$$

Having the mean and alternating stresses available, helical torsion springs are designed by following a procedure similar to that of helical compression springs.

SPIRAL TORSION SPRINGS

A spiral torsion spring (Figure 14.15) can also be analyzed by the foregoing procedure. Therefore, the highest stress occurring on the inner edge of the wire is given by Eqs. (14.39) and (14.40). Likewise, Eq. (14.41) can be applied directly to ascertain the angular deflection. Spiral springs are usually made of thin rectangular wire.

EXAMPLE 14.4   Spiral Torsion Spring: Design for Static Loading

For a torsional window-shade spring (Figure 14.15), determine the maximum operating moment and corresponding angular deflection.

Design Decisions: We select a music wire of $E = 207$ GPa; $d = 1.625$ mm, $D = 25$ mm, and $N_a = 350$. A safety factor of 1.5 is used.

Solution: By Eq. (14.12) and Table 14.2,

$$S_u = Ad^b = 2060(1.625^{-0.163}) = 1903 \text{ MPa}$$

From Eq. (7.23),

$$S_y = \frac{S_{ys}}{0.577} = 0.4\frac{1903}{0.577} = 1319 \text{ MPa}$$

Applying Eq. (14.36) with $C = 25/1.625 = 15.39$,

$$K_i = \frac{4(15.39)^2 - 15.39 - 1}{4(15.39)(15.39 - 1)} = 1.051$$

Through the use of Eq. (14.39), we have

$$M = Pa = \frac{\pi d^3 S_y/n}{32K_i} = \frac{\pi(1.625)^3(1319/1.5)}{32(1.051)}$$

$$= 352.5 \text{ N·mm}$$

The geometric properties of the spring are $L_w = \pi DN_a = \pi(25)(350) = 27{,}489$ mm and $I = \pi(1.625)^4/64 = 0.342$ mm⁴. Equation (14.41) results in

$$\theta_{\text{rad}} = \frac{ML_w}{EI} = \frac{352.5(27{,}489)}{(207 \times 10^3)(0.342)} = 136.9 \text{ rad}$$

Comment: The maximum moment winds the spring $136.9/2\pi = 21.8$ turns.

14.11 LEAF SPRINGS

A *leaf spring* is usually arranged as a cantilever or simply supported member. This thin beam or plate is also known as a flat spring, although it usually has some initial curvature. Springs in the form of a cantilever are often used as electrical contacts. For springs with uniform sections, we may use the results of Chapters 3 and 4. Recall from Sections 4.4 and 4.10 that, when the width b of the cross section is large compared with the depth h, it is necessary to multiply the deflection as given by the formula for a narrow beam section by $(1 - \nu^2)$, where ν is the Poisson's ratio.

A *cantilever spring* of *uniform stress* σ with a constant depth h, in a plan view looks like the triangle depicted in Figure 14.16 (see Section 3.8). However, near the free end, the wedge-shaped profile must be modified to have adequate strength to resist the shear force as depicted by the dashed lines in the figure. From the flexure formula, we have

$$\sigma = \frac{6PL}{bh^2} \qquad \text{(14.42)}$$

Figure 14.16
Cantilever spring of uniform stress.

As the cross section varies, end deflection δ may conveniently be obtained using Castigliano's theorem (see Section 5.6). It can be shown that

$$\delta = (1 - v^2)\frac{6P}{Eb}\left(\frac{L}{h}\right)^3$$

(14.43)

The corresponding spring rate is

$$k = \frac{P}{\delta} = \frac{Eb}{6(1 - v^2)}\left(\frac{h}{L}\right)^3$$

(14.44)

The quantity E represents modulus of elasticity.

MULTILEAF SPRINGS

Springs of varying width present a space problem. Multileaf springs are in widespread usage, particularly in automotive and railway service. An exact analysis of these springs is mathematically complex. For small deflections, an approximate solution can be obtained by the usual equations of beams, as shown in the following brief discussion.

A multileaf spring, approximating a triangular spring of uniform strength, is shown in Figure 14.17. Note that each half of the spring acts as a cantilever of length L. We observe from the figure that a constant strength triangle is cut into a series of leaves of equal width and rearranged in the form of a multileaf spring. A central bolt or clamp, used to hold the leaves together, causes a stress concentration. The triangular spring and equivalent

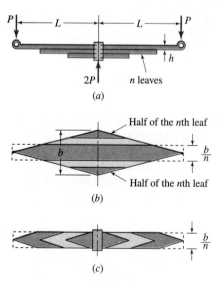

Figure 14.17   Multileaf spring: (a) front view of actual spring; (b) top view of approximation; (c) top view of equivalent spring.

Figure 14.18 Example 14.5.
Automotive-type leaf spring.

multipleaf spring have the *identical* stress and deflection characteristics, with the exception that the interleaf friction provides damping in the multileaf spring. Also, the multileaf spring can resist full load in only one direction; that is, leaves tend to separate when loaded in opposite direction. However, this is partially overcome by clips, as in vehicle suspension springs (Figure 14.18).

The Goodman criterion may be used in the design of leaf springs subject to cyclic loading, as illustrated in the solution of the following numerical problem.

Automotive-Type Multileaf Spring: Design for Fatigue Loading | **EXAMPLE 14.5**

A six-leaf spring is subjected to a load at the center that varies between P_{max} and P_{min} (Figure 14.18). Estimate the total length $2L$ and width of each leaf.

Given: $P_{min} = 80$ lb, $P_{max} = 400$ lb

Assumptions: Stress concentration at the center is such that $K_f = 1.2$. Use a survival rate of 50% and $C_f = C_s = 1$.

Design Decisions: We use a steel alloy spring of $S_u = 200$ ksi, $S'_e = 78$ ksi, $E = 30 \times 10^6$ psi, $v = 0.3$, $h = 0.25$ in., $k = 140$ lb/in. The material is shot peened. A safety factor of 1.4 is applied.

Solution: From Table 8.3, $C_r = 1$. The modified endurance limit, by Eq. (8.6), $S_e = (1)(1)(1)(/1.2)78 = 65$ ksi. Each half of a spring acts as a cantilever supporting half of the total load. The mean and the alternating loads are therefore

$$P_m = \frac{400 + 80}{2} = 240 \text{ lb}, \qquad P_a = \frac{400 - 80}{2} = 160 \text{ lb}$$

Inasmuch as bending stress is directly proportional to the load, we have $\sigma_a/\sigma_m = P_a/P_m = 2/3$. The mean stress, using Eq. (14.42), is

$$\sigma_m = \frac{6P_m L}{bh^2} = \frac{6(240)L}{b(0.25)^2} = 23,040\frac{L}{b} \qquad \text{(a)}$$

Substituting the given numerical values into Eq. (8.20), we have

$$\sigma_m = \frac{S_u/n}{\dfrac{\sigma_a}{\sigma_m}\dfrac{S_u}{S_e} + 1} = \frac{200/1.4}{\dfrac{2}{3}\dfrac{200}{65} + 1} = 46.82 \text{ ksi}$$

From Eq. (a),

$$46{,}820 = 23{,}040\frac{L}{b} \quad \text{or} \quad b = 0.492L \qquad \textbf{(b)}$$

Because the spring is loaded at the center with $2P$, Eq. (14.44) becomes $k = Ebh^3/3L^3(1 - \nu^2)$. Introducing the given data results in

$$140 = \frac{(30 \times 10^6)(0.492L)(0.25)^3}{3L^3(0.91)}$$

$$L = 24.56 \text{ in.}$$

Hence, Eq. (b) gives, $b = 0.492(24.56) = 12.08$ in. The *overall length* is $2L$, 49.12 in. The width of each of the six leaves is one-sixth of 12.08 in., or about 2.01 in.

14.12   MISCELLANEOUS SPRINGS

Many spring functions may also be acquired by the elastic bending of thin plates and shells of various shapes and by the blocks of rubber. Hence, there are spring washers, clips, constant-force springs, volute springs, rubber springs, and so on. A *volute spring* is a wide, thin strip of steel wound flat so that the coils fit inside one another, as shown in Figure 14.19. These springs have more lateral stability than helical compression springs, and rubbing of adjacent turns provides high damping. Here, we briefly discuss three commonly encountered types of miscellaneous springs.

CONSTANT-FORCE SPRINGS

The constant-force (Neg'ator) spring is a prestressed strip of flat spring stock that coils around a bushing or successive layers of itself (Figure 14.20). Usually, the inner coil is fastened to a flanged drum. When the spring is deflected by pulling on the outer end of the

Figure 14.19   Volute spring.

Figure 14.20   Constant-force spring.

Figure 14.21 Cross section through a Belleville spring.

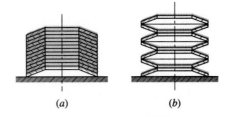

Figure 14.22 Belleville springs or washers: (a) in parallel stack; (b) in series stack.

coil, a nearly constant resisting force develops and there is a tendency for the material to recoil around itself. A uniform-force spring is widely employed for counterbalancing loads (such as in window sash), cable retractors, returning typewriter carriages, and making constant-torque spring motors. It provides very large deflection at about a constant pull force [1, 15].

BELLEVILLE SPRINGS

Belleville springs or *washers,* also known as coned-disk springs (Figure 14.21), patented by J. F. Belleville in 1867, are often used for supporting very *large* loads with *small* deflections. Some applications include various bolted connections, clutch plate supports, and gun recoil mechanisms. On loading, the disk tends to flatten out, spring action being obtained thus. The load-deflection characteristics are changed by varying the ratio h/t between cone height h and thickness t. Belleville springs are extremely compact and may be used singly or in combination of multiples of identical springs to meet needed characteristics. The forces associated with a coned-disk spring can be multiplied by stocking them in parallel (Figure 14.22a). On the other hand, the deflection corresponding to a given force can be increased by stacking the springs in series as shown in Figure 14.22b.

The theory of the Belleville springs is complicated. The following formulas are based on the simplifying assumption that radial cross sections of the spring do not distort during deflection. The results are in approximate agreement with available test data [1, 2, 21]. As is the case for a truncated cone shell, the upper edge of the spring is in compression and the lower edge is in tension [22].

The *load-deflection* relationship can be expressed in the form

$$P = \frac{E\delta}{(1-v^2)Kb^2}\left[\left(h - \frac{\delta}{2}\right)(h-\delta)t + t^3\right] \tag{14.45a}$$

where

$$K = \frac{6}{\pi \ln\alpha}\left(\frac{\alpha-1}{\alpha}\right)^2 \tag{14.45b}$$

The load at the *flat position* ($\delta = h$) is given by

$$P_{\text{flat}} = \frac{Eht^3}{(1-v^2)Kb^2} \tag{14.45c}$$

Here,

$$P = \text{load}$$
$$\delta = \text{deflection}$$
$$a = \text{inside radius}$$
$$b = \text{outside radius}$$
$$h = \text{cone height}$$
$$t = \text{thickness}$$
$$\alpha = \text{radius ratio} = b/a$$

Zero deflection and load ($\delta = 0$ and $P = 0$) are taken at the *free position* depicted in Figure 14.21.

Load-deflection characteristics are changed by varying the ratio between cone height and thickness, h/t. Figure 14.23 illustrates force-deflection curves for Belleville washers with four different h/t ratios. These curves are generated by applying Eqs. (14.45), where 1.0 deflection and 1.0 force refer to the deflection at the flat condition and the force at the flat condition, respectively [16, 17]. We see from the figure that coned-disk springs have nonlinear $P - \delta$ properties. For low values ($h/t = 0.4$), the spring acts almost linearly, and large h/t values result in prominent nonlinear behavior. At $h/t = \sqrt{2}$, the central portion of the curve approximates a horizontal line; that is, the load is nearly constant over a considerable deflection range. In the range $\sqrt{2} < h/2 \leq \sqrt{8}$, a prescribed force corresponds to more than one deflection. A phenomenon occurring at $h/t > \sqrt{2}$ is termed *snap-through buckling*, at which the spring deflection becomes unstable.

Interestingly, in snap-through buckling the spring quickly deflects or snaps to the next stable position. It can be shown that [6], if $h/t > \sqrt{8}$ the spring can snap into a deflection position for which the calculated force becomes negative. Then, a load in the direction opposite to the initial load will be required to return the spring to its unloaded configuration.

Stress distribution in the washer is nonuniform. The *largest* stress σ_A occurs at the upper inner edge A (convex side) at deflection δ and is compressive. The outside lower

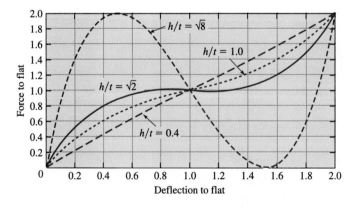

Figure 14.23   Force-deflection curves for Belleville springs.

edge B (concave side) has the largest tensile stress δ_B. The expressions for the foregoing stresses are

$$\sigma_A = -\frac{E\delta}{(1-v^2)Kb^2}\left[c_1\left(h-\frac{\delta}{2}\right)+c_2t\right] \qquad \text{(14.46a)}$$

$$\sigma_B = \frac{E\delta}{(1-v^2)Kb^2}\left[c_3\left(h-\frac{\delta}{2}\right)+c_4t\right] \qquad \text{(14.46b)}$$

where

$$c_1 = \frac{6}{\pi\,\ln\alpha}\left(\frac{\alpha-1}{\ln\alpha}-1\right)$$

$$c_2 = \frac{6}{\pi\,\ln\alpha}\left(\frac{\alpha-1}{2}\right)$$

$$c_3 = \left[\frac{\alpha\ln\alpha-(\alpha-1)}{\ln\alpha}\right]\left[\frac{\alpha}{(\alpha-1)^2}\right] \qquad \text{(14.46c)}$$

$$c_4 = \frac{\alpha}{2(\alpha-1)}$$

Stresses are highly concentrated at the edges of Belleville springs. When the yield strength is exceeded, a redistribution of the stresses occurs due to localized yielding. The compressive stress σ_A given by Eq. (14.46a) controls the design for static loading. If the spring is under *dynamic* loading, the alternating and mean stresses are determined from tensile stress σ_B defined by Eq. (14.46b). The factor of safety, according to the Goodman criterion, is found from Eq. (8.22).

RUBBER SPRINGS

A rubber spring and cushioning device is referred to as a *rubber mount* (Figure 14.24). Springs of this type are widely used due to their essentially shock and vibration damping

(a)   (b)

Figure 14.24 Cylindrical rubber mounts: (a) with shear loading; (b) with torsion loading.

qualities and low elastic moduli. The foregoing properties help dissipate energy and prevent sound transmission. Stresses and deformations in the rubber mounts for small deflections can be derived by the use of appropriate equations of mechanics of materials.

A cylindrical rubber spring with *direct shear loading* is shown in Figure 14.24a. The rubber is bonded to a steel ring on the outside and a steel shaft in the center. The shear stress τ at radius r is

$$\tau = \frac{P}{2\pi rh} \tag{14.47}$$

Maximum deflection δ occurs at inner edge ($r = d/2$):

$$\delta = \frac{P}{2\pi hG} \ln \frac{D}{a} \tag{14.48}$$

Here,

P = load

d = inner radius

h = depth of mount

D = outer radius

A cylindrical spring with *torsional shear loading* is depicted in Figure 14.24b. The maximum shear stress taking place at the inner edge ($r = d/2$) is given by

$$\tau_{max} = \frac{2T}{\pi d^2 h} \tag{14.49}$$

The angular rotation of the shaft, or maximum angle of twist ϕ, is

$$\phi = \frac{T}{\pi hG}\left(\frac{1}{d^2} - \frac{1}{D^2}\right) \tag{14.50}$$

The quantity T represents the torque.

Note that rubber does not follow Hooke's law but becomes increasingly stiff as the deformation is increased. The modulus of elasticity is contingent on the durameter hardness number of the rubber chosen for the mount [12]. The results of calculations must therefore be considered only approximate.

References

1. Associated Spring-Barnes Group. *Design Handbook*. Bristol, CN: Associated Spring-Barnes Group, 1987.
2. Wahl, A. M. *Mechanical Springs*, 2nd ed. New York: McGraw-Hill, 1963.
3. Shigley, J. E., and C. R. Mischke. *Standard Handbook of Machine Design*, 2nd ed. New York: McGraw-Hill, 1996.

4. Carlson, H. C. R. "Selection and Application of Spring Materials." *Mechanical Engineering* 78 (1956), pp. 331–34.

5. Carlson, H. C. R. *Spring Designer's Handbook.* New York: Marcel Dekker, 1978.

6. Burr, A. H., and J. B. Cheatham. *Mechanical Analysis and Design,* 2nd ed. Upper Saddle River, NJ: Prentice Hall, 1995.

7. *Metals Handbook,* vol.1, 9th ed. Metals Park, OH: ASM, 1978.

8. Rothbart, H. A., ed. *Mechanical Design and Systems Handbook,* 2nd ed. New York: McGraw-Hill, 1995.

9. Avallone, E. A., and T. Baumeister III, eds. *Mark's Standard Handbook for Mechanical Engineers,* 10th ed. New York: McGraw-Hill, 1996.

10. Samonov, C. "Computer-Aided Design of Helical Compression Springs." ASME paper no. 80-DET-69, 1980.

11. Dietrich, A. "Home Computers Aid Spring Design." *Design Engineering* (June 1981), pp. 31–35.

12. Spotts, M. F., and T. E. Shoup. *Machine Design,* 7th ed. Upper Saddle River, NJ: Prentice Hall, 1998.

13. Juvinall, R. C., and K. M. Marshak. *Fundamentals of Machine Component Design,* 3rd ed. New York: Wiley, 2000.

14. Hamrock, B. J., B. O., Jacobson, and S. R. Schmid. *Fundamentals of Machine Elements.* New York: McGraw-Hill, 1999.

15. Levinson, I. J. *Machine Design.* Reston, VA: Reston/Prentice Hall, 1978.

16. Mott, R. L. *Machine Elements in Mechanical Design,* 2nd ed. New York: Macmillan, 1992.

17. Norton, R. L. *Machine Design: An Integrated Approach,* 2nd ed. Upper Saddle River, NJ: Prentice Hall, 2000.

18. Shighley, J. E., and C. R. Mischke. *Mechanical Engineering Design,* 6th ed. New York: McGraw-Hill, 2001.

19. Dimarogonas, A. D. *Machine Design: A CAD Approach.* New York: Wiley, 2001.

20. Zimmerli, P. F. *Human Failures in Spring Design.* Mainspring, Associated Spring Corp., Bristol, CN, August–September 1957.

21. Ortwein, W. C. *Machine Component Design.* St. Paul, MN: West, 1990.

22. Ugural, A. C. *Stresses in Plates and Shells,* 2nd ed. New York: McGraw-Hill, 1999.

PROBLEMS

Sections 14.1 through 14.6

14.1 A steel torsion bar is used as a counterbalance spring for the trunk lid of an automobile (Figure 14.2a). Determine, when one end of the bar rotates 80° relative to the other end,

(*a*) The change in torque.

(*b*) The change in shear stress.

Given: $L = 1.25$ m, $d = 8$ mm, $G = 79$ Pa

14.2 A steel bar supports a load of 2 kN with a moment arm $R = 150$ mm (Figure 14.2a). Calculate

(*a*) The wire diameter.

(*b*) The length for a deflection of 40 mm.

Given: $n = 1.5$, $S_{ys} = 350$ MPa, $G = 79$ GPa

14.3 A helical spring must exert a force of 1 kN after being released 20 mm from its most highly compressed position. Determine the number of active coils.

Design Assumptions: The loading is static, $\tau_{all} = 450$ MPa, $G = 29$ GPa, $d = 7$ mm, and $C = 5$.

14.W1 Using the website at www.leesspring.com, rework Example 14.1.

14.W2 Check the site at www.acxesspring.com to review the common spring materials presented. List five commonly employed wire spring materials and their mechanical properties.

14.4 A helical compression spring used for static loading has $d = 3$ mm, $D = 15$ mm, $N_a = 10$, and squared ends. Determine

(a) The spring rate and the solid height.

(b) The maximum load that can be applied without causing yielding.

Design Decision: The spring is made of ASTM A227 hard-drawn steel wire of $G = 79$ GPa.

14.5 A helical compression spring is to support a 2 kN load. Determine

(a) The wire diameter.

(b) The free height.

(c) Whether buckling will occur in service.

Given: The spring has $r_c = 10\%$, $C = 5$, and $k = 90$ N/mm.
Assumptions: Both ends are squared and ground and constrained by parallel plates.
Design Decisions: The spring is made of steel of $S_{ys} = 500$ MPa, $S'_{es} = 280$ MPa, $G = 79$ GPa. Use a safety factor of 1.3.

14.6 A helical compression spring with ends squared and ground has $d = 1.8$ mm, $D = 15$ mm, $r_c = 15\%$, and $h_s = 21.6$ mm. Determine, using a safety factor of 2,

(a) The free height.

(b) Whether the spring will buckle in service, if one end is free to tip.

Design Decision: The spring is made of steel having $S_{ys} = 900$ MPa and $G = 79$ GPa.

14.7 Design a helical compression spring with squared and ground ends for a static load of 40 lb, $C = 8$, $k = 50$ lb/in., $r_c = 20\%$, and $n = 2.5$. Also check for possible buckling.

Assumption: The ends are constrained by parallel plates.
Design Decision: The spring is made of steel of $S_{ys} = 60$ ksi and $G = 11.5 \times 10^6$ psi.

14.8 A machine that requires a helical compression spring of $k = 120$ lb/in., $\tau_{all} = 75$ ksi, $r_c = 10\%$, $D = 3$ in., and is to support a static load of 400 lb. Determine

(a) The wire diameter.

(b) The free height.

(c) Whether the spring will buckle in service, if one end is free to tip.

Assumption: The ends are squared.
Design Decision: The spring is made of steel having $G = 11.5 \times 10^6$ psi.

Sections 14.7 and 14.8

14.9   Redo Problem14.5 for a load that varies between 2 kN and 4 kN, using the Soderberg relation. Also determine the surge frequency.

14.10   A helical compression spring for a cam follower supports a load that varies between 30 and 180 N. Determine

(a)   The factor of safety, according to the Goodman criterion.

(b)   The free height.

(c)   The surge frequency.

(d)   Whether the spring will buckle in service.

Design Decisions: The spring is made of music wire. Both ends are squared and ground; one end is free to tip.
Given: $d = 3$ mm, $D = 15$ mm, $N_a = 22$, $r_c = 10\%$, $G = 79$ GPa

14.11   A helical compression spring, made of 0.2-in. diameter music wire, carries a fluctuating load. The spring index is 8 and the factor of safety is 1.2. If the average load on the spring is 100 lb, determine the allowable values for the maximum and minimum loads. Employ the Goodman theory.

14.12   A helical compression spring made of a music wire has $d = 5$ mm, $D = 24$ mm, and $G = 79$ GPa. Determine

(a)   The factor of safety, according to the Goodman relation.

(b)   The number of active coils.

Requirements: The height of the spring varies between 65 and 72 mm with corresponding loads of 400 and 240 N.

14.13   A steel helical compression spring is to exert a force of 4 lb when its height is 3 in. and a maximum load of 18 lb when compressed to a height of 2.6 in. Determine, using the Soderberg criterion with a safety factor of 1.6,

(a)   The wire diameter.

(b)   The solid deflection.

(c)   The surge frequency.

(d)   Whether the spring will buckle in service, if ends are constrained by parallel plates.

Given: The spring has $C = 6$, $S_{ys} = 80$ ksi, $S'_{es} = 45$ ksi, $G = 11.5 \times 10^6$ psi, $r_c = 10\%$.
Design Assumption: Ends will be squared and ground.

14.14   Resolve Problem 14.13 for the case in which the helical spring is to exert a force of 2 lb at 5 in. height and a maximum load of 10 lb at 4.2 in. height.

14.15   An engine valve spring must exert a force of 300 N when the valve is closed (as shown in Figure P14.15) and 500 N when the valve is open. Apply the Goodman theory with a safety

Figure P14.15

factor of 1.6 to calculate

(*a*)   The wire diameter.

(*b*)   The number of active coils.

Given: The lift is 8 mm.
Design Decisions: The spring is made of steel having $S_{us} = 720\,\text{MPa}$, $S'_{es} = 330\,\text{MPa}$, $G = 79\,\text{GPa}$, and $C = 6$.

14.16   A helical spring, made of hard-drawn wire having $G = 29\,\text{GPa}$, supports a continuous load. Determine

(*a*)   The factor of safety based on the Soderberg criterion.

(*b*)   The free height.

(*c*)   The surge frequency.

(*d*)   Whether the spring will buckle in service.

Given: $d = 6\,\text{mm}$, $D = 30\,\text{mm}$, $r_c = 20\%$
Design Requirements: The ends are squared and ground; one end is free to tip. In the most-compressed condition, the force is 600 N; after 13 mm of release, the minimum force is 340 N.

Sections 14.9 through 14.12

14.17   A helical tension spring has $d_1 = 3\,\text{mm}$ and $D_1 = 30\,\text{mm}$. If a second spring is made of the same material and the same number of coils with $D_2 = 240$ mm, find the wire diameter d_2 that would be required to give the same spring rate as the first spring.

14.18   An extension coil spring is made of 0.02-in. music wire and has a mean diameter of coil of 0.2 in. The spring is wound with a pretension of 0.2 lb, and the load fluctuates from this value up to 1.0 lb. Determine the factor of safety guarding against a fatigue failure. Use the Goodman criterion.

Table 15.2 Basic dimensions of ISO (metric) screw threads

Nominal diameter, d (mm)	Coarse threads		Fine threads	
	Pitch, p (mm)	Tensile stress area, A_t (mm²)	Pitch, p (mm)	Tensile stress area, A_t (mm²)
2	0.4	2.07		
3	0.5	5.03		
4	0.7	8.78		
5	0.8	14.2		
6	1	20.1		
7	1	28.9		
8	1.25	36.6	1.25	39.2
10	1.5	58.0	1.25	61.2
12	1.75	84.3	1.25	92.1
14	2	115	1.5	125
16	2	157	1.5	167
18	2.5	192	1.5	216
20	2.5	245	1.5	272
24	3	353	2	384
30	3.5	561	2	621
36	4	817	2	915
42	4.5	1120	2	1260
48	5	1470	2	1670
56	5.5	2030	2	2300
64	6	2680	2	3030

SOURCE: [1].

Notes: Metric threads are specified by nominal diameter and pitch in millimeters; for example, M10 × 1.5. The letter M, which proceeds the diameter, is the clue to the metric designation.

Root or minor diameter $d_r \approx d - 1.227p$.

Figure 15.4 Typical power screw thread forms. All threads shown are external, $d_m = (d + d_r)/2$: (a) Acme; (b) square; (c) modified square.

Table 15.3   Standard sizes of power screw threads

Major diameter, d (in.)	Threads per inch	
	Acme, Acme stub	Square and modified square
$\frac{1}{4}$	16	10
$\frac{1}{2}$	10	$6\frac{1}{2}$
$\frac{5}{8}$	8	$5\frac{1}{2}$
$\frac{3}{4}$	6	5
$\frac{7}{8}$	6	$4\frac{1}{2}$
1	5	4
$1\frac{1}{4}$	5	$3\frac{1}{2}$
$1\frac{1}{2}$	4	2
$1\frac{3}{4}$	4	$2\frac{1}{2}$
2	4	$2\frac{1}{4}$
$2\frac{1}{4}$	3	$2\frac{1}{4}$
$2\frac{1}{2}$	3	2
$2\frac{3}{4}$	3	2
3	2	$1\frac{3}{4}$
4	2	$1\frac{1}{2}$
5	2	

SOURCE: [2].

15.3   MECHANICS OF POWER SCREWS

As noted previously, a power screw, sometimes called the *linear actuator* or *translation screw,* is in widespread usage in machinery to change angular motion into linear motion, to exert force, and to transmit power. Applications include the screws for vises, C-clamps, presses, micrometers, jacks (Figure 15.5), valve stems, and the lead screws for lathes and other equipment. In the usual configuration, the nut rotates in place, and the screw moves axially. In some designs, the screw rotates in place, and the nut moves axially. Forces may be large, but motion is usually slow and power is small. In all the foregoing cases, power screws operate on the same principle.

A simplified drawing of a screw jack having the Acme thread is shown in Figure 15.6. The load W can be lifted or lowered by the rotation of the nut that is supported by a washer, called a *thrust collar* (or a *thrust bearing*). It is, of course, assumed that the load and screw are prevented from turning when the nut rotates. Hence, there needs to be some friction at the load surface to prevent the screw from turning with the nut. Alternatively, the power screw could be turned against a nut that is prevented from turning to lift or lower the load. In either case, there is significant friction between the screw and nut as well as between the nut and the collar. Ordinarily, the screw is a hard steel, while the nut is made of a softer material (e.g., an alloy of aluminum, nickel, and bronze) to allow the parts to move smoothly.

Figure 15.5 Worm-gear screw jack. (Courtesy of Joyce/Dayton Corp.)

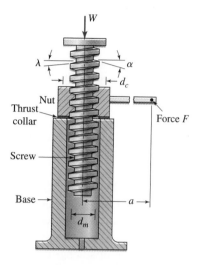

Figure 15.6 Schematic representation of power screw used as a screw jack. Note: Only the nut rotates in this model: d_m = mean thread diameter, d_c = mean collar diameter.

In this section, we develop expressions for ascertaining the values of the torque needed to lift and lower the load using a jack. We see from Figure 15.6 that turning the nut forces each portion of the nut thread to climb an inclined plane. This plane is depicted by unwrapping or developing one revolution of the helix in Figure 15.7a, which includes a small block representing the nut being slid up the inclined plane of an Acme thread. The forces acting on the nut as a free-body diagram are also noted in the figure. Clearly, one edge of the thread forms the hypotenuse of the right triangle, having a base as the circumference of the mean-thread-diameter circle and as the lead. Therefore,

$$\tan \lambda = \frac{L}{\pi d_m} \qquad (15.2)$$

where

λ = helix or lead angle

L = lead

d_m = mean diameter of thread contact surface

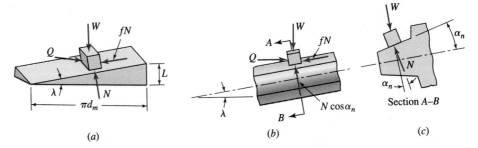

Figure 15.7   Forces acting on an Acme screw-nut interface when lifting load W: (a) a developed screw thread; (b) a segment of the thread; (c) thread angle measured in the plane normal to thread, α_n.

The preceding notation is the same as for worms (see Section 12.9) except that unnecessary subscripts are omitted.

TORQUE TO LIFT THE LOAD

The *sum* of all loads and normal forces acting on the entire thread surface in contact are denoted by W and N, respectively. To lift or rise the load, a tangential force Q acts to the right and the friction force fN acts to oppose the motion (Figure 15.7). The quantity f represents the coefficient of sliding friction between the nut and screw, or the coefficient of *thread friction*. The thread angle increases the frictional force by the wedging action of the threads. The conditions of equilibrium of the horizontal and vertical forces give

$$\sum F_h = 0: \quad Q - N(f\cos\lambda + \cos\alpha_n \sin\lambda) = 0$$

$$\sum F_v = 0: \quad W + N(f\sin\lambda - \cos\alpha_n \cos\lambda) = 0$$

(a)

where α_n is the normal thread angle and the other variables are defined in the figure. Inasmuch as we are not interested in the normal force N, we eliminate it from the foregoing equations and solve the result for Q. In so doing, we have

$$Q = W\frac{f + \cos\alpha_n \sin\lambda}{\cos\alpha_n \cos\lambda - f\sin\lambda}$$

(15.3)

The screw torque required to move the load up the inclined plane, after dividing the numerator and denominator by $\cos\lambda$, is then

$$T = \frac{1}{2}Qd_m = \frac{Wd_m}{2}\frac{f + \cos\alpha_n \tan\lambda}{\cos\alpha_n - f\tan\lambda}$$

(15.4)

But, the thrust collar also contributes a friction force. That is, the normal reactive force acting on contact surface due to W results in an additional force $f_c W$. Here, f_c is the sliding coefficient of the *collar friction* between the thrust collar and the surface that supports the screw. It is assumed that this frictional force acts at the mean collar diameter d_c (Figure 15.6). The torque needed to overcome collar friction is

$$T = \frac{W f_c d_c}{2} \tag{15.5}$$

The required total *torque T_u to lift the load* is found by addition of Eqs. (15.4) and (15.5):

$$T_u = \frac{W d_m}{2} \frac{f + \cos\alpha_n \tan\lambda}{\cos\alpha_n - f \tan\lambda} + \frac{W f_c d_c}{2} \tag{15.6}$$

TORQUE TO LOWER THE LOAD

The analysis of lowering a load is exactly the same as that just described, with the exception that the directions of Q and $f N$ (Figure 15.7b) are reversed. This leads to the equation for the total required *torque T_d to lower the load* as

$$T_d = \frac{W d_m}{2} \frac{f - \cos\alpha_n \tan\lambda}{\cos\alpha_n + f \tan\lambda} + \frac{W f_c d_c}{2} \tag{15.7}$$

VALUES OF FRICTION COEFFICIENTS

When a plain *thrust collar* is used, as shown in Figure 15.6, values of f and f_c vary customarily between 0.08 and 0.20 under conditions of ordinary service, lubrication, and the common materials of steel and cast iron or bronze. The lowest value applies for good workmanship, the highest value applies for poor workmanship, and some in between value for other work quality. The preceding range includes both starting and running friction. *Starting* friction can be about 4/3 times *running* friction. Should a rolling *thrust bearing* be used, f_c would usually be low enough (about 0.008 to 0.02) that collar friction can be omitted. For this case, the second term in Eqs. (15.6) and (15.7) is eliminated.

VALUES OF THREAD ANGLE IN THE NORMAL PLANE

A relationship between normal thread angle α_n, thread angle α, and helix angle λ can be obtained from a comparison of thread angles measured in axial plane and normal plane. Referring to Figures 15.6 and 15.7c, it can readily be verified that

$$\tan\alpha_n = \cos\lambda \tan\alpha \tag{15.8}$$

In most applications, λ is relatively small, and hence $\cos\lambda \approx 1$. So, we can set $\alpha_n \approx \alpha$ and Eq. (15.6) becomes

$$T_u = \frac{W d_m}{2} \frac{f + \cos\alpha \tan\lambda}{\cos\alpha - f \tan\lambda} + \frac{W f_c d_c}{2} \tag{15.9}$$

Obviously, for the case of the *square thread*, $\alpha = \alpha_n = 0$, and $\cos\alpha = 1$ in the preceding expressions.

15.4   OVERHAULING AND EFFICIENCY OF POWER SCREWS

A self-locking screw requires a positive torque to lower the load. This is a useful provision, particularly in screw jack applications. *Self-locking* refers to a condition in which the screw cannot be turned by applying an axial force of any magnitude to the nut. If collar friction is neglected, Eq. (15.7) shows that the *condition* for self-locking is

$$f \geq \cos\alpha_n \tan\lambda \tag{15.10}$$

For a square thread, the foregoing equation reduces to

$$f \geq \tan\lambda \tag{15.10a}$$

In other words, self-locking is obtained when the coefficient of thread friction is equal to or greater than the tangent of the thread helix angle. Note that Eq. (15.10) presumes a static situation and most power screws are self-locking.

Overhauling or back-driving screw is one that has low enough friction to enable the load to lower itself, by causing the screw to spin. In this situation, the inclined plane in Figure 15.7b moves to the right and force Q must act to the left to preserve uniform motion. It can be shown that the *torque T_o of overhauling* screw is

$$T_o = \frac{W d_m}{2} \frac{-f + \cos\alpha_n \tan\lambda}{\cos\alpha_n + f \tan\lambda} - \frac{W f_c d_c}{2} \tag{15.11}$$

A negative external lowering torque must now be maintained to keep the load from lowering.

SCREW EFFICIENCY

Screw efficiency is the ratio of the torque required to raise a load without friction to the torque required with friction. Using Eq. (15.6), *efficiency* is expressed in the form

$$e = \frac{d_m \tan \lambda}{d_m \dfrac{f + \cos\alpha_n \tan \lambda}{\cos\alpha_n - f \tan \lambda} + d_c f_c} \tag{15.12}$$

We observe from this equation that efficiency depends on only the screw geometry and the coefficient of friction. If the *collar friction* is *neglected*, the efficiency becomes

$$e = \frac{\cos\alpha_n - f \tan\lambda}{\cos\alpha_n + f \cot\lambda} \tag{15.13}$$

For a square thread, $\alpha_n = 0$ and Eq. (15.13) simplifies to

$$e = \frac{1 - f \tan\lambda}{1 + f \cot\lambda} \tag{15.13a}$$

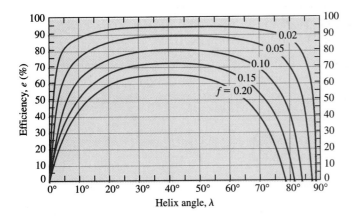

Figure 15.8 The efficiency of Acme screw threads (neglecting thrust collar friction).

Equation (15.13a) is plotted in Figure 15.8 for five values of f. We see from the curves that the power screws have very low mechanical efficiency when the helix angle is in the neighborhood of either $0°$ or $90°$. They generally have an efficiency of 30–90%, depending on the λ and f. We mention that values for square threads are higher by less than 1% over those for Acme screws in the figure.

The Torque and Efficiency of a Power Screw

EXAMPLE 15.1

A screw jack with an Acme thread of diameter d, similar to that illustrated in Figure 15.6, is used to lift a load of W. Determine

(a) The screw lead, mean diameter, and helix angle.

(b) The starting torque for lifting and for lowering the load.

(c) The efficiency of the jack when lifting the load, if collar friction is neglected.

(d) The length of a crank required, if $F = 150$ N is exerted by an operator.

Design Assumptions: The screw and nut are lubricated with oil. Coefficients of friction are estimated as $f = 0.12$ and $f_c = 0.09$.

Given: $d = 30$ mm and $W = 6$ kN. The screw is quadruple threaded having a pitch of $p = 4$ mm. The mean diameter of the collar is $d_c = 40$ mm.

Solution:

(a) From Figure 15.3, $d_m = d - p/2 = 30 - 2 = 28$ mm. Through the use of Eqs. (15.1) and (15.2), we have

$$L = np = 4(4) = 16 \text{ mm}$$

$$\lambda = \tan^{-1} \frac{16}{\pi(28)} = 10.31°$$

(b)   The coefficients of friction for starting are $f = \frac{4}{3}(0.12) = 0.16$ and $f_c = \frac{4}{3}(0.09) = 0.12$. For an Acme thread $\alpha = 14.5°$ (Figure 15.4a), by Eq. (15.8),

$$\alpha_n = \tan^{-1}(\cos\lambda \tan\alpha)$$

$$= \tan^{-1}(\cos10.31° \tan14.5°) = 14.28°$$

Then, application of Eqs. (15.6) and (15.7) results in

$$T_u = \frac{6(28)}{2} \frac{0.16 + \cos14.28° \tan10.31°}{\cos14.28° - (0.16)\tan10.31°} + \frac{6(0.12)40}{2}$$

$$= 30.05 + 14.4 = 44.45 \text{ N} \cdot \text{m}$$

$$T_d = \frac{6(28)}{2} \frac{0.16 - \cos14.28° \tan10.31°}{\cos14.28° + (0.16)\tan10.31°} + 14.4$$

$$= -1.37 + 14.4 = 13.03 \text{ N} \cdot \text{m}$$

Comments: The minus sign in the first term of T_d means that the screw alone is not self-locking and would rotate under the action of the load, except that the collar friction must be overcome too. Since T_d is positive, the screw does not overhaul.

(c)   The running torque needed to lift the load is based on $f = 0.12$. Using Eq. (15.13), we have

$$e = \frac{\cos14.28° - (0.12)\tan10.31°}{\cos14.28° + (0.12)\cot10.31°}$$

$$= 0.582 = 58.2\%$$

(d)   The length of the crank arm is

$$a = \frac{T_u}{F} = \frac{44.45}{150} = 0.296 \text{ m} = 296 \text{ mm}$$

15.5   BALL SCREWS

A *ball screw,* or so-called ball-bearing screw, is a linear actuator that transmits force or motion with minimum friction. A cutaway illustration of a ball screw and two of its precision assemblies supported by ball bearings at the ends are shown in Figure 15.9. Note that a circular groove is cut to proper conformity with the balls. The groove has a thread helix angle matching the thread angle of the groove within the nut. The *balls* are contained within the *nut* to produce an approximate rolling contact with the screw threads. The rotation of the screw (or nut) is converted into a linear motion and force with very little friction torque. During the motion, the balls are diverted from one end and middle of the nut and carried by two ball-return tubes (or ball guides) located outside of the nut to the

Return Tubes

Ball Nut   Bearing Balls

Ball Screw

(a)

(b)

Figure 15.9   Ball screw used as a positioning device: (a) cutaway of a ball screw;
(b) two assemblies. (Courtesy of Thomson Industries, Inc.)

middle and opposite end of the nut. Such recirculation allows accessible travel of the nut
relative to the screw.

A ball screw can support greater load than that of ordinary power screws of identical
diameter. The smaller size and lighter weight are usually an advantage. A thin film lubricant is required for these screws. Certain dimensions of ball screws have been standardized
by ANSI [3–5] but mainly for use in machine tools. Capacity ratings for ball screws are obtained by methods and equations identical to those for ball bearings, which can be found in
manufacturers' catalogs.

Efficiencies of 90% or greater are possible with ball screws over a wide range of helix
angles when converting rotary to axial motion. Ball screws may be preferred by the designers if higher screw efficiencies are required. As a positioning device, these screws are
used in many applications. Examples include the steering mechanism of automobiles, hospital bed mechanisms, automatic door closers, antenna drives, the aircraft control (e.g., a
ball or jack screw and gimbal nut assembly as an actuator on a linkage for extending and

retracting the wing flaps) and landing-gear actuator, jet aircraft engine thrust reverser actuators, and machine tool controls. Because of the low friction of ball screws, they are not self-locked. An auxiliary brake is required to hold a load driven by a ball screw for some applications.

15.6   THREADED FASTENER TYPES

The common element among screw fasteners used to connect or join two or more parts is their thread. Screws and *bolts* are the most-familiar threaded fastener types. The only difference between a screw and a bolt is that the bolt needs a *nut* to be used as a fastener (Figure 15.10a). On the other hand, a screw fits into a threaded hole. The same fastener is termed a *machine screw* or *cap screw* when it is threaded into a tapped hole rather than used with a nut, as shown in Figure 15.10b. *Stud* refers to a headless fastener, threaded on both ends, and screwed into the hole in one of the members being connected (Figure 15.10c).

Hexagon-head screws and bolts as well as hexagon nuts (see Figures 15.1 and 15.2) are commonly used for connecting machine components. Screws and bolts are also manufactured with round heads, square heads, oval heads, as well as various other head styles. Conventional bolts and nuts generally use standard threads, defined in Section 15.2. An almost endless number of threaded (and other) fasteners exist, many new types are constantly being developed [6–10]. Threaded fasteners must be designed so that they are lighter in weight, less susceptible to corrosion, and more resilient to loosening under vibration.

Flat or plain *washers* (Figure 15.1) are often used to increase the area of contact between the bolt head or nut and clamped part in a connection, as shown in Figure 15.10. They prevent stress concentration by the sharp edges of the bolt holes. Flat washer sizes are standardized to bolt size. A plain washer also forestalls marring of the clamped part surface by the nut when it is tightened. Belleville washers, discussed in Section 14.12, provide a controlled axial force over changes in bolt length.

Lock washers help prevent spontaneous loosening of standard nuts. The split lock washer acts as a spring under the nut. *Lock nuts* prevent too-spontaneous loosening of nuts due to vibration. Simply two nuts jammed together on the bolt or a nut with a cotter pin

(a)   (b)   (c)

Figure 15.10   Typical threaded fasteners: (a) bolt and nut; (b) cap screw; (c) stud.

serve for this purpose as well. The cotter pin is a wire that fits in diametrically opposite slots in the nut and passes through a drilled hole in the bolt. Lock nuts are considered to be more effective in preventing loosening than lock washers.

FASTENER MATERIALS AND STRENGTHS

A fastener is classified according to a grade or property class that defines its strength and material. Most fasteners are made from steel of specifications standardized by the SAE, ASTM, and ISO. The SAE grade (inch series) and SAE class (metric series) of steel threaded members are numbered according to tensile strength. The *proof strength* S_p corresponds to the axial stress at which the bolt or screw begins to develop a permanent set. It is close to but lower than the material yield strength. The *proof load* F_p is defined by

$$F_p = S_p A_t \qquad (15.14)$$

Here the *tensile stress area* A_t represents the minimum radial plane area for fracture through the threaded part of a bolt or screw. Numerical values of A_t are listed in Tables 15.1 and 15.2. The proof strength is obtained from Table 15.4 or 15.5. For other materials, an approximate value is about 10% less than for yield strength; that is, $S_p = 0.9S_y$ based on a 2% set.

Threads are generally formed by *rolling* and *cutting* or grinding. The former is stronger than the latter in fatigue and impact because of cold working. Hence, high-strength screws and bolts have rolled threads. The rolling should be done subsequent to hardening the bolt. The material of the nut must be selected carefully to match that of the bolt. The washers should be of hardened steel, where the bolt or nut compression load needs to be distributed over a large area of clamped part.

Table 15.4 SAE specifications and strengths for steel bolts

SAE grade	Size range diameter, d (in.)	Proof strength, S_p (ksi)	Yield strength,* S_y (ksi)	Tensile strength,** S_u (ksi)	Material carbon content
1	$\frac{1}{4}$–$1\frac{1}{2}$	33	36	60	Low or medium
2	$\frac{1}{4}$–$\frac{3}{4}$	55	57	74	Low or medium
2	$\frac{7}{8}$–$1\frac{1}{2}$	33	36	60	Low or medium
5	$\frac{1}{4}$–1	85	92	120	Medium, CD
5	$1\frac{1}{8}$–$1\frac{1}{2}$	74	81	105	Medium, CD
7	$\frac{1}{4}$–$1\frac{1}{2}$	105	115	133	Medium, alloy, Q&T
8	$\frac{1}{4}$–$1\frac{1}{2}$	120	130	150	Medium, alloy, Q&T

SOURCE: Society of Automotive Engineers Standard J429k (1979).

*Corresponds to permanent set not over 0.0001 in.

**Offset of 0.2%.

Q&T = quenched and tempered.

Table 15.5   Metric specifications and strengths for steel bolts

Class number	Size range diameter, d (mm)	Proof strength, S_p (MPa)	Yield strength, S_y (MPa)	Tensile strength, S_u (MPa)	Material carbon content
4.6	M5–M36	225	240	400	Low or medium
4.8	M1.6–M16	310	340	420	Low or medium
5.8	M5–M24	380	420	520	Low or medium
8.8	M3–M36	600	660	830	Medium, Q&T
9.8	M1.6–M16	650	720	900	Medium, Q&T
10.9	M5–M36	830	940	1040	Low, martensite, Q&T
12.9	M1.6–M36	970	1100	1220	Alloy, Q&T

| SOURCE: Society of Automotive Engineers Standard J1199 (1979).

A soft washer bends rather than uniformly distributes the load. Fasteners are also made of a variety of materials, including aluminum, brass, copper, nickel, Monel, stainless steel, titanium, berylium, and plastics. Appropriate coatings may be used in special applications in place of a more expensive material, for corrosion protection and to reduce thread friction and wear. Obviously, a designer has many options in selecting the fastener's material to suit the particular application.

15.7   STRESSES IN SCREWS

Stress distribution of the thread engagement between the screw and the nut is nonuniform. In reality, inaccuracies in thread spacing cause virtually all the load to be taken by the first pair of contacting threads and a large stress concentration is present here. While the stress concentration is to some extent relieved by the bending of the threads and the expansion of the nut, most bolt failures occur at this point. A concentration of stress also exists in the screw where the load is transferred through the nut to the adjoining member. Obviously, factors such as fillet radii at the thread roots, surface finish, and so on have significant affects on the actual stress values [2]. For ordinary threads, the *stress concentration factor* varies between 2 and 4 [11, 12].

Note that the screws should always have enough ductility to permit local yielding at thread roots without damage. For *static loading,* it is commonly assumed that the load carried by a screw and nut is about *uniformly* distributed throughout thread engagement. The stress distribution for threads with steady loads is usually determined by photoelastic analysis. A variety of methods are used to obtain a more nearly equal distribution of loads among the threads, including increasing the flexibility of the nut (or bolt), making the nut from a softer material than the bolt, and cutting the thread of the nut on a very small taper. A rule of thumb for the length of full thread engagement is $1.0d$ in steel, $1.5d$ in gray cast iron, and $2.0d$ in aluminum castings, where d is nominal thread size.

The following expressions for stresses in power screws and threaded fasteners are obtained through the use of the elementary formulas for stress. They enable the analyst to achieve a reasonable design for a static load. When bolts are subjected to fluctuating loads, stress concentration is very important.

AXIAL STRESS

Power screws may be under tensile or compressive stress; threaded fasteners normally carry only tension. The axial stress σ is then

$$\sigma = \frac{P}{A} \tag{15.15}$$

where

P = tensile or compressive load

$$A = \begin{cases} A_t \text{ from Tables 15.1 and 15.2} & \text{(threaded fasteners)} \\ \pi d_r^2 / 4, \quad d_r = \text{the root diameter} & \text{(power screws)} \end{cases}$$

TORSIONAL SHEAR STRESS

Power screws in operation and threaded fasteners during tightening are subject to torsion. The shear stress τ is given by

$$\tau = \frac{T_c}{J} = \frac{16T}{\pi d_r^3} \tag{15.16}$$

In the foregoing, we have

$$T = \begin{cases} \text{applied torque, for } f_c = 0 & \text{(power screws)} \\ \text{half the wrench torque} & \text{(threaded fasteners)} \end{cases}$$

$$d = \begin{cases} \text{from Figure 15.4} & \text{(power screws)} \\ \text{from Tables 15.1 and 15.2} & \text{(threaded screws)} \end{cases}$$

COMBINED TORSION AND AXIAL STRESS

The combined stress of Eqs. (15.15) and (15.16) can be treated as in Section 7.7, with the energy of distortion theory employed as a criterion for yielding.

BEARING STRESS

The direct compression or bearing stress σ_b is the pressure between the surface of the screw thread and the contacting surface of the nut:

$$\sigma_b = \frac{P}{\pi d_m h n_e} = \frac{Pp}{\pi d_m h L_n} \tag{15.17}$$

where

P = load

d_m = pitch or mean screw thread diameter

h = depth of thread (Figure 15.3)

n_e = number of threads in engagement = L_n/p

L_n = nut length

p = pitch

Exact values of σ_b are given in ANSI B1.1-1989 and various handbooks.

DIRECT SHEAR STRESS

The screw thread is considered to be loaded as a cantilevered beam [13]. The load is assumed to be uniformly distributed over the mean screw diameter. Hence, both the threads on the screw and the threads on the nut experience a transverse shear stress $\tau = 3P/2A$ at their roots. Here A is the cross-sectional area of the built-in end of the beam: $A = \pi d_r b n_e$ for the screw and $A = \pi d b n_e$ for the nut. Therefore, shear stress, for the *screw*, is

$$\tau = \frac{3P}{2\pi d_r b n_e} \tag{15.18}$$

and for the *nut*, is

$$\tau = \frac{3P}{2\pi d b n_e} \tag{15.19}$$

in which

d_r = the root diameter of the screw

d = the major diameter of the screw

b = thread thickness at the root (Figure 15.3)

The remaining terms are as defined earlier.

BUCKLING STRESS FOR POWER SCREWS

For the case in which the unsupported screw length is equal to or larger than about eight times the root diameter, the screw must be treated as a column. So, critical stresses are obtained as discussed in Sections 6.3 and 6.6.

Case Study 15-1 | DESIGN OF SCREWS FOR A WINCH CRANE HOOK

The steel crane hook supported by a trunnion or crosspiece as shown in Figure 15.11 is rated at $P = 3$ kN (refer to Case Study 1-1). Determine the necessary nut length L_n. Observe that a ball-thrust bearing permits rotation of the hook for positioning the load. The lower race of the bearing and a third (bottom) ring have matching spherical surfaces to allow self-alignment of the hook with the bearing load. Usually, bearing size selected for a given load and service has internationally standardized dimensions [3].

Assumptions: Both the threaded portion of the shank or bolt and the nut are made of $M12 \times 1.75$ class 5.8

rolled coarse threads. A stress concentration factor of $K_t = 3.5$ and a safety factor of $n = 5$ is used for threads.

Given: From Table 15.2,

$$p = 1.75 \text{ mm}, \qquad d = 12 \text{ mm},$$
$$d_m \approx 10.925 \text{ mm}, \quad d_r \approx 9.85 \text{ mm},$$
$$h = \frac{1}{2}(d - d_r) = 1.075 \text{ mm},$$
$$S_y = 420 \text{ MPa (from Table 15.5)}$$

Solution: See Figures 1.4 and 15.11.

Case Study (CONCLUDED)

Figure 15.11 Hook for the winch crane of Figure 1.4: (a) section of the trunnion with a thrust-ball bearing; (b) the critically stressed, modified trapezoidal section.

Bearing Strength. For the nut, incorporation of stress concentration factor into Eq. (15.17) gives the following design formula:

$$\frac{K_t P_p}{\pi d_m h L_n} = \frac{S_y}{n} \qquad \text{(15.17a)}$$

Substituting the given numerical values, we have

$$\frac{(3.5)(3000)(1.75)}{\pi(10.925)(1.075)L_n} = \frac{420}{5}$$

Solving,

$$L_n = 5.9 \text{ mm}$$

Shear Strength. Based on the energy of distortion theory of failure

$$S_{ys} = 0.577 S_y = 0.577 \times 420 = 242.3 \text{ MPa}$$

From Figure 15.3, thread thickness at the root,

$$b = \frac{p}{8} + 2h \tan 30° = 1.46 \text{ mm}$$

Equation (15.19) results in the design formula

$$\frac{3 K_t P_p}{2\pi d b L_n} = \frac{S_{ys}}{n} \qquad \text{(15.19a)}$$

Inserting the data given,

$$\frac{3(3.5)(3000)(1.75)}{2\pi(12)(1.46)L_n} = \frac{242.3}{5}$$

from which

$$L_n = 10.3 \text{ mm}$$

Comment: A standard nut length of 10 mm should be used.

15.8 BOLT TIGHTENING AND PRELOAD

Bolts are commonly used to hold parts together in opposing to forces likely to pull, or sometimes slide, them apart. Typical examples include connecting rod bolts and cylinder head bolts. *Bolt tightening* is prestressing at assembly. In general, bolted joints should be tightened to produce an *initial tensile force,* usually so-called the *preload* F_i. The advantages of an initial tension are especially noticeable in applications involving fluctuating loading, as demonstrated in Section 15.12, and in making a leakproof connection in pressure vessels. An increase of fatigue strength is obtained when initial tension is present in the bolt. The parts to be joined may or may not be separated by a gasket. In this section, we consider the situation when no gasket is used.

The *bolt strength* is the main factor in the design and analysis of bolted connections. Recall from Section 15.6 that the proof load F_p is the load that a bolt can carry without developing a permanent deformation. For both static and fatigue loading, the preload is often prescribed by

$$F_i = \begin{cases} 0.75F_p & \text{(reused connections)} \\ 0.9F_p & \text{(permanent connections)} \end{cases} \tag{15.20}$$

where the proof load F_p is obtained from Eq. (15.14). The amount of initial tension is clearly a significant factor in bolt design. It is usually maintained fairly constant in value.

TORQUE REQUIREMENT

The most important factor determining the preload in a bolt is the *torque required* to *tighten the bolt*. The torque may be applied manually by means of a wrench that has a dial attachment indicating the magnitude of the torque being enforced. Pneumatic or air wrenches give more consistent results than a manual torque wrench and are employed extensively.

An expression relating applied torque to initial tension can be obtained using Eq. (15.6) developed for power screws. Observe that load W of a screw jack is equivalent to F_i for a bolt and that collar friction in the jack corresponds to friction on the flat surface of the nut or under the screwhead. It can readily be shown that [9], for standard screw threads, Eq. (15.6) has the form

$$T = KdF_i \tag{15.21}$$

where

T = tightening torque

d = nominal bolt diameter

K = torque coefficient

F_i = initial tension or preload

For dry surface and *unlubricated* bolts or "average" condition of thread friction, taking $f = f_c = 0.15$, Eq. (15.6) results in $K = 0.2$. It is suggested that, for *lubricated* bolts, a value of 0.15 be used for torque coefficient. For various plated bolts see [14].

Note that Eq. (15.21) represents an approximate relationship between the induced initial tension and applied torque. Tests have shown that a typical joint loses about 5% or more of its preload owing to various relaxation effects. The exact tightening torque needed in a particular situation can likely be best ascertained experimentally through calibration. That is, a prototype can be built and accurate torque testing equipment used on it. Interestingly, bolts and washers are available with built-in sensors indicating a degree of tightness. Electronic assembly equipment is available [2, 9].

15.9 TENSION JOINTS UNDER STATIC LOADING

A principal utilization of bolts and nuts is clamping parts together in situations where the applied loads put the *bolts in tension.* Attention here is directed toward preloaded tension joints under static loading. We treat the case of two plates or parts fastened with a bolt and subjected to an external load P, as depicted in Figure 15.12a. The preload F_i, an initial tension, is applied to the bolt by tightening the nut prior to the load P. Clearly, the bolt axial load and the clamping force between the two parts F_p are both equal to F_i.

To determine what portion of the externally applied load is carried by the bolt and what portion by the connected parts in the assembly, refer the free-body diagram shown in Figure 15.12b. The *condition of equilibrium* of the forces gives

$$P = \Delta F_b + \Delta F_p \qquad \text{(a)}$$

The quantity ΔF_b is the increased bolt (tensile) force and ΔF_p represents the decreased clamping (compression) force between the parts. It is taken that the parts have not been separated by the application of the external load. The deformation of the bolt and the parts are defined by

$$\delta_b = \frac{\Delta F_b}{k_b}, \qquad \delta_p = \frac{\Delta F_p}{k_p} \qquad \text{(b)}$$

Here k_b and k_p represent the *stiffness constants* for the bolt and parts, respectively.

Because of the setup of the members in Figure 15.12a, the deformations given by Eqs. (b) are equal. The *compatibility condition* is then

$$\frac{\Delta F_b}{k_b} = \frac{\Delta F_p}{k_p} \qquad \text{(c)}$$

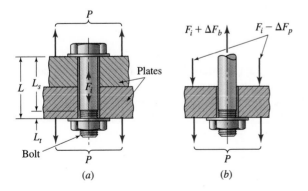

(a) *(b)*

Figure 15.12 A bolted connection: (a) complete joint with preload F_i and external load P; (b) isolated portion depicting increased bolt force ΔF_b and decreased force on parts or plates ΔF_p.

Combining Eqs. (a) and (c) yields

$$\Delta F_b = \frac{k_b}{k_b + k_p} P = CP, \qquad \Delta F_p = \frac{k_p}{k_b + k_p} P = (1 - C)P \tag{d}$$

The term C, called the *joint's stiffness factor* or simply the *joint constant,* is defined in Eq. (d) as

$$C = \frac{k_b}{k_b + k_p} \tag{15.22}$$

Note that, typically, k_b is small in comparison with k_p and C is a small fraction.

The total forces on the bolt and parts are $F_b = \Delta F_b + F_i$ and $F_p = \Delta F_p - F_i$, respectively. Therefore, we have

$$F_b = CP + F_i \qquad \text{(for } F_p < 0) \tag{15.23}$$

$$F_p = (1 - C)P - F_i \qquad \text{(for } F_p < 0) \tag{15.24}$$

where

$F_b =$ bolt axial tensile force

$F_p =$ clamping force on the two parts

$F_i =$ initial tension or preload

As indicated in the expressions, the foregoing results are valid only as long as some clamping force prevails on the parts: With no preload (loosened joint), $C = 1$, $F_i = 0$. We see that the ratios C and $1 - C$ in Eqs. (15.23) and (15.24) describe the proportions of the external load carried by the bolt and the parts, respectively. In all situations, the *parts* take a *greater portion* of the external load. This is significant when fluctuating loading is present.

FACTORS OF SAFETY FOR A JOINT

The tensile stress σ_b in the bolt can be found by dividing both terms of Eq. (15.24) by the tensile-stress area A_t:

$$\sigma_b = \frac{CP}{A_t} + \frac{F_i}{A_t} \tag{15.25}$$

A means of ensuring a safe joint requires that the external load be smaller than that needed to cause the joint separate. Let nP be the value of external load that would cause bolt failure and the limiting value of σ_b be the proof strength S_p. Substituting these, Eq. (15.25) becomes

$$\frac{CPn}{A_t} + \frac{F_i}{A_t} = S_p \tag{15.26}$$

It should be mentioned that the factor of safety is not applied to the preload. The foregoing can be rewritten to give the *bolt safety factor:*

$$n = \frac{S_p A_t - F_i}{CP} \tag{15.27}$$

As noted earlier, the tensile stress area A_t is furnished in Tables 15.1 and 15.2 and S_p is listed in Tables 15.4 and 15.5.

Equation (15.27) suggests that safety factor n is maximized by having no preload on the bolt. We also note that, for $n > 1$, the bolt stress is smaller than the proof strength. Separation occurs when in $F_p = 0$ in Eq. (15.24). Therefore, the *load safety factor* guarding against joint separation is

$$n_s = \frac{F_i}{P(1 - C)}$$ (15.28)

Here P is the *maximum* load applied to the joint.

15.10 GASKETED JOINTS

Sometimes, a sealing or gasketing material must be placed between the parts connected. Gaskets are made of materials that are soft relative to other joint parts. Obviously, the stiffer and thinner is the gasket, the better. The stiffness factor of a gasketed joint can be defined as

$$C = \frac{k_b}{k_b + k_c}$$ (15.29a)

The quantity k_c represents the combined constant found from

$$\frac{1}{k_c} = \frac{1}{k_g} + \frac{1}{k_p}$$ (15.29b)

where k_g and k_p are the spring rates of the gasket and connected parts, respectively.

When a full gasket extends over the entire diameter of a joint, the gasket pressure is

$$p = -\frac{F_p}{A_g}$$ (a)

in which A_g is the gasket area per bolt and F_p represents the clamping force on parts. For a load factor n_s, Eq. (15.24) becomes

$$F_p = (1 - C)n_s P - F_i$$ (b)

Carrying Eq. (a) into (b), *gasket pressure* may be expressed in the form

$$p = \frac{1}{A_g}[F_i - n_s P(1 - C)]$$ (15.30)

We point out that, to maintain the uniformity of pressure, bolts should not be spaced more than *six* bolt diameters apart.

15.11   DETERMINING THE JOINT STIFFNESS CONSTANTS

Application of the equations developed in Section 15.9 requires a determination of the spring rates of bolt and parts, or at least a reasonable approximation of their relative values. Recall from Chapter 4 that the axial deflection is found from the equation $\delta = PL/AE$ and spring rate by $k = P/\delta$. Thus, we have for the bolt and parts, respectively,

$$k_b = \frac{A_b E_b}{L} \tag{15.31}$$

$$k_p = \frac{A_p E_p}{L} \tag{15.32}$$

where

$\quad k_b$ = stiffness constant for bolt

$\quad k_p$ = stiffness constant for parts

$\quad A_b$ = cross-sectional area of bolt

$\quad A_p$ = effective cross-sectional area of parts

$\quad E$ = modulus of elasticity

$\quad L$ = grip, which represents approximate length of clamped zone

BOLT STIFFNESS

When the thread stops immediately above the nut as shown in Figure 15.12, the gross cross-sectional area of the bolt must be used in approximating k_b, since the unthreaded portion is stretched by the load. Otherwise, a bolt is treated as a spring in series when considering the threaded and unthreaded portions of the shank. For a bolt of axially loaded thread length L_t and the unthreaded shank length L_s (Figure 15.12a), the spring constant is

$$\frac{1}{k_b} = \frac{L_t}{A_t E_b} + \frac{L_s}{A_b E_b} \tag{15.33}$$

in which A_b is the gross cross-sectional area and A_t represents the tensile stress area of the bolt. Note that, ordinarily a bolt (or cap screw) has as little of its length threaded as practicable to maximize bolt stiffness. We then use Eq. (15.31) in calculating the bolt spring rate k_b.

STIFFNESS OF CLAMPED PARTS

The spring constant of clamped parts is seldom easy to ascertain and frequently approximated by employing an emprical procedure. Accordingly, the stress induced in the joint is assumed to be *uniform* throughout a region surrounding the bolt hole [15–20]. The region is often represented by a double-cone-shaped "barrel" geometry of a half-apex angle 30°, as depicted in Figure 15.13. The stress is taken to be 0 outside the region. The effective

Figure 15.13 A method for estimating the effective cross-sectional area of clamped parts A_p.

cross-sectional area A_p is equal to about the average area of the shaded section shown in the figure:

$$A_p = \frac{\pi}{4} \left[\left(\frac{d_w + d_2}{2} \right)^2 - d^2 \right]$$

The quantities $d_2 = d_w + L\tan30°$ and d_w represent the washer (or washer face) diameter. Note that $d_w = 1.5d$ for standard hexagon-headed bolts and cap screws. The preceding expression of A_p is used for estimating k_p from Eq. (15.32). It can be shown that [15], for connections using standard hexagon-headed bolts, the *stiffness constant for parts* is given by

$$k_p = \frac{0.58\pi E_p d}{2 \ln \left(5\frac{0.58L + 0.5d}{0.58L + 2.5d} \right)} \tag{15.34}$$

where d = bolt diameter, L = grip, and E = the modulus of elasticity of the single or two identical parts.

We should mention that the spring rate of clamped parts can be determined with good accuracy by experimentation or finite element analysis [21]. Various handbooks list rough estimates of the ratio k_p/k_b for typical gasketed and ungasketed joints. Sometimes $k_p = 3k_b$ is used for ungasketed ordinary joints.

Preloaded Fasteners in Static Loading

EXAMPLE 15.2

Figure 15.14 illustrates a portion of a cover plate bolted to the end of a thick-walled cylindrical pressure vessel. A total of N_b bolts are to be used to resist a separating force P. Determine

(a) The joint constant.

(b) The number N_b for a permanent connection.

(c) The tightening torque for an average condition of thread friction.

Given: The required joint dimensions and materials are shown in the figure. The applied load $P =$ 55 kips.

Figure 15.14   Example 15.2.

Design Assumptions: The effects of the flanges on the joint stiffness are omitted. The connection is permanent. A bolt safety factor of $n = 1.5$ is used.

Solution:

(a) Referring to Figure 15.14, Eq. (15.34) gives

$$k_p = \frac{0.58\pi(E_s/2)(0.75)}{2\ln\left[5\dfrac{0.58(2)+0.5(0.75)}{0.58(2)+2.5(0.75)}\right]} = 0.368E_s$$

Through the use of Eq. (15.31),

$$k_b = \frac{AE_s}{L} = \frac{\pi d^2 E_s}{4L} = \frac{\pi(0.75)^2 E_s}{4(2)} = 0.221E_s$$

Equation (15.22) is therefore

$$C = \frac{k_b}{k_b + k_p} = \frac{0.221}{0.221 + 0.368} = 0.375$$

(b) From Tables 15.1 and 15.4, we have $A_t = 0.334$ in.^{2} and $S_p = 105$ ksi. Applying Eq. (15.20),

$$F_i = 0.9S_p A_t = 0.90(105)(0.334) = 31.6 \text{ kips}$$

For N_b bolts, Eq. (15.26) can be written in the form

$$\frac{C(P/N_b)n}{A_t} + \frac{F_i}{A_t} = S_p$$

from which

$$N_b = \frac{CPn}{S_p A_t - F_i}$$

Substituting the numerical values, we have

$$N_b = \frac{(0.375)(55)(1.5)}{105(0.334) - 31.6} = 8.92$$

Comment: Nine bolts should be used.

(c) By Eq. (15.21),

$$T = 0.2dF_i = 0.2(0.75)(31.6) = 4.74 \text{ kip} \cdot \text{in.}$$

15.12 TENSION JOINTS UNDER DYNAMIC LOADING

Bolted joints with preload and subjected to fatigue loading can be analyzed directly by the methods discussed in Chapter 8. Since failure owing to fluctuating loading is more apt to occur to the bolt, our attention is directed toward the bolt in this section. As previously noted, the use of initial tension is important in problems for which the bolt carries cyclic loading. The maximum and minimum loads on the bolt are higher because of the initial tension. Consequently, the mean load is greater, but the alternating load component is reduced. Therefore, the fatigue effects, which depend primarily on the variations of the stress, are likewise reduced.

Reconsider the joint shown in Figure 15.12a, but let the applied force P vary between some minimum and maximum values, both positive. The mean and alternating loads are given by

$$P_m = \frac{1}{2}(P_{\max} + P_{\min}), \qquad P_a = \frac{1}{2}(P_{\max} + P_{\min})$$

Substituting P_m and P_a in place of P in Eq. (15.23), the mean and alternating forces felt by the bolt are

$$F_{bm} = CP_m + F_i \tag{15.35a}$$

$$F_{ba} = CP_a \tag{15.35b}$$

The mean and range stresses in the bolt are then

$$\sigma_{bm} = \frac{CP_m}{A_t} + \frac{F_i}{A_t} \tag{15.36a}$$

$$\sigma_{ba} = \frac{CP_a}{A_t} \tag{15.36b}$$

in which C represents the joint constant and A_t is the tensile-stress area. We observe from Eqs. (15.36) that, as long as separation does not occur, the alternating stress experienced by the bolt is reduced by the joint stiffness rate C. The mean stress is increased by the bolt preload.

For the bolted joints, the Goodman criterion given by Eq. (8.16), may be written as follows

$$\frac{\sigma_{ba}}{S_e} + \frac{\sigma_{bm}}{S_u} = 1$$

As before, the safety factor is not applied to the initial tension. Hence, introducing Eqs. (15.36) into this equation, we have

$$\frac{CP_a n}{A_t S_e} + \frac{CP_m n + F_i}{A_t S_u} = 1$$

The preceding is solved to give the *factor of safety* guarding against *fatigue failure* of the *bolt:*

$$n = \frac{S_u A_t - F_i}{C\left[P_a\left(\dfrac{S_u}{S_e}\right) + P_m\right]} \tag{15.37}$$

Alternatively,

$$n = \frac{S_u - \sigma_i}{C\left[\sigma_a\left(\dfrac{S_u}{S_e}\right) + \sigma_m\right]} \tag{15.38}$$

Here $\sigma_a = P_a/A_t$, $\sigma_m = P_m/A_t$, and $\sigma_i = F_i/A_t$. Recall from Section 8.9 that, these equation represents the Soderberg criterion when ultimate strength S_u is replaced by the yield strength S_y.

The modified endurance limit S_e is obtained from Eq. (8.6). For threaded finishes having good quality, a surface factor of $C_f = 1$ may be applicable. The size factor $C_s = 1$ (see Section 8.7) and by Eq. (8.3) we have $S'_e = 0.45 S_u$ for reversed axial loading. As a result,

$$S_e = C_r C_t \left(\frac{1}{K_f}\right)(0.45 S_u) \tag{15.39}$$

where C_r and C_t are the reliability and temperature factors. Table 15.6 gives average stress-concentration factors for the fillet under the bolt and also at the beginning of the threads on the shank. Cutting is the simplest method of producing threads. Rolling the threads provides a smoother thread finish than cutting. The fillet between the head and the shank reduces the K_f, as shown in the table. Unless otherwise specified, the threads are usually assumed to be rolled.

A very *common case* is that the fatigue loading fluctuates between 0 and some maximum value, such as in a bolted pressure vessel cycled from 0 to a maximum pressure. In this situation, the minimum tensile loading $P_{min} = 0$. The effect of initial tension with regard to fatigue loading is illustrated in the solution of the following sample problem.

Table 15.6   Fatigue stress concentration factors K_f for steel-threaded members

SAE grade (unified thread)	Metric grade (ISO thread)	Rolled threads	Cut threads	Fillet
0–2	3.6–5.8	2.2	2.8	2.1
4–8	6.6–10.9	3.0	3.8	2.3

| SOURCE: [11].

Preloaded Fasteners in Fatigue Loading

EXAMPLE 15.3

Figure 15.15a illustrates the connection of two steel parts with a single $\frac{5}{8}$-in.-11 UNC grade 5 bolt having rolled threads. Determine

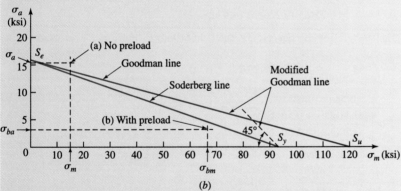

Figure 15.15 Example 15.3: (a) bolted parts carrying fluctuating loads; (b) fatigue diagram for bolts.

    (a) Whether the bolt fails when no preload is present.

    (b) If the bolt is safe with preload.

    (c) The fatigue factor of safety n when preload is present.

    (d) The static safety factors n and n_s.

Design Assumptions: The bolt may be reused when the joint is taken apart. Survival rate is 90%. Operating temperature is normal.

Given: The joint is subjected to a load P that varies continuously between 0 and 7 kips.

Solution: See Figure 15.15.

From Table 8.3 the reliability factor $C_r = 0.89$. Temperature factor is $C_t = 1$ (Section 8.7). Also,

$S_p = 85$ ksi, $S_y = 92$ ksi, $S_u = 120$ ksi   (from Table 15.4)

$K_f = 3$ (by Table 15.6)

$A_t = 0.226$ in.^{2} (from Table 15.1)

Equation (15.39) results in

$$S_e = (0.89)(1)\left(\frac{1}{3}\right)(0.45 \times 120) = 16 \text{ ksi}$$

The Soderberg and Goodman fatigue failure lines are shown in Figure 15.15b.

(a) For loosely held parts, when $F_i = 0$, load on the bolt equals the load on parts,

$$P_m = \frac{1}{2}(7 + 0) = 3.5 \text{ kips}, \qquad P_a = \frac{1}{2}(7 - 0) = 3.5 \text{ kips}$$

$$\sigma_a = \sigma_m = \frac{3.5}{0.226} = 15.5 \text{ ksi}$$

A plot of the stresses shown in Figure 15.15b indicates that failure will occur.

(b) Through the use of Eq. (15.20),

$$F_i = 0.75 S_p A_t = 0.75(85)(0.226) = 14.4 \text{ kips}$$

The grip is $L = 2.5$ in. By Eqs. (15.31) and (15.34) with $E_b = E_p = E$, we obtain

$$k_b = \frac{\pi d^2 E}{4L} = \frac{\pi (0.625)^2 E}{4(2.5)} = 0.123E$$

$$k_p = \frac{0.58\pi E(0.625)}{2 \ln\left[5\dfrac{0.58(2.5) + 0.5(0.625)}{0.58(2.5) + 2.5(0.625)}\right]} = 0.53E$$

The joint constant is then

$$C = \frac{k_b}{k_b + k_p} = \frac{0.123}{0.123 + 0.53} = 0.188$$

Comment: The foregoing means that only about 20% of the external load fluctuation is felt by the bolt and hence about 80% goes to decrease clamping pressure.

Applying Eqs. (15.35) and (15.36),

$$F_{bm} = C P_m + F_i$$

$$= 0.188(3.5) + 14.4 = 15.1 \text{ kips}$$

$$\sigma_{bm} = \frac{15.1}{0.226} = 66.8 \text{ ksi}$$

$$F_{ba} = C P_a = 0.118(3.5) = 0.66 \text{ kips}$$

$$\sigma_{ba} = \frac{0.66}{0.226} = 2.92 \text{ ksi}$$

A plot on the fatigue diagram shows that *failure* will *not occur* (Figure 15.15b).

(c) Equation (15.37) with $P_a = P_m$ becomes

$$n = \frac{S_u A_t - F_i}{C P_a \left[\left(\dfrac{S_u}{S_e} \right) + 1 \right]}$$

(15.40)

Introducing the given numerical values,

$$n = \frac{(120)(0.226) - 14.4}{(0.188)(3.5) \left[\left(\dfrac{120}{16} \right) + 1 \right]}$$

from which $n = 2.27$.

Comment: This is the factor of safety guarding against the fatigue failure. Observe from Figure 15.15b that the Goodman criteria led to a less conservative (higher) value for n.

(d) Substitution of the given data into Eqs. (15.27) and (15.28) gives

$$n = \frac{85(0.226) - 14.4}{(0.188)(7)} = 3.66$$

$$n_s = \frac{14.4}{7(1 - 0.188)} = 2.53$$

Comments: The factor of 3.66 prevents the bolt stress from becoming equal to proof strength. On the other hand, the factor of 2.53 guards against joint separation and the bolt taking the entire load.

15.13 RIVETED AND BOLTED JOINTS LOADED IN SHEAR

A rivet consists of a cylindrical body, known as the *shank*, usually with a rounded end called the *head*. The purpose of the rivet is to join together two plates while securing proper strength and tightness. If the rivet is heated prior to being placed in the hole, it is referred to as a hot-driven rivet, while if it is not heated, it is referred to as a cold-driven rivet. Rivets and bolts are ordinarily used in the construction of buildings, bridges, aircraft, and ships. The design of riveted and bolted connections is governed by construction codes formulated by such societies as the AISC [22] and the ASME.

Riveted and bolted joints loaded in shear are treated *exactly alike* in design and analysis. Figure 15.16 illustrates a simple riveted connection loaded in shear. It is obvious that the loading is eccentric and an unbalanced moment Pt exists. Hence, bending stress will be present. However, the usual procedure is to ignore the bending stress and compensate for its presence by a larger factor of safety. Table 15.7 lists various types of failure of the connection shown in the figure.

The *effective diameters* in a riveted joint are defined as follows. For a drilled hole $d_e = d + \frac{1}{16}$ in. (about 1.5 mm), and for a *punched hole* $d_e = d + \frac{1}{8}$ in. (about 3 mm). Here,

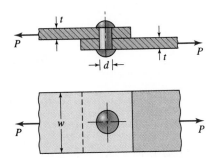

Figure 15.16   Riveted connection loaded in shear.

Table 15.7   Types of failure for riveted connections (Figure 15.16)

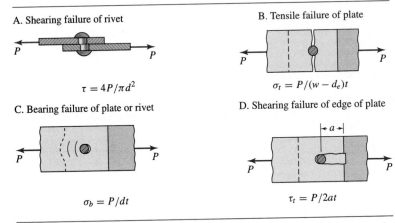

A. Shearing failure of rivet

$\tau = 4P/\pi d^2$

B. Tensile failure of plate

$\sigma_t = P/(w - d_e)t$

C. Bearing failure of plate or rivet

$\sigma_b = P/dt$

D. Shearing failure of edge of plate

$\tau_t = P/2at$

Notes: P = applied shear load; d = diameter of rivet; w = width of plate; t = thickness of the thinnest plate; d_e = effective hole diameter; a = closest distance from rivet to the edge of plate.

d represents the diameter of the rivet. Unless specified otherwise, we *assume* that the holes have been *punched*. Usually, shearing, or tearing, failure is avoided by spacing the rivet at least $1.5d$ away from the plate edge. To sum up, essentially three modes of failure must be considered in determining the capacity of a riveted or bolted connection: *shearing* failure of the rivet, *bearing* failure of the plate or rivet, and *tensile* failure of the plate. The associated normal and shear equations are given in the table.

EXAMPLE 15.4 | Determination of the Capacity of a Riveted Connection

The standard AISC connection for the W310 × 52 beam consists of two 102 × 102 × 6.4-mm angles, each 215 mm long, 22 mm rivets spaced 75 mm apart are used in 24 mm holes (Figure 15.17). Calculate the maximum load that the connection can carry.

Figure 15.17   Example 15.4.

Design Decisions: The allowable stresses are 100 MPa in shear and 335 MPa in bearing of rivets. Tensile failure cannot occur in this connection; only shearing and bearing capacities need be investigated.

Solution: The web thickness of the beam, $t_w = 7.6$ mm (from Table A.6), and the cross-sectional area of one rivet is $A_r = \pi(22)^2/4 = 380$ mm^2.
  Bearing on the web of the beam:

$$P_b = 3(7.6)(22)(335) = 168 \text{ kN} \qquad \text{(governs)}$$

Shear of six rivets:

$$P_b = 6(380)(100) = 228 \text{ kN}$$

Bearing of six rivets on angles:

$$P_b = 6(22)(6.4)(335) = 283 \text{ kN}$$

Comment: The capacity of this connection, the smallest of the forces obtained in the foregoing, is 168 kN.

JOINT TYPES AND EFFICIENCY

Most connections have many rivets or bolts in a variety of models. Riveted or bolted connections loaded in shear are of two types: *lap joints* and *butt joints.* In a lap joint, sometimes called a *single shear joint,* the two plates to be jointed overlap each other (Figure 15.18a).

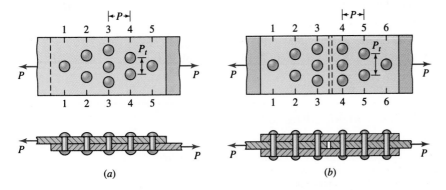

Figure 15.18   Types of riveted connections: (a) lap joint; (b) butt joint.

On the other hand, in a butt, also termed a *double shear joint,* the two plates to be connected (main plates) butt against one other (Figure 15.18b). *Pitch* is defined as the distance between adjacent rivet centers. It represents a significant geometric property of a joint. The *axial* pitch p for rivets is measured along a line parallel to the edge of the plate, while the corresponding distance along a line perpendicular to the edge of the plate is known as the *transverse* pitch p_t. Both kinds of pitch are depicted in the figure. The smallest symmetric group of rivets that repeats itself along the length of a joint is called a *repeating section*. The strength analysis of a riveted connection is based on its repeated section (see Example 15.5).

The *efficiency of joints* is defined as follows:

$$e = \frac{P_{all}}{P_t} \tag{15.41}$$

In the foregoing equation, P_{all} = the smallest of the allowable loads in shear, bearing, and tension; P_t = the static tensile yield load (strength) of plate with no hole. The most-efficient joint would be as strong in tension, shear, and bearing as the original plate to be joined is in tension. This can never be realized, since there must be at least one rivet hole in the plate: The allowable load of joint in tension therefore always is less than the strength of the plate with no holes.

For centrally applied loads, it is often assumed that the rivets are about equally stressed. In many cases, this cannot be justified by elastic analysis; however, ductile deformations permit an equal redistribution of the applied force, before the ultimate capacity of connection is reached. Also it is usually taken that the row of rivets immediately *adjacent* to the load carries the full load. Thus, the maximum load supported by such a row occurs when there is only one rivet in that row. The *actual load* carried by an interior row can be obtained from

$$P_i = \frac{n - n'}{n} P \tag{15.42}$$

where

P = externally applied load

P_i = the actual load, or portion of P, acting on a particular row i

n = total number of rivets in the joint

n' = total number of rivets in the row between the row being checked and the external load

For instance, load on row 3 of the joint in Figure 15.18a equals, $P_3 = (9 - 3)P/9 = 2P/3$. Likewise, load on row 2 or 5 of the joint in Figure 15.18b is $P_2 = P_5 = (12 - 1)P/12 = 11P/12$.

EXAMPLE 15.5 | Strength Analysis of a Multiple-Riveted Lap Joint

Figure 15.19a shows a multiple-riveted lap joint subjected to an axial load P. The dimensions are given in inches. Calculate the allowable load and efficiency of the joint.

Given: All rivets are $\frac{3}{4}$ in. in diameter.

Figure 15.19 Example 15.5: (a) a riveted lap joint; (b) enlarged view of a repeating group of rivets.

Design Assumptions: The allowable stresses are 20 ksi in tension, 15 ksi in shear, and 30 ksi in bearing.

Solution: The analysis is on the basis of the repeating section, which has four rivets and $L = 6$ in. (Figure 15.19b).

Plate in tension, without holes,

$$P_t = 20 \left(6 \times \frac{1}{2} \right) = 60 \text{ kips}$$

The rivet shear:

$$P_s = 4 \left[\frac{\pi}{4} \left(\frac{3}{4} \right)^2 \right] (15) = 26.51 \text{ kips} \qquad \text{(governs)}$$

The plate bearing,

$$P_b = 4 \left(\frac{1}{2} \times \frac{3}{4} \right) (30) = 45 \text{ kips}$$

The tension across sections 1-1 through 3-3 of the *bottom plate*, using Eq. (15.42):

$$\frac{4-3}{4} P_1 = \frac{1}{2} \left[6 - \left(\frac{3}{4} + \frac{1}{8} \right) \right] (20); \qquad P_1 = 205 \text{ kips}$$

$$\frac{4-1}{4} P_2 = \frac{1}{2} \left[6 - 2 \left(\frac{3}{4} + \frac{1}{8} \right) \right] (20); \qquad P_2 = 56.7 \text{ kips}$$

$$P_3 = \frac{1}{2} \left[6 - \left(\frac{3}{4} + \frac{1}{8} \right) \right] (20); \qquad P_3 = 51.25 \text{ kips}$$

The maximum allowable force that the joint can safely carry is the smallest of the force obtained in the preceding, $P_{all} = 26.51$ kips. The efficiency of this joint, from Eq. (15.41), is then

$$e = \frac{26.51}{60} \times 100 = 44.2\%$$

15.14   SHEAR OF RIVETS OR BOLTS DUE TO ECCENTRIC LOADING

For the case in which the load is applied eccentrically to a connection having a group of bolts or rivets, the effects of the torque or moment, as well as the direct force, must be considered. A typical structural problem is the situation that occurs when a horizontal beam is supported by a vertical column (Figure 15.20a). In this case, each bolt is subjected to a twisting moment $M = Pe$ and a direct shear force P. An enlarged view of bolt group with loading (P and M) acting at the centroid C of the group and the reactional shear forces acting at the cross section of each bolt are shown in Figure 15.20b.

Let us assume that the reactional *tangential force* due to moment, so-called moment load or secondary shear, on a bolt varies directly with the distance from the centroid C of the group of bolts and is directed perpendicular to the centroid. As a result,

$$\frac{F_1}{r_1} = \frac{F_2}{r_2} = \frac{F_3}{r_3} = \frac{F_4}{r_4}$$

In the preceding, F_i and r_i ($i = 1, \ldots, 4$) are the tangential force and radial distance from C to the center of each bolt, respectively. The externally applied moment and tangential forces are related as follows:

$$M = Pe = F_1 r_1 + F_2 r_2 + F_3 r_3 + F_4 r_4$$

Solving these equations simultaneously, we obtain

$$F_1 = \frac{Pe\,r_1}{r_1^2 + r_2^2 + r_3^2 + r_4^2}$$

(a)   (b)

Figure 15.20   (a) Bolted joint with eccentric load. (b) Bolt group with loading and reactional shear forces.

This expression can be written in the following general form:

$$F_i = \frac{Mr_j}{\sum_{j=1}^{n} r_j^2} \qquad (15.43)$$

where

F_i = tangential force

$M = Pe$, externally applied moment

n = number of bolts in the group

i = particular bolt whose load is to be found

It is customary to assume that the reactional *direct force F/n* is the same for all bolts of the joint. The vectorial sum of the tangential force and direct force is the resultant shear force on the bolt (Figure 15.20b). Clearly, only the bolt having the maximum resultant shear force need be considered. An inspection of the vector force diagram is often enough to eliminate all but two or three bolts as candidates for worst-loaded bolt.

Finding the Bolt Shear Forces Due to Eccentric Loading | **EXAMPLE 15.6**

A gusset plate is attached to a column by three identical bolts and vertically loaded, as shown in Figure 15.21a. The dimensions are in millimeters. Calculate the maximum bolt shear force and stress.

Figure 15.21 Example 15.6: (a) bolted connection; (b) bolt shear force and moment equilibrium.

Assumption: The bolt tends to shear across its major diameter.

Solution: For the bolt group, point C corresponds to the centroid of the triangular pattern, as shown in Figure 15.21b. This free-body diagram illustrates the bolt reactions and the external loading replaced at the centroid. Each bolt supports one-third of the vertical shear load, 4 kN, plus a tangential force F_i. The distances from the centroid to bolts are

$$r_1 = r_2 = \sqrt{(40)^2 + (75)^2} = 85.0 \text{ mm}, \qquad r_3 = 80 \text{ mm}$$

Equation (15.43) then results in

$$F_1 = F_2 = \frac{Mr_1}{r_1^2 + r_2^2 + r_3^2} = \frac{4500(85)}{2(85)^2 + (80)^2}$$

$$= \frac{382,500}{20,850} = 18.35 \text{ kN}$$

$$F_3 = \frac{4500(80)}{20,850} = 17.27 \text{ kN}$$

The vector sum of the two shear forces, obviously greatest for the bolt 2, can be obtained algebraically (or graphically):

$$V_2 = \left[\left(\frac{15}{17} \times 18.35 + 4 \right)^2 + \left(\frac{8}{17} \times 18.35 \right)^2 \right]^{1/2} = 21.96 \text{ kN}$$

The bolt shear-stress area is $A_s = \pi d^2 / 4 = \pi(14)^2 / 4 = 153.9 \text{ mm}^2$. Hence,

$$\tau = \frac{V_2}{A_s} = \frac{21,960}{153.9} = 142.7 \text{ MPa}$$

EXAMPLE 15.7

Shear Stress in Rivets Owing to Eccentric Loading

A riveted joint is under an inclined eccentric force P, as indicated in Figure 15.22a. Calculate the maximum shear stress in the rivets.

(a)   (b)

Figure 15.22 Example 15.7: (a) riveted connection; (b) enlarged view showing loading acting at the centroid and reactions on rivet 1.

Given: The rivets are 1 in. in diameter. $P = 10$ kips.

Solution: For simplicity in computations, the applied load P is first resolved into horizontal and vertical components. Each rivet carries one-half of the load. The centroid of the rivet group is between the top and bottom rivets at C. An inspection of Figure 15.22a shows that the top rivet 1 is under the highest stress (Figure 15.22b). Through the use of Eq. (15.43),

$$F_1 = \frac{Mr_1}{r_1^2 + r_2^2 + r_3^2 + r_4^2} = \frac{84(6)}{2(6)^2 + 2(2)^2} = 6.3 \text{ kip} \cdot \text{in}$$

Vector sum of the shear forces are

$$V_1 = [(1.5 + 6.3)^2 + 2^2]^{1/2} = 8.052 \text{ kips}$$

We then have

$$\tau = \frac{V_1}{\pi d^2/4} = \frac{4(8.052)}{\pi(1)^2} = 10.25 \text{ ksi}$$

15.15 WELDING

A *weld* is a joint between two surfaces produced by the application of localized heat. Here, we briefly discuss only welding between metal surfaces; thermoplastics can be welded much like metals. A *weldment* is fabricated by welding together a variety of metal forms cut to particular configuration. Nearly all welding is by fusion processes. Establishment of metallurgical bond between two parts by melting together the base metals with a filler metal is called the *fusion process*. Heat is brought about usually by an electric arc, electric current, or gas flame. Metals and alloys to arc and gas welding must be properly selected. Properties of welding filler material must be matched with those of base metal when possible. The joint strength would then be equal to the strength of the base metal, giving an efficiency of almost 100% for static loads.

WELDING PROCESSES AND PROPERTIES

Metallic arc welding, so-called shielded metal arc welding (SMAW), refers to a process where the heat is applied by an arc passing between an electrode and the work. The electrode is composed of suitable filler material with coating ordinarily similar to that of base metal. It is melted and fed into the joint as the weld is being formed. The coating is vaporized to provide a shielding gas-preventing oxidation at the weld as well as acts as a flux and directs the arc. Either direct or alternating current can be used with this process. A weld thickness greater than about $\frac{3}{8}$ in. is often produced on successive layers. In *metal-inert gas arc welding* or gas metal arc welding (GMAW), heat is applied by a gas flame. In this process, a bare or plated wire is continuously fed into the weld from a large spool. The wire serves as electrode and becomes the filler in the union. Uniform-quality welds are attainable with metal-gas welding.

Table 15.8   Typical weld-metal properties

AWS electrode number	Ultimate strength		Yield strength		Percent elongation
	ksi	(MPa)	ksi	(MPa)	
E6010	62	(427)	50	(345)	22
E6012	67	(462)	55	(379)	17
E6020	62	(427)	50	(345)	25
E7014	72	(496)	60	(414)	17
E7028	72	(496)	60	(414)	22

SOURCE: [24].

Resistance welding uses electric-current-generated heat that passes through the parts to be welded while they are clamped together firmly. Filler material is not ordinarily employed. Usually, thin metal parts may be connected by spot or continuous resistance welding. A spot weld is made by a pair of electrodes that apply pressure to either side of a lap joint and devise a complete circuit. Laser beam welding, plasma arc welding, and electron beam welding are utilized for special applications. The suitability of several metals and alloys to arc and gas welding is very important.

Materials and symbols for welding have been standardized by the ASTM and the American Welding Society (AWS). Numerous different kinds of electrodes have been standardized to fit a variety of conditions encountered in the welding of machinery and structures. Table 15.8 presents the characteristics for some E60 and E70 electrode classes. Note that the AWS *numbering system* is based on the use of an *E* prefix followed by four digits. The first two numbers on the left identify the approximate strength in ksi. The last digit denotes a group of welding technique variables, such as current supply. The next to last digit refers to a welding position number (1 for all and 2 for horizontal positions, respectively). Welding electrodes are available in diameters from $\frac{1}{16}$ to $\frac{5}{16}$ in. It should be mentioned that the *electrode* material is often the strongest material present in a joint [23–25].

THE STRENGTH OF WELDED JOINTS

Among numerous configurations of welds, we consider only two common butt and fillet types. The geometry of a typical *butt weld* loaded in tension and shear is shown in Figure 15.23. The equations for the stresses due to the loading are also given in the figure.

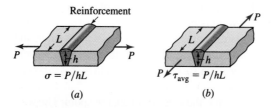

$$\sigma = P/hL \qquad\qquad \tau_{\text{avg}} = P/hL$$

(a)   (b)

Figure 15.23   Butt weld: (a) tension loading; (b) shear loading.

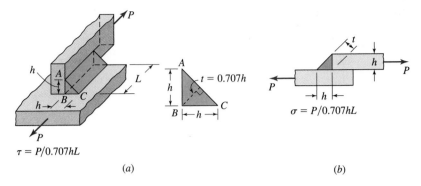

$\tau = P/0.707hL$

(a)

$\sigma = P/0.707hL$

(b)

Figure 15.24 Fillet weld: (a) shear loading; (b) transverse tension loading. Note: h = length of weld leg, t = throat length, L = weld length.

Note that the height h for a butt weld does not include the bulge or reinforcement used to compensate for voids or inclusions in the weld metal. Plates of $\frac{1}{4}$ in. and heavier should be beveled before welding as indicated.

Figure 15.24 illustrates two *fillet welds* loaded in shear and transverse tension. The corresponding average stress formula is written under each figure. Now the size of the fillet weld is defined as the leg length h. Normally, the two legs are of the same length h. In welding design, stresses are calculated for the *throat section*: minimum cross-sectional area A_t located at 45° to the legs. We have $A_t = tL = 0.707hL$, where t and L represent the throat length and length of weld, respectively (Figure 15.24a). We note that actual stress distribution in a weld is somewhat complicated and design depends on the stiffness of the base material and other factors that have been neglected. Particularly, for stress situation on the throat area in Figure 15.24b no exact solutions are available.

The foregoing average results are valid for design, however, because weld strengths are on the basis of tests on joints of these types. Having the material strengths available for a welded joint, the required weld size h can be obtained for a prescribed safety factor. The usual equation of the *factor of safety n* applies for *static loads*:

$$n = \frac{S_{ys}}{\tau} = \frac{0.5S_y}{\tau}$$ (15.44)

The quantities S_y and S_{ys} represent tensile yield and shear yield strengths of weld material, respectively.

STRESS CONCENTRATION AND FATIGUE IN WELDS

Abrupt changes in geometry take place in welds and hence stress concentrations are present. The weld and the plates at the base and reinforcement should be thoroughly blended together (Figure 15.23). The stresses are highest in the immediate vicinity of the weld. Sharp corners at the *toe* and *heel,* points A and B in Figure 15.24, should be rounded. Since welds are ductile materials, stress concentration effects are ignored for static loads. As has always been the case, when the loading fluctuates, a stress concentration factor is applied

Table 15.9   Fatigue stress concentration
factors K_f for welds

Type of weld	K_f
Reinforced butt weld	1.2
Toe of transverse fillet weld	1.5
End of parallel fillet weld	2.7
T-butt joint with sharp corners	2.0

| SOURCE: [25].

to alternating component. Approximate values for fatigue strength reduction factors are listed in Table 15.9.

Under *cyclic loading,* the welds fail long before the welded members. The fatigue factor of safety and working stresses in welds are defined by the AISC as well as AWS codes for buildings and bridges [2, 26, 27]. The codes allow the use of a variety of ASTM structural steels. For ASTM steels, tensile yield strength is one-half the ultimate strength in tension, $S_y = 0.5S_u$, for static or fatigue loads. Unless otherwise specified, an *as-forged surface* should always be used for weldments. Also, prudent design would suggest taking the size factor $C_s = 0.7$. Design calculations for fatigue loading can be made by the methods described in Section 8.11, as illustrated in following sample problem.

EXAMPLE 15.8   Design of a Butt Welding for Fatigue Loading

The tensile load P on a butt weld (Figure 15.23a) fluctuates continuously between 20 kN and 100 kN. Plates are 20 mm thick. Determine the required length L of the weld, applying the Goodman criterion.

Assumptions: Use an E6010 welding rod with a factor of safety of 2.5.

Solution: By Table 15.8 for E6010, $S_u = 427$ MPa. The endurance limit of the weld metal, from Eq. (8.6), is

$$S_e = C_f C_r C_s C_t (1/K_f) S'_e$$

Referring to Section 8.7, we have

$C_r = 1$ (based on 50% reliability)

$C_s = 0.7$ (lacking information)

$C_f = AS^b = 272(427)^{-0.995} = 0.657$ (by Eq. (8.7))

$C_t = 1$ (normal room temperature)

$K_f = 1.2$ (from Table 15.9)

$S'_e = 0.5S_u = 0.5(427) = 213.5$ MPa

Hence,

$$S_e = (1)(0.7)(0.657)(1)(1/1.2)(213.5) = 81.82 \text{ MPa}$$

The mean and alternating loads are given by

$$P_m = \frac{100 + 20}{2} = 60 \text{ kN}, \qquad P_a = \frac{100 - 20}{2} = 40 \text{ kN}$$

Corresponding stresses are

$$\sigma_m = \frac{60,000}{20L} = \frac{3000}{L}$$

$$\sigma_a = \frac{40,000}{20L} = \frac{2000}{L}$$

Through the use of Eq. (8.16), we have

$$\frac{S_u}{n} = \sigma_m + \frac{S_u}{S_e}\sigma_a; \qquad \frac{427}{2.5} = \frac{3000}{L} + \frac{427}{81.82}\left(\frac{2000}{L}\right)$$

Solving,

$$L = 78.67 \text{ mm}$$

Comment: A 79-mm long weld should be used.

15.16 WELDED JOINTS SUBJECTED TO ECCENTRIC LOADING

When a welded joint is under eccentrically applied loading, the effect of torque or moment must be taken into account as well as the direct load. The exact stress distribution in such a joint is complicated. A detailed study of both the rigidity of the parts being joined and the geometry of the weld is required [25, 26]. The following procedure, which is based on simplifying assumptions, leads to reasonably accurate results for most applications.

TORSION IN WELDED JOINTS

Figure 15.25 illustrates an eccentrically loaded joint, with the centroid of all the weld areas or weld group at point C. The load P is applied at a distance e from C, in the plane of the group. As a result, the welded connection is under torsion $T = Pe$ and the direct load P. The latter force causes a direct shear stress in the welds:

$$\tau_d = \frac{P}{A} \tag{15.45}$$

in which P is the applied load and A represents the throat area of all the welds. The preceding stress is taken to be uniformly distributed over the length of all welds. The torque

Figure 15.25   Welded joint in plane eccentric loading. Note: C is the centroid of the weld group.

causes the following torsional shear stress in the welds:

$$\tau_t = \frac{Tr}{J} \tag{15.46}$$

where

T = torque

r = distance from C to the point in the weld of interest

J = polar moment of inertia of the weld group about C (based on the throat area)

Resultant shear stress in the weld at radius r is given by the vector sum of the direct shear stress and torsional stress:

$$\tau = \left(\tau_d^2 + \tau_t^2\right)^{1/2} \tag{15.47}$$

Note that r usually represents the *farthest* distance from the centroid of the weld group.

BENDING IN WELDED JOINTS

Consider an angle welded to a column, as depicted in Figure 15.26. Load P acts at a distance e, out of plane of the weld group, producing bending in addition to direct shear. We again take a linear distribution of shear stress due to moment $M = Pe$ and a uniform distribution of direct shear stress. The latter stress τ_d is given by Eq. (15.45). The moment causes the shear stress

$$\tau_m = \frac{Mc}{I} \tag{15.48}$$

Here the distance c is measured from C to the farthest point on the weld. As in the previous case, the resultant shear stress τ in the weld is estimated by the vector sum of the direct

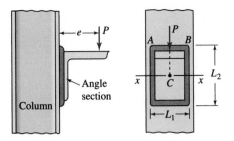

Figure 15.26   Welded joint under out-of-plane loading.

shear stress and the moment-induced stress:

$$\tau = \left(\tau_d^2 + \tau_m^2\right)^{1/2} \tag{15.49}$$

On the basis of the geometry and loading of Figure 15.26, we see that τ_d is downward and τ_m along edge AB is outward.

Centroid of the Weld Group

Let A_i denote the weld segment area and x_i and y_i the coordinates to the centroid of any (straight-line) segment of the weld group. Then, the centriod C of the weld group is located at

$$\bar{x} = \frac{\sum A_i x_i}{\sum A_i}, \qquad \bar{y} = \frac{\sum A_i y_i}{\sum A_i} \tag{15.50}$$

in which $i = 1, 2, \ldots, n$ for n welds. In the case of symmetric weld group, the location of the centroid is obvious.

Moments of Inertia of a Weld (Figure 15.27)

For simplicity, we assume that the effective weld width in the plane of the paper is the same as throat length $t = 0.707h$, shown in Figure 15.24a. The parallel axis theorem can be

Figure 15.27   Moments of inertia of a weld parallel to the y axis.

applied to find the moments of inertia about x and y axes through the centroid of the weld group:

$$I_x = I_{x'} + Ay_1^2 = \frac{tL^3}{12} + Lty_1^2$$

$$I_y = I_{y'} + Ax_1^2 = \frac{Lt^3}{12} + Ltx_1^2 = Ltx_1^2$$

(15.51)

Note that t is assumed to be *very small* in comparison with the other dimensions and hence $I_{y'} = Lt^3/12 = 0$ in the second of the preceding equations. The polar moment of inertia about an axis through C perpendicular to the plane of the weld is then

$$J = I_x + I_y = \frac{tL^3}{12} + Lt(x_1^2 + y_1^2)$$

(15.52)

The values of I and J for each weld about C should be calculated by using Eqs. (15.51) and (15.52); the results are added to obtain the moment and product of inertia of the entire joint. It should be mentioned that the moment and polar moment of inertias for the most common fillet welds encountered are listed in some publications [15]. The detailed procedure is illustrated in the following sample problems.

Case Study 15-2 | DESIGN OF A WELDED JOINT OF THE WINCH CRANE FRAME

The welded joint C, with identical fillets on both sides of the vertical frame of the winch crane frame, is under in-plane eccentric loading (refer to Case Study 1-1), as shown in Figure 15.28a. Determine the weld size h at the joint.

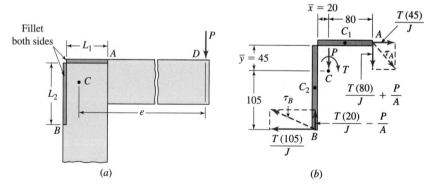

(a)   (b)

Figure 15.28 (a) Welded joint C of the winch crane shown in Fig. 1.4; (b) enlarged view of the weld group. Loading acts at the centroid C of the group and shear stresses at weld ends A and B.

Case Study (CONCLUDED)

Given: $L_1 = 100\,\text{mm}$, $\quad L_2 = 150\,\text{mm}$,

$\qquad e = 1.5\,\text{m}$, $\qquad P = 3\,\text{kN}$

Assumptions: An E6010 welding rod with a factor of safety $n = 2.2$ is used. The vertical frame of the crane is taken to be a rigid column.

Solution: See Figures 1.4 and 15.28.

Area Properties. The centroid of the weld group (Figure 15.27) is given by

$$\bar{x} = \frac{A_1 x_1 + A_2 x_2}{A_1 + A_2} = \frac{(100t)(50) + (150t)(10)}{100t + 150t}$$

$$= 20\,\text{mm}$$

$$\bar{y} = \frac{A_1 y_1 + A_2 y_2}{A_1 + A_2} = \frac{(100t)(0) + (150t)(75)}{100t + 150t} = 45\,\text{mm}$$

The torque equals

$$T = Pe = (3000)(1500) = 4.5\,\text{MN} \cdot \text{mm}$$

The centroidal moments of inertia are

$$I_x = \sum \frac{tL^3}{12} + Lty^2$$

$$= \frac{t(100)^3}{12} + (100)t(45)^2 + 0$$

$$+ (150)t(75 - 45)^2 = 420{,}833t$$

$$I_y = \sum \frac{tL^3}{12} + Ltx^2$$

$$= 0 + (100)t(50 - 20)^2$$

$$+ \frac{t(150)^3}{12} + (150)t(20)^2 = 431{,}250t$$

$$J = 852{,}083t \ \text{mm}^4$$

Since there are fillets at both sides of the column, the area properties are multiplied by 2.

Stresses. From Table 15.8, we have $S_y = 345$ MPa. By inspection of Figure 15.28b, either at point A or B, the combined torsional and direct shear stresses are greatest. At point A,

$$\tau_v = \frac{P}{A} + \frac{Tr_h}{J} = \frac{3(10)^3}{2(250t)} + \frac{4.5(10^6)(80)}{2(852{,}083t)}$$

$$= \frac{6}{t} + \frac{211.2}{t}$$

$$\tau_h = \frac{Tr_v}{J} = \frac{4.5(10^6)(45)}{2(852{,}083t)} = \frac{118.8}{t}$$

$$\tau_A = \left(\tau_v^2 + \tau_h^2\right)^{1/2} = \frac{247.6}{t}$$

Similarly, at point B,

$$\tau_v = -\frac{6}{t} + \frac{4.5(10^6)(20)}{2(852{,}083t)} = -\frac{6}{t} + \frac{52.8}{t}$$

$$\tau_h = \frac{4.5(10^6)(105)}{2(852{,}083t)} = -\frac{277.3}{t}$$

$$\tau_B = \frac{281.2}{t} \ \text{N/mm} \qquad \text{(governs)}$$

Weld Size. Therefore, by Eq. (15.44),

$$n\tau_B = 0.5S_y; \qquad 2.2\left(\frac{281.2}{t}\right) = 0.5(345)$$

from which $t = 3.59$ mm. Referring to Figure 15.25a, we obtain

$$h = \frac{t}{0.707} = \frac{3.59}{0.707} = 5.08 \ \text{mm}$$

Comment: A nominal size of 5-mm fillet welds should be used in the joint.

| **EXAMPLE 15.9** | Design of a Welded Joint under Out-of-Plane Eccentric Loading |

A welded joint is subjected to out-of-plane eccentric force P (Figure 15.26). What weld size is required?

Given: $L_1 = 60\,\text{mm}$, $L_2 = 90\,\text{mm}$, $e = 50\,\text{mm}$, $P = 15\,\text{kN}$

Assumption: An E6010 welding rod with factor of safety $n = 3$ is used.

Solution: By Table 15.8, for E6010, $S_y = 345$ MPa. The centriod lies at the intersection of the two axes of symmetry of the area enclosed by the weld group. The moment of inertia is

$$I_x = 2(60)t\,(45)^2 + \frac{2(90)^3 t}{12} = 364{,}500t \text{ mm}^4$$

The total weld area equals $A = 2(60t + 90t) = 300t$ mm^2. Moment is $M = 15(50) = 750$ kN $\cdot$ mm. The maximum shear stress, using Eq. (15.49):

$$\tau = \left[\left(\frac{15{,}000}{300t}\right)^2 + \left(\frac{750{,}000 \times 45}{364{,}500t}\right)^2 \right]^{1/2} = \frac{105.2}{t} \text{ N/mm}^2$$

Applying Eq. (15.44), we have

$$n\tau = 0.5S_y; \qquad 3\left(\frac{105.2}{t}\right) = 0.5(345) \quad \text{or} \quad t = 1.83\,\text{mm}$$

Hence,

$$h = \frac{t}{0.707} = \frac{1.83}{0.707} = 2.59\,\text{mm}$$

Comment: A nominal size of 3-mm fillet welds should be used throughout.

15.17   BRAZING AND SOLDERING

Brazing and soldering differ from welding essentially in that the temperatures are always below the melting point of the parts to be united, but the parts are heated above the melting point of the solder. It is important that the surfaces initially be clean. Soldering or brazing filler material acts somewhat similar to a molten metal glue or cement, which sets directly on cooling. Brazing or soldering can thus be categorized as bonding.

BRAZING PROCESS

Brazing starts with heating the workpieces to a temperature above 450°C. On contact with the parts to be united, the filler material melts and flows into the space between the workpieces. The filler materials are customarily alloys of copper, silver, or nickel. These may be handheld and fed into the joint (free of feeding) or preplaced as washers, shims, rings, slugs, and the like. Dissimilar metals, cast and wrought metals, as well as nonmetals and metals can be brazed together. Brazing is ordinarily accomplished by heating parts with a torch or in a furnace. Sometimes other brazing methods are used. A brief description of some processes of brazing follow. Note that, in all metals, either flux or an inert gas atmosphere is required.

Torch brazing utilizes acetylene, propane, and other fuel gas, burned with oxygen or air. It may be manual or mechanized. On the other hand, *furnace* brazing uses the heat of a gas-fired, electric, or other kind of furnace to raise the parts to brazing temperature. A technique that utilizes a high-frequency current to generate the required heat is referred to as *induction* brazing. As the name suggests, *dip* brazing involves the immersion of the parts in a molten bath. A method that utilizes resistance-welding machines to supply the heat is called *resistance* brazing. As currents are large, water cooling of electrodes is essential.

SOLDERING PROCESS

The procedure of soldering is identical to that of brazing. However, in soldering the filler metal has a melting temperature below 450°C and relatively low strength. Heating can be done with a torch or a high-frequency induction heating coil. Surfaces must be clean and covered with flux that is liquid at the soldering temperature. The flux is drawn into the joint and dissolves any oxidation present at the joint. When the soldering temperature is reached, the solder replaces the flux at the joint.

Cast iron, wrought-iron, and carbon steels can be soldered to each other or to brass, copper, nickel, silver, Monel, and other nonferrous alloys. Nearly all solders are tin-lead alloys, but alloys including antimony, zinc, and aluminum are also employed. The strength of a soldered union depends on numerous factors, such as the quality of the solder, thickness of the joint, smoothness of the surfaces, kind of materials soldered, and soldering temperature. Some common soldering applications involve electrical and electronic parts, sealing seams in radiators and in thin cans.

15.18   ADHESIVE BONDING

Adhesives are substances able to hold materials together by surface attachment. Nearly all structural adhesives are thermosetting as opposed to thermoplastic or heat-softening types, such as rubber cement and hot metals. Epoxies and urethanes are versatile and in widespread use as the structural adhesives [8]. Numerous other adhesive materials are used for various applications. Some remain liquid in the presence of oxygen, but they harden in restricted spaces, such as on bolt threads or in the spaces between a shaft and hub. *Adhesive bonding* is extensively utilized in the automotive and aircraft industries. Retaining compounds of adhesives can be employed to assemble cylindrical parts formerly needing press or shrink fits. In such cases, they eliminate press-fit stresses and reduce machining costs. Ordinary engineering adhesives have shear strengths varying from 25 to 40 MPa. The website at www.3m.com/bonding includes information and data on adhesives.

The *advantages* of adhesive bonding over mechanical fastening include the capacity to bond alike and dissimilar materials of different thickness; economic and rapid assembly; insulating characteristics; weight reduction; vibration dumping; and uniform stress distribution. On the other hand, examples of the *disadvantages* of the adhesive bonding are the preparation of surfaces to be connected, long cure times, possible need for heat and pressure for curing, service temperature sensitivity, service deterioration, tendency to creep under prolonged loading, and questionable long-term durability. The upper service temperature of most ordinarily employed adhesives is restricted to about 400°F. However, simpler, cheaper, stronger, and more easy-to-apply adhesives can be expected in the future.

DESIGN OF BONDED JOINTS

A design technique of rapidly growing significance is metal-to-metal adhesive bonding. Organic materials can be bonded as well. In cementing together metals, specific adhesion becomes important, inasmuch as the penetration of adhesive into the surface is insignificant. A number of metal-to-metal adhesives have been refined but their use has been confined mainly to lap or spot joints of relatively limited area. Metal-to-metal adhesives, as employed in making plymetal, have practical applications.

Three common methods of applying adhesive bonding are illustrated in Table 15.10. Here, based on an approximate analysis of joints, stresses are assumed to be uniform over the bonded surfaces. The actual stress distribution varies over the area with *aspect ratio* b/L. Highest and lowest stresses occur at the edges and in the center, respectively. Adhesive joints should be properly designed to support only shear or compression and very small tension. Connection geometry is most significant when relatively high-strength materials are united. Large bond areas are recommended, such as in a lap-joint (case *A* of the

Table 15.10   Some common types of adhesive joints

Configuration	Average stress
A. Lap	
	$\tau = \dfrac{P}{bL}$
B. Double lap	
	$\tau = \dfrac{P}{2bL}$
C. Scarf	Axial loading:
	$\sigma_{x'} = \dfrac{P}{bt}\cos^2\theta, \qquad \tau_{x'y'} = -\dfrac{P}{2bt}\sin 2\theta$
	Bending:
	$\sigma_{x'} = \dfrac{6M}{bt^2}\cos^2\theta, \qquad \tau_{x'y'} = -\dfrac{3M}{bt^2}\sin 2\theta$

⎮ Notes: P = centric load, M = moment, b = width of plate, t = thickness of thinnest plate, L = length of lap.

table), particularly connecting the metals. Nevertheless, this shear joint has noteworthy stress concentration of about 2 at the ends for an aspect ratio of 1.

It should be pointed out that the lap joints may be inexpensive because no preparation is required except, possibly, surface cleaning, while the machining of a scarf joint is impractical. The exact stress distribution depends on the thickness and elasticity of the joined members and adhesives. Stress concentration can arise because of the abrupt angles and changes in material properties. Load eccentricity is an important aspect in the state of stress of a single-lap joint. In addition, often the residual stresses associated with the mismatch in coefficient of thermal expansion between the adhesive and adherents may be significant [28, 29].

REFERENCES

1. *ANSI/ASME Standards,* B1.1-1989, B1.13-1983 (R1989). New York: American Standards Institute, 1989.
2. Horton, H. L., ed. *Machinery's Handbook,* 21st ed. New York: Industrial Press, 1974.
3. Burr, A. H., and J. B. Cheatham. *Mechanical Analysis and Design,* 2nd ed. Upper Saddle River, NJ: Prentice Hall, 1995.
4. Widmoyer, G. A. "Determine the Basic Design Specifications for a Ball Bearing Screw Assembly." *General Motors Engineering Journal* 2, no. 5 (1955), pp. 49–50.
5. ANSI B5.48. *Ball Screws.* New York: ASME, 1987.
6. Parmley, R. O., ed. *Standard Handbook of Fastening and Joining,* 2nd ed. New York: McGraw-Hill, 1989.
7. Fasteners and Joining Reference Issue. *Machine Design* (November 13, 1980).
8. Avallone, E. A., and T. Baumeister III, eds. *Mark's Standard Handbook for Mechanical Engineers,* 10th ed. New York: McGraw-Hill, 1996.
9. Kulak, G. I., J. W. Fisher, and H. A. Struik. *Guide to Design Criteria for Bolted and Riveted Joints,* 2nd ed. New York: Wiley, 1987.
10. Bickford, J. H. *An Introduction to the Design and Behavior of Bolted Joints,* 2nd ed. New York: Marcel Dekker, 1990.
11. Peterson, R. E. *Stress Concentration Factors.* New York: Wiley, 1974.
12. Dimarogones, A. D. *Machine Design: A CAD Approach.* New York: Wiley, 2001.
13. Deutchman, A. D., W. J. Michels, and C. E. Wilson. *Machine Design: Theory and Practice.* New York: Macmillan, 1975.
14. Fauppel, J. H., and F. E. Fisher. *Engineering Design,* 2nd ed. New York: Wiley, 1981.
15. Shigley, J. E., and C. R. Mischke. *Mechanical Engineering Design,* 6th ed. New York: McGraw-Hill, 2001.
16. Norton, R. L. *Machine Design: An Integrated Approach,* 2nd ed. Upper Saddle River, NJ: Prentice Hall, 2000.
17. Juvinall, R. C., and K. M. Marshak. *Fundamentals of Machine Component Design,* 3rd ed. New York: Wiley, 2000.
18. Hamrock, B. J., B. Jacobson, and S. R. Schmid. *Fundamentals of Machine Elements.* New York: McGraw-Hill, 1999.
19. Mott, R. L. *Machine Elements in Mechanical Design,* 2nd ed. New York: Macmillan, 1992.
20. Gould, H. H., and B. B. Minic. "Areas of Contact and Pressure Distribution in Bolted Joints." *Transactions of the ASME, Journal of Engineering for Industry,* 94 (1972), pp. 864–69.

21. Wileman, J., M. Choundury, and I. Green. "Computational Stiffness in Bolted Connections." *Transactions of the ASME, Journal of Mechanical Design* 113 (December 1991), pp. 432–37.

22. American Institute of Steel Construction. *Manual of Steel Construction,* 9th ed. New York: AISC, 1989.

23. Johnston, B. G., and F. J. Lin. *Basic Steel Design.* Upper Saddle River, NJ: Prentice Hall, 1974.

24. *American Welding Society Code AWSD.1.77.* Miami, FL: American Welding Society.

25. Norris, C. H. "Photoelastic Investigation of Stress Distribution in Transverse Fillet Welds." *Welding Journal* 24 (1945), p. 557s.

26. Jennings, C. H. "Welding Design." *Transactions of the ASME,* 58 (1936), p. 497, and 59 (1937), p. 462.

27. Osgood, C. C. *Fatigue Design.* New York: Wiley, 1970.

28. Pocius, A. V. *Adhesion and Adhesives Technology: An Introduction.* New York: Hanser, 1997.

29. Brinson, H. F., ed. *Engineering Materials Handbook,* vol. 3. *Adhesives and Sealents.* Metals Park, OH: ASM International, 1990.

PROBLEMS

Sections 15.1 through 15.7

15.1   A power screw is 75 mm in diameter and has a thread pitch of 15 mm. Determine the thread depth, the thread width or the width at pitch line, the mean and root diameters, and the lead, for the case in which

(a)   Square threads are used.

(b)   Acme threads are used.

15.2   A $1\frac{1}{2}$-in. diameter, double thread Acme screw is to be used in an application similar to that of Figure 15.6. Determine

(a)   The screw lead, mean diameter, and helix angle.

(b)   The starting torques for lifting and lowering the load.

(c)   The efficiency, if collar friction is negligible.

(d)   The force F to be exerted by an operator, for $a = 15$ in.

Given: $f = 0.1$, $f_c = 0.08$, $d_c = 2$ in., $W = 1.5$ kips

15.3   What helix angle would be required so that the screw of Problem 15.2 would just begin to overhaul? What would be the efficiency of a screw with this helix angle, for the case in which the collar friction is negligible.

15.4   A 32-mm diameter power screw has a double square thread with a pitch of 4 mm. Determine the power required to drive the screw.

Design Requirement: The nut is to move at a velocity of 40 mm/s and lift a load of $W = 6$ kN. Given: The mean diameter of the collar is 50 mm. Coefficients of friction are estimated as $f = 0.1$ and $f_c = 0.15$.

15.5 A square-thread screw has a mean diameter of $1\frac{3}{4}$ in. and a lead of $L = 1$ in. Determine the coefficient of thread friction.

Given: The screw consumes 5 hp when lifting a 2 kips weight at the rate of 25 fpm.
Design Assumption: The collar friction is negligible.

15.6 A $2\frac{3}{4}$-in. diameter square-thread screw is used to lift or lower a load of $W = 50$ kips at a rate of 2 fpm. Determine

(*a*) The revolutions per minute of the screw.

(*b*) The motor horsepower required to lift the weight, if screw efficiency $e = 85\%$ and $f = 0.15$.

Design Assumption: Because the screw is supported by a thrust ball bearing, the collar friction can be neglected.

15.7 A square-thread screw has an efficiency of 70% when lifting a weight. Determine the torque that a brake mounted on the screw must exert when lowering the load at a uniform rate.

Given: The coefficient of thread friction is estimated as $f = 0.12$ with collar friction negligible; the load is 50 kN and the mean diameter is 30 mm.

15.8 Determine the pitch that must be provided on a square-thread screw to lift a 2.5 kip weight at 40 fpm with power consumption of 5 hp.

Given: The mean diameter is 1.875 in. and $f = 0.15$.
Design Assumption: The collar friction is negligible.

15.9 A 1 in.-8 UNC screw supports a tensile of 12 kips. Determine

(*a*) The axial stress in the screw.

(*b*) The minimum length of nut engagement, if the allowable bearing stress is not to exceed 10 ksi.

(*c*) The shear stresses in the nut and screw.

15.10 A 50-mm diameter square-thread screw having a pitch of 8 mm carries a tensile load of 15 kN. Determine

(*a*) The axial stress in the screw.

(*b*) The minimum length of nut engagement needed, if the allowable bearing stress is not to exceed 10 MPa.

(*c*) The shear stresses in the nut and screw.

Sections 15.8 through 15.12

15.W1 Search the website at www.nutty.com. Perform a product search for various types of nuts, bolts, and washers. Review and list 15 commonly used configurations and descriptions of each of these elements.

15W2 Use the site at www.boltscience.com to review the current information related to bolted joint technology. List three usual causes of relative motion of threads.

15.11 The joint shown in Figure P15.11 has a 15-mm diameter bolt and a grip length of $L = 50$ mm. Calculate the maximum load that can be carried by the part without loosing all the initial compression in the part.

Given: The tightening torque of the nut for average condition of thread friction is 72 N·m by Eq. (15.21).

Aluminum part $E_a = \frac{1}{3} E_s$

Steel bolt E_s

P

Figure P15.11

15.12 The bolt of the joint shown in Figure P15.11 is $\frac{7}{8}$ in.-9 UNC, SAE grade 5, with a rolled thread. Apply the Goodman criterion to determine

(*a*) The permissible value of preload F_i if the bolt is to be safe for continuous operation with $n = 2$.

(*b*) The tightening torque for an average condition of thread friction.

Given: The value of load P on the part ranges continuously from 8 kips to 16 kips; the grip is $L = 2$ in. The survival rate is 95%.

15.13 The bolt of the joint depicted in Figure P15.11 is $M20 \times 2.5$-C, grade 7, with cut thread, $S_y = 620$ MPa, and $S_u = 750$ MPa. Calculate

(*a*) The maximum and minimum values of the fluctuating load P on the part, on the basis of the Soderberg theory.

(*b*) The tightening torque, if bolt is lubricated.

Given: The grip is $L = 50$ mm; the preload equals $F_i = 25$ kN; the average stress in the root of the screw is 160 MPa; the survival rate equals 90%.
Design Requirement: The safety factor is 2.2. The operating temperature is not elevated.

15.14 Figure P15.14 depicts a partial section from a permanent connection. Determine

(*a*) The total force and stress in each bolt.

(*b*) The tightening torque for an average condition of thread friction.

Given: A total of six bolts are used to resist an external load of $P = 18$ kips.

Cast iron
$E = 20 \times 10^6$ psi

$1\frac{1}{4}$ in.

Steel bolt
$\frac{5}{8}$ in. – 18 UNF
SAE grade 5
$E = 30 \times 10^6$ psi

Figure P15.14

15.15 A section of the connection illustrated in Figure P15.14 carries an external load that fluctuates between 0 and 4 kips. Using the Goodman criterion, determine the factor of safety n guarding against the fatigue failure of the bolt.

Given: The survival rate is 98%. The operating temperature is 900°F maximum.
Design Assumptions: All parts have rolled threads; each bolt has been preloaded to $F_i = 10$ kips.

15.16 The bolt of connection shown in Figure P15.16 is $M20 \times 2.5$, ISO course thread having $S_y = 630$ MPa. Determine

(a) The total force on the bolt, if the joint is reusable.

(b) The tightening torque, if the bolts are lubricated.

Given: The grip is $L = 60$ mm; the joint carries an external load of $P = 40$ kN.
Design Assumption: The bolt will be made of steel of modulus of elasticity E_s and the parts are cast iron with modulus of elasticity $E_c = E_s/2$.

Steel bolt

L

Parts

P

Figure P15.16

15.17 The connection shown in Figure P15.16 carries an external loading P that value ranges from 1 to 5 kips. Determine

(a) If the bolt fails without preload.

(b) Whether the bolt is safe when preload present.

(c)   The fatigue factor of safety n when preload is present.

(d)   The load factor n_s guarding against joint separation.

Design Assumptions: The bolt is made of steel (E_s) and the parts are cast iron with modulus of elasticity $E_c = E_s/2$. The operating temperature is normal. The bolt may be reused when the joint is taken apart. The survival rate is 90%.

Given: The steel bolt is $\frac{1}{2}$ in.-13 UNC, SAE grade 2, with rolled threads; the grip is $L = 2$ in.

15.18 The assembly shown in Figure P15.16 uses an $M14 \times 2$, ISO grade 8.8 course cut threads. Apply the Goodman criterion to determine the fatigue safety factor n of the bolt with and without initial tension.

Given: The joint constant is $C = 0.31$. The joint carries a load P varying from 0 to 10 kN. The operating temperature is 490°C maximum.

Design Assumptions: The bolt may be reused when the joint is taken apart. Survival probability is 95%.

15.19 Determine the maximum load P the joint described in Problem 15.18 can carry based on a static safety factor of 2.

Design Assumptions: The joint is reusable.

15.20 Figure P15.20 shows a portion of a high-pressure boiler accumulator having flat heads. The end plates are affixed using a number of bolts of $M16 \times 2$-C, grade 5.8, with rolled threads. Determine

(a)   The factor of safety n of the bolt against fatigue failure with and without preload.

(b)   The load factor n_s against joint separation.

Given: The fully modified endurance limit is $S_e = 100$ MPa. External load P varies from 0 to 12 kips/bolt.

Design Assumptions: Clamped parts have a stiffness k_p, five times the bolt stiffness k_b. The connection is permanent.

Figure P15.20

Sections 15.13 and 15.14

15.21 A double-riveted lap joint with plates of thickness t is to support a load P as shown in Figure P15.21. The rivets are 19 mm in diameter and spaced 50 mm apart in each row. Determine the shear, bearing, and tensile stresses.

Given: $P = 32$ kN, $t = 10$ mm

Figure P15.21

15.22 A double-riveted longitudinal lap joint (Figure P15.21) is made of plates of thickness t. Determine the efficiency of the joint.

Given: The $\frac{3}{4}$-in. diameter rivets have been drilled $2\frac{1}{2}$ in. apart in each row and $t = \frac{3}{8}$ in. Design Assumptions: The allowable stresses are 22 ksi in tension, 15 ksi in shear, and 48 ksi in bearing.

15.23 Figure P15.23 shows a bolted lap joint that uses $\frac{5}{8}$ in.-11 UNC, SAE grade 8 bolts. Determine the allowable value of the load P, for the following safety factors: 2, shear on bolts; 3, bearing of bolts; 2.5, bearing on members; and 3.5, tension of members.

Design Assumption: The members are made of cold-drawn AISI 1035 steel with $S_{ys} = 0.577S_y$.

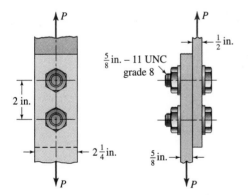

Figure P15.23

15.24 The bolted connection shown in Figure P15.24 uses $M14 \times 2$ course pitch thread bolts having $S_y = 640$ MPa and $S_{ys} = 370$ MPa. A tensile load $P = 20$ kN is applied to the connection. The dimensions are in millimeters. Determine the factor of safety n for all possible modes of failure.

Design Assumption: Members are made of hot-rolled 1020 steel.

15.25 A machine part is fastened to a frame by means of $\frac{1}{2}$ in.-13 UNC (Table 15.1) two rows of steel bolts, as shown in Figure P15.25. Each row also has two bolts. Determine the maximum allowable value of P.

Design Decisions: The allowable stresses for the bolts is 20 ksi in tension and 12 ksi in shear.

Figure P15.24   **Figure P15.25**

15.26 Three $M20 \times 2.5$ coarse-thread steel bolts (Table 15.2) are used to connect a part to a vertical column, as shown in Figure P15.26. Calculate the maximum allowable value of P.

Design Decisions: The allowable stresses for the bolt are 145 MPa in tension and 80 MPa in shear.

Figure P15.26

15.27 A riveted structural connection supports a load of 10 kN, as shown in Figure P15.27. What is the value of the force on the most heavily loaded rivet in the bracket? Determine the values of the shear stress for 20-mm rivets and the bearing stress if the Gusset plate is 15 mm thick.

Given: The applied loading is $P = 10$ kN.

15.28 The riveted connection shown in Figure P15.28 supports a load P. Determine the distance d.

Design Decision: The maximum shear stress on the most heavily loaded rivet are 100 MPa.
Given: The applied loading equals $P = 50$ kN.

15.29 Determine the value of the load P for the riveted joint shown in Figure P15.28.

Design Assumption: The allowable rivet stress in shear is 100 MPa.
Given: $d = 90$ mm

Figure P15.27

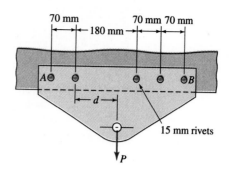

Figure P15.28

Sections 15.15 through 15.18

15.30 The plates in Figure 15.24a are 10 mm thick × 40 mm wide and made of steel having $S_y = 250$ MPa. They are welded together by a fillet weld with $h = 7$ mm leg, $L = 60$ mm long, $S_y = 350$ MPa, and $S_{ys} = 200$ MPa. Using a safety factor of 2.5 based on yield strength, determine the load P that can be carried by the joint.

15.31 Determine the lengths L_1 and L_2 of welds for the connection of a 75 × 10-mm steel plate ($\sigma_{all} = 140$ MPa) to a machine frame (Figure P15.31).

Given: 12-mm fillet welds having a strength of 1.2 kN per linear millimeter.

Figure P15.31

15.32 Calculate the required weld size for the bracket in Figure 15.26 if a load $P = 3$ kips is applied with eccentricity $e = 10$ in.

Design Assumptions: 8 ksi is allowed in shear; $L_1 = 4$ in., and $L_2 = 5$ in.

15.33 Resolve Problem 15.32 if the load P varies continuously from 2 to 4 kip. Apply the Goodman criterion.

Given: $S_u = 60$ ksi, $n = 2.5$

15.34 Determine the required length of weld L in Figure P15.34 if an E7014 electrode is used with a safety factor $n = 2.5$.

Given: $P = 100$ kN, $a = 60$ mm, $h = 12$ mm

Figure P15.34

15.35 Resolve Problem 15.34 if the load P varies continuously between 80 and 120 kN.

Design Decision: Use the Soderberg criterion.

15.36 Load P in Figure P15.34 varies continuously from 0 to P_{max}. Determine the value of P_{max} if an E6010 electrode is used, with a safety factor of 2. Apply the Goodman theory.

Given: $a = 3$ in., $L = 10$ in., $h = \frac{1}{4}$ in.

15.37 Calculate the size h of the two welds required to attach a plate to a frame as shown in Figure P15.37 if the plate supports an inclined force $P = 10$ kips.

Design Decisions: Use $n = 3$ and $S_y = 50$ ksi for the weld material.

Figure P15.37

15.38 The value of load P in Figure P15.37 ranges continuously between 2 and 10 kips. Using $S_u = 60$ ksi and $n = 1.5$, determine the required weld size. Employ the Goodman criterion.

AXISYMMETRIC PROBLEMS IN DESIGN

Outline

16.1   INTRODUCTION

In the class of axisymmetrically loaded members, the basic problem may be defined in terms of the radial coordinate. Typical examples are thick-walled cylinders, flywheels, press and shrink fits, curved beams subjected to pure bending, and thin-walled cylinders. This chapter concerns mainly "exact" stress distribution in this group of machine and structural members. The methods of the mechanics of materials and the theory of elasticity are applied. Consideration is given to thermal and plastic stresses, the material strength, and an appropriate theory of failure to obtain a safe and reliable design in Sections 16.6 and 16.7. We also discuss briefly symmetric bending of circular plates, axisymmetrically loaded shells, and filament-wound cylinders in Sections 16.9 through 16.14.

The buckling of thin-walled cylinders under axial compression and critical pressures in vessels are treated in the concluding section. There are several other problems of practical interest dealing with axisymmetric stress and deformation in a member. Among these are various situations involving rings reinforcing a juncture, hoses, semicircular barrel vaults, torsion of circular shafts of variable diameter, local stresses around a spherical cavity, and pressure between two spheres in contact (discussed in Section 3.14). For more detailed treatment of the members with axisymmetric loading, see, for example, [1–12].

16.2   BASIC RELATIONS

In the cases of axially loaded members, torsion of circular bars, and pure bending of beams, simplifying assumptions associated with deformation patterns are made so that strain (and stress) distribution for a cross section of each member can be ascertained. A basic hypothesis has been that plane sections remain plane subsequent to the loading. However, in axisymmetric and more complex problems, it is usually impossible to make similar assumptions regarding deformation. So, analysis begins with consideration of a general infinitesimal element, Hooke's law is stated, and the solution is found after stresses acting on any element and its displacements are known. At the boundaries of a member, the equilibrium of known forces (or prescribed displacement) must be satisfied by the corresponding infinitesimal elements.

Here, we present the basic relations of an axially symmetric two-dimensional problem referring to the geometry and notation of the thick-walled cylinder (Figure 16.1). The inside radius of the cylinder is a and the outside radius is b. The tangential stresses σ_θ and the radial stresses σ_r in the wall at a distance r from the center of the cylinder are caused by pressure. A typical infinitesimal element of unit thickness isolated from the cylinder is defined by two radii, r and $r + dr$, and an angle $d\theta$, as shown in Figure 16.2. The quantity F_r represents the radial body force per unit volume. The conditions of symmetry dictate that the stresses and deformations are independent of the angle θ and that the shear stresses must be 0. Note that the radial stresses acting on the parallel faces of the element differ by $d\sigma_r$, but the tangential stresses do not vary among the faces of the element. There can be no tangential displacement in an axisymmetrically loaded member of revolution; that is, $v = 0$. A point represented in the element has *radial* displacement u as a consequence of loading.

It can be demonstrated that [2] Eqs. (3.59a) in the absence of body forces, (3.61a), and (2.6) can be written in polar coordinates as given in the following outline.

Figure 16.1 Thick-walled cylinder.

Figure 16.2 Stress element of unit thickness.

Equations of Equilibrium

$$\frac{d\sigma_r}{dr} + \frac{\sigma_r - \sigma_\theta}{r} = 0 \tag{16.1}$$

Strain-Displacement Relations

$$\varepsilon_r = \frac{du}{dr}, \qquad \varepsilon_\theta = \frac{u}{r} \tag{16.2}$$

and the shear strain $\gamma_{r\theta} = 0$. Here ε_θ and ε_r are the tangential strain and radial strain, respectively. Substitution of the second into the first of Eqs. (16.2) gives a simple *compatibility condition* among the strains. This ensures the geometrically possible form of variation of strains from point to point within the member.

Hooke's Law

$$\varepsilon_r = \frac{1}{E}(\sigma_r - \nu\sigma_\theta), \qquad \varepsilon_\theta = \frac{1}{E}(\sigma_\theta - \nu\sigma_r) \tag{16.3}$$

The quantity E represents the modulus of elasticity and ν is the Poisson's ratio. The foregoing governing equations are sufficient to obtain a unique solution to a two-dimensional axisymmetric problem with specific *boundary conditions*. Applications to thick-walled cylinders, rotating disks, and pure bending of curved beams are illustrated in sections to follow.

16.3 THICK-WALLED CYLINDERS UNDER PRESSURE

The circular cylinder is usually divided into thin-walled and thick-walled classifications. In a thin-walled cylinder, the tangential stress may be regarded as constant with thickness. When the wall thickness exceeds the inner radius by more than 10%, the cylinder is usually considered *thick walled*. For this case, the variation of stress with radius can no longer be neglected. Thick-walled cylinders, which we deal with here, are used extensively in industry as pressure vessels, pipes, gun tubes, and the like.

SOLUTION OF THE BASIC RELATIONS

In a thick-walled cylinder subject to uniform internal or external pressure, the deformation is symmetrical about the axial (z) axis. The equilibrium condition and strain-displacement relations, Eqs. (16.1) and (16.2), apply to any point on a ring of unit length cut from the cylinder (Figure 16.1). When ends of the cylinder are open and unconstrained, so that $\sigma_z = 0$, the cylinder is in a condition of plane stress. Then, by Hooke's law (16.3), the strains are

$$\frac{du}{dr} = \frac{1}{E}(\sigma_r - v\sigma_\theta)$$

$$\frac{u}{r} = \frac{1}{E}(\sigma_\theta - v\sigma_r)$$

(16.4)

The preceding give the radial and tangential stresses, in terms of the radial displacement,

$$\sigma_r = \frac{E}{1 - v^2}(\varepsilon_r + v\varepsilon_\theta) = \frac{E}{1 - v^2}\left(\frac{du}{dr} + v\frac{u}{r}\right)$$

$$\sigma_\theta = \frac{E}{1 - v^2}(\varepsilon_\theta + v\varepsilon_r) = \frac{E}{1 - v^2}\left(\frac{u}{r} + v\frac{du}{dr}\right)$$

(16.5)

Introducing this into Eq. (16.1) results in the desired differential equation:

$$\frac{d^2u}{dr^2} + \frac{1}{r}\frac{du}{dr} - \frac{u}{r^2} = 0$$

(16.6)

The solution of this *equidimensional equation* is

$$u = c_1 r + \frac{c_2}{r}$$

(16.7)

The stresses may now be expressed in terms of the constants of integration c_1 and c_2 by inserting Eq. (16.7) into (16.5) as

$$\sigma_r = \frac{E}{1 - v^2}\left[c_1(1 + v) - c_2\left(\frac{1 - v}{r^2}\right)\right]$$

(a)

$$\sigma_\theta = \frac{E}{1 - v^2}\left[c_1(1 + v) + c_2\left(\frac{1 - v}{r^2}\right)\right]$$

(b)

STRESS AND RADIAL DISPLACEMENT FOR CYLINDER

For a cylinder under internal and external pressures p_i and p_o, respectively, the boundary conditions are

$$(\sigma_r)_{r=d} = -p_i, \qquad (\sigma_r)_{r=b} = -p_o$$

(16.8)

In the foregoing, the negative signs are used to indicate compressive stress. The constants are ascertained by introducing Eqs. (16.8) into (a); the resulting expressions are carried into Eqs. (16.7), (a), and (b). In so doing, the *radial* and *tangential stresses* and *radial displacement* are obtained in the forms:

$$\sigma_r = \frac{a^2 p_i - b^2 p_o}{b^2 - a^2} - \frac{(p_i - p_o)a^2 b^2}{(b^2 - a^2)r^2} \tag{16.9}$$

$$\sigma_\theta = \frac{a^2 p_i - b^2 p_o}{b^2 - a^2} + \frac{(p_i - p_o)a^2 b^2}{(b^2 - a^2)r^2} \tag{16.10}$$

$$u = \frac{1 - v}{E} \frac{(a^2 p_i - b^2 p_o)}{b^2 - a^2} + \frac{1 + v}{E} \frac{(p_i - p_o)a^2 b^2}{(b^2 - a^2)r^2} \tag{16.11}$$

These equations were first derived by French engineer G. Lame in 1833, for whom they are named. The maximum numerical value of σ_r occurs at $r = a$ to be p_i, provided that p_i exceeds p_o. When $p_o > p_i$, the maximum σ_r is found at $r = b$ and equals p_o. On the other hand, the maximum σ_θ occurs at either the inner or outer edge depending on the pressure ratio [2].

The *maximum shear stress* at any point in the cylinder, through the use of Eqs. (16.9) and (16.10), is found as

$$\tau_{max} = \frac{1}{2}(\sigma_\theta - \sigma_r) = \frac{(p_i - p_o)a^2 b^2}{(b^2 - a^2)r^2} \tag{16.12}$$

The largest value of this stress, corresponds to $p_o = 0$ and $r = a$:

$$\tau_{max} = \frac{p_i b^2}{b^2 - a^2} \tag{16.13}$$

that occurs on the planes making an angle of 45° with the planes of the principal stresses (σ_r and σ_θ). The pressure p_y that initiates *yielding* at the inner surface, by setting $\tau_{max} = S_y/2$ in Eq. (16.13), is

$$p_y = \frac{b^2 - a^2}{2b^2} S_y \tag{16.14}$$

where S_y is the tensile yield strength.

In the case of a pressurized *closed-ended cylinder*, the longitudinal stresses are in addition to σ_r and σ_θ. For a transverse section some distance from the ends, σ_z may be taken uniformly distributed over the wall thickness. The magnitude of the *longitudinal stress* is obtained by equating the net force acting on an end attributable to pressure loading to the internal z-directed force in the cylinder wall:

$$\sigma_z = \frac{p_i a^2 - p_o b^2}{b^2 - a^2} \tag{16.15}$$

where it is again assumed that the ends of the cylinder are not constrained. Also note that Eqs. (16.9) through (16.15) are applicable only away from the ends. The difficult problem of determining deformations and stresses near the junction of the thick-walled caps and the thick-walled cylinder lies outside of the scope of our analysis. This usually is treated by experimental approaches or by the finite element method, since its analytical solution depends on a general three-dimensional study in the theory of elasticity. For thin-walled cylinders, stress in the vicinity of the end cap junctions is presented in Section 16.12.

SPECIAL CASES

Internal Pressure Only

In this case, $p_o = 0$, and Eqs. (16.9) through (16.11) become

$$\sigma_r = \frac{a^2 p_i}{b^2 - a^2}\left(1 - \frac{b^2}{r^2}\right)$$

(16.16a)

$$\sigma_\theta = \frac{a^2 p_i}{b^2 - a^2}\left(1 + \frac{b^2}{r^2}\right)$$

(16.16b)

$$u = \frac{a^2 p_i r}{E(b^2 - r^2)}\left[(1 - \nu) + (1 + \nu)\frac{b^2}{r^2}\right]$$

(16.16c)

Since $b/r \geq 1$, σ_r is always compressive stress and is maximum at $r = a$. As for σ_θ, it is always a tensile stress and also has a maximum at $r = a$:

$$\sigma_{\theta,\max} = p_i \frac{b^2 + a^2}{b^2 - a^2}$$

(16.17)

To illustrate the variation of stress and radial distance for the case of no external pressure, dimensionless stress and displacement are plotted against dimensionless radius in Figure 16.3a for $b/a = 4$.

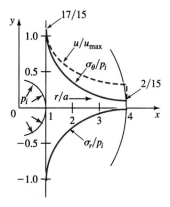

Figure 16.3   Distribution of stress and displacement in a thick-walled cylinder with $b/a = 4$ under internal pressure.

External Pressure Only

For this case, $p_i = 0$, and Eqs. (16.9) through (16.11) simplfy to

$$\sigma_r = -\frac{b^2 p_o}{b^2 - a^2}\left(1 - \frac{a^2}{r^2}\right) \tag{16.18a}$$

$$\sigma_\theta = -\frac{b^2 p_o}{b^2 - a^2}\left(1 + \frac{a^2}{r^2}\right) \tag{16.18b}$$

$$u = -\frac{b^2 p_o r}{E(b^2 - a^2)}(1 - v) + (1 + v)\frac{a^2}{r^2} \tag{16.18c}$$

Inasmuch as $a^2/r^2 \le 1$, the maximum σ_r occurs at $r = b$ and is always compressive. The maximum σ_θ is found at $r = a$ and is likewise always compressive:

$$\sigma_{\theta,\max} = -2p_o\frac{b^2}{b^2 - a^2} \tag{16.19}$$

Cylinder with an Eccentric Bore

The problem corresponding for cylinders having eccentric bore was solved by G. B. Jeffrey [1, 13]. For the case $p_o = 0$ and the eccentricity $e < a/2$ (Figure 16.4), the maximum tangential stress takes place at the internal surface at the thinnest part (point A). The result is as follows:

$$\sigma_{\theta,\max} = p_i\left[\frac{2b^2(b^2 + a^2 - 2ae - e^2)}{(a^2 + b^2)(b^2 - a^2 - 2ae - e^2)} - 1\right] \tag{16.20}$$

When $e = 0$, this coincides with Eq. (16.17).

Thick-Walled Spheres

Equations for thick-walled spheres may be derived following a procedure similar to that employed for thick-walled cylinders. Clearly, the notation of Figure 16.1 applies, with the

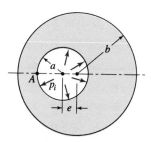

Figure 16.4 Thick-walled cylinder with eccentric bore (with $e < a/2$) under internal pressure.

sketch now representing a diametral cross section of a sphere. It can be shown that [1] the radial and tangential stresses are

$$\sigma_r = \frac{p_i a^3}{b^3 - a^3}\left(1 - \frac{b^3}{r^3}\right) - \frac{p_o b^3}{b^3 - a^3}\left(1 - \frac{a^3}{r^3}\right) \tag{16.21}$$

$$\sigma_\theta = \frac{p_i a^3}{b^3 - a^3}\left(1 + \frac{b^3}{2r^3}\right) - \frac{p_o b^3}{b^3 - a^3}\left(1 + \frac{a^3}{2r^3}\right) \tag{16.22}$$

Thick-walled spheres are used as vessels in high-pressure applications (e.g., in deep-sea vehicles). They yield lower stresses than other shapes and, under external pressure, the greatest resistance to buckling.

16.4   COMPOUND CYLINDERS: PRESS OR SHRINK FITS

A composite or *compound cylinder* is made by *shrink* or *press fitting* an outer cylinder on an inner cylinder. Recall from Section 9.6 that a press or shrink fit is also called interference fit. Contact pressure is caused by interference of metal between the two cylinders. Examples of compound cylinders are seen in various machine and structural members, compressors, extrusion presses, conduits, and the like. A fit is obtained by machining the hub hole to a slightly smaller diameter than that of the shaft. Figure 16.5 depicts a shaft and hub assembled by shrink fit; after the hub is heated, the contact comes through contraction on cooling. Alternatively, the two parts are forced slowly in press to form a press fit. The stresses and displacements resulting from the contact pressure p may readily be obtained from the equations of the preceding section.

Note from Figure 16.3 that most material is underutilized (i.e., only the innermost layer carries high stress) in a thick-walled cylinder subject to internal pressure. A similar conclusion applies to a cylinder under external pressure alone. The cylinders may be strengthened and the material used more effectively by shrink or press fits or by plastic flow, discussed in Section 16.7. Both cases are used in high-pressure technology. The technical literature contains an abundance of specialized information on multilayered cylinders in the form of graphs and formulas [3].

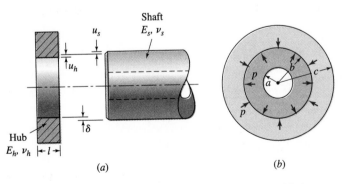

Figure 16.5   Notation for shrink and press fits: (a) unassembled parts; (b) after assembly.

In the unassembled stage (Figure 16.5a), the external radius of the shaft is larger than the internal radius of the hub by the amount δ. The increase u_h in the radius of the hub, using Eq. (16.16c):

$$u_h = \frac{bp}{E_h}\left(\frac{b^2 + c^2}{c^2 - b^2} + v_h\right) \tag{16.23}$$

The decrease u_s in the radius of the shaft, by Eq. (16.18c), is

$$u_s = -\frac{bp}{E_s}\left(\frac{a^2 + b^2}{b^2 - a^2} - v_s\right) \tag{16.24}$$

In the preceding, the subscripts h and s refer to the hub and shaft, respectively.

Radial interference or so-called shrinking allowance δ is equal to the sum of the absolute values of the expansion $|u_h|$ and of shaft contraction $|u_s|$:

$$\delta = \frac{bp}{E_h}\left(\frac{b^2 + c^2}{c^2 - b^2} + v_h\right) + \frac{bp}{E_s}\left(\frac{a^2 + b^2}{b^2 - a^2} - v_s\right) \tag{16.25}$$

When the hub and shaft are composed of the same material ($E_h = E_s = E$, $v_h = v_s$), the contact pressure from Eq. (16.25) may be obtained as

$$p = \frac{E\delta}{b}\frac{(b^2 - a^2)(c^2 - b^2)}{2b^2(c^2 - a^2)} \tag{16.26}$$

The stresses and displacements in the hub are then determined using Eqs. (16.16) by treating the contact pressure as p_i. Likewise, by regarding the contact pressure as p_o, the stresses and deformations in the shaft are calculated applying Eqs. (16.18).

An interference fit creates stress concentration in the shaft and hub at each end of the hub, owing to the abrupt change from uncompressed to compressed material. Some design modifications are often made in the faces of the hub close to the shaft diameter to reduce the stress concentrations at each sharp corner. Usually, for a press or shrink fit, a *stress concentration factor* K_t is used. The value of K_t, depending on the contact pressure, the design of the hub, and the maximum bending stress in the shaft, rarely exceeds 2 [14, 15]. Note that an approximation of the torque capacity of the assembly may be made on the basis of a *coefficient of friction* of about $f = 0.15$ between shaft and hub. The AGMA standard suggests a value of $0.15 < f < 0.20$ for shrink or press hubs, based on a ground finish on both surfaces.

Designing a Press Fit

EXAMPLE 16.1

A steel shaft of inner radius a and outer radius b is to be press fit in a cast iron disk having outer radius c and axial thickness or length of hub engagement of l (Figure 16.5). Determine

(a) The radial interference.

(b) The force required to press together the parts and the torque capacity of the joint.

Given: $a = 25$ mm, $b = 50$ mm, $c = 125$ mm, and $l = 100$ mm. The material properties are $E_s = 210$ GPa, $v_s = 0.3$, $E_c = 70$ GPa, and $v_c = 0.25$.

Assumptions: The maximum tangential stress in the disk is not to exceed 30 MPa; the contact pressure is uniform; and $f = 0.15$.

Solution:

(a) Through the use of Eq. (16.17), with $p_i = p$, $a = b$, and $b = c$, we have

$$p = \sigma_{\theta,\max} \frac{c^2 - b^2}{b^2 + c^2} = 30 \frac{125^2 - 50^2}{50^2 + 125^2} = 21.72 \text{ MPa}$$

From Eq. (16.25),

$$\delta = \frac{0.05(21.72)}{70 \times 10^3}\left(\frac{50^2 + 125^2}{125^2 - 50^2} + 0.25\right) + \frac{0.05(21.72)}{210 \times 10^3}\left(\frac{25^2 + 50^2}{50^2 - 25^2} - 0.3\right)$$

$$= 0.0253 + 0.0071 = 0.0324 \text{ mm}$$

(b) The *force* (axial or tangential) required for the assembly:

$$F = 2\pi bpfl \qquad\qquad\qquad (16.27a)$$

Introducing the required numerical values,

$$F = 2\pi(50)(21.72)(0.15)(100) = 102.4 \text{ kN}$$

The *torque capacity* or torque carried by the press fit is then

$$T = Fb = 2\pi b^2 fpl \qquad\qquad\qquad (16.27b)$$

Inserting the given data, we obtain

$$T = 102.4\ (0.05) = 5.12 \text{ kN} \cdot \text{m}$$

EXAMPLE 16.2   Design of a Duplex Hydraulic Conduit

A thick-walled concrete pipe (E_c, ν_c) with a thin-walled steel cylindrical liner or sleeve (E_s) of outer radius a is under internal pressure p_i, as shown in Figure 16.6. Develop an expression for the pressure p transmitted to the concrete pipe.

Figure 16.6   Example 16.2.

Design Decision: For practical purposes, we take

$$\frac{E_s}{E_c} = 15, \qquad \nu_c = 0.2, \qquad \frac{t}{a-t} = \frac{t}{a} \tag{a}$$

and $a \pm t = a$, since $a/t > 10$ for a thin-walled cylinder.

Solution: The sleeve is under internal pressure p_i and external pressure p:

$$\sigma_\theta = \frac{p_i - p}{t}(a - t) = (p_i - p)\left(\frac{a}{t} - 1\right) \tag{b}$$

Also, from Hooke's law and the second of Eqs. (16.2) with $r = a$,

$$\sigma_\theta = E_s \varepsilon_\theta = E_s \frac{u}{a} \tag{c}$$

The radial displacement at the bore ($r = a$) of pipe, using Eq. (16.16c), is

$$u = \frac{pa}{E_c}\left(\frac{a^2 + b^2}{b^2 - a^2} + \nu_c\right) \tag{d}$$

Evaluating u from Eqs. (b) and (c) and carrying into Eq. (d) lead to an expression from which the interface pressure can be obtained. In so doing, we obtain

$$p = \frac{p_i}{1 + \left(\dfrac{E_s}{E_c}\right)\left(\dfrac{t}{a-t}\right)\left(\dfrac{a^2 + b^2}{b^2 - a^2} + \nu_c\right)} \tag{16.28}$$

A design formula for the interface pressure is obtained on substitution of Eqs. (a) into the preceding equation:

$$p = \frac{p_i}{1 + 15\left(\dfrac{t}{a}\right)\left(\dfrac{R^2 + 1}{R^2 - 1} + 0.2\right)} \tag{16.29}$$

where the pipe radius ratio $R = b/a$. This formula can be used to prepare design curves for steel lined concrete conduits [3].

Comments: It is interesting to observe from Eq. (16.29) that, as the sleeve thickness t increases, the pressure p transmitted to the concrete decreases. But, for any given t/a ratio, the p increases as the R increases.

16.5   DISK FLYWHEELS

A *flywheel* is often used to smooth out changes in the speed of a shaft caused by torque fluctuations. Flywheels are therefore found in small and large machinery, such as compressors, punch presses (see Example 1.3), rock crushers, and internal combustion engines. Considerable stress may be induced in these members at high speed. Analysis of this effect is important, since failure of rotating disks is particularly hazardous. Designing of

Figure 16.7 A flywheel shrunk onto a shaft.

energy-storing flywheels for hybrid-electric cars is an active area of contemporary research. *Disk flywheels,* rotating annular disks of constant thickness, are made of high-strength steel plate. In this section, attention is directed to the design analysis of these flywheels using both equilibrium and energy approaches.

STRESS AND DISPLACEMENT

Figure 16.7 illustrates a flywheel of axial thickness or length of hub engagement l with inner radius a and outer radius b, shrunk onto a shaft. Let the contact pressure between the two parts be designated by p. An element of the disk is loaded by an outwardly directed *centrifugal force* $F_r = \rho\omega^2 r$ (Figure 16.2). Here, ρ is the mass density ($N \cdot s^2/m^4$ or $lb \cdot s^2/in.^4$) and ω represents the angular velocity or speed (rad/s). The condition of equilibrium, Eq. (16.1), becomes

$$\frac{d\sigma_r}{dr} + \frac{\sigma_r - \sigma_\theta}{r} + \rho\omega^2 r = 0 \tag{16.30}$$

The boundary conditions are $\sigma_r = -p$ at the inner surface ($r = a$) and $\sigma_r = 0$ at the outer surface ($r = b$).

The solution of Eq. (16.30) is obtained by following a procedure similar to that used in Section 16.2. It can be shown that [2] the combined radial stress (σ_r), tangential stress (σ_θ), and displacement (u) of a disk due to contact pressure p and angular speed ω are

$$\sigma_r = (\sigma_r)_p + \frac{3+v}{8}\left(a^2 + b^2 - \frac{a^2 b^2}{r^2} - r^2\right)\rho\omega^2 \tag{16.31a}$$

$$\sigma_\theta = (\sigma_\theta)_p + \frac{3+v}{8}\left(a^2 + b^2 - \frac{1+3v}{3+v}r^2 + \frac{a^2 b^2}{r^2}\right)\rho\omega^2 \tag{16.31b}$$

$$u = (u)_p + \frac{(3+v)(1-v)}{8E}\left(a^2 + b^2 - \frac{1+v}{3+v}r^2 + \frac{1+v}{1-v}\frac{a^2 b^2}{r^2}\right)\rho\omega^2 r \tag{16.31c}$$

Here, $(\sigma_r)_p$, $(\sigma_\theta)_p$, and $(u)_p$ are given by Eqs. (16.16) with $p_i = p$. The quantity v is the Poisson's ratio.

In most cases, *tangential stress* σ_θ *controls the design*. This stress is a maximum at the inner boundary ($r = a$) and is equal to

$$\sigma_{\theta,\max} = p\frac{a^2 + b^2}{b^2 - a^2} + \frac{\rho\omega^2}{4}[(1 - v)a^2 + (3 + v)b^2] \tag{16.32}$$

Clearly, the preceding problem in which pressure and rotation appears simultaneously could also be solved by superposition. Note that, due to *rotation* only, maximum radial stress occurs at $r = \sqrt{ab}$ and is given by

$$\sigma_{r,\max} = \frac{3 + v}{8}(b - a)^2\rho\omega^2 \tag{16.33}$$

Owing to the internal *pressure* alone, the largest radial stress is at the inner boundary and equals $\sigma_{r,\max} = -p$.

Customarily, inertial stress and displacement of a shaft are neglected. Therefore, for a shaft, we have approximately

$$\sigma_r = \sigma_\theta = -p$$

$$u = -\frac{1 - v}{E_s}pr \tag{16.34}$$

Note, however, that the contact pressure p depends on angular speed ω. For a given contact pressure p at angular speed ω, the required initial radial interference δ may be obtained using Eqs. (16.31c) and (16.34) for u. Hence, with $r = a$, we have

$$\delta = \frac{ap}{E_d}\left(\frac{a^2 + b^2}{b^2 - a^2} + v\right) + \frac{ap}{E_s}(1 - v) + \frac{a\rho\omega^2}{4E_d}[(1 - v)a^2 + (3 + v)b^2] \tag{16.35}$$

in which E_d and E_s represent the moduli of elasticity of the disk and shaft, respectively. The preceding equation is valid as long as a positive contact pressure is maintained.

Rotating Blade Design Analysis

EXAMPLE 16.3

A disk of uniform thickness, except where sharpened at the periphery, used at 12,000 rpm as a rotating blade for cutting blocks of paper or thin plywood. The disk is mounted on a shaft of 1-in. radius and clamped, as shown in Figure 16.8. Determine

(a) The factor of safety n according to the maximum shear stress criterion.

(b) The values of the maximum radial stress and displacement at outer edge.

Assumptions: The cutting forces are relatively small and speed is steady: loading is considered static. The disk outside radius is taken as 15 in. The stresses due to clamping are disregarded.

Design Decision: The disk material is a high-strength ASTM-A242 steel.

Figure 16.8   Example 16.3, a rotating blade (only a partial view shown) and shaft assembly.

Solution: The material properties are (Table B.1)

$$\rho = \frac{0.284}{386} = 7.358 \times 10^{-4}\,\mathrm{lb \cdot s^2/in.^4}, \qquad \nu = 0.3$$

$$E = 29 \times 10^6\,\mathrm{psi}, \qquad S_{ys} = 30\,\mathrm{ksi}$$

We have

$$\rho\omega^2 = 7.358 \times 10^{-4}\left(\frac{12{,}000 \times 2\pi}{60}\right)^2 = 1161.929$$

(a)   The tangential stress, expressed by Eq. (16.31a) with $p = 0$, has the form

$$\sigma_\theta = \frac{3+\nu}{8}\left(a^2 + b^2 - \frac{1+3\nu}{3+\nu}r^2 + \frac{a^2 b^2}{r^2}\right)\rho\omega^2$$

The stresses in the inner and outer edges of the blade are, from the preceding equation,

$$(\sigma_\theta)_{r=1} = \frac{3.3}{8}\left(1^2 + 5^2 - \frac{1.9 \times 1^2}{3.3} + \frac{1^2 \times 5^2}{1^2}\right)(1161.929) = 24.17\,\mathrm{ksi}$$

$$(\sigma_\theta)_{r=5} = \frac{3.3}{8}\left(1^2 + 5^2 - \frac{1.9 \times 5^2}{3.3} + 1^2\right)(1161.929) = 6.04\,\mathrm{ksi}$$

The maximum shear stress occurs at the inner surface ($r = 1$ in.), where $\sigma_r = 0$:

$$\tau_{max} = \frac{\sigma_\theta}{2} = \frac{24.17}{2} = 12.08\,\mathrm{ksi}$$

The factor of safety, based on the maximum shear stress theory, is then

$$n = \frac{S_{ys}}{\tau_{max}} = \frac{30}{12.08} = 2.48\,\mathrm{ksi}$$

Comment: Should there be starts and stops, the condition is one of fatigue failure and a lower value of n would be obtained by the techniques of Section 8.11.

(b) The largest radial stress in the disk, from Eq. (16.33), is given by

$$\sigma_{r,max} = \frac{3+v}{8}(b-a)^2 \rho \omega^2$$

$$= \frac{3.3}{8}(5-1)^2(1161.929) = 7669 \text{ ksi}$$

The radial displacement of the disk is expressed by Eq. (16.31c) with $p = 0$. Hence,

$$(u)_{r=5} = \frac{(3.3)(0.7)}{8 \times 29 \times 10^6}\left(1^2 + 5^2 - \frac{1.3 \times 1^2}{3.3} + \frac{1.3 \times 1^2}{0.7}\right)(1161.929)(5)$$

$$= 1.589 \times 10^{-3} \text{ in.}$$

is the radial displacement at the outer periphery.

Design of a Flywheel-Shaft Assembly

EXAMPLE 16.4

A 400-mm diameter flywheel is to be shrunk onto a 50-mm diameter shaft. Determine

(a) The required radial interference.
(b) The maximum tangential stress in the assembly.
(c) The speed at which the contact pressure becomes 0.

Requirement: At a maximum speed of $n = 5000$ rpm, a contact pressure of $p = 8$ MPa is to be maintained.

Design Decisions: Both the flywheel and shaft are made of steel having $\rho = 7.8$ kN · s²/m⁴, $E = 200$ GPa, and $v = 0.3$.

Solution:

(a) Applying Eq. (16.35), we have

$$\delta = \frac{25(10^{-3})p}{200(10^9)}\left(\frac{25^2 + 200^2}{200^2 - 25^2} + 1\right) + \frac{25(7.8)\omega^2}{4(200 \times 10^9)}[0.7(0.025)^2 + 3.3(0.2)^2]$$

$$= (0.254p + 32.282\omega^2)10^{-12} \tag{a}$$

For $p = 8$ MPa and $\omega = 5000(2\pi/60) = 523.6$ rad/s, Eq. (a) leads to $\delta = 0.011$ mm.

(b) Using Eq. (16.32),

$$\sigma_{\theta,max} = 8\frac{25^2 + 200^2}{200^2 - 25^2} + \frac{7800(523.6)^2}{4}[0.7(0.025)^2 + 3.3(0.2)^2]$$

$$= 8.254 + 70.802 = 79.06 \text{ MPa}$$

(c) Inserting $\delta = 0.011 \times 10^{-3}$ m and $p = 0$ into Eq. (a) results in

$$\omega = \left(\frac{0.011 \times 10^9}{32.282} \right)^{1/2} = 583.7 \text{ rad/s}$$

Therefore,

$$n = 583.7 \frac{60}{2\pi} = 5574 \text{ rpm}$$

Comment: At this speed, the shrink fit becomes completely ineffective.

The constant thickness disks discussed in the foregoing do not make optimum use of the material. Often disks are not flat, being thicker at the center than at the rim. Other types of rotating disks, offering many advantages over flat disks, are variable thickness and uniform stress disks. For these cases, the procedure outlined here must be modified. A number of problems of this type are discussed in [2].

ENERGY STORED

Heavy disks often serve as flywheels designed to store energy to maintain reasonably constant speed in a machine in spite of variations in input and output power. A flywheel absorbs and stores energy when speeded up and releases energy to the system when needed by slowing its rotational speed. The change in kinetic energy ΔE_k stored in a flywheel by a change in speed from ω_{max} to ω_{min}, by Eq. (1.10), is

$$\Delta E_k = \frac{1}{2} I \left(\omega_{\text{max}}^2 - \omega_{\text{min}}^2 \right) \tag{16.36}$$

The mass moment of inertia I about the axis of an annular disk flywheel of outer radius b and inner radius a (Figure 16.7) is given by

$$I = \frac{\pi}{2g} (b^4 - a^4) l \gamma \tag{b}$$

The weight of the disks is

$$W = \pi (b^2 - a^2) l \gamma \tag{c}$$

where

$l =$ the length of the hub engagement

$\gamma = \rho g$, specific weight

$g =$ acceleration of gravity

Substitution of Eq. (c) into (b) gives

$$I = \frac{W}{2g} (b^2 + a^2) \tag{16.37}$$

For a conservative system, the change in the kinetic energy is available as work output:

$$\Delta E_k = T \Delta \phi \tag{16.38}$$

in which T represents the torque and $\Delta \phi$ is the change in the angular rotation of the disk in radians. Ordinarily, there are two stages to the flywheel design [14]. First, the amount of energy needed for the required degree of smoothing must be found from Eq. (16.38) and the moment of inertia needed to absorb that energy calculated by Eq. (16.36). Then, a flywheel geometry must be defined by Eq. (16.37).

EXAMPLE 16.5

Flywheel Braking-Torque Requirement

A flywheel of outer diameter D, inner diameter d, and weight W, rotates at speed n. Determine the average braking torque required to stop the wheel in one-third revolution.

Given: $D = 10$ in., $\quad d = 2$ in., $\quad W = 30$ lb, $\quad n = 3600$ rpm

Solution: From Eqs. (16.36) and (16.38),

$$T\phi = \frac{1}{2} I \omega^2 \tag{16.39}$$

where

$$\phi = 2\pi/3 \text{ rad};$$

$$\omega = 3600(2\pi/60) = 377 \text{ rad/s};$$

$$I = \frac{30}{2(386)} (5^2 + 1^2) = 1.01 \text{ lb} \cdot \text{s}^2 \cdot \text{in.}^4$$

Hence,

$$T = \frac{1.01(377)^2}{2(2\pi/3)} = 34.27 \text{ kip} \cdot \text{in.}$$

16.6 THERMAL STRESSES IN CYLINDERS

Here, we are concerned with the stress and displacement associated with an axisymmetrical temperature $T(r)$ dependent on the radial dimension alone. Examples include heat exchanger tubes, chemical reaction vessels, clad reactor elements, nozzle sections of rockets, annular fins, and turbine disks. The deformation also is symmmetrical about the axis, and we may use the method developed in Section 16.3. The results of this section are restricted to the static, steady-state problem.

Consider a cylinder with ends assumed to be restrained so that the axial displacement $w = 0$. This exemplifies the case of plane strain (i.e., $\varepsilon_z = 0$). The applicable strain-stress relationships, from Eq. (16.5) with reference to Eq. (2.16), are assumed as follows:

$$\varepsilon_r = \frac{1}{E}[\sigma_r - \nu(\sigma_\theta + \sigma_z)] + \alpha T$$

$$\varepsilon_\theta = \frac{1}{E}[\sigma_\theta - \nu(\sigma_r + \sigma_z)] + \alpha T \qquad \text{(16.40)}$$

$$\varepsilon_z = \frac{1}{E}[\sigma_z - \nu(\sigma_r + \sigma_\theta)] + \alpha T$$

The third of the foregoing equations, setting $\varepsilon_z = 0$, leads to

$$\sigma_z = \nu(\sigma_r + \sigma_\theta) - \alpha E T \qquad \text{(a)}$$

Introducing this expression into the first two of Eqs. (16.40) results in

$$\varepsilon_r = \frac{1+\nu}{E}[(1-\nu)\sigma_r - \nu\sigma_\theta + \alpha E T]$$

$$\text{(16.41)}$$

$$\varepsilon_\theta = \frac{1+\nu}{E}[(1-\nu)\sigma_\theta - \nu\sigma_r + \alpha E T]$$

Equations (16.2) and (16.3) are unchanged for the case under discussion. Following a procedure similar to that outlined in Section 16.3, we obtain

$$u = \frac{(1+\nu)\alpha}{(1-\nu)r}\int_\alpha^r Tr\, dr + c_1 r + \frac{c_2}{r} \qquad \text{(b)}$$

$$\sigma_r = \frac{E}{1+\nu}\left[-\frac{(1+\nu)\alpha}{(1-\nu)r^2}\int_\alpha^r Tr\, dr + \frac{c_1}{1-2\nu} - \frac{c_2}{r^2}\right] \qquad \text{(c)}$$

$$\sigma_\theta = \sigma_r + r\frac{d\sigma_r}{dr} \qquad \text{(d)}$$

Equation (a) can then be written as

$$\sigma_z = -\frac{\alpha E T}{1-\nu} + \frac{2\nu E c_1}{(1+\nu)(1-2\nu)} \qquad \text{(e)}$$

Here c_1 and c_2 are constants of integration, to be determined from known boundary conditions at the edges of the cylinder.

CYLINDER WITH A CENTRAL HOLE

We now determine the expressions for stress components for a hollow cylinder subjected to a radial temperature change (Figure 16.9). The inner and outer edges are free of mechanical loads:

$$(\sigma_r)_{r=a} = 0, \qquad (\sigma_r)_{r=b} = 0 \qquad \text{(f)}$$

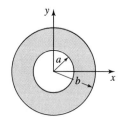

Figure 16.9   Cross section of a long circular cylinder under thermal loading.

Substituting these boundary conditions into Eq. (c), we have

$$c_1 = \frac{(1+v)(1-2v)\alpha}{(1-v)(b^2-a^2)} \int_a^b Tr\,dr$$

$$c_2 = \frac{(1+v)a^2\alpha}{(1-v)(b^2-a^2)} \int_a^b Tr\,dr$$

$$(g)$$

Equations (c), (d), and (e) provide the stresses as follows:

$$\sigma_r = \frac{E\alpha}{(1-v)r^2}\left(-\int_a^r Tr\,dr + \frac{r^2-a^2}{b^2-a^2}\int_a^b Tr\,dr\right)$$

$$(16.42)$$

$$\sigma_\theta = \frac{E\alpha}{(1-v)r^2}\left(-Tr^2 + \int_a^r Tr\,dr + \frac{r^2+a^2}{b^2-a^2}\int_a^b Tr\,dr\right)$$

$$\sigma_z = \frac{\alpha E}{1-v}\left(\frac{2}{b^2-a^2}\int_a^b Tr\,dr - T\right) \qquad (16.43a)$$

If the *ends* are *free*, it can be shown that the axial stress is given by

$$\sigma_z = \frac{\alpha E}{1-v}\left(\frac{2}{b^2-a^2}\int_a^b Tr\,dr - T\right) \qquad (16.43b)$$

Similarly, an expression for the displacement u may also be obtained by carrying Eqs. (g) into Eq. (b). For a given temperature distribution $T(r)$ over the cylinder, we can find the stresses using Eqs. (16.42) and (16.43) for each particular case.

STEADY-FLOW TEMPERATURE CHANGE $T(r)$

When the walls of the cylinder are at temperatures T_a and T_b at the inner ($r = a$) and outer ($r = b$) surfaces, respectively, the temperature distribution may be represented in the form [2]

$$T = \frac{T_a - T_b}{\ln(b/a)}\ln\frac{b}{r} \qquad (16.44)$$

This expression can be used with Eqs. (16.42) and (16.43) to determine stress components for steady-state temperature distribution in a thick-walled cylinder. The results are

$$\sigma_r = \frac{\alpha E(T_a - T_b)}{2(1-v)\ln(b/a)}\left[-\ln\frac{b}{r} - \frac{a^2(r^2-b^2)}{r^2(b^2-a^2)}\ln\frac{b}{a}\right]$$

$$\sigma_\theta = \frac{\alpha E(T_a - T_b)}{2(1-v)\ln(b/a)}\left[1 - \ln\frac{b}{r} - \frac{a^2(r^2+b^2)}{r^2(b^2-a^2)}\ln\frac{b}{a}\right] \qquad (16.45)$$

$$\sigma_z = \frac{\alpha E(T_a - T_b)}{2(1-v)\ln(b/a)}\left[1 - 2\ln\frac{b}{r} - \frac{2a^2}{b^2-a^2}\ln\frac{b}{a}\right]$$

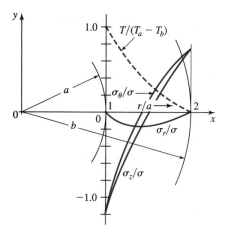

Figure 16.10   Thermal stress distribution in a thick-walled cylinder with $b/a = 2$ and $T_a > T_b$. Note: $\sigma = \alpha\, E(T_a - T_b)/2(1-\nu)$.

The dimensionless distribution of the temperature and stress over the cylinder wall for the particular case when $b/a = 2$ is shown in Figure 16.10. We see from the figure that the tangential σ_θ and axial σ_z stresses at the outer surface are equal and *tensile*. This is why internal heating may cause external cracks in materials weak in tension, such as in the chimneys and conduits of concrete masonry. On the contrary, the radial stress is compressive at all points and becomes 0 at the inner and outer edges of the cylinder.

Note that, in practice, a pressure loading is usually superimposed on the thermal stresses, as in chemical reaction pressure vessels. In this case, the internal pressure gives a tangential stress (Figure 16.3), causing a partial cancellation of compressive stress due to temperature. Also, when a cylinder (or disk) is rotating, stresses owing to the inertia may be superimposed over those due to temperature change and pressure.

SPECIAL CASES

In a *thin-walled cylinder*, as in the cylinder liner of an engine or compressor, we can simplify Eqs. (16.45). In this situation, it can be readily verified that the temperature distribution is nearly linear and the stresses have the values

$$(\sigma_\theta)_{r=a} = -\frac{E\alpha(T_a - T_b)}{2(1 - \nu)}$$

$$(\sigma_\theta)_{r=b} = \frac{E\alpha(T_a - T_b)}{2(1 - \nu)}$$

(16.46)

The preceding coincide with the stress expressions of an annular plate that is heated on sides and its edges are clamped [4]. Equations (16.46) can also be used with sufficient accuracy in the case of a *thin-walled spherical shell*.

*16.7 FULLY PLASTIC THICK-WALLED CYLINDERS

Because of variable nature of loading most machine components, plastic analysis is seldom pertinent to machine design. Nevertheless, in some cases, a saving of material can be effected by designing on the basis of complete yielding of a member, rather than predicating the design on loads at which yielding first occurs. The thick-walled cylinder exemplifies such a situation. The case of a thick-walled cylinder under internal pressure alone was treated in Section 16.2. Equation (16.14) was derived for the onset of yielding of the inner surface of the cylinder based on the maximum shear criterion.

As the internal pressure continues to increase, the plastic zone spreads toward the outer surface and an elastic-plastic state prevails in the cylinder. As the pressure increases further, eventually, the entire cross section becomes *fully plastic* at the ultimate pressure. When the pressure is reduced, the material unloads elastically. Elastic and plastic stresses are superimposed to produce a residual stress pattern (see Example 16.6). On reloading, the pressure needed to produce renewed yielding is greater than the pressure producing initial yielding, and the cylinder is strengthened. This section concerns only fully plastic action in a thick-walled cylinder made of a perfectly plastic material of yield strength S_y subjected to internal pressure. Two typical perfectly plastic materials are mild steel and nylon, which exhibit negligible elastic strains in comparison with large plastic deformations at practically constant strength S_y. Space limitations preclude including discussions of the perfectly yielded cylinders [2] and the stress-strain curves for loading and unloading.

Consider a cylinder of inner radius a and outer radius b subject to internal pressure p (Figure 16.11a). The ends of the cylinder are restrained so that axial strain $\varepsilon_z = 0$. Then, replacing ν by 1/2 (Section 2.5) in the third of Eqs. (2.8), the axial stress is found to be

$$\sigma_z = \frac{1}{2}(\sigma_r + \sigma_\theta) \tag{a}$$

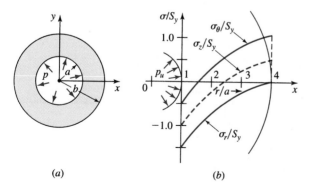

(a)   (b)

Figure 16.11 (a) Thick-walled cylinder of perfectly plastic material subjected to internal pressure. (b) Fully plastic stresses in the cylinder at the ultimate pressure for $b/a = 4$.

The equations of statics are applicable, regardless whether the elastic or plastic state is considered. A condition of equilibrium is given by Eq. (16.1),

$$\frac{d\sigma_r}{dr} + \frac{\sigma_r - \sigma_\theta}{r} = 0 \tag{b}$$

The boundary conditions are

$$(\sigma_r)_{r=a} = -p, \qquad (\sigma_r)_{r=b} = 0 \tag{c}$$

The *failure criterion,* on the basis of the maximum energy of distortion theory, Eq. (7.12), setting $\sigma_1 = \sigma_\theta$, $\sigma_2 = (\sigma_r + \sigma_\theta)/2$, $\sigma_3 = \sigma_r$, and $n = 1$, results in

$$\sigma_\theta - \sigma_r = kS_y, \qquad \text{where } k = \frac{2}{\sqrt{3}} \tag{16.47}$$

Similarly, according to the maximum shearing stress theory, Eq. (7.6),

$$\sigma_\theta - \sigma_r = kS_y, \qquad \text{where } k = 1 \tag{16.48}$$

Substituting Eq. (16.47) or (16.48) into Eq. (b), the *failure criteria* for *plastic design,* is

$$\frac{d\sigma_r}{dr} = k\frac{S_y}{r} \tag{16.49}$$

The preceding governing equation has the following solution

$$\sigma_r = kS_y \ln r + c \tag{d}$$

The constant of integration is obtained by applying the second of Eqs. (c): $c = -kS_y \ln b$. Equation (d) becomes

$$\sigma_r = -kS_y \ln\frac{b}{r} \tag{e}$$

The first of Eqs. (c) now gives the *ultimate pressure,* or the failure pressure for plastic design:

$$p_u = kS_y \ln\frac{b}{a} \tag{16.50}$$

This result, divided by a factor of safety n, applies satisfactorily for static loading.

An expression for σ_θ can be obtained by inserting Eq. (e) into Eq. (16.47) or (16.48). Then, Eq. (a) gives σ_z. In so doing, the *complete plastic stress* distribution is found as

$$\sigma_r = -kS_y \ln\frac{b}{r}$$

$$\sigma_\theta = kS_y\left(1 - \ln\frac{b}{r}\right) \tag{16.51}$$

$$\sigma_z = kS_y\left(\frac{1}{2} - \ln\frac{b}{r}\right)$$

The foregoing stresses are sketched in Figure 16.11b, whereas the distribution of the elastic radial and tangential stresses are illustrated in Figure 16.3. A comparison of the figures shows that the material is used more effectively by plastic action in a thick-walled cylinder.

| Plastic Design of a Thick-Walled Cylinder | **EXAMPLE 16.6** |

A perfectly plastic closed-ended cylinder of inner radius a and outer radius b is under internal pressure p (Figure 16.11a). Determine

(a) The fully plastic stresses at the inner surface.

(b) The residual tangential and axial stresses at the inner surface if the cylinder is unloaded from the ultimate pressure p_u.

Given: Material yield strength $S_y = 250$ MPa, $a = 50$ mm, and $b = 100$ mm.

Design Decision: Use the maximum shear stress failure criterion.

Solution: The magnitude of the ultimate pressure, using Eq. (16.50) with $k = 1$, is

$$p_u = 1(250)\ln 2 = 173.3 \text{ MPa}$$

(a) Applying Eqs. (16.51) at $r = a$,

$$\sigma_r = -p_u = -173.3 \text{ MPa}$$

$$\sigma_\theta = 1(250)(1 - \ln 2) = 76.7 \text{ MPa}$$

$$\sigma_z = 1(250)\left(\frac{1}{2} - \ln 2\right) = -48.3 \text{ MPa}$$

Note, as a check, that $\sigma_z = (\sigma_r + \sigma_\theta)/2$ gives the same result.

(b) Unloading is taken to be *linearly elastic*. Therefore, the elastic stresses at $r = a$, from Eqs. (16.17) and (16.15), give

$$\sigma_\theta = p_u \frac{a^2 + b^2}{b^2 - a^2} = (173.3)\frac{50^2 + 100^2}{100^2 - 50^2} = 288.8 \text{ MPa}$$

$$\sigma_z = p_u \frac{a^2}{b^2 - a^2} = (173.3)\frac{50^2}{100^2 - 50^2} = 57.8 \text{ MPa}$$

The residual stresses at the inner surface are

$$(\sigma_\theta)_{\text{res}} = 76.7 - 288.8 = -212.1 \text{ MPa}$$

$$(\sigma_z)_{\text{res}} = -48.3 - 57.8 = -106.1 \text{ MPa}$$

Comment: Complete plastic and residual stresses at any other location may be found in a like manner.

16.8 STRESSES IN CURVED BEAMS

Curved beams or bars in the form of hooks, C-clamps, press frames, chain links, and brackets are often used as machine or structural elements. Stresses in curved beams of rectangular cross sections already were discussed briefly in Section 3.7. Here, we are concerned with applications of both the theory of elasticity and the mechanics of materials approaches to initially curved bars or frames. Only elastic cases are treated.

EXACT SOLUTION

Figure 16.12a shows a beam of narrow rectangular cross section and circular axis subjected to equal end couples M such that pure bending occurs in the plane of the curvature. Since the bending moment is constant throughout the length of the bar, stress distribution is the same in all radial cross sections. This is the case of a plane stress problem with *axial symmetry* about θ. But, unlike the axisymmetrically loaded members of revolution treated in the preceding sections, there is a θ-dependent tangential displacement [1]. The condition of equilibrium is given by Eq. (16.1) as

$$\frac{d\sigma_r}{dr} + \frac{\sigma_r - \sigma_\theta}{r} = 0 \tag{a}$$

The conditions at the curved boundaries are

$$(\sigma_r)_{r=a} = (\sigma_r)_{r=b} = 0 \tag{b}$$

The conditions at the straight edges or ends are expressed as

$$t \int_a^b \sigma_\theta \, dr = 0, \qquad t \int_a^b r\sigma_\theta \, dr = M \tag{c}$$

Shear stress is also taken to be 0 throughout the beam.

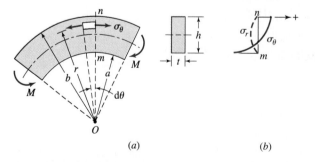

(a)   (b)

Figure 16.12 (a) A thin curved beam in pure bending.
(b) Distribution of stresses.

Solution of Eq. (a) is determined by following a procedure somewhat similar to that outlined in Section 16.2. It can be shown that the tangential and radial stress distributions in the beam are expressed as

$$\sigma_r = \frac{4M}{tb^2N}\left[\left(1-\frac{a^2}{b^2}\right)\ln\frac{r}{a}-\left(1-\frac{a^2}{r^2}\right)\ln\frac{b}{a}\right] \tag{16.52}$$

$$\sigma_\theta = \frac{4M}{tb^2N}\left[\left(1-\frac{a^2}{b^2}\right)\left(1+\ln\frac{r}{a}\right)-\left(1+\frac{a^2}{r^2}\right)\ln\frac{b}{a}\right] \tag{16.53}$$

where

$$N = \left(1-\frac{a^2}{b^2}\right)^2 - 4\frac{a^2}{b^2}\ln^2\frac{b}{a} \tag{16.54}$$

The bending moment is taken as *positive* when it tends to *decrease* the radius of curvature of the beam, as in Figure 16.12a. Using this sign convention, σ_r as determined from Eq. (16.52) is always negative, meaning that it is compressive. Similarly, when σ_θ is found to be positive, it is tensile; otherwise, it is compressive. A sketch of the stresses at section *mn* is presented in Figure 16.12b. Observe that the maximum stress magnitude is at the extreme fiber of the inner (concave) side.

TANGENTIAL STRESS IN A CURVED BEAM: WINKLER'S FORMULA

The approximate approach to the curved beams by E. Winkler is now explored. The fundamental assumptions of the elementary theory of straight beams are also valid for *Winkler's theory*. Only elastic bending is treated, with the usual condition that the modulus of elasticity is identical in tension and compression. Consider the pure bending of a curved beam of uniform cross section having a vertical (*y*) axis of symmetry (Figure 16.13a). An

Figure 16.13 (a) Pure bending of a beam with a cross-sectional axis of symmetry. (b) Stress distribution.

expression for the tangential stress is derived by applying the three principles of analysis based on the familiar hypothesis: Sections perpendicular to the axis of the beam remain subsequent to bending. This is depicted by the line *ef* in relation to a beam segment *abcd* subtended by the central angle θ.

Figure 16.13a shows that the deformation pattern of curved beams is the same as for straight beams. The initial length of a beam fiber such as *gh* depends on the distance *r* from the center of the curvature *O*. The total deformation of beam fibers, as the beam rotates through a small angle $d\theta$ follows a linear law. The tangential strain on the fiber *gh* may be expressed as

$$\varepsilon_\theta = \frac{hh'}{gh} = \frac{\varepsilon_c R d\theta + y d\theta}{(R+y)d\theta} \tag{d}$$

in which ε_c denotes the strain of the centroidal fiber. We see from this expression that ε_θ does *not* vary linearly over the depth of the beam as it does for straight beams. The tangential stress σ_θ on an element dA of the cross-sectional area is, using Hooke's law,

$$\sigma_\theta = E\varepsilon_\theta \tag{e}$$

Equations of equilibrium, $\sum F_x = 0$ and $\sum M_z = 0$, respectively, are

$$\int \sigma_\theta \, dA = 0, \qquad \int \sigma_\theta y \, dA = M \tag{f}$$

Integration in these equations extends over the entire cross-sectional area *A*.

We now substitute Eq. (e) together with (d) into Eqs. (f). After rearrangement and integration, we obtain the following equation for the tangential stress in a curved beam subject to pure bending:

$$\sigma_\theta = \frac{M}{AR}\left[1 + \frac{y}{Z(R+y)}\right] \tag{16.55}$$

This equation is often referred to as *Winkler's formula.* The quantity *Z* is an area property, known as the *curved beam factor:*

$$Z = -\frac{1}{A}\int \frac{y}{R+y} \, dA \tag{16.56}$$

Equation (16.55) indicates that the stress distribution in a curved bar follows a hyperbolic pattern, as sketched in Figure 16.13b. The maximum stress is always on the inner (concave) side of the beam. The *sign convention* applied to the bending moment is the same as that used in the exact theory. That is, the bending moment is positive when directed toward the concave side of the beam, as depicted in Figure 16.13a. If Eq. (16.55) results in a positive value, it is indicative of a tensile stress. The location of the neutral axis for a curved beam is obtained from Eq. (16.55) by setting $\sigma_\theta = 0$. As a result, we have

$$e = -\frac{ZR}{Z+1} \tag{16.57}$$

This denotes the distance between the centroidal axis ($y = 0$) and the neutral axis, as shown in the figure.

The tangential stress given by Eq. (16.55) may be superimposed to the stress produced by a centric normal load P. Hence, the *combined stress* in a curved beam is

$$\sigma_\theta = \frac{P}{A} + \frac{M}{AR}\left[1 + \frac{y}{Z(R+y)}\right] \tag{16.58}$$

As usual, a negative sign is associated with a compressive load. It is obvious that the conditions of axisymmetry do not apply for a beam subjected to combined loading.

On extensive *comparison of the results* of various theories, we find that Winkler's formula is adequate for practical applications. Its advantage lies in the relative ease with which it may be applied to any symmetric section. For curved beams in pure bending, there is a good agreement between the exact solutions given by Eq. (16.53) and Winkler results. Agreement between the two theories is not as good in situations of combined load, however. Tangential stress distribution predicted by the usual flexure formula, My/I, follows a linear pattern (Figure 16.13b). This provides, for beams with only slight curvature, reasonably good results. The stresses obtained by the Winkler's formula and the flexure formula are approximately the same for slender beams, as is usually assumed, $R/c > 20$. Here R represents the radius to the centroid, c is the distance from the centroid to the extreme fiber on the concave side of the beam, and R/c is called the index of curvature.

Note that evaluation of the integral for the Z over the cross-sectional area of ordinary geometrical form may readily be performed. Consider a curved frame of b by h rectangular cross section and mean radius R (Figure 16.14). Applying Eq. (16.56) with $c_A = c_B = c$,

$$Z = -\frac{1}{2tc}\int_{-c}^{c}\frac{y}{R+y}t\,dy = -\frac{1}{2c}\int_{-c}^{c}\frac{y}{R+y}\,dy \tag{g}$$

Usually, this expression may be obtained by direct integration by use of the binomial expansion and integration or by numerical techniques. By direct integration, Eq. (g) results in

$$Z = -\frac{1}{2c}\int_{-c}^{c}\left(1 - \frac{R}{R+y}\right)dy = -1 + \frac{R}{2c}\ln\left(\frac{R+c}{R-c}\right) \tag{h}$$

Observe that Z is a *dimensionless* quantity, it varies with the ratio of R to c and must be calculated with a reasonably high accuracy. Table 16.1 furnishes formulas for the curved beam factor Z for some typical cross sections [2]. Other methods of solution have also been developed [16–18]. Numerical evaluation of the integral can be done for complex shapes.

Table 16.1   Curved beam factor Z for various sections

Cross section	Formula for Z
A.	$$Z = -1 + 2\left(\frac{R}{c}\right)^2 - 2\left(\frac{R}{c}\right)\sqrt{\left(\frac{R}{c}\right)^2 - 1}$$
B.	$$Z = -1 + \frac{2R}{c^2 - c_1^2}\left[\sqrt{R^2 - c_1^2} - \sqrt{R^2 - c^2}\right]$$
C.	$$Z = -1 + \frac{R}{Ah}\left\{[b_1 h + (R + c_1)(b - b_1)]\right.$$ $$\times \ln \frac{R + c_1}{R - c} - (b - b_1)h\left.\right\}$$ $$A = (b_1 + b)h/2$$ For rectangular section: $c = c_1,\ b = b_1$ For triangular section: $b_1 = 0$
D.	$$Z = -1 + \frac{R}{A}[t \cdot \ln(R + c_1) + (b - t) \cdot \ln(R - c_2) - b \cdot \ln(R - c)]$$ $$A = bh - (b - t)(h - t)$$
E.	$$Z = -1 + \frac{R}{A}[b_1 \cdot \ln(R + c_1) + (t - b_1) \cdot \ln(R + c_3)$$ $$+ (b - t) \cdot \ln(R - c_2) - b \cdot \ln(R - c)]$$ $$A = (b + b_1 + c_2 + c_3)t$$

EXAMPLE 16.7   Determining Stresses in Curved Frame Using Various Methods

A circular frame of rectangular cross section and mean radius R is subjected to a load P as shown in Figure 16.14a. Compute the tangential stresses at points A and B, using

  (a) Winkler's theory.
  (b) The elementary theory.
  (c) The elasticity theory.

Given: $P = 200$ kN, $b = 100$ mm, $h = 200$ mm, $R = 300$ mm

Solution:

  (a) Applying Eq. (h) with $R = 300$ mm and $c_A = c_B = c = 100$ mm (Figure 16.14b), we have

$$Z = -1 + \frac{3}{2}\ln 2 = 0.0397$$

Figure 16.14 Example 16.7: (a) Curved frame with a vertical load at free end; (b) Rectangular cross section; (c) Stress resultants at a cross section A-B.

The tangential stresses are due to the compressive normal load $-P$ and the moment $M = PR$ acting at the centroid C of the cross section (Figure 16.14c). The maximum compression and tension values of σ_θ occur at points A and B, respectively. Substituting the given numerical values, Eq. (16.58) results in then

$$(\sigma_\theta)_A = -\frac{P}{A} + \frac{M}{AR}\left[1 + \frac{-c_A}{Z(R - c_A)}\right] = -\frac{Pc_A}{ZA(R - c_A)} \qquad \textbf{(16.59)}$$

$$= -\frac{200{,}000(0.1)}{0.0397 \times 0.02(0.3 - 0.1)} = -126 \text{ MPa}$$

$$(\sigma_\theta)_B = -\frac{P}{A} + \frac{M}{AR}\left[1 + \frac{c_B}{Z(R + c_B)}\right] = \frac{Pc_B}{ZA(R + c_B)} \qquad \textbf{(16.60)}$$

$$= \frac{200{,}000(0.1)}{0.0397 \times 0.02(0.3 + 0.1)} = 63 \text{ MPa}$$

(b) Through use of the flexure formula, with $M = PR = 200(300) = 60 \text{ kN} \cdot \text{m}$,

$$(\sigma_\theta)_B = -(\sigma_\theta)_A = \frac{Mc}{I} = \frac{60{,}000(0.1)}{(0.1)(0.2)^3/12} = 90 \text{ MPa}$$

(c) From Eq. (16.54), with $a = 300 - 100 = 200$ mm and $b = 300 + 100 = 400$ mm, we have

$$N = [1 - (0.5)^2]^2 - 4(0.5)^2 \ln^2 2 = 0.082$$

Superposition of $-P/A$ and Eq. (16.53) at $r = a$ gives

$$(\sigma_\theta)_A = -\frac{200{,}000}{0.02} + \frac{4(60{,}000)}{(0.1)(0.4)^2(0.082)}[(1 - 0.25)(1 + 0) - (1 + 1)\ln 2]$$

$$= -10 - 116.4 = -126.4 \text{ MPa}$$

Likewise, at $r = b$, we obtain $(\sigma_\theta)_B = -10 + 73.8 = 63.8$ MPa.

Comments: The foregoing shows that the results of the Winkler and elasticity theories are in good agreement. However, the usual flexure formula provides a result of unacceptable accuracy for the tangential stress in this nonslender curved beam.

Case Study 16-1 | DESIGN ANALYSIS OF A CRANE HOOK FOR THE WINCH CRANE

A crane hook for the winch crane, shown in Figure 15.11, repeated here, is rated at $P = 3$ kN (see Case Study 1-1). Determine the tangential stresses at points A and B using Winkler's formula. Note that, for a large number of manufactured crane hooks, the critical section AB can be closely approximated by a trapezoidal area with half an ellipse at the inner radius and an arc of a circle at the outer radius, as shown in Figure 15.11b. The solution for standardized crane hooks is expedited by readily available computer programs.

Assumptions: The critical section AB is taken to be trapezoidal. The hook is made of AISI 1020-HR steel with a safety factor of n against yielding.

Given: $r = 20$ mm, $b = 30$ mm, $b_1 = 10$ mm
      $h = 42$ mm, $n = 5$,
      $S_y = 210$ MPa   (from Table B.3)

(a)   (b)

Figure 15.11   Repeated. Hook for winch crane of Figure 1.4: (a) section of the trunnion with a thrust-ball bearing; (b) the critically stressed, modified trapezoidal section.

Solution: Referring to Figure 15.11b, we obtain the following quantities. The cross sectional area is

$$A = \frac{1}{2}(b + b_1)h = \frac{1}{2}(30 + 10)(42) = 840 \text{ mm}^2$$

The distance to the centroid C from the inner edge:

$$c_A = \frac{h(b + 2b_1)}{3(b + b_1)} = \frac{42(30 + 2 \times 10)}{3(30 + 10)} = 17.5 \text{ mm}$$

Hence,

$$c_B = h - c_A = 42 - 17.5 = 24.5 \text{ mm}$$

$$R = r + c_A = 20 + 17.5 = 37.5 \text{ mm}$$

By case C of Table 16.1, the curved beam factor, with $c = c_A$ and $c_1 = c_B$, is then

$$Z = -1 + \frac{37.5}{840(42)} \{[(10 \times 42)$$
$$+ (37.5 + 24.5)(30 - 10)] \ln \frac{37.5 + 24.5}{37.5 - 17.5}$$
$$- (30 - 10)(42)\} = -1 + 1.1034 = 0.1034$$

The circumferential stresses are determined through the use of Eqs. (16.59) and (16.60) with a tensile normal load P and bending moment $M = -PR$. Therefore

$$(\sigma_\theta)_A = \frac{Pc_A}{ZA(R - c_A)}$$

$$(\sigma_\theta)_B = -\frac{Pc_B}{ZA(R + c_B)}$$

where a minus sign means compression. Introducing the required values into the preceding expression, we have

$$(\sigma_\theta)_A = \frac{3000(0.0175)}{(0.1034)(840 \times 10^{-6})(0.0375 - 0.0175)}$$

$$= 30.22 \text{ MPa}$$

$$(\sigma_\theta)_B = -\frac{3000(0.0245)}{(0.1034)(840 \times 10^{-6})(0.0375 + 0.0245)}$$

$$= -13.65 \text{ MPa}$$

Comment: The allowable stress $\sigma_{\text{all}} = 210/5 = 42$ MPa is larger than the maximum stress of 30.22 MPa. That is, the crane hook can support a load of 3 kN with a factor of safety of 5 without yielding.

*16.9 AXISYMMETRICALLY LOADED CIRCULAR PLATES

The deflection w of a plate of circular shape depends on its radial position alone if the applied load and conditions of end support are independent of the angle θ. For this case, radial and tangential moments M_r and M_θ per unit length and force Q_r per unit length act on the circular plate element (Figure 16.15b). To develop the basic equations of a circular plate, we need only transform the appropriate formulations of Sections 4.10 and 4.11 from the Cartesian to polar coordinates.

The polar coordinate set (r, θ) and the Cartesian set (x, y) are related by the equations (Figure 16.15a):

$$x = r\cos\theta, \qquad r^2 = x^2 + y^2$$

$$y = r\sin\theta, \qquad \theta = \tan^{-1}\frac{y}{x}$$

Applying these relationships, the radial and tangential bending moments are found from Eqs. (4.48) as

$$M_r = -D\left(\frac{d^2w}{dr^2} + \frac{v}{r}\frac{dw}{dr}\right) \tag{16.61a}$$

$$M_\theta = -D\left(\frac{1}{r}\frac{dw}{dr} + v\frac{d^2w}{dr^2}\right) \tag{16.61b}$$

and the twisting moment $M_{r\theta} = 0$. The flexural rigidity, defined by Eq. (4.49), is $D = Et^3/12(1 - v^2)$. To obtain the maximum stresses in polar coordinates, it is necessary to replace subscripts x by r and y by θ in Eqs. (4.50):

$$\sigma_{r,\max} = \frac{6M_r}{t^2}, \qquad \sigma_{\theta,\max} = \frac{6M_\theta}{t^2} \tag{16.62}$$

in which t is the thickness of the plate.

The differential equation describing the surface deflection, as determined from Eq. (4.51) in a similar fashion [4], has the form

$$\nabla^4 w = \frac{1}{r}\frac{d}{dr}\left\{r\frac{d}{dr}\left[\frac{1}{r}\frac{d}{dr}\left(r\frac{dw}{dr}\right)\right]\right\} = \frac{p}{D} \tag{16.63}$$

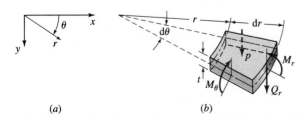

(a)   (b)

Figure 16.15 (a) Polar coordinates. (b) Axisymmetrically loaded circular plate element. (Only shear force and moments on the positive faces are shown.)

The deflection w is obtained by successive integration when $p(r)$ is given:

$$w = \int \frac{1}{r} \int r \int \frac{1}{r} \int \frac{pr}{D} \, dr \, dr \, dr \, dr \tag{16.64}$$

When the plate is under a *uniform loading* $p = p_o$, the general solution of Eq. (16.63) is

$$w = c_1 \ln r + c_2 r^2 \ln r + c_3 r^2 + c_4 + \frac{p_o r^4}{64D} \tag{16.65}$$

Here the c's are constants of integration.

EXAMPLE 16.8 | **Pump Diaphragm Stress Analysis**

Determine deflection and stress for a clamped circular plate of radius a, representing a pump diaphragm, subjected to uniformly distributed load p_o (Figure 16.16).

Figure 16.16   Example 16.8.

Assumptions: Presuppositions of the thin-plate bending theory given in Section 4.10 applies.

Design Decision: The boundary conditions are

$$w = 0, \qquad \frac{dw}{dr} = 0, \qquad (r = a)$$

Solution: The terms involving logarithms in Eq. (16.65) give an infinite displacement at $r = 0$ for all values of c_1 and c_2 except 0; so $c_1 = c_2 = 0$. Satisfying the preceding conditions, we have

$$c_3 = -\frac{p_o a^2}{32D}, \qquad c_4 = \frac{p_o a^4}{64D}$$

The deflection is then

$$w = \frac{p_o}{64D} (a^2 - r^2)^2 \tag{16.66}$$

The maximum displacement, occurring at the center of the plate, is

$$w_{max} = \frac{p_o a^4}{64D} \tag{16.67}$$

Expressions for the bending moments may be determined by Eqs. (16.66) in the form

$$M_r = \frac{p_o}{16}[(1+v)a^2 - (3+v)r^2]$$

$$\tag{16.68}$$

$$M_\theta = \frac{p_o}{16}[(1+v)a^2 - (1+3v)r^2]$$

Algebraically extreme values of the moments are found at the center and at the edge. At the edge ($r = a$), Eqs. (16.68) result in

$$M_r = -\frac{p_o a^2}{8}, \qquad M_\theta = -\frac{v p_o a^2}{8}$$

while at $r = 0$, $M_r = M_\theta = (1+v)p_o a^2/16$. We observe that the maximum moment occurs at the edge. Hence, we have the value of the maximum stress

$$\sigma_{r,max} = \frac{6M_r}{t^2} = -\frac{3p_o}{4}\left(\frac{a}{t}\right)^2$$

Comment: The minus sign means a compressive stress in the bottom half of the plate.

A similar procedure may be applied to symmetrically loaded circular plates subjected to different end conditions and loading. For reference purposes, Table 16.2 presents the

Table 16.2   Equations for maximum stresses σ_{max} and maximum deflections w_{max} that occur at the center of circular plates

Support and loading	σ_{max}	w_{max}
Edge simply supported, load uniform	$\frac{3}{8}(3+v)p_o\frac{a}{t^2}$	$\frac{3}{16}(1-v)(5+v)\frac{p_o a^2}{Et^3}$
Edge simply supported, load at center, $P = \pi r_c^2 p_o$ $r_c \to 0$, but $r_c > 0$	$\frac{3(1+v)P}{2\pi t^2}\left(\frac{1}{1+v} + \ln\frac{a}{r_c} - \frac{1-v}{1+v}\frac{r_c^2}{4a^2}\right)$	$\frac{3(1-v)(1+v)Pa^2}{4\pi Et^3}$
Fixed edge, load at center, $P = \pi r_c^2 p_o$ $r_c \to 0$, but $r_c > 0$	$\frac{3(1+v)P}{2\pi t^2}\left(\ln\frac{a}{r_c} + \frac{r_c^2}{4a^2}\right)$ (for $a > 1.7r_c$)	$\frac{3(1-v^2)Pa^2}{4\pi Et^2}$

Notes: p_o = uniform load per unit area, a = radius of plate, r_c = radius of central loaded area, t = thickness of plate, v = Poisson's ratio, E = modulus of elasticity.

maximum principal stress and maximum deflection for some commonly encountered cases [16]. Design calculations are facilitated by this type of compilation.

*16.10   THIN SHELLS OF REVOLUTION

Structural members resembling curved plates are known as *shells*. Examples include aircraft fuselages, pressure vessels, and a variety of containers. As was the case of plates, we limit our consideration to isotropic, homogeneous, elastic shells having a constant thickness that is small relative to the remaining dimensions. The surface bisecting the shell thickness is called the *midsurface*. To specify the geometry of a shell, we need to know only the form of the midsurface and the thickness of the shell at each point. In practice, if the ratio of thickness t to radius of curvature r equals or is less than $1/20$, the shell is classified as *thin*.

SHELL THEORIES

The stress analysis of shells normally embraces two distinct theories. The *membrane theory* is limited to moment-free membranes, which usually applies to a rather large proportion of the entire shell. The *bending theory* or general theory includes the affects of bending and enables us to treat discontinuities in the stress distribution occurring in a limited region in the vicinity of a load application, an abrupt change in the geometry, or in regions near boundaries. This method usually involves a membrane solution, corrected in those areas in which discontinuity effects are pronounced [4–6]. The following *assumptions* are generally made in the *small deflection analysis* of thin shells:

1. The ratio of the shell thickness to the radius of curvature of the midsurface is small compared to unity.
2. Displacements are very small compared with the shell thickness.
3. Sections perpendicular to the midsurface remain so to the deformed midsurface after bending.
4. The component of stress normal to the midsurface, σ_z, is negligible.

GEOMETRY OF SHELL OF REVOLUTION

Consider a particular type of shell having a surface of revolution, as shown in Figure 16.17. The midsurface of this shell is generated by rotation of a so-called meridian curve about an axis lying in the plane of the curve. The figure shows that a point on the shell is conveniently located by coordinates θ, ϕ, r and elemental surface *ABCD* is defined by two meridians and two parallel circles or parallels.

The planes associated with the principal radii of curvature r_ϕ and r_θ at any point on the midsurface of the shell are the *meridian plane* and the *parallel plane* at the point in question, respectively. The radii of curvature r_ϕ and r_θ are then related to sides *CD* and *AC*. The radius r_θ generates the shell surface in the direction perpendicular to the direction of the tangent to the meridian curve. The two radii r_ϕ and r_θ are related since $r_0 = r_\theta \sin \phi$ (Figure 16.17). Hence, lengths *AC* and *CD* of the curvilinear shell element are $r_0 \, d\theta = r_\theta \sin \phi \, d\phi$ and $r_\phi \, d\phi$, respectively.

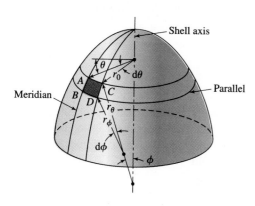

Figure 16.17 Geometry of shell of revolution.

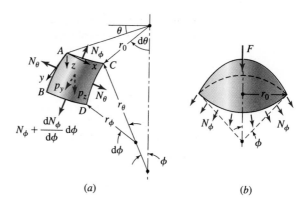

(a)

(b)

Figure 16.18 (a) Shell element with membrane forces per unit length and loading. (b) Truncated shell.

SYMMETRICALLY LOADED SHELLS OF REVOLUTION

In axisymmetrical problems involving shells of revolution, there are only two unknown *membrane forces per unit length,* the so-called *hoop force* per unit length N_θ and *meridional force* per unit length N_ϕ. The governing equations for these forces are developed from two conditions of equilibrium. Figure 16.18a shows the element $ABCD$ cut from the shell of Figure 16.17. Due to the condition of symmetry, the membrane forces and the loading display no variation with θ. We take the coordinate axes x and y tangent at A to the lines of principal curvature and the axis z normal to the midsurface (Figure 16.18a). Externally applied forces per unit area are represented by the components p_y and p_z in the y and z directions, respectively.

Equations of Equilibrium

The z-directed distributed load carried on the surface area of the shell element is given by

$$p_z r_0 r_\phi \, d\theta \, d\phi$$

The force acting on the top edge of the element, $N_\phi r_0 \, d\theta$, neglecting the higher-order terms, equals the force on the bottom edge. The z-directed component at each edge is then $N_\phi r_0 \, d\theta \sin(d\phi/2) \approx N_\theta r_0 \, d\theta \, d\phi/2$, and the resultant for both edges is

$$N_\phi r_0 \, d\theta \, d\phi$$

Inasmuch as the cross-sectional area along each of the two *lateral sides* of the shell element is $r_\phi \, d\phi$, the force on these areas equals $N_\theta r_\phi \, d\phi$. The resultant in the direction of the radius in the *parallel plane* for both such forces is $N_\theta r_\phi \, d\phi \, d\theta$. This produces the following component in the z direction:

$$N_\theta r_\phi \, d\phi \, d\theta \, \sin\phi$$

The equilibrium of z-directed forces given by the preceding three equations, $\sum F_z = 0$, after canceling $d\phi \, d\theta$, gives

$$N_\phi r_0 + N_\theta r_\phi \sin\phi + p_z r_0 r_\phi = 0$$

Dividing by $r_0 r_\phi$ and setting $r_0 = r_\phi \sin \phi$ results in a basic relation for the axisymmetrically loaded shell of revolution:

$$\frac{N_\phi}{r_\phi} + \frac{N_\theta}{r_\theta} = -p_z \tag{16.69a}$$

This is often referred to as the *membrane equation* of the shell.

An equation for the equilibrium of y-directed forces of the shell element may be derived similarly. But instead of solving the z and y equilibrium equations simultaneously, it is more convenient to determine N_ϕ from the equilibrium of the vertical forces acting on the portion of the shell subtended by the angle ϕ (Figure 16.18b):

$$N_\phi = -\frac{F}{2\pi r_0 \sin \phi} \tag{16.69b}$$

Then calculate N_θ from Eq. (16.69a). Here F represents the *resultant* of all *external loading* acting on the portion of the shell.

Compatibility of Deformations

In axisymmetrically loaded shells of revolution, due to their freedom of motion in the z direction, strains are produced to assure consistency with the stress distribution. These strains are compatible with one another. When a shell is under a concentrated surface loading or is constrained at its boundaries, membrane theory cannot satisfy the conditions on deformation everywhere. In such cases, however, the departure from membrane behavior is limited to a narrow region in the vicinity of the loading or the boundary. Membrane theory remains valid for the major portion of the shell, but the complete solution can be determined only by applying the bending theory [4–10].

16.11   SPECIAL CASES OF SHELLS OF REVOLUTION

Many types of shells are used as machine and structural components. The membrane stresses in any particular axisymmetrically loaded shell in the form of a surface of revolution may readily be found from Eqs. (16.69). Presented in this section is a brief discussion of three commonly encountered members. Note that the following groups do not relate only to the special types of axisymmetrical loading but are quite general.

SPHERICAL SHELL

Figure 16.19a shows a spherical shell. For this case, we need only set the mean radius $r = r_\theta = r_\phi$ and hence $r_0 = r \sin \phi$. Equations (16.69) then reduce to

$$N_\phi + N_\theta = -p_z r$$
$$N_\phi = -\frac{F}{2\pi r \sin^2 \phi} \tag{16.70}$$

Example 16.9 illustrates the application.

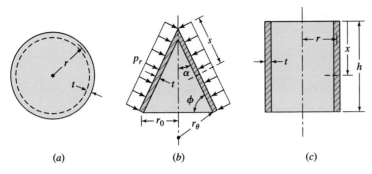

Figure 16.19 Common types of shells of revolution: (a) spherical; (b) conical; (c) vertical cylindrical.

The simplest case is that of a spherical shell under internal pressure. We have $p = -p_z$, $\phi = 90°$, and $F = -\pi r^2 p$. Owing to the symmetry of the spherical shell, $N_\phi = N_\theta = N$. Carrying these quantities into Eqs. (16.70) gives the membrane stress in a *spherical pressure vessel*:

$$\sigma = \frac{N}{t} = \frac{pr}{2t} \tag{16.71}$$

The quantity t is the thickness of the shell. Equation (16.71) represents uniform stresses in all directions of a pressurized sphere. As pointed out in Section 3.4, a sphere is the most favorably stressed shape for a vessel requiring the minimum wall thickness. It is used for extremely high-pressure operations and employed in space vehicles and missiles for the storage of liquidified gases at lower pressures but with lightweight thin walls.

CONICAL SHELLS

Figure 16.19b illustrates a conical shell under a radial loading p_r. We need only to set $r_\phi = \infty$ in Eq. (16.69a). This, together with Eq. (16.69b), gives the following pair of equations for obtaining the membrane forces per unit length under uniform pressure $p_z = p_r$:

$$N_\theta = p_r r_\theta = -\frac{p_r r_0}{\sin \theta}$$
$$\tag{16.72}$$
$$N_\phi = N_s = -\frac{F}{2\pi r_0 \sin \phi}$$

in which s represents the direction of the generator. The tangential or hoop and meridian stresses in a conical shell are found by dividing N_θ and N_s by the thickness t of the shell.

CIRCULAR CYLINDRICAL SHELLS

To obtain the membrane forces in a circular shell (Figure 16.19c), we set $\phi = \pi/2$, $p_z = p_r$, and mean radius $r = r_0 = $ constant in the cone expressions. Equations (16.72) are then

$$N_x = -\frac{F}{2\pi r}, \qquad N_\theta = -p_r r \tag{16.73}$$

where x represents the axial direction of the cylinder.

For a closed-ended *cylindrical vessel* subjected to internal pressure $p = -p_r$ and $F = -\pi r^2 p$. The preceding equations then give the tangential or hoop and the axial stresses:

$$\sigma_\theta = \frac{pr}{t}, \qquad \sigma_x = \frac{pr}{2t} \tag{16.74}$$

The quantity t is the thickness of the shell. We note that, for thin-walled cylindrical pressure vessels, $\sigma_\theta = 2\sigma_x$.

EXAMPLE 16.9 | Force Analysis of a Spherical Dome

Determine expressions for the meridianal and the hoop stresses in a spherical dome of radius a and thickness t, carrying only its own weight p per unit area (Figure 16.20).

Figure 16.20   Example 16.9.

Assumptions: The simple support shown in the figure is free to move as the shell deforms under loading: No bending is produced in the neighborhood of the edge.

Solution: The weight of that portion of the dome corresponding to the angle ϕ is obtained as

$$F = \int_0^\phi p(2\pi a \sin \phi \cdot a \, d\phi) = 2\pi a^2 p(1 - \cos \phi)$$

Also

$$p_z = p \cos \phi$$

Introducing into Eqs. (16.70) the preceding expressions for p_z and F, we have

$$N_\phi = -\frac{ap(1 - \cos \phi)}{\sin^2 \phi} = -\frac{ap}{1 + \cos \phi}$$

$$N_\theta = -ap\left(\cos \phi - \frac{1}{1 + \cos \phi}\right) \tag{16.75}$$

Division of the results by t gives the expressions for the membrane stresses.

Comments: In Eqs. (16.75), the negative signs mean compression. Obviously, N_ϕ is always compressive. The sign of N_θ, on the other hand, depends on ϕ. By the second equation, when $N_\theta = 0$, $\phi = 51°50'$. For ϕ smaller than this value, N_θ is compressive. If $\phi > 51°50'$, N_θ is tensile.

Design of a Parabolic Pressure Vessel

EXAMPLE 16.10

A parabolic shell is closed at the top by a thick plate and subjected to internal pressure p (Figure 16.21a). At level A-A, calculate the minimum permissible thickness t_{all} of the shell.

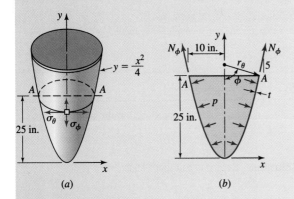

Figure 16.21   Example 16.10.

Given: $p = 200$ psi

Design Decision: The parabola is $y = x^2/4$, in which x and y are in inches.

Assumptions: At level A-A, the allowable membrane stress is $\sigma_{all} = 16.5$ ksi. Section A-A is away from the top.

Solution: Let the load resultant for the portion of the shell *below* plane A-A be F (Figure 16.21b). At level A-A, $y = 25$ in.:

$$x = r_0 = \sqrt{4(25)} = 10 \text{ in.}$$

and hence $dy/dx = x/2 = 5$. From geometry,

$$r_\theta = 10\left(\frac{\sqrt{26}}{5}\right) = 10.2 \text{ in.}$$

The familiar expression for the curvature then gives

$$r_\phi = \frac{[1 + (dy/dx)^2]^{1.5}}{d^2y/dx^2} = \frac{(1 + 5^2)^{1.5}}{1/2} = 265.15 \text{ in.}$$

The membrane forces per unit length at A-A can now be obtained by applying Eq. (16.69):

$$N_\phi = -\frac{F}{2\pi r_0 \sin\phi} = -\frac{-p\pi r_0^2}{2\pi r_0(5/\sqrt{26})} = \frac{\sqrt{26}\,pr_0}{10}$$

$$= \frac{\sqrt{26}(200)(10)}{10} = 1.02 \text{ kips}$$

$$\frac{N_\phi}{r_\phi} + \frac{N_\theta}{r_\theta} = -p_z: \qquad \frac{1.02}{265.15} + \frac{N_\theta}{10.2} = 0.2$$

Solving, $N_\theta = 2.0$ kips. Inasmuch as $N_\theta > N_\phi$, we have

$$t_{all} = \frac{N_\theta}{\sigma_{all}} = \frac{2.0}{16.5} = 0.121 \text{ in.}$$

Comment: The required shell thickness should be $\frac{1}{8}$ in. at level *A-A*.

Case Study 16-2 | DESIGN ANALYSIS OF CONICAL STORAGE TANK

A thin-walled container of conical shape supported from the top and filled with a heavy-liquid metal of specific weight γ is shown in Figure 16.22 (see Case Study 7-1). Determine

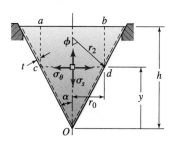

Figure 16.22   Conical tank.

(a)   The expressions for the tangential stress σ_θ and meridianal stress σ_ϕ.

(b)   The factor of safety *n* against yielding, using the maximum shear stress criterion.

Given: The container is made of ASTM-A36 structural steel.

$$h = 3 \text{ m}, \qquad \alpha = 45°,$$

$$t = 3 \text{ mm}, \qquad \gamma = 80 \text{ kN/m}^3$$

$$S_y = 250 \text{ MPa} \qquad \text{(from Table B.1)}$$

Assumption: The simple support depicted in the figure, free to move as the shell deforms under loading,

ensures that no bending is produced in the neighborhood of the edge.

Solution: Referring to Figure 16.22, we write

$$\phi = \frac{\pi}{2} - \alpha, \qquad r_0 = y \tan \alpha$$

where α is the half-cone angle. At any arbitrary level *y*, the pressure is

$$p = p_z = \gamma(h - y)$$

(a)   Substitution of the foregoing expressions into the first of Eqs. (16.72), after division by *t*, results in the hoop stress

$$\sigma_\theta = \frac{\gamma(h - y)y}{t} \frac{\tan\alpha}{\cos\alpha} \qquad (16.76a)$$

The load equals the weight of the liquid of volume *acOdb*. Then,

$$F = -\pi\gamma y^2 \left(h - y + \frac{1}{3}y \right) \tan^2 \alpha$$

Carrying this value into the second of Eqs. (16.72), and dividing the resulting expression by *t*, gives the meridianal stress:

$$\sigma_\phi = \sigma_s = \frac{\gamma(h - 2y/3)y}{2t} \frac{\tan\alpha}{\cos\alpha} \qquad (16.77a)$$

Comments: We see from these expressions that, at the bottom ($y = 0$), the pressure and the membrane stresses vanish. The exact stress distribution at the apex and edge is obtained by application of the bending theory [4].

Case Study (CONCLUDED)

(b) Differentiating with respect to y and equating to 0, it can be verified that the maximum values of the preceding membrane stresses and their location are as follows:

$$\sigma_{\theta,\max} = \frac{\gamma h^2}{4t} \frac{\tan\alpha}{\cos\alpha}, \qquad (\text{at } y = h/2) \qquad \textbf{(16.76b)}$$

$$\sigma_{s,\max} = \frac{3\gamma h^2}{16t} \frac{\tan\alpha}{\cos\alpha}, \qquad (\text{at } y = 3h/4) \qquad \textbf{(16.77b)}$$

Note that the largest stress occurs at midheight and is given by Eq. (16.76b). The design equation is given

by Eq. (7.19a) in Case Study 7-1 as

$$\frac{\gamma h^2}{4t} \frac{\tan\alpha}{\cos\alpha} = \frac{S_y}{n} \qquad \textbf{(16.78)}$$

Introducing the required values, we have

$$\frac{80(3)^2}{4(3)} \frac{\tan 45°}{\cos 45°} = \frac{250}{n}$$

from which $n = 2.95$.

16.12 PRESSURE VESSELS AND PIPING

The discussions of the last two sections are limited to the membrane stresses occurring over the entire wall thickness of thin shells. As noted previously, membrane theory must be augmented by bending theory in the vicinity of a discontinuity in load or geometry and in regions near boundaries. The latter approach considers membrane forces as well as shear forces and moments acting on the shell structure. Complete bending theory is mathematically intricate and the first solutions involving shell-bending stresses date back to only 1920. Since then great many problems in shell-like structures have been worked out.

Ever-broadening use of vessels for storage, industrial processing and power generation under unique conditions of temperature, pressure, and environment have given special emphasis to analytical, numerical, and experimental techniques for determining the appropriate working stresses. The finite element method has gained considerable favor in the design of vessels over other methods. A discontinuity of the membrane action in a vessel occurs at all points of external constraint or at the junction of the cylindrical shell and its head or end possessing different stiffness characteristics. Any incompatibility of deformation at the joint produces bending moments and shear forces. The stresses due to this bending and shear are called *discontinuity stresses*. Most of the situations involving pressure vessels and piping are covered in codes.

Since the bending is of a local character, the discontinuity stresses become negligibly small within a short distance. The narrow region at the edge of *spherical, elliptical,* and *conical* vessel heads can be assumed as *nearly cylindrical* in shape. Cylindrical shell equations can be used to obtain an approximate solution applicable at the juncture of vessels having spherical, elliptical, or conical ends. In the case of *flat-end* vessels, expressions for circular plates and cylindrical shells are utilized. The derivation of the discontinuity stresses in pressure vessels, using the bending and membrane theories of shells and the method of superposition is well known [3–11] and is not given here. As in thick-walled cylinders, the tangential or *hoop stress* is usually the *largest* and most critical in the design of the thin-walled vessels and piping.

A SHORT CATALOG OF RESULTS

Results of some common thin-walled vessel and pipe problems encountered in routine design are listed in the following paragraphs. Thin shell equations under uniform pressure apply to internal pressure p. They also pertain to cases of external pressure if the sign of p is changed. However, stresses so obtained are valid only if the pressure is not significant relative to that which causes failure by elastic instability. A degree of caution is necessary when applying the formulas for which there is uncertainty as to applicability and restriction of use. Particular emphasis should be given to the fact that high loading, extreme temperature, and rigorous performance requirements present difficult design challenges.

For a thin shell ($a/t > 10$) of radius a and thickness t, the geometric parameters are expressed by the following:

Cylinder

$$\beta^4 = \frac{Et}{4a^2 D} \tag{16.79}$$

It has the dimensions of L^{-1}, the reciprocal of length.

Sphere

$$\lambda = \beta a \tag{16.80}$$

The flexural rigidity of the shell is

$$D = \frac{Et^3}{12(1 - v^2)} \tag{16.81}$$

where E = modulus of elasticity and v = Poisson's ratio. Referring to the preceding two sections:

σ_x = axial bending stress in cylinder

σ_θ = circumferential bending stress

Δ = radial edge displacement or deflection of sphere

σ_ϕ = meridianal bending stress in sphere

w = radial displacement or deflection of cylinder

Moments M and forces V and P per unit length have units (in SI) N·m/m, or simply N, and N/m, respectively.

The circular cylindrical shell or thin-walled pipe, of special importance in engineering, is often classified into long and short. A *long cylindrical shell* loaded at one end or at the midspan is defined as of length

$$L > \frac{2\pi}{\beta} \tag{16.82}$$

Conditions at one end do not affect to the other end if the ends are far enough apart. Most cylindrical shells are mathematically long even when physically short. Cases 1 and 2 apply to "long" thin-walled pipes.

CASE 1. LONG PIPE, RADIAL CENTER LOAD P PER UNIT LENGTH

$$w = \frac{Pe^{-\beta x}}{8D\beta^3}\left[\sqrt{2}\sin\left(\beta x + \frac{\pi}{4}\right)\right] \qquad \text{(a)}$$

$$w_{max} = \frac{Pa^2\beta}{2Et} \qquad \text{(at } x = 0)$$

$$\sigma_{x,max} = \frac{3P}{2\beta t^2}, \qquad \sigma_{\theta,max} = \frac{P\beta}{2}\left(-\frac{a}{t} + \frac{3v}{\beta^2 t^2}\right) \qquad \text{(at } x = 0)$$

Equation (a) shows that the deflection (and moment) attenuates with distance as an exponentially damped sine wave of wave length, for $v = 0.3$, $2\pi/\beta \approx 4.89\sqrt{at}$; bending is of local character.

CASE 2. LONG PIPE, END MOMENT M PER UNIT LENGTH

$$w_{max} = -\frac{M}{2D\beta^2} \qquad \text{(at } x = 0)$$

$$\sigma_{x,max} = \frac{6M}{t^2}, \qquad \sigma_{\theta,max} = v\sigma_{x,max} \qquad \text{(at } x = 0)$$

CASE 3. SPHERICAL SEGMENT, RADIAL EDGE LOAD V PER UNIT LENGTH

$$\Delta = \frac{Va\lambda\sin^2\phi}{Etc_1}(1 + c_1 c_2) \qquad \text{(for } x = r)$$

$$\sigma_{\phi,max} = 0, \qquad \sigma_{\theta,max} = \frac{V\lambda^2\cos\phi}{ac_1} \qquad \text{(for } x = r)$$

where

$$c_1 = 1 - \frac{1-2v}{2\lambda}\cot\phi, \qquad c_2 = 1 - \frac{1+2v}{2\lambda}\cot\phi$$

For hemisphere ($\phi = \pi/2$), $\sin\phi = 1$, $\cos\phi = 0$, and $c_1 = c_2 = 1$.

CASE 4. SPHERICAL SEGMENT, EDGE MOMENT M PER UNIT LENGTH

$$\Delta = \frac{2M\lambda^2\sin\phi}{Etc_1} \qquad \text{(for } x = r)$$

$$\sigma_{\phi,max} = \frac{6M}{t^2} \qquad \text{(for } x = r)$$

$$\sigma_{\theta,max} = \frac{3M}{vtc_1}[(1 + v^2)(c_1 + c_2) - 2c_2] \qquad \text{(for } x = r)$$

The constants c_1 and c_2 are as defined in Case 3.

CASE 5. CYLINDRICAL VESSEL WITH HEMISPHERICAL HEADS, PRESSURE p

$$\sigma_{x,\max} = 1.293\frac{pa}{2t} \qquad (\text{at } x = \pi/4\beta)$$

$$\sigma_{\theta,\max} = 1.032\frac{pa}{t} \qquad (\text{at } x = 1.85/\beta)$$

The thickness t is uniform and $\nu = 0.3$. The foregoing maximum discontinuity stresses occur at the outer surface of the cylinder.

CASE 6. CYLINDRICAL VESSEL WITH ELLIPSOIDAL HEADS, PRESSURE p

$$\sigma_{x,\max} = 2.172\frac{pa}{2t}$$

$$\sigma_{\theta,\max} = 1.128\frac{pa}{t}$$

The thickness t is uniform, $\nu = 0.3$, and $b = a/2$. The preceding maximum discontinuity stresses take place at the same locations of the cylinder as in Case 5. Interestingly, the membrane stress in the spherical heads ($pa/2t$) is lower than in the cylinder (pa/t). But, for $a = 2b$, the largest membrane stress in the elliptical heads equals that in the cylinder.

16.13   THE ASME CODE FOR CONVENTIONAL PRESSURE VESSELS

With the advent of nuclear plants and outer space and undersea exploration, much attention is being given to the analysis and design of pressure vessels. We observe in the preceding section that the discontinuity stresses in the vicinity of the joint of a cylindrical shell and its head is quite a bit larger than the membrane stress in either portion. When a vessel is properly designed and constructed, however, these stresses are reduced greatly and it becomes unnecessary to consider them.

The ASME *code for pressure vessels* [19, 20] lists formulas for calculating the required minimum thickness of the shell and the ends. The following factors and a host of others contributing to an ideal vessel design are described by the code: approved techniques for joining the head to the shell, formulas for computing the thickness of shell and end, materials used in combination, temperature ranges, maximum allowable stress values, corrosion, types of closure, and so on. Table 16.3 gives the minimum required vessel thicknesses.

The required wall thickness for tubes and pipes under internal pressure are obtained according to the rules for a shell in the code. In its present status, the code applies when the pressure does not exceed 3000 psi. Pressures in excess of this amount may require special attention in the design and construction of the vessels, closures, and branch connections of piping systems [21–23]. Also, the preceding is only a partial description of the code specifications that, if complied with, provide assurance that discontinuity stresses may be neglected. For the complete requirements, reference should be made to the current edition of the code. The ASME publishes relevant books, conference papers, and a quarterly, the *Journal of Pressure Vessel Technology.*

Table 16.3 The ASME design code for pressure vessels

Geometry	Required thickness (in.)
Cylindrical shell	$t = \dfrac{pr}{\sigma e - 0.6p}$
Hemispherical head	$t = \dfrac{pr}{2\sigma e - 0.2p}$
Conical head	$t = \dfrac{pD}{2\cos\alpha(\sigma e - 0.6p)}$
Flat head	$t = d\sqrt{\dfrac{cp}{\sigma}}$
Ellipsoidal head	$t = \dfrac{pD}{2\sigma e - 0.2p}$
Wholly spherical shell	$t = \dfrac{pr}{2\sigma e - 0.2p}$

Notes: p = internal pressure (psi)

   r = inside radius of shell or hemispherical head (in.)

   σ = maximum allowable stress (psi)

   e = lowest joint efficiency ($e < 1.0$)

   D = inside diameter of conical head or inside length of the major axis of an ellipsoidal head (in.)

   α = half the apex angle of the conical head

   d = diameter of flat head (in.)

   c = a numerical coefficient depending on the method of attachment of head (such as $c = 0.5$ for circular plates welded to the end of the shell).

16.14 FILAMENT-WOUND PRESSURE VESSELS

A unique class of composites, formed by wrapping of high-strength filaments over a mandrel, followed by impregnation of the windings with a plastic binder and removal of the mandrel in pieces, is called *filament-wound cylinders*. A common system is the glass filament/epoxy resin combination. Filament structures of this type have an exceptional strength/weight ratio and reliability. They are in widespread use as lightweight vessels and thrust chambers in spacecraft, rockets, and airborne vehicles. Basic filament vessels contain longitudinal, circumferential, or helical windings. A combination of these windings is used if necessary.

Consider a *filament-wound vessel* with closed ends subjected to an internal pressure p (Figure 16.23). The tangential and axial stresses due to p are, from Eqs. (16.74),

$$\sigma_\theta = \frac{pa}{t}, \qquad \sigma_x = \frac{pa}{2t}$$

Hence,

$$\frac{\sigma_\theta}{\sigma_x} = 2 \tag{16.83}$$

Figure 16.23 Filament-wrapped cylindrical pressure vessel. Note: $F =$ tension force in the filament, $\psi =$ helix angle, $w =$ width of filaments, F_θ tangential component of filament force, $p =$ pressure.

The quantities a and t represent the average radius and wall thickness, respectively, of the vessel composed entirely of filament of tensile strength S_u and the binder.

The maximum tensile force carried by the filament may be expressed by

$$F = S_u wt$$

where w is the filament width wrapped at angle ψ. The corresponding circumferential force is

$$F_\theta = F \sin\psi$$

The filament cross-sectional area $A = wt/\sin\psi$. The tangential stress filament then can carry

$$\sigma_\theta = \frac{F_\theta}{A} = S_u \sin^2\psi \tag{16.84}$$

In a like manner, the axial stress may readily be ascertained in the form

$$\sigma_x = S_u \cos^2\psi \tag{16.85}$$

The preceding expressions give

$$\frac{\sigma_\theta}{\sigma_x} = \tan^2\psi \tag{16.86}$$

The *optimum helix angle of filament,* by Eqs. (16.83) and (16.86), is

$$\frac{\sigma_\theta}{\sigma_x} = 2 = \tan^2\psi \tag{16.87}$$

which yields $\psi = 54.7°$. This represents the condition of helical wrapping to support an internal pressure.

Note that, by additional use of circumferential filaments, the helix angle may be decreased for convenience in wrapping. Clearly, the preceding analysis applies only to the cylindrical portion of the vessel, away from the ends. Filament winding is also accomplished by laying down a pattern over a base material and forming a so-called filament-overlay composite [3, 24–26], for example, thin-walled (polyethylene) pipe overlaid with (nylon) cord or a wire of the same material of the shell.

16.15 BUCKLING OF CYLINDRICAL AND SPHERICAL SHELLS

The procedures of analysis and design of thin-walled curved structures or shells in a state of static equilibrium has been taken up in some detail in Sections 16.10 through 16.12. But, thin shell structures subjected to compression may fail by instability of the walls. The failure by *buckling* is similar to that which occurs in the compression of wing and fuselage structures, missile casings, long pipes, pressure vessels, and so forth. This section provides only a brief introduction to the buckling of a thin-walled homogeneous shell. Formulas for critical stresses for pressurized vessels are also furnished. Shell stability is an important area of engineering design [3–6, 27–29].

CYLINDRICAL SHELLS UNDER UNIAXIAL COMPRESSION

Consider a simply supported circular cylindrical shell of radius a, length L, and thickness t, subjected to uniform axial pressure (Figure 16.24). We note that the instability is particularly significant under this type of loading, because axial buckling is often synonymous with collapse of the shells. To determine the critical loading, the methods described in Chapter 6 may be applied.

It can be shown that [4], for $\nu = 0.3$, the length of half-sine waves L_w into which the shell buckles may be expressed in the form

$$L_w = \pi \sqrt[4]{\frac{a^2 t^2}{12(1 - \nu^2)}} \approx 1.72\sqrt{at} \qquad (16.88)$$

The corresponding minimum *buckling load* is

$$N_{cr} = \frac{Et^2}{a\sqrt{3(1 - \nu^2)}} = 0.605\frac{Et^2}{a} \qquad (16.89)$$

The *critical stress* is then $\sigma_{cr} = N_{cr}/t = 0.605E(t/a)$. Observe that the σ_{cr} depends on material properties, thickness, and radius and is independent of cylinder length. A thin-walled tube can wrinkle like tissue paper when subjected to axial compression. The value of the critical load defined by Eq. (16.89) often does not agree with experimental data. To relate the theoretical values introduced to actual test data, it is necessary to incorporate an empirical factor [29].

CYLINDRICAL AND SPHERICAL PRESSURE VESSELS

The determination of critical shell stresses is a very laborious process. The results are listed in Table 16.4 for four practical applications [3, 5, 18]. With them comes the off-repeated warning that they are valid only for pressure vessels of $a/t > 10$.

We point out that there are situations in which vessels under internal pressure can fail by buckling, such as hydraulic tubing loaded by rams. Various types of pressure vessel heads may also collapse when subjected to internal pressure owing to the existence of large tangential compressive stresses in the knuckle region. The problems involving buckling by external pressure have been of considerable interest, particularly for vacuum systems and

Figure 16.24
Buckled shape of a cylindrical shell under uniaxial pressure. The initial form is depicted by the dashed lines.

Table 16.4   Critical pressure p_{cr} formulas for buckling of thin-walled vessels

A. Long cylindrical shells under external pressure

$$p_{cr} = \frac{Et^3}{4a^3(1 - \nu^2)}$$

B. Long cylindrical shells under internal pressure from ram action

$$p_{cr} = \pi^2 \frac{Eta}{L^2}$$

C. Spherical shells under external pressure

$$p_{cr} = \frac{2Et^2}{a^2\sqrt{3(1 - \nu^2)}}$$

D. Spherical end caps (see Figure 16.22a) under external pressure

$$p_{cr} = \frac{2Et}{a\sqrt{3(1 - \nu^2)}}$$

| Notes: t = wall thickness, a = mean radius, L = length, E = modulus of elasticity, ν = Poisson's ratio.

structures, like submarines. Interestingly, *spheres*, in addition to the desirable features noted in Section 16.11, have the *greatest* buckling resistance. In single or tandem configuration, they form the pressure-carrying structure and living space in most deep-submergence oceanographic vehicles.

REFERENCES

1. Timoshenko, S. P., and J. N. Goodier. *Theory of Elasticity,* 3rd ed. New York: McGraw-Hill, 1970.
2. Ugural, A. C., and S. K. Fenster. *Advanced Strength and Applied Elasticity,* 4th ed. Upper Saddle River, NJ: Prentice Hall, 2003.
3. Faupel, J. H., and F. E. Fisher. *Engineering Design,* 2nd ed. New York: Wiley, 1981.
4. Ugural, A. C. *Stresses in Plates and Shells,* 2nd ed. New York: McGraw-Hill, 1999.
5. Timoshenko, S., and S. Woinowsky-Krieger. *Theory of Plates and Shells,* 2nd ed. McGraw-Hill, 1959.
6. Flügge, W. *Stresses in Shells,* 2nd ed. Berlin: Springer, 1990.
7. Flügge, W., ed. *Handbook of Engineering Mechanics.* New York: McGraw-Hill, 1962.
8. Gibson, J. E. *Thin Shells: Computing and Theory.* New York: Pergamon, 1980.
9. Pflüger, A. *Elementary Statics of Shells.* New York: McGraw-Hill, 1961.
10. Baker, E. H., L. Kovalesky, and F. L. Rish. *Structural Analysis of Shells.* Melbourne, FL: Krieger, 1981.
11. Krauss, H. *Thin Elastic Shells.* New York: Wiley, 1967.
12. Vinson, J. R. *Structural Mechanics: The Behavior of Plates and Shells.* New York: Wiley, 1974.
13. Jeffery, G. B. *Philosophical Transactions of the Royal Society,* London, 221 (series A), (1921), pp. 265–93.
14. Norton, R. L. *Machine Design: An Integrated Approach,* 2nd ed. Upper Saddle River, NJ: Prentice Hall, 2000.
15. Peterson, R. E., and A. M. Wahl. "Fatigue of Shafts at Fitted Members with Photoelastic Analysis." ASME *Journal of Applied Mechanics* 57 (1935), p. A173.

16. Boresi, A. P., and R. J. Schmidt. *Advanced Mechanics of Materials,* 6th ed. New York: Wiley, 2003.

17. Juvinall, R. C., and K. M. Marshak. *Fundamentals of Machine Component Design,* 2nd ed. New York: Wiley, 1991.

18. Young, W. C., and R. C. Budynas. *Roark's Formulas for Stress and Strain,* 7th ed. New York: McGraw-Hill, 2001.

19. Boiler and Pressure Vessel Code, Section VIII. New York: ASME.

20. Chuse, R., and S. M. Eber. *Pressure Vessels, the ASME Code Simplified,* 6th ed. New York: McGraw-Hill, 1984.

21. The M. W. Kellogg Co. *Design of Piping Systems.* New York: Wiley, 1956.

22. *International Design Criteria of Boilers and Pressure Vessels.* New York: ASME, 1958.

23. Harvey, J. F. *Theory and Design of Modern Pressure Vessels,* 2nd ed. New York: Litton, 1974.

24. Burr, A. H., and J. B. Cheatham. *Mechanical Analysis and Design,* 2nd ed. Upper Saddle River, NJ: Prentice Hall, 1995.

25. Stedfelt, R. L., and C. T. Hoover. "Design and Fabrication Technology for Filament Wound Vessels." *General Motors Engineering Journal,* Vol. 13, no. 4 (1966), pp. 24–31.

26. "Fiberglass-Reinforced Plastic Pressure Vessels," Section X. *ASME Boiler and Pressure Vessel Code.* New York: ASME, 1992.

27. Timoshenko, S. P., and J. M. Gere. *Theory of Elastic Stability.* New York: McGraw-Hill, 1961.

28. Brush, D. O., and B. O. Almroth. *Buckling of Bars, Plates, and Shells.* New York: McGraw-Hill, 1975.

29. "Buckling of Thin-Walled Cylinders," NASA-SP-807. NASA, September Honover, Maryland 1965.

PROBLEMS

Sections 16.1 through 16.7

16.1 A cylinder of inner radius a and the outer radius $3a$ is subjected to an internal pressure p_i. Determine the limiting values of the p_i, applying

(*a*) The maximum shear stress theory.

(*b*) The maximum energy of distortion theory.

Design Decision: The cylinder is made of steel of $S_y = 260$ MPa.

16.2 A solid steel shaft of radius b is pressed into a steel disk of outer radius $2b$ and the length of hub engagement $l = 3b$ (Figure 16.5). Determine the value of the radial interference in terms of b.

Given: The shearing stress in the shaft caused by the torque that the joint is to carry equals 100 MPa; $E = 210$ GPa, $f = 0.15$.

16.3 A cylinder of inner radius a and outer radius $2a$ is under internal pressure p_i. Calculate the allowable value of p_i, using

(*a*) The maximum principal stress theory.

(*b*) The Coulomb-Mohr theory.

Design Decisions: The cylinder is made of aluminum of $S_u = 350$ MPa and $S_{uc} = 650$ MPa.

16.4 A cast iron disk is to be shrunk on a 125-mm diameter steel shaft. Determine

(a) The contact pressure.

(b) The minimum allowable outside diameter of the disk.

Requirement: The tangential stress in the disk is not to exceed 60 MPa.
Given: The radial interference is 0.05 mm, $f = 0.15$, $E_c = 100$ GPa, $E_s = 200$ GPa, and $v = 0.3$.

16.5 A cast iron pinion with 100-mm dedendum diameter and $l = 50$ mm hub engagement length is to transmit a maximum torque of 150 N·m at low speeds. Calculate

(a) The required radial interference on a 25-mm diameter steel shaft.

(b) The maximum stress in the gear due to a press fit.

Given: $E_c = 100$ GPa, $E_s = 200$ GPa, $v = 0.3$, $f = 0.15$

16.6 A bronze bushing 50 mm in outer diameter and 30 mm in inner diameter is to be pressed into a hollow steel cylinder of 100-mm outer diameter. Determine the tangential stresses for the steel and bronze at the boundary between the two parts.

Given: $E_b = 105$ GPa, $E_s = 210$ GPa, $v = 0.3$
Design Requirement: The radial interference is $\delta = 0.025$ mm.

16.7 A cast iron cylinder of outer radius 150 mm is to be shrink fitted over a 50-mm radius steel shaft. Calculate the maximum tangential and radial stresses in both parts.

Given: $E_c = 120$ GPa, $v = 0.25$, $E_s = 210$ GPa, $v = 0.3$
Design Requirement: The radial interference is $\delta = 0.03$ mm.

16.8 When a steel disk of external diameter $4b$ is shrunk onto a steel shaft of diameter of $2b$, the internal diameter of the disk is increased by an amount λ. What reduction occurs in the diameter of the shaft?

Given: $v = 0.3$

16.9 A thick-walled disk flywheel has inner and outer radii of b and $4b$, respectively. Determine

(a) The radius b.

(b) The kinetic energy delivered for a 5% drop in speed.

Given: The maximum speed is 3600 rpm with a maximum stress from rotation equal to 75 MPa and $l = 50$ mm.
Design Decisions: The disk is made of steel having $\rho = 7.8$ kN·s^2/m^4 and $v = 0.3$.

16.10 A 200-mm diameter disk is shrunk onto a 40-mm diameter shaft. Determine

(a) The initial contact pressure required if the contact pressure is to be 3.2 MPa at 2400 rpm.

(b) The maximum stress when not rotating.

Design Decision: Both members are made of steel having $E = 210$ GPa, $v = 0.3$, and $\rho = 7.8$ kN·s^2/m^4.

16.11 A flywheel of 600-mm outer diameter and 100-mm inner diameter is to be press fit on a solid shaft with a radial interference of 0.02 mm. Calculate

(a) The maximum stress in the assembly at standstill.

(b) The speed n in rpm at which the press fit loosens as a result of rotation.

Design Decision: Both members are made of steel with $E = 200$ GPa, $\nu = 0.3$, and $\rho = 7.8$ kN $\cdot$ s^2/m^4.

16.12 A 60-mm thick steel flywheel has inner and outer radii, a and b, respectively. Determine the average braking torque required.

Given: $a = 50$ mm, $b = 200$ mm, $\rho = 7.8$ kN $\cdot$ s^2/m^4
Design Requirement: The flywheel speed must be reduced from 2400 to 1200 rpm in 2 revolutions.

16.13 A rolled steel disk flywheel has inner radius a, outer radius b, and length of hub engagement of l. It rotates on a shaft at a normal speed of 3000 rpm with a 10% drop during working cycle. Determine

(a) The maximum stress.

(b) The energy delivered per cycle.

Given: $a = 25$ mm, $b = 250$ mm, $l = 60$ mm, $\rho = 7.8$ kN $\cdot$ s^2/m^4, $\nu = 0.3$

16.14 A perfectly plastic cylinder ($S_y = 420$ MPa) with 40-mm and 60-mm inner and outer radii is under an internal pressure p. Determine

(a) The allowable pressure at the onset of yielding.

(b) The allowable ultimate pressure.

(c) The residual tangential stress at the outer surface if the cylinder is unloaded from the ultimate pressure.

Design Decisions: Use the maximum energy of distortion criterion with a safety factor of 3.

Section 16.8

16.15 The cross section of the circular cast iron frame of Figure 16.14a has a channel form, as shown in Figure P16.15. The dimensions are in millimeters. Determine the maximum load P.

Given: $R = 300$ mm
Design Decisions: Stress does not exceed 100 MPa on the critical section. Winkler's formula is used.

Section A–B

Figure P16.15

16.16 The allowable stress in compression for the clamp body shown in Figure P16.16 is 120 MPa. Calculate, applying Winkler's formula, the maximum permissible load the member can carry. The dimensions are in millimeters.

Figure P16.16

16.17 Calculate, using Winkler's formula, the maximum distance d for which tangential stress does not exceed 80 MPa on the cross section A-B of the frame shown in Figure P16.17.

Given: $P = 25$ KN

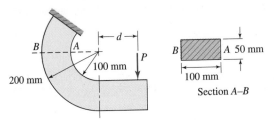

Figure P16.17

16.18 For the crane hook of circular cross section in Figure P16.18, determine

(a) The maximum load P that may be supported without exceeding a stress of 150 MPa at point A.

(b) The tangential stress at point B for the load found in part a.

Design Decision: Use Winkler's formula.

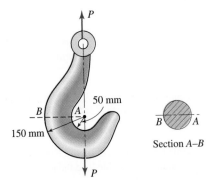

Figure P16.18

Sections 16.9 through 16.15

16.19 Determine maximum deflection and maximum tangential stress for a simply supported circular plate of radius a under uniformly distributed load p_o.

16.20 A circular clamped-edge window of an aircraft is under uniform pressure differential p_o between the cabin and the outside. Calculate the maximum value of the safety factor n, based on the following criteria:

(a) The maximum shear stress.

(b) The maximum energy of distortion.

Given: $v = 1/3$

Design Decisions: The plate is made of a material having yield strength S_y, thickness t, and radius a.

16.21 A pressure vessel control system includes a thin plate that is to close an electrical circuit by deflecting 0.05 in. at the center when the pressure reaches a value of 500 psi. Design the plate.

Given: $E = 30 \times 10^6$ psi, $v = 0.3$, $S_y = 80$ ksi; the plate has a radius of $a = 1.5$ in.
Assumption: The plate is simply supported at the edge.

16.22 An aluminum-alloy flat-clamped disk valve ($S_y = 150$ MPa, $v = 1/3$) of 10 mm thickness is under a liquid pressure of 200 kPa. Design the plate according to the maximum shear stress failure criterion.

Requirement: The factor of safety is 2.

16.23 A football of uniform skin thickness t is subjected to an internal pressure of p (Figure P16.23). Calculate the maximum stress at point A.

Given: $t = 2$ mm, $\qquad p = 100$ kPa

Figure P16.23

16.24 In the toroidal shell of Figure P16.24 under internal pressure p, the maximum membrane forces per unit length are

$$N_\theta = \frac{pa}{2}, \qquad N_\phi = \frac{pa}{b + a\sin\phi}\left(\frac{a}{2}\sin\phi + b\right)$$

Calculate the required minimum thickness t of the vessel.

Given: $p = 2.2$ MPa, $\qquad a = 50$ mm, $\qquad b = 250$ mm, $\qquad S_y = 240$ MPa, $\qquad n = 1.2$

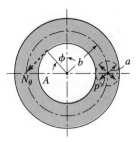

Figure P16.24

16.25 A circular cylindrical pressure vessel of radius a is under an internal pressure p. Determine the required wall thickness t, using a factor of safety of n and the following failure criteria:

(a) The maximum shear.

(b) The maximum energy of distortion.

(c) The Coulomb-Mohr theory.

Design Decision: The cylinder is made of a material with yield strength S_y and ultimate strength in compression equals twice the ultimate strength in tension, $S_{uc} = 2S_u$.

16.26 A closed-ended vertical cylindrical tank of radius r and height h (Figure 16.19c) is completely filled with a liquid of density γ and subjected to an additional gas pressure of p. Determine the wall thickness t required

(a) At the bottom.

(b) At $x = 3h/4$.

(c) At the ends, approximating these as clamped thin circular plates.

Given: $r = 5$ m, $h = 16$ m, $\gamma = 15$ kN/m^3, $p = 200$ kPa, and allowable stress is $\sigma_{\text{all}} = 150$ MPa.

16.27 A spherical tank is filled with a liquid of density γ and supported on a cylindrical pipe, as shown in Figure P16.27. Verify that the membrane force per unit length are given by

Figure P16.27

the expressions:

$$N_\phi = \frac{\gamma a^2}{6 \sin^2\phi}[1 - \cos^2\phi(3 - 2\cos\phi)] = \frac{\gamma a^2}{6}\left(1 - \frac{2\cos^2\phi}{1 + \cos\phi}\right)$$

$$N_\theta = \frac{\gamma a^2}{6}\left(5 - 6\cos\phi + \frac{2\cos^3\phi}{1 + \cos\phi}\right)$$

The preceding equations are valid for $\phi > \phi_0$.

16.28 During a stage of firing, a long cylindrical missile casing of diameter d and thickness t is subjected to axial compression. Calculate

(a) The half-length of the sine waves into which the shell buckles.

(b) The limiting stress for failure.

Given: $d = 1.5$ m, $t = 12$ mm
Design Decisions: The cylinder is made of steel of $S_y = 350$ MPa and $E = 210$ GPa.

16.29 A long pipe of diameter d and thickness t is to be used in a structure as a column. Determine the maximum axial load that can be applied without causing the shell to buckle.

Given: $d = 4$ ft, $t = \frac{1}{2}$ in.
Design Decision: The pipe is made of steel having $E = 30 \times 10^6$ psi.

17

FINITE ELEMENT ANALYSIS IN DESIGN

Outline

17.1 INTRODUCTION

In real design problems, generally structures are composed of a large assemblage of various members. In addition, the built-up structures or machines and their components involve complicated geometries, loadings, and material properties. Given these factors, it becomes apparent that the classical methods can no longer be used. For complex structures, the designer has to resort to more general approaches of analysis. The most widely used of these techniques is the finite element stiffness or displacement method. Unless otherwise specified, we refer to it as the *finite element method* (FEM).

Finite element analysis (FEA) is a numerical approach and well suited to digital computers. The method is based on the formulations of a simultaneous set of algebraic equations relating forces to corresponding displacements at discrete preselected points (called *nodes*) on the structure. These governing algebraic equations, also referred to as force-displacement relations, are expressed in matrix notation [1]. With the advent of high-speed, large-storage capacity digital computers, the finite element method gained great prominence throughout the industries in the solution of practical analysis and design problems of high complexity. The literature related to the FEA is extensive (for example, [2–20]). Numerous commercial FEA software programs are available, including some directed at the learning process. Most of the developments have now been coded into commercial programs.

The basic concept of the finite element approach is that the real structure can be discretized by a finite number of elements, connected not only at their nodes but along the interelement boundaries as well. Usually, triangular or rectangular shapes of elements are used in the finite element method. Figure 17.1 depicts how a real structure is modeled using triangular element shapes. The types of elements commonly employed in structural idealization are the truss, beam, two-dimensional elements, shell and plate bending, and three-dimensional elements. A solid model of an aircraft structure created using beam, plate, and shell elements is shown in Figure 17.2. Note that, in the analysis of large structural systems, such as ships, multistory buildings, or aircraft, the storage capacity of even general-purpose programs can sometimes be overrun. It may be necessary to use the *method of substructures* that divides the original system into smaller units. Once the stiffness of each substructure has been determined, the analysis of the system follows the familiar procedure of matrix methods used in structural mechanics.

 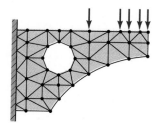

Figure 17.1 Tapered plate bracket and its triangular finite element model.

Figure 17.2 Finite element model of fuselage and a wing [4].

To adequately treat the subject of the FEA would require a far more lengthy presentation than could be justified here. Nevertheless, the subject is so important that any engineer concerned with the analysis and design of members should have at least an understanding of FEA. The fundamentals presented can clearly indicate the potential of the FEA as well as its complexities. It can be covered as an option, used as a "teaser" for a student's advance study of the topic, or as a professional reference. For simplicity, only four basic structural elements are discussed here: the one-dimensional axial element or truss element, the beam element or plane frame element, two-dimensional element, and the axisymmetric element. Sections 17.3, 17.5, and 17.8 present the formulation and general procedure for treating typical problems by the finite element method. Solutions of axial stress, plane stress, and axisymmetrical problems are demonstrated in various examples and case studies.

17.2   STIFFNESS MATRIX FOR AXIAL ELEMENTS

An axial element, also called a *truss bar* or simply *bar element,* can be considered as the simplest form of structural finite element. An element of this type with length L, modulus of elasticity E, and cross-sectional area A, is denoted by e (Figure 17.3). The two ends or joints or nodes are numbered 1 and 2, respectively. It is necessary to develop a set of two equations in matrix form to relate the *joint forces* ($\bar{F}_1$ and $\bar{F}_2$) to the *joint displacements* ($\bar{u}_1$ and $\bar{u}_2$).

DIRECT EQUILIBRIUM METHOD

The following derivation by the direct equilibrium approach is simple and clear. However, this method is practically applicable only for truss and frame elements. The equilibrium of the x-directed forces requires that $\bar{F}_1 = -\bar{F}_2$ (Figure 17.3). Because AE/L is the spring rate of the element, we have

$$\bar{F}_1 = \frac{AE}{L}(\bar{u}_1 - \bar{u}_2), \qquad \bar{F}_2 = \frac{AE}{L}(\bar{u}_2 - \bar{u}_1)$$

This may be written in matrix form

$$\left\{ \begin{matrix} \bar{F}_1 \\ \bar{F}_2 \end{matrix} \right\}_e = \frac{AE}{L} \begin{bmatrix} 1 & -1 \\ -1 & 1 \end{bmatrix} \left\{ \begin{matrix} \bar{u}_1 \\ \bar{u}_2 \end{matrix} \right\}_e \qquad \text{(17.1a)}$$

or symbolically

$$\{\bar{F}\}_e = [\bar{k}]_e \{\bar{u}\}_e \qquad \text{(17.1b)}$$

The quantity $[\bar{k}]_e$ is called the *stiffness matrix* of the element.

Figure 17.3   Axial (truss bar or bar) element.

ENERGY METHOD

The energy technique is more general and powerful than the direct approach just discussed, especially for sophisticated types of finite elements. To employ this method, it is necessary to first define a displacement function for the element (Figure 17.3):

$$\bar{u} = a_1 + a_2\bar{x} \tag{17.2}$$

in which a_1 and a_2 are constants. Clearly, Eq. (17.2) represents a linear displacement variation along the x axis of the element. The axial displacements of joints 1 (at $\bar{x} = 0$) and 2 (at $\bar{x} = L$), respectively, are therefore

$$\bar{u}_1 = a_1, \qquad \bar{u}_2 = a_1 + a_2L$$

Solving the preceding expressions, $a_1 = \bar{u}_1$ and $a_2 = -(\bar{u}_1 - \bar{u}_2)/L$. Carrying these into Eq. (17.2), we have

$$\bar{u} = \left(1 - \frac{\bar{x}}{L}\right)\bar{u}_1 + \frac{\bar{x}}{L}\bar{u}_2 \tag{17.3}$$

Then, by Eq. (3.60), the strain is

$$\varepsilon_x = \frac{d\bar{u}}{d\bar{x}} = \frac{1}{L}(-\bar{u}_1 + \bar{u}_2) \tag{17.4}$$

So, the element axial force:

$$\bar{F} = (E\varepsilon_x)A = \frac{AE}{L}(-\bar{u}_1 + \bar{u}_2) \tag{17.5}$$

The strain energy in the element is obtained by substituting Eq. (17.5) into Eq. (5.10) in the form

$$U = \int_0^L \frac{F^2 dx}{2AE} = \frac{AE}{2L}\left(\bar{u}_1^2 - 2\bar{u}_1\bar{u}_2 + \bar{u}_2^2\right) \tag{17.6}$$

Applying Castigliano's first theorem, Eq. (5.54), we obtain

$$\bar{F}_1 = \frac{\partial U}{\partial \bar{u}_1} = \frac{AE}{L}(\bar{u}_1 - \bar{u}_2)$$

$$\bar{F}_2 = \frac{\partial U}{\partial \bar{u}_2} = \frac{AE}{L}(-\bar{u}_1 + \bar{u}_2)$$

The matrix forms of the preceding equations are the same as those given by Eqs. (17.1).

GLOBAL STIFFNESS MATRIX

We now develop the global stiffness matrix for an element oriented arbitrarily in a two-dimensional plane. The *local coordinates* are chosen to conveniently represent the individual element, whereas the *global* or reference *coordinates* are chosen to be convenient for the whole structure. We designate the local and global coordinates systems for an axial element by $\bar{x}, \bar{y}$ and x, y, respectively (Figure 17.4).

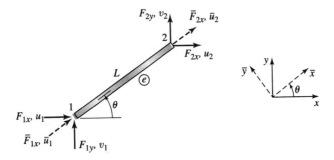

Figure 17.4   Local $(\bar{x}, \bar{y})$ and global (x, y) coordinates for a typical axial element e. All forces and displacements have a positive sense.

The figure depicts a typical axial element e lying along the $\bar{x}$ axis, which is oriented at an angle θ, measured *counterclockwise*, from the reference axis x. In the local coordinate system, each joint has an axial force $\bar{F}_x$, a transverse force $\bar{F}_y$, an axial displacement $\bar{u}$, and a transverse displacement $\bar{v}$. Referring to Figure 17.4, Eq. (17.1a) is expanded as

$$\begin{Bmatrix} \bar{F}_{1x} \\ \bar{F}_{1y} \\ \bar{F}_{2x} \\ \bar{F}_{2y} \end{Bmatrix}_e = \frac{AE}{L} \begin{bmatrix} 1 & 0 & -1 & 0 \\ 0 & 0 & 0 & 0 \\ -1 & 0 & 1 & 0 \\ 0 & 0 & 0 & 0 \end{bmatrix} \begin{Bmatrix} \bar{u}_1 \\ \bar{v}_1 \\ \bar{u}_2 \\ \bar{v}_2 \end{Bmatrix}_e \qquad \textbf{(17.7a)}$$

or

$$[\bar{F}]_e = [\bar{k}]_e \{\bar{\delta}\}_e \qquad \textbf{(17.7b)}$$

Clearly, $\{\bar{\delta}\}_e$ represents the nodal displacements in the local coordinate system.

We see from Figure 17.4 that the two local and global forces at joint 1 may be related by the following expressions:

$$\bar{F}_{1x} = F_{1x} \cos\theta + F_{1y} \sin\theta$$

$$\bar{F}_{1y} = -F_{1x} \sin\theta + F_{1y} \cos\theta$$

Similar expressions apply at joint 2. For brevity we designate

$$c = \cos\theta \quad \text{and} \quad s = \sin\theta$$

Thus, the local and global forces are related in the matrix form

$$\begin{Bmatrix} \bar{F}_{1x} \\ \bar{F}_{1y} \\ \bar{F}_{2x} \\ \bar{F}_{2y} \end{Bmatrix}_e = \begin{bmatrix} c & s & 0 & 0 \\ -s & c & 0 & 0 \\ 0 & 0 & c & s \\ 0 & 0 & -s & c \end{bmatrix} \begin{Bmatrix} F_{1x} \\ F_{1y} \\ F_{2x} \\ F_{2y} \end{Bmatrix}_e \qquad \textbf{(17.8a)}$$

or symbolically

$$\{\bar{F}\}_e = [T]\{F\}_e \qquad \textbf{(17.8b)}$$

In the foregoing, $[T]$ is the *coordinate transformation matrix:*

$$[T] = \begin{bmatrix} c & s & 0 & 0 \\ -s & c & 0 & 0 \\ 0 & 0 & c & s \\ 0 & 0 & -s & c \end{bmatrix} \qquad (17.9)$$

and $\{F\}_e$ represents the *global nodal force matrix:*

$$\{F\}_e = \begin{Bmatrix} F_{1x} \\ F_{1y} \\ F_{2x} \\ F_{2y} \end{Bmatrix}_e \qquad (17.10)$$

Inasmuch as the displacement transforms in the same manner as forces, we have

$$\begin{Bmatrix} \bar{u}_1 \\ \bar{v}_1 \\ \bar{u}_2 \\ \bar{v}_2 \end{Bmatrix}_e = [T] \begin{Bmatrix} u_1 \\ v_1 \\ u_2 \\ v_2 \end{Bmatrix}_e \qquad (17.11a)$$

or

$$\{\bar{\delta}\}_e = \{T\}\{\delta\}_e \qquad (17.11b)$$

Here, $\{\delta\}_e$ is the global nodal displacements. Carrying Eqs. (17.11b) and (17.8b) into (17.7b) leads to

$$[T]\{F\}_e = [\bar{k}]_e [T]\{\delta\}_e$$

or

$$\{F\}_e = [T]^{-1}[\bar{k}]_e [T]\{\delta\}_e$$

Note that the transformation matrix $[T]$ is an orthogonal matrix; that is, its inverse is the same as its transpose: $[T]^{-1} = [T]^T$, where the superscript T denotes the transpose. The global *force-displacement relations* for an element e is

$$\{F\}_e = [k]_e \{\delta\}_e \qquad (17.12)$$

where

$$[k]_e = [T]^T [\bar{k}]_e [T] \qquad (17.13)$$

Finally, to evaluate the *global stiffness matrix* for the element, we substitute Eq. (17.9) and $[k]_e$ from Eq. (17.7a) into Eq. (17.13):

$$[k]_e = \frac{AE}{L} \begin{bmatrix} c^2 & cs & -c^2 & -cs \\ cs & s^2 & -cs & -s^2 \\ -c^2 & -cs & c^2 & cs \\ -cs & -s^2 & cs & s^2 \end{bmatrix} = \frac{AE}{L} \begin{bmatrix} c^2 & cs & -c^2 & -cs \\ & s^2 & -cs & -s^2 \\ & & c^2 & cs \\ \text{Symmetric} & & & s^2 \end{bmatrix} \qquad (17.14)$$

This relationship shows that the element stiffness matrix depends on its dimensions, orientation, and material property.

AXIAL FORCE IN AN ELEMENT

Reconsider the general case of an axial element oriented arbitrarily in a two-dimensional plane, depicted in Figure 17.4. It can be shown that equation for the axial force is expressed in matrix form

$$F_{12} = \frac{AE}{L}[c \quad s]\begin{Bmatrix} u_2 - u_1 \\ v_2 - v_1 \end{Bmatrix}$$

(17.15)

This may be written for an element with nodes ij as follows:

$$F_{ij} = \left(\frac{AE}{L}\right)_{ij}[c \quad s]_{ij}\begin{Bmatrix} u_j - u_i \\ v_j - v_i \end{Bmatrix}$$

(17.16)

A positive (negative) value obtained for F_{ij} indicates that the element is in tension (compression). The axial stress in the element is given by $\sigma_{ij} = F_{ij}/A$.

17.3   FORMULATION OF THE FINITE ELEMENT METHOD AND ITS APPLICATION TO TRUSSES

Development of the governing equations appropriate to a truss demonstrates the formulation of the structural stiffness method, or the finite element method. As noted previously, a truss is an assemblage of axial elements that may be differently oriented. To derive truss equations, the global element relations given by Eq. (17.13) must be assembled. The preceding leads to the following force-displacement relations for the entire truss, the *system equations:*

$$\{F\} = [K]\{\delta\}$$

(17.17)

The *global nodal matrix* $\{F\}$ and the *global stiffness matrix* $[K]$ are

$$\{F\} = \sum_1^n \{F\}_e$$

(17.18a)

$$[K] = \sum_1^n [k]_e$$

(17.18b)

Here e designates an element and n is the number of elements making up the truss. It is noted that $[K]$ relates the global nodal force $\{F\}$ to the global displacement $\{\delta\}$ for the entire truss.

METHOD OF ASSEMBLAGE OF THE VALUES OF $[k]$

The element stiffness matrices in Eq. (17.18) cannot be directly added together or superimposed. To carry out proper summation, a convenient method is to label the columns and rows of each element stiffness matrix according to the displacement components associated with it. In so doing, the truss stiffness matrix $[K]$ is obtained simply by adding terms from the individual element stiffness matrix into their corresponding locations in $[K]$. This approach of assemblage of the element stiffness matrix is given in Case Study 17-1. An alternative way is to expand the $[k]_e$ for each element to the order of the truss stiffness matrix by adding rows and columns of zeros. However, for the problem involving a large number of elements, it becomes tedious to apply this approach.

PROCEDURE FOR SOLVING A PROBLEM

We now illustrate the use of the equations developed in the preceding paragraphs. The general procedure for solving a structural problem by application of the finite-element method may be summarized as shown in Figure 17.5. This outline is better understood when applied to planar structures, as shown in the solution of the following sample problem.

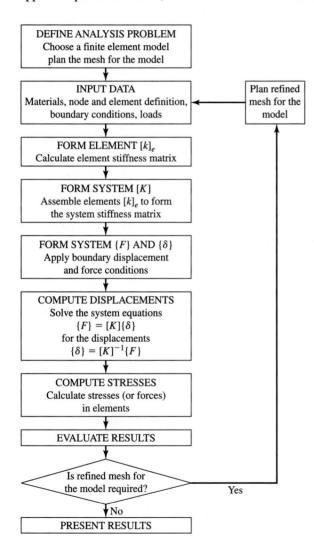

Figure 17.5   Finite element block diagram [13].

Case Study 17-1 | ANALYSIS AND DESIGN OF A TRUSS

A three-bar truss 123 (Figure 17.6a) is subjected to a horizontal force P acting at joint 2. By using the finite element method, analyze the truss and calculate the required cross-sectional area of each member.

Assumptions: All members will have the same length L and axial rigidity AE. Use a factor of safety of $n = 1.5$ on yielding.

Given: $S_y = 240$ MPa, $P = 200$ kN

Solution: The reactions are noted in Figure 17.6a. The node numbering is arbitrary for each element.

Input Data. At each node there are two displacements and two nodal force components (Figure 17.6b). Recall that θ is measured counterclockwise from the positive x axis to each element (Table 17.1). Inasmuch as the terms in $[k]_e$ involve c^2, s^2, and cs, a change in angle from θ to $\theta + \pi$, causing both c and s to change sign, does not affect the signs of the terms in the stiffness matrix. For

example, in the case of member 3, $\theta = 60°$ if measured counterclockwise at node 1 or 240° if measured counterclockwise at node 3. However, by substituting into Eq. (17.14), $[k]_e$ remains unchanged.

Element Stiffness Matrix. Using Eq. (17.14) and Table 17.1, we have for the elements 1, 2, and 3, respectively,

$$[k]_1 = \frac{AE}{L} \begin{array}{cccc} u_1 & v_1 & u_2 & v_2 \\ \begin{bmatrix} 1 & 0 & -1 & 0 \\ 0 & 0 & 0 & 0 \\ -1 & 0 & 1 & 0 \\ 0 & 0 & 0 & 0 \end{bmatrix} & \begin{array}{c} u_1 \\ v_1 \\ u_2 \\ v_2 \end{array} \end{array}$$

$$[k]_2 = \frac{AE}{4L} \begin{array}{cccc} u_2 & v_2 & u_3 & v_3 \\ \begin{bmatrix} 1 & -\sqrt{3} & -1 & \sqrt{3} \\ -\sqrt{3} & 3 & \sqrt{3} & -3 \\ -1 & \sqrt{3} & 1 & -\sqrt{3} \\ \sqrt{3} & -3 & -\sqrt{3} & 3 \end{bmatrix} & \begin{array}{c} u_2 \\ v_2 \\ u_3 \\ v_3 \end{array} \end{array}$$

$$[k]_3 = \frac{AE}{4L} \begin{array}{cccc} u_1 & v_1 & u_3 & v_3 \\ \begin{bmatrix} 1 & \sqrt{3} & -1 & -\sqrt{3} \\ \sqrt{3} & 3 & -\sqrt{3} & -3 \\ -1 & -\sqrt{3} & 1 & \sqrt{3} \\ -\sqrt{3} & -3 & \sqrt{3} & 3 \end{bmatrix} & \begin{array}{c} u_1 \\ v_1 \\ u_3 \\ v_3 \end{array} \end{array}$$

Note that the column and row of each stiffness matrix are labeled according to the nodal displacements associated with them.

Table 17.1 Data for the truss of Figure 17.6

Element	θ	c	s	c^2	cs	s^2
1	0°	1	0	1	0	0
2	120°	$-1/2$	$\sqrt{3}/2$	1/4	$-\sqrt{3}/4$	3/4
3	60°	1/2	$\sqrt{3}/2$	1/4	$\sqrt{3}/4$	3/4

(a)

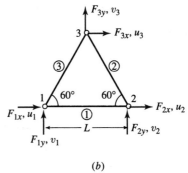

(b)

Figure 17.6 Basic plane truss.

Case Study (CONTINUED)

System Stiffness Matrix. There are a total of six components of displacement for the truss before boundary constraints are imposed. Therefore, the *order* of the truss stiffness matrix must be 6×6. Subsequent to addition of the terms from each element stiffness matrices into their corresponding locations in $[K]$, we readily obtain the global stiffness matrix for the truss:

$$[K] = \frac{AE}{4L} \begin{bmatrix} \overset{u_1}{4+1} & \overset{v_1}{0+\sqrt{3}} & \overset{u_2}{-4} & \overset{v_2}{0} & \overset{u_3}{-1} & \overset{v_3}{-\sqrt{3}} \\ 0+\sqrt{3} & 0+3 & 0 & 0 & -\sqrt{3} & -3 \\ -4 & 0 & 4+1 & 0-\sqrt{3} & -1 & \sqrt{3} \\ 0 & 0 & 0-\sqrt{3} & 0+3 & \sqrt{3} & -3 \\ -1 & -\sqrt{3} & -1 & \sqrt{3} & 1+1 & \sqrt{3}-\sqrt{3} \\ -\sqrt{3} & -3 & \sqrt{3} & -3 & \sqrt{3}-\sqrt{3} & 3+3 \end{bmatrix} \begin{matrix} u_1 \\ v_1 \\ u_2 \\ v_2 \\ u_3 \\ v_3 \end{matrix} \qquad \text{(a)}$$

System Force and Displacement Matrices. Accounting for the applied load and support constraints, with reference to Figure 17.6, the truss nodal force matrix is

$$\{F\} = \begin{Bmatrix} F_{1x} \\ F_{1y} \\ F_{2x} \\ F_{2y} \\ F_{3x} \\ F_{3y} \end{Bmatrix} = \begin{Bmatrix} R_1 \\ 0 \\ P \\ R_2 \\ R_{3x} \\ R_{3y} \end{Bmatrix} \qquad \text{(b)}$$

Similarly, accounting for the support conditions the truss nodal displacement matrix,

$$\{\delta\} = \begin{Bmatrix} u_1 \\ v_1 \\ u_2 \\ v_2 \\ u_3 \\ v_3 \end{Bmatrix} = \begin{Bmatrix} 0 \\ v_1 \\ u_2 \\ 0 \\ 0 \\ 0 \end{Bmatrix} \qquad \text{(c)}$$

Displacements. Substituting Eqs. (a), (b), and (c) into Eq. (17.18), the truss force-displacement relations are given by

$$\begin{Bmatrix} R_1 \\ 0 \\ P \\ R_2 \\ R_{3x} \\ R_{3y} \end{Bmatrix} = \frac{AE}{4L} \begin{bmatrix} 5 & \sqrt{3} & -4 & 0 & -1 & -\sqrt{3} \\ \sqrt{3} & 3 & 0 & 0 & -\sqrt{3} & -3 \\ -4 & 0 & 5 & -\sqrt{3} & -1 & \sqrt{3} \\ 0 & 0 & -\sqrt{3} & 3 & \sqrt{3} & -3 \\ -1 & -\sqrt{3} & -1 & \sqrt{3} & 2 & 0 \\ -\sqrt{3} & -3 & \sqrt{3} & -3 & 0 & 6 \end{bmatrix} \begin{Bmatrix} 0 \\ v_1 \\ u_2 \\ 0 \\ 0 \\ 0 \end{Bmatrix} \qquad \text{(d)}$$

To determine v_1 and u_2, only the part of Eqs. (d) relating to these displacements is considered. We then have

$$\begin{Bmatrix} 0 \\ P \end{Bmatrix} = \frac{AE}{4L} \begin{bmatrix} 3 & 0 \\ 0 & 5 \end{bmatrix} \begin{Bmatrix} v_1 \\ u_2 \end{Bmatrix}$$

Solving preceding equations simultaneously or by matrix inversion, the nodal displacements are obtained:

$$\begin{Bmatrix} v_1 \\ u_2 \end{Bmatrix} = \frac{4L}{15AE} \begin{bmatrix} 5 & 0 \\ 0 & 3 \end{bmatrix} \begin{Bmatrix} 0 \\ P \end{Bmatrix} = \frac{4PL}{5AE} \begin{Bmatrix} 0 \\ 1 \end{Bmatrix} \qquad \text{(e)}$$

(continued)

Case Study   (CONCLUDED)

Reactions. The values of v_1 and u_2 are used to determine reaction forces from Eq. (d) as follows:

$$\begin{Bmatrix} R_1 \\ R_2 \\ R_{3x} \\ R_{3y} \end{Bmatrix} = \frac{AE}{4L} \begin{bmatrix} \sqrt{3} & -4 \\ 0 & -\sqrt{3} \\ -\sqrt{3} & -1 \\ -3 & \sqrt{3} \end{bmatrix} \begin{Bmatrix} v_1 \\ u_2 \end{Bmatrix} = \frac{P}{5} \begin{Bmatrix} -4 \\ -\sqrt{3} \\ -1 \\ \sqrt{3} \end{Bmatrix}$$

The results may be verified by applying the equations of equilibrium to the free-body diagram of the entire truss, Figure 17.6a.

Axial Forces in Elements. Using Eqs. (17.16) and (e) and Table 17.1, we obtain

$$F_{12} = \frac{AE}{L} \begin{bmatrix} 1 & 0 \end{bmatrix} \begin{Bmatrix} \frac{4PL}{5AE} \\ 0 \end{Bmatrix} = \frac{4}{5} P$$

$$F_{23} = \frac{AE}{L} \begin{bmatrix} -\dfrac{1}{2} & \dfrac{\sqrt{3}}{2} \end{bmatrix} \begin{Bmatrix} -\frac{4PL}{5AE} \\ 0 \end{Bmatrix} = \frac{2}{5} P$$

$$F_{13} = \frac{AE}{L} \begin{bmatrix} \dfrac{1}{2} & \dfrac{\sqrt{3}}{2} \end{bmatrix} \begin{Bmatrix} 0 \\ 0 \end{Bmatrix} = 0$$

Stresses in Elements. Dividing the foregoing element forces by the cross-sectional area, we have $\sigma_{12} = 4P/5A$, $\sigma_{23} = 2P/5A$, and $\sigma_{13} = 0$.

Required Cross-Sectional Areas of Elements. The allowable stress is $\sigma_{all} = 240/1.5 = 160$ MPa. We then have $A_1 = 0.8(200 \times 10^3)/160 = 1000$ mm^2, $A_2 = 500$ mm^2, and $A_3 =$ any area.

17.4   BEAM AND FRAME ELEMENTS

Here, we formulate stiffness matrices for flexural or beam elements and axial-flexural or plane frame elements. Consider first an initially straight beam element of constant flexural rigidity EI and length L, as depicted in Figure 17.7. Such an element has a transverse deflection $\bar{v}$ and a slope $\bar{\theta}$ at each end or node. Corresponding to these displacements, a transverse shear force $\bar{F}_y$ and a bending moment $\bar{M}$ act at each node. The deflected configuration of the beam element is shown in Figure 17.8.

The linearly elastic behavior of a beam element is governed according to Eq. (4.16c) as $d^4v/dx^4 = 0$. The right-hand side of this equation is 0 because in the formulation of the stiffness matrix equations, we assume no loading between nodes. In the elements where there is a distributed load, the equivalent nodal load components are used. The solution is taken to be a cubic polynomial function of x,

$$\bar{v} = a_1 + a_2\bar{x} + a_3\bar{x}^2 + a_4\bar{x}^3 \tag{a}$$

Figure 17.7   Beam element; all forces and displacements have a positive sense.

Figure 17.8   Deformed beam element.

The constant values of a are obtained by using the conditions at both ends. The stiffness matrix can again be obtained by the procedure discussed in Section 17.2. It can be verified that [8] the nodal force-displacement relations in the matrix form are

$$\begin{Bmatrix} \bar{F}_{1y} \\ \bar{M}_1 \\ \bar{F}_{2y} \\ \bar{M}_2 \end{Bmatrix}_e = \frac{EI}{L^3} \begin{bmatrix} 12 & 6L & -12 & 6L \\ & 4L^2 & -6L & 2L^2 \\ & & 12 & -6L \\ \text{Symmetric} & & & 4L^2 \end{bmatrix} \begin{Bmatrix} \bar{v}_1 \\ \bar{\theta}_1 \\ \bar{v}_2 \\ \bar{\theta}_2 \end{Bmatrix}_e \qquad \textbf{(17.19a)}$$

or symbolically

$$\{\bar{F}\}_e = [\bar{k}]_e \{\bar{\delta}\}_e \qquad \textbf{(17.19b)}$$

The matrix $\{\bar{F}\}_e$ represents the force and moment components. Equation (17.19b) defines the stiffness matrix $[\bar{k}]_e$ for a beam element lying along a local coordinate axis $\bar{x}$. Having developed the stiffness matrix, formulation and solution of problems involving beam elements proceeds as discussed in Section 17.3.

Determining Displacements and Forces in a Statically Indeterminate Beam | **EXAMPLE 17.1**

A propped cantilevered beam of flexural rigidity EI is subjected to end load P as shown in Figure 17.9a. Using the finite element method, find

(a) The nodal displacements.

(b) The nodal forces and moments.

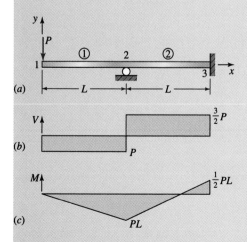

Figure 17.9 Example 17.2: (a) load diagram;
(b) shear diagram; (c) moment diagram.

Solution: We discretize the beam into elements with nodes 1, 2, and 3, as shown in Figure 17.9a. By Eq. (17.19),

$$[k]_1 = \frac{EI}{L^3} \begin{bmatrix} 12 & 6L & -12 & 6L \\ & 4L^2 & -6L & 2L^2 \\ & & 12 & -6L \\ \text{Symmetric} & & & 4L^2 \end{bmatrix} \begin{matrix} v_1 \\ \theta_1 \\ v_2 \\ \theta_2 \end{matrix}$$

with column headings $v_1 \quad \theta_1 \quad v_2 \quad \theta_2$

$$[k]_2 = \frac{EI}{L^3} \begin{bmatrix} 12 & 6L & -12 & 6L \\ & 4L^2 & -6L & 2L^2 \\ & & 12 & -6L \\ \text{Symmetric} & & & 4L^2 \end{bmatrix} \begin{matrix} v_2 \\ \theta_2 \\ v_3 \\ \theta_3 \end{matrix}$$

with column headings $v_2 \quad \theta_2 \quad v_3 \quad \theta_3$

(a) The global stiffness matrix of the beam can now be assembled: $[K] = [k]_1 + [k]_2$. The governing equations for the beam are then

$$\begin{Bmatrix} F_{1y} \\ M_1 \\ F_{2y} \\ M_2 \\ F_{3y} \\ M_3 \end{Bmatrix} = \frac{EI}{L^3} \begin{bmatrix} 12 & 6L & -12 & 6L & 0 & 0 \\ & 4L^2 & -6L & 2L^2 & 0 & 0 \\ & & 24 & 0 & -12 & 6L \\ & & & 8L^2 & -6L & 2L^2 \\ & & & & 12 & -6L \\ \text{Symmetric} & & & & & 4L^2 \end{bmatrix} \begin{Bmatrix} v_1 \\ \theta_1 \\ v_2 \\ \theta_2 \\ v_3 \\ \theta_3 \end{Bmatrix}$$ (17.20a)

or

$$\{F\} = [K]\{\delta\} \qquad (17.20b)$$

The boundary conditions are $v_2 = 0$, $\theta_3 = 0$, and $v_3 = 0$. Partitioning the first, second, and fourth of these equations associated with the unknown displacements:

$$\begin{Bmatrix} -P \\ 0 \\ 0 \end{Bmatrix} = \frac{EI}{L^3} \begin{bmatrix} 12 & 6L & 6L \\ 6L & 4L^2 & 2L^2 \\ 6L & 2L^2 & 8L^2 \end{bmatrix} \begin{Bmatrix} v_1 \\ \theta_1 \\ \theta_2 \end{Bmatrix}$$

Solving for nodal displacements, we obtain

$$v_1 = -\frac{7PL^3}{12EI}, \qquad \theta_1 = \frac{3PL^2}{4EI}, \qquad \theta_2 = \frac{PL^2}{4EI}$$

(b) Introducing these equations into Eq. (17.20a), after multiplying, the nodal forces and moments are found as

$$F_{1y} = -P, \qquad M_1 = 0, \qquad F_{2y} = \frac{5}{2}P,$$

$$M_2 = 0, \qquad F_{3y} = -\frac{3}{2}P, \qquad M_3 = \frac{1}{2}PL$$

Note that M_1 and M_2 are 0, since no reactive moments are present on the beam at nodes 1 and 2.

Comments: In general, it is necessary to determine the local nodal forces and moments associated with each element to analyze the entire structure. For the case under consideration, it may readily be observed from a free-body diagram of element 1 that $(M_2)_1 = -PL$. Hence, we obtain the shear and moment diagrams for the beam as shown in Figures 17.9b and 17.9c, respectively.

ARBITRARILY ORIENTED BEAM ELEMENT

In plane frame structures, the beam elements are no longer horizontal. They can be oriented in a two-dimensional plane as shown in Figure 17.10. So, it is necessary to expand $[k]_e$ to allow for the displacements transforming into u and v displacements in the global system. The moments are unaffected. Referring to the figure, the global force and displacement matrices are, respectively,

$$\{F\}_e = \begin{Bmatrix} F_{1x} \\ F_{1y} \\ M_1 \\ F_{2x} \\ F_{2y} \\ M_2 \end{Bmatrix}_e \tag{17.21a}$$

$$\{\delta\}_e = \begin{Bmatrix} u_1 \\ v_1 \\ \theta_1 \\ u_2 \\ v_2 \\ \theta_2 \end{Bmatrix}_e \tag{17.21b}$$

Following a procedure similar to that described in Section 17.2, the coordinate transformation matrix now becomes

$$[T] = \begin{bmatrix} c & s & 0 & 0 & 0 & 0 \\ -s & c & 0 & 0 & 0 & 0 \\ 0 & 0 & 1 & 0 & 0 & 0 \\ 0 & 0 & 0 & c & s & 0 \\ 0 & 0 & 0 & -s & c & 0 \\ 0 & 0 & 0 & 0 & 0 & 1 \end{bmatrix} \tag{17.22}$$

where, as before, $c = \cos\theta$ and $s = \sin\theta$. Substituting $[T]$ from Eq. (17.22) and $[k]_e$ from Eq. (17.19) into Eq. (17.13), the *global stiffness matrix* is formed:

$$[k]_e = \frac{EI}{L^3} \begin{bmatrix} 12s^2 & -12cs & -6Ls & -12s^2 & 12cs & -6Ls \\ & 12c^2 & 6Lc & 12cs & -12c^2 & 6Lc \\ & & 4L^2 & 6Ls & -6Lc & 2L^2 \\ & & & 12s^2 & -12cs & 6Ls \\ & & & & 12c^2 & -6Lc \\ \text{Symmetric} & & & & & 4L^2 \end{bmatrix} \tag{17.23}$$

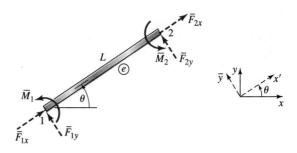

Figure 17.10 Global forces and displacements acting on an arbitrarily oriented beam element.

Figure 17.11 Local forces acting on arbitrarily oriented axial-flexural elements.

ARBITRARILY ORIENTED AXIAL-FLEXURAL BEAM OR FRAME ELEMENT

When a horizontal axial element (Figure 17.3) and a horizontal beam element (Figure 17.7) are combined, we obtain the axial-flexural beam element. In this case, the solution for the axial displacements and the transverse deflections and rotations can be carried out separately and independently. Local nodal forces acting on an axially flexural beam or frame element oriented in the two-dimensional plane with an angle θ with the x axis is shown in Figure 17.11. For this element, the stiffness matrix must undergo the routine coordinate transformation procedure described previously. In so doing, we obtain the *global stiffness matrix* for the element that contains the axial force, shear force, and bending moment effects [7]:

$$[k]_e = \frac{E}{L} \begin{bmatrix} Ac^2 + \frac{12I}{L^2}s^2 & \left(A - \frac{12I}{L^2}\right)cs & -\frac{6I}{L}s & -\left(Ac^2 + \frac{12I}{L^2}s^2\right) & -\left(A - \frac{12I}{L^2}\right)cs & -\frac{6I}{L}s \\ & As^2 + \frac{12I}{L^2}c^2 & \frac{6I}{L}c & -\left(A - \frac{12I}{L^2}\right)cs & -\left(As^2 + \frac{12I}{L^2}c^2\right) & \frac{6I}{L}c \\ & & 4I & \frac{6I}{L}s & -\frac{6I}{L}c & 2I \\ & & & Ac^2 + \frac{12I}{L^2}s^2 & \left(A - \frac{12I}{L^2}\right)cs & \frac{6I}{L}s \\ & & & & As^2 + \frac{12I}{L^2}c^2 & -\frac{6I}{L}c \\ \text{Symmetric} & & & & & 4I \end{bmatrix}$$

(17.24)

The global force and displacements are again given by Eqs. (17.21). Analysis and design of rigid-jointed frameworks can be undertaken by applying Eqs. (17.23) or (17.24). From the latter equation, we observe that the element stiffness matrix of a frame in general is a function of E, A, L, I, and the angle of orientation θ of the element with respect to the global-coordinate axes. With the element stiffness matrix developed, formulation and solution of a frame problem proceeds as discussed in Section 17.3. The following example illustrates the procedure.

Determination of Displacements in a Frame

EXAMPLE 17.2

A planar rectangular frame 1234 is fixed at both supports 1 and 4 (Figure 17.12). The load on the frame consists of a horizontal force P acting at joint 2 and a moment M applied at joint 3. By the finite element analysis, calculate the nodal displacements.

Figure 17.12 Example 17.3. Plane frame.

Table 17.2   Data for the frame of Figure 17.12

Element	1	2	3
θ	90°	0°	270°
c	0	1	0
s	1	0	−1
$12I/L^2$	0.4	0.2	0.4
$6I/L$	12	6	12
E/L	5×10^5	5×10^5	5×10^5

Given: $P = 4$ kips, $M = 2$ kip · in., $L = 5$ ft, $E = 30 \times 10^6$ psi, and $A = 5$ in.^{2} for all elements; $I = 120$ in.^{4} for elements 1 and 3 and $I = 60$ in.^{4} for element 2.

Solution: The global coordinate axes xy are indicated in Figure 17.12. Through the use of Eq. (17.24) and Table 17.2, the element stiffness matrices are obtained as

$$[k]_1 = 5(10^5) \begin{array}{c} \begin{array}{cccccc} u_1 & v_1 & \theta_1 & u_2 & v_2 & \theta_2 \end{array} \\ \begin{bmatrix} 0.4 & 0 & -12 & -0.4 & 0 & -12 \\ 0 & 5 & 0 & 0 & -5 & 0 \\ -12 & 0 & 480 & 12 & 0 & 240 \\ -0.4 & 0 & 12 & 0.4 & 0 & 12 \\ 0 & -5 & 0 & 0 & 5 & 0 \\ -12 & 0 & 240 & 12 & 0 & 480 \end{bmatrix} \begin{array}{c} u_1 \\ v_1 \\ \theta_1 \\ u_2 \\ v_2 \\ \theta_2 \end{array} \end{array}$$

$$[k]_2 = 5(10^5) \begin{array}{c} \begin{array}{cccccc} u_2 & v_2 & \theta_2 & u_3 & v_3 & \theta_3 \end{array} \\ \begin{bmatrix} 5 & 0 & 0 & -5 & 0 & 0 \\ 0 & 0.2 & 6 & 0 & -0.2 & 6 \\ 0 & 6 & 240 & 0 & -6 & 120 \\ -5 & 0 & 0 & 5 & 0 & 0 \\ 0 & -0.2 & -6 & 0 & 0.2 & -6 \\ 0 & 6 & 120 & 0 & -6 & 240 \end{bmatrix} \begin{array}{c} u_2 \\ v_2 \\ \theta_2 \\ u_3 \\ v_3 \\ \theta_3 \end{array} \end{array}$$

$$[k]_3 = 5(10^5) \begin{array}{c} \begin{array}{cccccc} u_3 & v_3 & \theta_3 & u_4 & v_4 & \theta_4 \end{array} \\ \begin{bmatrix} 0.4 & 0 & 12 & -0.4 & 0 & 12 \\ 0 & 5 & 0 & 0 & -5 & 0 \\ 12 & 0 & 480 & -12 & 0 & 240 \\ -0.4 & 0 & -12 & 0.4 & 0 & -12 \\ 0 & -5 & 0 & 0 & 5 & 0 \\ 12 & 0 & 240 & -12 & 0 & 480 \end{bmatrix} \begin{array}{c} u_3 \\ v_3 \\ \theta_3 \\ u_4 \\ v_4 \\ \theta_4 \end{array} \end{array}$$

We superpose the element stiffness matrices and apply the boundary conditions:

$$u_1 = v_1 = \theta_1 = 0, \qquad u_4 = v_4 = \theta_4 = 0$$

at nodes 1 and 4. This leads to the following reduced set of equations:

$$
\begin{Bmatrix} 4000 \\ 0 \\ 0 \\ 0 \\ 0 \\ 2000 \end{Bmatrix} = 5(10^5)
\begin{bmatrix}
5.4 & 0 & 12 & -5 & 0 & 0 \\
 & 5.2 & 6 & 0 & -0.2 & 6 \\
 & & 720 & 0 & -6 & 120 \\
 & & & 5.4 & 0 & 12 \\
 & & & & 5.2 & -6 \\
\text{Symmetric} & & & & & 720
\end{bmatrix}
\begin{Bmatrix} u_2 \\ v_2 \\ \theta_2 \\ u_3 \\ v_3 \\ \theta_3 \end{Bmatrix}
$$

Solving, the nodal deflections and rotations are

$$
\begin{Bmatrix} u_2 \\ v_2 \\ \theta_2 \\ u_3 \\ v_3 \\ \theta_3 \end{Bmatrix} =
\begin{Bmatrix} 18.208 \\ 0.582 \\ -0.271 \\ 17.408 \\ -0.582 \\ -0.248 \end{Bmatrix} (10^{-3})
\quad
\begin{matrix} \text{in.} \\ \text{in.} \\ \text{rad} \\ \text{in.} \\ \text{in.} \\ \text{rad} \end{matrix}
$$

The negative sign indicates a downward displacement or clockwise rotation.

17.5   PROPERTIES OF TWO-DIMENSIONAL ELEMENTS

So far we have dealt with only line elements connected at common nodes, forming trusses and frames. In this section, attention is directed toward the properties of two-dimensional finite elements of an isotropic elastic structure and general formulation of the finite element method for plane structures. To begin with, the plate shown in Figure 17.13a is discretized, as depicted in Figure 17.13b. The finite elements are connected not only at their nodes but along the interelement boundaries. All formulations are based on a counterclockwise labeling of the nodes i, j, and m. The simplest constant strain triangular finite element is used to clearly demonstrate the basic formulative method. The nodal displacements, represented by u and v in the x and y directions, respectively, are the primary unknowns.

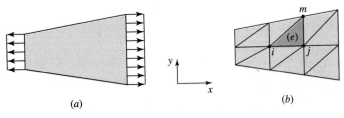

(a) (b)

Figure 17.13   Plate in tension: (a) before and (b) after division into finite elements.

DISPLACEMENT FUNCTIONS

Consider a typical finite element e with nodes i, j, and m (Fig. 17.13b). The nodal displacements are expressed in the convenient matrix form:

$$\{\delta\}_e = \begin{Bmatrix} u_i \\ v_i \\ u_j \\ v_j \\ u_m \\ v_m \end{Bmatrix} \tag{17.25}$$

The displacement functions, describing the displacements at any point within the element, $\{f\}_e$, is represented by

$$\{f\}_e = \begin{Bmatrix} u(x, y) \\ v(x, y) \end{Bmatrix} \tag{17.26a}$$

or

$$\{f\}_e = [N]\{\delta\}_e \tag{17.26b}$$

In the foregoing, the matrix $[N]$ is a function of position, to be obtained in the next section.

STRAIN, STRESS, AND DISPLACEMENT MATRICES

The strain and stress are defined in terms of displacement functions. The strain matrix may be written:

$$\{\varepsilon\}_e = \begin{Bmatrix} \varepsilon_x \\ \varepsilon_y \\ \gamma_{xy} \end{Bmatrix}_e = \begin{Bmatrix} \dfrac{\partial u}{\partial x} \\ \dfrac{\partial v}{\partial y} \\ \dfrac{\partial v}{\partial x} + \dfrac{\partial u}{\partial y} \end{Bmatrix} \tag{17.27a}$$

or

$$\{\varepsilon\}_e = [B]\{\delta\}_e \tag{17.27b}$$

in which $[B]$ is also obtained in the next section.

In a like manner, the stresses throughout the element are, by Hooke's law,

$$\{\sigma\}_e = \frac{E}{1 - v^2} \begin{bmatrix} 1 & v & 0 \\ v & 1 & 0 \\ 0 & 0 & (1 - v)/2 \end{bmatrix} \begin{Bmatrix} \varepsilon_x \\ \varepsilon_y \\ \gamma_{xy} \end{Bmatrix}_e \tag{17.28a}$$

or

$$\{\sigma\}_e = [D]\{\varepsilon\}_e \tag{17.28b}$$

Clearly, the *elasticity matrix* is

$$[D] = \frac{E}{1 - v^2} \begin{bmatrix} 1 & v & 0 \\ v & 1 & 0 \\ 0 & 0 & (1 - v)/2 \end{bmatrix} \tag{17.29a}$$

Table 17.3   Elastic constants for two-dimensional problems

Quantity	Plane strain	Plane stress
λ	$\dfrac{E}{1 - \nu^2}$	$\dfrac{E(1 - \nu)}{(1 + \nu)(1 - 2\nu)}$
D_{12}	ν	$\dfrac{\nu}{1 - \nu}$
D_{33}	$\dfrac{1 - \nu}{2}$	$\dfrac{1 - 2\nu}{2(1 - \nu)}$

In general, we write

$$[D] = \lambda \begin{bmatrix} 1 & D_{12} & 0 \\ D_{12} & 1 & 0 \\ 0 & 0 & D_{33} \end{bmatrix} \qquad \text{(17.29b)}$$

Recall from Section 3.18 that two-dimensional problems are of two classes: plane stress and plane strain. The constants λ, D_{12}, and D_{33} for a plane problem are defined in Table 17.3 [9, 10].

FINITE ELEMENT METHOD FOR TWO-DIMENSIONAL PROBLEMS

Through the use of the principle of minimum potential energy, we can develop the expressions for a *plane stress* and *plane strain* element. For this purpose, the total potential energy Π (see Section 5.8) is expressed in terms of two-dimensional element properties. Then, the minimizing condition, $\partial \Pi / \partial \{\delta\}_e = 0$, results in [10]

$$\{F\}_e = [k]_e \{\delta\}_e \qquad \text{(17.12)}$$

This is of the same form as obtained in Section 17.2 and $\{\delta\}_e$ represents element nodal displacement matrix. However, the element *stiffness matrix* $[k]_e$ and element *nodal force* matrix $\{F\}_e$ are now given by

$$[k]_e = \int_V [B]^T [D][B] \, dV \qquad \text{(17.30)}$$

$$\{F\}_e = \int_s [N]^T \{p\} \, ds \qquad \text{(17.31)}$$

where

p = boundary surface forces per unit area

s = boundary surface over which the forces p act

V = volume of the element

T = transpose of a matrix

We next assemble the element stiffness and nodal force matrices. This gives the following global governing equations for the entire member, the system equations:

$$\{F\} = [K]\{\delta\} \tag{17.17}$$

where

$$\{F\} = \sum_1^n (F)_e \qquad [K] = \sum_1^n [k]_e \tag{17.18}$$

as before. Now, n represents the number of finite elements making up the member. Note that, in the preceding formulations, the finite element stiffness matrix has been derived for a general orientation of global coordinates (x, y). Equation (17.17) is therefore applicable to all elements. Hence, no transformation from local to global equations is necessary. The general procedure for solving a problem by the finite element method is already shown in Figure 17.5. This outline is better understood when applied to a triangular element in the section to follow.

17.6 TRIANGULAR ELEMENT

We now develop the basic *constant strain triangular* (CST) plane stress and strain element. Boundaries of irregularly shaped members can be closely approximated and the expressions related to the triangular elements are simple. The treatment given here is brief. Various types of two-dimensional finite elements yield better solutions. Examples include *linear strain triangular* (LST) elements, triangular elements with additional side and interior nodes, rectangular elements with corner nodes, rectangular elements with additional side modes, and so on [4, 8, 14]. The LST element has six nodes: usual corner nodes and three additional nodes conveniently located at the midpoints of the sides. Hence, the element has 12 unknown displacements. The procedures for development of the equations for the LST element follow the same steps as that of the CST element.

DISPLACEMENT FUNCTION

Consider the triangular finite element i, j, m shown in Figure 17.14. The nodal displacement matrix $\{\delta\}_e$ is given by Eq. (17.25). The displacements u and v throughout the element can be assumed in the following linear form:

$$\{f\}_e = \begin{Bmatrix} u(x, y) \\ v(x, y) \end{Bmatrix} = \begin{Bmatrix} \alpha_1 + \alpha_2 x + \alpha_3 y \\ \alpha_4 + \alpha_5 x + \alpha_6 y \end{Bmatrix} \tag{17.32}$$

where the α represents constants. The foregoing expressions ensure that the compatibility of displacements on the boundaries of adjacent elements are satisfied.

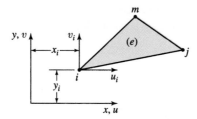

Figure 17.14   Basic triangular element.

The nodal displacements of the element are

$$u_i = \alpha_1 + \alpha_2 x_i + \alpha_3 y_i \qquad v_i = \alpha_4 + \alpha_5 x_i + \alpha_6 y_i$$
$$u_j = \alpha_1 + \alpha_2 x_j + \alpha_3 y_j \qquad v_j = \alpha_4 + \alpha_5 x_j + \alpha_6 y_j$$
$$u_m = \alpha_1 + \alpha_2 x_m + \alpha_3 y_m \qquad v_m = \alpha_4 + \alpha_5 x_m + \alpha_6 y_m$$

Solving these equations gives [7, 9]

$$\begin{Bmatrix} \alpha_1 \\ \alpha_2 \\ \alpha_3 \end{Bmatrix} = \frac{1}{2A} \begin{bmatrix} a_i & a_j & a_m \\ b_i & b_j & b_m \\ c_i & c_j & c_m \end{bmatrix} \begin{Bmatrix} u_i \\ u_j \\ u_m \end{Bmatrix}$$

$$\begin{Bmatrix} \alpha_4 \\ \alpha_5 \\ \alpha_6 \end{Bmatrix} = \frac{1}{2A} \begin{bmatrix} a_i & a_j & a_m \\ b_i & b_j & b_m \\ c_i & c_j & c_m \end{bmatrix} \begin{Bmatrix} v_i \\ v_j \\ v_m \end{Bmatrix}$$

(a)

The quantity A represents the area of the triangle:

$$A = \frac{1}{2}[x_i(y_j - y_m) + x_j(y_m - y_i) + x_m(y_i - y_j)] \tag{17.33}$$

and

$$a_i = x_j y_m - y_j x_m \qquad a_j = y_i x_m - x_i y_m \qquad a_m = x_i y_j - y_i x_j$$
$$b_i = y_j - y_m \qquad b_j = y_m - y_i \qquad b_m = y_i - y_j \tag{17.34}$$
$$c_i = x_m - x_j \qquad c_j = x_i - x_m \qquad c_m = x_j - x_i$$

Substituting Eq. (a) into Eq. (17.32), the displacement function is provided by

$$\{f\}_e = \begin{bmatrix} N_i & 0 & N_j & 0 & N_m & 0 \\ 0 & N_i & 0 & N_j & 0 & N_m \end{bmatrix} \begin{Bmatrix} u_i \\ v_i \\ u_j \\ v_j \\ u_m \\ v_m \end{Bmatrix} = [N]\{\delta\}_e \tag{17.35}$$

in which

$$N_i = \frac{1}{2A}(a_i + b_i x + c_i y)$$

$$N_j = \frac{1}{2A}(a_j + b_j x + c_j y) \qquad \text{(17.36)}$$

$$N_m = \frac{1}{2A}(a_m + b_m x + c_m y)$$

The *strain matrix* is obtained by carrying Eq. (17.35) into Eq. (17.27a):

$$\begin{Bmatrix} \varepsilon_x \\ \varepsilon_y \\ \gamma_{xy} \end{Bmatrix}_e = \frac{1}{2A} \begin{bmatrix} b_i & 0 & b_j & 0 & b_m & 0 \\ 0 & c_i & 0 & c_j & 0 & c_m \\ c_i & b_i & c_j & b_j & c_m & b_m \end{bmatrix} \{\delta\}_e \qquad \text{(17.37)}$$

Introducing Eq. (17.37) into Eq. (17.27b), we have

$$[B] = \frac{1}{2A} \begin{bmatrix} b_i & 0 & b_j & 0 & b_m & 0 \\ 0 & c_i & 0 & c_j & 0 & c_m \\ c_i & b_i & c_j & b_j & c_m & b_m \end{bmatrix} \qquad \text{(17.38a)}$$

or

$$[B] = [B_i][B_j][B_m] \qquad \text{(17.38b)}$$

where

$$[B_i] = \frac{1}{2A} \begin{bmatrix} b_i & 0 \\ 0 & c_i \\ c_i & b_i \end{bmatrix}, \qquad [B_j] = \frac{1}{2A} \begin{bmatrix} b_j & 0 \\ 0 & c_j \\ c_j & b_j \end{bmatrix}, \qquad [B_m] = \frac{1}{2A} \begin{bmatrix} b_m & 0 \\ 0 & c_m \\ c_m & b_m \end{bmatrix} \qquad \text{(17.39)}$$

Clearly matrix $[B]$ depends only on the element nodal coordinates, as seen from Eq. (17.34). Hence, the strain (and stress) is observed to be *constant* and, as already noted, the element of Figure 17.14 is called a constant strain triangle.

THE STIFFNESS MATRIX

For an element of constant thickness t the stiffness matrix can be obtained from Eq. (17.30) as follows:

$$[k]_e = [B]^T [D][B] t A \qquad \text{(17.40)}$$

This equation is assembled together with the elasticity matrix $[D]$ and $[B]$ given by Eqs. (17.29) and (17.38). Expanding the resulting expression, the stiffness matrix is usually written in a partitioned form of order 6×6. We point out that the element stiffness matrix is generally developed in most computer programs by performing the matrix triple products

Table 17.4   Nodal forces F_j and F_m of a CST element of thickness t due to some common loadings [10]

A. Linear load $p(y)$ per unit area

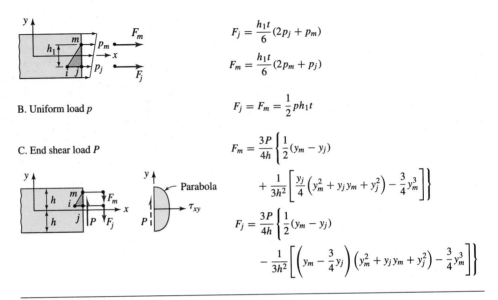

$$F_j = \frac{h_1 t}{6}(2p_j + p_m)$$

$$F_m = \frac{h_1 t}{6}(2p_m + p_j)$$

B. Uniform load p

$$F_j = F_m = \frac{1}{2}ph_1 t$$

C. End shear load P

$$F_m = \frac{3P}{4h}\left\{ \frac{1}{2}(y_m - y_j) \right.$$

$$\left. + \frac{1}{3h^2}\left[\frac{y_j}{4}\left(y_m^2 + y_j y_m + y_j^2 \right) - \frac{3}{4}y_m^3 \right] \right\}$$

$$F_j = \frac{3P}{4h}\left\{ \frac{1}{2}(y_m - y_j) \right.$$

$$\left. - \frac{1}{3h^2}\left[\left(y_m - \frac{3}{4}y_j \right)\left(y_m^2 + y_j y_m + y_j^2 \right) - \frac{3}{4}y_m^3 \right] \right\}$$

shown by Eq. (17.40). The explicit form of the stiffness matrix is rather lengthy and given in the specific publications on the subject, see, for example, [6].

ELEMENT NODAL FORCES DUE TO SURFACE LOADING

The nodal force attributable to applied external loading may be obtained either by evaluating the static resultants or applying Eq. (17.31). Some examples, found by the former approach, are given in Table 17.4. An expanded form of Eq. (17.40), together with those expressions given for the nodal forces, characterizes the CST element. The unknown displacements, strains, and stresses may now be determined applying the general outline given in Figure 17.5. The basic procedure employed in the finite method using CST or any other element is illustrated in the next section.

17.7   PLANE STRESS CASE STUDIES

Here, we present four case studies limited to plane stress situations and CST finite elements. A plate under tension, a deep beam or plate in pure bending, a plate with a hole subjected to an axial loading, and a disk carrying concentrated diametral compression are the members analyzed. There are very few elasticity or "exact" solutions to two-dimensional problems, especially for any but the simplest forms. As will become evident from the following discussion, the designer and stress analyst can reach a very accurate solution by applying proper techniques and modeling. Accuracy is usually limited by the willingness to model all the significant features of the problem and pursue the analysis until convergence is reached.

It should be mentioned that an "exact" solution is unattainable using the finite element method, and we seek instead an acceptable solution. The goal is then the establishment of a finite element that ensures convergence to the exact solution. The literature contains many comparisons among the various elements. The efficiency of a finite element solution can, in certain situations, be enhanced using a "mix" of elements. A denser mesh, for instance, within a region of severely changing or localized stress may save much time and effort.

Case Study 17-2 | ANALYSIS OF STRESSES AND DISPLACEMENTS IN A PLATE IN TENSION

A cantilever plate of depth h, length L, and thickness t supports a uniaxial tension load p as shown in Figure 17.15a. Outline the determination of deflections, strains, and stresses.

Given: $p = 4$ ksi, $E = 30 \times 10^6$ psi, $\nu = 0.3$, $t = \frac{1}{2}$ in., $L = 20$ in., $h = 10$ in.

Assumption: The plate is divided into two CST elements.

Solution: The discretized plate is depicted in Figure 17.15b. The origin of coordinates is placed at node 1, for convenience; however, it may be located at any point in the xy plane. The area of each element is

$$A = \frac{1}{2}hL = \frac{1}{2}(10)(20) = 100 \text{ in.}^2$$

The statically equivalent forces at nodes 2 and 3, $4(10 \times \frac{1}{2})/2 = 10$ kips, are shown in the figure. For plane stress, elasticity matrix $[D]$ is given by Eq. (17.29a).

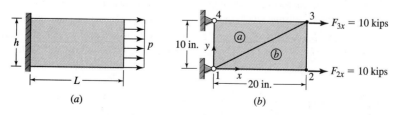

Figure 17.15 Cantilever plate: (a) before and (b) after being discretized.

(continued)

Case Study   (CONCLUDED)

Stiffness Matrix. For element *a*, on assigning $i = 1$, $j = 3$, and $m = 4$, Eq. (17.34) gives

$$b_1 = y_3 - y_4 = 10 - 10 = 0$$

$$b_3 = y_4 - y_1 = 10 - 0 = 10$$

$$b_4 = y_1 - y_3 = 0 - 10 = -10$$

(a)

$$c_1 = x_4 - x_3 = 0 - 20 = -20$$

$$c_3 = x_1 - x_4 = 0 - 0 = 0$$

$$c_4 = x_3 - x_1 = 20 - 0 = 20$$

Substitution of these and the given data into Eq. (17.40), after performing the matrix multiplications, results in stiffness matrix $[k]_a$. Similarly, for element *b*, assignment of $i = 1$, $j = 2$, and $m = 3$ into Eq. (17.34) leads to

$$b_1 = y_2 - y_3 = 0 - 10 = -10$$

$$b_2 = y_3 - y_1 = 10 - 0 = 10$$

$$b_3 = y_1 - y_2 = 0 - 0 = 0$$

$$c_1 = x_3 - x_2 = 20 - 20 = 0$$

$$c_2 = x_1 - x_3 = 0 - 20 = -20$$

$$c_3 = x_2 - x_1 = 20 - 0 = 20$$

and $[k]_b$ is determined. The displacements u_2, v_2, and u_4, v_4 are not involved in elements *a* and *b*, respectively. So, before summing $[k]_a$ and $[k]_b$ to form the system matrix, rows and columns of zeros must be added to each element matrix to account for the absence of these displacements, as mentioned in Section 17.3. Finally, superimposition of the resulting matrices gives the system matrix $[K]$.

Nodal Displacements. The boundary conditions are $u_1 = v_1 = u_4 = v_4 = 0$. The force-displacement relationship of the system is

$$\begin{Bmatrix} R_{1x} \\ R_{1y} \\ 10 \\ 0 \\ 10 \\ 0 \\ R_{4x} \\ R_{4y} \end{Bmatrix} = [K] \begin{Bmatrix} 0 \\ 0 \\ u_2 \\ v_2 \\ u_3 \\ v_3 \\ 0 \\ 0 \end{Bmatrix}$$

Next, to compare the quantities involved, we introduce the results without going through the computation of the $[K]$. It can be verified [8] that the preceding derivations yield

$$\begin{Bmatrix} 10 \\ 0 \\ 10 \\ 0 \end{Bmatrix} = \frac{187.5}{0.91} \begin{bmatrix} 48 & 0 & -28 & 14 \\ 0 & 87 & 12 & -80 \\ -28 & 12 & 48 & -26 \\ 14 & -80 & -26 & 87 \end{bmatrix} \begin{Bmatrix} u_2 \\ v_2 \\ u_3 \\ v_3 \end{Bmatrix}$$

Solving,

$$\begin{Bmatrix} u_2 \\ v_2 \\ u_3 \\ v_3 \end{Bmatrix} = \begin{Bmatrix} 2.4383 \\ 0.0163 \\ 2.6548 \\ 0.4261 \end{Bmatrix} (10^{-3}) \text{ in.}$$

(b)

Stresses. For element *a*, carrying Eqs. (a) and (b) into (17.37), we obtain the strain matrix $\{\varepsilon\}_a$. Equation (17.28), $[D]\{\varepsilon\}_a$, then result in

$$\begin{Bmatrix} \sigma_x \\ \sigma_y \\ \tau_{xy} \end{Bmatrix}_a = \begin{Bmatrix} 4020 \\ 1204 \\ 9.6 \end{Bmatrix} \text{ psi}$$

Element *b* is treated in a like manner.

Comments: Due to constant *x*-directed stress of 4000 psi applied on the edge of the plate, the normal stress is expected to be about 4000 psi in the element *a* (or *b*). The foregoing result for σ_x is therefore quite good. Interestingly, the support of the element *a* at nodes 1 and 4 causes a relatively high stress of $\sigma_y = 1204$ psi. Also note that the value of shear stress τ_{xy} is negligibly small, as anticipated. The effect of element size on solution accuracy will be demonstrated in the next example.

Case Study 17-3 | ANALYSIS OF A DEEP BEAM IN PURE BENDING BY THE ELASTICITY AND FINITE ELEMENT METHODS

A deep beam (or plate) is loaded by couples M, acting about its axis at the ends as shown in Figure 17.16a. Determine the stresses and displacements, using the methods of theory of elasticity and finite element.

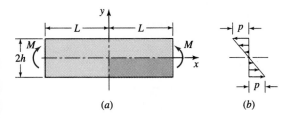

(a)   (b)

Figure 17.16   Beam in pure bending.

Given: $L = 76.2$ mm, $h = 50.8$ mm,   thickness $t = 25.4$ mm, $p = 6895$ kPa, $E = 207$ GPa, $\nu = 0.15$

Assumption: The weight of the member is negligible.

Elasticity or "Exact" Solution

To begin with we replace the end moments with the statically equivalent load per unit area $p = Mh/I$, as depicted in Figure 17.16b. The stresses are then

$$\sigma_x = -\frac{y}{h}p, \qquad \sigma_y = \tau_{xy} = 0 \qquad \textbf{(17.41)}$$

By Eqs. (2.6) and (3.61a),

$$\frac{\partial u}{\partial x} = -\frac{yp}{Eh}, \qquad \frac{\partial v}{\partial y} = \frac{\nu yp}{Eh}, \qquad \frac{\partial v}{\partial x} + \frac{\partial u}{\partial y} = 0$$

It can be shown that, satisfying the conditions $u(0,0) = 0$ and $u(L,0) = 0$, we have

$$u = -\frac{p}{Eh}xy, \qquad v = \frac{p}{2Eh}(x^2 + \nu y^2) \quad \textbf{(17.42)}$$

Introducing the data into Eqs. (17.41) and (17.42) gives

$$\sigma_x = -\frac{1}{0.0508}yp$$

$$u(0.0762, -0.0508) = 2.54 \times 10^{-6} \text{ m}$$

$$\sigma_{x,\max} = 6895 \text{ kPa}$$ (c)

$$v(0.0762, 0) = 1.904 \times 10^{-6} \text{ m}$$

Finite Element Solution

Based on symmetry and antisymmetry, only a quadrant (indictated by the shaded part in the figure) of the beam need be analyzed.

Boundary Conditions.   The discretized quarter-plate is composed of 12 triangular elements (Figure 17.17a). Since no deformation takes place along the x and y axes, the boundary conditions are

$$u_1 = u_2 = u_3 = u_6 = u_9 = u_{12} = 0, \qquad v_3 = 0$$

Nodal Forces.   Applying the equation in Case A of Table 17.4 to Figure 17.16b and inserting data given, we have

$$F_{10x} = \frac{0.0254 \times 0.0254}{6}(2 \times 6895 + 3447.5)$$

$$= 1853.5 \text{ N}$$

$$F_{11x} = \frac{0.0254 \times 0.0254}{6}(2 \times 3447.5 + 6895)$$

$$+ \frac{0.0254 \times 0.0254}{6}(2 \times 3447.5 + 0)$$

$$= 2224 \text{ N}$$

$$F_{12x} = \frac{0.0254 \times 0.0254}{6}(0 + 3447.5) = 370.7 \text{ N}$$

The other nodal forces are 0.

Results [10].   The values of the stresses, obtained by a procedure identical to that of the preceding example, are shown in Figure 17.17b. Note the considerable difference between the exact solution, Eq. (c), and that resulting from the coarse mesh arrangement used. To illustrate the effect of element size and orientation, the results corresponding to different grid configurations are given in Figures 17.17c and 17.17d. The displacements corresponding to Figures 17.17b through 17.17d and the exact solution are listed in Table 17.5.

Comments: Figures 17.17b and 17.17c show the effect of element orientation for the same number of elements

(continued)

Case Study (CONCLUDED)

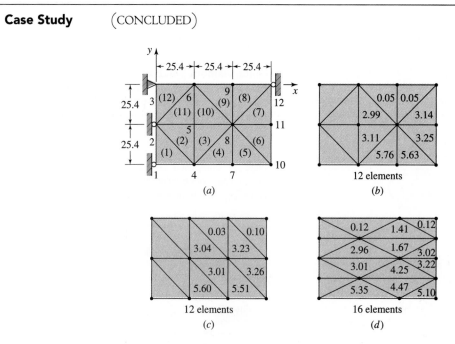

Figure 17.17   Effect of element size and orientation on stress in beam shown in Figure 17.16. Note: Stress (megapascals) obtained for an element is assigned to the centroid.

Table 17.5   Displacements

Case	Number of nodes	Displacement (10^6 m)	
		v_{12}	u_{10}
Figure 17.17b	12	1.547	2.133
Figure 17.17c	12	1.745	2.062
Figure 17.17d	15	1.572	1.976
Exact solution	—	1.905	2.540

and node locations. Observe that elements having large differences between their sides, or so-called weak elements, such as Figure 17.17d, yield less accurate results; even the number of nodes is larger than those of Figures 17.17b and 17.17c. Using *equilateral* or nearly equilateral, well-formed elements of finer mesh gives solutions approaching the exact values.

Case Study 17-4 | ANALYSIS OF STRESS CONCENTRATION IN A PLATE WITH A HOLE SUBJECTED TO UNIAXIAL TENSION

A thin plate containing a small circular hole of radius a is subjected to uniform tensile load of intensity σ_o at its edges, as shown in Figure 17.18a. Apply the finite element analysis to determine the theoretical stress concentration factor.

Case Study   (CONCLUDED)

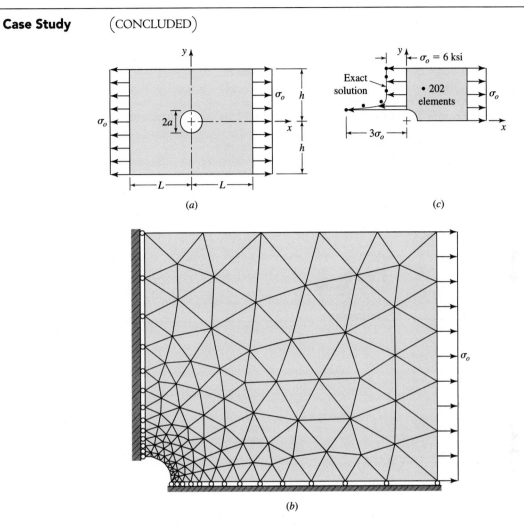

Figure 17.18   (a) Circular hole in a plate under uniaxial tension. (b) One-quadrant plate model. (c) Uniaxial stress (σ_x) distribution.

Given: $L = 24$ in., $a = 2$ in., $h = 20$ in., $\sigma_o = 6$ ksi, $E = 10 \times 10^6$ psi, $v = 0.3$

Solution: Owing to the symmetry, only any one-quarter of the plate need be analyzed, Figure 17.18b. The solution for the case in which the quarter plate discretized to contain 202 CST elements is given in [4]. The roller boundary conditions are also indicated in

the figure. The values of the normal edge stress σ_x, obtained by the finite element method and the theory of elasticity [9], are plotted in Figure 17.18c for comparison. We see from the figure that the agreement is reasonably good. The stress concentration factor for σ_x is $K_t \approx 3\sigma_o/\sigma_o = 3$.

Case Study 17-5 | ANALYSIS OF STRESS DISTRIBUTION IN A CIRCULAR DISK WITH A DIAMETRAL LOAD

A circular disk of constant thickness t is subjected to a pair of compression forces, as shown in Figure 17.19a. Demonstrate the accuracy of the finite element solution using well-formed (i.e., nearly equilateral) CST elements.

Given: The radius $a = 8t$ and $v = 0.3$.

Assumption: The disk is in a state of plane stress.

Solution: Because of symmetry, only a quarter of the disk need be considered (Figure 17.19b). The exact analytical as well as 30 and 109 CST element solutions of the normal stress σ_y are available [11, 12]. A plot of these results is shown in Figure 17.20. Note the considerable difference between the exact solution and the results for 30 elements but the particularly good agreement between the results for 109 elements and the exact solution.

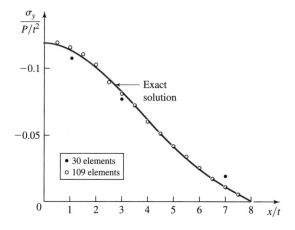

Figure 17.19   (a) Circular thin disk subjected to a diametral load. (b) One-quarter circular disk model with a coarse mesh.

Figure 17.20   Normal stress (σ_y) distribution along the x axis of the circular disk.

17.8   AXISYMMETRIC ELEMENT

This section can provide only a brief introduction to the finite element method for determination of strain and stress in an axisymmetrically loaded member, formed as a solid of revolution with material properties, support conditions, and loading, all of which are symmetrical about the z axis. The finite elements of the body of revolution or rings are used to discretize the axisymmetric member. We use a simple element e of triangular cross section and cylindrical coordinates (r, θ, z), shown in Figure 17.21. Note that a node now is a circle. Therefore, the elemental volume dV appearing in Eq. (17.30) is the volume of the ring element (i.e., $2\pi r\, dr\, dz$). Although the element lies in three-dimensional space, any of its arbitrarily selected vertical cross sections is a *plane triangle*. As observed earlier, no tangential displacement can exist in the axisymmetrical member, $v = 0$. Inasmuch as only the radial displacement u and axial displacement w in a (rz) plane are present, the expressions for displacements for plane strain may readily be extended to the axisymmetric analysis.

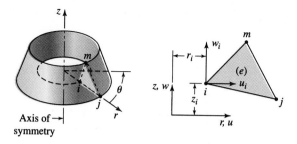

Figure 17.21 Symmetric solid finite element with triangular cross section.

Consider the triangular finite element i, j, m depicted in Figure 17.21. Properties belonging to the element are denoted by e. Referring to Eqs. (17.27) and (16.3), we define the strain-displacement matrix of the form

$$\{\varepsilon\}_e = \begin{Bmatrix} \varepsilon_r \\ \varepsilon_z \\ \varepsilon_\theta \\ \gamma_{rz} \end{Bmatrix} = \begin{Bmatrix} \dfrac{\partial u}{\partial r} \\[2mm] \dfrac{\partial w}{\partial z} \\[2mm] \dfrac{u}{r} \\[2mm] \dfrac{\partial u}{\partial z} + \dfrac{\partial w}{\partial r} \end{Bmatrix}_e \tag{17.43}$$

The nodal displacement matrix for the element is represented by

$$\{\delta\}_e = \begin{Bmatrix} u_i \\ u_j \\ u_m \\ w_i \\ w_j \\ w_m \end{Bmatrix}_e \tag{17.44}$$

The element strain-nodal displacement matrix may be expressed

$$\{\varepsilon\}_e = [B]\{\delta\}_e \tag{17.45}$$

In the preceding, $[B]$ is yet to be determined. The stress-strain relationship is as follows:

$$\{\sigma\}_e = \begin{Bmatrix} \sigma_r \\ \sigma_z \\ \sigma_\theta \\ \tau_{rz} \end{Bmatrix}_e = [D]\{\varepsilon\}_e \tag{17.46}$$

The *elasticity matrix* is given by

$$[D] = \frac{E}{(1+v)(1-2v)} \begin{bmatrix} 1-v & v & v & 0 \\ & 1-v & v & 0 \\ & & 1-v & 0 \\ \text{Symmetric} & & & (1-2v)/2 \end{bmatrix} \tag{17.47}$$

Following a procedure basically identical to that described in Section 17.6, it can be shown that [10]

$$[B] = \frac{1}{2A} \begin{bmatrix} b_i & b_j & b_m & 0 & 0 & 0 \\ 0 & 0 & 0 & c_i & c_j & c_m \\ d_i & d_j & d_m & 0 & 0 & 0 \\ c_i & c_j & c_m & b_i & b_j & b_m \end{bmatrix} \qquad \text{(17.48)}$$

The quantity A represents the area of the triangle defined by Eq. (17.33) and

$$b_i = z_j - z_m \qquad\qquad b_j = z_m - z_i \qquad b_m = z_i - z_j$$

$$c_i = r_m - r_j \qquad\qquad c_j = r_i - r_m \qquad c_m = r_j - r_i \qquad \text{(17.49)}$$

$$d_n = \frac{a_n}{r} + b_n + \frac{c_n z}{r} \qquad (n = i, j, m)$$

Note that the matrix $[B]$ includes the coordinates r and z. Therefore, the strains are *not constant,* as in the case of plane strain and plane stress. The *stiffness matrix* can now be obtained through the use of Eq. (17.30):

$$[k]_e = \int_V [B]^T [D][B] \, dV \qquad \text{(17.50)}$$

This may be rewritten, after integrating along the circumferential or ring boundary, as follows:

$$[k]_e = 2\pi \int r[B]^T [D][B] \, dr \, dz \qquad \text{(17.51)}$$

where the matrices $[D]$ and $[B]$ are defined by Eqs. (17.47) and (17.48), respectively.

Let the radial and axial components of force per unit length be denoted by f_r and f_z, respectively, of the circumferential boundary of a node or a radius r. The total radial and axial *nodal forces* are then

$$F_r = 2\pi r f_r, \qquad F_z = 2\pi r f_z \qquad \text{(17.52)}$$

Equations (17.51) and (17.52) characterize the triangular element for an axisymmetric problem. These are introduced into Eq. (17.18) and subsequently into Eq. (17.17) to evaluate the nodal displacements by satisfying the prescribed boundary conditions. The strains and stresses are then found using Eqs. (17.45) and (17.46). Most of what was learned in studying the two-dimensional problems in Sections 17.5 through 17.7 also applies to axisymmetrical problems.

Case Study 17-6 | STRESSES IN A THICK-WALLED HIGH-PRESSURE STEEL CYLINDER

A thick-walled pipe with the inner and outer radii a and b, respectively, is subjected to external pressure p_o (Figure 17.22a). Sketch the stress distribution, as obtained by the finite element and exact approaches of analysis, across the wall of the cylinder.

Case Study (CONCLUDED)

Section A–B

(a)

(b)

Figure 17.22 (a) Thick-walled cylinder under external pressure and modeling of its section A-B. (b) Distribution of tangential and radial stresses in the cylinder for $b = 2a$.

Given: $b = 2a$, $v = 0.3$

Solution: Only a slice of the cylinder needs be analyzed (Figure 17.22a). A total of 20 triangular elements, with gradually decreasing size toward the loaded region, are used. The boundary conditions for the mesh system may be represented by discrete rollers at the nodes along the top and bottom faces of the slice; that is, at these nodes, the w nodal displacements are to be constrained to vanish.

Using a general-purpose finite element program [4, 14–20], the tangential and radial stresses, σ_θ and σ_r are obtained. The nondimensional results are plotted in Figure 17.22b. Also shown are the exact results given by Eqs. (16.18a) and (16.18b). We see excellent agreement between the solutions determined by the two methods.

REFERENCES

1. Gere, J. M., and W. Weaver Jr. *Matrix Methods for Engineers*. New York: Van Nostrand, 1966.
2. Pao, Y. C. *Elements of Computer-Aided Design and Manufacturing*. New York: Wiley, 1984.
3. Szilard, R. *Theory and Analysis of Plates: Classical and Numerical Methods*. Upper Saddle River, NJ: Prentice Hall, 1974.

4. Yang, T. Y. *Finite Element Structural Analysis.* Upper Saddle River, NJ: Prentice Hall, 1986.

5. Weaver, W., Jr., and P. R. Johnston. *Finite Element for Structural Analysis.* Upper Saddle River, NJ: Prentice Hall, 1984.

6. Gallagher, R. H. *Finite Element Analysis—Fundamentals.* Upper Saddle River, NJ: Prentice Hall, 1975.

7. Martin, H. C., and G. F. Carey. *Introduction to Finite Element Analysis.* New York: McGraw-Hill, 1973.

8. Logan, D. L. *A First Course in the Finite Element Method.* Boston: PWS-Kent, 1986.

9. Knight, E. *The Finite Element Method in Mechanical Design.* Boston: PWS-Kent, 1993.

10. Ugural, A. C., and S. K. Fenster. *Advanced Strength and Applied Elasticity,* 4th ed. Upper Saddle River, NJ: Prentice Hall, 2003.

11. Boresi, A. P., and K. P. Chung. *Elasticity in Engineering,* 2nd ed. New York: Wiley, 2000.

12. Boresi, A. P., and D. M. Sidebottom. *Advanced Mechanics of Materials,* 3rd ed. New York: Wiley, 1985.

13. Ugural, A. C. *Stresses in Plates and Shells,* 2nd ed. New York: McGraw-Hill, 1999.

14. Zienkiewitcz, O. C., and R. I. Taylor. *The Finite Element Method,* 4th ed. vol. 2. *Solid and Fluid Mechanics, Dynamics and Non-Linearity.* London: McGraw-Hill, 1991.

15. Cook, R. D. *Concepts and Applications of Finite Element Analysis,* 2nd ed. New York: Wiley, 1980.

16. Segerlind, L. J. *Applied Finite Element Analysis,* 2nd ed. New York: Wiley, 1984.

17. Bathe, K. I. *Finite Element Procedures in Engineering Analysis.* Upper Saddle River, NJ: Prentice Hall, 1996.

18. Bernadou, M. *Finite Element Methods for Thin Shell Problems.* Chichester, UK: Wiley, 1996.

19. Dunham, R. S., and R. E. Nickell. "Finite Element Analysis of Axisymmetric Solids with Arbitrary Loadings" Report AD 655 253. Springfield, VA: National Technical Information Service, June 1967.

20. Utku, S. "Explicit Expressions for Triangular Torus Element Stiffness Matrix." *Journal of the American Aeronautics and Astronautics* 6, no. 6, (June 1968), pp. 1174–76.

PROBLEMS

Sections 17.1 through 17.3

17.1   A planar truss consisting of five members is supported at joints 1 and 4 as shown in Figure P17.1. Determine the global stiffness matrix for each element.

Assumption: All bars have the same axial rigidity *AE*.

Figure P17.1

17.2 through 17.4 The plane truss is loaded and supported as shown in Figures P17.2 through P17.4. Determine

(a) The global stiffness matrix for each element.

(b) The system matrix and the system force-displacement equations.

Assumption: The axial rigidity AE is the same for each element.

Figure P17.2

Figure P17.3

Figure P17.4

17.5 A vertical load 10 kN acts at joint 2 of the two-bar truss 123 shown in Figure P17.5. Determine

(a) The global stiffness matrix for each member.

(b) The system stiffness matrix.

(c) The nodal displacements.

(d) The reactions.

(e) The axial forces in each member and show the results on a sketch of each member.

Assumption: The axial rigidity $AE = 30$ MN is the same for each bar.

17.6 Redo Problem 17.5 for the structure shown in Figure P17.6, with $A = 1.8$ in.^{2} and $E = 30 \times 10^6$ psi.

17.7 Solve Problem 17.5 for the structure shown in Figure P17.7, with $AE = 10$ MN for each bar.

17.8 Resolve Problem 17.5 for the truss shown in Figure P17.8, with $AE = 125$ MN for each member.

Figure P17.5

Figure P17.6

Figure P17.7

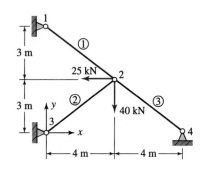

Figure P17.8

17.9 The two-bar plane structure shown in Figure P17.9, due to loading $P = 100$ kN, settles an amount of $u_1 = 25$ mm downward at support 1. Determine

(*a*) The global stiffness matrix for each member.

(*b*) The system matrix.

(*c*) The nodal displacements.

(*d*) The reactions.

(*e*) The axial forces in each member.

Given: $E = 210$ GPa, $A = 5 \times 10^{-4}$ m^2 for each bar.

17.10 A plane truss is loaded and supported as shown in Figure P17.10. Determine

(*a*) The global stiffness matrix for each member.

(*b*) The system stiffness matrix.

(*c*) The nodal displacements.

(*d*) The reactions.

(*e*) The axial forces in each member.

Assumption: The axial rigidity $AE = 20$ MN is the same for each bar.

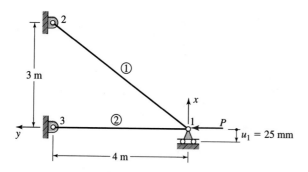

3 m

$u_1 = 25$ mm

4 m

Figure P17.9

80 kN

40 kN

3 m

4 m   4 m

Figure P17.10

Sections 17.4 through 17.8

17.11 A propped cantilever beam of constant flexural rigidity EI with a vertical load of 10 kips at its midspan is shown in Figure P17.11. Determine

(a) The stiffness matrix for each element.

(b) The system stiffness matrix and nodal displacements.

(c) The member end forces and moments.

(d) Sketch the shear and moment diagrams.

Given: $EI = 216 \times 10^6$ lb · in.²

10 kips

5 ft   5 ft

Figure P17.11

17.12 A plane frame 123 with hinged supports at joints 1 and 3 is subjected to a horizontal load of 30 kN (Figure P17.12). Determine

(a) The global stiffness matrix for each member.

(b) The system stiffness matrix.

(c) The displacements u_2, v_2, and θ_2.

Design Assumptions: Members 12 and 23 are identical with a square cross-sectional area of $A = b \times h = 900$ mm² and $E = 70$ GPa.

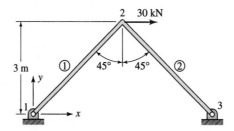

Figure P17.12

17.13 A frame 123 is fixed at supports 1 and 2 as shown in Figure P17.13. A horizontal load of 40 kN acts at joint 2. Determine

(a) The global stiffness matrix of each member.

(b) The system stiffness matrix.

(c) Displacements u_2, v_2, and θ_2.

Given: $E = 200$ GPa, $I_1 = 5\sqrt{2} \times 10^6$ mm^4, $I_2 = 5 \times 10^6$ mm^4, $A_1 = 2.5\sqrt{2} \times 10^3$ mm^2, $A_2 = 2.5 \times 10^3$ mm^2.

Figure P17.13

17.14 Redo Case Study 17-2, by discretizing the plate into four elements, as shown in Figure P17.14. Use a computer program with CST (or LST) elements.

17.15 Verify the results introduced in Case Study 17-5, using a computer program with CST (or LST) elements.

17.16 A steel plate with a hole is under a uniform axial tension loading P (Figure P17.16). The dimensions are in millimeters.

(a) Analyze the stresses using a computer program with the CST (or LST) elements.

(b) Compare the stress concentration factor K_t obtained in part a with that found from Figure C5.

Given: $P = 4$ kN and plate thickness $t = 10$ mm.

Figure P17.14

Figure P17.16

17.17 Redo Problem 17.16 for the plate shown in Figure P17.17.

Figure P17.17

17.18 A simply supported beam is subjected to a uniform loading of intensity w (Figure P17.18). Take $L = 10\,h$, $t = 1$, and $v = 0.3$. Refine the meshes to calculate the stress and deflection within 5% accuracy, by using a computer program with the CST (or LST) elements.

Given: The "exact" solution [9] is

$$\sigma_{x,\max} = \frac{3wL^2}{4th^2}, \qquad v_{\max} = -\frac{5wL^4}{16Eth^3} - \frac{3wL^2(1+v)}{5Eht}$$

in which t represents the thickness.

Figure P17.18

17.19 Redo Problem 17.18, for the case in which a cantilevered beam is under a uniform loading of intensity w (Figure P17.19).

Given: The "exact" solution [9] has the form

$$\sigma_{x,\max} = \frac{3wL^2}{4th^2}, \qquad v_{\max} = -\frac{3wL^4}{16Eth^3} - \frac{3wL^2(1+v)}{5Eht}$$

where t is the thickness.

Figure P17.19

17.20 Verify the results introduced in Case Study 17-6, using a finite element computer program with CST (or LST) elements.

17.21 A cast iron cylinder ($v = 0.2$) of a hydraulic device having inner radius a and outer radius $4a$ is subjected to an external pressure p_o. Using a finite element computer program with CST (or LST) elements, determine the distribution of tangential and radial stresses.

17.22 Redo Problem 17.21, if the cylinder is under an internal pressure p_i. Compare the results with the exact solution shown in Figure 16.3.

17.23 Resolve Problem 17.21 for the case in which the cylinder is subjected to internal pressure p_i and external pressure $p_o = 0.5p_i$.

Appendix A

UNITS, PROPERTIES OF SHAPES, AND BEAM DEFLECTIONS

Knowledge of the units of typical quantities and the characteristics of common areas and masses is essential in mechanical analysis and design. Quantities given in SI units can be converted to U.S. customary units by multiplying with the conversion factors furnished in Table A.1. To reverse the process, the number in customary units is divided by the factor. Prefixes can be attached to SI units to form multiples and submultiples (see Table A.2). Properties of most standard shapes encountered in practice are given in various handbooks. Tables A.3 through A.5 present several typical cases. Data for Tables A.4 and A.6 through A.8 were compiled from the listings found in the *AISC* Manual of Steel Construction (Chicago, American Institute of Steel Construction, 1989).

Representative expressions for deflection and slope for selected beams are given in Tables A.9 and A.10. Restrictions on the application of these equations include constancy of the flexural rigidity EI, symmetry of the cross section about the vertical y axis, and the magnitude of displacement v of the beam. In addition, equations apply to beams long in proportion to their depth and not disproportionally wide. Displacements are restricted to linearly elastic region, as shown by the presence of the elastic modulus E in the formulas.

Table A.1   Conversion factors: SI units to U.S. customary units

Quantity	SI unit	U.S. equivalent
Acceleration	m/s^2 (meter per square second)	3.2808 ft/s^2
Area	m^2 (square meter)	10.76 ft^2
Force	N (newton)	0.2248 lb
Intensity of force	N/m (newton per meter)	0.0685 lb/ft
Length	m (meter)	3.2808 ft
Mass	kg (kilogram)	2.2051 lb
Moment of a force	$N \cdot m$ (newton meter)	0.7376 lb $\cdot$ ft
Moment of inertia		
of a plane area	m^4 (meter to fourth power)	2.4025×10^6 in.^{4}
of a mass	$kg \cdot m^2$ (kilogram meter squared)	0.7376 ft $\cdot$ s^2
Power	W (watt)	0.7376 ft $\cdot$ lb/s
	kW (kilowatt)	1.3410 hp
Pressure or stress	Pa (pascal)	0.145×10^{-3} psi
Specific weight	kN/m^3 (kilonewton per cubic meter)	3.684×10^{-3} lb/in.^{3}
Velocity	m/s (meter per second)	3.2808 ft/s
Volume	m^3 (cubic meter)	35.3147 ft^3
Work or energy	J (joule, newton meter)	0.7376 ft $\cdot$ lb

Table A.2   SI prefixes

Prefix	Symbol	Factor	
tera	T	$10^{12} =$	1 000 000 000 000
giga	G	$10^9 =$	1 000 000 000
mega	M	$10^6 =$	1 000 000
kilo	k	$10^3 =$	1 000
hecto	h	$10^2 =$	100
deka	da	$10^1 =$	10
deci	d	$10^{-1} =$	0.1
centi	c	$10^{-2} =$	0.01
milli	m	$10^{-3} =$	0.001
micro	μ	$10^{-6} =$	0.000 001
nano	n	$10^{-9} =$	0.000 000 001
pico	p	$10^{-12} =$	0.000 000 000 001

Note: The use of the prefixes hecto, deka, and centi is not recommended. However, they are sometimes encountered in practice.

Table A.3   Properties of areas

1. Rectangle	**2. Circle**
$A = bh$ $$I_x = \frac{bh^3}{12}$$ $$J_c = \frac{bh(b^2 + h^2)}{12}$$	$A = \pi r^2$ $$I_x = \frac{\pi r^4}{4}$$ $$J_c = \frac{\pi r^4}{2}$$
3. Right triangle	**4. Semicircle**
$$A = \frac{bh}{2}$$ $$I_x = \frac{bh^3}{36} \qquad I_{xy} = -\frac{b^2 h^2}{72}$$ $$J_c = \frac{bh(b^2 + h^2)}{36}$$	$$A = \frac{\pi r^2}{2}$$ $$I_x = 0.110 r^2$$ $$I_y = \frac{\pi r^4}{8}$$
5. Ellipse	**6. Thin tube**
$A = \pi ab$ $$I_x = \frac{\pi ab^3}{4}$$ $$J_c = \frac{\pi ab(a^2 + b^2)}{4}$$	$A = 2\pi rt$ $$I_x = \pi r^3 t$$ $$J_c = 2\pi r^3 t$$
7. Isosceles triangle	**8. Half of thin tube**
$$A = \frac{bh}{2}$$ $$I_x = \frac{bh^3}{36} \qquad I_y = \frac{hb^3}{48}$$ $$J_c = \frac{bh}{144}(4h^2 + 3b^2)$$	$A = \pi rt$ $$I_x \approx 0.095 \pi r^3 t$$ $$I_y = 0.5 \pi r^3 t$$
9. Triangle 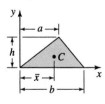	**10. Parabolic spandrel** ($y = kx^2$) 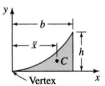
$$A = \frac{bh}{2}$$ $$\bar{x} = \frac{(a + b)}{3}$$	$$A = \frac{bh}{3}$$ $$\bar{x} = \frac{3b}{4}$$
11. Parabola ($y = kx^2$)	**12. General spandrel** ($y = kx^n$)
$$A = \frac{2bh}{3}$$ $$\bar{x} = \frac{3b}{8}$$	$$A = \frac{bh}{n + 1}$$ $$\bar{x} = \frac{n + 1}{n + 2} b$$

Notes: A = area, I = moment of inertia, J = polar moment of inertia.

Table A.4   Properties of some steel pipe and tubing

Standard weight pipe dimensions and properties

Nominal diameter (in.)	Outside diameter (in.)	Inside diameter (in.)	Wall thickness (in.)	Weight per foot (lb/ft) plain ends	A (in.^{2})	I (in.^{4})	S (in.^{3})	r (in.)
	Dimensions					Properties		
$\frac{1}{2}$	.840	.622	.109	.85	.250	.017	.041	.261
$\frac{3}{4}$	1.050	.824	.113	1.13	.333	.037	.071	.334
1	1.315	1.049	.133	1.68	.494	.087	.133	.421
$1\frac{1}{4}$	1.660	1.380	.140	2.27	.669	.195	.235	.540
$1\frac{1}{2}$	1.900	1.610	.145	2.72	.799	.310	.326	.623
2	2.375	2.067	.154	3.65	1.07	.666	.561	.787
$2\frac{1}{2}$	2.875	2.469	.203	5.79	1.70	1.53	1.06	.947
3	3.500	3.068	.216	7.58	2.23	3.02	1.72	1.16
4	4.500	4.026	.237	10.79	3.17	7.23	3.21	1.51

Square and rectangular structural tubing dimensions and properties

Nominal* size (in.)	Wall thickness (in.)	Weight per foot (lb/ft)	A (in.^{2})	I_x (in.^{4})	S_x (in.^{3})	r_x (in.)	I_y (in.^{4})	S_y (in.^{3})	r_y (in.)
	Dimensions				Properties**				
2 × 2	$\frac{3}{16}$	4.32	1.27	0.668	0.668	0.726			
	$\frac{1}{4}$	5.41	1.59	0.766	0.766	0.694			
2.5 × 2.5	$\frac{3}{16}$	5.59	1.64	1.42	1.14	0.930			
	$\frac{1}{4}$	7.11	2.09	1.69	1.35	0.899			
3 × 2	$\frac{3}{16}$	5.59	1.64	1.86	1.24	1.06	0.977	0.977	0.771
	$\frac{1}{4}$	7.11	2.09	2.21	1.47	1.03	1.15	1.15	0.742
3 × 3	$\frac{3}{16}$	6.87	2.02	2.60	1.73	1.13			
	$\frac{1}{4}$	8.81	2.59	3.16	2.10	1.10			
4 × 2	$\frac{3}{16}$	6.87	2.02	3.87	1.93	1.38	1.29	1.29	0.798
	$\frac{1}{4}$	8.81	2.59	4.69	2.35	1.35	1.54	1.54	0.770
4 × 4	$\frac{3}{16}$	9.42	2.77	6.59	3.30	1.54			
	$\frac{1}{4}$	12.21	3.59	8.22	4.11	1.51			

*Outside dimensions across flat sides ($h \times b$).

**Properties are based on a nominal outside corner radius equal to two times the wall thickness (t).

Notes: A = area, S = section modulus, I = moment of inertia, r = radius of gyration.

Table A.5   Properties of solids

---

1. Slender rod

$$m = \frac{\pi d^2 L \rho}{4}$$

$$I_y = I_z = \frac{mL^2}{12}$$

---

2. Thin disk

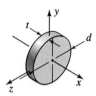

$$m = \frac{\pi d^2 t \rho}{4}$$

$$I_x = \frac{md^2}{8}$$

$$I_y = I_z = \frac{md^2}{16}$$

---

3. Rectangular prism

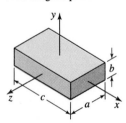

$$m = abc\rho$$

$$I_x = \frac{m}{12}(a^2 + b^2)$$

$$I_y = \frac{m}{12}(a^2 + c^2)$$

$$I_z = \frac{m}{12}(b^2 + c^2)$$

---

4. Cylinder

$$m = \frac{\pi d^2 L \rho}{4}$$

$$I_x = \frac{md^2}{8}$$

$$I_y = I_z = \frac{m}{48}(3d^2 + 4L^2)$$

---

5. Hollow cylinder

$$m = \frac{\pi L \rho}{4}\left(d_o^2 - d_i^2\right)$$

$$I_x = \frac{m}{8}\left(d_o^2 + d_i^2\right)$$

$$I_y = I_z = \frac{m}{48}\left(3d_o^2 + 3d_i^2 + 4L^2\right)$$

| Notes: ρ = mass density, m = mass, I = mass moment of inertia.

Table A.6   Properties of rolled-steel (*W*) shapes, wide-flange sections

| | | | Flange | | Web | Axis *x-x* | | | Axis *y-y* | |
| | | | | | | | | | | |
Designation*	Area, 10^3 mm^2	Depth, mm	Width, mm	Thickness, mm	thickness, mm	I, 10^6 mm^4	r, mm	S, 10^3 mm^3	I, 10^6 mm^4	r, mm
SI Units										
W 610 × 155	19.7	611	324	19.0	12.7	1290	256	4220	108	73.9
× 125	15.9	612	229	19.6	11.9	985	249	3220	39.3	49.7
W 460 × 158	20.1	476	284	23.9	15.0	795	199	3340	91.6	67.6
× 74	9.48	457	190	14.5	9.0	333	188	1457	16.7	41.9
× 52	6.65	450	152	10.8	7.6	212	179	942	6.4	31.0
W 410 × 114	14.6	420	261	19.3	11.6	462	178	2200	57.4	62.7
× 85	10.8	417	181	18.2	10.9	316	171	1516	17.9	40.6
× 60	7.61	407	178	12.8	7.7	216	168	1061	12	39.9
W 360 × 216	27.5	375	394	27.7	17.3	712	161	3800	282	101.1
× 122	15.5	363	257	21.7	13.0	367	154	2020	61.6	63.0
× 79	10.1	354	205	16.8	9.4	225	150	1271	24.0	48.8
W 310 × 107	13.6	311	306	17.0	10.9	248	135	1595	81.2	77.2
× 74	9.48	310	205	16.3	9.4	164	132	1058	23.4	49.8
× 52	6.65	317	167	13.2	7.6	119	133	748	10.2	39.1
W 250 × 80	10.2	256	255	15.6	9.4	126	111	985	42.8	65
× 67	8.58	257	204	15.7	8.9	103	110	803	22.2	51.1
× 49	6.26	247	202	11.0	7.4	70.8	106	573	15.2	49.3
W 200 × 71	9.11	216	206	17.4	10.2	76.6	91.7	709	25.3	52.8
× 59	7.55	210	205	14.2	9.1	60.8	89.7	579	20.4	51.8
× 52	6.65	206	204	12.6	7.9	52.9	89.2	514	17.7	51.6
W 150 × 37	4.74	162	154	11.6	8.1	22.2	69	274	7.12	38.6
× 30	3.79	157	153	9.3	6.6	17.2	67.6	219	5.54	38.1
× 24	3.06	160	102	10.3	6.6	13.4	66	167	1.84	24.6
× 18	2.29	153	102	7.1	5.8	9.2	63.2	120	1.25	23.3

*A wide-flange shape is designated by the letter *W* followed by the nominal depth in millimeters and the mass in kilogram per meter.
Notes: *I* = moment of inertia,   *S* = section modulus,   *r* = radius of gyration.

Table A.6 Properties of rolled-steel (W) shapes, wide-flange sections

			Flange		Web	Axis x-x			Axis y-y	
Designation*	Area, in.^{2}	Depth, in.	Width, in.	Thickness, in.	thickness, in.	I, in.^{4}	r, in.	S, in.^{3}	I, in.^{4}	r, in.
W 24 × 104	30.6	24.06	12.750	0.750	0.500	3100	10.1	258	259	2.91
× 84	24.7	24.10	9.020	0.770	0.470	2370	9.79	196	94.4	1.95
W 18 × 106	31.1	18.73	11.200	0.940	0.590	1910	7.84	204	220	2.66
× 50	14.7	17.99	7.495	0.570	0.355	800	7.38	88.9	40.1	1.65
× 35	10.3	17.70	6.000	0.425	0.300	510	7.04	57.6	15.3	1.22
W 16 × 77	22.6	16.52	10.295	0.760	0.455	1110	7.00	134	138	2.47
× 57	16.8	16.43	7.120	0.715	0.430	758	6.72	92.2	43.1	1.60
× 40	11.8	16.01	6.995	0.505	0.305	518	6.63	64.7	28.9	1.57
W 14 × 145	42.7	14.78	15.500	1.090	0.680	1710	6.33	232	677	3.98
× 82	24.1	14.31	10.130	0.855	0.510	882	6.05	123	148	2.48
× 53	15.6	13.92	8.060	0.660	0.370	541	5.89	77.8	57.7	1.92
W 12 × 72	21.1	12.25	12.040	0.670	0.430	597	5.31	97.4	195	3.04
× 50	14.7	12.19	8.080	0.640	0.370	394	5.18	64.7	56.3	1.96
× 35	10.3	12.50	6.560	0.520	0.300	285	5.25	45.6	24.5	1.54
W 10 × 54	15.8	10.09	10.030	0.615	0.370	303	4.37	60.0	103	2.56
× 45	13.3	10.10	8.020	0.620	0.350	248	4.33	49.1	53.4	2.01
× 33	9.71	9.73	7.960	0.435	0.290	170	4.19	35.0	36.6	1.94
W 8 × 48	14.1	8.50	8.110	0.685	0.400	184	3.61	43.3	60.9	2.08
× 40	11.7	8.25	8.070	0.560	0.360	146	3.53	35.5	49.1	2.04
× 35	10.3	8.12	8.020	0.495	0.310	127	3.51	31.2	42.6	2.03
W 6 × 25	7.34	6.38	6.080	0.455	0.320	53.4	2.70	16.7	17.1	1.52
× 20	5.88	6.20	6.020	0.365	0.260	41.4	2.66	13.4	13.3	1.50
× 16	4.74	6.28	4.030	0.405	0.260	32.1	2.60	10.2	4.43	0.967
× 12	3.55	6.03	4.000	0.280	0.230	22.1	2.49	7.31	2.99	0.918

U.S. Customary Units

SOURCE: The American Institute of Steel Construction, Chicago.

*A wide-flange shape is designated by the letter W followed by the nominal depth in inches and the weight in pounds per foot.

Table A.7   Properties of rolled-steel (S) shapes, American standard I beams

			SI Units							
			Flange		**Web**	**Axis x-x**			**Axis y-y**	
Designation*	Area, 10^3 mm^2	Depth, mm	Width, mm	Thickness, mm	thickness, mm	I, 10^6 mm^4	r, mm	S, 10^3 mm^3	I, 10^6 mm^4	r, mm
S 610 × 149	19.0	610	184	22.1	19.0	995	229	3260	19.9	32.3
× 119	15.2	610	178	22.1	12.7	878	241	2880	17.6	34.0
S 510 × 141	18.0	508	183	23.3	20.3	670	193	2640	20.7	33.8
× 112	14.3	508	162	20.1	16.3	533	193	2100	12.3	29.5
S 460 × 104	13.3	457	159	17.6	18.1	385	170	1685	10.0	27.4
× 81	10.4	457	152	17.6	11.7	335	180	1466	8.66	29.0
S 380 × 74	9.5	381	143	15.8	14.0	202	146	1060	6.53	26.2
× 64	8.13	381	140	15.8	10.4	186	151	977	5.99	27.2
S 310 × 74	9.48	305	139	16.8	17.4	127	116	833	6.53	26.2
× 52	6.64	305	129	13.8	10.9	95.3	120	625	4.11	24.9
S 250 × 52	6.64	254	126	12.5	15.1	61.2	96	482	3.48	22.9
× 38	4.81	254	118	12.5	7.9	51.6	103	406	2.83	24.2
S 200 × 34	4.37	203	106	10.8	11.2	27	78.7	266	1.79	20.3
× 27	3.5	203	102	10.8	6.9	24	82.8	236	1.55	21.1
S 150 × 26	3.27	152	90	9.1	11.8	11.0	57.9	144	0.96	17.2
× 19	2.36	152	84	9.1	5.8	9.20	62.2	121	0.76	17.9
S 100 × 14	1.80	102	70	7.4	8.3	2.83	39.6	55.5	0.38	14.5
× 11	1.45	102	67	7.4	4.8	2.53	41.6	49.6	0.32	14.8

| *An American standard beam is designated by the letter S followed by the nominal depth in millimeters and the mass in kilograms per meter.

Table A.7 Properties of rolled-steel (S) shapes, American standard I beams

			Flange		**Web**	**Axis x-x**			**Axis y-y**	
Designation*	**Area, in.²**	**Depth, in.**	**Width, in.**	**Thickness, in.**	**thickness, in.**	**I, in.⁴**	**r, in.**	**S, in.³**	**I, in.⁴**	**r, in.**
S 24 × 100	29.4	24.00	7.247	0.871	0.747	2390	9.01	199	47.8	1.27
× 79.9	23.5	24.00	7.001	0.871	0.501	2110	9.47	175	42.3	1.34
S 20 × 95	27.9	20.00	7.200	0.916	0.800	1610	7.60	161	49.7	1.33
× 75	22.1	20.00	6.391	0.789	0.641	1280	7.60	128	29.6	1.16
S 18 × 70	20.6	18.00	6.251	0.691	0.711	926	6.71	103	24.1	1.08
× 54.7	16.1	18.00	6.001	0.691	0.461	804	7.07	89.4	20.8	1.14
S 15 × 50	14.7	15.00	5.640	0.622	0.550	486	5.75	64.8	15.7	1.03
× 42.9	12.6	15.00	5.501	0.622	0.411	447	5.95	59.6	14.4	1.07
S 12 × 50	14.7	12.00	5.477	0.659	0.687	305	4.55	50.8	15.7	1.03
× 35	10.3	12.00	5.078	0.544	0.428	229	4.72	38.2	9.87	0.980
S 10 × 35	10.3	10.00	4.944	0.491	0.594	147	3.78	29.4	8.36	0.901
× 25.4	7.46	10.00	4.661	0.491	0.311	124	4.07	24.7	6.79	0.954
S 8 × 23	6.77	8.00	4.171	0.425	0.441	64.9	3.10	16.2	4.31	0.798
× 18.4	5.41	8.00	4.001	0.425	0.271	57.6	3.26	14.4	3.73	0.831
S 6 × 17.25	5.07	6.00	3.565	0.359	0.465	26.3	2.28	8.77	2.31	0.675
× 12.5	3.67	6.00	3.332	0.359	0.232	22.1	2.45	7.37	1.82	0.705
S 4 × 9.5	2.79	4.00	2.796	0.293	0.326	6.79	1.56	3.39	0.903	0.569
× 7.7	2.26	4.00	2.663	0.293	0.193	6.08	1.64	3.04	0.764	0.581

SOURCE: The American Institute of Steel Construction, Chicago.

*An American standard beam is designated by the letter S followed by the nominal depth in inches and the weight in pounds per foot.

Table A.8   Properties of rolled-steel (L) shapes, angles with equal legs

Size and thickness, mm	Mass, kg/m	Area, 10^3 mm^2	Axis x-x or y-y				Axis z-z, r, mm
			I, 10^6 mm^4	r, mm	S, 10^3 mm^3	$\bar{x}$ or $\bar{y}$, mm	r, mm
L 203 × 203 × 25.4	75.9	9.68	37	61.8	259	60.2	39.6
× 19	57.9	7.36	29	62.8	200	57.9	40.1
× 12.7	39.3	5.0	20.2	63.6	137	55.6	40.4
L 152 × 152 × 25.4	55.7	7.1	14.8	45.6	140.4	47.2	29.7
× 15.9	36	4.59	10.1	46.8	92.8	43.9	30.0
× 9.5	22.2	2.8	6.41	47.8	57.8	41.7	30.2
L 127 × 127 × 19	35.1	4.48	6.53	38.2	74.2	38.6	24.8
× 12.7	24.1	3.07	4.70	39.2	51.8	36.3	25.0
× 9.5	18.3	2.33	3.64	39.5	39.7	35.3	25.1
L 102 × 102 × 19	27.5	3.51	3.19	30.1	46.0	32.3	19.8
× 12.7	19	2.42	2.31	30.9	32.3	30.0	19.9
× 6.4	9.8	1.25	1.27	31.8	17.2	27.7	20.2
L 89 × 89 × 9.5	12.6	1.6	1.20	27.3	18.9	25.7	17.5
× 6.4	8.6	1.09	0.84	27.7	13.0	24.6	17.6
L 76 × 76 × 12.7	14	1.77	0.92	22.8	17.5	23.7	14.8
× 6.4	7.3	0.93	0.52	23.6	9.46	21.4	15.0

Table A.8   Properties of rolled-steel (L) shapes, angles with equal legs

			Axis x-x or y-y				Axis z-z,
Size and thickness, in.	Weight, lb/ft	Area, in.^{2}	I, in.^{4}	r, in.	S, in.^{3}	$\bar{x}$ or $\bar{y}$, in.	r, in.
$L\,8 \times 8 \times 1$	51.0	15.0	89.0	2.44	15.8	2.37	1.56
$\times \frac{3}{4}$	38.9	11.4	69.7	2.47	12.2	2.28	1.58
$\times \frac{1}{2}$	26.4	7.75	48.6	2.50	8.36	2.19	1.59
$L\,6 \times 6 \times 1$	37.4	11.0	35.5	1.80	8.57	1.86	1.17
$\times \frac{5}{8}$	24.2	7.11	24.2	1.84	5.66	1.73	1.18
$\times \frac{3}{8}$	14.9	4.36	15.4	1.88	3.53	1.64	1.19
$L\,5 \times 5 \times \frac{3}{4}$	23.6	6.94	15.7	1.51	4.53	1.52	0.975
$\times \frac{1}{2}$	16.2	4.75	11.3	1.54	3.16	1.43	0.983
$\times \frac{3}{8}$	12.3	3.61	8.74	1.56	2.42	1.39	0.990
$L\,4 \times 4 \times \frac{3}{4}$	18.5	5.44	7.67	1.19	2.81	1.27	0.778
$\times \frac{1}{2}$	12.8	3.75	5.56	1.22	1.97	1.18	0.782
$\times \frac{1}{4}$	6.6	1.94	3.04	1.25	1.05	1.09	0.795
$L\,3\frac{1}{2} \times 3\frac{1}{2} \times \frac{3}{8}$	8.5	2.48	2.87	1.07	1.15	1.01	0.687
$\times \frac{1}{4}$	5.8	1.69	2.01	1.09	0.794	0.968	0.694
$L\,3 \times 3 \times \frac{1}{2}$	9.4	2.75	2.22	0.898	1.07	0.932	0.584
$\times \frac{1}{4}$	4.9	1.44	1.24	0.930	0.577	0.842	0.592

U.S. Customary Units

| SOURCE: The American Institute of Steel Construction, Chicago.

Table A.9　Deflections and slopes of beams

Load and support	Maximum deflection	Slope at end	Equation of elastic curve
1.	$-\dfrac{PL^3}{3EI}$	$-\dfrac{PL^2}{2EI}$	$v = \dfrac{Px^2}{6EI}(x - 3L)$
2.	$-\dfrac{ML^2}{2EI}$	$-\dfrac{ML}{EI}$	$v = -\dfrac{Mx^2}{2EI}$
3.	$-\dfrac{wL^4}{8EI}$	$-\dfrac{wL^3}{6EI}$	$v = -\dfrac{wx^2}{24EI}(x^2 - 4Lx + 6L^2)$
4.	$-\dfrac{w_o L^4}{30EI}$	$-\dfrac{w_o L^3}{24EI}$	$v = \dfrac{w_o x^2}{120EIL}(x^3 - 5Lx^2 + 10L^2x - 10L^3)$
5.	$-\dfrac{PL^3}{48EI}$	$\pm\dfrac{PL^2}{16EI}$	$v = \dfrac{Px}{48EI}(4x^2 - 3L^2) \qquad (x \le L/2)$
6.	For $a > b$: $-\dfrac{Pb(L^2 - b^2)^{3/2}}{9\sqrt{3}EIL}$ $x_m = \sqrt{\dfrac{L^2 - b^2}{3}}$	$\theta_A = -\dfrac{Pb(L^2 - b^2)}{6EIL}$ $\theta_B = \dfrac{Pa(L^2 - a^2)}{6EIL}$	$v = \dfrac{Pbx}{6EIL}(x^2 - L^2 + b^2) \qquad (x \le a)$

(continued)

Table A.9 (continued)

Load and support	Maximum deflection	Slope at end	Equation of elastic curve
7.	$-\dfrac{ML^2}{9\sqrt{3}EI}$	$\theta_A = -\dfrac{ML}{6EI}$ $\theta_B = \dfrac{ML}{3EI}$	$v = \dfrac{Mx}{6EIL}(x^2 - L^2)$
8.	$-\dfrac{5wL^4}{384EI}$	$\pm\dfrac{wL^3}{24EI}$	$v = -\dfrac{wx}{24EI}(x^3 - 2Lx^2 + L^3)$
9.	$\pm\dfrac{ML^2}{36\sqrt{12}EI}$	$\pm\dfrac{ML}{24EI}$	$v = \dfrac{Mx}{24EIL}(4x^2 - L^2) \qquad (x \le L/2)$
10.	$-\dfrac{Pb^2L}{3EI}$	$\theta_A = \dfrac{Pab}{6EI}$ $\theta_B = -\dfrac{Pb}{6EI}(2L + b)$	$v = \dfrac{Pbx}{6aEI}(a^2 - x^2) \qquad (0 \le x \le a)$

Table A.10   Reactions and deflections of statically indeterminate beams

Load and support	Reactions*		Deflections
1.	$R_A = R_B = \dfrac{P}{2}$ $M_A = M_B = \dfrac{PL}{8}$		$v_{max} = v_C = -\dfrac{PL^3}{192}$
2.	$R_A = \dfrac{Pb^2}{L^3}(3a+b)$ $M_A = \dfrac{Pab^2}{L^2}$	$R_B = \dfrac{Pa^2}{L^3}(a+3b)$ $M_B = \dfrac{Pa^2b}{L^2}$	For $a > b$: $v_C = -\dfrac{Pb^2}{48EI}(3L-4b)$
3.	$R_A = \dfrac{5}{16}P,$ $M_B = \dfrac{3}{16}PL$	$R_B = \dfrac{11}{16}P$	$v_C = -\dfrac{7PL^3}{768EI}$
4. 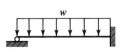	$R_A = R_B = \dfrac{wL}{2}$ $M_A = M_B = \dfrac{wL^2}{12}$		$v_{max} = v_C = -\dfrac{wL^4}{384EI}$
5. 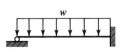	$R_A = \dfrac{3}{32}wL,$ $M_A = \dfrac{5}{192}wL^2,$	$R_B = \dfrac{13}{32}wL$ $M_B = \dfrac{11}{192}wL^2$	$v_C = -\dfrac{wL^4}{768EI}$
6. 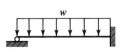	$R_A = \dfrac{3}{8}wL,$ $M_B = \dfrac{1}{8}wL^2$	$R_B = \dfrac{5}{8}wL$	$v_C = -\dfrac{wL^4}{192EI}$

| *For all the cases tabulated, the senses of the reactions and the notations are the same as those shown in case 1.

Appendix B

Material Properties

The properties of materials vary widely, depending on numerous factors, including chemical composition, manufacturing processes, internal defects, heat treatment, temperature, and dimensions of test specimens. Hence, the values furnished in Tables B.1 through B.8 are representative but are not necessarily suitable for a specific application. In some cases, a range of values given in the listings show the possible variations in characteristics.

Unless otherwise indicated, the modulus of elasticity E and other properties are for materials in tension. The specific data were compiled from broad tabulations listed in the references cited. For details, see, for example, [1–8] of Chapter 2. Note that the reference issues of *Machine Design Materials* (Cleveland: Penton/IPC) also constitute an excellent source of data on a great variety of materials.

Table B.1 Average properties of common engineering materials*

SI Units

Material	Density, Mg/m³	Ultimate strength, MPa			Yield strength,† MPa		Modulus of elasticity, GPa	Modulus of rigidity, GPa	Coefficient of thermal expansion, 10⁻⁶/°C	Elongation in 50 mm, %	Poisson's ratio
		Tension	Compression**	Shear	Tension	Shear					
Steel											0.27–0.3
Structural, ASTM-A36	7.86	400	—	—	250	145	200	79	11.7	30	
High strength, ASTM-A242	7.86	480	—	—	345	210	200	79	11.7	21	
Stainless (302), cold rolled	7.92	860	—	—	520	—	190	73	17.3	12	
Cast iron											0.2–0.3
Gray, ASTM A-48	7.2	170	650	240	—	—	70	28	12.1	0.5	
Malleable, ASTM A-47	7.3	340	620	330	230	—	165	64	12.1	10	
Wrought iron	7.7	350	—	240	210	130	190	70	12.1	35	0.3
Aluminum											0.33
Alloy 2014-T6	2.8	480	—	290	410	220	72	28	23	13	
Alloy 6061-T6	2.71	300	—	185	260	140	70	26	23.6	17	
Brass, yellow											0.34
Cold rolled	8.47	540	—	300	435	250	105	39	20	8	
Annealed	8.47	330	—	220	105	65	105	39	20	60	
Bronze, cold rolled (510)	8.86	560	—	—	520	275	110	41	17.8	10	0.34
Copper, hard drawn	8.86	380	—	165	260	160	120	40	16.8	4	0.33
Magnesium alloys	1.8	140–340	—	—	80–280	—	45	17	27	2–20	0.35
Nickel	8.08	310–760	—	—	140–620	—	210	80	13	2–50	0.31
Titanium alloys	4.4	900–970	—	—	760–900	—	100–120	39–44	8–10	10	0.33
Zinc alloys	6.6	280–390	—	—	210–320	—	83	31	27	1–10	0.33
Concrete											0.1–0.2
Medium strength	2.32	—	28	—	—	—	24	—	10	—	
High strength	2.32	—	40	—	—	—	30	—	10	—	
Timber‡ (air dry)											
Douglas fir	0.54	—	55	7.6	—	—	12	—	4	—	
Southern pine	0.58	—	60	10	—	—	11	—	4	—	
Glass, 98% silica	2.19	—	50	—	—	—	65	28	80	—	0.2–0.27
Graphite	0.77	20	240	35	—	—	70	—	7	—	
Rubber	0.91	14	—	—	—	—	—	—	162	600	0.45–0.5

*Properties may vary widely with changes in composition, heat treatment, and method of manufacture.
**For ductile metals the compression strength is assumed to be the same as that in tension.
† Offset of 0.2%.
‡ Loaded parallel to the grain.

Table B.1 Average properties of common engineering materials*

		U.S. Customary Units									
	Specific weight, lb/in.³	Ultimate strength, ksi			Yield strength,† ksi		Modulus of elasticity, 10⁶ psi	Modulus of rigidity, 10⁶ psi	Coefficient of thermal expansion, 10⁻⁶/°F	Elongation in 2 in., %	Poisson's ratio
Material		Tension	Compression**	Shear	Tension	Shear					
Steel											0.27–0.3
Structural, ASTM-A36	0.284	58	—	—	36	21	29	11.5	6.5	30	
High strength, ASTM-A242	0.284	70	—	—	50	30	29	11.5	6.5	21	
Stainless (302), cold rolled	0.286	125	—	—	75	—	28	10.6	9.6	12	
Cast iron											0.2–0.3
Gray, ASTM A-48	0.260	25	95	35	—	—	10	4.1	6.7	0.5	
Malleable, ASTM A-47	0.264	50	90	48	33	—	24	9.3	6.7	10	
Wrought iron	0.278	50	—	35	30	18	27	10	6.7	35	0.3
Aluminum											0.33
Alloy 2014-T6	0.101	70	—	42	60	32	10.6	4.1	12.8	13	
Alloy 6061-T6	0.098	43	—	27	38	20	10.0	3.8	13.1	17	
Brass, yellow											0.34
Cold rolled	0.306	78	—	43	63	36	15	5.6	11.3	8	
Annealed	0.306	48	—	32	15	9	15	5.6	11.3	60	
Bronze, cold rolled (510)	0.320	81	—	—	75	40	16	5.9	9.9	10	0.34
Magnesium alloys	0.065	20–49	—	24	11–40	—	6.5	2.4	15	2–20	0.35
Copper, hard drawn	0.320	55	—	—	38	23	17	6	9.3	4	0.33
Nickel	0.320	45–110	—	—	20–90	—	30	11.4	7.2	2–50	0.31
Titanium alloys	0.160	130–140	—	—	110–130	—	15–17	5.6–6.4	4.5–5.5	10	0.33
Zinc alloys	0.240	40–57	—	—	30–46	—	12	4.5	15	1–10	0.33
Concrete											0.1–0.2
Medium strength	0.084	—	4	—	—	—	3.5	—	5.5	—	
High strength	0.084	—	6	—	—	—	4.3	—	5.5	—	
Timber‡ (air dry)											
Douglas fir	0.020	—	7.9	1.1	—	—	1.7	—	2.2	—	
Southern pine	0.021	—	8.6	1.4	—	—	1.6	—	2.2	—	
Glass, 98% silica	0.079	7	—	—	—	—	9.6	4.1	44	—	0.2–0.27
Graphite	0.028	3	35	5	—	—	10	—	3.9	—	
Rubber	0.033	2	—	—	—	—	—	—	90	600	0.45–0.5

*Properties may vary widely with changes in composition, heat treatment, and method of manufacture.

**For ductile metals the compression strength is assumed to be the same as that in tension.

†Offset of 0.2%.

‡Loaded parallel to the grain.

Table B.2   Typical mechanical properties of gray cast iron

ASTM class*	Ultimate strength S_u, MPa	Compressive strength S_{uc}, MPa	Modulus of elasticity, GPa		Brinell hardness H_B	Fatigue stress concentration factor K_f
			Tension	Torsion		
20	150	575	66–97	27–39	156	1.00
25	180	670	79–102	32–41	174	1.05
30	215	755	90–113	36–45	201	1.10
35	250	860	100–120	40–48	212	1.15
40	295	970	110–138	44–54	235	1.25
50	365	1135	130–157	50–54	262	1.35
60	435	1295	141–162	54–59	302	1.50

*Minimum values of S_u (in ksi) are given by the class number.

Note: To convert from MPa to ksi, divide given values by 6.895.

Table B.3   Mechanical properties of some hot-rolled (HR) and cold-drawn (CD) steels

UNS number	AISI/SAE number	Processing	Ultimate strength* S_u, MPa	Yield strength* S_y, MPa	Elongation in 50 mm, %	Reduction in area, %	Brinell hardness H_B
G10060	1006	HR	300	170	30	55	86
		CD	330	280	20	45	95
G10100	1010	HR	320	180	28	50	95
		CD	370	300	20	40	105
G10150	1015	HR	340	190	28	50	101
		CD	390	320	18	40	111
G10200	1020	HR	380	210	25	50	111
		CD	470	390	15	40	131
G10300	1030	HR	470	260	20	42	137
		CD	520	440	12	35	149
G10350	1035	HR	500	270	18	40	143
		CD	550	460	12	35	163
G10400	1040	HR	520	290	18	40	149
		CD	590	490	12	35	170
G10450	1045	HR	570	310	16	40	163
		CD	630	530	12	35	179
G10500	1050	HR	620	340	15	35	179
		CD	690	580	10	30	197
G10600	1060	HR	680	370	12	30	201
G10800	1080	HR	770	420	10	25	229
G10950	1095	HR	830	460	10	25	248

SOURCE: 1986 SAE Handbook, p. 2.15.

*Values listed are estimated ASTM minimum values in the size range 18 to 32 mm.

Note: To convert from MPa to ksi, divide given values by 6.895.

Table B.4 Mechanical properties of selected heat-treated steels

AISI number	Treatment	Temperature, °C	Ultimate strength S_u, MPa	Yield strength S_y, MPa	Elongation in 50 mm, %	Reduction in area, %	Brinell hardness H_B
1030	WQ&T	205	848	648	17	47	495
	WQ&T	425	731	579	23	60	302
	WQ&T	650	586	441	32	70	207
	Normalized	925	521	345	32	61	149
	Annealed	870	430	317	35	64	137
1040	OQ&T	205	779	593	19	48	262
	OQ&T	425	758	552	21	54	241
	OQ&T	650	634	434	29	65	192
	Normalized	900	590	374	28	55	170
	Annealed	790	519	353	30	57	149
1050	WQ&T	205	1120	807	9	27	514
	WQ&T	425	1090	793	13	36	444
	WQ&T	650	717	538	28	65	235
	Normalized	900	748	427	20	39	217
	Annealed	790	636	365	24	40	187
1060	OQ&T	425	1080	765	14	41	311
	OQ&T	540	965	669	17	45	277
	OQ&T	650	800	524	23	54	229
	Normalized	900	776	421	18	37	229
	Annealed	790	626	372	22	38	179
1095	OQ&T	315	1260	813	10	30	375
	OQ&T	425	1210	772	12	32	363
	OQ&T	650	896	552	21	47	269
	Normalized	900	1010	500	9	13	293
	Annealed	790	658	380	13	21	192
4130	WQ&T	205	1630	1460	10	41	467
	WQ&T	425	1280	1190	13	49	380
	WQ&T	650	814	703	22	64	245
	Normalized	870	670	436	25	59	197
	Annealed	865	560	361	28	56	156
4140	OQ&T	205	1770	1640	8	38	510
	OQ&T	425	1250	1140	13	49	370
	OQ&T	650	758	655	22	63	230
	Normalized	870	870	1020	18	47	302
	Annealed	815	655	417	26	57	197

SOURCE: *ASM Metals Reference Book*, 2nd ed. Metals Park, OH: American Society for Metals, 1983.

Notes: To convert from MPa to ksi, divide given values by 6.895.

Values tabulated for 25-mm round sections and of gage length 50-mm. The properties for quenched and tempered steel are from a single heat: OQ&T = oil-quenched and tempered; WQ&T = water-quenched and tempered.

Table B.5   Mechanical properties of some annealed (An.) and cold-worked (CW) wrought stainless steels

AISI type	Ultimate strength S_u, (MPa) An.	CW	Yield strength S_y, (MPa) An.	CW	Elongation in 50 mm, % An.	CW	Izod impact J (N · m) An.	CW
Austentic								
302	586	758	241	517	60	35	149	122
303	620	758	241	552	50	22	115	47
304	586	758	241	517	60	55	149	122
347, 348	620	758	241	448	50	40	149	—
Martensitic								
410	517	724	276	586	35	17	122	102
414	793	896*	620*	862	20	15*	68	—
431	862	896*	655*	862*	20	15*	68	—
440 A, B, C	724	796*	414	620*	14	7*	3	3*
Ferritic								
430, 430F	517	572	296	434	27	20	—	—
446	572	586	365	483	23	20	3	—

SOURCES: *Metal Progress Databook 1980,* Vol. 118, no. 1, Metals Park, OH: American Society for Metals (June 1980); *ASME Handbook Metal Properties,* New York: McGraw-Hill, 1954.

Note: To convert from MPa to ksi, divide given values by 6.895.

*Annealed and cold drawn.

Table B.6   Mechanical properties of some aluminum alloys

Alloy	Ultimate strength S_u MPa	(ksi)	Yield strength S_y MPa	(ksi)	Elongation in 50 mm, %	Brinell hardness H_B
Wrought:						
1100-H14	125	(18)	115	(17)	20	32
2011-T3	380	(55)	295	(43)	15	95
2014-T4	425	(62)	290	(42)	20	105
2024-T4	470	(68)	325	(47)	19	120
6061-T6	310	(45)	275	(40)	17	95
6063-T6	240	(35)	215	(31)	12	73
7075-T6	570	(83)	505	(73)	11	150
Cast:						
201-T4*	365	(53)	215	(31)	20	—
295-T6*	250	(36)	165	(24)	5	—
355-T6*	240	(35)	175	(25)	3	—
-T6**	290	(42)	190	(27)	4	—
356-T6*	230	(33)	165	(24)	2	—
-T6**	265	(38)	185	(27)	5	—
520-T4*	330	(48)	180	(26)	16	—

SOURCES: *ASM Metals Reference Book,* Metals Park, OH: American Society for Metals, 1981; *1981 Materials Selector,* vol. 92, no. 6, Cleveland: Materials Engineering, Penton/IPC, (December 1980).

*Sand casting.

**Permanent-mold casting.

Table B.7   Mechanical properties of some copper alloys

Alloy	UNS number	Ultimate strength S_u, MPa	Yield strength S_y, MPa	Elongation in 50 mm, %
Wrought:				
Leaded				
Beryllium copper	C17300	469–1379	172–1227	43–3
Phos bronze	C54400	469–517	393–434	20–15
Aluminum				
Silicon-bronze	C64200	517–703	241–469	32–22
Silicon bronze	C65500	400–745	152–414	60–13
Manganese bronze	C67500	448–579	207–414	33–19
Cast:				
Leaded				
Red brass	C83600	255	117	30
Yellow brass	C85200	262	90	35
Manganese bronze	C86200	655	331	20
Bearing bronze	C93200	241	124	20
Aluminum bronze	C95400	586–724	241–372	18–8
Copper nickel	C96200	310	172	20

SOURCE: 1981 materials reference issue, *Machine Design*, 53, no. 6 (March 19, 1981).
Note: To convert from MPa to ksi, divide given values by 6.895.

Table B.8   Selected mechanical properties of some common plastics

Plastic	Ultimate strength S_u MPa	(ksi)	Elongation in 50 mm, %	Izod impact strength J	(ft · lb)
Acrylic	72	(10.5)	6	0.5	(0.4)
Cellulose accetate	14–18	(2–7)	—	1.4–9.5	(1–7)
Epoxy (glass-filled)	69–138	(10–20)	4	2.7–41	(2–30)
Fluorocarbon	23	(3.4)	300	4.1	(3)
Nylon (6/6)	83	(12)	60	1.4	(1)
Phenolic (wood-flour filled)	48	(7)	0.4–0.8	0.4	(0.3)
Polycarbonate	62–72	(9–10.5)	110–125	16–22	(12–16)
Polyester (25% glass filled)	110–90	(16–23)	1–3	1.4–2.6	(1.0–1.9)
Polypropylene	34	(5)	10–20	0.7–3.0	(0.5–2.2)

SOURCES: 1981 materials reference issue, *Machine Design*, 53, no. 6 (March 19, 1981); 1981 materials selector issue, *Materials Engineering*, 92, no. 6 (December 1980).

Appendix C

STRESS CONCENTRATION FACTORS

In the following charts, the theoretical or geometric stress concentration factors K_t for some common cases are presented as an aid to the reader in the solution of practical problems. These graphs were selected from the extensive charts found in [8–13] of Chapter 3. Equations to estimate most of these curves have been included to allow automatic generation of the K_t during calculations. Figures C.1 through C.6 are for flat bars and C.7 through C.13 relate to cylindrical members. Note that the results pertain to an isotropic material and for use in Eqs. (3.40).

C.1　Theoretical stress-concentration factor K_t for a filleted bar in axial tension.

C.2　Theoretical stress-concentration factor K_t for a filleted bar in bending.

C.3　Theoretical stress-concentration factor K_t for a notched bar in axial tension.

C.4　Theoretical stress-concentration factor K_t for a notched bar in bending.

C.5　Theoretical stress-concentration factor K_t: A—for a flat bar loaded in tension by a pin through the transverse hole; B—for a flat bar with a transverse hole in axial tension.

C.6　Theoretical-stress concentration factor K_t for a flat bar with a transverse hole in bending.

C.7　Theoretical stress-concentration factor K_t for a shaft with a shoulder fillet in axial tension.

C.8　Theoretical stress-concentration factor K_t for a shaft with a shoulder fillet in torsion.

C.9　Theoretical stress-concentration factor K_t for a shaft with a shoulder fillet in bending.

C.10　Theoretical stress-concentration factor K_t for a grooved shaft in axial tension.

C.11　Theoretical stress-concentration factor K_t for a grooved shaft in torsion.

C.12　Theoretical stress-concentration factor K_t for a grooved shaft in bending.

C.13　Theoretical stress-concentration factor K_t for a shaft with a transverse hole in axial tension, bending, and torsion.

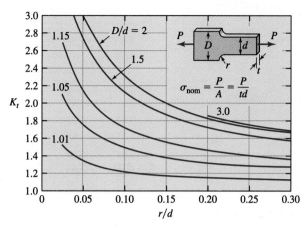

Approximate formula
$$K_t \approx B \left(\frac{r}{d}\right)^a, \text{ where:}$$

D/d	B	a
2.00	1.100	−0.321
1.50	1.077	−0.296
1.15	1.014	−0.239
1.05	0.998	−0.138
1.01	0.977	−0.107

Figure C.1 Theoretical stress-concentration factor K_t for a filleted bar in axial tension [9 and 12, Chapter 3].

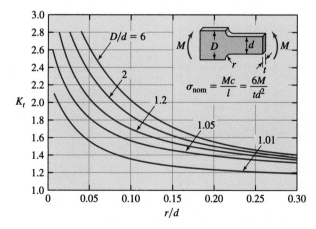

Approximate formula
$$K_t \approx B \left(\frac{r}{d}\right)^a, \text{ where:}$$

D/d	B	a
6.00	0.896	−0.358
2.00	0.932	−0.303
1.20	0.996	−0.238
1.05	1.023	−0.192
1.01	0.967	−0.154

Figure C.2 Theoretical stress-concentration factor K_t for a filleted bar in bending [9 and 12, Chapter 3].

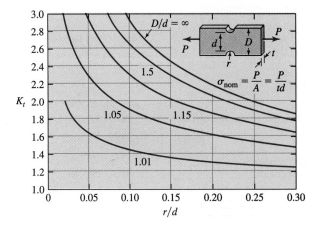

D/d	B	a
∞	1.110	−0.417
1.50	1.133	−0.366
1.15	1.095	−0.325
1.05	1.091	−0.242
1.01	1.043	−0.142

Approximate formula
$$K_t \approx B\left(\frac{r}{d}\right)^a, \text{ where:}$$

$$\sigma_{nom} = \frac{P}{A} = \frac{P}{td}$$

Figure C.3   Theoretical stress-concentration factor K_t for a notched bar in axial tension [9 and 12, Chapter 3].

D/d	B	a
∞	0.971	−0.357
1.50	0.983	−0.334
1.15	0.993	−0.303
1.05	1.025	−0.240
1.01	1.061	−0.134

Approximate formula
$$K_t \approx B\left(\frac{r}{d}\right)^a, \text{ where:}$$

$$\sigma_{nom} = \frac{Mc}{I} = \frac{6M}{td^2}$$

Figure C.4   Theoretical stress-concentration factor K_t for a notched bar in bending [9 and 12, Chapter 3].

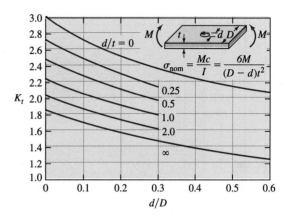

Figure C.5 Theoretical stress-concentration factor K_t: A—for a flat bar loaded in tension by a pin through the transverse hole; B—for a flat bar with a transverse hole in axial tension [13, Chapter 3].

Figure C.6 Theoretical-stress concentration factor K_t for a flat bar with a transverse hole in bending [9, Chapter 3].

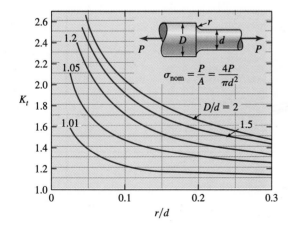

Approximate formula
$$K_t \approx B \left(\frac{r}{d}\right)^a, \text{ where:}$$

D/d	B	a
2.00	1.015	−0.300
1.50	1.000	−0.282
1.20	0.963	−0.255
1.05	1.005	−0.171
1.01	0.984	−0.105

Figure C.7 Theoretical stress-concentration factor K_t for a shaft with a shoulder fillet in axial tension [9 and 12, Chapter 3].

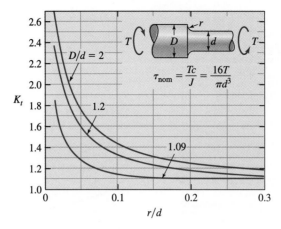

Figure C.8   Theoretical stress-concentration factor K_t for a shaft with a shoulder fillet in torsion [9 and 12, Chapter 3].

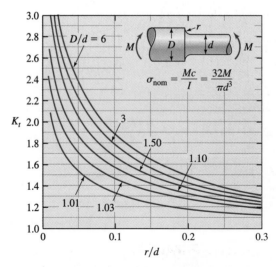

Figure C.9   Theoretical stress-concentration factor K_t for a shaft with a shoulder fillet in bending [9 and 12, Chapter 3].

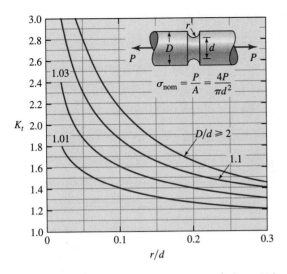

Figure C.10 Theoretical stress-concentration factor K_t for a grooved shaft in axial tension [9, Chapter 3].

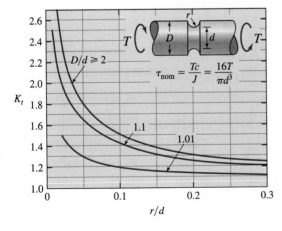

Approximate formula

$$K_t \approx B \left(\frac{r}{d}\right)^a, \text{ where:}$$

D/d	B	a
2.00	0.890	−0.241
1.10	0.923	−0.197
1.01	0.972	−0.102

Figure C.11 Theoretical stress-concentration factor K_t for a grooved shaft in torsion [9 and 12, Chapter 3].

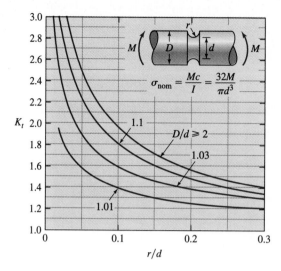

D/d	B	a
2.00	0.936	−0.331
1.10	0.955	−0.283
1.03	0.990	−0.215
1.01	0.994	−0.152

Approximate formula
$$K_t \approx B \left(\frac{r}{d}\right)^a, \text{ where:}$$

$$\sigma_{\text{nom}} = \frac{Mc}{I} = \frac{32M}{\pi d^3}$$

Figure C.12   Theoretical stress-concentration factor K_t for a grooved shaft in bending [9 and 12, Chapter 3].

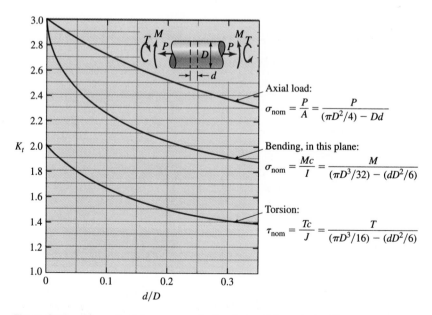

Axial load:
$$\sigma_{\text{nom}} = \frac{P}{A} = \frac{P}{(\pi D^2/4) - Dd}$$

Bending, in this plane:
$$\sigma_{\text{nom}} = \frac{Mc}{I} = \frac{M}{(\pi D^3/32) - (dD^2/6)}$$

Torsion:
$$\tau_{\text{nom}} = \frac{Tc}{J} = \frac{T}{(\pi D^3/16) - (dD^2/6)}$$

Figure C.13   Theoretical stress-concentration factor K_t for a shaft with a transverse hole in axial tension, bending, and torsion [9, Chapter 3].

Appendix D

SOLUTION OF THE STRESS CUBIC EQUATION

PRINCIPAL STRESSES

Numerous methods for solving a cubic equation are in common use. The following is a practical approach for calculating the roots of stress cubic equation (see Section 3.16):

$$\sigma_i^3 - I_1\sigma_i^2 + I_2\sigma_i - I_3 = 0 \qquad (i = 1, 2, 3) \tag{3.54}$$

where

$$I_1 = \sigma_x + \sigma_y + \sigma_z$$

$$I_2 = \sigma_x\sigma_y + \sigma_x\sigma_z + \sigma_y\sigma_z - \tau_{xy}^2 - \tau_{yz}^2 - \tau_{xz}^2 \tag{3.55}$$

$$I_3 = \sigma_x\sigma_y\sigma_z + 2\tau_{xy}\tau_{yz}\tau_{xz} - \sigma_x\tau_{yz}^2 - \sigma_y\tau_{xz}^2 - \sigma_z\tau_{xy}^2$$

In accordance with the method, expressions that provide direct means for solving both two- and three-dimensional stress problems are (see [2] of Chapter 3)

$$\sigma_a = 2S[\cos(\alpha/3)] + \tfrac{1}{3}I_1$$

$$\sigma_b = 2S\{\cos[(\alpha/3) + 120°]\} + \tfrac{1}{3}I_1 \tag{D.1}$$

$$\sigma_c = 2S\{\cos[(\alpha/3) + 240°]\} + \tfrac{1}{3}I_1$$

Here, the constants are expressed by

$$S = \left(\tfrac{1}{3}R\right)^{1/2}$$

$$\alpha = \cos^{-1}\left(-\frac{Q}{2T}\right)$$

$$R = \tfrac{1}{3}I_1^2 - I_2 \tag{D.2}$$

$$Q = \tfrac{1}{3}I_1I_2 - I_3 - \tfrac{2}{27}I_1^3$$

$$T = \left(\tfrac{1}{27}R^3\right)^{1/2}$$

The invariants I_1, I_2, and I_3 are represented in terms of the given stress components by Eq. (3.55). The principal stresses found from Eqs. (D.1) are *redesignated* using numerial subscripts so that algebraically $\sigma_1 > \sigma_2 > \sigma_3$.

DIRECTION COSINES

The values of the direction cosines of a principal stress are determined using Eqs. (3.52) and (3.50), as already discussed in Section 3.16. However, the following simpler method is preferred:

$$
\begin{bmatrix}
(\sigma_x - \sigma_i) & \tau_{xy} & \tau_{xz} \\
\tau_{xy} & (\sigma_y - \sigma_i) & \tau_{yz} \\
\tau_{xz} & \tau_{yz} & (\sigma_z - \sigma_i)
\end{bmatrix}
\begin{Bmatrix}
l_i \\
m_i \\
n_i
\end{Bmatrix} = 0
\tag{3.52}
$$

The *cofactors* of the determinant of the preceding matrix on the elements of the first row are given by

$$
a_i =
\begin{vmatrix}
(\sigma_y - \sigma_i) & \tau_{yz} \\
\tau_{yz} & (\sigma_z - \sigma_i)
\end{vmatrix}
$$

$$
b_i = -
\begin{vmatrix}
\tau_{xy} & \tau_{yz} \\
\tau_{xz} & (\sigma_z - \sigma_i)
\end{vmatrix}
\tag{D.3}
$$

$$
c_i =
\begin{vmatrix}
\tau_{xy} & (\sigma_y - \sigma_i) \\
\tau_{xz} & \tau_{yz}
\end{vmatrix}
$$

Let us introduce the notation

$$
k_i = \frac{1}{\left(a_i^2 + b_i^2 + c_i^2\right)^{1/2}}
\tag{D.4}
$$

the *direction cosines* are then expressed in the form

$$
l_i = a_i k_i \qquad m_i = b_i k_i \qquad n_i = c_i k_i
\tag{D.5}
$$

Clearly, Eqs. (D.5) give $l_i^2 + m_i^2 + n_i^2 = 1$.

The foregoing procedures are well adopted to a quality scientific calculator or digital computer.

Answers to Selected Problems

Chapter 1. Introduction to Design

1.1 $V_D = 4$ kips, $M_D = 4$ kip·ft., $V_E = 2.5$ kips, $M_E = 6$ kip·ft.

1.3 (a) $R_{Cx} = 75$ kN $\leftarrow$, $R_{Cy} = 50$ kN $\uparrow$, $R_{Bx} = 75$ kN $\rightarrow$, $R_{By} = 10$ kN $\downarrow$.

(b) $F_D = 85$ kN, $V_D = 30$ kN, $M_D = 37.5$ kN·m.

1.5 $R_B = 12.58$ kips.

1.6 (a) $T = 0.6$ kN·m.

(b) $R_A = 4.427$ kN, $R_E = 8.854$ kN.

1.8 $V_y = 64$ N, $M_z = 31.2$ N·m.

1.10 $T_d = 0.3$ kN·m.

1.13 $e = 90\%$.

1.16 (a) $\varepsilon_{c,max} = 2000\mu$.

(b) $\varepsilon_r = 1000\mu$.

1.19 $\Delta L_{BD} = 0.232$ mm.

1.21 $\varepsilon_x = \varepsilon_y = -363\mu$, $\gamma_{xy} = 1651\mu$.

Chapter 2. Materials

2.2 (a) $E = 53$ GPa.

(b) $v = 0.25$.

(c) $G = 21.2$ GPa.

2.3 (a) $v = 0.25$.

(b) $E = 8.335 \times 10^6$ psi.

(c) $a' = 3.0015$ in.

(d) $G = 3.334 \times 10^6$ psi.

2.5 $\varepsilon_x = 1327\mu$.

2.7 $L' = 99.96$ mm, $a' = 49.98$ mm, $b' = 9.996$ mm.

2.9 $n = 3.16$.

2.11 $d = 0.667$ in.

Chapter 3. Stress and Strain

3.2 $a_{min} = 73$ mm.

3.4 (a) $\sigma_{BD} = -1.018$ MPa.

(b) $\tau_A = 9.697$ MPa.

3.5 $\alpha = 54.7°$.

3.8 $b = \dfrac{d}{\sqrt{3}}$, $h = d\sqrt{\dfrac{2}{3}}$.

3.11 $b = 56.5$ mm.

3.13 $h = h_1 \left(\dfrac{x}{L}\right)^{3/2}$.

3.16 $t = 0.399$ in.

3.19 $M = 38.45$ kip·in.

3.21 (a) $\sigma_x = -37.1$ kPa, $\sigma_y = -2.9$ kPa, $\tau_{xy} = 47$ kPa.

(b) $\tau_{max} = 50$ kPa.

3.24 (a) $\sigma_x = 25$ ksi, $\sigma_y = -5$ ksi, $\tau_{xy} = -8.66$ ksi.

(b) $\sigma_1 = 27.32$ ksi, $\sigma_2 = -7.32$ ksi, $\theta'_p = 15°$.

3.26 $\sigma_{x'} = 140.3$ MPa, $\sigma_{y'} = 1.07$ MPa, $\tau_{x'y'} = 12.28$ MPa.

3.29 Point A: (a) $\sigma_1 = 864$ psi, $\sigma_2 = 234$ psi.

(b) $\tau_{max} = 315$ psi.

3.31 $\sigma_{x'} = 19.6$ ksi, $\tau_{x'y'} = 2.87$ ksi.

3.35 (a) $\gamma_{max} = 566\mu$.

(b) $\Delta L_{AC} = 1.41 \times 10^{-4}$ in.

3.37 $p_{all} = 1.281$ MPa.

3.39 $P_{all} = 29.3$ kN.

3.41 (a) $a = 0.135$ mm.

(b) $p_o = 943.1$ MPa.

3.44 $p_o = 414.9$ MPa.

3.48 $\sigma = 24.98$ MPa,
$\tau = 21.27$ MPa.

Chapter 4.   Deflection and Impact

4.1 (a) $d = 8.74$ mm.

(b) $k = 2000$ kN/m.

4.2 $R_A = 7.041$ kips,
$R_B = 0.959$ kips.

4.4 (a) $P = 178.2$ kN.

(b) $\delta_a = 0.0653$ mm.

4.5 (a) $\phi_D = 6.82°$.

(b) $\tau_{AB} = 41.92$ MPa.

4.7 $h = 197$ mm.

4.11 $v_1 = \dfrac{M_o L}{6EIL} (-6aL + 3a^2 + 2L^2 + x^2)$.

4.14 $R = \dfrac{5P(E_2 I_2)}{2(E_1 I_1 + E_2 I_2)}$.

4.18 $R_A = R_B = \dfrac{P}{2}$, $M_A = -M_B = \dfrac{PL^4}{8}$ ↻,

$v = -\dfrac{Px^2}{48EI} (3L - 4x)$.

4.22 $R_A = \dfrac{3}{4}P↓$, $R_B = \dfrac{7}{4}P↑$, $M_A = \dfrac{1}{2}Pa$ ↻.

4.25 (a) $\delta_{max} = 7.38$ mm.

(b) $\sigma_{max} = 198$ MPa.

4.27 $d = 1.943$ in.

4.30 (a) $\phi_{max} = 4.24°$.

(b) $\tau_{max} = 243.3$ MPa.

4.32 $M_{max} = 46.875$ lb · in.

4.34 (b) $\sigma_{max} = 0.75 \, p_o \left(\dfrac{b}{t}\right)^2$.

(c) $w_{max} = 0.071$ in.,
$\sigma_{max} = 9.375$ ksi.

Chapter 5.   Energy Methods in Design

5.2 $U_s = \dfrac{1}{20} \dfrac{w^2 L^3}{AG}$.

5.5 $v_A = 2Pa (a + L) \left[\dfrac{a}{6EI} + \dfrac{3}{5AGL}\right]$.

5.7 $\delta_B = \dfrac{P}{12EI}(4L^3 + 6\pi RL^2 + 24R^2 L + 3\pi R^3)$.

5.9 $R_A = \dfrac{2}{3}\dfrac{M_o}{L}↑$, $R_B = 2\dfrac{M_o}{L}↓$,

$R_C = \dfrac{4}{3}\dfrac{M_o}{L}↑$.

5.11 $\delta_h = 60\dfrac{wa^4}{EI}→$.

5.13 $\delta_B = \dfrac{PR^2}{2EI}$.

5.15 (b) $\phi_B = \dfrac{1}{GJ}\left(\dfrac{T_o L}{2} + PaL\right)$.

5.18 $(\delta_C)_v = 2.828\dfrac{PL}{AE}↓$,

$(\delta_C)_h = 3.828\dfrac{PL}{AE}→$.

5.22 $F = \dfrac{4P}{\pi}↑$.

5.24 $\delta_C = \dfrac{PL}{3E}\left(\dfrac{11}{A} + \dfrac{8L^2}{I}\right)↓$.

5.30 $v = \dfrac{w_o}{EI}\left(\dfrac{L}{\pi}\right)^4 \sin\dfrac{\pi x}{L}$.

5.32 (a) $v = \dfrac{Px^2}{6EI}(3L - x)$.

(b) $v_{max} = \dfrac{PL^3}{3EI}$, $\theta_{max} = \dfrac{PL^2}{2EI}$.

5.34 $v_A = \dfrac{Pc^2(L - c)^2}{4EIL}$.

Chapter 6.   Buckling Design of Members

6.1 $d = 29$ mm.

6.2 $d = 25.7$ mm.

6.4 $P_{all} = 4.5$ kN.

6.6 $F_{all} = 129.6$ kN.

6.7 $P_{all} = 13.83$ kips.

6.12 $d = 121$ mm.

6.13 $P_{all} = 637.5$ kips.

6.16 $P_{all} = 341.3$ kN.

6.18 $L_e = 131.9$ in.

6.22 $P_{cr} = 12\dfrac{EI}{L^2}$.

6.23 $P_{cr} = \dfrac{9}{4}\dfrac{EI_o}{L^2}$.

Chapter 7.   Failure Criteria and Reliability

7.1 $\sigma = 231.6$ MPa.

7.3 $P = 256$ kN,
$\sigma = 97.5$ MPa.

7.5 $M = 2.76$ kN·m.

7.7 $P = 490$ lb.

7.9 (a) $n = 2.86$.
(b) $n = 2.61$.

7.13 (a) $t = 0.208$ in.
(b) $t = 0.18$ in.

7.15 (a) $n = 1.94$.
(b) $n = 1.82$.

7.17 (a) $n = 1.21$.
(b) $n = 1.06$.

7.20 $T = 14.11$ kips·in.

7.23 $\tau = 111.1$ MPa.

7.26 $R \approx 99.94\%$.

7.29 (a) $\sigma = 7.645$ ksi.
(b) $R \approx 76\%$.

7.31 10%.

Chapter 8.   Fatigue

8.5 $S_e = 38.5$ MPa.

8.8 (a) $n = 4.64$.
(b) $n = 1.57$.

8.12 (a) $T = 624.9$ N·m.

8.15 $n = 1.98$.

8.17 $t = 0.736$ in.

8.19 $h = 1.45$ mm.

8.21 $h = 0.021$ in.

8.22 $P_o = 30.23$ N.

8.24 $n = 1.63$.

8.27 $n = 1.4$.

Chapter 9.   Shafts and Associated Parts

9.2 (a) $D_{AC} = 14.22$ mm, $D_{BC} = 20.52$ mm.
(b) $\phi_{AB} = 49.25°$.

9.3 $W_a/W_s = 0.598$.

9.5 (a) $D = 42.71$ mm.
(b) $D = 42.3$ mm.

9.7 $D = 63.5$ mm.

9.9 $n = 2.07$.

9.11 $n = 1.93$.

9.14 $D = 2.24$ in.

9.16 $n_{cr} = 594$ rpm.

9.18 $n_{cr} = 966$ rpm.

9.20 $n = 1.205$.

Chapter 10. Bearings and Lubrication

10.2 (a) $T_f = 42.64$.
  (b) hp $= 16.24$.
  (c) $f = 0.057$.

10.4 $\eta = 25.95$ mPa $\cdot$ s.

10.6 $W = 563$ lb.

10.8 (a) $f = 0.02$.
  (b) hp $= 0.714$.

10.10 (a) $h_0 = 0.008$ mm.
  (b) kW $= 0.017$.

10.11 (a) $h_0 = 0.013$ mm.
  (b) $p_{max} = 4.808$ MPa.

10.13 (a) $\eta = 52.8$ mPa $\cdot$ s.
  (b) kW $= 0.377$.

10.14 $t = 85.2°C$.

10.16 $L_{10} = 344.8$ hr.

10.18 $L_{10} = 119.5$ hr.

10.20 18.8%.

10.22 $L_{10} = 949.3$ hr.   (for 03 series)

10.26 $L_5 = 267$ hr.

Chapter 11. Spur Gears

11.1 $h = 0.562$ in., $h_k = 0.5$ in.,
  $r_b = 3.759$ in., $r_o = 4.25$ in.

11.2 $N_1 = 60$, $N_2 = 180$.

11.4 $N_g = 88$, $d_p = 88$ mm, $c = 220$ mm.

11.6 (a) $F_{t1} = 210$ lb, $F_{r1} = 76.43$ lb.
  (b) $R_C = 223.5$ lb, $T_C = 630$ lb $\cdot$ in.

11.8 (a) $F_{t1} = 8.843$ kN, $F_{r1} = 3.219$ kN.
  (b) $R_C = 9.411$ kN, $T_C = 1.326$ kN $\cdot$ m.

11.10 (a) $F_{t2} = 280$ lb, $F_{r2} = 130.6$ lb,
  $F_{t3} = 490$ lb, $F_{r3} = 278.5$ lb.
  (b) $R_C = 540.7$ lb,
  $T_C = 1960$ lb $\cdot$ in.

11.12 (a) $F_b = 692.2$ lb.
  (b) $F_w = 340.2$ lb.
  (c) $F_t = 114.8$ lb.

11.13 (a) $F_b = 2.75$ kN.
  (b) $F_w = 1.81$ kN.
  (c) $F_t = 617.5$ kN.

11.15 (a) $F_b = 6.76$ kN.
  (b) $F_w = 2.35$ kN.

11.16 No.

11.18 hp $= 6.95$.

11.20 hp $= 19.59$.

11.23 hp $= 36.33$.

Chapter 12. Helical, Bevel, and Worm Gears

12.1 (a) $p_n = 0.524$ in., $p = 0.605$ in.,
  $p_a = 1.048$ in.
  (b) $P = 5.196$, $\phi = 28.3°$.
  (c) $d_p = 3.849$ in., $d_g = 7.698$ in.

12.3 kW $= 75.58$.

12.5 (a) $F_{t1} = F_{t2} = F_{t3} = 263.1$ lb.
  (b) $T_1 = 840$ lb $\cdot$ in., $T_2 = 0$, $T_3 = 1680$ lb $\cdot$ in.

12.7 $n = 1.21$.

12.9 (a) hp $= 17.63$.

(b) hp $= 31.1$.

12.10 (a) $d_p = 2.5$ in., $d_g = 5.25$ in.

(b) $\alpha_p = 25.46°$, $\alpha_g = 64.54°$.

(c) $b = 0.969$ in.

(d) $c = 0.026$ in.

12.12 The gears are safe.

12.15 kW $= 29.25$.

12.17 $F_t = 10.08$ kips.

12.19 (a) $\lambda = 10.39°$.

(b) $F_{wt} = F_{ga} = 420.2$ lb.

(c) $(hp)_m = 8.74$.

12.22 $(hp)_d = 1.392$.   No.

Chapter 13.   Belts, Chains, Clutches, and Brakes

13.1 (a) $F_1 = 296.6$ lb, $F_2 = 128.5$ lb.

(b) $L = 151.8$ in.

13.3 kW $= 34.6$.

13.5 $F_{max} = 1.143$ kN.

13.6 $c = 13.132$ in.

13.8 (a) $p_{max} = 254.6$ kPa, $T = 180$ N $\cdot$ m.

(b) $p_{max} = 191$ kPa, $T = 183.8$ N $\cdot$ m.

13.10 (a) $d = 17.64$ in.

(b) $F_a = 1.833$ kips.

13.12 $w = 1.451$ in.

13.15 $T = 602$ N $\cdot$ m.

13.17 $F_1 = 14$ kN, $F_2 = 3.987$ kN,
kW $= 31.46$.

13.19 $F_1 = 3,085$ N, $F_2 = 538.6$ N.

13.22 hp $= 12.76$.

13.24 $F_a = 366.04$ N.   No.

13.25 (a) $F_a = 1.542$ kN.   No.

(b) $R_A = 2.632$ kN.

13.28 $b = 1.414r$.

Chapter 14.   Springs

14.1 (a) $T = 35.48$ N $\cdot$ m.

(b) $\tau = 353$ MPa.

14.3 $N_a = 7.49$.

14.4 (a) $h_s = 39$ mm.

(b) $P_{max} = 320.4$ N.

14.6 (a) $h_f = 45.53$ mm.

(b) The spring is safe.

14.9 (a) $d = 14.94$ mm.

(b) $h_f = 274.6$ mm.

(c) The spring is safe.
$f_n = 4370$ cpm.

14.11 $P_{min} = 72.4$ lb,
$P_{max} = 127.6$ lb.

14.12 (a) $n = 2.49$.

(b) $N_a = 17.3$.

14.13 (a) $d = 0.103$ in.

(c) $f_n = 9270$ cpm.

(d) The spring is safe.

14.15 (a) $d = 5.41$ mm.

(b) $N_a = 9.89$.

14.18 $n = 1.30$.

14.20 (a) $M = 6.016$ lb $\cdot$ in.

(b) $\theta = 64.6°$.

Chapter 15.   Power Screws, Fasteners, and Connections

15.4 kW $= 1.23$.

15.6 (a) $n = 48$ rpm.

(b) $(hp)_{req} = 12.1$.

15.7 $T_o = 145.3$ N $\cdot$ m.

15.10 (a) $\sigma = 10.8$ MPa.

(b) $L_{ne} = 20.8$ mm.

(c) Nut: $\tau = 13.8$ MPa,
screw: $\tau = 16.4$ MPa.

15.11 $P = 54.67$ kN.

15.13 (a) $P_{max} = 37.27$ kN, $P_{min} = 21.53$ kN.

(b) $T = 75$ N $\cdot$ m.

15.15 $n = 2.77$.

15.16 (a) $P_b = 118.5$ kN.

(b) $T = 312.6$ N $\cdot$ m.

15.18 $n = 2.07$ (with preload),
$n = 1.40$ (without preload).

15.20 (a) $n = 4.5$ (with preload),
$n = 2.19$ (no preload).

(b) $n_s = 5.37$.

15.22 $e = 64.3\%$.

15.25 $P_{all} = 4.57$ kips.

15.27 $V_B = 2.15$ kN, $\tau_B = 6.843$ MPa,
$\sigma_B = 7.167$ MPa.

15.28 $d = 54.3$ mm.

15.30 $P = 23.76$ kN.

15.32 $h = 0.19$ in.

15.34 $L = 199.6$ mm.

15.37 $h = 0.22$ in.

Chapter 16. Axisymmetric Problems in Design

16.4 (a) $p = 30.71$ MPa.

(b) $2c = 220$ mm.

16.6 Steel: $\sigma_{\theta,max} = 62.2$ MPa,
bronze: $\sigma_{\theta,max} = -116.8$ MPa.

16.8 $\Delta d_s = 0.356\lambda$.

16.10 (a) $p = 5.167$ MPa.

(b) $\sigma_\theta = 5.596$ MPa.

16.11 (a) $\sigma_{\theta,max} = 41.11$ MPa.

(b) $\omega = 3539$ rpm.

16.14 (a) $p_y = 38.89$ MPa.

(b) $p_u = 65.55$ MPa.

(c) $(\sigma_\theta)_{res} = 56.8$ MPa.

16.16 $P = 1.949$ kN.

16.18 (a) $P = 84.59$ kN.

(b) $(\sigma_\theta)_B = -50$ MPa.

16.21 $t = 0.108$ in.

16.22 $a = 238.1$ mm.

16.23 $\sigma_\theta = 2.93$ MPa.

16.26 (a) $t = 14.67$ mm.

(b) $t = 12.67$ mm.

(c) Top end: $t = 158$ mm,
bottom end: $t = 235$ mm.

Chapter 17. Finite Element Analysis in Design

17.5 (c) $\begin{Bmatrix} u_2 \\ v_2 \end{Bmatrix} = \begin{Bmatrix} 1.0 \\ -3.9 \end{Bmatrix}$ mm.

(d) $\begin{Bmatrix} F_{1x} \\ F_{1y} \\ F_{3x} \\ F_{3y} \end{Bmatrix} = \begin{Bmatrix} 7632 \\ 10176 \\ -7500 \\ 0 \end{Bmatrix}$ N.

(e) $F_{12} = 22.32$ kN (T),
$F_{23} = -7.5$ kN (C).

17.6 (c) $\begin{Bmatrix} u_1 \\ v_1 \end{Bmatrix} = \begin{Bmatrix} 2.30 \\ -8.81 \end{Bmatrix} (10^{-3})$ in.

(d) $\begin{Bmatrix} R_{2y} \\ R_{3x} \\ R_{3y} \\ R_{4x} \end{Bmatrix} = \begin{Bmatrix} 5964.5 \\ 1037 \\ 1037 \\ -1035 \end{Bmatrix}$ lb.

(e) $F_{12} = 3964.5$ lb (T),
$F_{13} = 1464.5$ lb (T),
$F_{14} = -1035$ lb (C).

17.7 (c) $\begin{Bmatrix} u_2 \\ v_2 \end{Bmatrix} = \begin{Bmatrix} 18 \\ -60.4 \end{Bmatrix}$ mm.

(d) $\begin{Bmatrix} R_{1x} \\ R_{1y} \\ R_{3x} \end{Bmatrix} = \begin{Bmatrix} 45.024 \\ 60.032 \\ -45 \end{Bmatrix}$ kN.

(e) $F_{12} = -404$ kN (C),
$F_{23} = 180$ kN (T).

17.9 (c) $v_1 = 8.87$ mm.

(d) $\begin{Bmatrix} F_{2x} \\ F_{2y} \\ F_{3x} \\ F_{3y} \end{Bmatrix} = \begin{Bmatrix} 99.59 \\ 132.79 \\ 0 \\ -232.84 \end{Bmatrix}$ kN.

(e) $F_{12} = 166$ kN (T),
$F_{13} = -232.8$ kN (C).

17.11 (b) $\begin{Bmatrix} v_2 \\ \theta_2 \\ \theta_3 \end{Bmatrix} = \begin{Bmatrix} -0.729 \text{ in.} \\ 0.0052 \text{ rad} \\ 0.02083 \text{ rad} \end{Bmatrix}$

(c) $\begin{Bmatrix} F_{1y} \\ M_1 \\ F_{3y} \end{Bmatrix} = \begin{Bmatrix} 6.876 \text{ kips} \\ 225 \text{ kip} \cdot \text{in} \\ 3.125 \text{ kips} \end{Bmatrix}$

Index

Properties of areas

1. Rectangle

$$A = bh$$

$$I_x = \frac{bh^3}{12}$$

$$J_c = \frac{bh(b^2 + h^2)}{12}$$

2. Circle

$$A = \pi r^2$$

$$I_x = \frac{\pi r^4}{4}$$

$$J_c = \frac{\pi r^4}{2}$$

3. Right triangle

$$A = \frac{bh}{2}$$

$$I_x = \frac{bh^3}{36} \qquad I_{xy} = -\frac{b^2 h^2}{72}$$

$$J_c = \frac{bh(b^2 + h^2)}{36}$$

4. Semicircle

$$A = \frac{\pi r^2}{2}$$

$$I_x = 0.110 r^2$$

$$I_y = \frac{\pi r^4}{8}$$

5. Ellipse

$$A = \pi ab$$

$$I_x = \frac{\pi ab^3}{4}$$

$$J_c = \frac{\pi ab(a^2 + b^2)}{4}$$

6. Thin tube

$$A = 2\pi rt$$

$$I_x = \pi r^3 t$$

$$J_c = 2\pi r^3 t$$

7. Isosceles triangle

$$A = \frac{bh}{2}$$

$$I_x = \frac{bh^3}{36} \qquad I_y = \frac{hb^3}{48}$$

$$J_c = \frac{bh}{144}(4h^2 + 3b^2)$$

8. Half of thin tube

$$A = \pi rt$$

$$I_x \approx 0.095 \pi r^3 t$$

$$I_y = 0.5 \pi r^3 t$$

9. Triangle

$$A = \frac{bh}{2}$$

$$\bar{x} = \frac{(a + b)}{3}$$

10. Parabolic spandrel ($y = kx^2$)

$$A = \frac{bh}{3}$$

$$\bar{x} = \frac{3b}{4}$$

Vertex

11. Parabola ($y = kx^2$)

Vertex

$$A = \frac{2bh}{3}$$

$$\bar{x} = \frac{3b}{8}$$

12. General spandrel ($y = kx^n$)

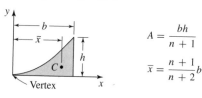

$$A = \frac{bh}{n + 1}$$

$$\bar{x} = \frac{n + 1}{n + 2} b$$

Vertex

| Notes: A = area, I = moment of inertia, J = polar moment of inertia.